THEORY, TABLES, AND DATA
FOR
COMPRESSIBLE FLOW

THEORY, TABLES, AND DATA FOR COMPRESSIBLE FLOW

Compiled by

William B. Brower, Jr.
Troy, New York

●HEMISPHERE PUBLISHING CORPORATION
A member of the Taylor & Francis Group

New York   Washington   Philadelphia   London

THEORY, TABLES, AND DATA FOR COMPRESSIBLE FLOW

Copyright © 1990 by Hemisphere Publishing Corporation. All rights reserved. Printed in the United States of America. Except as permitted under the United States Copyright Act of 1976, no part of this publication may be reproduced or distributed in any form or by any means, or stored in a data base or retrieval system, without the prior written permission of the publisher.

1 2 3 4 5 6 7 8 9 0   B R B R   9 8 7 6 5 4 3 2 1 0

Cover design by Renée Winfield.
A CIP catalog record for this book is available from the British Library.

Library of Congress Cataloging-in-Publication Data

Brower, William B.
   Theory, tables, and data for compressible flow / compiled by
William B. Brower, Jr.
     p. cm.
   Includes bibliographical references.
   1. Compressibility.   I. Title.
TL574.C4B76   1990
629.132′323—dc20   89-78372
                                                     CIP

ISBN 1-56032-065-6

CONTENTS

PREFACE

Perhaps 20 years ago that most valued reference, NACA Report 1135, a product of the research staff of Ames Aeronautical Laboratory, went out of print. Fortunately, by the time of its demise the Xerox machine had come along and, no doubt, like many other teachers of compressible-flow courses, I have been able to make it available to my own students at something like its original cost of $0.75!

Gradually, however, I found that I was xeroxing for use in my classes increasingly more material which had been excerpted from a wide variety of sources. Over the same period of time the main-frame computer, accessed by terminal, had come along providing the opportunity for various computing projects relating to compressible flow for undergraduates. Several useful tables resulted from these efforts.

It thus seemed natural to assemble these items in a form, and within one cover, that would serve the needs for quick reference by working engineers as well. This work represents the amalgamation of these documents.

No claim is made for completeness. The choice of topics in the THEORY section—obviously influenced by NACA Rep. 1135—has been expanded according to my interests but limited to what can be treated in reasonably concise form. The blending of thermodynamics and fluid mechanics is based upon the original approach of my friend and former colleague, J. V. Foa.

Informed criticism is welcomed, as are suggestions for additional topics for inclusion in future editions. Although reasonable diligence in proofing the manuscript has been observed, I am resigned to the probability of overlooked errors and will be grateful to have them pointed out.

W. B. Brower, Jr.
Troy, New York

ACKNOWLEDGEMENT

I wish to express appreciation to several of my former students in putting together certain of the tables included herein: John Brooks for the tables on oblique shocks, Debi Beebe for the Fanno-flow and isothermal-flow tables, Jack Simon for certain editing work, in particular on the cone-flow tables, and Doug Krehbiel for carrying out the calculations on the thermophysical properties of air.

My sincere thanks go to Bonnie J. McBride of NASA Lewis for her cooperation in providing access to the CETA program which was used for the tabulation of thermophysical values for air, as part of on-going hypersonic research at RPI.

I want to thank Dwight Sangrey, former Dean of Engineering at Rensselaer Polytechnic Institute, for encouraging me to pursue this project and providing access to the RPI IBM 3086D computer on which all the calculations were performed. In the same regard I wish to express thanks to Joseph V. Smith, Assistant Dean of Engineering, for his cooperation over an extended period of time.

In preparation of the THEORY manuscript Mary Ellen Frank typed preliminary versions of almost every section. Betty Alix agonized through an incredible number of drafts in putting the manuscript into the MTS *Textform*. The final editing of the tables was provided by Ajay Divakaran.

I also wish to thank Dr. B.A. Younglove of NIST (formerly NBS) and the American Institute of Physics for permission to publish selected values of thermophysical properties of argon, hydrogen, nitrogen and oxygen. My wife Yolanda undertook the onerous task of transcribing these data, and the data from *U.S. Standard Atmosphere, 1976* in a form suitable for inclusion in this work. For this and other invaluable contributions I want to express my gratitude.

SYMBOLS

The equation number following each symbol, corresponds to the first equation in which that symbol appears for the particular use.

Roman

a	lapse rate in troposphere (21-1)
a	non-dimensional sound speed (19-9)
a	speed of sound (5-1)
a^o	total speed of sound (6-3)
$a*$	critical speed of sound (6-5)
$a_0, a_1, a_2, \ldots\, b_2, b_3, b_4$	coefficients in Busemann's theory (14-3)
a, b	coefficients in Busemann's theory (14-12)
A	cross-sectional area of duct (3-4)
$A*$	cross-sectional area of duct where $M = 1$ (9-5)
A, B, C, D	coefficients in Molenbroek's equation (18-33)
$B_0, B_1, B_2,\ C_0, C_1, C_2, C_3, C_0', C_1', C_2', C_3'$	coefficients in Busemann's theory (14-4), (14-10), (15-7)
c	airfoil chord length (16-5)
c	shock propagation speed (4-1)
c_d, c_ℓ	section drag, and lift coefficients (16-10), (16-6)
$c_f, \bar{c}_f$	local, and mean skin-friction coefficients (10-1), (10-12)
$c_{m0.5c}$	section moment coefficient (16-15)
c_p, c_v	coefficients of specific heat (1-2)
C	coefficient in nozzle flow (9-5)
C	mean free-speed (20-3)
$C.P.$	center of pressure (16-21)
C_p	pressure coefficient (7-14)
d, d_b	airfoil section drag, base drag (16-9), (16-12)
$\dot{d}$	denotes path differential (1-1)
$\bar{dq}, dq$	differential heat added, (1-1)
$d\psi$	differential heat added due to frictional dissipation (1-1)
ds	differential arc length on airfoil surface (16-5)
D	duct diameter (10-1a)
D/Dt	substantial derivative operator (3-1)

e	specific internal energy (1-1)
f	friction factor (10-1)
f_{ij}	denotes ratio of flow variable between two states (4-4a)
$F(x,t)$	arbitrary flow function (3-11b)
$F(M;\gamma)$	Mach number function for momentum calculation (6-16)
$F_1(M;\gamma)$	Mach number function in Busemann's theory (14-11)
$\mathbf{F}$	force of duct on fluid contained within (6-13)
g	uniform gravitational-field constant (21-3)
$G(M;\gamma)$	Mach number function for Fanno flow (10-11)
G	Newton's universal gravitational constant (21-10)
h	airfoil thickness function (16-1)
h	specific enthalpy (1-1)
h^o	total enthalpy (3-6a)
$H(M;\gamma)$	Mach number function in isothermal flow (11-6)
H	vertical coordinate of atmosphere in geometric units (21-12)
k	heat conduction coefficient (20-15)
k	molar mass (2-2)
Kn	Knudsen number (20-8)
ℓ	lift force on airfoil (16-5)
L	characteristic body length in Knudsen number (20-8)
L	duct length in Fanno flow (10-14)
m	molecular mass (20-2)
$\dot{m}$	mass rate-of-flow (3-14)
$\dot{\overline{m}}$	mass flux (3-16)
$\overline{m}$	molecular weight (2-2)
$m_{0.5c}$	aerodynamic moment of airfoil about mid chord (16-13)
m,n	natural coordinates (18-32)
M	Mach number (6-1)
M	mass of the earth (21-10)
n	number of degrees of freedom of a molecule (2-6a)
n	polytropic exponent in atmosphere (21-6)
N	number of molecules (20-1)
O	denotes the order of a quantity (13-11)
p	static, or thermodynamic pressure (1-1)
P	duct perimeter (3-5)

Pr	Prandtl number (20-16)
q	dynamic pressure (6-8)
q	heat flux in Fourier's equation (20-15)
$\dot{q}$	path time-derivative of heat-addition function (3-3)
R,R^*	specific and universal gas constants (2-1),(2-2)
R_0	radius of the earth (21-10)
Re	Reynolds number (10-15)
R,θ,ϕ	spherical coordinates (19-1)
s	specific entropy function (1-5)
S	cross-sectional area of airfoil profile (16-20)
S	parameter in Sutherland's law (20-14)
t	time variable (3-1)
T	static temperature (1-2)
T°	total temperature (6-1)
u	horizontal flow velocity (3-1)
u	unified atomic mass unit (20-2)
u,v	velocity components parallel to x,y-axes (7-17)
$\overline{u},\overline{v},\overline{w}$	molecular mean-free-velocity components (20-3)
$\overline{u},\overline{v}$	non-dimensional velocity components in hodograph plane (13-5)
$\overline{u'},\overline{v'}$	non-dimensional perturbation velocity components in hodograph plane (13-10)
$\hat{u},\hat{v}$	non-dimensional scaled, perturbation velocity components in transonic flow (13-13)
u_R,u_θ,u_ϕ	velocity components in spherical coordinates (19-1)
U	free-stream speed (7-14)
v	specific volume (1-1)
V	flow speed in three-dimensions (18-6)
$\mathbf{V}$	vector velocity (18-1)
V	volume (20-1)
w	flow speed in two-dimensional or axi-symmetric flow (12-1), (19-15)
W	potential energy function of earth's gravitational field (21-12)
x	space coordinate in one-dimensional horizontal flow (3-1)
x,y	space coordinates in two-dimensions, Section 12
y	coordinate normal to chord-line of airfoil (16-1)
y_c	camber function for airfoil (16-1)
z	vertical coordinate of atmosphere in geopotential units (21-1)

z^* tropopause (21-1)

z^{**} upper limit of stratosphere constant-temperature region (21-9)

Greek

α	angle-of-attack of airfoil (16-2)
α	$(\gamma+1)/(\gamma-1)$ (8-2)
β	shock angle (12-2)
γ	ratio of specific heats (1-4)
Δ	indicates a finite difference, or jump, in a quantity (4-2)
ϵ	non-dimensional small parameter (17-1)
ϵ	symbol which takes on numerical value of 0 or 1 (18-31)
θ	denotes flow deviation from upstream uniform value, or flow deflection across shock (12-2)
λ	mean-free-path length (20-7)
μ	dynamic viscosity (20-10)
μ	Mach angle (12-9)
ν	molecular number density (20-1)
$\nu(M;\gamma)$	Prandtl-Meyer function (15-3)
ξ	airfoil chordwise coordinate (16-1)
ρ	mass density (2-1)
σ	molecular diameter (20-7)
τ_w	wall shear stress (3-5)
Φ	velocity potential (18-10)
ψ	flow direction behind shock in axi-symmetric flow (19-16)
$\dot{\psi}$	path time-derivative of heat added due to frictional work on particle boundary (3-8)
ω	vector vorticity (18-9)
$\omega_R, \omega_\theta, \omega_\phi$	vorticity components in spherical coordinates (19-5)

Subscript

a	denotes standard atmospheric value, Section 21
c	denotes value on cone surface, Section 20
f	denotes station in isothermal flow where $M_f \to 1/\gamma$, Section 11
m	denotes value at extremum, Section 12
n	denotes component of Mach number normal to oblique shock (12-2)
P	denotes piston value, Section 4
r	denotes reference value, Section 21
w	denotes wedge value in oblique-shock flow, Section 12
∞	denotes value where $M \to \infty$, Section 12

Superscript

o	denotes total condition in a flow, Section 3
$*$	denotes local value where $M \to 1$ in nozzle flow, or Fanno flow, or flow behind shock
$*$	denotes value at the tropopause altitude
$**$	denotes value at upper limit of the constant-temperature region in troposphere

THEORY

1. Thermodynamics of General Substances

The *first law of thermodynamics*, written for an observer motionless with respect to the mass center of a fluid particle, is

$$\bar{d}q = \bar{d}q + \bar{d}\psi = de + pdv = dh - vdp , \qquad (1\text{-}1)$$

where the symbol $\bar{d}X$ denotes the differential of a path function and where the thermodynamic variables such as *internal energy e, enthalpy h, specific volume* $v = 1/\rho$, *entropy s*, etc., are written in the intensive form — i.e. per-unit-mass. The quantity $\bar{d}q$ refers to *heat addition* per-unit-mass in the form of conduction, combustion, and/or radiation. The quantity $\bar{d}\psi$ refers to heat added due to the work of viscous forces on the boundary of the particle, computed in the frame of the same observer. Since this is dissipative, $\bar{d}\psi \geq 0$.

The *specific heats at constant pressure* and *constant volume*, respectively, are defined as

$$c_p \equiv (\bar{d}q/dT)_p = (\partial h/\partial T)_p , \quad c_v \equiv (\bar{d}q/dT)_v = (\partial e/\partial T)_v . \qquad (1\text{-}2)$$

Expressing $e = e(v,T)$ it follows, with (1-1) and (1-2), that

$$c_p - c_v = [(\partial e/\partial v)_T + p](\partial v/\partial T)_p . \qquad (1\text{-}3)$$

The *ratio of specific heats* is defined as

$$\gamma \equiv c_p/c_v , \qquad (1\text{-}4)$$

which may vary with pressure and temperature, for a general substance.

Consideration of the *second law of thermodynamics* leads to the introduction of a new variable, the entropy s, which must be expressible as a function of any two, independent, state variables, e.g. $s = s(T,v)$, or $s = s(p,T)$, such that

$$Tds = \bar{d}q + \bar{d}\psi = de + pdv = dh - vdp . \qquad (1\text{-}5)$$

It should be evident that in (1-5) the entropy relation has been introduced into the first law of thermodynamics.

The requirement that ds be an exact differential leads to a number of relations, called *exactness criteria*, two of which follow:

$$(\partial e/\partial v)_T = T(\partial p/\partial T)_v - p , \qquad (1\text{-}6a)$$

$$(\partial h/\partial p)_T = v - T(\partial v/\partial T)_p . \qquad (1\text{-}6b)$$

1

$$(\partial h/\partial p)_T = v - T(\partial v/\partial T)_p \quad . \qquad (1\text{-}6b)$$

If the first of equations (1-6) is substituted into (1-3) the alternative form is obtained:

$$c_p - c_v = T(\partial p/\partial T)_v (\partial v/\partial T)_p \quad . \qquad (1\text{-}7)$$

2. Thermally and Calorically Perfect Gases

A *thermally perfect gas* is defined as one whose *thermal equation of state* is

$$p = RT/v = \rho RT \quad , \qquad (2\text{-}1)$$

where the *specific volume* $v = 1/\rho$, and where the *specific gas-constant R* is related to the *universal gas-constant R^** by

$$R = R^*/\overline{m}k \quad , \qquad (2\text{-}2)$$

where $\overline{m}$ is the *molecular weight* (which is actually dimensionless), and k is a multiplying constant of unit magnitude which specifies the *molar mass* in units consistent with those of the universal gas-constant. That is, k is expressed in *gm/mol, kg/kmol,* or *slug/mol,* and so forth. Tables of gas constants can be found in the data section of this work.

For a thermally perfect gas it follows from (1-7) that

$$c_p - c_v = R \quad , \qquad (2\text{-}3)$$

and that e, h, c_p, c_v are all functions only of T; thus

$$e = \int c_v dT \quad , \quad h = \int c_p dT \quad . \qquad (2\text{-}4)$$

For a *calorically perfect gas* we put

$$c_p = const. \quad , \quad c_v = const. \quad , \quad \gamma = const. \qquad (2\text{-}5)$$

Then, for a thermally and calorically perfect gas,

$$c_p = \gamma R/(\gamma - 1), \quad c_v = R/(\gamma - 1), \quad \gamma = (n+2)/n, \qquad (2\text{-}6a)$$

$$e = c_v T, \quad h = c_p T, \qquad (2\text{-}6b)$$

where, from the kinetic theory of gases, n is the number of *classical degrees of freedom* of a molecule, i.e. excluding vibrational effects. For a monatomic molecule $n = 3$; for a diatomic molecule $n = 5$; and for triatomic molecules $n = 5$ or 6, for linear, or for polar, molecules, respectively.

For a thermally perfect gas, the entropy relation, equation (1-5), becomes

$$d(s/R) \quad = \quad (c_v/R)d(\ell nT) + d(\ell nv) \quad , \tag{2-7a}$$

$$= \quad (c_p/R)d(\ell nT) - d(\ell np) \quad , \tag{2-7b}$$

$$= \quad (c_v/R)d(\ell np) + (c_p/R)d(\ell nv) \quad . \tag{2-7c}$$

If also calorically perfect, equations (2-7) can be integrated in closed form between any state, designated by the subscript 1, to an arbitrary state (no subscript). For example, (2-7c) becomes

$$pv^\gamma \quad = \quad p_1 v_1^\gamma exp[(\gamma - 1)\Delta s/R] \quad , \tag{2-8a}$$

where

$$\Delta s \quad \equiv \quad s - s_1 \quad . \tag{2-8b}$$

For an isentropic transformation, defined by

$$s \quad = \quad constant \quad , \tag{2-9}$$

the familiar

$$pv^\gamma \quad = \quad p_1 v_1^\gamma \quad = \quad constant \tag{2-10}$$

results.

3. The Equations of One-Dimensional Duct Flow

We now shift from an observer motionless with respect to the center of mass of a particle to an observer located at an arbitrary point but, unless otherwise specified, at rest with respect to the wall of the duct. Initially we suppose that all variables, including the duct cross-sectional area, may be time dependent. For simplicity, however, we restrict considerations to a straight, horizontal duct. A particle is considered as an arbitrarily thin disc of fluid occupying station x. The pressure, temperature, flow velocity, etc., are considered to be uniform across the disc. The particle velocity for horizontal flow is

$$u \quad \equiv \quad Dx/Dt \quad , \tag{3-1}$$

where we have employed the *substantial-derivative operator*[1]

$$D/Dt \quad \equiv \quad \partial/\partial t + u\,\partial/\partial x \quad , \tag{3-2}$$

to indicate differentiation following a particle.

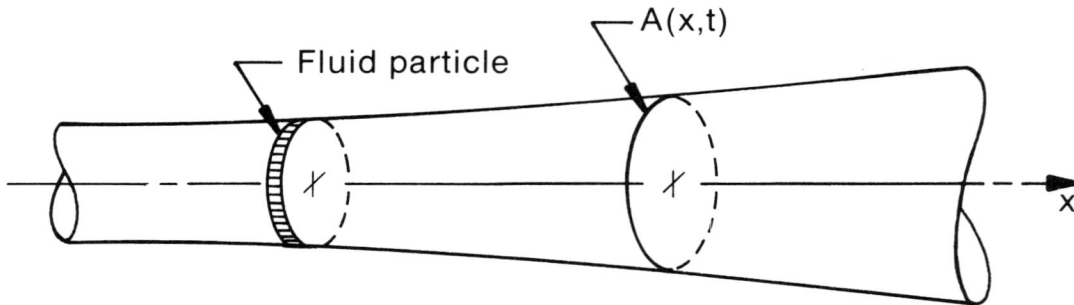

Figure 3-1. Schematic for horizontal-duct flow.

Consequently, the first-law relations can be taken over directly into the new frame of reference. For example, we divide (1-5) by dt. Then, where in each term the combination d/dt operating on a state variable appears, we replace d/dt by D/Dt. However, to denote the fact that the derivatives dq/dt and $d\psi/dt$ are path functions, not expressible as substantial derivatives, we replace them by $\dot{q}$ and $\dot{\psi}$, respectively. Therefore, the first law of thermodynamics for an arbitrary observer becomes

$$TDs/Dt \quad = \quad \dot{q} + \dot{\psi} \quad = \quad De/Dt + p Dv/Dt \quad = \quad Dh/Dt - v Dp/Dt \quad . \tag{3-3}$$

The <u>continuity equation</u>. For a flow without mass injection or extraction, *conservation of mass* is expressed as

$$\frac{\partial}{\partial t}(\rho A) + \frac{\partial}{\partial x}(\rho u A) \quad = \quad 0 \quad , \tag{3-4}$$

where the cross-sectional area of the duct is $A = A(x,t)$.

The <u>dynamical equation</u>. For gas flows gravitational effects are negligible except for cases of very large height changes. For horizontal flow they are truly zero. The

---

[1]In fluid mechanics it is assumed, unless stated otherwise, that the independent variables are the *space* and the *time* coordinates; in one-dimensional horizontal flow, these are x and t.

dynamical equation of motion, under this restriction,[2] and for an inertial observer, is

$$\frac{Du}{Dt} = -\frac{1}{\rho}\frac{\partial p}{\partial x} - \frac{\tau_w P}{\rho A} \quad , \tag{3-5}$$

where τ_w is the *wall viscous stress* and P is the *duct perimeter*.

The energy equation. The *energy equation* is viewed herein as the result of combining the first law of thermodynamics, equation (3-3), with the dynamical equation, by eliminating the derivative $\partial p/\partial x$, which appears in both, since $Dp/Dt = \partial p/\partial t + u\partial p/\partial x$. This results in the alternative forms, valid for a general substance, in the absence of shaft work:

$$\frac{Dh^o}{Dt} = \frac{1}{\rho}\frac{\partial p}{\partial t} + T\frac{Ds}{Dt} - u\frac{\tau_w P}{\rho A} \quad , \tag{3-6a}$$

$$= \frac{1}{\rho}\frac{\partial p}{\partial t} + \dot{q} + \dot{\psi} - u\frac{\tau_w P}{\rho A} \quad . \tag{3-6b}$$

We have introduced in (3-6) the *total enthalpy* of a particle

$$h^o \equiv e + pv + \tfrac{1}{2}u^2 = h + \tfrac{1}{2}u^2 \quad . \tag{3-7}$$

In (3-6b), the rate-of-work per-unit-mass due to the presence of wall viscous stresses, computed in the frame of an observer moving with a particle, is precisely

$$\dot{\psi} = u\frac{\tau_w P}{\rho A} \quad . \tag{3-8}$$

Therefore, (3-6b) reduces to

$$\frac{Dh^o}{Dt} = \frac{1}{\rho}\frac{\partial p}{\partial t} + \dot{q} \quad . \tag{3-9}$$

Thus, in the absence of mechanical work, the only way to change the total enthalpy of a particle is by heat transfer or by non-steady pressure forces. Equation (3-9) is restricted to an inertial observer who is motionless with respect to the duct wall. The viscous terms do not appear explicitly, since the wall does no work on the fluid in the frame of the inertial observer, and the viscous effects manifest themselves by redistributing the elements making up the total enthalpy $h^o = e + pv + \tfrac{1}{2}u^2$, among themselves.

---

[2]It is also implied that we are neglecting any normal viscous stresses in (3-5).

The definition of an *adiabatic flow* is one in which $đq \equiv 0$, or

$$\dot{q} \equiv 0 \; . \tag{3-10}$$

Note that in an adiabatic flow, not only is the heat addition, due to conductic.. from the wall or by combustion or radiation, required to be negligible, but heat conduction from particle to particle is also taken as zero in the streamwise direction.

The <u>equations</u> <u>of</u> <u>steady</u> <u>flow</u>. The restriction of *steady flow* is expressed by

$$\partial/\partial t \equiv 0 \; . \tag{3-11a}$$

Consequently, where $F = F(x,t)$ is any flow function, such as pressure, or velocity, it follows that

$$F(x,t) \;\rightarrow\; F(x) \quad \text{only,} \quad \frac{\partial}{\partial x} \;\rightarrow\; \frac{d}{dx} \; . \tag{3-11b}$$

Therefore, from (3-9), for steady, adiabatic flow,

$$Dh^o/Dt \;\rightarrow\; udh^o/dx \equiv 0 \; , \tag{3-12a}$$

or,

$$h^o \;=\; h + \tfrac{1}{2} u^2 \;=\; constant \; . \tag{3-12b}$$

Since (3-12b) applies to every particle we see that in an adiabatic, workless, steady flow the total enthalpy is the same for every particle. Such a flow is said to be *isoenergetic*.[3]

Similarly, the dynamical equation for steady flow is

$$u \frac{du}{dx} \;=\; -\frac{1}{\rho}\frac{dp}{dx} - \frac{\tau_w P}{\rho A} \; . \tag{3-13}$$

From (3-4) the *mass flow-rate* is

$$\dot{m} \equiv \rho u A \;=\; constant \; . \tag{3-14}$$

Equations (3-12b), (3-13), and (3-14), which are valid for arbitrary substances, together with the appropriate state relations, comprise our set of working relations for isoenergetic

---

[3] A three-dimensional flow subject to the same restrictions is said to be *homenergetic*.

flow.

The equations of constant-area flow. The definition of *constant-area flow* is simply

$$A \equiv constant \ . \tag{3-15}$$

Under this additional restriction (3-14) becomes a statement about the *mass flux*

$$\dot{\overline{m}} \equiv \dot{m}/A = \rho u = \rho_1 u_1 = \rho_2 u_2 = constant \ , \tag{3-16}$$

where the subscripts *1,2* refer to an arbitrary pair of stations. Equation (3-16) can be combined with (3-13) to produce the *steady-state momentum equation* for a constant-area duct:

$$d(p + \dot{\overline{m}}u) = - \frac{\tau_w P}{A} dx \ , \tag{3-17a}$$

which can be integrated between *1* and *2*. Thus

$$p_2 + \rho_2 u_2^2 = p_1 + \rho_1 u^2 - \int_1^2 \frac{\tau_w P}{A} dx \ . \tag{3-17b}$$

If we additionally restrict considerations to an arbitrarily short duct-segment, the flow may be treated as though it were *inviscid*, i.e. such that

$$\tau_w \equiv 0 \ . \tag{3-18}$$

Then, (3-17b) becomes

$$p_1 + \rho_1 u_1^2 = p_2 + \rho_2 u_2^2 \ . \tag{3-19}$$

Equations (3-16), (3-19), and (3-12b) provide the basis for treating the flow due to a finite disturbance, i.e. a shock wave, propagating through a constant-area duct.

4. On the Propagation of Finite Disturbances

Consider a fluid within a constant-area duct to be initially (i.e. for $t < 0$) at pressure p_1, density ρ_1, and static enthalpy h_1. The fluid initial velocity with respect to the wall is $u'_1 \equiv 0$ everywhere. A piston, starting at $t = 0$, moves impulsively to the left at velocity $u'_P < 0$. Since the fluid in contact with the piston must move at the piston velocity, a finite disturbance (a wave front of infinitesimal thickness) is created which moves to the left at velocity $u'_s = -c$ with respect to the wall. Because an increasingly larger mass of fluid is forced to move at the piston velocity we must have $c > |u'_P|$. Such a flow is inherently non-steady in the frame of an observer motionless with respect to the wall.

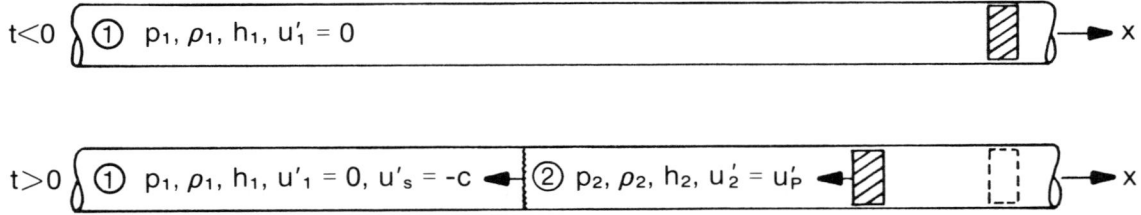

t<0 ① p_1, ρ_1, h_1, $u'_1 = 0$ → x

t>0 ① p_1, ρ_1, h_1, $u'_1 = 0$, $u'_s = -c$ ◄ ② p_2, ρ_2, h_2, $u'_2 = u'_P$ ◄ → x

Figure 4-1.   Propagation of a finite disturbance.

However, if we transform to an observer moving with the disturbance we can use the previously derived equations for steady flow. Denoting velocities in the frame of an observer fixed on the wall with primes, the velocities in the frame of the shock (unprimed) are given by:

$$u_{1,2} = u'_{1,2} + c \; , \quad or \quad u_1 = c \; , \quad u_2 = u'_2 + c \; . \tag{4-1}$$

Hence, the *speed of propagation* c of a finite disturbance relative to an observer embedded in the undisturbed fluid is equal to the speed of the flow u_1 ahead of the shock (i.e. in region 1) with respect to an observer fixed on the shock.

Substitution of (4-1) into (3-16) and (3-19), and the elimination of u_2 therefrom, gives an expression for the propagation speed in terms of the *pressure* and *density jumps* $\Delta p \equiv p_2 - p_1$, $\Delta \rho \equiv \rho_2 - \rho_1$, across the shock:

$$c^2 = \frac{\Delta p}{\Delta \rho}(1 + \frac{\Delta \rho}{\rho_1}) \; . \tag{4-2}$$

Similarly, in *any* frame of reference, the *velocity jump* is

$$\Delta u \equiv u_1 - u_2 = u'_1 - u'_2 = -c \, \Delta \rho / (\rho_1 + \Delta \rho) \; . \tag{4-3}$$

Alternatively, if we introduce the energy equation (3-12b) (for isoenergetic flow) we can eliminate the propagation speed in (4-3) to produce an expression involving only ratios of state variables and the undisturbed flow conditions in *1*. Introducing the special notation $f_{ij} \equiv f_i / f_j$, where f can be any flow function p, ρ, h, T, u, etc., we obtain

$$\frac{(h_{21}-1)\rho_{21}}{(p_{21}-1)(\rho_{21}+1)} = \frac{p_1}{2\rho_1 h_1} \; . \tag{4-4a}$$

Equation (4-4a) is referred to herein as the *Rankine-Hugoniot equation for general substances*. It is valid for gases, liquids, or solids, and is independent of any state relation. Alternative and equivalent forms of (4-4a) are:

$$h_2 - h_1 = \tfrac{1}{2}(p_2 - p_1)(v_2 + v_1) \; , \tag{4-4b}$$

$$e_2 - e_1 = -\tfrac{1}{2}(p_1 + p_2)(v_2 - v_1) \; . \tag{4-4c}$$

5. On the Speed of Sound

By the *speed of sound* a we mean the limiting speed of a vanishingly small disturbance; thus, from (4-2),

$$a^2 \equiv \lim_{\Delta\rho\to 0} c^2 = \lim_{\Delta\rho\to 0} \frac{\Delta p}{\Delta \rho} \; . \tag{5-1}$$

However, (5-1) is not uniquely defined, and an additional restriction must be imposed, namely, that the propagation of the infinitesimal disturbance shall not of itself create any modification of the fluid entropy. Thus

$$a^2 \equiv \lim_{\substack{\Delta\rho\to 0 \\ s=const.}} c^2 = (\partial p/\partial \rho)_s = \gamma(\partial p/\partial \rho)_T \; , \tag{5-2}$$

where the right-hand side of (5-2) follows routinely from the preceding thermodynamical relations for an isentropic process. Note that this additional restriction is not equivalent to requiring that every particle of the flow shall have the same entropy. Equation (5-2) is valid for an arbitrary substance.

For a thermally perfect gas (5-2) and (2-1) combine to yield

$$a^2 = \gamma R T = \gamma p/\rho \; . \tag{5-3}$$

6. Relations for Isoenergetic and Isentropic Duct Flow of Perfect Gases

The following relations apply to ducts of arbitrary cross-sectional areas $A = A(x)$. Combining (3-12b) and (2-6), and dividing by $c_p T$, we obtain

$$T^o/T \;=\; 1 + u^2/2c_p T \;=\; 1 + \tfrac{1}{2}(\gamma-1)(u/a)^2 \;=\; 1 + \tfrac{1}{2}(\gamma-1)M^2 \;, \qquad (6\text{-}1)$$

where we have introduced the dimensionless ratio

$$M \;\equiv\; u/a \;, \qquad (6\text{-}2)$$

which is the *Mach number*, and which is the most important dimensionless parameter in compressible flows. T^o is called the *total temperature* (sometimes called the *stagnation temperature*, also). For an observer who sees a particle at temperature T in relative motion at speed u, T^o is the temperature that the particle would acquire were it decelerated to rest isoenergetically in his/her frame. Note that T^o is a useful reference quantity including flows (in contrast to the hypothetical deceleration process) which are non-isoenergetic, but in which T^o may vary from particle to particle.

There are other reference flow speeds often encountered. The *total speed-of-sound* for an isoenergetic flow of total temperature T^o is given by

$$a^o \;\equiv\; (\gamma R T^o)^{\tfrac{1}{2}} \;. \qquad (6\text{-}3)$$

Then, with (2-6), (5-3) and (6-3), the energy equation can be written in the following equivalent forms:

$$\frac{(a^o)^2}{\gamma-1} \;=\; c_p T^o \;=\; c_p T + \frac{u^2}{2} \;=\; \frac{a^2}{\gamma-1} + \frac{u^2}{2}$$

$$=\; \frac{\gamma}{\gamma-1}\frac{p}{\rho} + \frac{u^2}{2} \;=\; \text{const} \;. \qquad (6\text{-}4)$$

At a point where the flow speed and the local speed-of-sound are equal we put $T \to T^*$, $a \to a^*$, thus $u^* = a^*$, or $M = 1$, and

$$T^o/T^* \;=\; (a^o/a^*)^2 \;=\; (\gamma+1)/2 \;. \qquad (6\text{-}5)$$

The quantity $a*$ is called the *critical speed of sound*. Alternatively, from (6-4) and (6-5), the theoretical *maximum flow-speed* u_m for a gas expanded into a vacuum (where $p \rightarrow 0$, $T \rightarrow 0$, $a \rightarrow 0$, $M \rightarrow \infty$) can be expressed as:

$$u_m \quad = \quad a^o [2/(\gamma - 1)]^{\frac{1}{2}} \quad = \quad a*[(\gamma + 1)/(\gamma - 1)]^{\frac{1}{2}} \quad . \tag{6-6}$$

If the hypothetical process previously referred to is also isentropic then we can also define a *total pressure* p^o and a *total density* ρ^o. The principal isentropic and isoenergetic relations are summarized as:

$$\frac{p^o}{p} \quad = \quad (\frac{\rho^o}{\rho})^\gamma \quad = \quad (\frac{T^o}{T})^{\frac{\gamma}{\gamma - 1}} \quad = \quad (\frac{a^o}{a})^{\frac{2\gamma}{\gamma - 1}}$$

$$= \quad (\frac{\gamma + 1}{2} \frac{a*}{a})^{\frac{2\gamma}{\gamma - 1}} \quad = \quad [1 + \tfrac{1}{2}(\gamma - 1)M^2]^{\frac{\gamma}{\gamma - 1}} \quad . \tag{6-7}$$

In many fluid-mechanical applications it is convenient to introduce a reference *dynamic pressure* q such that

$$q \quad \equiv \quad \tfrac{1}{2}\rho u^2 \quad = \quad \tfrac{1}{2}(\rho u^2/a^2)(\gamma p/\rho) \quad .$$

In terms of the Mach number

$$q/p \quad = \quad \gamma M^2/2 \quad ; \tag{6-8}$$

and,

$$q/p^o \quad = \quad \gamma M^2/2 \; [1 + \tfrac{1}{2}(\gamma - 1)M^2]^{\frac{\gamma}{\gamma - 1}} \quad . \tag{6-9}$$

<u>The momentum equation for steady flow</u>. If we multiply (3-13) by $\rho A dx$ then

$$\rho u A du \quad = \quad -A dp \; - \; \tau_w P dx \quad . \tag{6-10}$$

Then, adding $d(pA) = p dA + A dp$ to both sides, and noting that $\dot{m} = \rho u A$, we can solve for

$$p dA \; - \; \tau_w P dx \quad = \quad d[pA + \dot{m}u] \quad . \tag{6-11}$$

Now, the left-hand side of (6-11) represents the algebraic sum of the differential pressure-force and the viscous force exerted by an element of the duct dx in length, on the fluid within the element. Thus, we put

$$d\mathbf{F} \equiv pdA - \tau_w Pdx \quad , \tag{6-12}$$

and integrate between stations 1 and 2. This results in the following expression for the force $\mathbf{F}_{1,2}$ exerted by the duct wall on the fluid:

$$\mathbf{F}_{1,2} = (pA + \dot{m}u)_2 - (pA + \dot{m}u)_1 \quad . \tag{6-13}$$

The quantity $pA + \dot{m}u$ is variously referred to as the *impulse* or *stream force*. This simple relation is very powerful since it is in integrated form and can be applied to any flow, constant-density or compressible, viscous or inviscid, and is applicable whether or not the flow is truly one-dimensional, even involving separated regions, as long as the flow is essentially uniform and steady at stations 1 and 2. Equation (6-13) states that the force of the wall upon the flow between 1 and 2 is equal to the change of stream force between the two stations.

We can convert (6-13) to a Mach-number relation by putting

$$pA + \dot{m}u = \dot{m}[pA/\dot{m} + u] = \dot{m}[\frac{p}{\rho u} + u] \quad . \tag{6-14}$$

Therefore, with $u = aM$, $p/\rho = a^2/\gamma$,

$$\frac{p}{\rho u} + u = \frac{a^2}{\gamma}\frac{1}{aM} + aM = a^o(a/a^o)\frac{1 + \gamma M^2}{\gamma M} \quad . \tag{6-15}$$

With (6-15) and (6-7) equation (6-14) reduces to

$$pA + \dot{m}u = \dot{m}a^o F(M; \gamma) \quad , \tag{6-16a}$$

where we define a Mach-number function

$$F(M; \gamma) \equiv \frac{1 + \gamma M^2}{\gamma M[1 + \frac{1}{2}(\gamma - 1)M^2]^{\frac{1}{2}}} \quad . \tag{6-16b}$$

Consequently, for isoenergetic flow the momentum equation (6-13) can be stated as

$$\mathbf{F}_{1,2} = \dot{m}a^o [F_2(M; \gamma) - F_1(M; \gamma)] \tag{6-17}$$

For flows with heat addition between stations *1* and *2* the appropriate form is simply

$$\mathbf{F}_{1,2} \;\; = \;\; \dot{m} \, [a_2^o F_2(M; \; \gamma) - a_1^o F_1(M; \; \gamma)] \;\; . \tag{6-18}$$

A tabulation of $F(M; \; \gamma)$ is included in Tables A.1 to A.3.

7. Formulas Involving Alternative Speed Ratios for Isoenergetic and Isentropic Flow

In some applications it is more convenient to non-dimensionalize the flow speed with respect to one of the reference speeds a^o, a^*, or u_m. These are readily related to the local Mach-number through the energy equation (6-4). Thus, expressing each speed ratio in terms of the Mach number,

$$(u/a^o)^2 \;\; = \;\; 2(u/a^*)^2/(\gamma + 1) \;\; = \;\; 2(u/u_m)^2/(\gamma - 1) \;\; = \;\; M^2/[1 + \tfrac{1}{2}(\gamma - 1)M^2] \;\; , \tag{7-1}$$

Therefore, we can express the principal flow-ratios in terms of the various speed-ratios as given in the following expressions:

- **Parameter u/a^***

$$T/T^o \;\; = \;\; (a/a^o)^2 \;\; = \;\; (p/p^o)^{\tfrac{\gamma-1}{\gamma}} \;\; = \;\; (\rho/\rho^o)^{\gamma - 1} \;\; = \;\; 1 - \tfrac{\gamma-1}{\gamma+1}(u/a^*)^2 \;\; , \tag{7-2}$$

$$M^2 \;\; = \;\; \tfrac{2}{\gamma+1}(u/a^*)^2/[1 - \tfrac{\gamma-1}{\gamma+1}(u/a^*)^2] \;\; , \tag{7-3}$$

$$q/p \;\; = \;\; \tfrac{\gamma}{\gamma+1}(u/a^*)^2/[1 - \tfrac{\gamma-1}{\gamma+1}(u/a^*)^2] \;\; , \tag{7-4}$$

$$q/p^o \;\; = \;\; \tfrac{\gamma}{\gamma+1}(u/a^*)^2[1 - \tfrac{\gamma-1}{\gamma+1}(u/a^*)^2]^{\tfrac{1}{\gamma-1}} \;\; . \tag{7-5}$$

- **Parameter u/a^o**

$$T/T^o \;\; = \;\; (a/a^o)^2 \;\; = \;\; (p/p^o)^{\tfrac{\gamma-1}{\gamma}} \;\; = \;\; (\rho/\rho^o)^{\gamma - 1} \;\; = \;\; 1 - \tfrac{\gamma-1}{2}(u/a^o)^2 \;\; , \tag{7-6}$$

$$M^2 \;\; = \;\; (u/a^o)^2/[1 - \tfrac{\gamma-1}{2}(u/a^o)^2] \;\; , \tag{7-7}$$

$$q/p \;\; = \;\; \tfrac{1}{2}\gamma(u/a^o)^2/[1 - \tfrac{\gamma-1}{2}(u/a^o)^2] \;\; , \tag{7-8}$$

$$q/p^o \;\; = \;\; \tfrac{1}{2}\gamma(u/a^o)^2[1 - \tfrac{\gamma-1}{2}(u/a^o)^2]^{\tfrac{1}{\gamma-1}} \;\; . \tag{7-9}$$

● <u>Parameter u/u_m</u>

$$T/T^o \;=\; (a/a^o)^2 \;=\; (p/p^o)^{\frac{\gamma-1}{\gamma}} \;=\; (\rho/\rho^o)^{\gamma-1} \;=\; 1 - (u/u_m)^2 \;, \qquad (7\text{-}10)$$

$$M^2 \;=\; \frac{2}{\gamma-1}(u/u_m)^2/[1-(u/u_m)^2]\;, \qquad (7\text{-}11)$$

$$q/p \;=\; \frac{\gamma}{\gamma-1}(u/u_m)^2/[1-(u/u_m)^2]\;, \qquad (7\text{-}12)$$

$$q/p^o \;=\; \frac{\gamma}{\gamma-1}(u/u_m)^2[1-(u/u_m)^2]^{\frac{1}{\gamma-1}}\;. \qquad (7\text{-}13)$$

<u>Pressure coefficient in a two-dimensional flow.</u> In a two-dimensional flow let u and v be the components of velocity parallel to, and normal to, respectively, the velocity at a reference station (subscript 1) far upstream which is aligned with the x-axis. Thus, with $u_1 \equiv U,\; v_1 \equiv 0$, the *pressure coefficient* at an arbitrary station (subscript 2) is defined as

$$C_p \;\equiv\; \frac{p_2 - p_1}{\tfrac{1}{2}\rho_1 U^2} \;=\; \frac{(p_{21}-1)p_1}{q_1}\;, \qquad p_{21} \sim \frac{p_2}{p_1} \qquad (7\text{-}14)$$

where $q_1 = \rho_1 U^2/2$ is the free-stream dynamic pressure. If we introduce the free-stream Mach number $M_1 \equiv U/a_1$ for a perfect gas, with $a_1^2 = \gamma p_1/\rho_1$, then (7-15) becomes

$$C_p \;=\; (2/\gamma M_1^2)\,(p_{21}-1)\;. \qquad (7\text{-}15)$$

Equation (7-15) is the standard form for the pressure coefficient in a compressible flow.

Frequently, it is important, for analytical purposes, to express the pressure coefficient in terms of the velocities at station 1 and 2, on the premise that the flow is isoenergetic and isentropic between. For isentropic flow

$$p_{21} \;=\; T_{21}^{\frac{\gamma}{\gamma-1}} \;=\; a_{21}^{\frac{2\gamma}{\gamma-1}}\;. \qquad (7\text{-}16)$$

Now, from the energy equation (6-4) for isoenergetic flow written between 1 and 2, we have

$$a_1^2/(\gamma-1) + \tfrac{1}{2}U^2 \;=\; a_2^2/(\gamma-1) + \tfrac{1}{2}(u_2^2 + v_2^2)\;,$$

or

$$a_{21}^2 = 1 + \tfrac{1}{2}(\gamma-1)(U^2/a_1^2)[1-(u_2^2+v_2^2)/U^2] \; ,$$

$$= 1 + \tfrac{1}{2}(\gamma-1)M_1^2[1-(u_2^2+v_2^2)/U^2] \; . \tag{7-17}$$

Combining (7-15) through (7-17) we have the desired result:

$$C_p = (2/\gamma M_1^2)\left[\left[1 + \tfrac{1}{2}(\gamma-1)M_1^2[1-(u_2^2+v_2^2)/U^2]\right]^{\frac{\gamma}{\gamma-1}} - 1\right] \; . \tag{7-18}$$

Equation (7-18) can readily be extended to other coordinate systems, or to three-dimensional flow.

8. Relations for Flow through a Normal Shock

For a thermally and calorically perfect gas it is possible to obtain explicit relations for the ratio of state variables across a normal shock from (4-4a) when combined with the perfect-gas relation. Thus, the thermal and caloric equations of state, (2-1) and (2-6b) produce the ratios

$$p_{21} = \rho_{21}T_{21} \; , \quad h_{21} = T_{21} \; . \tag{8-1}$$

With (8-1) and (2-6b), $p_1/2\rho_1 h_1 = (\gamma-1)/2\gamma,$ thus (4-4a) reduces to

$$\rho_{21} = \frac{1+a p_{21}}{a+p_{21}} \; , \quad a \equiv \frac{\gamma+1}{\gamma-1} \; . \tag{8-2}$$

Equation (8-2) is the Rankine-Hugoniot equation for a perfect gas.

It is particularly convenient to express the state ratios as a function of M_1, the Mach number of the flow immediately ahead of (i.e. upstream of) the shock. Thus (3-19), (3-16), (8-1), (8-2) and (5-3) combine to produce

$$p_{21} = \frac{2\gamma}{\gamma+1}M_1^2 - \frac{\gamma-1}{\gamma+1} \; , \tag{8-3a}$$

$$\rho_{21} = \frac{(\gamma-1)M_1^2}{(\gamma-1)M_1^2+2} = u_{12} \; , \tag{8-3b}$$

and

$$a_{21}^2 \;=\; T_{21} \;=\; \frac{[2\gamma M_1^2-(\gamma-1)][(\gamma-1)M_1^2+2]}{(\gamma+1)^2 M_1^2} \; . \tag{8-3c}$$

By interchanging the subscripts in (8-3a) we can obtain a similar expression in p_{12} and M_2. These can be multiplied together to give a relation for M_2, the flow Mach-number behind the shock, in terms of M_1, i.e.

$$M_2^2 \;=\; \frac{(\gamma-1)M_1^2+2}{2\gamma M_1^2-(\gamma-1)} \; . \tag{8-4}$$

If we write the appropriate form of the energy equation (6-4) on either side of the shock, and eliminate M_1 and M_2 by means of (8-4), an especially simple relation is obtained:

$$u_1 u_2 \;=\; a^{*2} \, , \tag{8-5}$$

which is known as *Prandtl's relation for a normal shock*.

Another sometimes useful result relates the speed of propagation c, of the shock to the velocity jump Δu, by eliminating ρ, $\Delta\rho$ and Δp from (4-3). Thus

$$\frac{c}{\Delta u} \;=\; \frac{\gamma+1}{4}\left[1 + [1 + 4a_1/(\gamma+1)\Delta u]^2\right]^{\frac{1}{2}} . \tag{8-6}$$

We can also obtain the ratio of the total pressures across the shock:

$$\frac{p_2^o}{p_1^o} \;=\; \frac{p_2^o}{p_2}\frac{p_2}{p_1}\frac{p_1}{p_1^o} \;=\; \left[\frac{(\gamma+1)M_1^2}{(\gamma-1)M_1^2+2}\right]^{\frac{\gamma}{\gamma-1}} \left[\frac{\gamma+1}{2M_1^2-(\gamma-1)}\right]^{\frac{1}{\gamma-1}} . \tag{8-7}$$

The first and the third factors above are the ratios of static-to-total pressure for isoenergetic and isentropic flow, whereas the middle ratio p_{21} is the Rankine-Hugoniot relation in the form of (8-3a), which is an isoenergetic relation. Equation (8-7) is a variation of the *Rayleigh pitot formula*, the original form of which gives the ratio p_1/p_2^o .

Entropy variation across shock. For a thermally and calorically perfect gas the entropy change between any two states, denoted *1, 2*, can be obtained by integrating (2-7b):

$$\Delta s/R \;=\; \ell n\, T_{21}^{\frac{\gamma}{\gamma-1}} - \ell n\, p_{21} \; . \tag{8-8}$$

For a flow, however, it is more convenient to write (8-8) in terms of the corresponding total conditions. Thus, since there is no entropy change involved in the hypothetical process of decelerating a particle to rest isoenergetically and isentropically, i.e. from (p_1, T_1) to $(p_1^{\,o}, T_1^{\,o})$, and from (p_2, T_2) to $(p_2^{\,o}, T_2^{\,o})$, we can write the same entropy change in (8-8) in terms of total conditions. That is,

$$\Delta s/R \quad = \quad \ell n (T_{21}^o)^{\frac{\gamma}{\gamma-1}} \; - \; \ell n \, p_{21}^o \quad . \tag{8-9}$$

Equation (8-9) is valid, in general, for any flow of a perfect gas; i.e. viscous or inviscid, flows with shaft work, flows with heat addition. For isoenergetic flows of a perfect gas — in which T^o is a constant — (8-9) becomes

$$\Delta s/R \quad = \quad \ell n \, (p_1^o/p_2^o) \quad = \quad \ell n \, p_{12}^o \quad . \tag{8-10}$$

Equation (8-10), in conjunction with (8-7), provides the most convenient form for numerical computations of the entropy change across a shock. However, to examine the general variation of entropy for a shock of arbitrary strength, the preceeding combination is not convenient, since the first bracket in (8-7) increases with increasing M_1, whereas the second decreases. Thus, we use (2-7c) instead which, after integrating, becomes, with $v_{21} = 1/\rho_{21}$,

$$\Delta s/R \quad = \quad \frac{1}{\gamma - 1} \, [\ell n \, p_{21} - \gamma \ell n \, \rho_{21}] \quad . \tag{8-11}$$

We then employ the series expansion

$$\ell n \, \rho_{21} \quad = \quad 2 \left[\left[\frac{\rho_{21} - 1}{\rho_{21} + 1} \right] + \frac{1}{3} \left[\frac{\rho_{21} - 1}{\rho_{21} + 1} \right]^3 + \frac{1}{5} \left[\frac{\rho_{21} - 1}{\rho_{21} + 1} \right]^5 + ... \right] \quad , \tag{8-12}$$

which is uniformly convergent for $\rho_{21} \geq 1$. A similar equation can be written for $\ell n \, p_{21}$. With the Rankine-Hugoniot equation,

$$\frac{\rho_{21} - 1}{\rho_{21} + 1} \quad = \quad \frac{1}{\gamma} \frac{p_{21} - 1}{p_{21} + 1} \quad . \tag{8-13}$$

If (8-13) is combined with (8-12), and if both series expansions are introduced into (8-11), the following form is obtained:

$$\Delta s/R \quad = \quad \frac{2}{\gamma - 1} \left[\frac{1}{3}(1 - 1/\gamma^2) \left[\frac{p_{21} - 1}{p_{21} + 1} \right]^3 + \frac{1}{5} (1 - 1/\gamma^4) \left[\frac{p_{21} - 1}{p_{21} + 1} \right]^5 + ... \right] \quad . \tag{8-14}$$

Equation (8-14) is an odd function in the *shock strength* $p_{21} - 1$. Thus we see that

$$\Delta s/R \ =\ \gtreqless\ 0\ ,\quad if\ p_{21} \gtreqless 1\ . \tag{8-15}$$

It is remarkable that the first-order term involving $p_{21} - 1$ does not appear in (8-14) and thus, to the lowest order (i.e. in the largest term), the entropy change is proportional to the cube of the shock strength. If p_{21} is eliminated from (8-14) by use of (8-3a), then the lowest-order term is

$$\frac{\Delta s}{R} \ =\ \frac{2\gamma}{3(\gamma+1)^2} \ (M_1^2 - 1)^3 \ -\ \frac{2\gamma^2}{(\gamma+1)^3}\ (M_1^2 - 1)^4 \ +\ ...\ . \tag{8-16}$$

Now (8-11) through (8-16) are written for a particle between the two states *1* and *2*. If the particle is viewed as an isolated system undergoing an adiabatic transformation between *1* and *2*, then the second law of thermodynamics requires that there shall be no entropy decrease. Thus, the possibility of an expansion shock (i.e. $p_{21} < 1$) is ruled out. This result is restricted to a perfect gas.

With the rearrangement of (8-3a) and (8-5) we obtain

$$p_{21} \ =\ M_1^2 \ +\ \frac{\gamma-1}{\gamma+1}\ (M_1^2 - 1)\ , \tag{8-17a}$$

$$M_2^2 \ =\ 1 \ -\ \frac{\gamma+1}{\gamma-1}\ \frac{(M_1^2 - 1)}{2\gamma M_1^2/(\gamma-1)-1}\ . \tag{8-17b}$$

Thus, if

$$M_1 \geq 1,\qquad p_{21} \geq 1,\qquad M_2 \leq 1,\qquad \Delta s/R\ \geq\ 0\ . \tag{8-18}$$

Consequently, under the stated restrictions, for a standing normal-shock the upstream flow must always be supersonic, and the downstream subsonic.

A number of the shock relations are tabulated as a function of M_1 for $\gamma = 5/3,\ 7/5,$ and $4/3$ in Tables A.1 to A.3.

9. Flow in a De Laval, or Convergent-Divergent, Nozzle.

For the isoenergetic flow of a perfect gas in a horizontal duct of fixed geometry (but of non-constant area), the equations can be integrated if viscous effects are treated as negligible. For well designed, short nozzles (the nozzle contour cannot be designed on the basis of one-dimensional theory) this restriction enables the derivation of elementary, but highly useful results which, in some cases, can later be modified on the basis of boundary layer theory to account for viscous effects.

The <u>Hugoniot</u> velocity-area relation. Consider a flow from a reservoir into a duct which is initially convergent, Figure 9-1. If the flow is adiabatic and inviscid it is then also isentropic. The dynamical equation (3-13) can be put in the form

$$du/u \; = \; -dp/\rho u^2 \; = \; -(dp/\rho u^2)(dp/d\rho) \; = \; -(1/M^2)(d\rho/\rho) \; , \qquad (9\text{-}1)$$

where $dp/d\rho = (dp/d\rho)_s = a^2$. Taking the logarithmic differential of (3-14), and then combining with (9-1) by eliminating $d\rho/\rho$ we have

$$\frac{du}{u} \; = \; -\frac{dA/A}{1-M^2} \; . \qquad (9\text{-}2)$$

Thus, in order to accelerate $(du > 0)$ a perfect gas in a duct we must have

$$dA/A \; \lessgtr \; 0, \quad for\, M \; \lessgtr \; 1 \; . \qquad (9\text{-}3)$$

Starting from a reservoir where $u \simeq 0$, the only way to attain a supersonic flow is through a convergent-divergent duct with a continuously decreasing pressure. At the duct minimum-area, or throat, $dA = 0$, and the flow Mach number is unity. Such a geometry is named after its inventor de Laval.

<u>Mach</u> <u>number</u> <u>relations.</u> We can use the Mach number relations already derived in conjunction with the continuity equation to obtain the working relations for flow in a duct. The *mass flux* is obtained from (3-16):

$$\dot{m}/A \; = \; \rho u \; ,$$

$$= \; (\rho/\rho^o)\rho^o\,(u/a)\,(a/a^o)a^o \; ,$$

$$= \; \left[M[1+\tfrac{1}{2}(\gamma-1)M^2]^{-\frac{\gamma+1}{2(\gamma-1)}}\right]\rho^o a^o \; , \qquad (9\text{-}4a)$$

$$= \; \left[M[1+\tfrac{1}{2}(\gamma-1)M^2]^{-\frac{\gamma+1}{2(\gamma-1)}}\right](\gamma/R)^{\frac{1}{2}}\,(p^o/\sqrt{T^o}) \; . \qquad (9\text{-}4b)$$

In (9-4) ρ^o, a^o, p^o, and T^o are not only the reservoir values, but they are also the total conditions everywhere in the flow in the absence of shocks.

It is straightforward to show that $\dot{m}/A$ has an extremum at the throat where $M = 1$, and that it is a maximum. The area at which this occurs is denoted as A^*. Putting

$M = 1$ in (9-4b) thus results in

$$\dot{m}/A^* = [2/(\gamma+1)]^{\frac{\gamma+1}{2(\gamma-1)}} (\gamma/R)^{\frac{1}{2}} (p^0/\sqrt{T^0}) = C(p^0/\sqrt{T^0}), \quad (9\text{-}5a)$$

where

$$C \equiv [2/(\gamma+1)]^{\frac{\gamma+1}{2(\gamma-1)}} (\gamma/R)^{\frac{1}{2}}. \quad (9\text{-}5b)$$

For air, with $\gamma = 7/5$, we have for the engineering system of units (also variously referred to as British, or U.S. customary units), and for SI:

	Units			
	$\dot{m}/A^*$	p^0	T^0	C
Engineering System	$\dfrac{slug}{sec\ ft^2}$	lb/ft^2	$^\circ R$	0.01653
SI	$kg/(s \cdot m^2)$	N/m^2	K	0.04041

Dividing (9-5a) by (9.4b) yields the standard *area-Mach number relation* for nozzle flow.

$$A/A^* = M^{-1} \left[\frac{2}{\gamma+1} [1 + \tfrac{1}{2}(\gamma-1)M^2] \right]^{\frac{\gamma+1}{2(\gamma-1)}}. \quad (9\text{-}6)$$

Tabulations of A/A^* for $\gamma = 5/3$, $7/5$, and $4/3$ are included in Tables A.1 to A.3. If the Mach number is eliminated from (9-6) by use of (6-7) we can obtain the area ratio as a function of the static-to-total pressure-ratio:

$$A/A^* = \frac{[(\gamma-1)/2]^{\frac{1}{2}} [2/(\gamma+1)]^{\frac{\gamma+1}{2(\gamma-1)}}}{(p/p^0)^{1/\gamma} \left[1 - (p/p^0)^{\frac{\gamma-1}{\gamma}} \right]^{\frac{1}{2}}}. \quad (9\text{-}7)$$

<u>Critical conditions in a de Laval nozzle</u>. When the Mach number is unity (at the throat) the flow is said to be choked. Conditions at the throat are said to be critical because any further changes of the pressure downstream of the throat are not communicated to the reservoir, and the mass-flow-rate remains unchanged. The throat conditions are

designated as p^*, T^*, ρ^*, a^* where

$$T^*/T^o \;=\; (a^*/a^o)^2 \;=\; (p^*/p^o)^{\frac{\gamma-1}{\gamma}} \;=\; (\rho^*/\rho^o)^{\gamma-1} \;=\; \frac{2}{\gamma+1} \;. \tag{9-8}$$

The pressure gradient and the Mach-number gradient at the critical condition can be shown to be given by

$$(dM/dx)^* \;=\; -[(\gamma+1)/2\gamma p^*](dp/dx)^* \;, \tag{9-9a}$$

$$(dp/dx)^* \;=\; \pm[(\gamma p^*)/(\gamma+1)^{\frac{1}{2}}]\,[(d^2A/dx^2)^*/A^*]^{\frac{1}{2}} \;, \tag{9-9b}$$

the choice of the plus, or minus sign depending, of course, on whether or not the flow is subsonic or supersonic, respectively, downstream of the throat.

$\underline{\text{Flow}}$ $\underline{\text{downstream}}$ $\underline{\text{of a}}$ $\underline{\text{choked}}$ $\underline{\text{throat.}}$ Plots of (9-6) and (9-7), given in Figure 9-1, show that A/A^* is a double-valued function, whereas p/p^o is singled-valued. Upstream of the throat the flow is subsonic with $p > p^*$. Downstream of the throat the flow may be either subsonic or supersonic depending on the pressure in the downstream dump-tank. For a given exit area-ratio there are two discrete pressures for isentropic choked flow, one for subsonic, and one for supersonic flow in the divergent portion of the nozzle.

For dump-tank pressures less than the value corresponding to the subsonic choked-flow solution shock waves are generated (by unsteady processes) which may appear in the nozzle or in the jet downstream of the nozzle. For a discussion of the various nozzle solutions see Liepmann and Roshko (1957, p. 127).

For the case of a standing normal-shock in the divergent portion of the duct we know that M_2 behind the shock is subsonic; then, since the area is increasing downstream we know from (9-2) that the flow is decelerating. Furthermore, for isoenergetic flow we can show that

$$\frac{dM}{M} \;=\; -\frac{1+\frac{1}{2}(\gamma-1)M^2}{1-M^2}\,\frac{dA}{A} \;. \tag{9-10}$$

Thus, the Mach number must decrease downstream of a shock which stands in a divergent duct.

We can also apply (9-5a) to either side of the shock. Since $A_1 = A_2$ and $T^o = $ *constant*, we find that

$$A_2^*/A_1^* \;=\; p_1^o/p_2^o \;. \tag{9-11}$$

Consequently, the appropriate reference throat-area in the region following the shock is A_2^* as given by (9-11). If a second convergent-divergent nozzle is employed downstream of the shock to generate a second throat, (9-11) indicates that the area of the second throat must be greater than that of the first. Furthermore, by (8-10) and (9-11) we know that the flow downstream of the shock has a higher level of entropy than upstream, and that the total pressure downstream is less than that upstream.

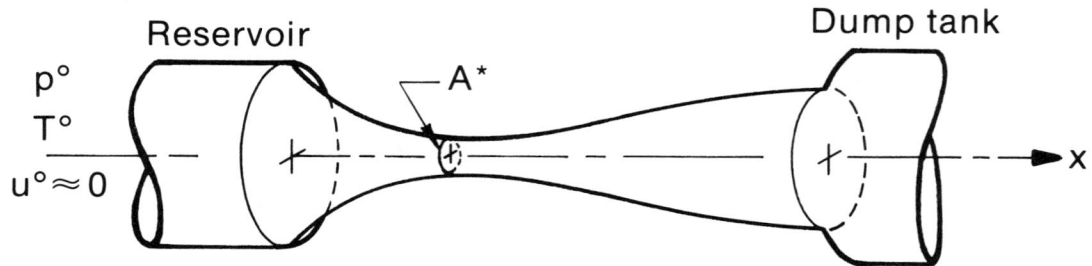

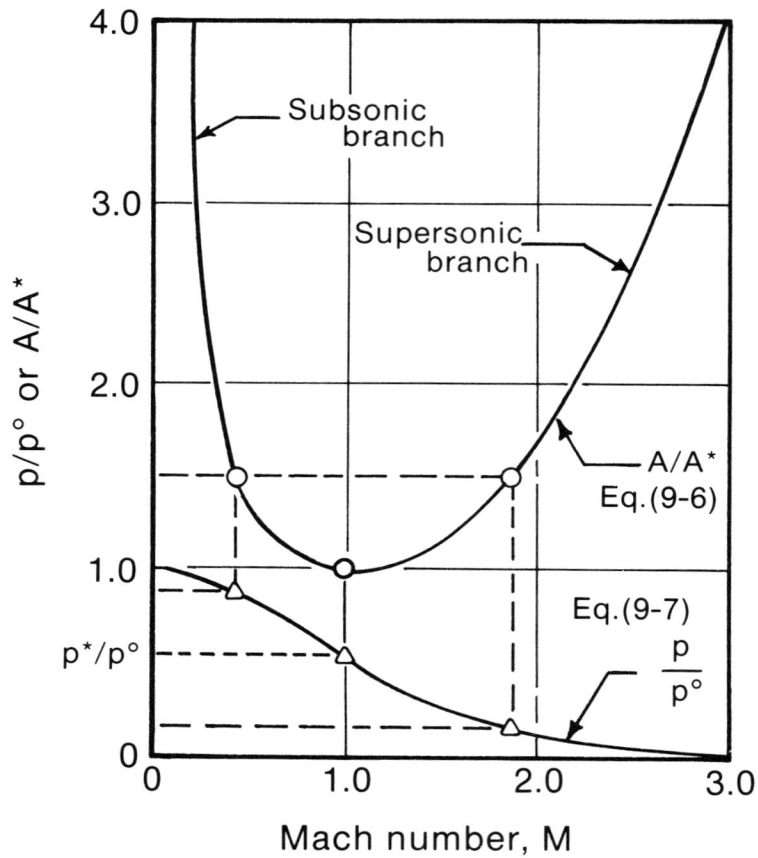

Figure 9-1. Isentropic, isoenergetic flow through a convergent-divergent nozzle for $\gamma = 7/5$.

10. <u>Flows</u> <u>in</u> <u>Ducts</u> <u>with</u> <u>Friction</u> — <u>Fanno</u> <u>Flow</u>

There are few general results for ducts of varying cross-sectional area in which viscosity plays an important role. The reader may consult Crocco (1958) for a comprehensive one-dimensional flow treatment with important emphasis on frictional effects.

For steady flow of a perfect gas in constant-area ducts there are two flows — so-called *Fanno flow* and *isothermal flow* — which yield useful results. A Fanno flow is an isoenergetic flow in which the frictional effects are handled by the same correlations used in constant-density pipe flows. Requiring that $A = constant$, the dynamical equation (3-5) can be reworked into the following form:

$$du/u = -(1/\gamma M^2)dp/p - 2c_f dx/D , \qquad (10\text{-}1a)$$

where the *skin-friction coefficient* c_f is defined as

$$c_f \equiv \tau_w/(\tfrac{1}{2}\rho u^2) . \qquad (10\text{-}1b)$$

The skin-friction coefficient is related to the friction factor f of pipe flow by $f = 4c_f$. In modifying the dynamical equation we have put, for a circular cross-section duct of diameter D, $P/A = 4/D$. For non-circular ducts D should be interpreted as the hydraulic diameter, $D_h = 4A/P$.

Then, combining (2-1) and (5-3), the thermal equation of state can be written in logarithmic differential form:

$$dp/p = d\rho/\rho + 2da/a . \qquad (10\text{-}2)$$

With $u = Ma$

$$du/u = dM/M + da/a . \qquad (10\text{-}3)$$

Differentiating the energy equation (6-4), and introducing the Mach number, we have

$$\frac{2}{\gamma-1} \frac{da}{a} + M^2 \frac{du}{u} = 0 ; \qquad (10\text{-}4)$$

and from the definition of the total pressure (6-7),

$$dp^o/p^o = dp/p + \frac{\gamma M^2(dM/M)}{1+\tfrac{1}{2}(\gamma-1)M^2} . \qquad (10\text{-}5)$$

Following the scheme developed by Shapiro (1953, p. 162) we see that there are 6 dimensionless ratios du/u, dp/p, $c_f dx/D$, dM/M, da/a and dp^o/p^o. These six ratios are related by five equations where the multiplying coefficients (referred to as *influence coefficients*) of the ratios are functions exclusively of the local Mach-number and the parameter γ. Thus, we can eliminate any four to get a relation between any two differential ratios. For example,

$$du/u \;=\; \frac{dM/M}{1+\tfrac{1}{2}(\gamma-1)M^2} \; , \tag{10-6}$$

$$dp/p \;=\; -\,\frac{[1+(\gamma-1)M^2](dM/M)}{1+\tfrac{1}{2}(\gamma-1)M^2} \; . \tag{10-7}$$

We deduce from these two equations that in an accelerating/decelerating flow the Mach number increases/decreases and that the pressure decreases/increases.

On the other hand we can solve for

$$du/u \;=\; \frac{2\gamma M^2}{1-M^2}\, c_f\,\frac{dx}{D} \; ; \tag{10-8}$$

thus, if the flow is initially subsonic/supersonic the flow accelerates/decelerates. Consequently, every Fanno flow tends asymptotically to a Mach number of unity.

Furthermore, since

$$dp^o/p^o \;=\; -\,2\gamma M^2 c_f\, dx/D \; , \tag{10-9}$$

the total pressure always decreases in the direction of the flow. Then, since the flow is isoenergetic, it follows from (8-10) that the entropy increases in the flow direction.

The integrated Mach-number distribution in a duct. If (10-6) and (10-7) are substituted into (10-1a) the following expression is obtained:

$$c_f\, dx/D \;=\; \frac{(1-M^2)dM^2}{4\gamma M^4[1+\tfrac{1}{2}(\gamma-1)M^2]} \; . \tag{10-10}$$

The right-hand side, which depends purely on the Mach number, with γ as parameter, allows definition of a Mach-number function $G(M;\ \gamma)$ involving an integral with limits from a station of Mach number M to the (hypothetical) downstream station where the flow (theoretically) becomes sonic. Thus, we put

$$
G(M; \ \gamma) \ \equiv \ (1/4\gamma) \int_{M^2}^{1} \frac{(1-M^2)dM^2}{M^4[1+\tfrac{1}{2}(\gamma-1)M^2]} \ ,
$$

$$
= \ (1/4\gamma) \left[\frac{1-M^2}{M^2} \ + \ \frac{\gamma+1}{2} \ell n \ \frac{(\gamma+1)M^2}{2[1+\tfrac{1}{2}(\gamma-1)M^2)]} \right] \tag{10-11}
$$

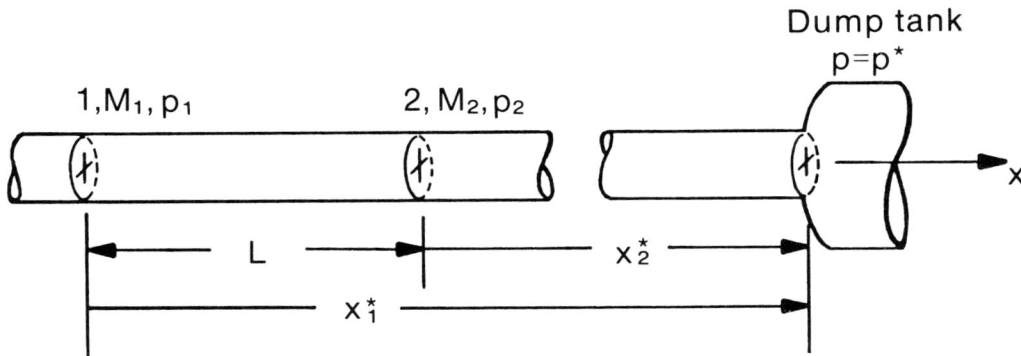

Figure 10-1. Geometry of Fanno flow.

The left-hand side of (10-10) depends on c_f which is truly a variable, and which is a function of the unknown shear-stress, velocity, and density. This difficulty is surmounted by employing a mean value of the skin-friction coefficient

$$
\overline{c}_f \ \equiv \ (1/x^*) \int_{0}^{x^*} c_f dx \ , \tag{10-12}
$$

where x^* is the length of duct (see Figure 10-1) required for the flow at the upstream station, at Mach number M, to accelerate/decelerate to unity. Thus, the solution of (10-10) is given by

$$
\overline{c}_f x^*/D \ = \ G(M; \ \gamma) \ , \tag{10-13}
$$

where $G(M; \ \gamma)$ is tabulated in Table B for $\gamma = 5/3, \ 7/5, \ 4/3$.

The length of duct L required to go from a station where the Mach number is M_1 to a station at M_2 is the difference in x^* obtained by two successive applications of (10-13), i.e.

$$\overline{c}_f L/D \;=\; \overline{c}_f(x_1^* - x_2^*)/D \;=\; G(M_1;\,\gamma) \;-\; G(M_2;\,\gamma)\;. \tag{10-14}$$

For smooth-wall ducts the practice is to use *Prandtl's logarithmic law of friction* for the skin-friction coefficient, according to which

$$1/\sqrt{c_f} \;=\; 4.0\,\log_{10}(Re\,\sqrt{c_f}) - 0.396\;, \tag{10-15}$$

where $Re \equiv \rho u D/\mu$ is the local Reynolds number. Experience shows that any reasonable value of the mean Reynolds number of the flow (for example, the average of the values at stations *1* and *2*) gives adequate accuracy. For rough walls a Moody chart can be employed to determine c_f. A tabulation of values of c_f vs Re from Prandtl's relation is given in Table D.

Integrated expressions for the flow functions. We select as reference the station where the Mach number approaches unity. The local conditions are designated as u^*, a^*, p^*, ρ^*, T^*, etc.; but it should be noted that these values are not all identical to the critical-flow conditions in a de Laval nozzle, which employ the same symbols. It can be verified that the following expressions obtain:

$$T/T^* \;=\; \frac{(\gamma+1)/2}{1+\tfrac{1}{2}(\gamma-1)M^2}\;, \tag{10-16}$$

and

$$T/T^* \;=\; (a/a^*)^2 \;=\; (Mp/p^*)^2 \;=\; (\rho^*/\rho M)^2 \;=\; (u/Mu^*)^2 \;=\; (p^{o^*}/Mp^o)^{\frac{2(\gamma-1)}{\gamma+1}}$$

$$=\; \left[\frac{1}{M}\exp\left(\frac{s-s^*}{R}\right)\right]^{\frac{2(\gamma-1)}{\gamma+1}} \;=\; \left[\frac{(\gamma+1)M^2 F}{(1+\gamma M^2)F^*}\right]^2\;. \tag{10-17}$$

In (10-17) the function F is the Mach-number function which was defined in (6-16b) for the momentum equation.

To obtain the ratio of a given variable between any two stations, (10-16) or (10-17) can be employed twice. For example,

$$T_{21} \;=\; \frac{T_2/T^*}{T_1/T^*} \;=\; \frac{1+\tfrac{1}{2}(\gamma-1)M_1^2}{1+\tfrac{1}{2}(\gamma-1)M_2^2}\;. \tag{10-18}$$

11. Isothermal Flow

Isothermal flow is one in which the static temperature T is constant. Under the restrictions of a constant-area duct, steady flow, and a perfect gas, the analysis employs a procedure analogous to that in Fanno flow. Thus,

$$\frac{du}{u} = \frac{dM}{M} = -\frac{dp}{p} = -\frac{d\rho}{\rho} = \frac{2\gamma M^2}{1-\gamma M^2} \, c_f \frac{dx}{D} \quad . \tag{11-1}$$

Equation (11-1) delineates a subsonic, limiting Mach number $M_f \equiv 1/\sqrt{\gamma}$ towards which isothermal flow tends asymptotically. From the definition of T^o, equation (6-1) yields

$$\frac{dT^o}{T^o} = \frac{(\gamma-1)M^2}{1+\frac{1}{2}(\gamma-1)M^2} \, \frac{dM}{M} \quad . \tag{11-2}$$

Thus, T^o increases/decreases for $M \gtrless M_f$. Under the restrictions cited the energy equation (3-9) reduces to

$$\frac{Dh^o}{Dt} \rightarrow c_p u \frac{dT^o}{dx} = \dot{q} \quad , \tag{11-3}$$

where $\dot{q}$ is the rate of heat-addition per-unit-mass due to conduction from the pipe wall. If (11-2) and (11-3) are combined with the definition of Mach number, then

$$\dot{q} = \frac{2\gamma a^3 M^5}{1-\gamma M^2} \, \frac{c_f}{D} \quad . \tag{11-4}$$

Thus, heat is transferred to/from the gas if $M \gtrless M_f$. However, even for low subsonic values of the Mach number (say $M = 0.1$), the heat-transfer rate required to maintain isothermal flow starts to grow so large that the surroundings cannot provide the energy specified by (11-4), except under extraordinary circumstances. Thus, in practical cases isothermal flow is restricted to very small Mach numbers.

The differential relation for the total pressure is

$$\frac{dp^o}{p^o} = -\frac{2\gamma M^2[1-\frac{1}{2}(\gamma+1)M^2]}{(1-\gamma M^2)[1+\frac{1}{2}(\gamma-1)M^2]} \, c_f \frac{dx}{D} \quad . \tag{11-5}$$

We see that the total pressure always decreases in the flow direction except for the narrow range of Mach number $1/\sqrt{\gamma} < M < [2/(\gamma+1)]^{\frac{1}{2}}$, where it increases.

Integration of the appropriate combination in (11-1) enables the calculation of x_f, which is the theoretical duct length required to attain M_f at the dump-tank inlet from a specified upstream Mach-number M.

Defining

$$H(M; \gamma) = (1/4) [(1-\gamma M^2)/\gamma M^2 + \ln\gamma M^2] , \qquad (11\text{-}6)$$

then

$$c_f x_f/D = H(M; \gamma) , \qquad (11\text{-}7)$$

where $H(M; \gamma)$ is tabulated in Table C for $\gamma = 5/3, 7/5, 4.3$. For isothermal flow of a perfect gas $c_f = \overline{c}_f = constant$, since $\rho u, D, \mu$, and hence Re, are all constant.

<u>Integrated</u> <u>expressions</u> <u>for</u> <u>the</u> <u>flow</u> variables. From the preceding relations, using f to denote values at the station where $M_f = 1/\sqrt{\gamma}$, then

$$M/M_f = u/u_f = p_f/p = \rho_f/\rho , \qquad (11\text{-}8)$$

$$p^o/p_f^o = \frac{1}{M\sqrt{\gamma}} \left[\frac{2\gamma}{3\gamma-1} [1+\tfrac{1}{2}(\gamma-1)M^2] \right]^{\frac{\gamma}{\gamma-1}} , \qquad (11\text{-}9)$$

$$T^o/T_f^o = \frac{2\gamma}{3\gamma-1} [1+\tfrac{1}{2}(\gamma-1)M^2] . \qquad (11\text{-}10)$$

The entropy relation, for a non-isoenergetic flow, when combined with (11-9) and (11-10) produces

$$(s-s_f)/R = \ln \left[\frac{(T^o/T_f^o)^{\frac{\gamma}{\gamma-1}}}{p^o/p_f^o} \right] = \ln(M\sqrt{\gamma}) . \qquad (11\text{-}11)$$

<u>Approximate</u> <u>forms</u> <u>for</u> <u>low-Mach-number</u> <u>flow.</u> A good example of an actual isothermal flow is that found in natural-gas pipelines. To reduce pumping losses the flow speeds are kept small; the corresponding Mach numbers are also small, the order of 0.01. The right-hand side of (11-6) can be expanded as a series in terms of the small quantity $(M-M_1)/M_1$, where M_1 is the initial Mach number at the station $x = 0$, to yield the first-order approximation

$$M/M_1 = 1 + 2\gamma M_1^2 c_f(L/D)(x/L) = u/u_1 , \qquad (11\text{-}12)$$

where the station $x/L = 1$ is chosen to correspond to some small increment to the initial Mach number M_1, for example, such that at $x/L = 1, M = 1.1M_1$.

The corresponding first-order expression for the pressure distribution is obtained by use of (11-8), i.e.,

$$p/p_1 = (M/M_1)^{-1} = [1 + 2\gamma M_1^2 c_f(L/D)(x/L)]^{-1}$$
$$= 1 - 2\gamma M_1^2 c_f(L/D)(x/L) = \rho/\rho_1 \ . \tag{11-13}$$

Corresponding relations for the total pressure and total temperature are readily obtained from (11-9) and (11-10).

12. Relations for Oblique Shock Flow

The simplest procedure to analyze the flow through an oblique shock is to modify the equations for the flow through a normal shock by transforming to an observer, see Figure 12-1, who moves parallel to the shock at velocity $-v$ downward. Thus, in the frame of the second observer, the particle velocities on either side of the shock acquire a component, positive upward, of $v_1 = v_2 \equiv v$.

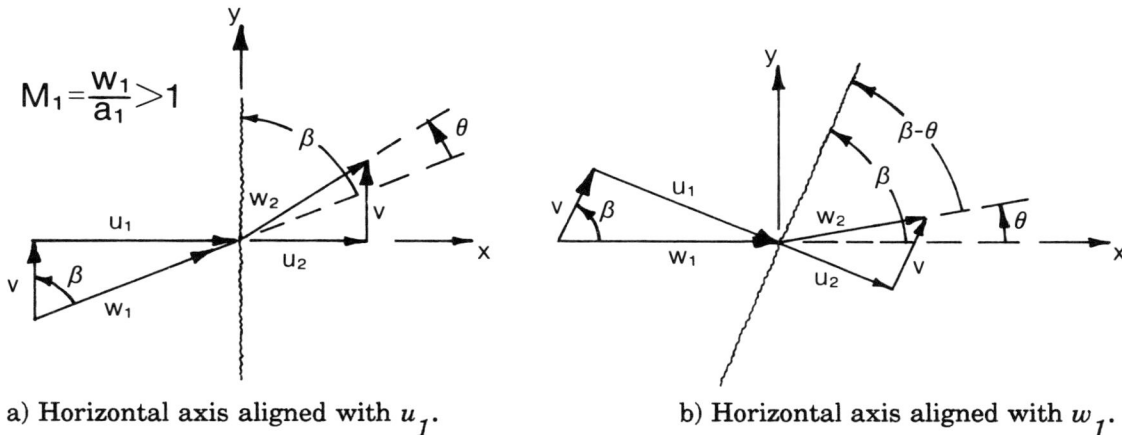

a) Horizontal axis aligned with u_1.

b) Horizontal axis aligned with w_1.

Figure 12-1. Velocity field for oblique shock.

If, then, the coordinate axes are also rotated so that the new x-axis is aligned with the velocity vector ahead of the shock, the path of a particle is seen to deviate from the original direction by the angle θ.

$\rho u = \rho_1 u_1 = \rho_2 u_2$

In this transformation the only quantity appearing in the conservation equations for steady flow [(3-16), (3-19) and (3-12b)], which is thereby modified, is the total enthalpy, which becomes

$(\rho + \rho u^2)_1 = (\rho + \rho u^2)_2$

$$h^o = h_1^o = h_1 + \tfrac{1}{2}(u_1^2 + v^2) = h_2 + \tfrac{1}{2}(u_2^2 + v^2)$$
$$= h_1 + \tfrac{1}{2}w_1^2 = h_2 + \tfrac{1}{2}w_2^2 = h_2^o = constant \ . \tag{12-1}$$

Of course, the new Mach numbers are defined as $M_1 \equiv w_1/a_1$ and $M_2 \equiv w_2/a_2$. Since, in terms of the new shock geometry, $u_1 = w_1 \sin\beta$, $u_2 = w_2 \sin(\beta-\theta)$, and since the state variables p_1, ρ_1, h_1, a_1, p_2, ρ_2, h_2, a_2, etc., remain unaltered in the change of observer, we can utilize the relations of Section 8 directly by the simple artifice of replacing M_1, M_2 of that section by

$$M_1 \rightarrow M_{1n} = M_1 \sin\beta \quad, \quad M_2 \rightarrow M_{2n} = M_2 \sin(\beta-\theta) \quad, \qquad (12\text{-}2)$$

respectively, where M_{1n} and M_{2n} are called the normal Mach-number components. Of course, for a shock to exist, we must still have $M_{1n} > 1$, or $w_1 \sin\beta > a_1$.

A preferred objective of the analysis of an oblique shock would be to obtain an explicit expression for the shock angle β, measured with respect to the flow direction ahead of the shock, as a function of the turning angle θ and the flow variables ahead of the shock. This does not seem to be possible and we must settle for an implicit relation. The shock geometry is introduced by putting

see (7-14), p.14

$$\frac{\rho_2}{\rho_1} = \rho_{21} = u_{12} = \frac{u_1/v}{u_2/v} = \frac{\tan\beta}{\tan(\beta-\theta)} \quad . \qquad (12\text{-}3)$$

This equation can be solved to produce

$$\tan\beta = \pm (\rho_{21} - 1)\cot\theta \left[1 \pm \left[1 - \frac{4\rho_{21}\tan^2\theta}{(\rho_{21}-1)^2} \right]^{\frac{1}{2}} \right] \quad . \qquad (12\text{-}4)$$

Equation (12-4) indicates the possibility of multiple solutions. Furthermore, since it is independent of any equation of state it has sometimes been used to investigate shocks within a fluid of variable specific-heat.

When the radicand in (12-4) is put equal to zero a unique value for θ is thereby determined, namely

$$\tan\theta_{max} = (\rho_{21} - 1)/2 \sqrt{\rho_{21}} \quad . \qquad (12\text{-}5)$$

According to Liepmann and Roshko (1957, p. 391) this corresponds to θ_m, the maximum possible turning-angle through an oblique shock, also called the shock detachment-angle for wedge flow. In fact, a few sample computations for a perfect gas verify that the angle determined by (12-5) is not the detachment angle, and that it is of no physical significance. The same computations disclose that only the positive sign in (12-4) corresponds to a physical flow. A relation to determine β_m, and thereby θ_m, follows as (12-10).

The rest of this section is restricted to perfect gases. An equation relating θ and β is obtained by combining (12-3) and (8-3b). This leads to the following equivalent expressions:

$$\cot\theta = \tan\beta \left[\frac{(\gamma+1)M_1^2}{2(M_1^2 \sin^2\beta - 1)} - 1 \right] \quad, \qquad (12\text{-}6a)$$

or

$$\tan\theta = 2\cot\beta \left[\frac{M_1^2 \sin^2\beta - 1}{M_1^2(\gamma + \cos 2\beta) + 2} \right] . \tag{12-6b}$$

Equations (12-6) can also be written as a cubic in $sin^2\beta$, namely

$$\sin^6\beta + b\sin^4\beta + c\sin^2\beta + d = 0 , \tag{12-7a}$$

where

$$b = -(M_1^2 + 2)/M_1^2 - \gamma \sin^2\theta ,$$

$$c = (2M_1^2 + 1)/M_1^4 + [(\gamma+1)^2 M_1^2 + 4(\gamma-1)]\sin^2\theta/4M_1^2 , \tag{12-7b}$$

$$d = -\cos^2\theta/M_1^4 .$$

The two positive solutions of β correspond to the strong-, and weak-shock solutions discussed in the following paragraphs. The third, which corresponds to a negative turning angle, and to a pressure ratio $p_{21} < 1$, has previously been ruled out by the second law of thermodynamics. Cardan's solution of the cubic [e.g. see Spiegel (1968, p. 32)] would yield explicit expressions for $sin^2\beta = F(\theta; M_1, \gamma)$ but they are not convenient to use and are, therefore, omitted.

To obtain expressions for the ratios of the state variables across the shock, in terms of upstream flow conditions and the shock geometry, taking into account the fact that the flow is isoenergetic, we introduce (12-2) into (8-3), (8-7) and (8-10); then

shock angle

$$p_{21} = \frac{2\gamma}{\gamma+1} M_1^2 \sin^2\beta - \frac{\gamma-1}{\gamma+1} , \tag{12-8a}$$

$$\rho_{21} = \frac{(\gamma+1)M_1^2\sin^2\beta}{(\gamma-1)M_1^2\sin^2\beta + 2} = u_{12} , \tag{12-8b}$$

$$a_{21}^2 = T_{21} = \frac{[2\gamma M_1^2\sin^2\beta - (\gamma-1)][(\gamma-1)M_1^2\sin^2\beta + 2]}{(\gamma+1)^2 M_1^2\sin^2\beta} , \tag{12-8c}$$

and

$$p_{21}^{o} = \left[\frac{(\gamma+1)M_1^2 sin^2 \beta}{(\gamma-1)M_1^2 sin^2 \beta+2} \right]^{\frac{\gamma}{\gamma-1}} \left[\frac{\gamma+1}{2\gamma M_1^2 sin^2 \beta-(\gamma-1)} \right]^{\frac{1}{\gamma-1}} = \rho_{21}^{o} = e^{-\Delta s/R} \quad .$$

(12-8d)

Combining (7-14) and (12-8a), we have for the pressure coefficient,

$$C_p = 4(sin^2 \beta - 1/M_1^2)/(\gamma+1) \quad .$$

(12-8e)

An expression for M_2 can be obtained by a procedure similar to that followed in deriving (8-4); thus

$$M_2^2 = \frac{(\gamma+1)^2 M_1^4 sin^2 \beta - 4(M_1^2 sin^2 \beta - 1)(\gamma M_1^2 sin^2 \beta + 1)}{[2\gamma M_1^2 sin^2 \beta - (\gamma-1)][(\gamma+1)M_1^2 sin^2 \beta + 2]} \quad .$$

(12-8f)

The flow speeds on either side of an oblique shock are given by $w_1^2 = u_1^2 + v^2$, and $w_2^2 = u_2^2 + v^2$. The following equation gives the flow-speed ratio:

$$w_{21}^2 = 1 - \frac{4(M_1^2 sin^2 \beta - 1)(\gamma M_1^2 sin^2 \beta + 1)}{(\gamma+1)^2 M_1^4 sin^2 \beta} \quad .$$

(12-8g)

Due to the fact that the total enthalpy in the oblique-shock model has been modified in transforming from that of the associated normal-shock, Prandtl's relation for an oblique shock takes the form

$$u_1 u_2 = a^{*2} - \frac{\gamma-1}{\gamma+1} v^2 \quad .$$

(12-8h)

Equation (4-2) for the propagation speed of a normal shock remains valid. It is emphasized that an oblique shock propagates normal to itself so that, as for a normal shock, $u_1 = c$. Equation (4-3) for the jump in normal velocity across a shock in a perfect gas also applies to the oblique shock.

Mapping the solutions of (12-6). We seek the values of β for which $\theta = 0$. Setting the two factors on the right, in turn, equal to zero leads, first, to $\beta = \pi/2$ which is the normal-shock solution. Putting the second factor in (12-6b) equal to zero yields the second solution, namely, the case for which $sin\beta = 1/M_1$. The negative root is ignored since it corresponds to an identical flow which is a reflection about the x-axis.

The value of β obtained in the second limiting case is of sufficient importance that it is assigned a special symbol μ, and is called the *Mach angle*. That is

$$\mu_1 \equiv \lim_{\substack{\theta \to 0 \\ p_{21} \to 1}} \beta = \sin^{-1}(1/M_1), \qquad \mu_1 \leq \beta \leq \pi/2 \ . \tag{12-9}$$

The additional restriction in (12-9), that $p_{21} \to 1$, differentiates the Mach-angle solution from the normal-shock solution for which $p_{21} > 1$. The Mach angle represents the limiting case of a vanishingly weak shock. The corresponding lines $dy/dx = \pm tan\mu_1$, extended downstream, form a Mach wedge which bounds the region of flow affected by an infinitesimal disturbance located at the (upstream) tip of the wedge.

Since β is a double-valued function of θ, the plot of θ versus β for fixed values of the parameters M_1 and γ must exhibit a maximum in the interval $\mu_1 < \beta \leq \pi/2$. Denoting the maximum value as θ_m and the corresponding shock angle as β_m, then

$$\sin^2\beta_m = \frac{1}{4\gamma M_1^2}\left[(\gamma+1)M_1^2 - 4 + \left[(\gamma+1)[(\gamma+1)M_1^4 + 8(\gamma-1)M_1^2 + 16]\right]^{\frac{1}{2}}\right] \ . \tag{12-10}$$

Figure 12-2 is a plot of the oblique-shock solutions for $\gamma = 7/5$, with M_1 as parameter. The solutions for $\mu_1 \leq \beta < \beta_m$ and for $\beta_m < \beta \leq \pi/2$ are classified as weak-, and strong-shock solutions, respectively, since, for a specified $\theta < \theta_m$, the corresponding $p_{21weak} < p_{21strong}$.

The flow downstream of a weak oblique-shock is supersonic except for a very limited region designated as $\theta^* < \theta < \theta_m$. The location of the sonic point is obtained by setting $M_2 = 1$ in equation (12-8f), and solving the resulting quadratic for β^*; thus

$$\sin^2\beta^* = \frac{1}{4\gamma M_1^2}\left[(\gamma+1)M_1^2 - (3-\gamma) + \left[(\gamma+1)[(\gamma+1)M_1^4 - 2(3-\gamma)M_1^2 + (\gamma+9)]\right]^{\frac{1}{2}}\right] \ . \tag{12-11}$$

The flow downstream of a strong shock is always subsonic. Strong shocks in steady flow have been observed experimentally primarily for two-, or three-dimensional bodies of finite length, e.g. in connection with a supersonic flow past a symmetrical double-wedge body where the half-angle θ_w at the wedge-nose exceeds the theoretical maximum turning angle θ_m for the specified values of M_1 and γ, i.e. for $\theta_w > \theta_m$. In such cases the shock is always curved and stands in the flow ahead of the wedge tip, and where, on the centerline, the shock is locally a normal shock. The shock is said to be "detached" from the body. The computation of the flow involving detached shocks is notoriously difficult.

An extensive tabulation of oblique-shock functions for $\gamma = 5/3$, $7/5$ and $4/3$ can be found in Tables E.

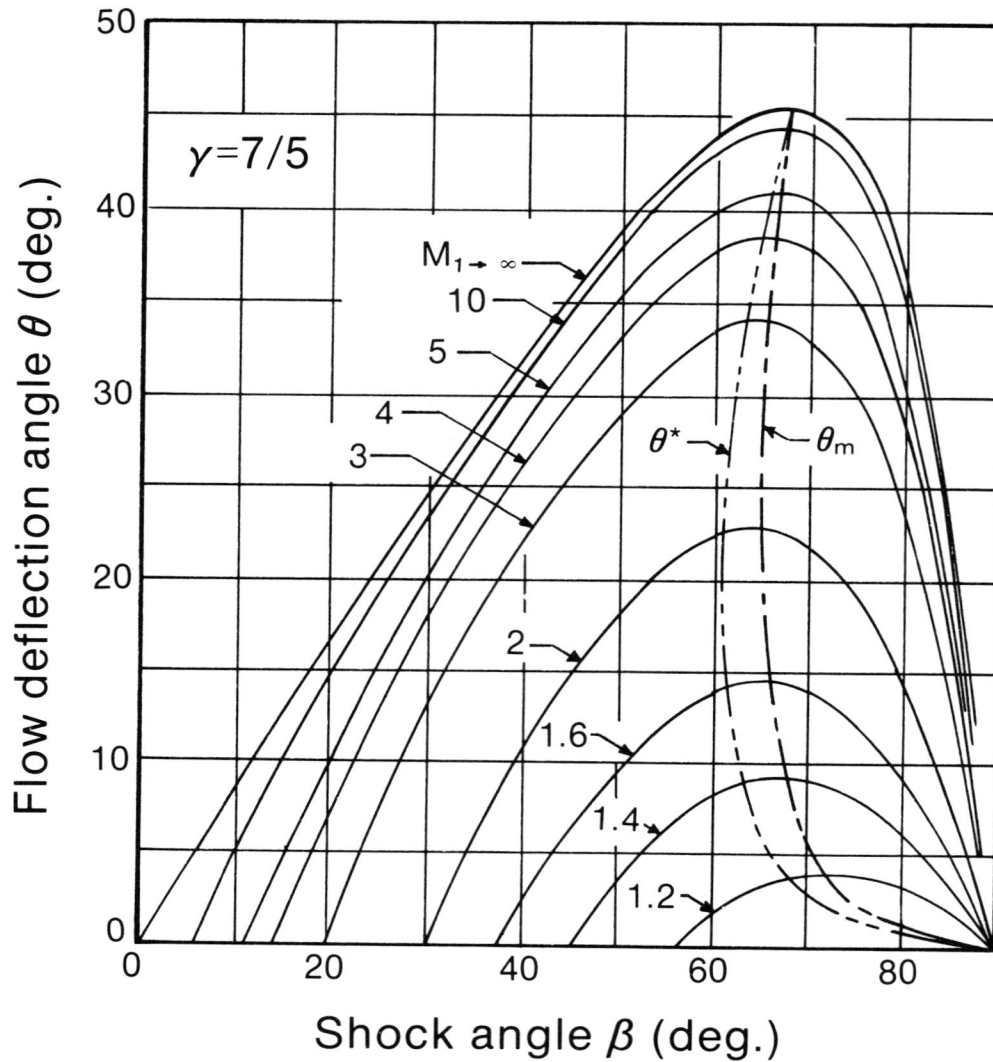

Figure 12-2. Mapping of the oblique-shock solution, $\gamma = 7/5$.

Relations for the limiting case as the upstream Mach-number approaches infinity.

We denote for any variable F,

$$\lim_{M_1 \to \infty} F \equiv F_\infty \; ;$$

then, from (12-6),

$$\tan\theta_\infty = \frac{\sin 2\beta_\infty}{\gamma + \cos 2\beta_\infty} \; , \tag{12-12a}$$

or, equivalently

$$cos2\beta_\infty = -\gamma sin^2\theta_\infty \pm cos\theta_\infty(1-\gamma^2 sin^2\theta_\infty)^{\frac{1}{2}} \quad , \qquad (12\text{-}12b)$$

where the $\pm$ signs refer to the weak-, and strong-shock solutions, respectively.

For the ratios of the state variables across the shock we have, from (12-8):

$$p_{21\infty} = T_{21\infty} = a_{21\infty} = \infty \quad ; \qquad (12\text{-}12c)$$

$$\rho_{21\infty} = \frac{\gamma+1}{\gamma-1} = u_{12\infty} \quad ; \qquad (12\text{-}12d)$$

$$M_{2\infty}^2 = \frac{(\gamma+1)^2 - 4\gamma sin^2\beta_\infty}{2\gamma(\gamma+1)sin^2\beta_\infty} \quad ; \qquad (12\text{-}12e)$$

$$w_{21\infty}^2 = 1 - \frac{4\gamma}{(\gamma+1)^2} sin^2\beta_\infty \quad . \qquad (12\text{-}12f)$$

The last relation can be combined with (12-12b), resulting in

$$w_{21\infty}^2 = \frac{(\gamma-1)^2 - 4\gamma sin^2\theta_\infty \pm 4\gamma cos\theta_\infty(1-\gamma^2 sin^2\theta_\infty)^{\frac{1}{2}}}{(\gamma+1)^2} \quad . \qquad (12\text{-}12g)$$

for the weak-, and strong-shock solutions, respectively. The pressure coefficients for these two cases are:

$$C_{p\infty} = \frac{4}{\gamma+1} sin^2\beta_\infty = \frac{2}{\gamma+1} [1 + \gamma sin^2\theta_\infty \mp (1-\gamma^2 sin^2\theta_\infty)^{\frac{1}{2}}] \quad . \qquad (12\text{-}12h)$$

From (12-10) and (12-11)

$$sin^2\beta_{m\infty} = sin^2\beta_\infty^* = \frac{\gamma+1}{2\gamma} \quad . \qquad (12\text{-}12i)$$

and from (12-8a)

$$p_{21\infty}^o = \rho_{21\infty}^o = e_\infty^{-\Delta s/R} = 0 \quad . \qquad (12\text{-}12j)$$

13. The Shock Polar

The *shock polar* is an alternative representation of the oblique shock which is due to Busemann (1937), and which employs the velocity components downstream of the shock parallel to, and normal to the velocity ahead of the shock as the principal variables. It is a special case of the *hodograph* representation of two-dimensional flows.

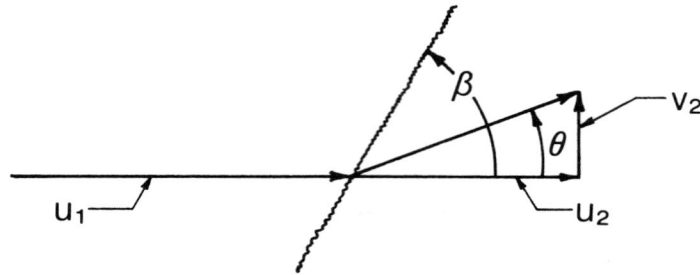

Figure 13-1. Geometry of the shock polar.

Denoting the velocity components parallel to the x,y-coordinates as u and v, the conservation equations for a perfect gas — at a specified total temperature or, equivalently, at a specified value of $a*$ — are:

- Mass
$$\rho_1 u_1 \sin\beta \;=\; \rho_2(u_2 \sin\beta - v_2 \cos\beta) \;; \tag{13-1}$$

- Momentum
$$p_1 + \rho_1^2 u_1^2 \sin^2\beta \;=\; p_2 + \rho_2(u_2 \sin\beta - v_2 \cos\beta)^2 \;; \tag{13-2}$$

- Energy
$$\frac{\gamma}{\gamma-1}\frac{p_1}{\rho_1} + \tfrac{1}{2}u_1^2 \;=\; \frac{\gamma}{\gamma-1}\frac{p_2}{\rho_2} + \tfrac{1}{2}(u_2^2 + v_2^2) \;=\; \frac{\gamma+1}{2(\gamma-1)}\,a*^2 \;. \tag{13-3}$$

Since the velocity components parallel to the shock must be equal,

$$u_1 \cos\beta \;=\; u_2 \cos\beta + v_2 \sin\beta \;,$$

or

$$\tan\beta \;=\; (u_1 - u_2)/v_2 \;. \tag{13-4}$$

Equations (13-1) through (13-4) involve 10 variables in 6 independent equations. However, as can be verified by dividing (13-2) by (13-1), the pressure terms appear only in the combination p/ρ. Thus, there are really only 9 independent variables and it is therefore possible to reduce the system to one independent equation involving u_1, u_2, v_2, and a^*. We introduce the speed ratios $\bar{u}_1 = u_1/a^*$, $\bar{u}_2 = u_2/a^*$, $\bar{v}_2 = v_2/a^*$ in place of the velocities. The resulting equation can be written as

$$\bar{v}_2^2 = \frac{(\bar{u}_1 - \bar{u}_2)^2 (\bar{u}_1\bar{u}_2 - 1)}{1 + \frac{2}{\gamma+1}\bar{u}_1^2 - \bar{u}_1\bar{u}_2} \quad . \tag{13-5}$$

Therefore, if we regard $\bar{u}_1$ and γ as parameters, $\bar{v}_2 = \bar{v}_2(\bar{u}_2; \bar{u}_1, \gamma)$.

The <u>shock</u> <u>polar</u>. The *shock polar* is a plot, for fixed $\bar{u}_1$ and γ, of $\bar{v}_2$ versus $\bar{u}_2$. Note that there are two possibilities for which $\bar{v}_2 = 0$. The first, when $\bar{u}_1 = \bar{u}_2$, corresponds to the no-shock condition. When the second factor in (13-5) is zero, then $\bar{u}_1\bar{u}_2 = 1$, or

$$u_1 u_2 = a^{*2} \quad , \tag{13-6}$$

which is Prandtl's relation, corresponding to the normal-shock solution.

From Figure 13-1 the turning angle is given by

$$tan\theta = v_2/u_2 = \bar{v}_2/\bar{u}_2 \quad . \tag{13-7}$$

It can be shown that the maximum value of the turning angle θ_m, which corresponds to the shock-polar coordinate designated as $\bar{u}_{2m}$ is given by

$$\bar{u}_{2m} = \frac{1}{2\gamma\bar{u}_1}\left[\bar{u}_1^2 + 2(\gamma+1) - [\bar{u}_1^4 - 4(\gamma-1)\bar{u}_1^2 + 4(\gamma+1)]^{\frac{1}{2}}\right] \quad . \tag{13-8}$$

Similarly, the sonic point behind the shock designated by $\bar{u}_2^*$, which corresponds to θ^*, is given by

$$\bar{u}_2^* = \frac{\gamma+1}{4\gamma\bar{u}_1}\left[\bar{u}_1^2 + 3 - \left[\bar{u}_1^4 - \frac{2(\gamma^2 + 6\gamma - 3)}{(\gamma+1)^2}\bar{u}_1^2 + \frac{\gamma+9}{\gamma+1}\right]^{\frac{1}{2}}\right] \quad . \tag{13-9}$$

Equations (13-8) and (13-9) could, individually, be substituted into (13-5) and then combined with (13-7) and (13-4) to obtain explicit expressions for the corresponding values $\bar{v}_{2m}$, θ_m and β_m, or $\bar{v}_2^*$, θ^*, β^*. The resulting expressions appear to be extremely cumbersome so that no effort has been made to complete the analysis for these quantities.

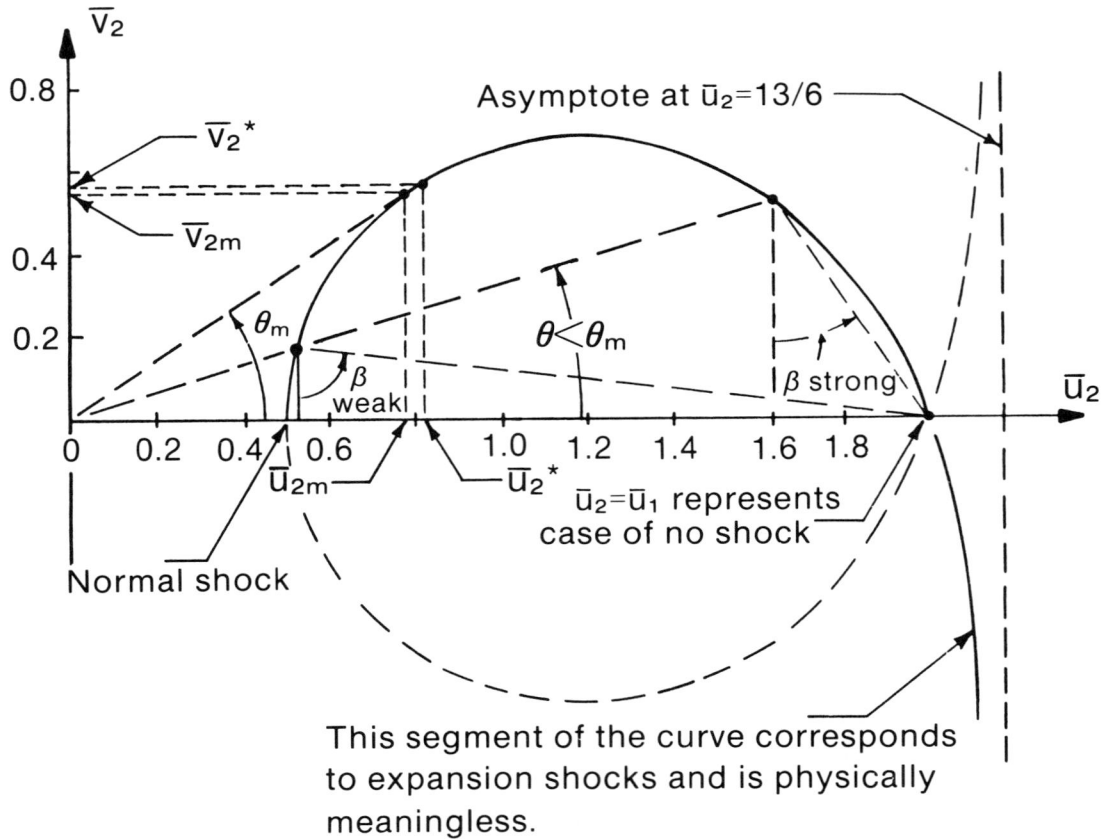

Figure 13-2. The shock polar for $\gamma = 7/5$, $\overline{u}_1 = 2.0$, for which:

$$\overline{u}_{2m} = 0.789, \quad \overline{v}_{2m} = 0.555, \quad \theta_m = 35.11^o,$$

$$\overline{u}_2^* = 0.819, \quad \overline{v}_2^* = 0.574, \quad \theta^* = 35.50^o \quad .$$

The corresponding free-stream Mach number is $M_1 = 3.56$.

A plot of the shock polar for $\gamma = 7/5$, $\overline{u}_1 = 2.0$, is shown in Figure 13-2. Only the portion of the curve shown as a solid line is physically meaningful. The dotted line for $\overline{u}_2 > 2$ corresponds to negative values of the turning angle and a negative entropy-change. It is interesting to compare the graphical methods by which the weak-, and strong-shock solutions are illustrated on the shock polar with the representation of Figure 12-2. The weak- and strong-shock solutions are delineated by putting $\overline{u}_2 \lessgtr \overline{u}_{2m}$, respectively.

Cole's scaled shock-polar for transonic flow. For the transonic-flow regime Cole and Cook (1986, p. 105) have presented an approximate representation which collapses the neighborhood of the shock polar near the sonic point (i.e. the neighborhood of $\overline{u}_1 = 1$) onto a single curve. Introducing (dimensionless) perturbation velocities such that

$$\bar{u}_1' \equiv (u_1 - a^*)/a^* = \bar{u}_1 - 1 \quad , \quad \bar{u}_2' \equiv \bar{u}_2 - 1 \quad , \quad \bar{v}_2' \equiv \bar{v}_2 \quad , \qquad (13\text{-}10)$$

and substituting into (13-5) we obtain

$$(\bar{v}_2')^2 = \frac{(\bar{u}_2' - \bar{u}_1')^2 \left[\bar{u}_1' + \bar{u}_2' + \bar{u}_1'\bar{u}_2'\right]}{\frac{2}{\gamma+1}[1 + 0(\bar{u}_1')]} \quad . \qquad (13\text{-}11a)$$

where

$$0(\bar{u}_1') = \frac{3-\gamma}{\gamma+1} \bar{u}_1' - \frac{\gamma+1}{2} \bar{u}_2' + \bar{u}_1'^2 - \frac{\gamma+1}{2} \bar{u}_1'\bar{u}_2' \quad . \qquad (13\text{-}11b)$$

If, we neglect the second-order term $\bar{u}_1' \, \bar{u}_2'$, with respect to the first-order quantity $\bar{u}_1' + \bar{u}_2'$, and the terms $0(\bar{u}_1')$ with respect to unity, (13-11a) becomes

$$(\bar{v}_2')^2 = \tfrac{1}{4} (\gamma+1)(\bar{u}_2' - \bar{u}_1')^2 (\bar{u}_1' + \bar{u}_2') \quad . \qquad (13\text{-}12)$$

Equation (13-12) is the shock-polar relation in terms of the perturbation velocities.

Equation (13-12) is then scaled by introducing new variables of order unity such that

$$\hat{u} \equiv \frac{(\gamma+1)\bar{u}_2'}{(\gamma+1)\bar{u}_1'} = \frac{\bar{u}_2'}{\bar{u}_1'} \quad , \qquad \hat{v} \equiv \frac{(\gamma+1)\bar{v}_2'}{[(\gamma+1)\bar{u}_1']^{3/2}} \quad . \qquad (13\text{-}13)$$

With (13-13), the final form of the scaled shock-polar equation is

$$\hat{v}^2 = \tfrac{1}{4} (\hat{u}-1)^2 (1+\hat{u}) = \tfrac{1}{4} (\hat{u}^3 - \hat{u}^2 - \hat{u} + 1) \quad . \qquad (13\text{-}14)$$

A plot of (13-14) is shown in Figure 13-3. The curve has the following distinctive features:

1) The scaled shock-polar is independent of the ratio of specific heats.

2) When $\hat{v} = 0$, (13-14) has three roots. For $\hat{u} = 1$ we have a double root corresponding to $\bar{u}_1' = \bar{u}_2'$ which is the no-shock case.

3) The third root, corresponds to $\hat{u} = -1$, or $\overline{u}_1' = -\overline{u}_2'$; hence the flow downstream of the shock is subsonic and the case is that of the normal shock.

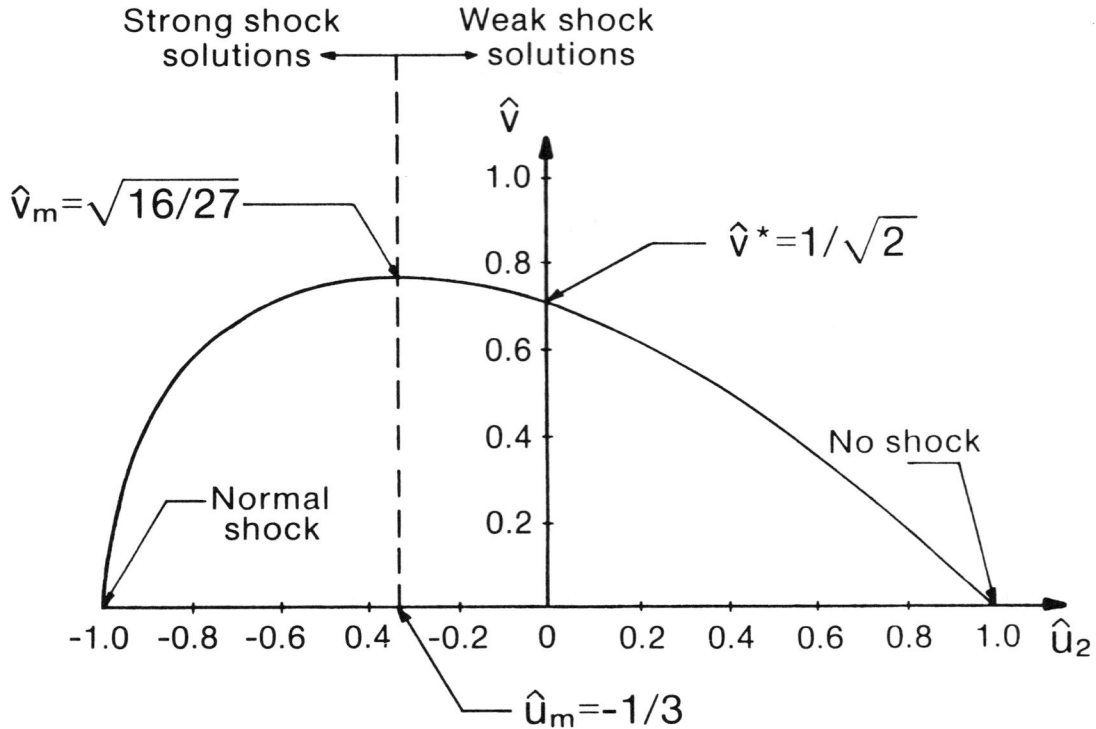

Strong shock solutions ←⎯⎯⎯⎯→ Weak shock solutions

$\hat{V}_m = \sqrt{16/27}$

$\hat{v}^* = 1/\sqrt{2}$

Normal shock

No shock

$\hat{u}_m = -1/3$

Figure 13-3. The scaled shock-polar for transonic flow.

4) To determine the maximum turning angle we can combine (13-7), (13-10) and (13-13) to obtain the approximate form

$$\tan^2\theta \;=\; \left[\frac{\overline{v}_2}{\overline{u}_2}\right]^2 \;=\; \frac{(\gamma+1)(\overline{u}_1')^3(\hat{v}_2)^2}{(1+\overline{u}_1'\hat{u})^2} \;=\; \tfrac{1}{4}(\gamma+1)(\overline{u}_1')^3(\hat{u}-1)^2(1+\hat{u}) \quad ,(13\text{-}15)$$

where the right-hand side has been approximated to the same order as the scaled shock-polar. However, since the turning angle in the transonic regime is small we can also put $\tan^2\theta \rightarrow \theta^2$; consequently, to the same order of approximation,

$$\theta^2 \;=\; \tfrac{1}{4}(\gamma+1)(\overline{u}_1')^3(\hat{u}-1)^2(\hat{u}+1) \quad . \tag{13-16}$$

Equation (13-16) exhibits a maximum at $\hat{u}_m = -1/3$, producing

$$\theta_m^2 = (16/27)(\gamma+1)(\bar{u}_1')^3 \quad . \tag{13-17}$$

The values for $\hat{u} \gtrless -1/3$ correspond to the strong-, and weak-shock solutions, respectively. The corresponding approximations for the free-stream Mach number and shock angle at θ_m are:

$$M_1 = 1 + (3/2)[(\gamma+1)\theta_m/4]^{2/3} \quad , \tag{13-18}$$

$$\beta_m = \pi/2 - [(\gamma+1)\theta_m/4]^{1/3} \quad . \tag{13-19}$$

5) For sonic flow downstream of the shock, the exact expression

$$(\bar{u}_2^*)^2 + (v_2^*)^2 = 1 \quad ,$$

translates to

$$1 + 2\bar{u}_2'^* + (\bar{u}_2'^*)^2 + (\bar{v}_2'^*)^2 = 1 \quad .$$

The corresponding first-order approximation is

$$\bar{u}_2'^* = 0 \quad , \text{ or } \quad \hat{u}^* = 0 \quad , \tag{13-20a}$$

for which

$$\hat{v}_2^* = 1/\sqrt{2} \quad , \quad \text{ or } \bar{v}_2'^* = [(\gamma+1)/2]^{1/2}(\bar{u}_1')^{3/2} \quad . \tag{13-20b}$$

Therefore, from (13-16)

$$\theta^{*2} = \tfrac{1}{2}(\gamma+1)(\bar{u}_1')^3 \quad . \tag{13-21}$$

All of these transonic-flow relations are valid only for free-stream Mach numbers close to unity and, hence, only for very small turning angles. For example, for $\gamma = 7/5$, $M_1 = 1.10$, exact theory yields $\theta_m = 1.5152°$, $\beta_m = 76.2695°$, $\theta^* = 1.4062°$,

$\beta^* = 73.2502°$, and $\overline{u}_1 = 1.08124$, $\overline{u}_2 = 0.99966$. The approximate relations, using $\overline{u}'_{1m} = 0.08124$, produce $\theta_m = 1.53°$, $\beta_m = 75.39°$, $\theta^* = 1.45°$ and $M_1 = 1.0975$.

14. Busemann's Approximate Theory for the Pressure Coefficient across an Oblique Shock

Busemann (1935) undertook to obtain an approximate, but explicit, solution for the pressure coefficient in the flow behind an oblique shock. Substitution of (7-15) for the pressure coefficient into the oblique-shock relation (12-8a) for the static-pressure ratio yields

$$M_1^2 sin^2\beta - 1 = [(\gamma+1)/4]M_1^2 C_p \ , \tag{14-1a}$$

or

$$sin^2\beta = 1/M_1^2 + [(\gamma+1)/4]C_p \ . \tag{14-1b}$$

Equation (14-1a) is substituted into the denominator of (12-6a). Squaring the result produces

$$cot^2\theta = \frac{sin^2\beta}{1-sin^2\beta} (2/C_p - 1)^2 \ . \tag{14-2}$$

Eliminating β by use of (14-1b) and rearranging, the following cubic in C_p results:

$$a_3 C_p^3 + a_2 C_p^2 + a_1 C_p + a_o = 0 \ , \tag{14-3a}$$

where

$$a_3 = (\gamma+1)M_1^4/4 \ , \qquad a_1 = sin^2\theta[(\gamma+1)M_1^2 - 4] \ ,$$

$$a_2 = 1 - M_1^2 - \gamma M_1^2 sin^2\theta \ , \qquad a_0 = 4 sin^2\theta \ . \tag{14-3b}$$

The exact solution of (14-3) being inconvenient, Busemann assumed a series expansion of the form

$$C_p = C'_0 + C'_1\theta + C'_2\theta^2 + C'_3\theta^3 + ... \ , \tag{14-4}$$

and replaced $sin\theta$ by its own series expansion. The reason for the primes on the coefficients in (14-4) is that Busemann also developed a series for C_p corresponding to the Prandtl-Meyer expansion (to be discussed in Section 15). The coefficients for the two series turn out to be identical only in the first two terms, i.e. $C'_1 = C_2$, $C'_2 = C_2$.

The weak-shock solution. In order to ensure that the weak-shock solution is obtained it is necessary to put $C_0' \equiv 0$. Substitution of (14-4) and (14-3b) into (14-3a) produces an infinite series of the form

$$0 = b_2 \theta^2 + b_3 \theta^3 + b_4 \theta^4 + \dots , \qquad (14\text{-}5)$$

where the coefficients b_m depend on M_1, γ and the unknown C_{m-1}'. For example,

$$b_2 = (1-M_1^2)C_1'^2 + 4 . \qquad (14\text{-}6)$$

It is elementary to prove that for a solution of (14-5) to be valid for all θ, it is necessary that each $b_m \equiv 0$. This yields a sequence of algebraic equations for C_{m-1}'.

We list below the first three non-zero coefficients:

$$C_1' = \frac{2}{(M_1^2-1)^{\frac{1}{2}}} = C_1 , \qquad (14\text{-}7a)$$

$$C_2' = \frac{(\gamma+1)M_1^4 - 4(M_1^2-1)}{2(M_1^2-1)^2} = C_2 , \qquad (14\text{-}7b)$$

$$C_3' = \frac{1}{(M_1^2-1)^{7/2}} \left[\frac{(\gamma+1)^2}{16} M_1^8 + \frac{3\gamma^2-12\gamma-7}{12} M_1^6 + \frac{3(\gamma+1)}{2} M_1^4 - 2M_1^2 + \frac{4}{3} \right]. (14\text{-}7c)$$

In fact, the solution for C_1' allows for both positive and negative values. However, since for $\theta > 0$ the negative sign would correspond to an expansion, only the positive value has been retained.[4]

The real objective in replacing the exact result by a series expansion in θ is based on the expectation that for θ not too large it should be possible to truncate the series after a few terms and still get reasonable values for C_p. A comparison between the first-, second-, and third-order approximations with the exact theory for $\gamma = 7/5$, $M_1 = 2.0$ is made in Table 14-1.

---

[4] The originally published values for C_3 and C_3' (the latter in a slightly different form) were both erroneous. Laitone (1947, p. 25) first gave the correct expression for C_3 but his value for C_3' was wrong. Although Puckett and Li (1947, p. 336) later obtained the correct expression for C_3' their published expression involved a transcription error. The correct form was finally contained in a note by Kahane and Lees (1947, p. 600). The expression of (14-7c) is taken directly from NACA Report 1135, equation (152). The first-order coefficient C_1 had earlier been derived by Ackeret (1925), by a different method.

Since $p_{21} = 1 + \gamma M_1^2 C_p/2$, we can combine (14-7) with (8-14) to obtain an approximate expression for the entropy change across a weak oblique shock:

$$\Delta s/R \;=\; \ell n p_{12}^0 \;=\; \frac{2(\gamma+1)}{3\gamma}\left[\frac{p_{21}-1}{p_{21}+1}\right]^3 + \frac{2(\gamma+1)(\gamma^2+1)}{5\gamma^4}\left[\frac{p_{21}-1}{p_{21}+1}\right]^5 + \dots$$

$$= \frac{\gamma(\gamma+1)M_1^6}{12(M_1^2-1)^{3/2}}\,\theta^3 - \frac{\gamma(\gamma+1)M_1^6}{16(M_1^2-1)^3}\,[(5\gamma-1)M_1^4 - 2(3\gamma-2)M_1^2-4]\theta^4 + \dots \quad (14\text{-}8)$$

Equation (14-8), due the necessity to expand the term $(p_{21} + 1)^3$ in a series in $p_{21} - 1$, is subject to the additional restriction that $p_{21} - 1 < 1$.

The preceding expansions can also be combined with (12-8g) to produce a relation for the speed ratio across a weak shock, namely

$$w_{21} \;=\; 1 - \tfrac{1}{2} C_1 \theta - \tfrac{1}{2}(C_2-1)\theta^2$$

$$- \left[\frac{(\gamma+1)M_1^4 - 2(M_1^2-1)}{32}\,C_1^3 - \frac{(M_1^2-1)}{4}\,C_1 C_2 + \tfrac{1}{2}C_3'\right]\theta^3 + 0(\theta^4) \quad . \quad (14\text{-}9)$$

The strong-shock solution. The procedure differs from that of the weak-shock case in that we require (14-4) to reduce to the normal-shock solution as $\theta \to 0$. To avoid confusion with the weak-shock solution the series is denoted

$$C_p \;=\; B_0 + B_1\theta + B_2\theta^2 + \dots \quad . \quad (14\text{-}10)$$

The first three non-zero coefficients are:

$$B_0 \;=\; 4(M_1^2-1)/(\gamma+1)M_1^2 \;, \qquad B_2 = -[\gamma-1)M_1^2+2]^2/(\gamma+1)(M_1^2-1)^2 \;,$$

$$B_4 \;=\; -\,[(\gamma-1)M_1^2+2]^2 F_1(M_1;\ \gamma)/12(\gamma+1)(M_1^2-1)^5 \;, \quad (14\text{-}11)$$

where

$$F_1(M_1;\ \gamma) \;=\; (3\gamma^2+6\gamma-1)M_1^6 + 3(\gamma-1)^2 M_1^4 + 12\gamma M_1^2 + 4 \quad .$$

Since strong shocks are encountered only on blunt bodies where they are detached, or in transonic flows where the shock is nearly orthogonal to the surface from which it originates, equation (14-10) must be valid for negative, as well as positive values of θ. This fact is confirmed by the finding that the odd-numbered B_n are zero.

The preceding expressions provide a simple means for computing the complement of the shock angle $\pi/2 - \beta$ as a series in θ. Putting

$$\pi/2 - \beta = a\theta + b\theta^3 + O(\theta^5) , \tag{14-12}$$

and substituting (14-12) and (14-10) into (14-1a), we find that

$$a = [(\gamma-1)M_1^2 + 2]/2(M_1^2-1) , \tag{14-13}$$

$$b = \left[(M_1^2-1)[(\gamma-1)M_1^2+2]^3 + [(\gamma-1)M_1^2 + 2]F_1(M_1 ; \gamma) \right]/48(M_1^2-1)^4 .$$

Equations (14-12) and (14-13) provide good accuracy for values of θ not exceeding $\frac{1}{2}\theta_m$ for the specified M_1 and γ. The approximate theory is compared with the exact strong-shock theory in Table 14-1.

Table 14-1. Comparison of exact and approximate expressions for pressure coefficient across weak-, and strong-shocks.

	C_p at $M_1 = 2.0$, for $\gamma = 7/5$						
	Weak shock				Strong shock		
θ(Deg)	Exact Theory	First-Order	Second-Order	Third-Order	Exact Theory	Second Order	Fourth Order
0	0.0	0.0	0.0	0.0	1.2500	1.2500	1.2500
2	0.0421	0.0403	0.0421	0.0421	1.2493	1.2493	1.2493
4	0.0881	0.0806	0.0878	0.0881	1.2470	1.2470	1.2470
6	0.1383	0.1209	0.1370	0.1382	1.2432	1.2432	1.2432
8	0.1929	0.1612	0.1898	0.1927	1.2376	1.2383	1.2376
10	0.2523	0.2015	0.2462	0.2516	1.2299	1.2317	1.2301
12	0.3173	0.2418	0.3062	0.3155	1.2198	1.2237	1.2204
14	0.3884	0.2821	0.3697	0.3845	1.2063	1.2142	1.2060
16	0.4670	0.3225	0.4368	0.4590	1.1885	1.2032	1.1928
18	0.5552	0.3628	0.5075	0.5390	1.1639	1.1908	1.1740
20	0.6581	0.4031	0.5818	0.6250	1.1275	1.1769	1.1514
22	0.8702	0.4433	0.6596	0.7171	1.0612	1.1615	1.1242

15. The Prandtl-Meyer Theory for Continuous Turning in Supersonic Flow

Equation (14-9) gives the expression for the ratio of the flow speeds across a shock to third-order in the turning angle. If we keep only the zeroth and first-order terms, and if we replace θ by $\Delta\theta \equiv \theta_2 - \theta_1$, $w_{21} - 1$ by $(w_2 - w_1)/w_1 \equiv \Delta w/w_1$, then

$$\frac{\Delta w}{w_1} = -\tfrac{1}{2}C_1\Delta\theta \;, \quad where \; C_1 = 1/(M_1^2 - 1)^{\frac{1}{2}} \;;$$

thus

$$\Delta w/\Delta\theta = -w_1/(M_1^2 - 1)^{\frac{1}{2}} \;. \tag{15-1}$$

Recalling that both the first- and second-order Busemann coefficients correspond to isentropic theory, if we take the limit of (15-1) as $\Delta\theta \to 0$, we obtain the exact expression for a vanishingly weak isentropic-wave, i.e.

$$dw/d\theta = -w/(M^2 - 1)^{\frac{1}{2}} \;, \tag{15-2}$$

where M and w represent the local values. Since the flow is isentropic, then, for $d\theta > 0$, (15-2) is equally valid for a compression $(dw < 0)$ or an expansion $(dw > 0)$. Rearranging, we next we define a new function $\nu = \nu(M; \gamma)$ such that

$$d\nu \equiv (M^2 - 1)^{\frac{1}{2}} \, dw/w = -d\theta \;. \tag{15-3}$$

By use of the energy equation (12-1), combined with the perfect-gas relations and the Mach-number definition $w = Ma$, (15-3) becomes

$$d\nu = \frac{(M^2 - 1)^{\frac{1}{2}} \, dM/M}{1 + \tfrac{1}{2}(\gamma - 1)M^2} \;. \tag{15-4}$$

Since $d\nu$ does not exist for $M < 1$, we arbitrarily set $\nu = 0$ at $M = 1$; thus

$$\nu = \int_0^\nu d\nu \equiv \int_1^M \frac{(M^2 - 1)^{\frac{1}{2}} \, dM/M}{1 + \tfrac{1}{2}(\gamma - 1)M^2} \;,$$

which has the integral

$$\nu(M; \gamma) = a^{\frac{1}{2}} arctan[(M^2 - 1)/a]^{\frac{1}{2}} - arctan(M^2 - 1)^{\frac{1}{2}} \;, \tag{15-5a}$$

where

$$a = (\gamma + 1)/(\gamma - 1) \;. \tag{15-5b}$$

The Mach-number function $\nu(M;\ \gamma)$ is called the *Prandtl-Meyer function*, and was first employed by Prandtl (1907). A thorough discussion of its properties is given by Ferri (1954). The maximum possible value of ν occurs as $M \to \infty$, and is $\nu_{max} = \pi(a^{\frac{1}{2}} - 1)/2$, where $\nu_{max} = 130.45^{o}$ for $\gamma = 7/5$. Tabulations of ν versus M for $\gamma = 5/3,\ 7/5$ and $4/3$ are given in the data section.

Note on use of the Prandtl-Meyer function. Equation (15-3) can be written as $d(\nu +\ \theta) = 0$, or

$$\nu +\ \theta\ =\ \nu_1 +\ \theta_1 =\ constant\ , \qquad (15\text{-}6)$$

where ν_1 and θ_1 are the values for the flow in a uniform region prior to the initiation of an expansion or compression.

Prandtl first pointed out, as shown in Figure 15-1, that it is possible to join two regions of uniform supersonic flow where $M_2 > M_1$, by a centered Prandtl-Meyer expression, where the total temperature and total pressure remain the same throughout. The flow properties along any radial line (which are Mach lines, or characteristics) emanating from the corner are invariant. Such a flow has no characteristic length (or reference length) and thus is dependent only on the turning angle θ or, equivalently, the function ν. A flow lacking a characteristic length is said to be *self-similar*.

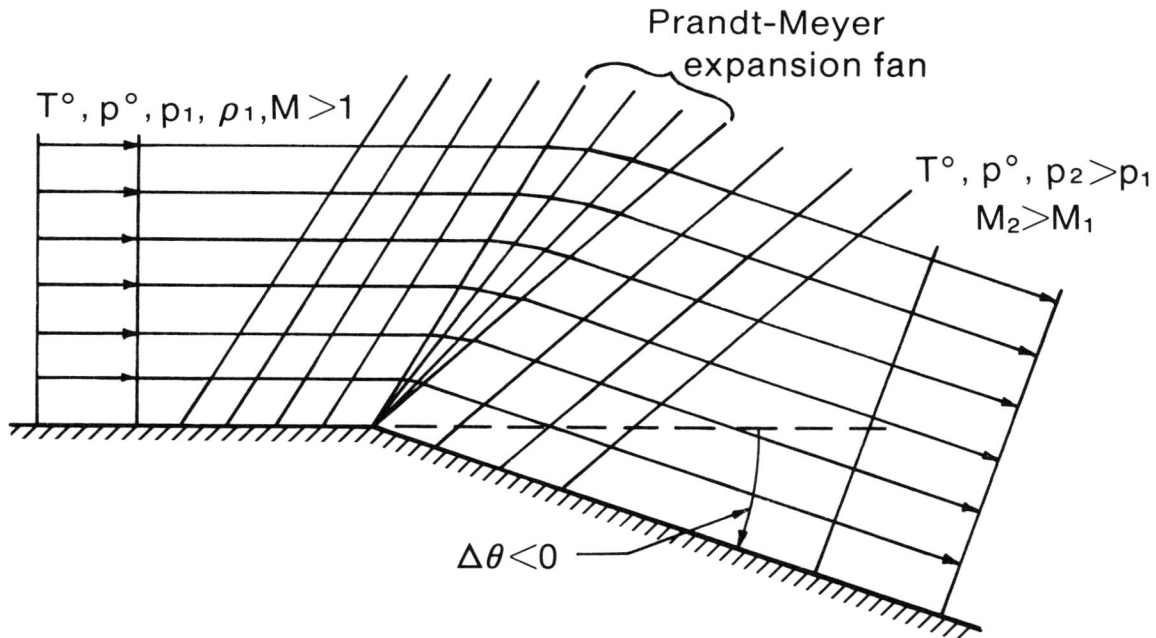

Figure 15-1. Sketch of centered Prandtl-Meyer expansion.

Supersonic compression using continuous turning (i.e. in a reverse Prandtl-Meyer expansion) can also be obtained, as shown in Figure 15-2. Since the flow-angle θ and the Mach angle μ both increase in the flow direction, the Mach lines tend to converge. In a practical sense this means that the lateral extent of the flow must be restricted. Intersection of the Mach lines implies that at the intersection the Prandtl-Meyer function (say) at a point may take on more than one value, which is physically impossible in a continuous flow. In fact, the intersection of any pair of Mach lines of the same family (i.e. either left-, or right-running) indicates the formation of a shock wave whose analysis must be handled by special techniques, c.f. Courant and Friedrichs (1945, p. 171). To avoid this, the lateral extent of the flow field is limited by introducing a wall, as shown in Figure 15-2, lying on a theoretical streamline. The flow of Figure 15-2 is also reversible in direction and can be used as an expansion from M_2 to $M_1 > M_2$ where the turning is initiated on the upper wall. In these flows, the experimental realization would be modified by effects of viscosity on the wall. Experiments reveal that the case of supersonic compression is inherently more susceptible to drastic modification (than an expansion) due to the tendency of the wall boundary-layer to separate, or to generate disturbances resulting in the formation of oblique shocks.

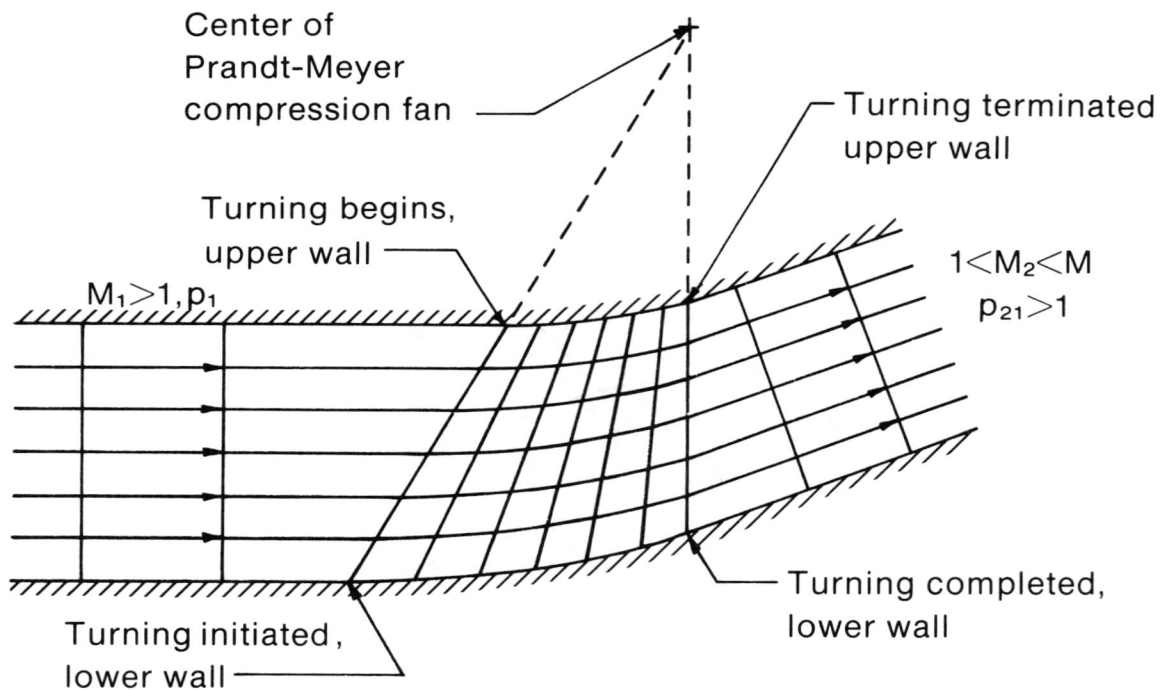

Figure 15-2. Sketch of an isentropic compression.

Busemann's theory applied to the Prandtl-Meyer function. By procedures similar to those of Section 14 Busemann obtained the pressure coefficient in the uniform region downstream of a centered Prandtl-Meyer flow expressed as a series expansion about the uniform conditions upstream designated as p_1, M_1, v_1, etc. Thus, where $\theta < 0$ corresponds to an expansion, the appropriate series, analogous to the approximate expression for a shock (i.e. a compression where $\theta > 0$), is

$$C_p = C_1\theta + C_2\theta^2 + C_3\theta^3 + 0(\theta^4) \;, \tag{15-7a}$$

where C_1 and C_2 are identical to the values given in (14-7). The third coefficient is

$$C_3 = \frac{1}{(M_1^2-1)^{7/2}} \left[\frac{\gamma+1}{6}M_1^8 + \frac{2\gamma^2-7\gamma-5}{6}M_1^6 + \frac{5(\gamma+1)}{3}M_1^4 - 2M_1^2 + \frac{4}{3} \right] \tag{15-7b}$$

Table 15-1. Comparison of the Prandtl-Meyer (exact) theory in an expansion for the pressure coefficient with the Busemann theory.

$-C_p$			
Expansion from $M_1 = 2.0$ for $\gamma = 7/5$.			
$-\theta(\text{Deg})$	Exact Theory	First Order	Second Order
0	0.0	0.0	0.0
2	0.0386	0.0403	0.0385
4	0.0738	0.0806	0.0735
6	0.1059	0.1209	0.1048
8	0.1350	0.1612	0.1326
10	0.1614	0.2015	0.1569
12	0.1853	0.2418	0.1775
14	0.2068	0.2821	0.1946
16	0.2262	0.3225	0.2081
18	0.2434	0.3627	0.2180
20	0.2589	0.4030	0.2244
22	0.2726	0.4434	0.2271

Since the first two coefficients are identical, we can truncate the series after the second-order term, resulting in

$$C_p = C_1\theta + C_2\theta^2 \;, \tag{15-8}$$

which gives excellent results for surface pressures on thin, supersonic airfoils undergoing either weak compressions or expansions starting at the leading edge. The turning angle θ is measured from the free-stream direction, with the caution that it is necessary for the user to specify, based on the flow geometry, whether an expansion (θ negative) or compression (θ positive) is attained. Equation (15-8) is usually known as the *Busemann second-order theory for surface pressure*.

Table 15-1 presents a comparison of the exact values of the pressure coefficient in an expansion from M_1 = 2.0 for γ = 7/5 with the first-order, and second-order approximations.

16. Busemann's Theory Applied to Thin, Two-Dimensional, Supersonic Airfoils

Airfoil nomenclature. The geometry of an airfoil, see Figure 16-1, is described, first, by specifying a *chord line* of length c (the chord), the extremities of which are referred to as the *leading*, and *trailing edges*, respectively, and which are fixed on the ξ-axis. A camber line $y_c = y_c(\xi)$ of arbitrary shape passes through the leading- and trailing-edge points. A thickness function $h = h(\xi) > 0$, is "wrapped around" the camber line thus defining the upper, and lower surfaces:

$$y_u(\xi) \;=\; y_c + h \;,\quad y_\ell(\xi) \;=\; y_c - h \;. \tag{16-1}$$

According to (16-1) the thickness at station ξ is measured perpendicular to the chord line. Furthermore, to avoid a detached shock at the leading edge we require that $h(0) = 0$, and $|dy/d\xi| << 1$ everywhere on both the upper and lower surface. On the other hand the thickness at the trailing edge is not necessarily zero.

The angle-of-attack a is defined such that a nose-up pitch motion corresponds to a increasing.

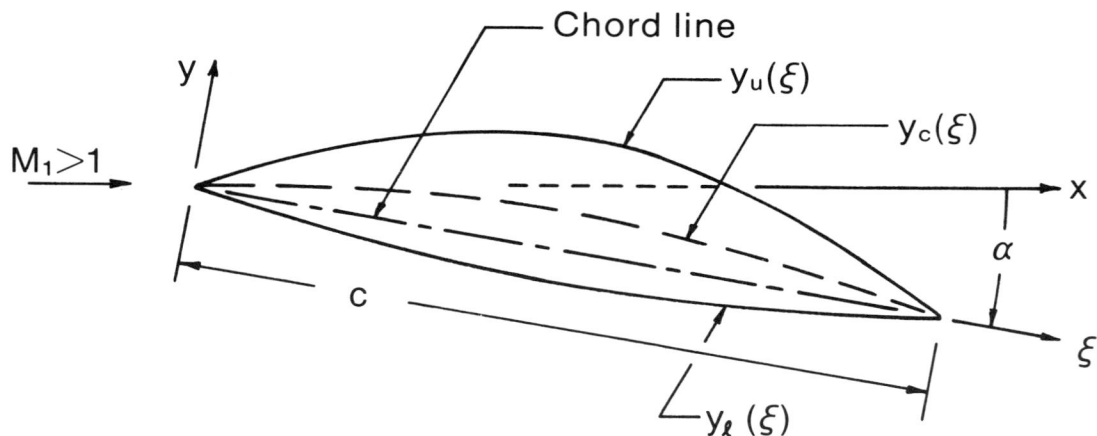

Figure 16-1. Geometry of a supersonic airfoil.

Pressure distribution. If we start with an airfoil of some specified thickness $h(\xi)$, but of zero camber $(y_c \equiv 0)$ and zero angle-of-attack, we see that the upstream surfaces near the leading edge would both be in compression   Then local, positive camber would increase the flow deviation on the upper surface and decrease it on the lower. On the other hand, a positive angle-of-attack would have the opposite effect. Thus, for a slender shape, where we can replace the sine and tangent of the local flow-angle on the surface by the angle itself, we have, approximately,

$$\theta_u = \frac{dh}{d\xi} + \frac{dy_c}{d\xi} - a \;, \qquad \theta_\ell = \frac{dh}{d\xi} - \frac{dy_c}{d\xi} + a \;. \tag{16-2}$$

Then, according to the Busemann theory, equation (15-8), the lower- and upper-surface pressures are given by

$$C_{p\ell} = C_1 \theta_\ell + C_2 \theta_\ell^2 \;, \qquad C_{pu} = C_1 \theta_u + C_2 \theta_u^2 \;. \tag{16-3}$$

For future use, the *pressure-coefficient jump* ΔC_p on the airfoil surface, becomes

$$\Delta C_p \equiv C_{p\ell} - C_{pu} \;,$$

$$= 2C_1 \left[a - \frac{dy_c}{d\xi} \right] + 4C_2 \left[a\frac{dh}{d\xi} - \frac{dh}{d\xi}\frac{dy_c}{d\xi} \right] \;. \tag{16-4}$$

The lift force. Aerodynamic *lift* is defined to be the component of the aerodynamic force perpendicular to the relative wind, i.e. perpendicular to the x-axis. However, it is more convenient to integrate with respect to the ξ-variable. Thus, for a differential segment $d\xi$, the arc length along the lower surface is ds_ℓ. If we, arbitrarily, require that the airfoil interior pressure be free-stream at p_1, then the local normal-force per-unit-span is $df_\ell = (p_\ell - p_1)ds_\ell$, of which the lift component is

$$d\ell_\ell = df_\ell \cos\theta_\ell = (p_\ell - p_1)ds_\ell \cos\theta_\ell \;.$$

If we define the angle made by the ξ-axis and the local tangent-line as $\sigma_\ell = \theta_\ell - a$, then, referring to Figure 16-2 we see that $d\xi = ds_\ell \cos\sigma_\ell$, hence

$$d\ell_\ell = (p_\ell - p_1)d\xi \cdot (\cos\theta_\ell / \cos\sigma_\ell) \;.$$

Similarly, the local lift on the upper surface is

$$d\ell_u = -(p_\ell - p_1)d\xi \cdot (\cos\theta_u / \cos\sigma_u) \;.$$

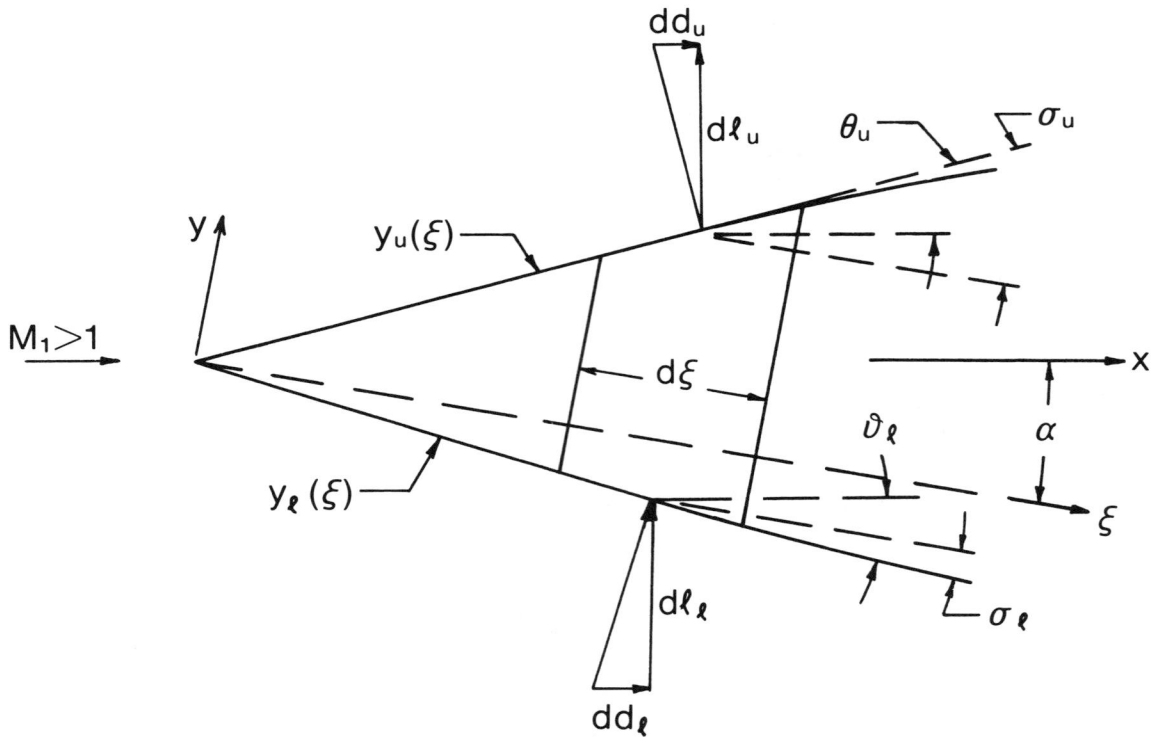

Figure 16-2. Sketch for computing the pressure forces.

Therefore, retaining two orders of magnitude in the process, for thin airfoils we can set the parentheses above involving the cosines of the angles equal to unity, by the thin-airfoil restriction. Thus

$$d\ell = d\ell_{\ell} + d\ell_{u} \simeq [(p_{\ell}-p_{1}) - (p_{u}-p_{1})]d\xi \quad .$$

The section lift is, therefore

$$\ell = \int_{0}^{c} [(p_{\ell}-p_{1}) - (p_{u}-p_{1})]d\xi \quad . \tag{16-5}$$

The section *lift-coefficient* is defined as

$$c_{\ell} \equiv \ell/q_{1}c \quad ,$$

yielding

$$c_{\ell} = \frac{1}{c} \int_{0}^{c} \Delta C_{p}d\xi \quad , \tag{16-6}$$

where q_1 is the free-stream dynamic pressure, and where we have introduced the definition of C_p to obtain the pressure-coefficient jump. It is convenient to regard ξ as a non-dimensional coordinate along a chord line of unit length. Thus (16-6) becomes

$$c_\ell \;=\; \int_0^1 \Delta C_p \, d\xi \;.$$
(16-7)

Combining (16-4) and (16-7), then integrating where possible,

$$c_\ell \;=\; 2C_1 a + 4C_2 \left[a \int_0^1 \frac{dh}{d\xi} \, d\xi - \int_0^1 \frac{dh}{d\xi} \frac{dy_c}{d\xi} \, d\xi \right] \;.$$
(16-8)

We note that the first integral above is zero if the thickness vanishes at the trailing edge. The term $dy_c/d\xi$ in the first parenthesis of (16-4) makes no contribution to the lift whatever.

The drag force. For the computation of the *drag force*, which is the force component parallel to the flow, the procedure is similar to the computation of the lift. The sum of the differential drag contributions is

$$dd \;=\; (p_\ell - p_1) d\xi \cdot \frac{\sin\theta_\ell}{\cos\sigma_\ell} + (p_u - p_1) d\xi \cdot \frac{\sin\theta_u}{\cos\sigma_u} \;,$$

and the section drag under the thin-airfoil approximation becomes

$$d \;\simeq\; \int_0^c \left[(p_\ell - p_1)\theta_\ell + (p_u - p_1)\theta_u \right] d\xi \;.$$
(16-9)

The *section drag-coefficient* is given by

$$c_d \;\equiv\; d/q_1 c \;=\; \frac{1}{c} \int_0^c \left[C_{p\ell}\theta_\ell + C_{pu}\theta_u \right] d\xi \;,$$
(16-10)

and, in non-dimensional terms,

$$c_d \;=\; C_1 \int_0^1 \left[\theta_\ell^2 + \theta_u^2 \right] d\xi + C_2 \int_0^1 \left[\theta_\ell^3 + \theta_u^3 \right] d\xi \;,$$
(16-11)

which becomes, with (16-2),

$$
c_d = 2C_1\left[a^2 + \int_0^1 \left[(\frac{dh}{d\xi})^2 + (\frac{dy_c}{d\xi})\right]d\xi\right] + 2C_2\int_0^1 (\frac{dh}{d\xi})^3 d\xi
$$

$$
+ 6C_2\left[a^2\int_0^1 \frac{dh}{d\xi}d\xi + \int_0^1 \frac{dh}{d\xi}(\frac{dy_c}{d\xi})^2 d\xi - 2a\int_0^1 \frac{dh}{d\xi}\frac{dy_c}{d\xi}d\xi\right] .
$$

$$(16\text{-}12)$$

It should be noted, for an airfoil terminating at $\xi = 1$ in a finite thickness, there is an additional contribution called the base drag $d_b = 2(p_1-p_b)h(1)$ in which the base pressure cannot be theoretically computed. Furthermore, of course, the current theory does not account for the viscous drag.

The **pitching moment**. In supersonic flow it is convenient to compute the pitching moment about the mid-chord, positive for nose-up motion. The component of the differential surface-forces normal to the chord-line is given by

$$
df_\ell \cos\sigma_\ell - df_u \cos\sigma_u = [(p_\ell - p_1) - (p_u - p_1)]d\xi ,
$$

exactly, which is equal to the approximate expression for the differential lift. The integrated moment due to this force, about the mid-chord, is

$$
m_{0.5c} = \int_0^c (\tfrac{1}{2}c - \xi)[(p_\ell - p_1) - (p_u - p_1)]d\xi .
$$

$$(16\text{-}13)$$

Then, with the definition of the *section pitching-moment coefficient*,

$$
c_{m0.5c} \equiv m_{0.5c}/q_1 c^2 ;
$$

$$(16\text{-}14)$$

in the non-dimensional scheme,

$$
c_{m0.5c} = \int_0^1 (\tfrac{1}{2} - \xi)\Delta C_p d\xi = \tfrac{1}{2}c_\ell - \int_0^1 \xi \Delta C_p d\xi .
$$

$$(16\text{-}15)$$

With (16-4) and (16-8) this reduces to

$$c_{m0.5c} = 2C_1 \int_0^1 \xi \frac{dy_c}{d\xi} d\xi + C_2 \left[2a \int_0^1 \frac{dh}{d\xi} d\xi - 2 \int_0^1 \frac{dh}{d\xi} d\xi \right.$$

$$\left. - 4a \int_0^1 \xi \frac{dh}{d\xi} d\xi + 4 \int_0^1 \xi \frac{dh}{d\xi} \frac{dy_c}{d\xi} d\xi \right] . \qquad (16\text{-}16)$$

Lighthill's ideal[5] supersonic airfoil. The results embodied in equations (16-8), (16-12) and (16-16) are equivalent to the expressions given by Lighthill (1954, pp. 388-392). If we put $C_2 = 0$ in these equations the theory of Ackeret (1925) is obtained.

Lighthill examined, within the confines of this inviscid theory, restrictions on the allowable shape in the sense of producing lift at the cost of the least drag. The resulting airfoil is referred to herein as "ideal." In the first place, we see from (16-12) that thickness, camber, and angle-of-attack all produce positive drag to second order. For structural reasons thickness cannot be eliminated, and for a fixed shape the only possible way to vary lift is through angle-of-attack variation. However, from the last term in (16-8) we see that positive camber tends to interact with the thickness distribution to produce negative lift in supersonic flow, which would be a reversal of its effect in subsonic flow. Thus, we see that it is useful to eliminate camber on two accounts and to choose an airfoil which is symmetrical about the chord line. For simplicity we require the trailing-edge thickness to be zero. Under these restrictions

$$c_\ell = 2C_1 a , \qquad (16\text{-}17)$$

which is the same as the Ackeret theory.

The only remaining third-order term in (16-12) is the integral

$$\int_0^1 \left(\frac{dh}{d\xi} \right)^3 d\xi .$$

If we require the thickness distribution to be symmetrical about the mid-chord this integral is zero. This is not, however, the optimum case since, for a properly chosen thickness-distribution, this integral can be made negative, though its magnitude is small compared to the second-order term. Thus, simplicity suggests a doubly symmetrical

---

[5]An ideal airfoil is not in any sense an optimal design but merely one which exhibits certain characteristics while eliminating drag due to camber.

thickness-distribution. The drag coefficient then becomes

$$c_d = 2C_1 \left[a^2 + \int_0^1 (\frac{dh}{d\xi})^2 \, d\xi \right] ,$$

(16-18)

which also corresponds to the Ackeret theory, where the drag is of second order. In two-dimensional supersonic flow the pressure drag, called *wave drag*, is unavoidable, in spite of the fact that the theory is isentropic to second order. There is no counterpart of wave drag in subsonic flow.

Under the preceding restrictions (16-16) reduces to

$$c_{m0.5c} = -4C_2 a \int_0^1 \xi \frac{dh}{d\xi} \, d\xi .$$

(16-19)

It is important to note that (16-19) embodies the principal distinction between the Busemann and Ackeret theories. Measurements have verified that Ackeret's expression for the pressure distribution $C_p = C_1 \theta$, is subject to significant error but still gives reasonable values for the lift and the drag coefficients, which are identical to those of the 2nd-order Busemann theory for an ideal airfoil. The reason is that the second-order contribution to the pressure is positive on both surfaces and these terms cancel each other in both the lift and drag integrations. The second-order expression for the pressure compares very favorably with experiment. The Ackeret theory, however, fails in the computation of the pitching moment. Lighthill has additional illuminating remarks on these matters in his article.

Equation (16-19) can be further simplified. Integrating by parts, and keeping in mind that $h(0) = h(1) = 0$ on an ideal airfoil,

$$\int_0^1 \xi \frac{dh}{d\xi} d\xi = -\int_0^1 h(\xi) d\xi = -\tfrac{1}{2}S ,$$

(16-20)

where S is the cross-sectional area of the airfoil profile. Furthermore, if we denote the *center-of-pressure* as $(C.P.)c$, where $C.P.$ is the fraction forward of the mid-chord designating the location where the concentrated lift-force must act to produce the same pitching moment, then

$$m_{0.5c} = c_{m0.5c}(q_1 c^2) = \ell(C.P.)c = c_\ell(q_1 c)(C.P.)c .$$

(16-21)

Hence, combining (16-19), (16-20) and (16-17),

$$C.P. = \frac{C_2}{C_1} \frac{S}{c^2} = \frac{(\gamma+1)M_1^4 - 4(M_1^2-1)}{4(M_1^2-1)^{3/2}} \frac{S}{c^2} .$$

(16-22)

For $\gamma = 7/5$ the ratio C_2/C_1 varies slowly between *1.2* and *2.0* over the range *1.25* $\leq M_1$ $\leq$ *3.0* (the minimum value of $C_2/C_1 = 1.2$ occurs at $M_1 \simeq 1.6$). The ratio $S/c^2 << 1$ for thin airfoils. For example, for a doubly symmetric diamond-airfoil of thickness ratio $t/c = 0.05$, at $M_1 = \sqrt{2}$, $C_2/C_1 = 1.4$, C.P. = *0.056* forward of the mid-chord.

17. Hayes's Rule for Subsonic Surface Pressures

The derivation of the Prandtl-Glauert similarity law relating the pressure and aerodynamic forces — to first-order in a small parameter — on a subsonic airfoil, to a related airfoil shape in incompressible flow can be found, for example, in Liepmann and Roshko (1957, pp. 253-256). Hayes (1955) developed a second-order expression (which encompasses the Prandtl-Glauert law) relating the subsonic-flow pressure distribution on an airfoil to that of the same airfoil in incompressible flow. Van Dyke (1964, p. 76) formalized the rule in convenient form.

The rule can be stated as follows: for an airfoil, (actually, for any thin shape, such as a wavy wall) if ϵ is any non-dimensional small parameter, such as the angle of attack, or camber parameter, or thickness parameter, suppose that at $M_1 = 0$ the pressure distribution is given by

$$C_p = \epsilon C_{p1}(x) + \epsilon^2 C_{p2}(x) \dots \quad ; \qquad (17\text{-}1a)$$

then for $0 \leq M_1 < 1$,

$$C_p = \frac{\epsilon}{(1-M_1^2)^{\frac{1}{2}}} C_{p1}(x) + \epsilon^2 \frac{(\gamma+1)M_1^4 + 4(1-M_1^2)}{4(1-M_1^2)^2} C_{p2}(x) + \dots \qquad (17\text{-}1b)$$

The expressions $C_{p1}(x)$ and $C_{p2}(x)$ are the first- and second-order contributions, respectively, due to the effect of the parameter ϵ. A complete exposition of the procedure to obtain the higher-order contributions due to the thickness effect in incompressible flow can be found in Van Dyke (1964). For a treatment of the effects of camber and angle-of-attack, see Van Dyke (1956).

The subsonic Mach-number functions multiplying $C_{p1}(x)$ and $C_{p2}(x)$ are obviously similar to the supersonic coefficients C_1 and C_2 of Sections 14 and 15. It should be noted that both the subsonic and supersonic relations break down in the limit as $M_1 \rightarrow 1$.

18. The Equations of Inviscid Three-Dimensional Flow

The continuity equation for source-free flow is

$$\frac{\partial \rho}{\partial t} + \nabla \cdot (\rho \mathbf{V}) = 0 \quad . \qquad (18\text{-}1)$$

Written in an inertial frame of reference, the dynamical equation for inviscid flow with negligible body-forces is

$$\frac{D\mathbf{V}}{Dt} = -\frac{1}{\rho}\nabla p \quad, \tag{18-2}$$

where the substantial-derivative operator is

$$D/Dt = \partial/\partial t + \mathbf{V}\cdot\nabla \quad. \tag{18-3}$$

The state relations for a thermally and calorically perfect-gas are the same as in one-dimensional flow:

$$p = \rho RT \quad, \tag{18-4a}$$

and

$$h = c_p T \quad, \tag{18-4b}$$

$$c_p = \gamma/(\gamma-1) \quad, \quad \gamma = c_p/c_v \quad, \quad c_p \text{ and } c_v \text{ are constants} \quad. \tag{18-4c}$$

Similarly, the first law of thermodynamics when combined with the entropy relation for a perfect gas is unchanged:

$$pv^\gamma = p_1 v_1^\gamma \exp[(\gamma-1)\Delta s/R] \quad. \tag{18-5}$$

Equation (18-5) refers to a fixed particle between any two states.

In a transformation which is *workless* (absence of shaft work) and inviscid, the energy equation for negligible body forces is the same as (3-9), viz.

$$\frac{Dh^o}{Dt} = \frac{1}{\rho}\frac{\partial p}{\partial t} + \dot{q} \quad, \tag{18-6a}$$

where

$$h^o = h + \tfrac{1}{2}V^2 \quad, \quad V \equiv |\mathbf{V}| \quad. \tag{18-6b}$$

If the flow is also adiabatic (18-6a) becomes

$$\frac{Dh^o}{Dt} = \frac{1}{\rho}\frac{\partial p}{\partial t} \quad. \tag{18-6c}$$

A thermodynamic transformation of a particle of fixed identity, which is both adiabatic and inviscid, is isentropic, more precisely is *particle isentropic*; that is

$$\frac{Ds}{Dt} \equiv \frac{\partial s}{\partial t} + \mathbf{V} \cdot \nabla s = 0 \ . \tag{18-7}$$

The relations for the speed of sound, derived in Section 5, remain valid in three dimensions. Thus

$$a^2 \equiv (\partial p / \partial \rho)_s = \gamma (\partial p / \partial \rho)_T \ , \tag{18-8a}$$

generally; and, for a thermally perfect gas,

$$a^2 = \gamma R T = \gamma p / \rho \ . \tag{18-8b}$$

Vorticity (ω) is a vector, kinematical flow-property which is utilized to describe the rotation of a fluid particle about its own axis. It is equal to twice the angular velocity. In vector notation it is given by the operation

$$\omega = \nabla \mathbf{x} \mathbf{V} \ . \tag{18-9}$$

In an *irrotational flow* (definition: $\omega \equiv 0$) it follows that a scalar, point function Φ, called the *velocity potential*, can be defined such that

$$\mathbf{V} = \nabla \Phi \ . \tag{18-10}$$

Crocco's equation. The left-hand side of (18-2) can be expanded by a vector theorem. Thus

$$\begin{aligned}
\frac{D\mathbf{V}}{Dt} &= \frac{\partial \mathbf{V}}{\partial t} + (\mathbf{V} \cdot \nabla)\mathbf{V} \ , \\
&= \frac{\partial \mathbf{V}}{\partial t} + \nabla(\tfrac{1}{2}V^2) + (\nabla \mathbf{x} \mathbf{V}) \mathbf{x} \mathbf{V} \ , \\
&= \frac{\partial \mathbf{V}}{\partial t} + \nabla(\tfrac{1}{2}V^2) + \omega \mathbf{x} \mathbf{V} \ .
\end{aligned} \tag{18-11}$$

Then, the first law of thermodynamics (1-5), written for an observer stationary with respect to the mass center, can be written as a field equation (for an arbitrary, fixed instant) as

$$T\nabla s = \nabla h - \upsilon \nabla p \ . \tag{18-12}$$

Substitution of (18-12) and (18-11) into (18-2) leads to

$$\frac{\partial \mathbf{V}}{\partial t} + \nabla(\tfrac{1}{2}V^2) + \omega \mathbf{x} \mathbf{V} = T\nabla s - \nabla h \ ,$$

or

$$\frac{\partial \mathbf{V}}{\partial t} + \nabla(h + \tfrac{1}{2}V^2) = T\nabla s + \mathbf{V} \mathbf{x} \omega \quad,$$

$$\frac{\partial \mathbf{V}}{\partial t} + \nabla h^o = T\nabla s + \mathbf{V} \mathbf{x} \omega \quad. \tag{18-13}$$

Equation (18-13) is *Crocco's equation* subject to the restrictions of inviscid flow with negligible body forces.

The pressure-velocity relation for inviscid, adiabatic flow. As previously noted, for inviscid, adiabatic flow the flow must be particle isentropic. If, in addition, all particles originate in a region of uniform conditions (e.g. at p_1, ρ_1, h_1 or T_1, s_1, and $\mathbf{V}_1$) then it follows (barring the presence of shock waves) that $s =$ constant throughout. Such a flow is said to be *homentropic*. Therefore, under these restrictions, (18-5) can be substituted into the right side of (18-2) to produce

$$\frac{1}{\rho}\nabla p = \frac{1}{Cp^{1/\gamma}}\nabla p = \frac{\gamma}{C(\gamma-1)}\nabla(p)^{\frac{\gamma-1}{\gamma}} = \frac{\gamma}{\gamma-1}\nabla(\tfrac{p}{\rho}) \quad.$$

Thus (18-2) becomes

$$\frac{D\mathbf{V}}{Dt} = -\frac{\gamma}{\gamma-1}\nabla(\tfrac{p}{\rho}) \quad. \tag{18-14}$$

We next take the curl of (18-14); then, since the curl of a vector point-function is identically zero, we have

$$\nabla \mathbf{x} \frac{D\mathbf{V}}{Dt} = \frac{D\omega}{Dt} = -\frac{\gamma}{\gamma-1}\nabla \mathbf{x} \nabla(\tfrac{p}{\rho}) = 0 \quad.$$

Thus, for a particle, ω is at most constant. However, since the flow originated in a region where $\omega_1 = 0$, then $\omega \equiv 0$ and a velocity potential must exist.

If the left-hand side of (18-14) is replaced by (18-11) and the velocity-potential relation (18-10) introduced, equation (18-14) becomes

$$\frac{\partial}{\partial t}(\nabla\Phi) + \nabla(\tfrac{1}{2}V^2) = -\frac{\gamma}{\gamma-1}\nabla(\tfrac{p}{\rho}) \quad,$$

or

$$\nabla(\frac{\partial \Phi}{\partial t} + \frac{\gamma}{\gamma-1}\frac{p}{\rho} + \tfrac{1}{2}V^2) = 0 \quad,$$

which has the integral

$$\frac{\partial \Phi}{\partial t} + \frac{\gamma}{\gamma - 1}\frac{p}{\rho} + \tfrac{1}{2}V^2 = F(t) \ , \tag{18-15}$$

where $F(t)$ is determined by the boundary conditions. Consequently, under the stated restrictions, the theoretical problem is reduced to solving for the velocity field since the pressure follows immediately from (18-15).

$\underline{\text{The equations of steady flow}}$. If we further restrict considerations to a steady flow ($\partial/\partial t \equiv 0$) in which all particles are considered to originate from a region of uniform conditions as previously noted, then from (18-13), and under the steady-flow restriction,

$$\nabla h^o = T\nabla s = 0 \ . \tag{18-16}$$

Consequently, all particles have the same total enthalpy h^o and, in the absence of shocks, all particles have the same entropy. Such flows are said to be *homenergetic* and *homentropic*, respectively.

Therefore, the energy equation (18-6a) for a perfect gas[6] reduces to $Dh^o/Dt = 0$, or

$$h^o = c_p T^o = c_p T + \tfrac{1}{2}V^2 = constant \ , \tag{18-17a}$$

throughout. If (18-17a) is combined with the expression for sound speed we have the familiar

$$T^o/T = 1 + \tfrac{1}{2}(\gamma - 1)M^2 \ , \tag{18-17b}$$

where

$$M \equiv V/a \ . \tag{18-18}$$

Thus from (18-16) and (18-17a) the requirement of homenergetic, perfect-gas flow guarantees that the total temperature T^o = constant throughout.

It also follows from (18-5), for homentropic flow, that

$$pv^\gamma = constant, \tag{18-19}$$

throughout. In such case the hypothetical total-pressure and total-density are defined,

---

[6] Alternatively, since $Dh^o/Dt = \partial h^o/\partial t + \mathbf{V}\cdot\nabla h^o$, in steady flow the velocity vector $\mathbf{V}$ could be parallel to the gradient of the total enthalpy (or total temperature). This possibility is excluded from further consideration.

which are constants throughout, and are related to the local conditions by

$$T^o/T \;=\; (p^o/p)^{\frac{\gamma-1}{\gamma}} \;=\; (\rho^o/\rho)^{(\gamma-1)} = 1 + \tfrac{1}{2}(\gamma-1)M^2 \;. \qquad (18\text{-}20)$$

Flows with shock waves. An extremely important class of flows exists which may involve shocks, straight or curved, near, or originating on a body which is subject to a uniform irrotational flow (therefore, homenergetic and homentropic) upstream of the shocks. Behind (downstream of) the shock the flow is often treated (as a first approximation) as inviscid.

From (18-13) the appropriate form of Crocco's equation under the homenergetic restriction is

$$T\nabla s \;=\; \omega \mathbf{x} \mathbf{V} \;. \qquad (18\text{-}21)$$

For a perfect gas, equation (8-9), the general expression for the entropy change of a particle as it traverses a shock remains valid, i.e.

$$(s\text{-}s_1)/R \;=\; \ell n (T^o/T_1^o)^{\frac{\gamma}{\gamma-1}} - \ell n (p^o/p_1^o) \;. \qquad (18\text{-}22)$$

Since the flow across a shock is homenergetic, T^o is constant, therefore

$$\nabla(s/R) \;=\; -\nabla(\ell n p^o) \;=\; -(1/p^o)\nabla p^o \;. \qquad (18\text{-}23)$$

With (18-23), (18-21) becomes

$$\omega \mathbf{x} \mathbf{V} \;=\; -(RT/p^o)\nabla p^o \;. \qquad (18\text{-}24)$$

For a straight shock (e.g. a weak shock generated by a wedge, or a conical shock generated by a cone aligned with the flow) the total-pressure decrease is the same for all particles passing through the shock because the decrease depends (for a perfect gas) only on the upstream Mach-number component normal to the shock. Consequently, the region downstream of the shock is also irrotational.[7]

For a curved, detached shock, the flow behind the shock involves curved streamlines with a non-uniform total-pressure field and, according to (18-24), must be rotational. To make things worse there are regions of mixed subsonic and supersonic flow whose boundaries are not known in advance. Such geometries invariable preclude analytical

---

[7]An alternative possibility is that ω and $\mathbf{V}$ are everywhere parallel to each other. Such a flow is called a *Beltrami flow*, a category which is not further considered herein.

solutions and resort must be made to numerical computations.

The equation of Molenbroek. We now derive the fundamental equation which governs the velocity distribution in homenergetic, particle-isentropic flow. With the steady-flow restriction the continuity equation can be expanded as follows:

$$\nabla \cdot (\rho \mathbf{V}) = \rho \nabla \cdot \mathbf{V} + \mathbf{V} \cdot \nabla \rho = 0 \quad . \tag{18-25}$$

From (18-2) and (18-11) we can solve for

$$\nabla p = -\rho \nabla (\tfrac{1}{2} V^2) - \rho \omega \mathbf{x} \mathbf{V} \quad . \tag{18-26}$$

If we consider that $p = p(s,\rho)$, the terms such as $\partial p / \partial x$, which appear in ∇p can be written as

$$\frac{\partial p}{\partial x} = \left(\frac{\partial p}{\partial \rho}\right)_s \frac{\partial \rho}{\partial x} + \left(\frac{\partial p}{\partial s}\right)_\rho \frac{\partial s}{\partial x} = a^2 \frac{\partial \rho}{\partial x} + \left(\frac{\partial p}{\partial s}\right)_\rho \frac{\partial s}{\partial x} \quad .$$

Thus

$$\nabla p = a^2 \nabla \rho + \left(\frac{\partial p}{\partial s}\right)_\rho \nabla s \quad ,$$

and

$$\mathbf{V} \cdot \nabla p = a^2 \mathbf{V} \cdot \nabla \rho + \left(\frac{\partial p}{\partial s}\right)_\rho \mathbf{V} \cdot \nabla s = a^2 \mathbf{V} \cdot \nabla \rho \quad , \tag{18-27}$$

where we have utilized the fact that for steady, particle-isentropic flow (18-7) requires that $\mathbf{V} \cdot \nabla s = 0$. Equation (18-27) gives us one expression for $\mathbf{V} \cdot \nabla p$. A second can be obtained by taking the dot product of $\mathbf{V}$ with equation (18-26). Equating these we have

$$\mathbf{V} \cdot \nabla p = a^2 \mathbf{V} \cdot \nabla \rho = -\rho \mathbf{V} \cdot \nabla (\tfrac{1}{2} V^2) - \rho \mathbf{V} \cdot (\omega \mathbf{x} \mathbf{V}) \quad . \tag{18-28}$$

Now, since the vector $\omega \mathbf{x} \mathbf{V}$ is orthogonal to $\mathbf{V}$, the last term in (18-28) is zero, and we can eliminate $\mathbf{V} \cdot \nabla \rho$ from (18-28) by means of (18-25). Thus

$$a^2 \mathbf{V} \cdot \nabla \rho = -\rho a^2 \nabla \cdot \mathbf{V} = -\rho \mathbf{V} \cdot \nabla (\tfrac{1}{2} V^2) \quad , \tag{18-29}$$

or

$$\nabla \cdot \mathbf{V} - (1/a^2) \, \mathbf{V} \cdot \nabla (\tfrac{1}{2} V^2) = 0 \quad . \tag{18-30}$$

It appears that (18-30) was first derived by Molenbroek (1890) and, in the writer's view, it is appropriate after a century to refer to it as *Molenbroek's equation*.

In two-dimensional, or axi-symmetric flow (axis of symmetry aligned with the x-axis), (18-30) becomes

$$(1 - \frac{u^2}{a^2}) \frac{\partial u}{\partial x} - \frac{uv}{a^2} (\frac{\partial u}{\partial y} + \frac{\partial v}{\partial x}) + (1 - \frac{v^2}{a^2}) \frac{\partial v}{\partial y} + \epsilon \frac{v}{y} = 0 . \qquad (18\text{-}31)$$

In (18-31) u and v are the velocity components parallel to the x- and y-axes, respectively, and $\epsilon \equiv 0,1$ in two-dimensional flow, and axi-symmetric flow, respectively.

The additional equation needed to complete the governing equations comes from (18-21); namely

$$T \frac{ds}{dn} = - V\omega_m = - V(\frac{V}{R} - \frac{\partial V}{\partial n}) . \qquad (18\text{-}32)$$

Equation (18-32) is written in path (or natural) coordinates where n is the direction normal to a streamline directed towards the local center-of-curvature, $\omega_m = (V/R - \partial V/\partial n)$ is the vorticity, m is the bi-normal direction perpendicular to the x,y-plane, and R is the local radius of curvature. For a streamline concave up, m coincides with the $+z$-axis in a two-dimensional flow. When concave down m coincides with the negative z-axis. The appropriate interpretation for axi-symmetric flow follows readily.

If the flow is irrotational then $\omega_m \equiv 0$, and a velocity potential $\Phi = \Phi(x,y)$ exists. Expressed in terms of the velocity potential, the velocity components are $u = \Phi_x$, $v = \Phi_y$. Equation (18-31) then becomes

$$(1 - \frac{\Phi_x^2}{a^2})\Phi_{xx} - \frac{2\Phi_x \Phi_y}{a^2} \Phi_{xy} + (1 - \frac{\Phi_y^2}{a^2})\Phi_{yy} + \epsilon\frac{\Phi_y}{y} = 0 . \qquad (18\text{-}33)$$

Equation (18-33) has the form $A\Phi_{xx} + 2B\Phi_{xy} + C\Phi_{yy} + D\Phi_y = 0$. The theory of partial differential equations[1] shows that the nature of solution satisfying (18-33) depends on whether the *discriminant*

$$B^2 - AC = \frac{\Phi_x^2 + \Phi_y^2}{a^2} - 1 = \frac{u^2 + v^2}{a^2} - 1 = M^2 - 1 \lessgtr 0 . \qquad (18\text{-}34)$$

That is, if $M > 1$ the flow is locally supersonic and the equation is said to be *hyperbolic*. If $M = 1$ the flow is sonic and the equation is *parabolic*. If $M < 1$ the flow is subsonic and the equation is *elliptical*. The analytical difficulties in solving (18-33) in applications where the flow type changes from subsonic to supersonic, or the reverse, are substantial.

---

[1]For example, see Carrier and Pearson (1976, pp. 75-83).

19. Axi-symmetric Supersonic Flow About a Circular Cone.

One of the few exact, three-dimensional solutions of Molenbroek's equation in supersonic flow — and one which is of major practical aerodynamic significance — is due to Taylor and Maccoll (1933). For a steady, uniform, supersonic free-stream over an infinite, circular cone aligned with the flow, Figure 19-1, a solution can be found (as long as the vertex half-angle is not too large) which has a co-axial conical shock surrounding the cone. In the region between the cone and the shock there exists a non-uniform, homentropic — therefore irrotational — supersonic flow, though at a higher entropy level than upstream of the shock.

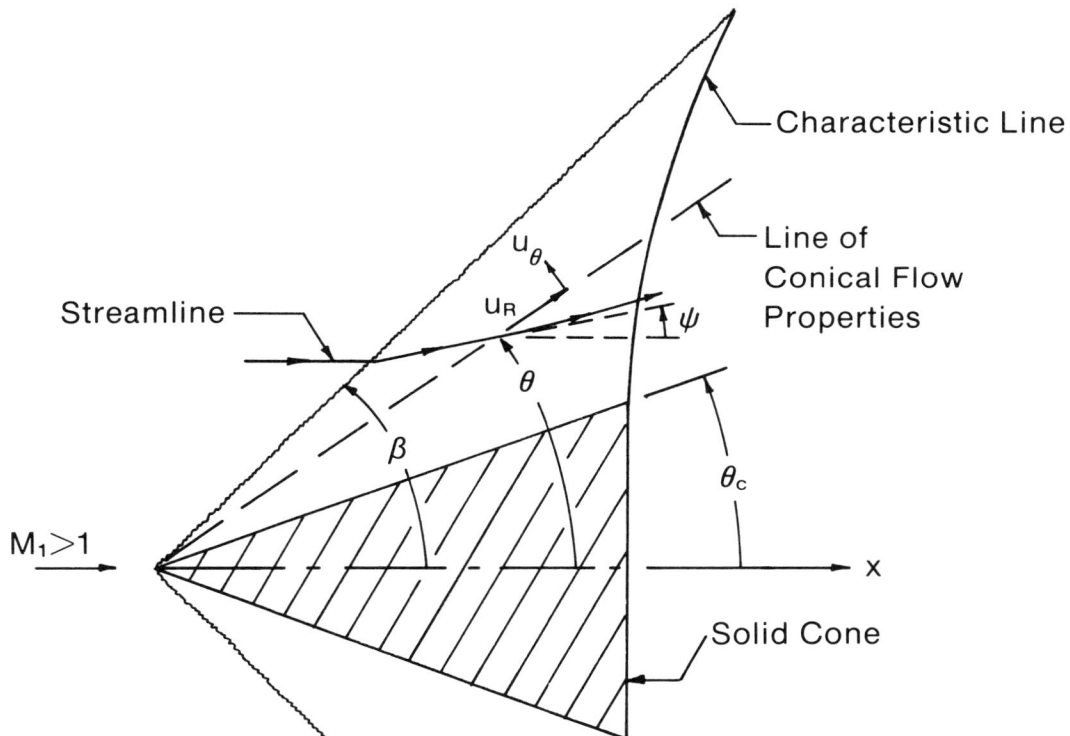

Figure 19-1. Axi-symmetric supersonic flow about a cone.

Taylor and Maccoll postulated that along any radial line emanating from the vertex, flow conditions (e.g. p, ρ, T, $\mathbf{V}$) are invariant.[9] Such a flow is said to be *conical*, and

---

[9]This does not imply that the velocity vector along a ray is locally tangent, however.

is a member of the class of self-similar flows, of which the Prandtl-Meyer flow of Section 15 is also an example. Since there is no characteristic length associated with the flow, which is axi-symmetric, the independent variables can be reduced to only one space variable, namely the angle between the flow axis and the ray called the azimuthal angle.

Consequently, the flow is most easily analyzed in terms of spherical coordinates R, θ, ϕ, where θ is the azimuthal angle, and ϕ is the meridional angle (rotation about the x-axis). Referring to (18-30) we write the two vector terms in Molenbroek's equation as follows:

$$\nabla \cdot \mathbf{V} = \frac{\partial u_R}{\partial R} + \frac{2u_R}{R} + \frac{\cot\theta \cdot u_\theta}{R} + \frac{1}{R}\frac{\partial u_\theta}{\partial \theta} + \frac{1}{R\sin\theta}\frac{\partial u_\phi}{\partial \phi} \quad , \tag{19-1a}$$

$$(\mathbf{V} \cdot \nabla)(\tfrac{1}{2}V^2) = \tfrac{1}{2}\left[u_R \frac{\partial}{\partial R} + \frac{u_\theta}{R}\frac{\partial}{\partial \theta} + \frac{u_\phi}{R\sin\theta}\frac{\partial}{\partial \phi} \right](u_R^2 + u_\theta^2 + u_\phi^2) \quad . \tag{19-1b}$$

We now impose on (19-1) the axi-symmetric-flow restrictions and the conical-flow restriction, respectively:

$$\partial/\partial\phi \equiv 0 \ , \quad u_\phi \equiv 0 \ ; \quad \partial/\partial R \equiv 0 \ . \tag{19-2}$$

In consequence of (19-2) $\partial/\partial\theta \to d/d\theta$; thus

$$\nabla \cdot \mathbf{V} = \frac{1}{R}(2u_R + \cot\theta \cdot u_\theta + du_\theta/d\theta) \quad , \tag{19-3a}$$

and

$$\mathbf{V} \cdot \nabla(\tfrac{1}{2}V^2) = \frac{1}{R}(u_R u_\theta du_R/d\theta + u_\theta^2 du_\theta/d\theta) \quad . \tag{19-3b}$$

Combining (19-3) and (18-30) produces

$$2u_R + \cot\theta \cdot u_\theta - (u_R u_\theta/a^2)du_R/d\theta + (1 - u_\theta^2/a^2)du_\theta/d\theta = 0 \quad . \tag{19-4}$$

We have not yet made use of the irrotationality condition. In spherical coordinates the vorticity components are given by

$$\omega_R = \frac{1}{R\sin\theta} \left[\frac{\partial}{\partial\theta} (\sin\theta \cdot u_\phi) - \frac{\partial u_\theta}{\partial\phi} \right] , \qquad (19\text{-}5a)$$

$$\omega_\theta = \frac{1}{R\sin\theta} \left[\frac{\partial u_R}{\partial\phi} - \sin\theta \frac{\partial}{\partial R} (Ru_\phi) \right] , \qquad (19\text{-}5b)$$

$$\omega_\phi = \frac{1}{R} \left[\frac{\partial}{\partial R} (Ru_\theta) - \frac{\partial u_R}{\partial\theta} \right] . \qquad (19\text{-}5c)$$

In putting each of these equal to zero only the third yields new information, viz.

$$du_R/d\theta = u_\theta , \qquad (19\text{-}6)$$

where we note that the conical-flow restriction does not apply to R itself, i.e. $\partial R/\partial R = 1$. Introducing (19-6) into the two right-hand terms of (19-4) produces the following simplification:

$$\frac{d^2 u_R}{d\theta^2} + u_R = - \frac{u_R + u_\theta \cot\theta}{1 - u_\theta^2/a^2} . \qquad (19\text{-}7)$$

Equations (19-6) and (19-7), along with the energy equation

$$u_R^2 + u_\theta^2 + \frac{2}{\gamma-1} a^2 = u_m^2 , \qquad (19\text{-}8)$$

are the governing equations of the flow, where u_m is the theoretical maximum flow-speed introduced in Section 6. The first two form a set of coupled, non-linear ordinary differential equations which can be solved numerically. This reduction represents an enormous analytical simplification over the original, non-linear, partial-differential equations.

Non-dimensionalization of the equations. If we introduce, new, non-dimensional variables such that $u \equiv u_R/u_m$, $v \equiv u_\theta/u_m$, and if we replace[10] a/u_m by a, the set then

---

[10]The reason for the dual use of the symbol a is to achieve agreement (with several minor deviations) with the notation of Sims (1964a), which is the most comprehensive source of numerical results.

becomes:

$$\frac{d^2u}{d\theta^2} + u = \frac{a^2(u + v\cot\theta)}{v^2 - a^2} \quad , \tag{19-9}$$

$$v = du/d\theta \quad , \tag{19-10}$$

$$a^2 = \tfrac{1}{2}(\gamma - 1)\left[1 - (u^2 + v^2)\right] \quad . \tag{19-11}$$

Boundary conditions. On the cone surface the flow must be tangent. Thus, for

$$\theta = \theta_c : \quad v = 0 \quad . \tag{19-12}$$

The resulting surface-speed is denoted as u_c. The solution requires that for a specified θ_c a guess be made for the initial surface speed u_c and the computation proceeds towards the shock wave whose location is unknown. Reverting, temporarily, to the (dimensional) notation of Section 12, we have from Prandtl's relation (12-8h), and from the shock geometry, respectively,

$$u_1 u_2 = a^{*2} - \frac{\gamma - 1}{\gamma + 1} v^2 = \frac{\gamma - 1}{\gamma + 1}(u_m^2 - v^2) \quad , \tag{19-13a}$$

$$\tan\beta = u_1/v \quad . \tag{19-13b}$$

where the subscript 2 denotes flow conditions immediately downstream of the shock. Noting that, in spherical coordinates, $u_2 = -u_\theta$, $v = u_R$, then elimination of u_1 from (19-13) yields

$$\tan\beta = \frac{\gamma - 1}{\gamma + 1} \frac{u_R^2 - u_m^2}{u_R u_\theta} \quad , \tag{19-14a}$$

written in the dimensional spherical coordinates. In the non-dimensional scheme, therefore, the boundary condition behind the shock becomes for

$$\theta = \beta : \quad \tan\beta = \frac{\gamma - 1}{\gamma + 1} \frac{u^2 - 1}{uv} \quad , \tag{19-14b}$$

where $v < 0$. Since the Rankine-Hugoniot equation is built into (19-14) some writers refer to it by the same appelation.

At an arbitrary point within the flow field, the flow speed is given by

$$w^2 = u_R^2 + u_\theta^2$$

in dimensional terms. The corresponding non-dimensional version is

$$(w/u_m)^2 = u^2 + v^2 \ . \tag{19-15}$$

Consequently, the local flow direction ψ, with respect to the x-axis, is given by

$$\psi = \theta + tan^{-1}(u_\theta/u_R) = \theta + tan^{-1}(v/u) \ . \tag{19-16}$$

where, for emphasis, we note that u_θ, v, and $tan^{-1}(v/u)$ are all negative.

Free-stream Mach number. Again, in the notation of Section 12, the free-stream Mach-number is given by

$$M_1^2 = w_1^2/a_1^2 = v^2/a_1^2 \cos^2\beta = u_R^2/a_1^2 \cos^2\beta \ , \tag{19-17}$$

the last equation being written in spherical coordinates. The energy equation, written ahead of the shock, is

$$w_1^2 + \frac{2}{\gamma-1} a_1^2 = u_m^2 \ ;$$

Thus, since $w_1 = u_R/\cos\beta$

$$a_1^2 = \frac{\gamma-1}{2} (u_m^2 - u_R^2/\cos^2\beta) \ . \tag{19-18}$$

Combining (19-17) and (19-18), and then returning to the non-dimensional notation, we have

$$M_1^2 = \frac{2}{\gamma-1} \left[\frac{u_R^2}{(u_m^2 - u_R^2/\cos^2\beta)\cos^2\beta} \right] = \frac{2}{\gamma-1} \frac{u^2}{\cos^2\beta - u^2} \ . \tag{19-19}$$

Note on existing computations. The first results, those of Taylor and Maccoll (1933) were restricted to calculations on cones of semi-vertex angles of 10°, 20° and 30°. It is a tribute to these authors that they identified all the crucial questions and obtained results — presumably using hand-operated state-of-the-art mechanical calculators — unmodified by later calculations.

The first extensive electronic-machine computations were done at MIT's Center for Analysis under the direction of Kopal (1947) a work which required a large staff and which was completed over a period of several years. These computations employed $\gamma = 1.405$ as the most appropriate value for air. As previously remarked, for a specified cone-angle and surface-speed, the computation proceeds from the cone surface towards the shock until a station is found where the Rankine-Hugoniot equation is satisfied. This identifies the shock location. A drawback of Kopal's computations is that the free-stream Mach numbers thus obtained are given as un-rounded 5-digit numbers which appear at irregular intervals, and are thus not in the most convenient format for typical use. This deficiency should not obscure the fact that this study stands as one of the landmarks of early, extensive, computational projects in applied aerodynamics.

The rapid development of the high-speed computer in the 17 years following Kopal's work is demonstrated by the publication of Sims (1964a) tables for the same flow. Apparently undertaken as a one-person effort, he essentially repeated the computations of Kopal, except that he used $\gamma = 7/5$,[11] and that, for each case computed, an iteration was performed by modifying the initial-value of the surface speed until a specified value of the free-stream Mach number was obtained. A principal purpose of reproducing Kopal's work was to obtain numerical data in the conical-flow field for use as initial values in an even more extensive computation of the flow about yawed cones. The results of the second study can be found in Sims (1964b).

Table F-2 in this work summarizes the principal results of Sims (1964a) for the shock angle, surface pressure, temperature, density, etc. as a function of the free-stream Mach number with cone semi-vertex angle as the principal parameter, for the case of $\gamma = 7/5$. Computations are made for cone angles between 2.5° and 30° in increments of 2.5°, and for $M_1 = 1.5, 1.75, 2.0, 2.5, 3.0, 3.5, 4.0, 4.5, 5.0, 6.0, 7.0, 8.0, 10.0, 12.0, 15.0$ and 20.0.

Tables F-1a and F.1b reproduce Sims's Tables 1 and 10 for the flow properties on the cone surface, and immediately behind the shock, respectively, for the minimum free-stream Mach number for a specified cone angle θ_c for which a solution of the equations could be found which is supersonic everywhere in the flow downstream of the shock. This corresponds, ideally, to the case where sonic flow $M_c = 1$ is attained on the cone surface. For computational reasons the actual cone converged surface-value was chosen to be $M_c = 1.0000050$.

20. On Thermophysical Properties of Gases

One objective of the kinetic theory of gases is to demonstrate that the macroscopic concepts of thermodynamics, e.g. pressure, temperature, internal energy, can all be derived from the principles of mechanics. In Jeans (1952) he gives what is perhaps the clearest and most elegant introduction to the subject. In the following paragraphs we list, along with some explanatory material, most of the derived relations which refer to an ensemble of a single species of monatonic molecules (unless otherwise specified), confined within a fixed volume. The volume is in isolation from its environment, meaning that no mass, momentum, or energy is transported across its boundaries.

---

[11]Experience has shown that, for engineering purposes, the differences in using $\gamma = 7/5$ for air, instead of $\gamma = 1.405$, are insignificant.

Molecular density. If the volume V contains N molecules the molecular density ν is given by

$$\nu \equiv N/V \ . \tag{20-1}$$

At sea-level pressure and temperature (see Section 21) the molecular density of helium, for example, would be $\nu = 2.547 \times 10^{25}$ per cubic meter.

Mass density. If $\overline{m}$ is the *molecular weight* ($\overline{m}$ is dimensionless, in fact) and u is the *unified atomic-mass unit*,[12] the (macroscopic) mass density is

$$\rho \equiv \overline{m} u \nu = m \nu \ , \tag{20-2}$$

where $m = \overline{m} u$ is the *molecular mass*.

Mean-free-speed. One of the earlier fundamental triumphs in kinetic theory was the demonstration by Maxwell in which he showed that the expression for the distribution of the molecular velocities in a gas ensemble is of negative exponential form, the fraction of molecules exceeding a certain value rapidly becoming smaller as the specified value increases.

The *mean-free-speed*, denoted as C, is the value of the speed, at any instant, at which the sum of the individual kinetic energies of random motion of all molecules with speeds greater than C just equals that of all molecules with speeds below.

The principle *equi-partition of energy* denies the possibility of the existence of any preferred direction such that the components of velocity of all molecules in that direction might involve a total kinetic energy greater than for any other. Thus if $\overline{u}, \overline{v}, \overline{w}$ are the mean velocities in rectangular cartesian coordinates, which produce the same, total kinetic-energy, we must have

or

$$\overline{u}^2 = \overline{v}^2 = \overline{w}^2 = C^2/3 \ , \tag{20-3a}$$

$$C^2 = \overline{u}^2 + \overline{v}^2 + \overline{w}^2 \ . \tag{20-3b}$$

Pressure. Pressure is conceived to be the effect of the bombardment of a wall by the molecules contained thereby. Each collision is assumed to occur in such a way that the component of velocity normal to the wall is reversed, the particle kinetic-energy remaining constant. Thus pressure is the time-averaged force per-unit-area over all colliding molecules, and over all possible velocities. The resulting relation is

$$p = \frac{2}{3} \frac{N(\tfrac{1}{2}mC^2)}{V} = \frac{2}{3} \frac{Kinetic\ Energy}{Volume} \ . \tag{20-4}$$

---

[12] In SI, $1\ u = 1.660\ 53 \times 10^{-27}\ kg.$

Indtoducing the definition of density from (20-2), (20-4) becomes

$$p = \frac{1}{3} \rho C^2 .$$

(20-5)

Equations (20-4) and (20-5) are valid strictly for a monotonic gas.

Temperature. Comparing (20-5) with the equation of state for a thermally perfect gas we see that

$$C^2 = 3RT .$$

(20-6)

Thus, temperature is a measure of the average kinetic-energy of random motion of all particles.

Mean-free-path length. The ceaseless motion of an ensemble of particles naturally results in inter-molecular collisions — more properly called encounters — between pairs of molecules. The probability of triple encounters in a dilute gas[13] is so small as to be negligible. If molecules are considered to be perfectly elastic spheres of diameter σ, then the mean-free-path length, λ, between collisions, can be shown to be

$$\lambda = \frac{1}{(\pi\sqrt{2})\nu\sigma^2} .$$

(20-7)

Thus, as we would expect, the mean-free-path length is inversely proportional to the number density.

There are many techniques for measuring the diameter of a molecule, most of them indirect, in which the effective molecular diameter is inferred. For example, Hirschfelder, Curtiss, and Bird (1954) indicate that the molecular diameters for argon, neon, and nitrogen, respectively, as inferred from measurements of viscosity, are $10^8\sigma = 3.64, 2.58,$ and $3.75\ cm$. The corresponding values from diffusion measurements are $3.47, 2.42$ and 3.48. The notion of a molecule as a sphere, of course, is strictly for macroscopic computational convenience, and disappears on the level of quantum mechanics.

Knudsen number. Workers in the dynamics of rarefied gases[14] are often confronted with the problem of deciding whether or not continuum theory applies. The answer hinges on the magnitude of a dimensionless parameter called the *Knudsen number Kn*, which is defined as

$$Kn \equiv \lambda/L ,$$

(20-8)

where L is a reference, or characteristic, length associated with the physical body under study. For example, in the flow about a sphere the reference length is evidently the sphere diameter.

---

[13]In contrast, in dense gases triple encounters must be taken into account.
[14]c.f. Patterson (1971).

The criterion used by the aerodynamicist divides the flow into regimes thus, if:

$$Kn < 0.1, \quad \text{continuum theory applies;}$$
$$0.1 < Kn < 1.0, \quad \text{this specifies a transition regime involving slip flow;}$$
$$Kn > 1.0, \quad \text{free-molecule flow governs.}$$

The ranges designated thereby should be regarded only as approximate.

This approach can be extended to provide a criterion to distinguish a dilute gas from a dense gas in which we expect that the average spacing between molecules (were they arranged in a cubical grid) would start to approach the molecular diameter. Thus, to specify a dilute gas we require that

$$\sigma/\lambda \ll 1 \ . \tag{20-9}$$

Dynamic viscosity. The *dynamic viscosity* μ is a macroscopic coefficient which arises in the equations of three-dimensional flow, e.g. Howarth (1956). It comes in as a multiplying coefficient in the assumed *stress-strain relation*, the most elementary expression of which is found in *Newton's law of friction*

$$\tau = \mu \frac{d\mu}{dy} \ , \tag{20-10}$$

where, in a two-dimensional flow with straight, horizontal streamlines, the shear stress is linearly proportional to the gradient of the macroscopic velocity-component $u(y)$ (sometimes called the directed velocity) normal to the direction of the shearing stress.

When the same phenomenon is viewed from the kinetic viewpoint, the diffusion of molecules between adjacent layers is accompanied by a transport of molecular momentum, the molecules arriving from the faster layer tending to speed up the slower, and conversely. The shearing stress thus predicted by the most elementary molecular model is

$$\tau = \frac{1}{3} \rho C \lambda \frac{du}{dy} \ . \tag{20-11}$$

Equating (20-10) to (20-11) leads to the relation $\mu = \frac{1}{3}\rho C\lambda$. A more sophisticated analysis due to S. Chapman, in which the distribution of molecular velocities is accounted for, leads to

$$\mu = 0.499 \ \rho C \lambda \ , \tag{20-12}$$

a result which is restricted to dilute gases.

If we eliminate λ from equation (20-7) by use of (20-2), then

$$\mu = \frac{0.499mC}{\pi\sqrt{2}\sigma^2} \sim T^{\frac{1}{2}} . \tag{20-13}$$

Thus, we see that in a dilute gas viscosity depends on the temperature but not on the density. This fact was demonstrated experimentally by Robert Boyle in 1660.

Sutherland's law for viscosity. As already noted, molecules are not elastic spheres, in fact. To overcome this deficiency Sutherland proposed a more sophisticated model[15] in which each molecule carries with it a force field which attracts other molecules at large distances and repels those close by. The result is a law of the form

$$\frac{\mu}{\mu_r} = \frac{T_r + S}{T + S} (\frac{T}{T_r})^{3/2} , \tag{20-14}$$

where the parameter S depends on the gas. The reference viscosity μ_r corresponds to a measured value at temperature T_r which lies within the range at which (20-14) applies. If measurements are available at two reference temperatures then the need to know the value of S can be bypassed. Although (20-9) represents a substantial improvement over (20-7) it is still limited to dilute gases, i.e. it applies in the limit as $\rho \to 0$.

Heat conduction coefficient. Heat conduction in a gas is the result of the transport (by molecular diffusion) of kinetic energy of random motion due to the existence of a temperature gradient in the gas. An analysis, which is similar to that for viscosity, leads to a comparison of (the macroscopic) Fourier's law of heat conduction [$q = -k(dT/dy)$, where q is the heat flux term], with that derived from kinetic theory. The form for k, according to Sutherland, is identical to that for the coefficient of viscosity, namely,

$$\frac{k}{k_r} = \frac{\mu}{\mu_r} = \frac{T_r + S}{T + S} (\frac{T}{T_r})^{3/2} . \tag{20-15}$$

The Prandtl number. Application of the techniques of dimensional analysis to heat-transfer problems reveals that the Prandtl number

$$Pr \equiv \mu c_p / k , \tag{20-16}$$

is one of the important, governing dimensionless parameters. It was shown by Eucken[16] that for a perfect gas

$$Pr = \frac{4\gamma}{9\gamma - 5} . \tag{20-17}$$

That is, under the restrictions cited, the Prandtl number is constant, for which $Pr = 0.667$,

---

[15]See the treatise by Chapman and Cowling (1953) for the original references.
[16]See Jeans (1952, p. 190) for details.

0.737, 0.762, for $\gamma = 5/3$, *7/5*, and *4/3*, respectively. Experiments show that for a dilute gas, at temperatures sufficiently higher than the critical temperature, equation (20-17) is reasonably accurate, although the error increases with the complexity of the molecule.

<u>On real gases</u>. For most applications in aerodynamics, where the air stagnation-temperature remains below, say, 1 000 K, perfect-gas relations and Sutherland's law for the viscosity and heat-conduction coefficients, provide sufficient analytical tools to predict most gas properties.

However, starting in the 1950's, demands of the military and space programs resulted in requirements for knowledge of the behavior of air and pure gases (such as H_2, N_2, O_2, etc.), and a variety of fuels at higher temperatures. Above 1 000 K the vibration of the atoms of a molecule comes into play, followed by dissociation as the temperature continues to increase; ultimately, at extremely high temperatures ionization results. The question is further complicated — particularly for air at high temperatures and low pressures — by the phenomenon of recombination, a process which may produce more complicated molecules such as CO_2, NO_2, N_2O_3, etc. The need for more precise information stimulated a number of theoretical and experimental investigations to pinpoint the properties of gases at high temperatures, and for a wide range of pressures.

During the same period extensive investigations of gas properties were undertaken at very low temperatures over a wide range of pressures, including the neighborhood of the critical point (for example, for nitrogen, the critical values are $T_c = 126.4~K$, $p_c = 33.98$ *bar*). To predict state and transport properties over the vast range of pressures involved requires the use of extremely sophisticated equations of state in conjunction with relations governing the transport properties. Furthermore, in order to establish with sufficient accuracy the large number of constants in these semi-empirical relations, a significant sampling of experimental data over the entire range must be available.

<u>Data sources for real gases</u>. The most comprehensive single source of references known to the writer is found in Sauerwein and Dalton (1985). For example, under the sub-title of *Viscosity* they list 50 references, three quarters of which refer to gases. The dates of publication are almost all later than 1970.

The earliest extensive tabulation of data for air and its components is that of Hilsenrath et al. (1955) which tabulates values up to 3 000 K and to 100 atm for the state and transport properties. The actual temperature-range tabulated for a given quantity is subject to significant variation.

This tabulation has been superseded by a series of Russian publications, the earliest of which by Stupochenko et al. (1957) tabulates thermodynamic functions for air, over the range 1 000 to 12 000 K. As far as the writer can ascertain it is available only in the original Russian version.

In two highly useful volumes by Predvoditelev et al. (1962 a,b); the molecular weight and eight state-functions ($h, e, s, c_p, c_v, \gamma, a$, and ρ), and the mole fractions of the principal molecules and ions are tabulated versus the temperature, for a wide selection of isobars. Extensive tables of the gas in the ionized state are also given. Temperatures range from 6 000 to 20 000 K and pressures from 0.001 to 1 000 atm. Unfortunately, no computations for the transport properties of viscosity and heat conduction seem to have been made as part of this study.

As part of a comprehensive study, for a large selection of gases and liquids, Vargaftik (1975), in an English translation, tabulates properties of air to 6 000 K. See also Vasserman, et al. (1971), for an earlier study, restricted to air up to 1 300 K.

An excellent source of tabulated data in the cryogenic range (below 600 K) is given by Stephan and Lucas (1979) where they present smoothed data down to the critical point for 16 different gases.

Based on the work of McCarty (1975) and others at the National Bureau of Standards, Younglove (1982) has published extensive tabulations for the thermophysical properties argon, ethylene, parahydrogen, nitrogen, nitrogen floride, and oxygen. Younglove and Ely (1987) extend the coverage to methane, ethane, propane, isobutane, and normal butane. Tables G.1, G.3 and G.4.[17] are excerpted from Younglove (1982). Table G.2, for parahydrogen, is based on the more extensive computations from McCarty (1975) which include transport properties.

Recently the thermodynamic properties of air have been tabulated by Sychev et al. (1987) from 70 K to 1500 K for pressures from 0.01 to 100 MPa. This work is the 6th volume of a 10-volume series, edited by Selover (1987), which covers helium, nitrogen, methane, ethane, oxygen, air, ethylene, freon 1, freon 2, and the noble gases. Volumes on hydrogen, propane, and n-hexane are understood to be in preparation.

As part of a massive study to evaluate the consistency of experimental measurements in the literature Touloukian, Saxena and Hestermans (1975) have devoted an entire volume to viscosity. Although this work cannot fail to impress the reader by its vast scope, one is left with a feeling of dismay to recognize that there remain extensive gaps where the data is insufficient or non-existent.

In Table G.5 of this work the thermophysical properties of air are tabulated for a range of temperatures from 300 K to 6 000 K, and for 11 isobars from 0.01 MPa to 1 000 MPa. The computations were made using the CETA program of NASA Lewis by Gordon and McBride (1976). See also Gordon et al. (1984).

21. On the U.S. Standard Atmosphere, 1976

For the purpose of calculations it is essential to have a reference atmosphere, based on measured values, which represents mean values between fluctuations, daily, seasonal and, in fact, between the extremes encountered over decades. The *U.S. Standard Atmosphere, 1976*, produced under the editorship of Dubin, Hull & Champion (1976), is the latest of a series of standard atmospheres, the first of which is due to Diehl (1925). In the following a brief description of the earth's atmosphere is given, followed by a summary of the principal constants which define atmospheric air and the analytical relations which govern the distribution. In the data section of this work a number of the tabulated values from Dubin, et al. are excerpted.

---

[17]A computer program (MIPROPS) to calculate the thermophysical properties for 12 fluids (those listed above plus nitrogen trifloride) is available from the National Institute of Standards and Technology (formerly the National Bureau of Standards). For information write to Joan Sauerwein, Standard Reference Data, NIST, Physics A320, Gaithersburg, Maryland, 20899, or call 301-975-2208.

The atmospheric shells. The atmosphere can be visualized as a series of four concentric shells which surround the earth, and which blend together at their contiguous boundaries. In fact these shells overlap and the altitude designating a boundary between two shells is arbitrary, but representative of the mean value of the blending region.

The *troposphere* is the shell adjacent to the earth and is distinguished by an almost linearly decreasing temperature variation. The troposphere, the domain of weather, is in convective equilibrium with the sun-warmed surface of the earth. Near an altitude of 11 km the temperature becomes, and remains, essentially constant (about 227 K) over a finite height-change. The altitude at which this first occurs is designated as the *tropopause*.

The second shell, the *stratosphere*, begins at the tropopause. The region of (almost) constant temperature is thicker at the poles and is almost non-existent at the equator. The appearance of ozone occurs in the stratosphere. Eventually the temperature starts to increase and continues to rise until a second (almost) constant-temperature region is encountered near 47 km. The beginning of the second constant-temperature (about 270 K) region is called the *stratopause*. As in all layers the pressure and density decrease with height.

The stratopause marks the beginning of the third shell which is the *mesosphere*. In the lower portion of the mesosphere the temperature is (almost) constant, but with increasing height the temperature starts to fall again, reaching a low of about 187 K near 86 km. The mesophere is in radiative equilibrium between the ultraviolet-ozone heating by the upper fringe of the ozone region, and the infared-ozone and carbon-dioxide cooling by radiation to space. The upper limit of the mesosphere is designated as the *mesopause*. For an extensive discussion of the upper-atmosphere physics see Jursa (1985).

The shell beyond the mesopause is called the *thermosphere*. It is a region where pressure and density are so low, and where the mean-free-path length (see Section 21) is so large, that the concept of air as a continuum is inapplicable and, in fact, is where most aerodynamics effects (such as frictional heating, or fluid-dynamic drag) become negligible, even from a free-molecule viewpoint. In this region the pressure and density asymptotically approach zero, whereas the temperature (more precisely, the *kinetic temperature*) asymptotically approaches the order of 1 000 K at 10^3 km.

Supplementary shells. There are alternative ways of classifying the atmosphere other than by temperature distribution. These alternative shells necessarily overlap the primary shells just described. In the *homosphere* (up to about 100 km) there is a continual mixing process which keeps the relative proportions of the constituents approximately constant. Surrounding the homosphere is the heterosphere in which the chemical process of diffusive separation dominates. Exterior to the heterosphere is the *exosphere* from which particles may escape the earth's gravitational field and be transported to outer space.

The *ionosphere* is the domain of charged particles and consists of a number of layers lying between 80 km and 300 km. Ionization is dominated by the interaction of solar ultraviolet- and x-radiation with that portion of the atmosphere receiving sunlight.

The composition of air. Air is a mixture of gases of differing molecular weights whose composition may vary even on the surface of the earth due to changes primarily in moisture content and carbon dioxide. Based on careful measurements Table 21-1 presents the normal composition of clean, dry air near sea level, which has been adopted as standard by international agreement. Molecular weights are based on the carbon isotope $C^{12} \equiv 12.000\ 0$.

Table 21-1. Fractional-volume composition and molecular weights for sea-level dry air, *U.S. Standard Atmosphere, 1976.*

Gas	Fractional Volume	Molecular Weight ($\overline{m}$) (Dimensionless)
Nitrogen (N_2)	0.780 84	28.013 4
Oxygen (O_2)	0.209 476	31.998 8
Argon (Ar)	0.009 34	39.948
Carbon Dioxide (CO_2)	0.000 314	44.009 95
Total Fraction = 0.999 97		Mean Molecular = 28.964 4 Weight

The remaining fraction of 0.000 03 is made up of neon, helium, krypton, xenon, methane, and hydrogen in extremely small proportions. There are also present, in varying, low concentrations, a number of pollutants including nitrous oxide, nitric oxide, nitrogen dioxide, nitric-acid vapor, hydrogen sulfide, ammonia, sulfur dioxide, and carbon monoxide. There is also an ozone concentration which varies with altitude. The most highly variable constituent of ordinary air is, of course, water vapor.

<u>Sea-level reference conditions</u>. Based on decades of measurements the mean sea-level reference conditions for *40° north latitude* are specified in Table 21-2.

<u>Temperature-height relations</u>. For the purpose of defining a temperature profile, up to 86 km the temperature function is represented by a series of straight-line segments with discontinuous slopes at the junctions. A complete tabulation of these values is given in Table 4 of *U.S. Standard Atmosphere, 1976.* Above 86 km, where the particles are not in thermal equilibrium, the relations used to specify temperature variation are more complicated.

<u>Temperature-height relation in the troposphere</u>[18]. Based on thousands of measurements the tropopause is taken to occur at $z^* = 11\ km$, where the tropopause temperature is $T^* = 216.65\ K = 389.97\ °R$. The *lapse-rate a* in the tropopause is defined to be the negative of the mean temperature-gradient between sea level and 11 km. Thus

$$a \equiv (T_a - T^*)/z^* = 6.5 \times 10^{-3}\ K/m = 3.5662 \times 10^{-3}\ °R/ft \ . \qquad (21\text{-}1)$$

[18]The variable z stands for height in geopotential rather than geometric units. The relation between the two is explained at the end of this section.

Table 21-2. Sea-level reference conditions for dry air. U.S. Standard Atmosphere, 1976.

Quantity	Symbol	SI (kg, m, s, K)	Engineering System (slug, ft, sec, °R)
Uniform gravitational constant	g	$9.806\ 65$ m/s^2	32.174 ft/sec^2
Universal gas constant	R^*	$8.314\ 32\times10^3\ \dfrac{N\cdot m}{kmol\cdot K}$ $8.314\ 32\times10^3\ \dfrac{m^2\cdot kg}{kmol\cdot s^2\cdot K}$	$4.97190\times10^4\ \dfrac{ft^2\cdot slug}{slug\ mol\cdot sec^2\cdot R}$
Mean molecular weight	m	$28.964\ 4$	28.9644
Specific gas constant	$R = R^*/mk$	287.053 m^2/s$^2\cdot$K	1716.6 ft^2/sec$^2\cdot$R
Pressure	p_a	$1.013\ 250\times10^5$ N/m^2	2116.2 lb/ft^2
Temperature	T_a	288.150 K	518.67 °R
Density	ρ_a	$1.225\ 0$ kg/m^3	0.0023769 slug/ft^3
Speed of sound	a_a	340.294 m/s	1116.45 ft/sec
Dynamic viscosity	ν_a	$1.789\ 4\times10^{-5}$ kg/m$\cdot$s	3.7372×10^{-7} lb sec/ft^2
Coefficient of heat conduction	k_a	$2.532\ 6\times10^{-2}\ \dfrac{J}{s\cdot m\cdot K}$	$3.1631\times10^{-3}\ \dfrac{ft\ lb_f}{sec\cdot ft\cdot R} = 1.463\times10^{-2}\ \dfrac{Btu}{hr\cdot ft\cdot F}$

For practical purposes the ratio of specific heats is taken as $\gamma = 1.400$.

Therefore, the temperature distribution in the atmosphere is given by

$$T = T_a - az , \quad z < z^* .$$ (21-2)

Pressure-height relation in the troposphere. The fundamental equation of fluid statics is

$$dp/dz = -\rho g .$$ (21-3)

With the perfect-gas relation this becomes

$$dp/p = -(g/R)(dz/dT)(dT/T) .$$ (21-4)

Thus, with (21-2), integrating from sea-level, see Table 21-1,

$$p/p_a = (T/T_a)^{g/aR} ,$$ (21-5)

which has the form of a polytropic relation $pv^n = constant$, where $n = (1 - aR/g)^{-1} = 1.233$. Combining (21-2) and (21-5) with the polytropic expression the pressure- and density-height relations become

$$p/p_a = (1-az/T_a)^{\frac{n}{n-1}} , \quad z \le z^* ;$$ (21-6)

$$\rho/\rho_a = (1-az/T_a)^{\frac{1}{n-1}} , \quad z \le z^* .$$ (21-7)

The pressure p^* and density ρ^*, at the tropopause, are obtained from (21-5):

$$(p^*/p_a)^{\frac{n-1}{n}} = (\rho^*/\rho_a)^{n-1} = (T^*/T_a) .$$ (21-8)

Thus, $p^* = 0.226\ 32 \times 10^5\ N/m^2$, $\rho^* = 0.363\ 92\ kg/m^3$.

Conditions in the lower stratosphere. According to our atmospheric model the temperature remains constant at T^* over the range $z^* \le z < z^{**} = 20\ km$. Thus, in the isothermal portion of the stratosphere,

$$p/p^* = \rho/\rho^* = exp[-g(z-z^*)/RT^*] , \quad z^* \le z < z^{**} .$$ (21-9)

For the portion of the stratosphere above z^{**} and below the stratopause the temperature profile is approximated by a series of linear temperature profiles, and this process of piecewise representation to obtain analytic expressions is extended into the mesosphere except that at appropriate altitudes the appearance of real gas effects (dissociation and ionization), and other chemical effects such as the presence of ozone, become of importance and must be taken into account.

Note on geopotential, versus geometric, height. In transatmospheric applications one must account for the fact that gravitational forces must be calculated by Newton's inverse-square, universal law-of-gravitation. The differences between the uniform-gravitational-model and the true law are minor, if not negligible, for heights up to 100 km.

The gravitational forces on an object of mass m located on the earth's surface (the earth is treated as a sphere of radius $R_o = 6\ 356.766\ km = 3949.427\ mi.$), according to the *uniform gravitational law*, or *the inverse-square law*, must both yield the same result, i.e.

$$F \quad = \quad -mg \quad = -\ G\ \frac{mM}{R_o^2}\ , \tag{21-10}$$

where $G = 667\ 0 \times 10^{-11}\ N \cdot m/kg^2$, and M is the mass of the earth. Thus, we see that

$$GM \quad = \quad gR_o^2\ . \tag{21-11}$$

For an object at *geometric height H* above the earth's surface the gravitational potential-energy function for the inverse-square law is, therefore, given by

$$dW/dH \quad = \quad gR_o^2/(R_o+H)^2\ ,$$

or

$$W \quad = \quad gR_oH/(R_o+H)\ . \tag{21-12}$$

The *geopotential height z* is defined by equating its potential-energy function $W = gz$, to that in (21-12); thus[19].

$$z/H \quad = \quad 1/(1+H/R_o) \quad or \quad H/z \quad = \quad 1(1-z/R_o)\ , \tag{21-13}$$

is the relation between the two heights. For $z < 63.6\ km$ the difference is only one percent, and a 10% difference requires an altitude of 636 km.

Tables H.1 and H.2 tabulate selected values for the *U.S. Standard Atmosphere, 1976*, in SI units, and engineering units, respectively, up to 85.5 km. Table H.3 tabulates selected values from 86 to 1,000 km.

---

[19]The use of the symbols z and H in this work has been interchanged with that employed in the *U.S. Standard Atmosphere, 1976*

REFERENCES AND AUTHOR INDEX

The numbers in square brackets show the pages on which the references appear.

Anonymous (1953), *Equations, Tables, and Charts for Compressible Flow*, NACA Rep. 1135 by the Ames Research Staff. [43]

J. Ackeret (1925), *Luftkraft auf Flügel, die mit grösserer als Schallgeschwindigkeit bewegt werden*, Z.F.M., 16, 72-74. [43,55]

A. Busemann (1935), *Aerodynamischer Auftrieb bei Überschallgeschwindigkeit*. Luftfahrtforschung, 12, 210-220. [42]

A. Busemann (1937), *Hodengraphenmethode der Gasdynamik*, Z.A.M.M., 12, 73-79. [36]

G.F. Carrier and C.E. Pearson (1976), *Partial Differential Equations*, Academic Press. [64]

S. Chapman and T.G. Cowling (1953), *The Mathematical Theory of Non-Uniform Gases*, 2nd edition, Cambridge University Press, reprint of the 1939 edition. [74]

J.D. Cole and L.P. Cook (1986), *Transonic Aerodynamics*, North-Holland Publishing Co., New York. [38]

R. Courant and K.O. Friedrichs (1948), *Supersonic Flow and Shock Waves*, Interscience Publishers, New York. [48]

L. Crocco (1958), *One-Dimensional Treatment of Steady Gas Dynamics*, Section B in *Fundamentals of High Speed Gas Dynamics*, edited by Howard W. Emmons, Volume III of High Speed Aerodynamics and Jet Propulsion, Princeton University Press. [23]

W.S. Diehl (1925), *Standard Atmosphere Tables and Data*, NACA Rep. 218. [76]

M. Dubin, A.R. Hall, K.S.W. Champion (1976), *U.S. Standard Atmosphere, 1976*, U.S Government Printing Office. The authors listed are the co-chairmen of the U.S. Committee on Extension to the Standard Atmosphere. [76]

A. Ferri (1954), *The Method of Characteristics*, Section G of *General Theory of High Speed Aerodynamics*, vol. VI of High Speed Aerodynamics and Jet Propulsion, edited by W.R. Sears, Princeton University Press. [57]

S. Gordon and B.J. McBride (1976), *Computer Program for Calculations of Complex Chemical Equilibrium Compositions, Rocket Performance, Incident and Reflected Shocks, and Chapman-Jouget Detonations*, NASA SP-273. [76]

S. Gordon and B.J. McBride and F.J. Zeleznik (1984), *Computer Programs for Calculations of Complex Chemical Equilibrium Compositions and Applications, Supplement I−Transport Properties*, NASA TM 86885. [76]

W.D. Hayes (1955), *Second-order Pressure-Law for Two-dimensional Compressible Flow*, Journal of Aeronautical Sciences, vol. 22, pp. 284-286. [57]

J.O. Hirschfelder, C.F. Curtiss, and R.B. Bird (1954), *Molecular Theory of Gases and Liquids*, John Wiley and Sons, N. Y. [72]

J. Hilsenrath, C.W. Beckett, W.S. Benedict, L. Fano, H.J. Hoge, J.F. Masi, R.L. Nuttall, Y.S. Touloukian, and H.W. Wooley (1955), *Tables of Thermal Properties of Gases*, NBS Circular 564. [75]

L. Howarth (1956), *Modern Developments in Fluid Dynamics*, vol. 1, Clarendon Press, Oxford. [73]

J. Jeans (1952), *An Introduction to the Kinetic Theory of Gases*, Cambridge University Press, reprint of the 1940 edition. [70,74]

A. Jursa (1985), Scientific Editor, *Handbook of Geophysics and the Space Environment*, U.S. Air Force Geophysics Laboratory, NTIS Document Accession Number: ADA 167000. [77]

A. Kahane and L. Lees (1947), *Letter to the Editor*, Journal of the Aeronautical Sciences, vol. 14, p. 600. [43]

Z. Kopal (1947), *Tables of Supersonic Flow Around Cones*, Center of Analysis, Department of Electrical Engineering, Massachusetts Institute of Technology, Technical Report No. 1. [70]

E.V. Laitone (1947), *Exact and Approximate Solutions of Two-Dimensional Oblique Shock Flow*, Journal of the Aeronautical Sciences, vol. 14. [43]

H. Liepmann and A. Roshko (1957), *Elements of Gas Dynamics*, John Wiley and Sons. [30,57]

M.J. Lighthill (1954), *Higher Approximations*, Section E in *General Theory of High Speed Aerodynamics*, edited by W.R. Sears, Volume VI of High Speed Aerodynamics and Jet Propulsion, Princeton University Press. [55]

R.D. McCarty (1975), *Hydrogen Technological Survey - Thermophysical Properties*, NASA SP-3089. [76]

P. Molenbroek (1890), *Über einige Bewegungen eines Gases bei Annahme eines Geschwindigkeitspotentials*, Archiv der Mathematik und Physik (Grunert Hoppe), vol. 2, reprinted in *Foundations of High Speed Aerodynamics*, facsimile of 19 fundamental papers from scientific journals, collected by G.F. Carrier, Dover Publications, 1951. [63]

G.N. Patterson (1971), *Introduction to the Kinetic Theory of Gas Flows*, University of Toronto Press. [72]

L. Prandtl (1907), *Neue Untersuchungen über die Strömende Bewegung der Gase und Dämpfe*, Physicalische Zeitschrift, 8, p. 23. [47]

A.S. Predvoditelev, E.V. Stupochenko, E. Samuilov, I.P. Stakhanov, A.S. Pleshanov, and I.B. Rozhdestvenskii (1962), *Tables of Thermodynamic Functions of Air, a) 6,000 to 12,000 K and 0.001 to 1000 atm., b) 12,000 to 20,000 K and 0.001 to 1000 atm.*, translated from the 1957 Russian version, Associated Technical Services, Inc., Glen Ridge, N.J. This same work was also translated by Infosearch, Ltd., London in 1959. [75]

A.E. Puckett and T.Y. Li (1947), *Letter to the Editor*, Journal of the Aeronautical Sciences, vol. 14, p. 336. [43]

J.C. Sauerwein and G.R. Dalton (1985), *Standard Reference Data Publications*, NBS Special Publication 708, Office of Standard Reference Data, National Bureau of Standards, Gaithersburg, Maryland. [75]

Theodore B. Selover, Jr. (1987), Editor of *National Standard Reference Data of the USSR: A Series of Property Tables*, 10 volumes. 1: Methane; 2: Nitrogen; 3: Methane; 4: Ethane; 5: Oxygen; 6: Air; 7: Ethylene; 8: Freons, Part 1; 9: Freons, Part 2; 10: Neon, Argon, Krypton and Xenon. [76]

A.H. Shapiro (1953), *The Dynamics and Thermodynamics of Compressible Fluid Flow*, vol. 1, Ronald Press. [24]

J.L. Sims (1964a), *Tables for Supersonic Flow around Right Circular Cones at Zero Angle of Attack*, NASA SP-3004. [67,70]

J.L. Sims (1964b), *Tables for Supersonic Flow around Right Circular Cones at Small Angle of Attack*, NASA SP-3007. [70]

M. Spiegel (1968), *Mathematical Handbook of Formulas and Tables*, Schaum's Outline Series, McGraw-Hill Book Co. [31]

K. Stephan and K. Lucas (1979), *Viscosity of Dense Fluids*, Plenum Press, N. Y. [76]

E.V. Stupochenko, E. Samuilov, I.P. Stakhanov, A.S. Pleshanov, and I.B. Rozhdestvenskii, and I.B. Razhdestvenskii (1957), *Thermodynamic Functions of Air in the Temperature Range 1000 K to 12,000 K*, collected works of the Combustion Physics Laboratory of the Energetic Institute of the U.S.S.R. Academy of Sciences, Moscow. [75]

V.V. Sychov, A.A. Vasserman, A.D. Kozlov, G.A. Spiridonov, and V.A. Tsymarny (1987), *Thermodynamic Properties of Air*, National Standard Reference Data Service of the USSR, vol. 6, Hemisphere Publishing Corp., New York. [76]

G.I. Taylor and J.W. Maccoll (1933), *The Air Pressure on a Cone Moving at High Speeds*, Proc. Roy. Soc., A, 139, p. 278. [65,69]

Y.S. Touloukian, S.C. Saxena, and P. Hestermans (1975), *Viscosity*, vol. 11 of *Thermophysical Properties of Matter*, IFI/Plenum, New York/Washington. [76]

M.D. Van Dyke (1956), *Second-order Subsonic Airfoil Theory Including Edge Effects*, NACA Rep. 1274. [57]

M.D. Van Dyke (1964), *Perturbation Methods in Fluid Mechanics*, Academic Press, an annoted edition was printed in 1975 by Parabolic Press. [57]

N.B. Vargaftik (1975), *Handbook of Physical Properties of Liquids and Gases: Pure Substances and Mixtures*, 2nd edition, translated from the Russian, Hemisphere Publishing Corp. [76]

A.A. Vasserman, Ya. Z. Kazavchinskii and V.A. Rabinovich (1971), *Thermophysical Properties of Air and Air Components*, Israel Program for Scientific Translations, from a 1966 Russian publication. [84]

B.A. Younglove (1982), *Thermophysical Properties of Fluids*, Journal of Physical and Chemical Reference Data, vol. 11, Supplement No. 1, ACS, AIP, NBS. [76]

B.A. Younglove and J.F. Ely (1987), *Thermophysical Properties of Fluids II*, Journal of Physical and Chemical Reference Data, vol. 16, Supplement No. 4, ACS, AIP, NBS. pp. 577-798. [76]

TABLES AND DATA

TABLE A.1

ISENTROPIC FLOW FUNCTIONS

AND

NORMAL SHOCK FUNCTIONS

$\gamma = 5/3$

ISENTROPIC FUNCTIONS FOR SUBSONIC COMPRESSIBLE FLOW

$$\gamma = 5/3$$

M	T/T°		p/p°		ρ/ρ°		A/A^*		q/p°		u/u_m	u/a^*	$F(\gamma;M)$
0.00	1.000	+0	1.000	+0	1.000	+0	∞		0.000		0.0000	0.0000	∞
0.01	1.000		1.000		1.000		5.625		8.333	−5	0.0058	0.0115	60.01
0.02	9.997	−1	9.998	−1	9.999	−1	2.813		3.332	−4	0.0115	0.0231	30.02
0.03	9.993		9.996		9.997		1.876		7.494		C.0173	0.0346	20.03
0.04	9.987		9.992		9.995		1.408		1.332	−3	0.0231	0.0462	15.04
0.05	9.979		9.988		9.992		1.127		2.079		0.0289	0.0577	12.05
0.06	9.970		9.982		9.988		9.398	+0	2.991		Ս.0346	0.0692	10.05
0.07	9.959		9.976		9.984		8.062		4.067		0.0404	0.0808	8.634
0.08	9.947		9.968		9.979		7.061		5.305		0.0461	0.0923	7.572
0.09	9.933		9.960		9.973		6.284		6.705		0.0519	0.1038	6.748
0.10	9.917		9.950		9.967		5.663		8.264		0.0576	0.1153	6.090
0.11	9.900		9.940		9.960		5.155		9.982		0.0634	0.1268	5.553
0.12	9.881		9.928		9.952		4.733		1.186	−2	0.0691	0.1382	5.108
0.13	9.861		9.916		9.944		4.376		1.389		0.0748	0.1497	4.732
0.14	9.839		9.903		9.935		4.071		1.607		0.0806	0.1611	4.411
0.15	9.815		9.889		9.926		3.806		1.840		0.0863	0.1726	4.135
0.16	9.790		9.873		9.915		3.576		2.088		0.0920	0.1840	3.893
0.17	9.763		9.857		9.905		3.373		2.351		0.0977	0.1954	3.682
0.18	9.735		9.840		9.893		3.193		2.628		0.1034	0.2067	3.495
0.19	9.705		9.822		9.881		3.032		2.920		0.1090	0.2181	3.328
0.20	9.674		9.803		9.868		2.888		3.225		0.1147	0.2294	3.179
0.21	9.642		9.783		9.855		2.758		3.543		0.1204	0.2407	3.045
0.22	9.608		9.763		9.841		2.640		3.875		0.1260	0.2520	2.924
0.23	9.572		9.741		9.827		2.533		4.220		0.1316	0.2633	2.814
0.24	9.536		9.719		9.812		2.435		4.577		0.1373	0.2745	2.714
0.25	9.498		9.695		9.796		2.345		4.947		0.1429	0.2857	2.623
0.26	9.458		9.671		9.780		2.262		5.328		0.1484	0.2969	2.539
0.27	9.417		9.646		9.763		2.186		5.721		0.1540	0.3080	2.462
0.28	9.375		9.620		9.745		2.115		6.125		0.1596	0.3192	2.392
0.29	9.332		9.594		9.727		2.050		6.540		0.1651	0.3303	2.327
0.30	9.288		9.566		9.709		1.989		6.966		0.1707	0.3413	2.266
0.31	9.242		9.538		9.690		1.933		7.401		0.1762	0.3524	2.210
0.32	9.195		9.509		9.670		1.880		7.847		0.1817	0.3634	2.158
0.33	9.147		9.479		9.650		1.831		8.301		0.1872	0.3743	2.110
0.34	9.098		9.449		9.629		1.784		8.764		0.1926	0.3852	2.065
0.35	9.048		9.417		9.608		1.741		9.236		0.1981	0.3961	2.023
0.36	8.997		9.385		9.586		1.700		9.716		0.2035	0.4070	1.984
0.37	8.944		9.353		9.564		1.662		1.020	−1	0.2089	0.4178	1.948
0.38	8.891		9.319		9.541		1.626		1.070		0.2143	0.4286	1.913
0.39	8.837		9.285		9.517		1.592		1.120		0.2197	0.4393	1.881
0.40	8.782		9.250		9.494		1.560		1.171		0.2250	0.4500	1.851
0.41	8.726		9.215		9.469		1.530		1.222		0.2303	0.4607	1.823
0.42	8.669		9.179		9.445		1.501		1.274		0.2357	0.4713	1.797
0.43	8.611		9.142		9.419		1.474		1.327		0.2409	0.4819	1.772
0.44	8.553		9.105		9.394		1.449		1.380		0.2462	0.4924	1.748
0.45	8.493		9.067		9.368		1.424		1.433		0.2515	0.5029	1.726
0.46	8.433		9.028		9.341		1.401		1.487		0.2567	0.5134	1.705
0.47	8.373		8.989		9.314		1.380		1.541		0.2619	0.5238	1.686
0.48	8.311		8.949		9.287		1.359		1.596		0.2671	0.5341	1.667
0.49	8.249		8.909		9.259		1.339		1.651		0.2722	0.5444	1.650
0.50	8.186		8.869		9.231		1.320		1.706		0.2773	0.5547	1.633

ISENTROPIC FUNCTIONS FOR SUBSONIC COMPRESSIBLE FLOW

$\gamma = 5/3$

M	$T/T°$	$p/p°$	$\rho/\rho°$	A/A^*	$q/p°$	u/u_m	u/a^*	$F(\gamma;M)$
0.50	8.186 -1	8.869 -1	9.231 -1	1.320 +0	1.706 -1	0.2773	0.5547	1.633
0.51	8.123	8.827	9.202	1.302	1.761	0.2825	0.5649	1.618
0.52	8.059	8.786	9.173	1.286	1.816	0.2875	0.5751	1.603
0.53	7.995	8.744	9.144	1.269	1.872	0.2926	0.5852	1.589
0.54	7.930	8.701	9.114	1.254	1.927	0.2976	0.5953	1.576
0.55	7.865	8.658	9.084	1.239	1.983	0.3026	0.6053	1.564
0.56	7.799	8.615	9.054	1.225	2.038	0.3076	0.6153	1.552
0.57	7.733	8.571	9.023	1.212	2.094	0.3126	0.6252	1.541
0.58	7.667	8.526	8.992	1.200	2.149	0.3175	0.6351	1.531
0.59	7.600	8.482	8.960	1.187	2.205	0.3224	0.6449	1.521
0.60	7.533	8.437	8.929	1.176	2.260	0.3273	0.6547	1.512
0.61	7.465	8.391	8.897	1.165	2.315	0.3322	0.6644	1.503
0.62	7.398	8.346	8.864	1.155	2.370	0.3370	0.6740	1.495
0.63	7.330	8.300	8.832	1.145	2.424	0.3418	0.6836	1.487
0.64	7.262	8.253	8.799	1.135	2.479	0.3466	0.6932	1.480
0.65	7.194	8.207	8.766	1.126	2.533	0.3514	0.7027	1.473
0.66	7.125	8.160	8.732	1.118	2.586	0.3561	0.7122	1.466
0.67	7.057	8.113	8.698	1.110	2.640	0.3608	0.7215	1.460
0.68	6.988	8.065	8.665	1.102	2.693	0.3654	0.7309	1.454
0.69	6.919	8.018	8.630	1.094	2.745	0.3701	0.7402	1.449
0.70	6.851	7.970	8.596	1.088	2.797	0.3747	0.7494	1.444
0.71	6.782	7.922	8.561	1.081	2.849	0.3793	0.7586	1.439
0.72	6.713	7.873	8.527	1.075	2.900	0.3838	0.7677	1.434
0.73	6.645	7.825	8.492	1.069	2.951	0.3884	0.7768	1.430
0.74	6.576	7.776	8.456	1.063	3.001	0.3929	0.7858	1.426
0.75	6.508	7.728	8.421	1.058	3.050	0.3974	0.7947	1.422
0.76	6.439	7.679	8.386	1.053	3.099	0.4018	0.8036	1.419
0.77	6.371	7.630	8.350	1.048	3.148	0.4062	0.8125	1.416
0.78	6.303	7.581	8.314	1.043	3.195	0.4106	0.8212	1.413
0.79	6.235	7.532	8.278	1.039	3.242	0.4150	0.8300	1.410
0.80	6.167	7.482	8.242	1.035	3.289	0.4193	0.8386	1.407
0.81	6.099	7.433	8.205	1.031	3.335	0.4236	0.8472	1.405
0.82	6.032	7.383	8.169	1.028	3.380	0.4279	0.8558	1.402
0.83	5.964	7.334	8.133	1.025	3.424	0.4321	0.8643	1.400
0.84	5.897	7.284	8.096	1.022	3.468	0.4364	0.8727	1.398
0.85	5.831	7.235	8.059	1.019	3.511	0.4406	0.8811	1.397
0.86	5.764	7.185	8.022	1.016	3.553	0.4447	0.8894	1.395
0.87	5.698	7.136	7.985	1.014	3.594	0.4489	0.8977	1.394
0.88	5.632	7.086	7.948	1.012	3.635	0.4530	0.9059	1.392
0.89	5.567	7.037	7.911	1.010	3.675	0.4570	0.9141	1.391
0.90	5.502	6.987	7.874	1.008	3.714	0.4611	0.9222	1.390
0.91	5.437	6.938	7.837	1.006	3.752	0.4651	0.9302	1.389
0.92	5.372	6.888	7.800	1.005	3.789	0.4691	0.9382	1.388
0.93	5.308	6.839	7.762	1.004	3.826	0.4731	0.9461	1.388
0.94	5.245	6.789	7.725	1.003	3.862	0.4770	0.9540	1.387
0.95	5.181	6.740	7.687	1.002	3.897	0.4809	0.9618	1.387
0.96	5.119	6.691	7.650	1.001	3.931	0.4848	0.9695	1.386
0.97	5.056	6.642	7.612	1.001	3.964	0.4886	0.9772	1.386
0.98	4.994	6.593	7.575	1.000	3.997	0.4924	0.9849	1.386
0.99	4.933	6.544	7.538	1.000	4.029	0.4962	0.9925	1.386
1.00	4.871	6.495	7.500	1.000	4.059	0.5000	1.000	1.386

ISENTROPIC FUNCTIONS FOR SUPERSONIC COMPRESSIBLE FLOW

$$\gamma = 5/3$$

M	T/T°	p/p°	ρ/ρ°	A/A^{*}	q/p°	u/u_{m}	u/a^{*}	$F(\gamma;M)$
1.00	4.871 −1	6.495 −1	7.500 −1	1.000 +0	4.059 −1	0.5000	1.000	1.386
1.01	4.811	6.447	7.463	1.000	4.090	0.5037	1.007	1.386
1.02	4.751	6.398	7.425	1.000	4.119	0.5074	1.015	1.386
1.03	4.691	6.350	7.388	1.001	4.147	0.5111	1.022	1.386
1.04	4.632	6.301	7.350	1.001	4.175	0.5148	1.029	1.386
1.05	4.573	6.253	7.313	1.002	4.201	0.5184	1.037	1.387
1.06	4.515	6.205	7.275	1.003	4.227	0.5220	1.044	1.387
1.07	4.457	6.158	7.238	1.004	4.252	0.5256	1.051	1.387
1.08	4.400	6.110	7.200	1.005	4.276	0.5291	1.058	1.388
1.09	4.343	6.063	7.163	1.006	4.300	0.5326	1.065	1.388
1.10	4.287	6.015	7.126	1.007	4.322	0.5361	1.072	1.389
1.11	4.231	5.968	7.089	1.008	4.344	0.5396	1.079	1.390
1.12	4.176	5.921	7.052	1.010	4.365	0.5430	1.086	1.390
1.13	4.121	5.875	7.014	1.012	4.385	0.5464	1.093	1.391
1.14	4.067	5.828	6.977	1.014	4.404	0.5498	1.100	1.392
1.15	4.013	5.782	6.940	1.015	4.423	0.5531	1.106	1.393
1.16	3.960	5.736	6.904	1.017	4.440	0.5565	1.113	1.394
1.17	3.907	5.690	6.867	1.020	4.457	0.5598	1.120	1.394
1.18	3.855	5.645	6.830	1.022	4.473	0.5630	1.126	1.395
1.19	3.804	5.599	6.793	1.024	4.489	0.5663	1.133	1.396
1.20	3.753	5.554	6.757	1.027	4.503	0.5695	1.139	1.397
1.21	3.702	5.509	6.720	1.029	4.517	0.5727	1.145	1.398
1.22	3.652	5.465	6.684	1.032	4.530	0.5759	1.152	1.399
1.23	3.603	5.420	6.648	1.035	4.542	0.5790	1.158	1.401
1.24	3.554	5.376	6.611	1.038	4.554	0.5821	1.164	1.402
1.25	3.506	5.332	6.575	1.041	4.565	0.5852	1.170	1.403
1.26	3.458	5.288	6.539	1.044	4.575	0.5883	1.177	1.404
1.27	3.411	5.245	6.504	1.047	4.585	0.5913	1.183	1.405
1.28	3.364	5.202	6.468	1.051	4.593	0.5943	1.189	1.406
1.29	3.318	5.159	6.432	1.054	4.601	0.5973	1.195	1.408
1.30	3.273	5.116	6.397	1.057	4.609	0.6003	1.201	1.409
1.31	3.227	5.074	6.361	1.061	4.615	0.6032	1.206	1.410
1.32	3.183	5.031	6.326	1.065	4.621	0.6061	1.212	1.411
1.33	3.139	4.990	6.291	1.069	4.627	0.6090	1.218	1.413
1.34	3.095	4.948	6.256	1.073	4.632	0.6119	1.224	1.414
1.35	3.052	4.907	6.221	1.077	4.636	0.6147	1.229	1.415
1.36	3.010	4.866	6.186	1.081	4.639	0.6176	1.235	1.417
1.37	2.968	4.825	6.152	1.085	4.642	0.6204	1.241	1.418
1.38	2.927	4.784	6.117	1.089	4.644	0.6231	1.246	1.419
1.39	2.886	4.744	6.083	1.094	4.646	0.6259	1.252	1.421
1.40	2.845	4.704	6.048	1.098	4.647	0.6286	1.257	1.422
1.41	2.805	4.664	6.014	1.103	4.648	0.6313	1.263	1.423
1.42	2.766	4.625	5.980	1.108	4.647	0.6340	1.268	1.425
1.43	2.727	4.586	5.947	1.112	4.647	0.6367	1.273	1.426
1.44	2.689	4.547	5.913	1.117	4.646	0.6393	1.279	1.428
1.45	2.651	4.508	5.880	1.122	4.644	0.6419	1.284	1.429
1.46	2.613	4.470	5.846	1.127	4.642	0.6445	1.289	1.431
1.47	2.576	4.432	5.813	1.132	4.639	0.6471	1.294	1.432
1.48	2.540	4.394	5.780	1.138	4.636	0.6496	1.299	1.433
1.49	2.504	4.357	5.747	1.143	4.632	0.6521	1.304	1.435
1.50	2.468	4.320	5.714	1.148	4.628	0.6546	1.309	1.436

NORMAL SHOCK FUNCTIONS FOR SUPERSONIC COMPRESSIBLE FLOW

$$\gamma = 5/3$$

M	p_{21}	T_{21}	ρ_{21}	M_2	p^o_{21}	μ	ν
1.00	1.000 +0	1.000 +0	1.000	1.000	1.000 +0	90.00	0.000
1.01	1.025	1.010	1.015	0.9901	1.000	81.93	0.040
1.02	1.050	1.020	1.030	0.9806	1.000	78.64	0.113
1.03	1.076	1.030	1.045	0.9713	1.000	76.14	0.206
1.04	1.102	1.040	1.060	0.9623	1.000	74.06	0.315
1.05	1.128	1.049	1.075	0.9535	9.999 -1	72.25	0.436
1.06	1.154	1.059	1.090	0.9450	9.998	70.63	0.570
1.07	1.181	1.069	1.105	0.9367	9.996	69.16	0.712
1.08	1.208	1.079	1.120	0.9286	9.995	67.81	0.864
1.09	1.235	1.088	1.135	0.9207	9.992	66.55	1.024
1.10	1.262	1.098	1.150	0.9131	9.990	65.38	1.191
1.11	1.290	1.108	1.165	0.9056	9.987	64.28	1.364
1.12	1.318	1.118	1.179	0.8983	9.983	63.24	1.543
1.13	1.346	1.127	1.194	0.8912	9.979	62.25	1.727
1.14	1.374	1.137	1.209	0.8843	9.974	61.31	1.917
1.15	1.403	1.146	1.224	0.8776	9.969	60.41	2.111
1.16	1.432	1.156	1.239	0.8710	9.963	59.55	2.310
1.17	1.461	1.166	1.253	0.8646	9.957	58.73	2.512
1.18	1.490	1.175	1.268	0.8583	9.950	57.94	2.718
1.19	1.520	1.185	1.283	0.8522	9.942	57.18	2.927
1.20	1.550	1.195	1.297	0.8463	9.933	56.44	3.140
1.21	1.580	1.204	1.312	0.8404	9.924	55.74	3.355
1.22	1.610	1.214	1.326	0.8347	9.915	55.05	3.573
1.23	1.641	1.224	1.341	0.8291	9.904	54.39	3.794
1.24	1.672	1.234	1.355	0.8237	9.893	53.75	4.017
1.25	1.703	1.243	1.370	0.8184	9.882	53.13	4.242
1.26	1.734	1.253	1.384	0.8132	9.869	52.53	4.469
1.27	1.766	1.263	1.399	0.8081	9.856	51.94	4.698
1.28	1.798	1.273	1.413	0.8031	9.842	51.38	4.928
1.29	1.830	1.282	1.427	0.7982	9.828	50.82	5.160
1.30	1.862	1.292	1.441	0.7934	9.813	50.29	5.394
1.31	1.895	1.302	1.455	0.7888	9.797	49.76	5.628
1.32	1.928	1.312	1.470	0.7842	9.781	49.25	5.864
1.33	1.961	1.322	1.484	0.7797	9.764	48.75	6.101
1.34	1.994	1.332	1.498	0.7753	9.746	48.27	6.340
1.35	2.028	1.342	1.512	0.7710	9.727	47.80	6.579
1.36	2.062	1.352	1.526	0.7668	9.708	47.33	6.818
1.37	2.096	1.362	1.539	0.7627	9.689	46.88	7.059
1.38	2.130	1.372	1.553	0.7586	9.669	46.44	7.300
1.39	2.165	1.382	1.567	0.7547	9.648	46.01	7.542
1.40	2.200	1.392	1.581	0.7508	9.626	45.59	7.784
1.41	2.235	1.402	1.594	0.7470	9.604	45.17	8.027
1.42	2.270	1.412	1.608	0.7432	9.581	44.77	8.270
1.43	2.306	1.422	1.621	0.7395	9.558	44.37	8.513
1.44	2.342	1.433	1.635	0.7359	9.534	43.98	8.757
1.45	2.378	1.443	1.648	0.7324	9.510	43.60	9.001
1.46	2.414	1.453	1.662	0.7289	9.485	43.23	9.245
1.47	2.451	1.463	1.675	0.7255	9.460	42.87	9.489
1.48	2.488	1.474	1.688	0.7222	9.434	42.51	9.733
1.49	2.525	1.484	1.701	0.7189	9.407	42.16	9.977
1.50	2.562	1.495	1.714	0.7157	9.380	41.81	10.22

ISENTROPIC FUNCTIONS FOR SUPERSONIC COMPRESSIBLE FLOW

$$\gamma = 5/3$$

M	$T/T°$	$p/p°$	$\rho/\rho°$	A/A^*	$q/p°$	u/u_m	u/a^*	$F(\gamma;M)$
1.50	2.468 −1	4.320 −1	5.714 −1	1.148 +0	4.628 −1	0.6546	1.309	1.436
1.51	2.433	4.283	5.682	1.154	4.624	0.6571	1.314	1.438
1.52	2.399	4.246	5.649	1.160	4.618	0.6596	1.319	1.439
1.53	2.365	4.210	5.617	1.165	4.613	0.6620	1.324	1.441
1.54	2.331	4.174	5.585	1.171	4.607	0.6644	1.329	1.442
1.55	2.298	4.138	5.553	1.177	4.600	0.6668	1.334	1.443
1.56	2.265	4.103	5.521	1.183	4.594	0.6692	1.338	1.445
1.57	2.233	4.067	5.490	1.189	4.586	0.6716	1.343	1.446
1.58	2.201	4.033	5.458	1.195	4.579	0.6739	1.348	1.448
1.59	2.170	3.998	5.427	1.201	4.571	0.6762	1.353	1.449
1.60	2.139	3.964	5.396	1.208	4.562	0.6785	1.357	1.451
1.61	2.108	3.929	5.365	1.214	4.553	0.6808	1.362	1.452
1.62	2.078	3.896	5.334	1.220	4.544	0.6831	1.366	1.454
1.63	2.048	3.862	5.303	1.227	4.535	0.6853	1.371	1.455
1.64	2.019	3.829	5.273	1.234	4.525	0.6875	1.375	1.457
1.65	1.990	3.796	5.243	1.240	4.515	0.6897	1.379	1.458
1.66	1.962	3.763	5.212	1.247	4.504	0.6919	1.384	1.459
1.67	1.934	3.731	5.182	1.254	4.493	0.6941	1.388	1.461
1.68	1.906	3.699	5.153	1.261	4.482	0.6962	1.393	1.462
1.69	1.879	3.667	5.123	1.268	4.471	0.6984	1.397	1.464
1.70	1.852	3.635	5.094	1.275	4.459	0.7005	1.401	1.465
1.71	1.825	3.604	5.064	1.283	4.447	0.7026	1.405	1.467
1.72	1.799	3.573	5.035	1.290	4.435	0.7046	1.409	1.468
1.73	1.773	3.542	5.006	1.297	4.422	0.7067	1.413	1.469
1.74	1.748	3.511	4.977	1.305	4.409	0.7087	1.417	1.471
1.75	1.723	3.481	4.949	1.313	4.396	0.7107	1.422	1.472
1.76	1.698	3.451	4.920	1.320	4.383	0.7127	1.426	1.474
1.77	1.674	3.421	4.892	1.328	4.369	0.7147	1.429	1.475
1.78	1.650	3.392	4.864	1.336	4.356	0.7167	1.433	1.476
1.79	1.626	3.363	4.836	1.344	4.342	0.7186	1.437	1.478
1.80	1.603	3.334	4.808	1.352	4.327	0.7206	1.441	1.479
1.81	1.580	3.305	4.780	1.360	4.313	0.7225	1.445	1.481
1.82	1.557	3.277	4.753	1.368	4.298	0.7244	1.449	1.482
1.83	1.535	3.248	4.725	1.377	4.283	0.7263	1.453	1.483
1.84	1.513	3.220	4.698	1.385	4.268	0.7281	1.456	1.485
1.85	1.491	3.193	4.671	1.393	4.253	0.7300	1.460	1.486
1.86	1.470	3.165	4.644	1.402	4.238	0.7318	1.464	1.487
1.87	1.449	3.138	4.618	1.411	4.222	0.7336	1.467	1.489
1.88	1.428	3.111	4.591	1.419	4.207	0.7354	1.471	1.490
1.89	1.408	3.084	4.565	1.428	4.191	0.7372	1.475	1.491
1.90	1.388	3.058	4.539	1.437	4.175	0.7390	1.478	1.493
1.91	1.368	3.032	4.513	1.446	4.159	0.7408	1.482	1.494
1.92	1.349	3.006	4.487	1.455	4.142	0.7425	1.485	1.495
1.93	1.329	2.980	4.461	1.464	4.126	0.7442	1.489	1.497
1.94	1.310	2.954	4.436	1.474	4.110	0.7459	1.492	1.498
1.95	1.292	2.929	4.410	1.483	4.093	0.7476	1.495	1.499
1.96	1.273	2.904	4.385	1.493	4.076	0.7493	1.499	1.501
1.97	1.255	2.879	4.360	1.502	4.059	0.7510	1.502	1.502
1.98	1.237	2.854	4.335	1.512	4.042	0.7526	1.505	1.503
1.99	1.220	2.830	4.310	1.521	4.025	0.7543	1.509	1.504
2.00	1.203	2.806	4.286	1.531	4.008	0.7559	1.512	1.506

NORMAL SHOCK FUNCTIONS FOR SUPERSONIC COMPRESSIBLE FLOW

$\gamma = 5/3$

M	p_{21}	T_{21}	ρ_{21}	M_2	p_{21}^o	μ	ν
1.50	2.562 +0	1.495 +0	1.714	0.7157	9.380 -1	41.81	10.22
1.51	2.600	1.505	1.727	0.7125	9.353	41.47	10.47
1.52	2.638	1.516	1.740	0.7094	9.325	41.14	10.71
1.53	2.676	1.526	1.753	0.7064	9.296	40.81	10.95
1.54	2.714	1.537	1.766	0.7034	9.268	40.49	11.20
1.55	2.753	1.548	1.779	0.7004	9.238	40.18	11.44
1.56	2.792	1.558	1.791	0.6975	9.208	39.87	11.68
1.57	2.831	1.569	1.804	0.6947	9.178	39.57	11.92
1.58	2.870	1.580	1.817	0.6919	9.148	39.27	12.17
1.59	2.910	1.591	1.829	0.6891	9.117	38.97	12.41
1.60	2.950	1.602	1.842	0.6864	9.085	38.68	12.65
1.61	2.990	1.613	1.854	0.6838	9.053	38.40	12.89
1.62	3.030	1.624	1.866	0.6812	9.021	38.12	13.13
1.63	3.071	1.635	1.879	0.6786	8.989	37.84	13.37
1.64	3.112	1.646	1.891	0.6761	8.956	37.57	13.61
1.65	3.153	1.657	1.903	0.6736	8.923	37.31	13.85
1.66	3.194	1.668	1.915	0.6712	8.889	37.04	14.09
1.67	3.236	1.679	1.927	0.6687	8.856	36.79	14.33
1.68	3.278	1.691	1.939	0.6664	8.821	36.53	14.57
1.69	3.320	1.702	1.951	0.6641	8.787	36.28	14.80
1.70	3.362	1.713	1.963	0.6618	8.752	36.03	15.04
1.71	3.405	1.725	1.974	0.6595	8.717	35.79	15.27
1.72	3.448	1.736	1.986	0.6573	8.682	35.55	15.51
1.73	3.491	1.748	1.998	0.6551	8.647	35.31	15.74
1.74	3.534	1.759	2.009	0.6530	8.611	35.08	15.98
1.75	3.578	1.771	2.021	0.6508	8.575	34.85	16.21
1.76	3.622	1.782	2.032	0.6488	8.539	34.62	16.44
1.77	3.666	1.794	2.043	0.6467	8.503	34.40	16.67
1.78	3.710	1.806	2.055	0.6447	8.466	34.18	16.91
1.79	3.755	1.818	2.066	0.6427	8.429	33.96	17.14
1.80	3.800	1.830	2.077	0.6407	8.392	33.75	17.36
1.81	3.845	1.841	2.088	0.6388	8.355	33.54	17.59
1.82	3.890	1.853	2.099	0.6369	8.318	33.33	17.82
1.83	3.936	1.865	2.110	0.6350	8.281	33.12	18.05
1.84	3.982	1.878	2.121	0.6332	8.243	32.92	18.27
1.85	4.028	1.890	2.131	0.6314	8.206	32.72	18.50
1.86	4.074	1.902	2.142	0.6296	8.168	32.52	18.73
1.87	4.121	1.914	2.153	0.6278	8.130	32.33	18.95
1.88	4.168	1.926	2.163	0.6261	8.092	32.14	19.17
1.89	4.215	1.939	2.174	0.6243	8.054	31.95	19.39
1.90	4.262	1.951	2.184	0.6227	8.016	31.76	19.62
1.91	4.310	1.964	2.195	0.6210	7.977	31.57	19.84
1.92	4.358	1.976	2.205	0.6193	7.939	31.39	20.06
1.93	4.406	1.989	2.216	0.6177	7.900	31.21	20.28
1.94	4.454	2.001	2.226	0.6161	7.862	31.03	20.49
1.95	4.503	2.014	2.236	0.6145	7.823	30.85	20.71
1.96	4.552	2.027	2.246	0.6130	7.785	30.68	20.93
1.97	4.601	2.039	2.256	0.6115	7.746	30.51	21.14
1.98	4.650	2.052	2.266	0.6099	7.708	30.34	21.36
1.99	4.700	2.065	2.276	0.6085	7.669	30.17	21.57
2.00	4.750	2.078	2.286	0.6070	7.630	30.00	21.79

ISENTROPIC FUNCTIONS FOR SUPERSONIC COMPRESSIBLE FLOW

$\gamma = 5/3$

M	$p/p°$	$T/T°$	$\rho/\rho°$	A/A^*	$q/p°$	u/u_m	u/a^*	$F(\gamma;M)$
2.00	1.203 -1	2.806 -1	4.286 -1	1.531 +0	4.008 -1	0.7559	1.512	1.506
2.01	1.186	2.782	4.261	1.541	3.991	0.7575	1.515	1.507
2.02	1.169	2.758	4.237	1.551	3.974	0.7591	1.518	1.508
2.03	1.152	2.735	4.213	1.561	3.956	0.7607	1.521	1.509
2.04	1.136	2.711	4.189	1.571	3.939	0.7623	1.525	1.511
2.05	1.120	2.688	4.165	1.582	3.921	0.7638	1.528	1.512
2.06	1.104	2.666	4.142	1.592	3.904	0.7654	1.531	1.513
2.07	1.088	2.643	4.118	1.602	3.886	0.7669	1.534	1.514
2.08	1.073	2.620	4.095	1.613	3.868	0.7684	1.537	1.516
2.09	1.058	2.598	4.072	1.623	3.851	0.7699	1.540	1.517
2.10	1.043	2.576	4.049	1.634	3.833	0.7714	1.543	1.518
2.11	1.028	2.554	4.026	1.645	3.815	0.7729	1.546	1.519
2.12	1.014	2.533	4.003	1.656	3.797	0.7744	1.549	1.520
2.13	9.997 -2	2.511	3.981	1.667	3.779	0.7758	1.552	1.522
2.14	9.857	2.490	3.958	1.678	3.761	0.7773	1.555	1.523
2.15	9.719	2.469	3.936	1.689	3.743	0.7787	1.557	1.524
2.16	9.583	2.448	3.914	1.700	3.725	0.7801	1.560	1.525
2.17	9.449	2.428	3.892	1.712	3.707	0.7815	1.563	1.526
2.18	9.317	2.407	3.870	1.723	3.689	0.7829	1.566	1.527
2.19	9.187	2.387	3.848	1.734	3.671	0.7843	1.569	1.528
2.20	9.059	2.367	3.827	1.746	3.653	0.7857	1.571	1.530
2.21	8.933	2.347	3.805	1.758	3.635	0.7871	1.574	1.531
2.22	8.808	2.328	3.784	1.770	3.617	0.7884	1.577	1.532
2.23	8.686	2.308	3.763	1.781	3.599	0.7898	1.579	1.533
2.24	8.565	2.289	3.742	1.793	3.581	0.7911	1.582	1.534
2.25	8.447	2.270	3.721	1.806	3.563	0.7924	1.585	1.535
2.26	8.330	2.251	3.700	1.818	3.545	0.7937	1.587	1.536
2.27	8.214	2.232	3.680	1.830	3.527	0.7950	1.590	1.537
2.28	8.101	2.214	3.659	1.842	3.509	0.7963	1.593	1.538
2.29	7.989	2.195	3.639	1.855	3.491	0.7975	1.595	1.539
2.30	7.879	2.177	3.619	1.867	3.473	0.7988	1.598	1.541
2.31	7.771	2.159	3.599	1.880	3.455	0.8001	1.600	1.542
2.32	7.664	2.141	3.579	1.893	3.437	0.8013	1.603	1.543
2.33	7.559	2.124	3.559	1.906	3.419	0.8025	1.605	1.544
2.34	7.455	2.106	3.540	1.919	3.401	0.8038	1.607	1.545
2.35	7.353	2.089	3.520	1.932	3.383	0.8050	1.610	1.546
2.36	7.252	2.071	3.501	1.945	3.366	0.8062	1.612	1.547
2.37	7.153	2.054	3.482	1.958	3.348	0.8074	1.615	1.548
2.38	7.055	2.038	3.463	1.971	3.330	0.8085	1.617	1.549
2.39	6.959	2.021	3.444	1.985	3.312	0.8097	1.619	1.550
2.40	6.864	2.004	3.425	1.998	3.295	0.8109	1.622	1.551
2.41	6.771	1.988	3.406	2.012	3.277	0.8120	1.624	1.552
2.42	6.679	1.972	3.388	2.026	3.259	0.8132	1.626	1.553
2.43	6.589	1.956	3.369	2.039	3.242	0.8143	1.629	1.554
2.44	6.499	1.940	3.351	2.053	3.224	0.8154	1.631	1.555
2.45	6.412	1.924	3.333	2.067	3.207	0.8165	1.633	1.556
2.46	6.325	1.908	3.315	2.081	3.189	0.8176	1.635	1.557
2.47	6.240	1.893	3.297	2.096	3.172	0.8187	1.637	1.558
2.48	6.156	1.877	3.279	2.110	3.155	0.8198	1.640	1.559
2.49	6.073	1.862	3.261	2.124	3.137	0.8209	1.642	1.559
2.50	5.991	1.847	3.243	2.139	3.120	0.8220	1.644	1.560

NORMAL SHOCK FUNCTIONS FOR SUPERSONIC COMPRESSIBLE FLOW

$$\gamma = 5/3$$

M	p_{21}	T_{21}	ρ_{21}	M_2	p_{21}^o	μ	ν
2.00	4.750 +0	2.078 +0	2.286	0.6070	7.630 −1	30.00	21.79
2.01	4.800	2.091	2.295	0.6055	7.591	29.84	22.00
2.02	4.850	2.104	2.305	0.6041	7.553	29.67	22.21
2.03	4.901	2.117	2.315	0.6027	7.514	29.51	22.42
2.04	4.952	2.130	2.324	0.6013	7.475	29.35	22.63
2.05	5.003	2.144	2.334	0.5999	7.436	29.20	22.84
2.06	5.054	2.157	2.343	0.5986	7.398	29.04	23.05
2.07	5.106	2.170	2.353	0.5972	7.359	28.89	23.25
2.08	5.158	2.184	2.362	0.5959	7.320	28.74	23.46
2.09	5.210	2.197	2.371	0.5946	7.282	28.59	23.66
2.10	5.262	2.210	2.380	0.5933	7.243	28.44	23.87
2.11	5.315	2.224	2.390	0.5921	7.205	28.29	24.07
2.12	5.368	2.238	2.399	0.5908	7.166	28.15	24.27
2.13	5.421	2.251	2.408	0.5896	7.128	28.00	24.48
2.14	5.474	2.265	2.417	0.5883	7.089	27.86	24.68
2.15	5.528	2.279	2.426	0.5871	7.051	27.72	24.88
2.16	5.582	2.293	2.434	0.5859	7.013	27.58	25.08
2.17	5.636	2.307	2.443	0.5848	6.975	27.44	25.27
2.18	5.690	2.321	2.452	0.5836	6.936	27.31	25.47
2.19	5.745	2.335	2.461	0.5825	6.898	27.17	25.67
2.20	5.800	2.349	2.469	0.5813	6.860	27.04	25.86
2.21	5.855	2.363	2.478	0.5802	6.823	26.90	26.06
2.22	5.910	2.377	2.486	0.5791	6.785	26.77	26.25
2.23	5.966	2.391	2.495	0.5780	6.747	26.64	26.45
2.24	6.021	2.405	2.503	0.5769	6.710	26.52	26.64
2.25	6.078	2.420	2.512	0.5759	6.672	26.39	26.83
2.26	6.134	2.434	2.520	0.5748	6.635	26.26	27.02
2.27	6.191	2.449	2.528	0.5738	6.597	26.14	27.21
2.28	6.247	2.463	2.536	0.5728	6.560	26.02	27.40
2.29	6.305	2.478	2.544	0.5717	6.523	25.89	27.59
2.30	6.362	2.493	2.552	0.5707	6.486	25.77	27.77
2.31	6.420	2.507	2.560	0.5698	6.449	25.65	27.96
2.32	6.477	2.522	2.568	0.5688	6.413	25.53	28.15
2.33	6.536	2.537	2.576	0.5678	6.376	25.42	28.33
2.34	6.594	2.552	2.584	0.5669	6.340	25.30	28.51
2.35	6.653	2.567	2.592	0.5659	6.303	25.19	28.70
2.36	6.711	2.582	2.600	0.5650	6.267	25.07	28.88
2.37	6.771	2.597	2.607	0.5641	6.231	24.96	29.06
2.38	6.830	2.612	2.615	0.5631	6.195	24.85	29.24
2.39	6.889	2.627	2.623	0.5622	6.159	24.74	29.42
2.40	6.949	2.642	2.630	0.5614	6.123	24.63	29.60
2.41	7.009	2.658	2.638	0.5605	6.088	24.52	29.78
2.42	7.070	2.673	2.645	0.5596	6.053	24.41	29.95
2.43	7.130	2.688	2.652	0.5587	6.017	24.30	30.13
2.44	7.191	2.704	2.660	0.5579	5.982	24.20	30.31
2.45	7.252	2.719	2.667	0.5571	5.947	24.09	30.48
2.46	7.314	2.735	2.674	0.5562	5.912	23.99	30.66
2.47	7.375	2.751	2.681	0.5554	5.878	23.88	30.83
2.48	7.437	2.766	2.689	0.5546	5.843	23.78	31.00
2.49	7.499	2.782	2.696	0.5538	5.809	23.68	31.17
2.50	7.562	2.798	2.703	0.5530	5.775	23.58	31.34

ISENTROPIC FUNCTIONS FOR SUPERSONIC COMPRESSIBLE FLOW

$\gamma = 5/3$

M	p/p°	T/T°	ρ/ρ°	A/A^*	q/p°	u/u_m	u/a^*	$F(\gamma;M)$
2.50	5.991 −2	1.847 −1	3.243 −1	2.139 +0	3.120 −1	0.8220	1.644	1.560
2.51	5.911	1.832	3.226	2.154	3.103	0.8230	1.646	1.561
2.52	5.832	1.818	3.209	2.168	3.086	0.8241	1.648	1.562
2.53	5.754	1.803	3.191	2.183	3.069	0.8251	1.650	1.563
2.54	5.677	1.788	3.174	2.198	3.052	0.8262	1.652	1.564
2.55	5.601	1.774	3.157	2.213	3.035	0.8272	1.654	1.565
2.56	5.527	1.760	3.140	2.228	3.018	0.8282	1.657	1.566
2.57	5.453	1.746	3.124	2.243	3.001	0.8292	1.659	1.567
2.58	5.381	1.732	3.107	2.259	2.984	0.8302	1.661	1.568
2.59	5.309	1.718	3.090	2.274	2.968	0.8312	1.663	1.569
2.60	5.239	1.704	3.074	2.290	2.951	0.8322	1.665	1.569
2.61	5.170	1.691	3.058	2.305	2.934	0.8332	1.666	1.570
2.62	5.101	1.677	3.041	2.321	2.918	0.8342	1.668	1.571
2.63	5.034	1.664	3.025	2.337	2.902	0.8351	1.670	1.572
2.64	4.968	1.651	3.009	2.353	2.885	0.8361	1.672	1.573
2.65	4.903	1.638	2.993	2.369	2.869	0.8371	1.674	1.574
2.66	4.838	1.625	2.978	2.385	2.853	0.8380	1.676	1.575
2.67	4.775	1.612	2.962	2.401	2.836	0.8389	1.678	1.575
2.68	4.712	1.599	2.946	2.418	2.820	0.8399	1.680	1.576
2.69	4.651	1.587	2.931	2.434	2.804	0.8408	1.682	1.577
2.70	4.590	1.574	2.916	2.451	2.788	0.8417	1.683	1.578
2.71	4.530	1.562	2.900	2.468	2.772	0.8426	1.685	1.579
2.72	4.472	1.550	2.885	2.484	2.757	0.8435	1.687	1.579
2.73	4.414	1.538	2.870	2.501	2.741	0.8444	1.689	1.580
2.74	4.356	1.526	2.855	2.518	2.725	0.8453	1.691	1.581
2.75	4.300	1.514	2.840	2.535	2.710	0.8461	1.692	1.582
2.76	4.244	1.502	2.826	2.553	2.694	0.8470	1.694	1.583
2.77	4.190	1.490	2.811	2.570	2.679	0.8479	1.696	1.583
2.78	4.136	1.479	2.797	2.587	2.663	0.8487	1.698	1.584
2.79	4.082	1.467	2.782	2.605	2.648	0.8496	1.699	1.585
2.80	4.030	1.456	2.768	2.623	2.633	0.8504	1.701	1.586
2.81	3.978	1.445	2.753	2.640	2.618	0.8513	1.703	1.586
2.82	3.927	1.434	2.739	2.658	2.602	0.8521	1.704	1.587
2.83	3.877	1.423	2.725	2.676	2.587	0.8529	1.706	1.588
2.84	3.828	1.412	2.711	2.694	2.573	0.8537	1.707	1.589
2.85	3.779	1.401	2.697	2.713	2.558	0.8546	1.709	1.589
2.86	3.731	1.390	2.684	2.731	2.543	0.8554	1.711	1.590
2.87	3.684	1.380	2.670	2.749	2.528	0.8562	1.712	1.591
2.88	3.637	1.369	2.656	2.768	2.514	0.8569	1.714	1.592
2.89	3.591	1.359	2.643	2.787	2.499	0.8577	1.716	1.592
2.90	3.545	1.348	2.629	2.806	2.485	0.8585	1.717	1.593
2.91	3.501	1.338	2.616	2.824	2.470	0.8593	1.719	1.594
2.92	3.457	1.328	2.603	2.843	2.456	0.8601	1.720	1.595
2.93	3.413	1.318	2.590	2.863	2.442	0.8608	1.722	1.595
2.94	3.370	1.308	2.577	2.882	2.427	0.8616	1.723	1.596
2.95	3.328	1.298	2.564	2.901	2.413	0.8623	1.725	1.597
2.96	3.286	1.288	2.551	2.921	2.399	0.8631	1.726	1.597
2.97	3.245	1.279	2.538	2.940	2.385	0.8638	1.728	1.598
2.98	3.205	1.269	2.525	2.960	2.371	0.8646	1.729	1.599
2.99	3.165	1.260	2.513	2.980	2.358	0.8653	1.731	1.599
3.00	3.126	1.250	2.500	3.000	2.344	0.8660	1.732	1.600

NORMAL SHOCK FUNCTIONS FOR SUPERSONIC COMPRESSIBLE FLOW

$$\gamma = 5/3$$

M	p_{21}	T_{21}	ρ_{21}	M_2	p^o_{21}	μ	ν
2.50	7.562 +0	2.798 +0	2.703	0.5530	5.775 -1	23.58	31.34
2.51	7.624	2.814	2.710	0.5522	5.741	23.48	31.51
2.52	7.687	2.830	2.717	0.5514	5.707	23.38	31.68
2.53	7.750	2.846	2.723	0.5507	5.673	23.28	31.85
2.54	7.814	2.862	2.730	0.5499	5.639	23.19	32.02
2.55	7.877	2.878	2.737	0.5491	5.606	23.09	32.18
2.56	7.941	2.894	2.744	0.5484	5.573	22.99	32.35
2.57	8.005	2.910	2.751	0.5477	5.540	22.90	32.52
2.58	8.070	2.927	2.757	0.5469	5.507	22.81	32.68
2.59	8.134	2.943	2.764	0.5462	5.474	22.71	32.84
2.60	8.199	2.960	2.770	0.5455	5.441	22.62	33.01
2.61	8.264	2.976	2.777	0.5448	5.409	22.53	33.17
2.62	8.330	2.993	2.783	0.5441	5.377	22.44	33.33
2.63	8.395	3.009	2.790	0.5434	5.344	22.35	33.49
2.64	8.461	3.026	2.796	0.5427	5.312	22.26	33.65
2.65	8.527	3.043	2.803	0.5420	5.281	22.17	33.81
2.66	8.594	3.059	2.809	0.5414	5.249	22.08	33.97
2.67	8.660	3.076	2.815	0.5407	5.218	22.00	34.13
2.68	8.727	3.093	2.821	0.5401	5.186	21.91	34.28
2.69	8.794	3.110	2.828	0.5394	5.155	21.82	34.44
2.70	8.862	3.127	2.834	0.5388	5.124	21.74	34.60
2.71	8.929	3.144	2.840	0.5381	5.093	21.66	34.75
2.72	8.997	3.161	2.846	0.5375	5.063	21.57	34.90
2.73	9.065	3.179	2.852	0.5369	5.032	21.49	35.06
2.74	9.134	3.196	2.858	0.5363	5.002	21.41	35.21
2.75	9.202	3.213	2.864	0.5357	4.972	21.32	35.36
2.76	9.271	3.231	2.870	0.5351	4.942	21.24	35.51
2.77	9.340	3.248	2.876	0.5345	4.912	21.16	35.66
2.78	9.410	3.266	2.881	0.5339	4.883	21.08	35.81
2.79	9.479	3.283	2.887	0.5333	4.853	21.00	35.96
2.80	9.549	3.301	2.893	0.5327	4.824	20.93	36.11
2.81	9.619	3.319	2.899	0.5321	4.795	20.85	36.26
2.82	9.690	3.336	2.904	0.5316	4.766	20.77	36.41
2.83	9.760	3.354	2.910	0.5310	4.737	20.69	36.55
2.84	9.831	3.372	2.915	0.5304	4.708	20.62	36.70
2.85	9.902	3.390	2.921	0.5299	4.680	20.54	36.84
2.86	9.973	3.408	2.927	0.5294	4.652	20.47	36.99
2.87	1.005 +1	3.425	2.932	0.5288	4.624	20.39	37.13
2.88	1.012	3.444	2.937	0.5283	4.596	20.32	37.27
2.89	1.019	3.462	2.943	0.5277	4.568	20.25	37.42
2.90	1.026	3.481	2.948	0.5272	4.540	20.17	37.56
2.91	1.033	3.499	2.954	0.5267	4.513	20.10	37.70
2.92	1.041	3.517	2.959	0.5262	4.486	20.03	37.84
2.93	1.048	3.536	2.964	0.5257	4.458	19.96	37.98
2.94	1.055	3.554	2.969	0.5252	4.431	19.89	38.12
2.95	1.063	3.573	2.975	0.5247	4.405	19.82	38.26
2.96	1.070	3.591	2.980	0.5242	4.378	19.75	38.40
2.97	1.078	3.610	2.985	0.5237	4.352	19.68	38.53
2.98	1.085	3.629	2.990	0.5232	4.325	19.61	38.67
2.99	1.092	3.648	2.995	0.5227	4.299	19.54	38.81
3.00	1.100	3.666	3.000	0.5222	4.273	19.47	38.94

A.1-11

ISENTROPIC FUNCTIONS FOR SUPERSONIC COMPRESSIBLE FLOW

$\gamma = 5/3$

M	p/p^o	T/T^o	ρ/ρ^o	A/A^*	q/p^o	u/u_m	u/a^*	$F(\gamma;M)$
3.00	3.126 -2	1.250 -1	2.500 -1	3.000 +0	2.344 -1	0.8660	1.732	1.600
3.02	3.049	1.232	2.475	3.040	2.317	0.8674	1.735	1.601
3.04	2.974	1.213	2.451	3.081	2.290	0.8689	1.738	1.603
3.06	2.901	1.195	2.427	3.122	2.263	0.8702	1.741	1.604
3.08	2.830	1.178	2.403	3.163	2.237	0.8716	1.743	1.605
3.10	2.761	1.161	2.379	3.206	2.211	0.8730	1.746	1.606
3.12	2.694	1.144	2.356	3.248	2.185	0.8743	1.749	1.608
3.14	2.629	1.127	2.333	3.291	2.160	0.8756	1.751	1.609
3.16	2.566	1.111	2.310	3.335	2.135	0.8769	1.754	1.610
3.18	2.504	1.094	2.288	3.379	2.110	0.8782	1.756	1.611
3.20	2.444	1.079	2.266	3.423	2.086	0.8794	1.759	1.612
3.22	2.386	1.063	2.244	3.468	2.061	0.8807	1.761	1.614
3.24	2.329	1.048	2.223	3.514	2.038	0.8819	1.764	1.615
3.26	2.274	1.033	2.202	3.560	2.014	0.8831	1.766	1.616
3.28	2.221	1.018	2.181	3.607	1.991	0.8843	1.769	1.617
3.30	2.168	1.004	2.160	3.654	1.968	0.8854	1.771	1.618
3.32	2.118	9.897 -2	2.140	3.701	1.945	0.8866	1.773	1.619
3.34	2.068	9.758	2.119	3.749	1.922	0.8877	1.775	1.620
3.36	2.020	9.621	2.100	3.798	1.900	0.8888	1.778	1.621
3.38	1.973	9.486	2.080	3.847	1.878	0.8899	1.780	1.622
3.40	1.927	9.354	2.061	3.897	1.857	0.8910	1.782	1.623
3.42	1.883	9.224	2.041	3.947	1.835	0.8921	1.784	1.624
3.44	1.840	9.096	2.023	3.997	1.814	0.8932	1.786	1.625
3.46	1.798	8.971	2.004	4.049	1.793	0.8942	1.788	1.626
3.48	1.757	8.847	1.986	4.100	1.773	0.8952	1.791	1.627
3.50	1.717	8.726	1.967	4.152	1.752	0.8962	1.793	1.628
3.52	1.678	8.607	1.949	4.205	1.732	0.8973	1.795	1.629
3.54	1.640	8.490	1.932	4.259	1.712	0.8982	1.797	1.630
3.56	1.603	8.375	1.914	4.312	1.693	0.8992	1.798	1.631
3.58	1.567	8.262	1.897	4.367	1.674	0.9002	1.800	1.632
3.60	1.532	8.150	1.880	4.422	1.655	0.9011	1.802	1.633
3.62	1.498	8.041	1.863	4.477	1.636	0.9021	1.804	1.634
3.64	1.465	7.934	1.846	4.533	1.617	0.9030	1.806	1.635
3.66	1.432	7.828	1.830	4.590	1.599	0.9039	1.808	1.636
3.68	1.401	7.724	1.814	4.647	1.581	0.9048	1.810	1.637
3.70	1.370	7.622	1.798	4.705	1.563	0.9057	1.811	1.637
3.72	1.340	7.521	1.782	4.763	1.545	0.9065	1.813	1.638
3.74	1.311	7.422	1.766	4.822	1.528	0.9074	1.815	1.639
3.76	1.282	7.325	1.751	4.881	1.511	0.9083	1.817	1.640
3.78	1.255	7.229	1.735	4.941	1.494	0.9091	1.818	1.641
3.80	1.227	7.135	1.720	5.002	1.477	0.9099	1.820	1.642
3.82	1.201	7.043	1.705	5.063	1.460	0.9107	1.822	1.642
3.84	1.175	6.952	1.691	5.125	1.444	0.9116	1.823	1.643
3.86	1.150	6.862	1.676	5.187	1.428	0.9124	1.825	1.644
3.88	1.126	6.774	1.662	5.250	1.412	0.9131	1.826	1.645
3.90	1.102	6.688	1.648	5.314	1.396	0.9139	1.828	1.645
3.92	1.079	6.602	1.634	5.378	1.381	0.9147	1.829	1.646
3.94	1.056	6.518	1.620	5.442	1.366	0.9154	1.831	1.647
3.96	1.034	6.436	1.606	5.508	1.351	0.9162	1.832	1.648
3.98	1.012	6.355	1.592	5.574	1.336	0.9169	1.834	1.648
4.00	9.908 -3	6.275	1.579	5.640	1.321	0.9177	1.835	1.649

NORMAL SHOCK FUNCTIONS FOR SUPERSONIC COMPRESSIBLE FLOW

$$\gamma = 5/3$$

M	P_{21}	T_{21}	ρ_{21}	M_2	P^o_{21}	μ	ν
3.00	1.100 +1	3.666 +0	3.000	0.5222	4.273 -1	19.47	38.94
3.02	1.115	3.704	3.010	0.5213	4.222	19.34	39.21
3.04	1.130	3.742	3.020	0.5204	4.171	19.21	39.48
3.06	1.145	3.781	3.029	0.5195	4.121	19.08	39.74
3.08	1.161	3.819	3.039	0.5186	4.071	18.95	40.00
3.10	1.176	3.858	3.048	0.5177	4.022	18.82	40.26
3.12	1.192	3.897	3.058	0.5168	3.974	18.69	40.52
3.14	1.207	3.937	3.067	0.5160	3.926	18.57	40.77
3.16	1.223	3.976	3.076	0.5152	3.879	18.45	41.02
3.18	1.239	4.016	3.085	0.5144	3.833	18.33	41.27
3.20	1.255	4.056	3.094	0.5136	3.787	18.21	41.52
3.22	1.271	4.097	3.102	0.5128	3.741	18.09	41.77
3.24	1.287	4.137	3.111	0.5120	3.697	17.98	42.01
3.26	1.303	4.178	3.119	0.5113	3.652	17.86	42.25
3.28	1.320	4.219	3.128	0.5105	3.609	17.75	42.49
3.30	1.336	4.261	3.136	0.5098	3.566	17.64	42.73
3.32	1.353	4.302	3.144	0.5091	3.524	17.53	42.96
3.34	1.369	4.344	3.152	0.5084	3.482	17.42	43.20
3.36	1.386	4.386	3.160	0.5077	3.440	17.32	43.43
3.38	1.403	4.428	3.168	0.5070	3.400	17.21	43.66
3.40	1.420	4.471	3.176	0.5063	3.359	17.11	43.88
3.42	1.437	4.514	3.183	0.5056	3.320	17.00	44.11
3.44	1.454	4.557	3.191	0.5050	3.280	16.90	44.33
3.46	1.471	4.600	3.198	0.5044	3.242	16.80	44.55
3.48	1.489	4.644	3.206	0.5037	3.204	16.70	44.77
3.50	1.506	4.687	3.213	0.5031	3.166	16.60	44.99
3.52	1.524	4.732	3.220	0.5025	3.129	16.51	45.20
3.54	1.541	4.776	3.227	0.5019	3.092	16.41	45.41
3.56	1.559	4.820	3.234	0.5013	3.056	16.31	45.63
3.58	1.577	4.865	3.241	0.5007	3.020	16.22	45.84
3.60	1.595	4.910	3.248	0.5002	2.985	16.13	46.04
3.62	1.613	4.955	3.255	0.4996	2.951	16.04	46.25
3.64	1.631	5.001	3.261	0.4990	2.916	15.95	46.45
3.66	1.649	5.047	3.268	0.4985	2.883	15.86	46.66
3.68	1.668	5.093	3.275	0.4980	2.849	15.77	46.86
3.70	1.686	5.139	3.281	0.4974	2.816	15.68	47.06
3.72	1.705	5.186	3.287	0.4969	2.784	15.59	47.25
3.74	1.723	5.232	3.294	0.4964	2.752	15.51	47.45
3.76	1.742	5.279	3.300	0.4959	2.720	15.42	47.64
3.78	1.761	5.327	3.306	0.4954	2.689	15.34	47.84
3.80	1.780	5.374	3.312	0.4949	2.658	15.26	48.03
3.82	1.799	5.422	3.318	0.4944	2.628	15.18	48.22
3.84	1.818	5.470	3.324	0.4940	2.598	15.10	48.40
3.86	1.837	5.518	3.330	0.4935	2.569	15.02	48.59
3.88	1.857	5.567	3.335	0.4930	2.539	14.94	48.78
3.90	1.876	5.615	3.341	0.4926	2.511	14.86	48.96
3.92	1.896	5.664	3.347	0.4921	2.482	14.78	49.14
3.94	1.915	5.714	3.352	0.4917	2.454	14.70	49.32
3.96	1.935	5.763	3.358	0.4913	2.427	14.63	49.50
3.98	1.955	5.813	3.363	0.4908	2.400	14.55	49.68
4.00	1.975	5.863	3.368	0.4904	2.373	14.48	49.85

ISENTROPIC FUNCTIONS FOR SUPERSONIC COMPRESSIBLE FLOW

$$\gamma = 5/3$$

M	$p/p°$	$T/T°$	$\rho/\rho°$	A/A^*	$q/p°$	u/u_m	u/a^*	$F(\gamma;M)$
4.00	9.908 −3	6.275 −2	1.579 −1	5.640 +0	1.321 −1	0.9177	1.835	1.649
4.02	9.702	6.196	1.566	5.707	1.306	0.9184	1.837	1.650
4.04	9.501	6.119	1.553	5.775	1.292	0.9191	1.838	1.650
4.06	9.305	6.043	1.540	5.843	1.278	0.9198	1.840	1.651
4.08	9.113	5.968	1.527	5.912	1.264	0.9205	1.841	1.652
4.10	8.926	5.894	1.514	5.982	1.250	0.9212	1.842	1.652
4.12	8.744	5.821	1.502	6.052	1.237	0.9218	1.844	1.653
4.14	8.566	5.750	1.490	6.123	1.223	0.9225	1.845	1.654
4.16	8.392	5.679	1.478	6.194	1.210	0.9232	1.846	1.654
4.18	8.222	5.610	1.465	6.266	1.197	0.9238	1.848	1.655
4.20	8.056	5.542	1.454	6.339	1.184	0.9245	1.849	1.656
4.22	7.894	5.475	1.442	6.412	1.171	0.9251	1.850	1.656
4.24	7.736	5.409	1.430	6.486	1.159	0.9257	1.851	1.657
4.26	7.581	5.344	1.419	6.561	1.146	0.9264	1.853	1.658
4.28	7.430	5.280	1.407	6.636	1.134	0.9270	1.854	1.658
4.30	7.283	5.216	1.396	6.712	1.122	0.9276	1.855	1.659
4.32	7.139	5.154	1.385	6.788	1.110	0.9282	1.856	1.659
4.34	6.998	5.093	1.374	6.866	1.098	0.9288	1.858	1.660
4.36	6.860	5.033	1.363	6.943	1.087	0.9293	1.859	1.660
4.38	6.726	4.973	1.352	7.022	1.075	0.9299	1.860	1.661
4.40	6.595	4.915	1.342	7.101	1.064	0.9305	1.861	1.662
4.42	6.467	4.857	1.331	7.181	1.053	0.9311	1.862	1.662
4.44	6.341	4.801	1.321	7.261	1.042	0.9316	1.863	1.663
4.46	6.219	4.745	1.311	7.343	1.031	0.9322	1.864	1.663
4.48	6.099	4.690	1.300	7.424	1.020	0.9327	1.865	1.664
4.50	5.982	4.635	1.290	7.507	1.009	0.9333	1.867	1.664
4.52	5.867	4.582	1.280	7.590	9.988 −2	0.9338	1.868	1.665
4.54	5.755	4.529	1.271	7.674	9.885	0.9343	1.869	1.665
4.56	5.646	4.478	1.261	7.759	9.782	0.9348	1.870	1.666
4.58	5.539	4.426	1.251	7.844	9.681	0.9353	1.871	1.666
4.60	5.434	4.376	1.242	7.930	9.582	0.9359	1.872	1.667
4.62	5.332	4.326	1.232	8.017	9.483	0.9364	1.873	1.667
4.64	5.232	4.278	1.223	8.104	9.386	0.9369	1.874	1.668
4.66	5.134	4.229	1.214	8.192	9.290	0.9373	1.875	1.668
4.68	5.038	4.182	1.205	8.281	9.195	0.9378	1.876	1.669
4.70	4.945	4.135	1.196	8.370	9.101	0.9383	1.877	1.669
4.72	4.853	4.089	1.187	8.460	9.009	0.9388	1.878	1.670
4.74	4.763	4.043	1.178	8.551	8.918	0.9393	1.879	1.670
4.76	4.676	3.999	1.169	8.643	8.828	0.9397	1.879	1.671
4.78	4.590	3.954	1.161	8.735	8.739	0.9402	1.880	1.671
4.80	4.506	3.911	1.152	8.828	8.651	0.9406	1.881	1.672
4.82	4.424	3.868	1.144	8.922	8.564	0.9411	1.882	1.672
4.84	4.343	3.826	1.135	9.017	8.478	0.9415	1.883	1.673
4.86	4.265	3.784	1.127	9.112	8.393	0.9420	1.884	1.673
4.88	4.188	3.743	1.119	9.208	8.310	0.9424	1.885	1.673
4.90	4.112	3.702	1.111	9.304	8.227	0.9428	1.886	1.674
4.92	4.038	3.662	1.103	9.402	8.146	0.9433	1.887	1.674
4.94	3.966	3.623	1.095	9.500	8.065	0.9437	1.887	1.675
4.96	3.895	3.584	1.087	9.599	7.985	0.9441	1.88 8	1.675
4.98	3.826	3.545	1.079	9.699	7.907	0.9445	1.889	1.676
5.00	3.758	3.507	1.072	9.799	7.829	0.9449	1.890	1.676

NORMAL SHOCK FUNCTIONS FOR SUPERSONIC COMPRESSIBLE FLOW

$$\gamma = 5/3$$

M	p_{21}	T_{21}	ρ_{21}	M_2	p_{21}^o	μ	ν
4.00	1.975 +1	5.863 +0	3.368	0.4904	2.373 -1	14.48	49.85
4.02	1.995	5.913	3.374	0.4900	2.346	14.40	50.03
4.04	2.015	5.964	3.379	0.4896	2.320	14.33	50.20
4.06	2.035	6.014	3.384	0.4892	2.294	14.26	5C.37
4.08	2.056	6.065	3.389	0.4888	2.269	14.19	50.54
4.10	2.076	6.117	3.394	0.4884	2.244	14.12	50.71
4.12	2.097	6.168	3.399	0.4880	2.219	14.05	5C.88
4.14	2.117	6.220	3.404	0.4876	2.195	13.98	51.05
4.16	2.138	6.272	3.409	0.4873	2.170	13.91	51.21
4.18	2.159	6.324	3.414	0.4869	2.147	13.84	51.37
4.20	2.180	6.376	3.419	0.4865	2.123	13.77	51.54
4.22	2.201	6.429	3.423	0.4862	2.100	13.71	51.70
4.24	2.222	6.482	3.428	0.4858	2.077	13.64	51.86
4.26	2.243	6.535	3.433	0.4855	2.055	13.58	52.02
4.28	2.265	6.589	3.437	0.4851	2.032	13.51	52.17
4.30	2.286	6.642	3.442	0.4848	2.010	13.45	52.33
4.32	2.308	6.696	3.446	0.4844	1.989	13.38	52.48
4.34	2.329	6.751	3.450	0.4841	1.967	13.32	52.64
4.36	2.351	6.805	3.455	0.4838	1.946	13.26	52.79
4.38	2.373	6.860	3.459	0.4834	1.925	13.20	52.94
4.40	2.395	6.915	3.463	0.4831	1.905	13.14	53.09
4.42	2.417	6.970	3.467	0.4828	1.884	13.08	53.24
4.44	2.439	7.025	3.472	0.4825	1.864	13.02	53.39
4.46	2.461	7.081	3.476	0.4822	1.845	12.96	53.54
4.48	2.484	7.137	3.480	0.4819	1.825	12.90	53.68
4.50	2.506	7.193	3.484	0.4816	1.806	12.84	53.83
4.52	2.529	7.250	3.488	0.4813	1.787	12.78	53.97
4.54	2.551	7.306	3.492	0.4810	1.768	12.73	54.11
4.56	2.574	7.363	3.496	0.4807	1.749	12.67	54.25
4.58	2.597	7.421	3.499	0.4804	1.731	12.61	54.40
4.60	2.620	7.478	3.503	0.4801	1.713	12.56	54.53
4.62	2.643	7.536	3.507	0.4799	1.695	12.50	54.67
4.64	2.666	7.594	3.511	0.4796	1.678	12.45	54.81
4.66	2.689	7.652	3.514	0.4793	1.660	12.39	54.95
4.68	2.713	7.710	3.518	0.4791	1.643	12.34	55.08
4.70	2.736	7.769	3.522	0.4788	1.626	12.29	55.22
4.72	2.760	7.828	3.525	0.4785	1.610	12.23	55.35
4.74	2.783	7.887	3.529	0.4783	1.593	12.18	55.48
4.76	2.807	7.947	3.532	0.4780	1.577	12.13	55.61
4.78	2.831	8.006	3.536	0.4778	1.561	12.08	55.74
4.80	2.855	8.066	3.539	0.4775	1.545	12.03	55.87
4.82	2.879	8.126	3.543	0.4773	1.529	11.97	56.00
4.84	2.903	8.187	3.546	0.4770	1.514	11.92	56.13
4.86	2.927	8.248	3.549	0.4768	1.498	11.87	56.26
4.88	2.952	8.309	3.552	0.4766	1.483	11.83	56.38
4.90	2.976	8.370	3.556	0.4763	1.468	11.78	56.51
4.92	3.001	8.431	3.559	0.4761	1.454	11.73	56.63
4.94	3.025	8.493	3.562	0.4759	1.439	11.68	56.76
4.96	3.050	8.555	3.565	0.4756	1.425	11.63	56.88
4.98	3.075	8.617	3.568	0.4754	1.411	11.58	57.00
5.00	3.100	8.679	3.571	0.4752	1.397	11.54	57.12

ISENTROPIC FUNCTIONS FOR SUPERSONIC COMPRESSIBLE FLOW

$$\gamma = 5/3$$

M	$p/p°$	$T/T°$	$\rho/\rho°$	A/A^*	$q/p°$	u/u_m	u/a^*	$F(\gamma;M)$
5.00	3.758 −3	3.507 −2	1.072 −1	9.799 +0	7.829 −2	0.9449	1.890	1.676
5.05	3.595	3.415	1.053	1.005 +1	7.639	0.9459	1.892	1.677
5.10	3.440	3.326	1.034	1.031	7.455	0.9469	1.894	1.678
5.15	3.292	3.240	1.016	1.058	7.276	0.9478	1.896	1.679
5.20	3.152	3.156	9.987 −2	1.085	7.103	0.9487	1.898	1.680
5.25	3.019	3.076	9.817	1.112	6.934	0.9496	1.899	1.681
5.30	2.893	2.998	9.650	1.140	6.771	0.9505	1.901	1.682
5.35	2.773	2.922	9.488	1.168	6.613	0.9514	1.903	1.682
5.40	2.658	2.849	9.329	1.197	6.459	0.9522	1.904	1.683
5.45	2.549	2.779	9.174	1.226	6.310	0.9530	1.906	1.684
5.50	2.446	2.710	9.023	1.256	6.165	0.9538	1.908	1.685
5.55	2.347	2.644	8.876	1.287	6.024	0.9546	1.909	1.686
5.60	2.253	2.580	8.732	1.318	5.887	0.9553	1.911	1.686
5.65	2.163	2.518	8.591	1.349	5.754	0.9561	1.912	1.687
5.70	2.078	2.458	8.454	1.381	5.625	0.9568	1.914	1.688
5.75	1.996	2.400	8.319	1.413	5.500	0.9575	1.915	1.689
5.80	1.919	2.343	8.188	1.447	5.378	0.9582	1.916	1.689
5.85	1.844	2.288	8.060	1.480	5.260	0.9589	1.918	1.690
5.90	1.774	2.235	7.935	1.514	5.145	0.9595	1.919	1.691
5.95	1.706	2.184	7.813	1.549	5.033	0.9601	1.920	1.691
6.00	1.641	2.134	7.693	1.584	4.924	0.9608	1.922	1.692
6.05	1.580	2.085	7.576	1.620	4.818	0.9614	1.923	1.692
6.10	1.521	2.038	7.461	1.656	4.715	0.9620	1.924	1.693
6.15	1.464	1.992	7.349	1.693	4.615	0.9626	1.925	1.694
6.20	1.410	1.948	7.240	1.731	4.517	0.9631	1.926	1.694
6.25	1.359	1.905	7.133	1.769	4.423	0.9637	1.927	1.695
6.30	1.309	1.863	7.028	1.808	4.330	0.9642	1.928	1.695
6.35	1.262	1.822	6.925	1.847	4.241	0.9648	1.930	1.696
6.40	1.217	1.783	6.825	1.887	4.153	0.9653	1.931	1.696
6.45	1.173	1.745	6.727	1.928	4.068	0.9658	1.932	1.697
6.50	1.132	1.707	6.630	1.969	3.985	0.9663	1.933	1.697
6.55	1.092	1.671	6.536	2.010	3.904	0.9668	1.934	1.698
6.60	1.054	1.636	6.444	2.053	3.826	0.9672	1.935	1.698
6.65	1.017	1.601	6.353	2.096	3.749	0.9677	1.935	1.699
6.70	9.823 −4	1.568	6.265	2.139	3.675	0.9682	1.936	1.699
6.75	9.487	1.536	6.178	2.183	3.602	0.9686	1.937	1.700
6.80	9.164	1.504	6.093	2.228	3.531	0.9691	1.938	1.700
6.85	8.854	1.473	6.010	2.274	3.462	0.9695	1.939	1.701
6.90	8.556	1.443	5.928	2.320	3.394	0.9699	1.940	1.701
6.95	8.270	1.414	5.848	2.367	3.329	0.9703	1.941	1.702
7.00	7.996	1.386	5.770	2.414	3.265	0.9707	1.941	1.702
7.05	7.732	1.358	5.693	2.462	3.202	0.9711	1.942	1.702
7.10	7.479	1.331	5.617	2.511	3.141	0.9715	1.943	1.703
7.15	7.235	1.305	5.543	2.560	3.082	0.9719	1.944	1.703
7.20	7.000	1.280	5.471	2.610	3.024	0.9723	1.945	1.703
7.25	6.775	1.255	5.400	2.661	2.967	0.9726	1.945	1.704
7.30	6.558	1.230	5.330	2.713	2.912	0.9730	1.946	1.704
7.35	6.350	1.207	5.261	2.765	2.858	0.9733	1.947	1.705
7.40	6.149	1.184	5.194	2.817	2.806	0.9737	1.947	1.705
7.45	5.956	1.161	5.128	2.871	2.754	0.9740	1.948	1.705
7.50	5.770	1.139	5.064	2.925	2.704	0.9744	1.949	1.706

NORMAL SHOCK FUNCTIONS FOR SUPERSONIC COMPRESSIBLE FLOW

$$\gamma = 5/3$$

M	p_{21}	T_{21}	ρ_{21}	M_2	p^o_{21}	μ	ν
5.00	3.100 +1	8.679 +0	3.571	0.4752	1.397 -1	11.54	57.12
5.05	3.163	8.837	3.579	0.4747	1.362	11.42	57.42
5.10	3.226	8.995	3.586	0.4741	1.329	11.31	57.71
5.15	3.290	9.156	3.594	0.4735	1.297	11.20	58.00
5.20	3.355	9.317	3.601	0.4731	1.266	11.09	58.28
5.25	3.420	9.481	3.607	0.4726	1.236	10.98	58.56
5.30	3.486	9.646	3.614	0.4722	1.206	10.88	58.84
5.35	3.553	9.812	3.620	0.4717	1.178	10.77	59.10
5.40	3.620	9.980	3.627	0.4713	1.150	10.67	59.37
5.45	3.688	1.015 +1	3.633	0.4708	1.123	10.57	59.63
5.50	3.756	1.032	3.639	0.4704	1.097	10.48	59.89
5.55	3.825	1.049	3.645	0.4700	1.072	10.38	60.14
5.60	3.895	1.067	3.651	0.4696	1.048	10.29	60.39
5.65	3.965	1.084	3.656	0.4692	1.024	10.19	60.63
5.70	4.036	1.102	3.662	0.4689	1.001	10.10	60.87
5.75	4.108	1.120	3.667	0.4685	9.783 -2	10.02	61.11
5.80	4.180	1.138	3.672	0.4681	9.565	9.929	61.34
5.85	4.252	1.156	3.678	0.4678	9.353	9.843	61.57
5.90	4.326	1.175	3.683	0.4674	9.147	9.759	61.79
5.95	4.400	1.193	3.688	0.4671	8.946	9.676	62.02
6.00	4.475	1.212	3.692	0.4668	8.752	9.594	62.24
6.05	4.550	1.231	3.697	0.4665	8.562	9.514	62.45
6.10	4.626	1.250	3.702	0.4661	8.378	9.436	62.66
6.15	4.702	1.269	3.706	0.4658	8.199	9.358	62.87
6.20	4.780	1.288	3.710	0.4656	8.025	9.282	63.08
6.25	4.857	1.308	3.715	0.4653	7.856	9.207	63.28
6.30	4.936	1.327	3.719	0.4650	7.691	9.134	63.48
6.35	5.015	1.347	3.723	0.4647	7.531	9.061	63.68
6.40	5.095	1.367	3.727	0.4644	7.374	8.990	63.88
6.45	5.175	1.387	3.731	0.4642	7.222	8.919	64.07
6.50	5.256	1.407	3.735	0.4639	7.075	8.850	64.26
6.55	5.337	1.428	3.739	0.4637	6.930	8.782	64.44
6.60	5.420	1.448	3.742	0.4634	6.790	8.715	64.63
6.65	5.502	1.469	3.746	0.4632	6.653	8.649	64.81
6.70	5.586	1.490	3.749	0.4629	6.520	8.584	64.99
6.75	5.670	1.511	3.753	0.4627	6.391	8.520	65.16
6.80	5.755	1.532	3.756	0.4625	6.264	8.457	65.34
6.85	5.840	1.553	3.760	0.4623	6.141	8.395	65.51
6.90	5.926	1.575	3.763	0.4621	6.021	8.333	65.68
6.95	6.012	1.596	3.766	0.4619	5.904	8.273	65.84
7.00	6.100	1.618	3.769	0.4616	5.790	8.213	66.01
7.05	6.187	1.640	3.772	0.4614	5.679	8.155	66.17
7.10	6.276	1.662	3.775	0.4612	5.570	8.097	66.33
7.15	6.365	1.685	3.778	0.4611	5.464	8.040	66.49
7.20	6.455	1.707	3.781	0.4609	5.361	7.984	66.65
7.25	6.545	1.730	3.784	0.4607	5.261	7.928	66.80
7.30	6.636	1.752	3.787	0.4605	5.162	7.874	66.95
7.35	6.727	1.775	3.790	0.4603	5.066	7.820	67.10
7.40	6.820	1.798	3.792	0.4601	4.973	7.767	67.25
7.45	6.912	1.822	3.795	0.4600	4.881	7.714	67.40
7.50	7.006	1.845	3.797	0.4598	4.792	7.662	67.54

ISENTROPIC FUNCTIONS FOR SUPERSONIC COMPRESSIBLE FLOW

$$\gamma = 5/3$$

M	$p/p°$	$T/T°$	$\rho/\rho°$	$A/A*$	$q/p°$	u/u_m	$u/a*$	$F(\gamma;M)$
7.50	5.770 −4	1.139 −2	5.064 −2	2.925 +1	2.704 −2	0.9744	1.949	1.706
7.55	5.590	1.118	5.000	2.980	2.655	0.9747	1.949	1.706
7.60	5.418	1.097	4.938	3.036	2.608	0.9750	1.950	1.706
7.65	5.252	1.077	4.877	3.092	2.561	0.9753	1.951	1.707
7.70	5.091	1.057	4.816	3.149	2.515	0.9756	1.951	1.707
7.75	4.937	1.038	4.757	3.207	2.471	0.9759	1.952	1.707
7.80	4.788	1.019	4.700	3.265	2.427	0.9762	1.952	1.708
7.85	4.644	1.000	4.643	3.325	2.385	0.9765	1.953	1.708
7.90	4.506	9.823 −3	4.587	3.385	2.343	0.9768	1.954	1.708
7.95	4.372	9.647	4.532	3.445	2.303	0.9771	1.954	1.708
8.00	4.243	9.476	4.478	3.507	2.263	0.9774	1.955	1.709
8.05	4.119	9.308	4.425	3.569	2.224	0.9776	1.955	1.709
8.10	3.999	9.144	4.373	3.632	2.186	0.9779	1.956	1.709
8.15	3.883	8.984	4.322	3.696	2.149	0.9782	1.956	1.710
8.20	3.771	8.828	4.271	3.760	2.113	0.9784	1.957	1.710
8.25	3.662	8.675	4.222	3.825	2.077	0.9787	1.957	1.710
8.30	3.558	8.525	4.173	3.891	2.042	0.9789	1.958	1.710
8.35	3.457	8.379	4.126	3.958	2.008	0.9792	1.958	1.711
8.40	3.359	8.237	4.079	4.026	1.975	0.9794	1.959	1.711
8.45	3.265	8.097	4.032	4.094	1.943	0.9796	1.959	1.711
8.50	3.174	7.961	3.987	4.163	1.911	0.9799	1.960	1.711
8.55	3.086	7.827	3.942	4.233	1.880	0.9801	1.960	1.712
8.60	3.001	7.697	3.898	4.304	1.849	0.9803	1.961	1.712
8.65	2.918	7.569	3.855	4.376	1.819	0.9805	1.961	1.712
8.70	2.838	7.445	3.813	4.448	1.790	0.9808	1.961	1.712
8.75	2.761	7.322	3.771	4.521	1.762	0.9810	1.962	1.712
8.80	2.686	7.203	3.730	4.595	1.734	0.9812	1.962	1.713
8.85	2.614	7.086	3.689	4.670	1.706	0.9814	1.963	1.713
8.90	2.544	6.972	3.649	4.746	1.679	0.9816	1.963	1.713
8.95	2.476	6.860	3.610	4.822	1.653	0.9818	1.964	1.713
9.00	2.411	6.750	3.572	4.900	1.627	0.9820	1.964	1.713
9.05	2.347	6.643	3.534	4.978	1.602	0.9822	1.964	1.714
9.10	2.286	6.538	3.496	5.057	1.577	0.9824	1.965	1.714
9.15	2.226	6.435	3.460	5.137	1.553	0.9826	1.965	1.714
9.20	2.168	6.334	3.423	5.217	1.529	0.9827	1.966	1.714
9.25	2.112	6.235	3.388	5.299	1.506	0.9829	1.966	1.714
9.30	2.058	6.138	3.353	5.382	1.483	0.9831	1.966	1.715
9.35	2.005	6.044	3.318	5.465	1.461	0.9833	1.967	1.715
9.40	1.954	5.951	3.284	5.549	1.439	0.9834	1.967	1.715
9.45	1.905	5.860	3.250	5.634	1.417	0.9836	1.967	1.715
9.50	1.857	5.771	3.217	5.720	1.396	0.9838	1.968	1.715
9.55	1.810	5.684	3.185	5.807	1.376	0.9839	1.968	1.715
9.60	1.765	5.598	3.153	5.895	1.355	0.9841	1.968	1.716
9.65	1.721	5.514	3.121	5.984	1.336	0.9843	1.969	1.716
9.70	1.679	5.432	3.090	6.073	1.316	0.9844	1.969	1.716
9.75	1.637	5.351	3.059	6.164	1.297	0.9846	1.969	1.716
9.80	1.597	5.272	3.029	6.255	1.278	0.9847	1.970	1.716
9.85	1.558	5.195	2.999	6.348	1.260	0.9849	1.970	1.716
9.90	1.520	5.119	2.970	6.441	1.242	0.9850	1.970	1.717
9.95	1.484	5.044	2.941	6.535	1.224	0.9852	1.970	1.717
10.00	1.448	4.971	2.913	6.630	1.207	0.9853	1.971	1.717

NORMAL SHOCK FUNCTIONS FOR SUPERSONIC COMPRESSIBLE FLOW

$$\gamma = 5/3$$

M	p_{21}	T_{21}	ρ_{21}	M_2	p^o_{21}	μ	ν
7.50	7.006 +1	1.845 +1	3.797	0.4598	4.792 −2	7.662	67.54
7.55	7.100	1.868	3.800	0.4596	4.705	7.611	67.68
7.60	7.195	1.892	3.802	0.4595	4.620	7.561	67.83
7.65	7.290	1.916	3.805	0.4593	4.537	7.511	67.97
7.70	7.386	1.940	3.807	0.4592	4.456	7.462	68.10
7.75	7.482	1.964	3.810	0.4590	4.377	7.414	68.24
7.80	7.580	1.988	3.812	0.4589	4.300	7.366	68.37
7.85	7.677	2.013	3.814	0.4587	4.224	7.319	68.51
7.90	7.776	2.037	3.817	0.4586	4.150	7.272	68.64
7.95	7.875	2.062	3.819	0.4584	4.078	7.226	68.77
8.00	7.974	2.087	3.821	0.4583	4.007	7.181	68.89
8.05	8.075	2.112	3.823	0.4582	3.938	7.136	69.02
8.10	8.176	2.137	3.825	0.4580	3.871	7.092	69.15
8.15	8.277	2.163	3.827	0.4579	3.805	7.048	69.27
8.20	8.379	2.188	3.829	0.4578	3.740	7.005	69.39
8.25	8.482	2.214	3.831	0.4576	3.677	6.962	69.51
8.30	8.586	2.240	3.833	0.4575	3.616	6.920	69.63
8.35	8.690	2.266	3.835	0.4574	3.555	6.878	69.75
8.40	8.794	2.292	3.837	0.4573	3.496	6.837	69.87
8.45	8.900	2.318	3.839	0.4572	3.439	6.797	69.98
8.50	9.006	2.345	3.841	0.4570	3.382	6.757	70.10
8.55	9.112	2.372	3.842	0.4569	3.327	6.717	70.21
8.60	9.219	2.398	3.844	0.4568	3.273	6.678	70.32
8.65	9.327	2.425	3.846	0.4567	3.220	6.639	70.43
8.70	9.436	2.452	3.847	0.4566	3.168	6.601	70.54
8.75	9.545	2.480	3.849	0.4565	3.117	6.563	70.65
8.80	9.654	2.507	3.851	0.4564	3.068	6.525	70.75
8.85	9.765	2.535	3.852	0.4563	3.019	6.488	70.86
8.90	9.876	2.562	3.854	0.4562	2.971	6.452	70.96
8.95	9.987	2.590	3.856	0.4561	2.925	6.415	71.07
9.00	1.010 +2	2.618	3.857	0.4560	2.879	6.380	71.17
9.05	1.021	2.647	3.859	0.4559	2.834	6.344	71.27
9.10	1.033	2.675	3.860	0.4558	2.790	6.309	71.37
9.15	1.044	2.703	3.862	0.4557	2.747	6.275	71.47
9.20	1.055	2.732	3.863	0.4556	2.705	6.240	71.57
9.25	1.067	2.761	3.864	0.4555	2.664	6.206	71.67
9.30	1.079	2.790	3.866	0.4554	2.623	6.173	71.76
9.35	1.090	2.819	3.867	0.4553	2.584	6.140	71.86
9.40	1.102	2.848	3.869	0.4553	2.545	6.107	71.95
9.45	1.114	2.878	3.870	0.4552	2.507	6.075	72.04
9.50	1.126	2.907	3.871	0.4551	2.469	6.043	72.14
9.55	1.137	2.937	3.873	0.4550	2.433	6.011	72.23
9.60	1.149	2.967	3.874	0.4549	2.397	5.979	72.32
9.65	1.161	2.997	3.875	0.4548	2.362	5.948	72.41
9.70	1.174	3.027	3.876	0.4548	2.327	5.917	72.49
9.75	1.186	3.058	3.878	0.4547	2.293	5.887	72.58
9.80	1.198	3.088	3.879	0.4546	2.260	5.857	72.67
9.85	1.210	3.119	3.880	0.4545	2.227	5.827	72.76
9.90	1.223	3.150	3.881	0.4545	2.195	5.797	72.84
9.95	1.235	3.181	3.882	0.4544	2.164	5.768	72.92
10.00	1.247	3.212	3.883	0.4543	2.133	5.739	73.01

ISENTROPIC FUNCTIONS FOR SUPERSONIC COMPRESSIBLE FLOW

$\gamma = 5/3$

M	$p/p°$	$T/T°$	$\rho/\rho°$	$A/A*$	$q/p°$	u/u_m	$u/a*$	$F(\gamma;M)$
10.0	1.448 −4	4.971 −3	2.913 −2	6.630 +1	1.207 −2	0.9853	1.971	1.717
10.1	1.380	4.829	2.857	6.823	1.173	0.9856	1.971	1.717
10.2	1.315	4.692	2.803	7.020	1.140	0.9859	1.972	1.717
10.3	1.254	4.561	2.750	7.221	1.109	0.9862	1.972	1.718
10.4	1.197	4.434	2.699	7.425	1.079	0.9864	1.973	1.718
10.5	1.142	4.312	2.649	7.634	1.049	0.9867	1.973	1.718
10.6	1.091	4.194	2.601	7.846	1.021	0.9869	1.974	1.719
10.7	1.042	4.081	2.554	8.062	9.941 −3	0.9871	1.974	1.719
10.8	9.958 −5	3.971	2.508	8.283	9.679	0.9874	1.975	1.719
10.9	9.520	3.865	2.463	8.507	9.425	0.9876	1.975	1.719
11.0	9.105	3.763	2.419	8.736	9.181	0.9878	1.976	1.719
11.1	8.712	3.665	2.377	8.968	8.945	0.9880	1.976	1.720
11.2	8.339	3.570	2.336	9.205	8.716	0.9883	1.977	1.720
11.3	7.985	3.478	2.296	9.446	8.496	0.9885	1.977	1.720
11.4	7.648	3.389	2.256	9.691	8.282	0.9887	1.977	1.720
11.5	7.328	3.304	2.218	9.941	8.076	0.9888	1.978	1.721
11.6	7.025	3.221	2.181	1.019 +2	7.877	0.9890	1.978	1.721
11.7	6.736	3.141	2.145	1.045	7.683	0.9892	1.978	1.721
11.8	6.461	3.063	2.109	1.072	7.497	0.9894	1.979	1.721
11.9	6.200	2.988	2.075	1.098	7.316	0.9896	1.979	1.721
12.0	5.951	2.916	2.041	1.125	7.140	0.9897	1.980	1.721
12.1	5.714	2.845	2.008	1.153	6.971	0.9899	1.980	1.722
12.2	5.488	2.777	1.976	1.181	6.806	0.9901	1.980	1.722
12.3	5.272	2.711	1.944	1.210	6.647	0.9902	1.981	1.722
12.4	5.067	2.648	1.914	1.239	6.492	0.9904	1.981	1.722
12.5	4.871	2.586	1.884	1.268	6.343	0.9905	1.981	1.722
12.6	4.685	2.526	1.855	1.298	6.197	0.9907	1.981	1.722
12.7	4.506	2.468	1.826	1.328	6.057	0.9908	1.982	1.723
12.8	4.336	2.411	1.798	1.359	5.920	0.9910	1.982	1.723
12.9	4.174	2.357	1.771	1.390	5.787	0.9911	1.982	1.723
13.0	4.018	2.304	1.744	1.422	5.659	0.9912	1.983	1.723
13.1	3.870	2.252	1.718	1.455	5.534	0.9914	1.983	1.723
13.2	3.728	2.202	1.693	1.487	5.412	0.9915	1.983	1.723
13.3	3.592	2.154	1.668	1.521	5.295	0.9916	1.983	1.723
13.4	3.462	2.107	1.643	1.554	5.180	0.9917	1.984	1.724
13.5	3.338	2.061	1.620	1.589	5.069	0.9919	1.984	1.724
13.6	3.219	2.017	1.596	1.623	4.961	0.9920	1.984	1.724
13.7	3.105	1.973	1.573	1.659	4.856	0.9921	1.984	1.724
13.8	2.996	1.931	1.551	1.695	4.754	0.9922	1.984	1.724
13.9	2.891	1.891	1.529	1.731	4.655	0.9923	1.985	1.724
14.0	2.791	1.851	1.508	1.768	4.558	0.9924	1.985	1.724
14.1	2.695	1.813	1.487	1.805	4.464	0.9925	1.985	1.724
14.2	2.602	1.775	1.466	1.843	4.373	0.9926	1.985	1.724
14.3	2.514	1.739	1.446	1.882	4.284	0.9927	1.986	1.725
14.4	2.429	1.703	1.426	1.921	4.197	0.9928	1.986	1.725
14.5	2.348	1.669	1.407	1.960	4.113	0.9929	1.986	1.725
14.6	2.269	1.635	1.388	2.000	4.031	0.9930	1.986	1.725
14.7	2.194	1.602	1.369	2.041	3.951	0.9931	1.986	1.725
14.8	2.122	1.571	1.351	2.082	3.873	0.9932	1.986	1.725
14.9	2.053	1.540	1.333	2.124	3.798	0.9933	1.987	1.725
15.0	1.986	1.509	1.316	2.166	3.724	0.9934	1.987	1.725

NORMAL SHOCK FUNCTIONS FOR SUPERSONIC COMPRESSIBLE FLOW

$$\gamma = 5/3$$

M	p_{21}	T_{21}	ρ_{21}	M_2	p^o_{21}	μ	ν
10.0	1.247 +2	3.212 +1	3.883	0.4543	2.133 -2	5.739	73.01
10.1	1.273	3.275	3.886	0.4542	2.073	5.682	73.17
10.2	1.298	3.338	3.888	0.4541	2.016	5.626	73.33
10.3	1.324	3.402	3.890	0.4539	1.960	5.572	73.49
10.4	1.349	3.467	3.892	0.4538	1.906	5.518	73.65
10.5	1.376	3.532	3.894	0.4537	1.855	5.465	73.80
10.6	1.402	3.598	3.896	0.4535	1.805	5.413	73.95
10.7	1.429	3.665	3.898	0.4534	1.757	5.363	74.10
10.8	1.455	3.732	3.900	0.4533	1.710	5.313	74.24
10.9	1.483	3.800	3.901	0.4532	1.666	5.264	74.38
11.0	1.510	3.868	3.903	0.4531	1.622	5.216	74.52
11.1	1.538	3.937	3.905	0.4530	1.580	5.169	74.66
11.2	1.565	4.007	3.907	0.4529	1.540	5.123	74.79
11.3	1.594	4.077	3.908	0.4528	1.501	5.077	74.93
11.4	1.622	4.148	3.910	0.4527	1.463	5.033	75.06
11.5	1.651	4.220	3.911	0.4526	1.427	4.989	75.18
11.6	1.679	4.292	3.913	0.4525	1.391	4.946	75.31
11.7	1.709	4.365	3.914	0.4524	1.357	4.903	75.43
11.8	1.738	4.438	3.916	0.4523	1.324	4.862	75.55
11.9	1.768	4.512	3.917	0.4522	1.292	4.821	75.67
12.0	1.797	4.587	3.918	0.4522	1.261	4.780	75.79
12.1	1.828	4.662	3.920	0.4521	1.231	4.741	75.91
12.2	1.858	4.738	3.921	0.4520	1.202	4.702	76.02
12.3	1.889	4.815	3.922	0.4519	1.174	4.663	76.13
12.4	1.919	4.892	3.923	0.4518	1.147	4.626	76.24
12.5	1.951	4.970	3.925	0.4518	1.120	4.589	76.35
12.6	1.982	5.048	3.926	0.4517	1.094	4.552	76.46
12.7	2.014	5.127	3.927	0.4516	1.070	4.516	76.56
12.8	2.045	5.207	3.928	0.4516	1.045	4.481	76.67
12.9	2.078	5.287	3.929	0.4515	1.022	4.446	76.77
13.0	2.110	5.368	3.930	0.4514	9.992 -3	4.412	76.87
13.1	2.143	5.450	3.931	0.4514	9.771	4.378	76.97
13.2	2.175	5.532	3.932	0.4513	9.557	4.345	77.06
13.3	2.209	5.615	3.933	0.4512	9.348	4.312	77.16
13.4	2.242	5.698	3.934	0.4512	9.146	4.280	77.25
13.5	2.276	5.782	3.935	0.4511	8.950	4.248	77.35
13.6	2.309	5.867	3.936	0.4511	8.759	4.217	77.44
13.7	2.344	5.952	3.937	0.4510	8.573	4.186	77.53
13.8	2.378	6.038	3.938	0.4510	8.392	4.156	77.62
13.9	2.413	6.125	3.939	0.4509	8.217	4.126	77.71
14.0	2.447	6.212	3.940	0.4509	8.046	4.096	77.79
14.1	2.483	6.300	3.941	0.4508	7.880	4.067	77.88
14.2	2.518	6.388	3.941	0.4508	7.719	4.038	77.96
14.3	2.554	6.477	3.942	0.4507	7.562	4.010	78.05
14.4	2.589	6.567	3.943	0.4507	7.409	3.982	78.13
14.5	2.626	6.657	3.944	0.4506	7.260	3.955	78.21
14.6	2.662	6.748	3.944	0.4506	7.115	3.928	78.29
14.7	2.699	6.840	3.945	0.4505	6.974	3.901	78.37
14.8	2.735	6.932	3.946	0.4505	6.837	3.874	78.45
14.9	2.773	7.025	3.947	0.4504	6.703	3.848	78.52
15.0	2.810	7.118	3.947	0.4504	6.573	3.823	78.60

ISENTROPIC FUNCTIONS FOR SUPERSONIC COMPRESSIBLE FLOW

$$\gamma = 5/3$$

M	$p/p°$	$T/T°$	$\rho/\rho°$	A/A^*	$q/p°$	u/u_m	u/a^*	$F(\gamma;M)$
15.0	1.986 −5	1.509 −3	1.316 −2	2.166 +2	3.724 −3	0.9934	1.987	1.725
15.1	1.922	1.480	1.299	2.209	3.652	0.9935	1.987	1.725
15.2	1.860	1.451	1.282	2.252	3.582	0.9936	1.987	1.725
15.3	1.801	1.423	1.265	2.296	3.514	0.9937	1.987	1.725
15.4	1.744	1.396	1.249	2.341	3.447	0.9937	1.988	1.726
15.5	1.689	1.370	1.233	2.386	3.382	0.9938	1.988	1.726
15.6	1.637	1.344	1.218	2.431	3.319	0.9939	1.988	1.726
15.7	1.586	1.319	1.203	2.478	3.257	0.9940	1.988	1.726
15.8	1.537	1.294	1.188	2.525	3.197	0.9940	1.988	1.726
15.9	1.490	1.270	1.173	2.572	3.138	0.9941	1.988	1.726
16.0	1.444	1.247	1.158	2.620	3.081	0.9942	1.988	1.726
16.1	1.400	1.224	1.144	2.669	3.025	0.9943	1.989	1.726
16.2	1.358	1.202	1.130	2.718	2.970	0.9943	1.989	1.726
16.3	1.317	1.180	1.117	2.768	2.917	0.9944	1.989	1.726
16.4	1.278	1.159	1.103	2.819	2.865	0.9945	1.989	1.726
16.5	1.240	1.138	1.090	2.870	2.814	0.9945	1.989	1.726
16.6	1.204	1.118	1.077	2.921	2.764	0.9946	1.989	1.726
16.7	1.169	1.098	1.064	2.974	2.716	0.9947	1.989	1.727
16.8	1.135	1.079	1.052	3.027	2.668	0.9947	1.990	1.727
16.9	1.102	1.060	1.040	3.080	2.622	0.9948	1.990	1.727
17.0	1.070	1.041	1.027	3.134	2.577	0.9948	1.990	1.727
17.1	1.039	1.023	1.016	3.189	2.533	0.9949	1.990	1.727
17.2	1.010	1.006	1.004	3.245	2.490	0.9950	1.990	1.727
17.3	9.813 −6	9.887 −4	9.925 −3	3.301	2.447	0.9950	1.990	1.727
17.4	9.537	9.719	9.812	3.358	2.406	0.9951	1.990	1.727
17.5	9.270	9.555	9.701	3.415	2.366	0.9951	1.990	1.727
17.6	9.012	9.395	9.593	3.473	2.326	0.9952	1.990	1.727
17.7	8.763	9.238	9.485	3.532	2.288	0.9952	1.991	1.727
17.8	8.522	9.085	9.380	3.592	2.250	0.9953	1.991	1.727
17.9	8.289	8.935	9.277	3.652	2.213	0.9954	1.991	1.727
18.0	8.063	8.788	9.175	3.713	2.177	0.9954	1.991	1.727
18.1	7.845	8.645	9.075	3.774	2.142	0.9955	1.991	1.727
18.2	7.633	8.504	8.976	3.836	2.107	0.9955	1.991	1.727
18.3	7.429	8.367	8.879	3.899	2.073	0.9956	1.991	1.727
18.4	7.231	8.232	8.784	3.962	2.040	0.9956	1.991	1.727
18.5	7.039	8.101	8.690	4.027	2.008	0.9956	1.991	1.728
18.6	6.854	7.972	8.597	4.091	1.976	0.9957	1.991	1.728
18.7	6.674	7.846	8.507	4.157	1.945	0.9957	1.992	1.728
18.8	6.500	7.722	8.417	4.223	1.914	0.9958	1.992	1.728
18.9	6.331	7.601	8.329	4.290	1.884	0.9958	1.992	1.728
19.0	6.168	7.483	8.242	4.358	1.855	0.9959	1.992	1.728
19.1	6.009	7.367	8.157	4.426	1.827	0.9959	1.992	1.728
19.2	5.856	7.253	8.073	4.496	1.799	0.9960	1.992	1.728
19.3	5.707	7.142	7.990	4.565	1.771	0.9960	1.992	1.728
19.4	5.562	7.033	7.909	4.636	1.744	0.9960	1.992	1.728
19.5	5.422	6.926	7.828	4.707	1.718	0.9961	1.992	1.728
19.6	5.286	6.822	7.749	4.779	1.692	0.9961	1.992	1.728
19.7	5.154	6.719	7.671	4.852	1.667	0.9962	1.992	1.728
19.8	5.027	6.619	7.595	4.926	1.642	0.9962	1.992	1.728
19.9	4.903	6.520	7.519	5.000	1.618	0.9962	1.993	1.728
20.0	4.782	6.423	7.445	5.075	1.594	0.9963	1.993	1.728

NORMAL SHOCK FUNCTIONS FOR SUPERSONIC COMPRESSIBLE FLOW

$\gamma = 5/3$

M	p_{21}	T_{21}	ρ_{21}	M_2	p^o_{21}	μ	ν
15.0	2.810 +2	7.118 +1	3.947	0.4504	6.573 -3	3.823	78.60
15.1	2.848	7.212	3.948	0.4503	6.446	3.797	78.67
15.2	2.885	7.307	3.949	0.4503	6.322	3.772	78.75
15.3	2.924	7.402	3.949	0.4503	6.201	3.748	78.82
15.4	2.962	7.498	3.950	0.4502	6.083	3.723	78.89
15.5	3.000	7.595	3.951	0.4502	5.969	3.699	78.96
15.6	3.039	7.692	3.951	0.4501	5.857	3.675	79.03
15.7	3.078	7.790	3.952	0.4501	5.748	3.652	79.10
15.8	3.118	7.888	3.953	0.4501	5.641	3.629	79.17
15.9	3.157	7.987	3.953	0.4500	5.538	3.606	79.24
16.0	3.197	8.087	3.954	0.4500	5.436	3.583	79.31
16.1	3.237	8.187	3.954	0.4500	5.338	3.561	79.37
16.2	3.278	8.288	3.955	0.4499	5.241	3.539	79.44
16.3	3.318	8.390	3.955	0.4499	5.147	3.517	79.50
16.4	3.359	8.492	3.956	0.4499	5.055	3.496	79.56
16.5	3.400	8.595	3.956	0.4498	4.965	3.475	79.63
16.6	3.442	8.698	3.957	0.4498	4.878	3.454	79.69
16.7	3.483	8.802	3.957	0.4498	4.792	3.433	79.75
16.8	3.525	8.907	3.958	0.4497	4.708	3.413	79.81
16.9	3.567	9.012	3.958	0.4497	4.627	3.392	79.87
17.0	3.610	9.118	3.959	0.4497	4.547	3.372	79.93
17.1	3.652	9.225	3.959	0.4497	4.469	3.353	79.99
17.2	3.695	9.332	3.960	0.4496	4.393	3.333	80.05
17.3	3.738	9.440	3.960	0.4496	4.318	3.314	80.10
17.4	3.782	9.548	3.961	0.4496	4.245	3.295	80.16
17.5	3.825	9.657	3.961	0.4495	4.174	3.276	80.21
17.6	3.869	9.767	3.962	0.4495	4.104	3.257	80.27
17.7	3.913	9.877	3.962	0.4495	4.036	3.239	80.32
17.8	3.958	9.988	3.962	0.4495	3.970	3.221	80.38
17.9	4.002	1.010 +2	3.963	0.4494	3.904	3.203	80.43
18.0	4.047	1.021	3.963	0.4494	3.841	3.185	80.48
18.1	4.092	1.032	3.964	0.4494	3.778	3.167	80.54
18.2	4.138	1.044	3.964	0.4494	3.717	3.150	80.59
18.3	4.183	1.055	3.964	0.4493	3.657	3.133	80.64
18.4	4.229	1.067	3.965	0.4493	3.599	3.116	80.69
18.5	4.275	1.078	3.965	0.4493	3.542	3.099	80.74
18.6	4.322	1.090	3.966	0.4493	3.486	3.082	80.79
18.7	4.368	1.101	3.966	0.4493	3.431	3.066	80.84
18.8	4.415	1.113	3.966	0.4492	3.377	3.049	80.89
18.9	4.462	1.125	3.967	0.4492	3.324	3.033	80.93
19.0	4.510	1.137	3.967	0.4492	3.273	3.017	80.98
19.1	4.557	1.149	3.967	0.4492	3.222	3.001	81.03
19.2	4.605	1.161	3.968	0.4492	3.173	2.986	81.08
19.3	4.653	1.173	3.968	0.4491	3.125	2.970	81.12
19.4	4.702	1.185	3.968	0.4491	3.077	2.955	81.17
19.5	4.750	1.197	3.969	0.4491	3.031	2.940	81.21
19.6	4.799	1.209	3.969	0.4491	2.985	2.925	81.26
19.7	4.848	1.221	3.969	0.4491	2.940	2.910	81.30
19.8	4.898	1.234	3.970	0.4490	2.897	2.895	81.34
19.9	4.947	1.246	3.970	0.4490	2.854	2.880	81.39
20.0	4.997	1.259	3.970	0.4490	2.812	2.866	81.43

ISENTROPIC FUNCTIONS FOR SUPERSONIC COMPRESSIBLE FLOW

$$\gamma = 5/3$$

M	p/p°	T/T°	ρ/ρ°	A/A^*	q/p°	u/u_m	u/a^*	$F(\gamma;M)$
20.0	4.782 −6	6.423 −4	7.445 −3	5.075 +2	1.594 −3	0.9963	1.993	1.728
21.0	3.753	5.555	6.757	5.867	1.379	0.9966	1.993	1.729
22.0	2.979	4.835	6.161	6.737	1.201	0.9969	1.994	1.729
23.0	2.388	4.235	5.639	7.690	1.053	0.9972	1.994	1.729
24.0	1.933	3.730	5.182	8.730	9.276 −4	0.9974	1.995	1.729
25.0	1.577	3.302	4.777	9.859	8.216	0.9976	1.995	1.730
26.0	1.298	2.937	4.418	1.108 +3	7.310	0.9978	1.996	1.730
27.0	1.075	2.624	4.099	1.240	6.533	0.9979	1.996	1.730
28.0	8.973 −7	2.354	3.812	1.382	5.862	0.9981	1.996	1.730
29.0	7.534	2.119	3.555	1.535	5.279	0.9982	1.996	1.730
30.0	6.363	1.915	3.322	1.699	4.772	0.9983	1.997	1.730
31.0	5.403	1.736	3.112	1.873	4.327	0.9984	1.997	1.730
32.0	4.612	1.579	2.921	2.060	3.936	0.9985	1.997	1.731
33.0	3.956	1.440	2.747	2.258	3.590	0.9986	1.997	1.731
34.0	3.409	1.317	2.589	2.469	3.284	0.9987	1.997	1.731
35.0	2.950	1.208	2.443	2.693	3.012	0.9988	1.998	1.731
36.0	2.563	1.110	2.310	2.929	2.768	0.9988	1.998	1.731
37.0	2.236	1.023	2.187	3.180	2.551	0.9989	1.998	1.731
38.0	1.957	9.441 −5	2.073	3.444	2.355	0.9990	1.998	1.731
39.0	1.719	8.734	1.969	3.722	2.179	0.9990	1.998	1.731
40.0	1.515	8.097	1.872	4.015	2.020	0.9991	1.998	1.731
41.0	1.340	7.519	1.782	4.323	1.877	0.9991	1.998	1.731
42.0	1.188	6.996	1.698	4.646	1.746	0.9992	1.998	1.731
43.0	1.056	6.520	1.620	4.985	1.627	0.9992	1.998	1.731
44.0	9.417 −8	6.086	1.547	5.340	1.519	0.9992	1.999	1.731
45.0	8.417	5.690	1.479	5.712	1.420	0.9993	1.999	1.731
46.0	7.542	5.327	1.416	6.101	1.330	0.9993	1.999	1.731
47.0	6.774	4.995	1.356	6.506	1.247	0.9993	1.999	1.731
48.0	6.098	4.690	1.300	6.930	1.171	0.9993	1.999	1.731
49.0	5.502	4.409	1.248	7.371	1.101	0.9994	1.999	1.731
50.0	4.974	4.150	1.199	7.831	1.036	0.9994	1.999	1.731
51.0	4.505	3.911	1.152	8.310	9.765 −5	0.9994	1.999	1.731
52.0	4.089	3.689	1.108	8.807	9.214	0.9994	1.999	1.731
53.0	3.718	3.485	1.067	9.324	8.703	0.9995	1.999	1.731
54.0	3.386	3.295	1.028	9.861	8.229	0.9995	1.999	1.732
55.0	3.090	3.119	9.908 −4	1.042 +4	7.789	0.9995	1.999	1.732
56.0	2.824	2.955	9.557	1.100	7.380	0.9995	1.999	1.732
57.0	2.585	2.802	9.225	1.160	6.999	0.9995	1.999	1.732
58.0	2.370	2.660	8.910	1.222	6.643	0.9996	1.999	1.732
59.0	2.176	2.527	8.611	1.286	6.312	0.9996	1.999	1.732
60.0	2.001	2.403	8.327	1.352	6.002	0.9996	1.999	1.732
65.0	1.341	1.890	7.096	1.719	4.722	0.9996	1.999	1.732
70.0	9.261 −9	1.514	6.119	2.146	3.782	0.9997	1.999	1.732
75.0	6.561	1.231	5.331	2.639	3.075	0.9997	1.999	1.732
80.0	4.752	1.014	4.685	3.203	2.534	0.9998	1.999	1.732
85.0	3.510	8.456 −6	4.151	3.841	2.113	0.9998	2.000	1.732
90.0	2.638	7.124	3.702	4.560	1.780	0.9998	2.000	1.732
95.0	2.013	6.058	3.323	5.362	1.514	0.9998	2.000	1.732
100.0	1.558	5.194	2.999	6.254	1.298	0.9998	2.000	1.732
∞	0.000	0.000	0.000	∞	0.000	1.000	2.000	1.732

NORMAL SHOCK FUNCTIONS FOR SUPERSONIC COMPRESSIBLE FLOW

$$\gamma = 5/3$$

M	P_{21}	T_{21}	ρ_{21}	M_2	P^o_{21}	μ	ν
20.0	4.997 +2	1.259 +2	3.970	0.4490	2.812 -3	2.866	81.43
21.0	5.510	1.387	3.973	0.4488	2.433	2.729	81.84
22.0	6.047	1.521	3.975	0.4487	2.119	2.605	82.21
23.0	6.610	1.662	3.977	0.4486	1.857	2.492	82.54
24.0	7.197	1.809	3.979	0.4485	1.636	2.388	82.85
25.0	7.810	1.962	3.981	0.4484	1.449	2.293	83.14
26.0	8.447	2.121	3.982	0.4483	1.289	2.204	83.40
27.0	9.110	2.287	3.984	0.4482	1.152	2.123	83.64
28.0	9.797	2.459	3.985	0.4481	1.034	2.047	83.87
29.0	1.051 +3	2.637	3.986	0.4481	9.309 -4	1.976	84.08
30.0	1.125	2.821	3.987	0.4480	8.414	1.910	84.28
31.0	1.201	3.012	3.988	0.4480	7.629	1.849	84.46
32.0	1.280	3.209	3.988	0.4479	6.939	1.791	84.63
33.0	1.361	3.412	3.989	0.4479	6.330	1.737	84.80
34.0	1.445	3.621	3.990	0.4478	5.790	1.685	84.95
35.0	1.531	3.837	3.990	0.4478	5.310	1.637	85.09
36.0	1.620	4.059	3.991	0.4478	4.881	1.592	85.23
37.0	1.711	4.287	3.991	0.4477	4.497	1.549	85.36
38.0	1.805	4.521	3.992	0.4477	4.152	1.508	85.48
39.0	1.901	4.762	3.992	0.4477	3.842	1.469	85.60
40.0	2.000	5.009	3.993	0.4477	3.562	1.433	85.71
41.0	2.101	5.262	3.993	0.4476	3.308	1.398	85.81
42.0	2.205	5.521	3.993	0.4476	3.078	1.364	85.91
43.0	2.311	5.787	3.994	0.4476	2.869	1.333	86.01
44.0	2.420	6.059	3.994	0.4476	2.678	1.302	86.10
45.0	2.531	6.337	3.994	0.4476	2.504	1.273	86.18
46.0	2.645	6.621	3.994	0.4476	2.345	1.246	86.27
47.0	2.761	6.912	3.995	0.4475	2.198	1.219	86.34
48.0	2.880	7.209	3.995	0.4475	2.064	1.194	86.42
49.0	3.001	7.512	3.995	0.4475	1.940	1.169	86.49
50.0	3.125	7.821	3.995	0.4475	1.827	1.146	86.56
51.0	3.251	8.137	3.995	0.4475	1.721	1.124	86.63
52.0	3.380	8.459	3.996	0.4475	1.624	1.102	86.70
53.0	3.511	8.787	3.996	0.4475	1.534	1.081	86.76
54.0	3.645	9.121	3.996	0.4475	1.451	1.061	86.82
55.0	3.781	9.462	3.996	0.4475	1.373	1.042	86.88
56.0	3.920	9.808	3.996	0.4474	1.301	1.023	86.93
57.0	4.061	1.016 +3	3.996	0.4474	1.234	1.005	86.99
58.0	4.205	1.052	3.996	0.4474	1.171	0.988	87.04
59.0	4.351	1.089	3.997	0.4474	1.113	0.971	87.09
60.0	4.500	1.126	3.997	0.4474	1.058	0.955	87.14
65.0	5.281	1.321	3.997	0.4474	8.324 -5	0.882	87.36
70.0	6.125	1.532	3.998	0.4474	6.666	0.819	87.55
75.0	7.031	1.759	3.998	0.4473	5.421	0.764	87.71
80.0	8.000	2.001	3.998	0.4473	4.467	0.716	87.85
85.0	9.031	2.259	3.998	0.4473	3.725	0.674	87.98
90.0	1.012 +4	2.532	3.999	0.4473	3.138	0.637	88.09
95.0	1.128	2.821	3.999	0.4473	2.669	0.603	88.19
100.0	1.250	3.126	3.999	0.4473	2.288	0.573	88.28
∞	∞	∞	4.000	0.4472	0.000	0.000	90.00

TABLE A.2

ISENTROPIC FLOW FUNCTIONS

AND

NORMAL SHOCK FUNCTIONS

$$\gamma = 7/5$$

ISENTROPIC FUNCTIONS FOR SUBSONIC COMPRESSIBLE FLOW

$\gamma = 7/5$

M	$p/p°$	$T/T°$	$\rho/\rho°$	A/A^*	$q/p°$	u/u_m	u/a^*	$F(\gamma;M)$
0.00	1.000 +0	1.000 +0	1.000 +0	∞	0.000 +0	0.0000	0.0000	90.00
0.01	1.000	1.000	1.000	5.787	7.000	0.0045	0.0110	71.43
0.02	9.997 −1	9.998 −1	1.000	2.894	2.799 −4	0.0089	0.0219	35.73
0.03	9.994	9.996	9.998 −1	1.930	6.296	0.0134	0.0329	23.84
0.04	9.989	9.992	9.997	1.448	1.119 −3	0.0179	0.0438	17.89
0.05	9.983	9.988	9.995	1.159	1.747	0.0224	0.0548	14.33
0.06	9.975	9.982	9.993	9.666 +0	2.514	0.0268	0.0657	11.96
0.07	9.966	9.976	9.990	8.292	3.418	0.0313	0.0766	10.27
0.08	9.955	9.968	9.987	7.262	4.460	0.0358	0.0876	9.003
0.09	9.944	9.960	9.984	6.461	5.638	0.0402	0.0985	8.020
0.10	9.930	9.950	9.980	5.822	6.951	0.0447	0.1094	7.236
0.11	9.916	9.940	9.976	5.299	8.399	0.0491	0.1204	6.596
0.12	9.900	9.928	9.971	4.864	9.979	0.0536	0.1313	6.064
0.13	9.883	9.916	9.966	4.497	1.169 −2	0.0580	0.1422	5.615
0.14	9.864	9.903	9.961	4.182	1.353	0.0625	0.1531	5.232
0.15	9.844	9.888	9.955	3.910	1.550	0.0669	0.1639	4.901
0.16	9.823	9.873	9.949	3.673	1.760	0.0714	0.1748	4.612
0.17	9.800	9.857	9.943	3.464	1.983	0.0758	0.1857	4.359
0.18	9.776	9.840	9.936	3.278	2.217	0.0802	0.1965	4.135
0.19	9.751	9.822	9.928	3.112	2.464	0.0847	0.2074	3.935
0.20	9.725	9.803	9.921	2.964	2.723	0.0891	0.2182	3.756
0.21	9.697	9.783	9.913	2.829	2.994	0.0935	0.2290	3.596
0.22	9.668	9.762	9.904	2.708	3.276	0.0979	0.2398	3.450
0.23	9.638	9.740	9.895	2.597	3.569	0.1023	0.2506	3.318
0.24	9.607	9.718	9.886	2.496	3.874	0.1067	0.2614	3.198
0.25	9.575	9.694	9.877	2.403	4.189	0.1111	0.2722	3.088
0.26	9.541	9.670	9.867	2.317	4.515	0.1155	0.2829	2.987
0.27	9.506	9.645	9.856	2.238	4.851	0.1199	0.2936	2.894
0.28	9.470	9.619	9.846	2.166	5.197	0.1242	0.3043	2.809
0.29	9.433	9.592	9.835	2.098	5.553	0.1286	0.3150	2.730
0.30	9.395	9.564	9.823	2.035	5.919	0.1330	0.3257	2.657
0.31	9.355	9.535	9.811	1.977	6.293	0.1373	0.3364	2.589
0.32	9.315	9.506	9.799	1.922	6.677	0.1417	0.3470	2.526
0.33	9.274	9.476	9.787	1.871	7.069	0.1460	0.3576	2.468
0.34	9.231	9.445	9.774	1.823	7.470	0.1503	0.3682	2.413
0.35	9.188	9.413	9.761	1.778	7.878	0.1546	0.3788	2.362
0.36	9.143	9.380	9.747	1.736	8.295	0.1589	0.3893	2.314
0.37	9.098	9.347	9.734	1.696	8.719	0.1632	0.3999	2.270
0.38	9.052	9.313	9.719	1.659	9.149	0.1675	0.4104	2.228
0.39	9.004	9.278	9.705	1.623	9.587	0.1718	0.4209	2.188
0.40	8.956	9.243	9.690	1.590	1.003 −1	0.1761	0.4313	2.152
0.41	8.907	9.207	9.675	1.559	1.048	0.1804	0.4418	2.117
0.42	8.857	9.170	9.659	1.529	1.094	0.1846	0.4522	2.084
0.43	8.807	9.132	9.643	1.501	1.140	0.1888	0.4626	2.054
0.44	8.755	9.094	9.627	1.474	1.186	0.1931	0.4729	2.025
0.45	8.703	9.055	9.611	1.449	1.234	0.1973	0.4833	1.997
0.46	8.650	9.016	9.594	1.425	1.281	0.2015	0.4936	1.972
0.47	8.596	8.976	9.577	1.402	1.329	0.2057	0.5038	1.947
0.48	8.541	8.935	9.560	1.380	1.378	0.2099	0.5141	1.924
0.49	8.486	8.894	9.542	1.359	1.426	0.2141	0.5243	1.903
0.50	8.430	8.852	9.524	1.340	1.475	0.2182	0.5345	1.882

ISENTROPIC FUNCTIONS FOR SUBSONIC COMPRESSIBLE FLOW

$\gamma = 7/5$

M	$p/p°$	$T/T°$	$\rho/\rho°$	A/A^*	$q/p°$	u/u_m	u/a^*	$F(\gamma;M)$
0.50	8.430 -1	8.852 -1	9.524 -1	1.340 +0	1.475 -1	0.2182	0.5345	1.882
0.51	8.374	8.809	9.506	1.321	1.525	0.2224	0.5447	1.863
0.52	8.317	8.766	9.487	1.303	1.574	0.2265	0.5548	1.844
0.53	8.259	8.723	9.468	1.286	1.624	0.2306	0.5649	1.827
0.54	8.201	8.679	9.449	1.270	1.674	0.2347	0.5750	1.811
0.55	8.142	8.634	9.430	1.255	1.724	0.2388	0.5851	1.795
0.56	8.082	8.589	9.410	1.240	1.774	0.2429	0.5951	1.781
0.57	8.022	8.544	9.390	1.226	1.825	0.2470	0.6051	1.767
0.58	7.962	8.498	9.370	1.213	1.875	0.2511	0.6150	1.754
0.59	7.901	8.451	9.349	1.200	1.925	0.2551	0.6249	1.741
0.60	7.840	8.405	9.328	1.188	1.976	0.2592	0.6348	1.729
0.61	7.778	8.357	9.307	1.177	2.026	0.2632	0.6447	1.718
0.62	7.716	8.310	9.286	1.166	2.076	0.2672	0.6545	1.708
0.63	7.654	8.262	9.265	1.155	2.127	0.2712	0.6643	1.698
0.64	7.591	8.213	9.243	1.145	2.177	0.2752	0.6740	1.688
0.65	7.528	8.164	9.221	1.136	2.226	0.2791	0.6837	1.679
0.66	7.465	8.115	9.199	1.127	2.276	0.2831	0.6934	1.671
0.67	7.401	8.066	9.176	1.118	2.326	0.2870	0.7031	1.663
0.68	7.338	8.016	9.153	1.110	2.375	0.2909	0.7127	1.656
0.69	7.274	7.966	9.131	1.102	2.424	0.2949	0.7223	1.648
0.70	7.209	7.916	9.107	1.094	2.473	0.2988	0.7318	1.642
0.71	7.145	7.865	9.084	1.087	2.521	0.3026	0.7413	1.636
0.72	7.080	7.814	9.061	1.081	2.569	0.3065	0.7508	1.630
0.73	7.016	7.763	9.037	1.074	2.617	0.3103	0.7602	1.624
0.74	6.951	7.712	9.013	1.068	2.664	0.3142	0.7696	1.619
0.75	6.886	7.660	8.989	1.062	2.711	0.3180	0.7789	1.614
0.76	6.821	7.609	8.964	1.057	2.758	0.3218	0.7883	1.609
0.77	6.756	7.557	8.940	1.052	2.804	0.3256	0.7975	1.605
0.78	6.691	7.505	8.915	1.047	2.849	0.3294	0.8068	1.601
0.79	6.625	7.452	8.890	1.043	2.894	0.3331	0.8160	1.597
0.80	6.560	7.400	8.865	1.038	2.939	0.3369	0.8251	1.594
0.81	6.495	7.347	8.840	1.034	2.983	0.3406	0.8343	1.591
0.82	6.430	7.295	8.815	1.030	3.026	0.3443	0.8433	1.588
0.83	6.365	7.242	8.789	1.027	3.069	0.3480	0.8524	1.585
0.84	6.300	7.189	8.763	1.024	3.112	0.3517	0.8614	1.582
0.85	6.235	7.136	8.737	1.021	3.153	0.3553	0.8704	1.580
0.86	6.170	7.083	8.711	1.018	3.195	0.3590	0.8793	1.578
0.87	6.106	7.030	8.685	1.015	3.235	0.3626	0.8882	1.576
0.88	6.041	6.977	8.659	1.013	3.275	0.3662	0.8970	1.574
0.89	5.977	6.924	8.632	1.011	3.314	0.3698	0.9058	1.573
0.90	5.913	6.870	8.606	1.009	3.352	0.3734	0.9146	1.571
0.91	5.849	6.817	8.579	1.007	3.390	0.3769	0.9233	1.570
0.92	5.785	6.764	8.552	1.006	3.427	0.3805	0.9320	1.569
0.93	5.721	6.711	8.525	1.004	3.464	0.3840	0.9406	1.568
0.94	5.658	6.658	8.498	1.003	3.499	0.3875	0.9493	1.567
0.95	5.595	6.604	8.471	1.002	3.534	0.3910	0.9578	1.566
0.96	5.532	6.551	8.444	1.001	3.569	0.3945	0.9663	1.566
0.97	5.469	6.498	8.416	1.001	3.602	0.3980	0.9748	1.565
0.98	5.407	6.445	8.389	1.000	3.635	0.4014	0.9832	1.565
0.99	5.345	6.392	8.361	1.000	3.667	0.4048	0.9916	1.565
1.00	5.283	6.339	8.333	1.000	3.698	0.4082	1.000	1.565

ISENTROPIC FUNCTIONS FOR SUPERSONIC COMPRESSIBLE FLOW

$\gamma = 7/5$

M	$p/p°$	$T/T°$	$\rho/\rho°$	A/A^*	$q/p°$	u/u_m	u/a^*	$F(\gamma;M)$
1.00	5.283 -1	6.339 -1	8.333 -1	1.000 +0	3.698 -1	0.4082	1.000	1.565
1.01	5.221	6.287	8.306	1.000	3.728	0.4116	1.008	1.565
1.02	5.160	6.234	8.278	1.000	3.758	0.4150	1.017	1.565
1.03	5.099	6.181	8.250	1.001	3.787	0.4184	1.025	1.565
1.04	5.039	6.129	8.222	1.001	3.815	0.4217	1.033	1.566
1.05	4.979	6.077	8.193	1.002	3.842	0.4250	1.041	1.566
1.06	4.919	6.024	8.165	1.003	3.869	0.4284	1.049	1.567
1.07	4.860	5.972	8.137	1.004	3.895	0.4316	1.057	1.567
1.08	4.801	5.920	8.108	1.005	3.919	0.4349	1.065	1.568
1.09	4.742	5.869	8.080	1.006	3.944	0.4382	1.073	1.569
1.10	4.684	5.817	8.052	1.008	3.967	0.4414	1.081	1.570
1.11	4.626	5.766	8.023	1.010	3.989	0.4446	1.089	1.571
1.12	4.568	5.714	7.994	1.011	4.011	0.4478	1.097	1.572
1.13	4.511	5.663	7.966	1.013	4.032	0.4510	1.105	1.573
1.14	4.455	5.612	7.937	1.015	4.052	0.4542	1.113	1.574
1.15	4.398	5.562	7.908	1.017	4.072	0.4574	1.120	1.575
1.16	4.343	5.511	7.880	1.020	4.090	0.4605	1.128	1.576
1.17	4.287	5.461	7.851	1.022	4.108	0.4636	1.136	1.578
1.18	4.232	5.411	7.822	1.025	4.125	0.4667	1.143	1.579
1.19	4.178	5.361	7.793	1.028	4.141	0.4698	1.150	1.580
1.20	4.124	5.312	7.764	1.030	4.157	0.4729	1.158	1.582
1.21	4.070	5.262	7.735	1.033	4.171	0.4759	1.166	1.583
1.22	4.017	5.213	7.706	1.037	4.185	0.4789	1.173	1.585
1.23	3.965	5.164	7.677	1.040	4.198	0.4820	1.181	1.587
1.24	3.912	5.115	7.648	1.043	4.211	0.4850	1.188	1.588
1.25	3.861	5.067	7.619	1.047	4.222	0.4879	1.195	1.590
1.26	3.809	5.019	7.590	1.050	4.233	0.4909	1.203	1.592
1.27	3.759	4.971	7.561	1.054	4.244	0.4939	1.210	1.593
1.28	3.708	4.924	7.532	1.058	4.253	0.4968	1.217	1.595
1.29	3.659	4.876	7.503	1.062	4.262	0.4997	1.224	1.597
1.30	3.609	4.829	7.474	1.066	4.270	0.5026	1.231	1.599
1.31	3.560	4.782	7.445	1.071	4.277	0.5055	1.238	1.601
1.32	3.512	4.736	7.416	1.075	4.283	0.5083	1.245	1.603
1.33	3.464	4.690	7.387	1.080	4.289	0.5112	1.252	1.605
1.34	3.417	4.644	7.358	1.084	4.294	0.5140	1.259	1.607
1.35	3.370	4.598	7.329	1.089	4.299	0.5168	1.266	1.609
1.36	3.323	4.553	7.300	1.094	4.303	0.5196	1.273	1.611
1.37	3.277	4.508	7.271	1.099	4.306	0.5224	1.280	1.613
1.38	3.232	4.463	7.242	1.104	4.308	0.5252	1.286	1.615
1.39	3.187	4.419	7.213	1.109	4.310	0.5279	1.293	1.617
1.40	3.143	4.374	7.184	1.115	4.311	0.5307	1.300	1.619
1.41	3.099	4.331	7.155	1.120	4.312	0.5334	1.307	1.621
1.42	3.055	4.287	7.126	1.126	4.312	0.5361	1.313	1.623
1.43	3.012	4.244	7.097	1.132	4.311	0.5388	1.320	1.626
1.44	2.969	4.201	7.069	1.138	4.310	0.5414	1.326	1.628
1.45	2.927	4.158	7.040	1.144	4.308	0.5441	1.333	1.630
1.46	2.886	4.116	7.011	1.150	4.306	0.5467	1.339	1.632
1.47	2.845	4.074	6.982	1.156	4.303	0.5493	1.346	1.634
1.48	2.804	4.032	6.954	1.163	4.299	0.5519	1.352	1.637
1.49	2.764	3.991	6.925	1.169	4.295	0.5545	1.358	1.639
1.50	2.724	3.950	6.897	1.176	4.290	0.5571	1.365	1.641

NORMAL SHOCK FUNCTIONS FOR SUPERSONIC COMPRESSIBLE FLOW

$$\gamma = 7/5$$

M	p_{21}	T_{21}	ρ_{21}	M_2	p_{21}^0	μ	ν
1.00	1.000 +0	1.000 +0	1.000	1.0000	1.000 +0	90.00	0.000
1.01	1.023	1.007	1.017	0.9901	1.000	81.93	0.045
1.02	1.047	1.013	1.033	0.9805	1.000	78.64	0.126
1.03	1.071	1.020	1.050	0.9712	1.000	76.14	0.229
1.04	1.095	1.026	1.067	0.9620	1.000	74.06	0.351
1.05	1.120	1.033	1.084	0.9531	9.999 -1	72.25	0.487
1.06	1.144	1.039	1.101	0.9444	9.997	70.63	0.637
1.07	1.169	1.046	1.118	0.9360	9.996	69.16	0.797
1.08	1.194	1.052	1.135	0.9277	9.994	67.81	0.968
1.09	1.219	1.059	1.152	0.9197	9.992	66.55	1.148
1.10	1.245	1.065	1.169	0.9118	9.989	65.38	1.336
1.11	1.271	1.071	1.186	0.9041	9.986	64.28	1.532
1.12	1.297	1.078	1.203	0.8966	9.982	63.24	1.735
1.13	1.323	1.084	1.221	0.8892	9.978	62.25	1.944
1.14	1.350	1.090	1.238	0.8820	9.973	61.31	2.160
1.15	1.376	1.097	1.255	0.8750	9.967	60.41	2.381
1.16	1.403	1.103	1.272	0.8682	9.961	59.55	2.607
1.17	1.430	1.109	1.290	0.8615	9.953	58.73	2.838
1.18	1.458	1.115	1.307	0.8549	9.946	57.94	3.074
1.19	1.485	1.122	1.324	0.8485	9.937	57.18	3.314
1.20	1.513	1.128	1.342	0.8422	9.928	56.44	3.558
1.21	1.541	1.134	1.359	0.8360	9.918	55.74	3.806
1.22	1.570	1.141	1.376	0.8300	9.907	55.05	4.057
1.23	1.598	1.147	1.394	0.8241	9.896	54.39	4.311
1.24	1.627	1.153	1.411	0.8183	9.884	53.75	4.569
1.25	1.656	1.159	1.429	0.8126	9.871	53.13	4.829
1.26	1.685	1.166	1.446	0.8071	9.857	52.53	5.093
1.27	1.715	1.172	1.463	0.8017	9.842	51.94	5.358
1.28	1.745	1.178	1.481	0.7963	9.827	51.38	5.627
1.29	1.775	1.185	1.498	0.7911	9.811	50.82	5.897
1.30	1.805	1.191	1.516	0.7860	9.794	50.29	6.170
1.31	1.835	1.197	1.533	0.7809	9.776	49.76	6.444
1.32	1.866	1.204	1.551	0.7760	9.758	49.25	6.721
1.33	1.897	1.210	1.568	0.7712	9.738	48.75	6.999
1.34	1.928	1.216	1.585	0.7664	9.718	48.27	7.279
1.35	1.959	1.223	1.603	0.7618	9.697	47.80	7.560
1.36	1.991	1.229	1.620	0.7572	9.676	47.33	7.843
1.37	2.023	1.235	1.638	0.7527	9.654	46.88	8.127
1.38	2.055	1.242	1.655	0.7483	9.630	46.44	8.412
1.39	2.087	1.248	1.672	0.7440	9.607	46.01	8.699
1.40	2.120	1.255	1.690	0.7397	9.582	45.59	8.986
1.41	2.153	1.261	1.707	0.7355	9.557	45.17	9.275
1.42	2.186	1.268	1.724	0.7314	9.531	44.77	9.564
1.43	2.219	1.274	1.742	0.7274	9.504	44.37	9.854
1.44	2.252	1.281	1.759	0.7235	9.477	43.98	10.15
1.45	2.286	1.287	1.776	0.7196	9.448	43.60	10.44
1.46	2.320	1.294	1.793	0.7158	9.420	43.23	10.73
1.47	2.354	1.300	1.811	0.7120	9.390	42.87	11.02
1.48	2.389	1.307	1.828	0.7083	9.360	42.51	11.32
1.49	2.423	1.314	1.845	0.7047	9.329	42.16	11.61
1.50	2.458	1.320	1.862	0.7011	9.298	41.81	11.90

ISENTROPIC FUNCTIONS FOR SUPERSONIC COMPRESSIBLE FLOW

$\gamma = 7/5$

M	$p/p°$	$T/T°$	$\rho/\rho°$	$A/A*$	$q/p°$	u/u_m	$u/a*$	$F(\gamma;M)$
1.50	2.724 −1	3.950 −1	6.897 −1	1.176 +0	4.290 −1	0.5571	1.365	1.641
1.51	2.685	3.909	6.868	1.183	4.285	0.5596	1.371	1.643
1.52	2.646	3.869	6.840	1.190	4.279	0.5622	1.377	1.646
1.53	2.608	3.829	6.811	1.197	4.273	0.5647	1.383	1.648
1.54	2.570	3.789	6.783	1.204	4.267	0.5672	1.389	1.650
1.55	2.533	3.750	6.755	1.212	4.259	0.5697	1.395	1.653
1.56	2.496	3.711	6.726	1.219	4.252	0.5722	1.402	1.655
1.57	2.459	3.672	6.698	1.227	4.243	0.5746	1.408	1.657
1.58	2.423	3.633	6.670	1.234	4.235	0.5771	1.414	1.660
1.59	2.388	3.595	6.642	1.242	4.226	0.5795	1.420	1.662
1.60	2.353	3.557	6.614	1.250	4.216	0.5819	1.425	1.664
1.61	2.318	3.520	6.586	1.258	4.206	0.5843	1.431	1.667
1.62	2.284	3.483	6.558	1.267	4.196	0.5867	1.437	1.669
1.63	2.250	3.446	6.530	1.275	4.185	0.5891	1.443	1.671
1.64	2.217	3.409	6.502	1.284	4.174	0.5914	1.449	1.674
1.65	2.184	3.373	6.475	1.292	4.162	0.5937	1.454	1.676
1.66	2.152	3.337	6.447	1.301	4.150	0.5961	1.460	1.678
1.67	2.120	3.302	6.420	1.310	4.138	0.5984	1.466	1.681
1.68	2.088	3.267	6.392	1.319	4.125	0.6007	1.471	1.683
1.69	2.057	3.232	6.365	1.328	4.112	0.6029	1.477	1.685
1.70	2.026	3.197	6.337	1.338	4.099	0.6052	1.482	1.688
1.71	1.996	3.163	6.310	1.347	4.085	0.6075	1.488	1.690
1.72	1.966	3.129	6.283	1.357	4.071	0.6097	1.493	1.692
1.73	1.936	3.095	6.256	1.366	4.056	0.6119	1.499	1.695
1.74	1.907	3.062	6.229	1.376	4.042	0.6141	1.504	1.697
1.75	1.878	3.029	6.202	1.386	4.027	0.6163	1.510	1.700
1.76	1.850	2.996	6.175	1.397	4.011	0.6185	1.515	1.702
1.77	1.822	2.964	6.148	1.407	3.996	0.6206	1.520	1.704
1.78	1.795	2.932	6.121	1.417	3.980	0.6228	1.526	1.707
1.79	1.767	2.900	6.095	1.428	3.964	0.6249	1.531	1.709
1.80	1.741	2.868	6.068	1.439	3.947	0.6270	1.536	1.711
1.81	1.714	2.837	6.042	1.450	3.931	0.6292	1.541	1.714
1.82	1.688	2.806	6.015	1.461	3.914	0.6312	1.546	1.716
1.83	1.662	2.776	5.989	1.472	3.897	0.6333	1.551	1.718
1.84	1.637	2.745	5.963	1.484	3.879	0.6354	1.556	1.721
1.85	1.612	2.716	5.937	1.495	3.862	0.6374	1.561	1.723
1.86	1.588	2.686	5.911	1.507	3.844	0.6395	1.566	1.725
1.87	1.563	2.656	5.885	1.519	3.826	0.6415	1.571	1.727
1.88	1.539	2.627	5.859	1.531	3.808	0.6435	1.576	1.730
1.89	1.516	2.599	5.833	1.543	3.790	0.6455	1.581	1.732
1.90	1.493	2.570	5.807	1.555	3.771	0.6475	1.586	1.734
1.91	1.470	2.542	5.782	1.568	3.753	0.6495	1.591	1.737
1.92	1.447	2.514	5.756	1.580	3.734	0.6514	1.596	1.739
1.93	1.425	2.486	5.731	1.593	3.715	0.6534	1.601	1.741
1.94	1.403	2.459	5.706	1.606	3.696	0.6553	1.605	1.743
1.95	1.381	2.432	5.680	1.619	3.677	0.6572	1.610	1.746
1.96	1.360	2.405	5.655	1.633	3.657	0.6591	1.615	1.748
1.97	1.339	2.379	5.630	1.646	3.638	0.6610	1.619	1.750
1.98	1.319	2.352	5.605	1.660	3.618	0.6629	1.624	1.752
1.99	1.298	2.326	5.580	1.673	3.599	0.6648	1.628	1.755
2.00	1.278	2.301	5.556	1.687	3.579	0.6667	1.633	1.757

NORMAL SHOCK FUNCTIONS FOR SUPERSONIC COMPRESSIBLE FLOW

$$\gamma = 7/5$$

M	p_{21}	T_{21}	ρ_{21}	M_2	p^o_{21}	μ	ν
1.50	2.458 +0	1.320 +0	1.862	0.7011	9.298 -1	41.81	11.90
1.51	2.493	1.327	1.879	0.6976	9.266	41.47	12.20
1.52	2.529	1.334	1.896	0.6941	9.233	41.14	12.49
1.53	2.564	1.340	1.913	0.6907	9.200	40.81	12.79
1.54	2.600	1.347	1.930	0.6874	9.166	40.49	13.08
1.55	2.636	1.354	1.947	0.6841	9.132	40.18	13.38
1.56	2.672	1.361	1.964	0.6809	9.097	39.87	13.68
1.57	2.709	1.367	1.981	0.6777	9.062	39.57	13.97
1.58	2.746	1.374	1.998	0.6746	9.026	39.27	14.27
1.59	2.783	1.381	2.015	0.6715	8.989	38.97	14.56
1.60	2.820	1.388	2.032	0.6685	8.952	38.68	14.86
1.61	2.857	1.395	2.048	0.6655	8.915	38.40	15.15
1.62	2.895	1.402	2.065	0.6625	8.877	38.12	15.45
1.63	2.933	1.409	2.082	0.6596	8.838	37.84	15.75
1.64	2.971	1.416	2.099	0.6568	8.799	37.57	16.04
1.65	3.009	1.423	2.115	0.6540	8.760	37.31	16.34
1.66	3.048	1.430	2.132	0.6512	8.720	37.04	16.63
1.67	3.087	1.437	2.148	0.6485	8.680	36.79	16.93
1.68	3.126	1.444	2.165	0.6458	8.640	36.53	17.22
1.69	3.165	1.451	2.181	0.6432	8.599	36.28	17.51
1.70	3.205	1.458	2.198	0.6406	8.557	36.03	17.81
1.71	3.245	1.465	2.214	0.6380	8.516	35.79	18.10
1.72	3.285	1.473	2.230	0.6355	8.474	35.55	18.39
1.73	3.325	1.480	2.247	0.6330	8.431	35.31	18.69
1.74	3.365	1.487	2.263	0.6305	8.389	35.08	18.98
1.75	3.406	1.495	2.279	0.6281	8.346	34.85	19.27
1.76	3.447	1.502	2.295	0.6257	8.303	34.62	19.56
1.77	3.488	1.509	2.311	0.6234	8.259	34.40	19.85
1.78	3.530	1.517	2.327	0.6211	8.215	34.18	20.14
1.79	3.571	1.524	2.343	0.6188	8.171	33.96	20.43
1.80	3.613	1.532	2.359	0.6165	8.127	33.75	20.72
1.81	3.655	1.539	2.375	0.6143	8.083	33.54	21.01
1.82	3.698	1.547	2.391	0.6121	8.038	33.33	21.30
1.83	3.740	1.554	2.407	0.6099	7.993	33.12	21.59
1.84	3.783	1.562	2.422	0.6078	7.948	32.92	21.88
1.85	3.826	1.569	2.438	0.6057	7.903	32.72	22.16
1.86	3.869	1.577	2.454	0.6036	7.857	32.52	22.45
1.87	3.913	1.585	2.469	0.6016	7.812	32.33	22.73
1.88	3.957	1.592	2.485	0.5996	7.766	32.14	23.02
1.89	4.000	1.600	2.500	0.5976	7.720	31.95	23.30
1.90	4.045	1.608	2.516	0.5956	7.674	31.76	23.58
1.91	4.089	1.616	2.531	0.5937	7.628	31.57	23.87
1.92	4.134	1.624	2.546	0.5918	7.582	31.39	24.15
1.93	4.179	1.631	2.561	0.5899	7.535	31.21	24.43
1.94	4.224	1.639	2.577	0.5880	7.489	31.03	24.71
1.95	4.269	1.647	2.592	0.5862	7.442	30.85	24.99
1.96	4.315	1.655	2.607	0.5844	7.396	30.68	25.27
1.97	4.361	1.663	2.622	0.5826	7.349	30.51	25.55
1.98	4.407	1.671	2.637	0.5808	7.302	30.34	25.82
1.99	4.453	1.679	2.652	0.5791	7.256	30.17	26.10
2.00	4.500	1.687	2.667	0.5774	7.209	30.00	26.38

ISENTROPIC FUNCTIONS FOR SUPERSONIC COMPRESSIBLE FLOW

$$\gamma = 7/5$$

M	$p/p°$	$T/T°$	$\rho/\rho°$	A/A^*	$q/p°$	u/u_m	u/a^*	$F(\gamma;M)$
2.00	1.278 -1	2.301 -1	5.556 -1	1.687 +0	3.579 -1	0.6667	1.633	1.757
2.01	1.258	2.275	5.531	1.702	3.559	0.6685	1.638	1.759
2.02	1.239	2.250	5.507	1.716	3.539	0.6703	1.642	1.761
2.03	1.220	2.225	5.482	1.730	3.519	0.6722	1.646	1.764
2.04	1.201	2.201	5.458	1.745	3.498	0.6740	1.651	1.766
2.05	1.182	2.176	5.434	1.760	3.478	0.6758	1.655	1.768
2.06	1.164	2.152	5.409	1.775	3.458	0.6775	1.660	1.770
2.07	1.146	2.128	5.385	1.790	3.437	0.6793	1.664	1.772
2.08	1.128	2.105	5.361	1.805	3.417	0.6811	1.668	1.774
2.09	1.111	2.081	5.337	1.821	3.396	0.6828	1.673	1.777
2.10	1.094	2.058	5.314	1.837	3.376	0.6846	1.677	1.779
2.11	1.077	2.035	5.290	1.853	3.355	0.6863	1.681	1.781
2.12	1.060	2.013	5.266	1.869	3.335	0.6880	1.685	1.783
2.13	1.044	1.990	5.243	1.885	3.314	0.6897	1.689	1.785
2.14	1.027	1.968	5.220	1.902	3.293	0.6914	1.694	1.787
2.15	1.011	1.946	5.196	1.918	3.273	0.6931	1.698	1.789
2.16	9.958 -2	1.925	5.173	1.935	3.252	0.6948	1.702	1.791
2.17	9.803	1.903	5.150	1.952	3.231	0.6964	1.706	1.793
2.18	9.651	1.882	5.127	1.970	3.210	0.6981	1.710	1.796
2.19	9.501	1.861	5.104	1.987	3.189	0.6997	1.714	1.798
2.20	9.353	1.841	5.082	2.005	3.169	0.7013	1.718	1.800
2.21	9.208	1.820	5.059	2.023	3.148	0.7029	1.722	1.802
2.22	9.065	1.800	5.036	2.041	3.127	0.7045	1.726	1.804
2.23	8.924	1.780	5.014	2.059	3.106	0.7061	1.730	1.806
2.24	8.786	1.760	4.991	2.078	3.086	0.7077	1.734	1.808
2.25	8.649	1.741	4.969	2.096	3.065	0.7093	1.737	1.810
2.26	8.515	1.721	4.947	2.115	3.044	0.7108	1.741	1.812
2.27	8.383	1.702	4.925	2.134	3.023	0.7124	1.745	1.814
2.28	8.253	1.683	4.903	2.154	3.003	0.7139	1.749	1.816
2.29	8.125	1.665	4.881	2.173	2.982	0.7155	1.753	1.818
2.30	7.998	1.646	4.859	2.193	2.962	0.7170	1.756	1.820
2.31	7.874	1.628	4.838	2.213	2.941	0.7185	1.760	1.822
2.32	7.752	1.610	4.816	2.233	2.920	0.7200	1.764	1.824
2.33	7.632	1.592	4.795	2.254	2.900	0.7215	1.767	1.826
2.34	7.513	1.574	4.773	2.274	2.880	0.7230	1.771	1.828
2.35	7.397	1.557	4.752	2.295	2.859	0.7244	1.775	1.829
2.36	7.282	1.539	4.731	2.316	2.839	0.7259	1.778	1.831
2.37	7.169	1.522	4.710	2.337	2.819	0.7273	1.782	1.833
2.38	7.058	1.505	4.689	2.359	2.798	0.7288	1.785	1.835
2.39	6.949	1.489	4.668	2.381	2.778	0.7302	1.789	1.837
2.40	6.841	1.472	4.647	2.403	2.758	0.7316	1.792	1.839
2.41	6.735	1.456	4.626	2.425	2.738	0.7330	1.796	1.841
2.42	6.631	1.440	4.606	2.448	2.718	0.7345	1.799	1.843
2.43	6.528	1.424	4.585	2.470	2.698	0.7358	1.802	1.844
2.44	6.427	1.408	4.565	2.493	2.678	0.7372	1.806	1.846
2.45	6.328	1.392	4.545	2.517	2.659	0.7386	1.809	1.848
2.46	6.230	1.377	4.524	2.540	2.639	0.7400	1.813	1.850
2.47	6.134	1.362	4.504	2.564	2.619	0.7413	1.816	1.852
2.48	6.039	1.347	4.484	2.588	2.600	0.7427	1.819	1.854
2.49	5.946	1.332	4.464	2.612	2.580	0.7440	1.822	1.855
2.50	5.854	1.317	4.445	2.636	2.561	0.7453	1.826	1.857

NORMAL SHOCK FUNCTIONS FOR SUPERSONIC COMPRESSIBLE FLOW

$$\gamma = 7/5$$

M	P_{21}	T_{21}	ρ_{21}	M_2	P^o_{21}	μ	ν
2.00	4.500 +0	1.687 +0	2.667	0.5774	7.209 −1	30.00	26.38
2.01	4.546	1.696	2.681	0.5757	7.162	29.84	26.65
2.02	4.593	1.704	2.696	0.5740	7.116	29.67	26.93
2.03	4.641	1.712	2.711	0.5723	7.069	29.51	27.20
2.04	4.688	1.720	2.725	0.5707	7.022	29.35	27.47
2.05	4.736	1.728	2.740	0.5691	6.975	29.20	27.75
2.06	4.784	1.737	2.754	0.5675	6.929	29.04	28.02
2.07	4.832	1.745	2.769	0.5659	6.882	28.89	28.29
2.08	4.880	1.754	2.783	0.5643	6.835	28.74	28.56
2.09	4.929	1.762	2.798	0.5628	6.789	28.59	28.83
2.10	4.978	1.770	2.812	0.5613	6.742	28.44	29.09
2.11	5.027	1.779	2.826	0.5598	6.696	28.29	29.36
2.12	5.076	1.787	2.840	0.5583	6.650	28.15	29.63
2.13	5.126	1.796	2.854	0.5568	6.603	28.00	29.89
2.14	5.176	1.805	2.868	0.5554	6.557	27.86	30.16
2.15	5.226	1.813	2.882	0.5540	6.511	27.72	30.42
2.16	5.276	1.822	2.896	0.5526	6.465	27.58	30.69
2.17	5.327	1.830	2.910	0.5512	6.419	27.44	30.95
2.18	5.377	1.839	2.924	0.5498	6.373	27.31	31.21
2.19	5.428	1.848	2.937	0.5484	6.327	27.17	31.47
2.20	5.480	1.857	2.951	0.5471	6.282	27.04	31.73
2.21	5.531	1.866	2.965	0.5457	6.236	26.90	31.99
2.22	5.583	1.874	2.978	0.5444	6.191	26.77	32.25
2.23	5.635	1.883	2.992	0.5431	6.146	26.64	32.50
2.24	5.687	1.892	3.005	0.5418	6.101	26.52	32.76
2.25	5.739	1.901	3.019	0.5406	6.056	26.39	33.02
2.26	5.792	1.910	3.032	0.5393	6.011	26.26	33.27
2.27	5.845	1.919	3.045	0.5381	5.966	26.14	33.52
2.28	5.898	1.928	3.058	0.5368	5.922	26.02	33.78
2.29	5.951	1.938	3.071	0.5356	5.878	25.89	34.03
2.30	6.004	1.947	3.084	0.5344	5.833	25.77	34.28
2.31	6.058	1.956	3.097	0.5332	5.789	25.65	34.53
2.32	6.112	1.965	3.110	0.5321	5.746	25.53	34.78
2.33	6.167	1.974	3.123	0.5309	5.702	25.42	35.03
2.34	6.221	1.984	3.136	0.5298	5.659	25.30	35.28
2.35	6.276	1.993	3.149	0.5286	5.615	25.19	35.52
2.36	6.331	2.002	3.162	0.5275	5.572	25.07	35.77
2.37	6.386	2.012	3.174	0.5264	5.529	24.96	36.01
2.38	6.441	2.021	3.187	0.5253	5.487	24.85	36.26
2.39	6.497	2.031	3.199	0.5242	5.444	24.74	36.50
2.40	6.553	2.040	3.212	0.5231	5.402	24.63	36.74
2.41	6.609	2.050	3.224	0.5221	5.360	24.52	36.99
2.42	6.665	2.059	3.237	0.5210	5.318	24.41	37.23
2.43	6.722	2.069	3.249	0.5200	5.276	24.30	37.47
2.44	6.779	2.079	3.261	0.5189	5.235	24.20	37.71
2.45	6.836	2.088	3.273	0.5179	5.194	24.09	37.94
2.46	6.893	2.098	3.285	0.5169	5.153	23.99	38.18
2.47	6.950	2.108	3.297	0.5159	5.112	23.88	38.42
2.48	7.008	2.118	3.309	0.5149	5.071	23.78	38.65
2.49	7.066	2.128	3.321	0.5140	5.031	23.68	38.89
2.50	7.124	2.137	3.333	0.5130	4.991	23.58	39.12

ISENTROPIC FUNCTIONS FOR SUPERSONIC COMPRESSIBLE FLOW

$\gamma = 7/5$

M	$p/p°$	$T/T°$	$\rho/\rho°$	A/A^*	$q/p°$	u/u_m	u/a^*	$F(\gamma;M)$
2.50	5.854 -2	1.317 -1	4.445 -1	2.636 $+0$	2.561 -1	0.7453	1.826	1.857
2.51	5.763	1.302	4.425	2.661	2.541	0.7467	1.829	1.859
2.52	5.675	1.288	4.405	2.686	2.522	0.7480	1.832	1.861
2.53	5.587	1.274	4.386	2.711	2.503	0.7493	1.835	1.862
2.54	5.501	1.260	4.366	2.737	2.484	0.7506	1.839	1.864
2.55	5.416	1.246	4.347	2.763	2.465	0.7519	1.842	1.866
2.56	5.333	1.232	4.328	2.789	2.446	0.7531	1.845	1.868
2.57	5.251	1.219	4.309	2.815	2.427	0.7544	1.848	1.869
2.58	5.170	1.205	4.290	2.842	2.409	0.7557	1.851	1.871
2.59	5.091	1.192	4.271	2.869	2.390	0.7569	1.854	1.873
2.60	5.012	1.179	4.252	2.896	2.372	0.7582	1.857	1.874
2.61	4.936	1.166	4.233	2.923	2.353	0.7594	1.860	1.876
2.62	4.860	1.153	4.215	2.951	2.335	0.7606	1.863	1.878
2.63	4.785	1.140	4.196	2.979	2.317	0.7618	1.866	1.879
2.64	4.712	1.128	4.177	3.007	2.299	0.7631	1.869	1.881
2.65	4.640	1.116	4.159	3.036	2.281	0.7643	1.872	1.883
2.66	4.569	1.103	4.141	3.064	2.263	0.7655	1.875	1.884
2.67	4.499	1.091	4.123	3.093	2.245	0.7666	1.878	1.886
2.68	4.430	1.079	4.105	3.123	2.227	0.7678	1.881	1.888
2.69	4.362	1.068	4.086	3.153	2.209	0.7690	1.884	1.889
2.70	4.296	1.056	4.069	3.183	2.192	0.7702	1.887	1.891
2.71	4.230	1.044	4.051	3.213	2.175	0.7713	1.889	1.892
2.72	4.166	1.033	4.033	3.244	2.157	0.7725	1.892	1.894
2.73	4.102	1.022	4.015	3.274	2.140	0.7736	1.895	1.896
2.74	4.040	1.011	3.998	3.306	2.123	0.7747	1.898	1.897
2.75	3.978	9.995 -2	3.980	3.337	2.106	0.7759	1.901	1.899
2.76	3.918	9.887	3.963	3.369	2.089	0.7770	1.903	1.900
2.77	3.858	9.779	3.946	3.401	2.072	0.7781	1.906	1.902
2.78	3.800	9.673	3.928	3.434	2.056	0.7792	1.909	1.903
2.79	3.742	9.568	3.911	3.467	2.039	0.7803	1.911	1.905
2.80	3.686	9.464	3.894	3.500	2.022	0.7814	1.914	1.906
2.81	3.630	9.361	3.877	3.533	2.006	0.7825	1.917	1.908
2.82	3.575	9.260	3.861	3.567	1.990	0.7835	1.919	1.909
2.83	3.521	9.160	3.844	3.601	1.974	0.7846	1.922	1.911
2.84	3.468	9.061	3.827	3.635	1.958	0.7857	1.925	1.912
2.85	3.415	8.963	3.810	3.670	1.942	0.7867	1.927	1.914
2.86	3.364	8.866	3.794	3.705	1.926	0.7878	1.930	1.915
2.87	3.313	8.771	3.778	3.741	1.910	0.7888	1.932	1.917
2.88	3.263	8.676	3.761	3.777	1.894	0.7899	1.935	1.918
2.89	3.214	8.583	3.745	3.813	1.879	0.7909	1.937	1.920
2.90	3.166	8.490	3.729	3.849	1.864	0.7919	1.940	1.921
2.91	3.118	8.399	3.713	3.886	1.848	0.7929	1.942	1.923
2.92	3.072	8.309	3.697	3.923	1.833	0.7939	1.945	1.924
2.93	3.025	8.220	3.681	3.961	1.818	0.7949	1.947	1.925
2.94	2.980	8.132	3.665	3.999	1.803	0.7959	1.950	1.927
2.95	2.936	8.044	3.649	4.037	1.788	0.7969	1.952	1.928
2.96	2.892	7.958	3.634	4.076	1.773	0.7979	1.955	1.930
2.97	2.848	7.873	3.618	4.115	1.759	0.7989	1.957	1.931
2.98	2.806	7.789	3.602	4.154	1.744	0.7998	1.959	1.932
2.99	2.764	7.706	3.587	4.194	1.730	0.8008	1.962	1.934
3.00	2.723	7.624	3.572	4.234	1.715	0.8018	1.964	1.935

NORMAL SHOCK FUNCTIONS FOR SUPERSONIC COMPRESSIBLE FLOW

$$\gamma = 7/5$$

M	P_{21}	T_{21}	ρ_{21}	M_2	P^o_{21}	μ	ν
2.50	7.124 +0	2.137 +0	3.333	0.5130	4.991 −1	23.58	39.12
2.51	7.183	2.147	3.345	0.5120	4.951	23.48	39.35
2.52	7.241	2.157	3.357	0.5111	4.911	23.38	39.59
2.53	7.300	2.167	3.368	0.5102	4.872	23.28	39.82
2.54	7.360	2.177	3.380	0.5092	4.832	23.19	40.05
2.55	7.419	2.187	3.392	0.5083	4.793	23.09	40.28
2.56	7.478	2.197	3.403	0.5074	4.754	22.99	40.51
2.57	7.538	2.208	3.415	0.5065	4.716	22.90	40.73
2.58	7.598	2.218	3.426	0.5056	4.678	22.81	40.96
2.59	7.659	2.228	3.438	0.5048	4.640	22.71	41.19
2.60	7.719	2.238	3.449	0.5039	4.602	22.62	41.41
2.61	7.780	2.249	3.460	0.5030	4.564	22.53	41.64
2.62	7.841	2.259	3.471	0.5022	4.527	22.44	41.86
2.63	7.902	2.269	3.482	0.5013	4.490	22.35	42.08
2.64	7.964	2.280	3.494	0.5005	4.453	22.26	42.30
2.65	8.025	2.290	3.505	0.4997	4.416	22.17	42.53
2.66	8.087	2.300	3.516	0.4988	4.380	22.08	42.75
2.67	8.150	2.311	3.526	0.4980	4.343	22.00	42.97
2.68	8.212	2.322	3.537	0.4972	4.307	21.91	43.18
2.69	8.275	2.332	3.548	0.4964	4.272	21.82	43.40
2.70	8.338	2.343	3.559	0.4956	4.236	21.74	43.62
2.71	8.401	2.353	3.570	0.4949	4.201	21.66	43.83
2.72	8.464	2.364	3.580	0.4941	4.166	21.57	44.05
2.73	8.528	2.375	3.591	0.4933	4.131	21.49	44.26
2.74	8.591	2.386	3.601	0.4926	4.097	21.41	44.48
2.75	8.655	2.396	3.612	0.4918	4.063	21.32	44.69
2.76	8.720	2.407	3.622	0.4911	4.029	21.24	44.90
2.77	8.784	2.418	3.633	0.4903	3.995	21.16	45.11
2.78	8.849	2.429	3.643	0.4896	3.961	21.08	45.32
2.79	8.914	2.440	3.653	0.4889	3.928	21.00	45.53
2.80	8.979	2.451	3.663	0.4882	3.895	20.93	45.74
2.81	9.045	2.462	3.674	0.4875	3.862	20.85	45.95
2.82	9.110	2.473	3.684	0.4868	3.830	20.77	46.16
2.83	9.176	2.484	3.694	0.4861	3.797	20.69	46.36
2.84	9.242	2.495	3.704	0.4854	3.765	20.62	46.57
2.85	9.309	2.507	3.714	0.4847	3.733	20.54	46.78
2.86	9.375	2.518	3.724	0.4840	3.702	20.47	46.98
2.87	9.442	2.529	3.733	0.4834	3.670	20.39	47.18
2.88	9.509	2.540	3.743	0.4827	3.639	20.32	47.38
2.89	9.576	2.552	3.753	0.4820	3.608	20.25	47.59
2.90	9.644	2.563	3.763	0.4814	3.578	20.17	47.79
2.91	9.712	2.574	3.772	0.4807	3.547	20.10	47.99
2.92	9.780	2.586	3.782	0.4801	3.517	20.03	48.19
2.93	9.848	2.597	3.792	0.4795	3.487	19.96	48.39
2.94	9.917	2.609	3.801	0.4788	3.457	19.89	48.58
2.95	9.985	2.620	3.810	0.4782	3.428	19.82	48.78
2.96	1.005 +1	2.632	3.820	0.4776	3.399	19.75	48.98
2.97	1.012	2.644	3.829	0.4770	3.370	19.68	49.17
2.98	1.019	2.655	3.839	0.4764	3.341	19.61	49.37
2.99	1.026	2.667	3.848	0.4758	3.312	19.54	49.56
3.00	1.033	2.679	3.857	0.4752	3.284	19.47	49.75

ISENTROPIC FUNCTIONS FOR SUPERSONIC COMPRESSIBLE FLOW

$\gamma = 7/5$

M	p/p°	T/T°	ρ/ρ°	A/A^*	q/p°	u/u_m	u/a^*	$F(\gamma;M)$
3.00	2.723 −2	7.624 −2	3.572 −1	4.234 +0	1.715 −1	0.8018	1.964	1.935
3.02	2.643	7.462	3.541	4.315	1.687	0.8037	1.969	1.938
3.04	2.565	7.305	3.511	4.398	1.659	0.8055	1.973	1.940
3.06	2.489	7.151	3.481	4.483	1.631	0.8074	1.973	1.943
3.08	2.416	7.000	3.452	4.569	1.604	0.8092	1.982	1.946
3.10	2.345	6.853	3.423	4.657	1.578	0.8110	1.987	1.948
3.12	2.277	6.709	3.394	4.746	1.551	0.8128	1.991	1.951
3.14	2.210	6.569	3.365	4.837	1.525	0.8146	1.995	1.953
3.16	2.146	6.431	3.337	4.930	1.500	0.8163	2.000	1.956
3.18	2.084	6.297	3.309	5.024	1.475	0.8180	2.004	1.958
3.20	2.023	6.166	3.281	5.120	1.450	0.8197	2.008	1.961
3.22	1.965	6.038	3.254	5.218	1.426	0.8214	2.012	1.963
3.24	1.908	5.913	3.227	5.318	1.402	0.8230	2.016	1.966
3.26	1.853	5.791	3.200	5.419	1.378	0.8246	2.020	1.968
3.28	1.800	5.672	3.173	5.523	1.355	0.8262	2.024	1.970
3.30	1.748	5.555	3.147	5.628	1.332	0.8278	2.028	1.973
3.32	1.698	5.441	3.121	5.735	1.310	0.8294	2.032	1.975
3.34	1.650	5.330	3.095	5.844	1.288	0.8310	2.035	1.977
3.36	1.603	5.221	3.070	5.955	1.266	0.8325	2.039	1.979
3.38	1.557	5.114	3.044	6.068	1.245	0.8340	2.043	1.981
3.40	1.513	5.010	3.020	6.183	1.224	0.8355	2.047	1.984
3.42	1.470	4.908	2.995	6.300	1.203	0.8370	2.050	1.986
3.44	1.428	4.809	2.970	6.419	1.183	0.8384	2.054	1.988
3.46	1.388	4.712	2.946	6.540	1.163	0.8399	2.057	1.990
3.48	1.349	4.617	2.922	6.663	1.144	0.8413	2.061	1.992
3.50	1.311	4.524	2.899	6.789	1.124	0.8427	2.064	1.994
3.52	1.275	4.433	2.875	6.916	1.105	0.8441	2.068	1.996
3.54	1.239	4.345	2.852	7.046	1.087	0.8454	2.071	1.998
3.56	1.205	4.258	2.829	7.178	1.069	0.8468	2.074	2.000
3.58	1.171	4.173	2.807	7.312	1.051	0.8481	2.078	2.002
3.60	1.139	4.090	2.784	7.449	1.033	0.8495	2.081	2.004
3.62	1.107	4.009	2.762	7.588	1.016	0.8508	2.084	2.006
3.64	1.077	3.930	2.740	7.729	9.986 −2	0.8521	2.087	2.008
3.66	1.047	3.852	2.718	7.873	9.818	0.8533	2.090	2.010
3.68	1.018	3.776	2.697	8.019	9.653	0.8546	2.093	2.012
3.70	9.905 −3	3.702	2.675	8.168	9.491	0.8558	2.096	2.014
3.72	9.635	3.630	2.654	8.319	9.332	0.8571	2.099	2.015
3.74	9.372	3.559	2.633	8.473	9.176	0.8583	2.102	2.017
3.76	9.118	3.490	2.613	8.629	9.022	0.8595	2.105	2.019
3.78	8.871	3.422	2.592	8.788	8.872	0.8607	2.108	2.021
3.80	8.631	3.355	2.572	8.949	8.723	0.8618	2.111	2.022
3.82	8.398	3.291	2.552	9.113	8.578	0.8630	2.114	2.024
3.84	8.172	3.227	2.532	9.280	8.435	0.8642	2.117	2.026
3.86	7.953	3.165	2.513	9.450	8.294	0.8653	2.120	2.028
3.88	7.740	3.104	2.493	9.622	8.156	0.8664	2.122	2.029
3.90	7.534	3.045	2.474	9.797	8.021	0.8675	2.125	2.031
3.92	7.333	2.987	2.455	9.975	7.887	0.8686	2.128	2.033
3.94	7.139	2.930	2.436	1.016 +1	7.756	0.8697	2.130	2.034
3.96	6.949	2.874	2.418	1.034	7.628	0.8708	2.133	2.036
3.98	6.766	2.820	2.399	1.053	7.502	0.8718	2.136	2.037
4.00	6.588	2.767	2.381	1.072	7.378	0.8729	2.138	2.039

NORMAL SHOCK FUNCTIONS FOR SUPERSONIC COMPRESSIBLE FLOW

$\gamma = 7/5$

M	P_{21}	T_{21}	ρ_{21}	M_2	P^o_{21}	μ	ν
3.00	1.033 +1	2.679 +0	3.857	0.4752	3.284 -1	19.47	49.75
3.02	1.047	2.702	3.875	0.4740	3.228	19.34	50.14
3.04	1.061	2.726	3.893	0.4729	3.173	19.21	50.52
3.06	1.076	2.750	3.911	0.4717	3.118	19.08	50.90
3.08	1.090	2.774	3.929	0.4706	3.065	18.95	51.27
3.10	1.104	2.798	3.946	0.4695	3.013	18.82	51.65
3.12	1.119	2.823	3.964	0.4685	2.961	18.69	52.02
3.14	1.134	2.847	3.981	0.4674	2.910	18.57	52.38
3.16	1.148	2.872	3.998	0.4664	2.860	18.45	52.75
3.18	1.163	2.897	4.015	0.4654	2.811	18.33	53.11
3.20	1.178	2.922	4.031	0.4644	2.763	18.21	53.47
3.22	1.193	2.947	4.048	0.4634	2.715	18.09	53.82
3.24	1.208	2.972	4.064	0.4624	2.668	17.98	54.18
3.26	1.223	2.998	4.080	0.4615	2.623	17.86	54.53
3.28	1.238	3.023	4.096	0.4605	2.577	17.75	54.87
3.30	1.254	3.049	4.112	0.4596	2.533	17.64	55.22
3.32	1.269	3.075	4.127	0.4587	2.490	17.53	55.56
3.34	1.285	3.101	4.143	0.4578	2.447	17.42	55.90
3.36	1.300	3.127	4.158	0.4569	2.405	17.32	56.24
3.38	1.316	3.153	4.173	0.4561	2.363	17.21	56.57
3.40	1.332	3.180	4.188	0.4552	2.323	17.11	56.90
3.42	1.348	3.207	4.203	0.4544	2.283	17.00	57.23
3.44	1.364	3.233	4.218	0.4535	2.243	16.90	57.56
3.46	1.380	3.260	4.232	0.4527	2.205	16.80	57.89
3.48	1.396	3.288	4.247	0.4519	2.167	16.70	58.21
3.50	1.412	3.315	4.261	0.4512	2.130	16.60	58.53
3.52	1.429	3.342	4.275	0.4504	2.093	16.51	58.84
3.54	1.445	3.370	4.289	0.4496	2.057	16.41	59.16
3.56	1.462	3.398	4.302	0.4489	2.022	16.31	59.47
3.58	1.478	3.425	4.316	0.4481	1.987	16.22	59.78
3.60	1.495	3.453	4.330	0.4474	1.953	16.13	60.09
3.62	1.512	3.482	4.343	0.4467	1.920	16.04	60.39
3.64	1.529	3.510	4.356	0.4460	1.887	15.95	60.70
3.66	1.546	3.539	4.369	0.4453	1.855	15.86	61.00
3.68	1.563	3.567	4.382	0.4446	1.823	15.77	61.30
3.70	1.580	3.596	4.395	0.4440	1.792	15.68	61.59
3.72	1.598	3.625	4.407	0.4433	1.762	15.59	61.89
3.74	1.615	3.654	4.420	0.4426	1.732	15.51	62.18
3.76	1.633	3.683	4.432	0.4420	1.702	15.42	62.47
3.78	1.650	3.713	4.445	0.4414	1.673	15.34	62.76
3.80	1.668	3.742	4.457	0.4407	1.645	15.26	63.04
3.82	1.686	3.772	4.469	0.4401	1.617	15.18	63.32
3.84	1.703	3.802	4.481	0.4395	1.590	15.10	63.61
3.86	1.721	3.832	4.492	0.4389	1.563	15.02	63.89
3.88	1.740	3.862	4.504	0.4383	1.536	14.94	64.16
3.90	1.758	3.893	4.515	0.4377	1.510	14.86	64.44
3.92	1.776	3.923	4.527	0.4372	1.485	14.78	64.71
3.94	1.794	3.954	4.538	0.4366	1.460	14.70	64.98
3.96	1.813	3.985	4.549	0.4361	1.436	14.63	65.25
3.98	1.831	4.015	4.560	0.4355	1.411	14.55	65.52
4.00	1.850	4.047	4.571	0.4350	1.388	14.48	65.78

ISENTROPIC FUNCTIONS FOR SUPERSONIC COMPRESSIBLE FLOW

$$\gamma = 7/5$$

M	$p/p°$	$T/T°$	$\rho/\rho°$	A/A^*	$q/p°$	u/u_m	u/a^*	$F(\gamma;M)$
4.00	6.588 −3	2.767 −2	2.381 −1	1.072 +1	7.378 −2	0.8729	2.138	2.039
4.02	6.415	2.714	2.363	1.091	7.256	0.8739	2.141	2.040
4.04	6.246	2.663	2.345	1.111	7.136	0.8749	2.143	2.042
4.06	6.083	2.614	2.328	1.131	7.018	0.8759	2.146	2.044
4.08	5.924	2.565	2.310	1.151	6.903	0.8769	2.148	2.045
4.10	5.770	2.517	2.293	1.171	6.789	0.8779	2.150	2.046
4.12	5.621	2.470	2.276	1.192	6.678	0.8789	2.153	2.048
4.14	5.475	2.424	2.259	1.213	6.568	0.8799	2.155	2.049
4.16	5.334	2.379	2.242	1.235	6.461	0.8808	2.158	2.051
4.18	5.197	2.335	2.225	1.257	6.355	0.8818	2.160	2.052
4.20	5.063	2.292	2.209	1.279	6.252	0.8827	2.162	2.054
4.22	4.934	2.250	2.192	1.302	6.150	0.8836	2.164	2.055
4.24	4.808	2.209	2.176	1.324	6.050	0.8845	2.167	2.056
4.26	4.685	2.169	2.160	1.348	5.951	0.8854	2.169	2.058
4.28	4.566	2.129	2.144	1.371	5.855	0.8863	2.171	2.059
4.30	4.450	2.091	2.129	1.395	5.760	0.8872	2.173	2.060
4.32	4.338	2.053	2.113	1.420	5.667	0.8881	2.175	2.062
4.34	4.229	2.016	2.098	1.444	5.575	0.8889	2.178	2.063
4.36	4.122	1.979	2.083	1.469	5.485	0.8898	2.180	2.064
4.38	4.019	1.944	2.068	1.495	5.397	0.8906	2.182	2.066
4.40	3.919	1.909	2.053	1.521	5.310	0.8915	2.184	2.067
4.42	3.821	1.875	2.038	1.547	5.225	0.8923	2.186	2.068
4.44	3.726	1.841	2.023	1.574	5.141	0.8931	2.188	2.069
4.46	3.633	1.809	2.009	1.601	5.059	0.8939	2.190	2.071
4.48	3.544	1.777	1.995	1.628	4.978	0.8947	2.192	2.072
4.50	3.456	1.745	1.980	1.656	4.899	0.8955	2.194	2.073
4.52	3.371	1.714	1.966	1.684	4.821	0.8963	2.196	2.074
4.54	3.288	1.684	1.952	1.713	4.744	0.8971	2.197	2.075
4.56	3.208	1.655	1.939	1.742	4.669	0.8979	2.199	2.077
4.58	3.129	1.626	1.925	1.772	4.595	0.8986	2.201	2.078
4.60	3.053	1.597	1.911	1.801	4.522	0.8994	2.203	2.079
4.62	2.979	1.570	1.898	1.832	4.451	0.9001	2.205	2.080
4.64	2.907	1.542	1.885	1.863	4.381	0.9008	2.207	2.081
4.66	2.837	1.516	1.872	1.894	4.312	0.9016	2.208	2.082
4.68	2.768	1.489	1.859	1.926	4.244	0.9023	2.210	2.083
4.70	2.702	1.464	1.846	1.958	4.178	0.9030	2.212	2.084
4.72	2.637	1.439	1.833	1.991	4.112	0.9037	2.214	2.086
4.74	2.574	1.414	1.820	2.024	4.048	0.9044	2.215	2.087
4.76	2.513	1.390	1.808	2.057	3.985	0.9051	2.217	2.088
4.78	2.453	1.366	1.796	2.091	3.923	0.9058	2.219	2.089
4.80	2.395	1.343	1.783	2.126	3.862	0.9065	2.220	2.090
4.82	2.338	1.320	1.771	2.161	3.802	0.9071	2.222	2.091
4.84	2.283	1.298	1.759	2.197	3.743	0.9078	2.224	2.092
4.86	2.229	1.276	1.747	2.233	3.686	0.9085	2.225	2.093
4.88	2.177	1.255	1.735	2.269	3.629	0.9091	2.227	2.094
4.90	2.126	1.233	1.724	2.306	3.573	0.9097	2.228	2.095
4.92	2.076	1.213	1.712	2.344	3.518	0.9104	2.230	2.096
4.94	2.028	1.193	1.701	2.382	3.464	0.9110	2.232	2.097
4.96	1.981	1.173	1.689	2.421	3.411	0.9116	2.233	2.098
4.98	1.935	1.153	1.678	2.460	3.359	0.9123	2.235	2.099
5.00	1.890	1.134	1.667	2.500	3.308	0.9129	2.236	2.100

NORMAL SHOCK FUNCTIONS FOR SUPERSONIC COMPRESSIBLE FLOW

$$\gamma = 7/5$$

M	p_{21}	T_{21}	ρ_{21}	M_2	p^o_{21}	μ	ν
4.00	1.850 +1	4.047 +0	4.571	0.4350	1.388 −1	14.48	65.78
4.02	1.869	4.078	4.582	0.4344	1.365	14.40	66.05
4.04	1.887	4.109	4.593	0.4339	1.342	14.33	66.31
4.06	1.906	4.141	4.603	0.4334	1.319	14.26	66.57
4.08	1.925	4.173	4.614	0.4329	1.297	14.19	66.82
4.10	1.944	4.205	4.624	0.4324	1.276	14.12	67.08
4.12	1.964	4.237	4.635	0.4319	1.255	14.05	67.33
4.14	1.983	4.269	4.645	0.4314	1.234	13.98	67.59
4.16	2.002	4.301	4.655	0.4309	1.213	13.91	67.84
4.18	2.022	4.334	4.665	0.4304	1.193	13.84	68.08
4.20	2.041	4.366	4.675	0.4299	1.174	13.77	68.33
4.22	2.061	4.399	4.685	0.4295	1.154	13.71	68.58
4.24	2.081	4.432	4.694	0.4290	1.135	13.64	68.82
4.26	2.100	4.465	4.704	0.4286	1.116	13.58	69.06
4.28	2.120	4.498	4.713	0.4281	1.098	13.51	69.30
4.30	2.140	4.532	4.723	0.4277	1.080	13.45	69.54
4.32	2.160	4.565	4.732	0.4272	1.063	13.38	69.77
4.34	2.181	4.599	4.741	0.4268	1.045	13.32	70.01
4.36	2.201	4.633	4.750	0.4264	1.028	13.26	70.24
4.38	2.221	4.667	4.759	0.4260	1.011	13.20	70.47
4.40	2.242	4.701	4.768	0.4255	9.950 −2	13.14	70.70
4.42	2.262	4.736	4.777	0.4251	9.788	13.08	70.93
4.44	2.283	4.770	4.786	0.4247	9.630	13.02	71.16
4.46	2.304	4.805	4.795	0.4243	9.474	12.96	71.38
4.48	2.325	4.840	4.803	0.4239	9.321	12.90	71.61
4.50	2.346	4.875	4.812	0.4236	9.171	12.84	71.83
4.52	2.367	4.910	4.820	0.4232	9.024	12.78	72.05
4.54	2.388	4.945	4.829	0.4228	8.879	12.73	72.27
4.56	2.409	4.981	4.837	0.4224	8.737	12.67	72.49
4.58	2.430	5.016	4.845	0.4220	8.597	12.61	72.70
4.60	2.452	5.052	4.853	0.4217	8.460	12.56	72.92
4.62	2.473	5.088	4.861	0.4213	8.325	12.50	73.13
4.64	2.495	5.124	4.869	0.4210	8.193	12.45	73.34
4.66	2.517	5.160	4.877	0.4206	8.063	12.39	73.55
4.68	2.538	5.197	4.885	0.4203	7.935	12.34	73.76
4.70	2.560	5.233	4.893	0.4199	7.810	12.29	73.97
4.72	2.582	5.270	4.900	0.4196	7.687	12.23	74.17
4.74	2.604	5.307	4.908	0.4192	7.566	12.18	74.38
4.76	2.627	5.344	4.915	0.4189	7.447	12.13	74.58
4.78	2.649	5.381	4.923	0.4186	7.330	12.08	74.78
4.80	2.671	5.418	4.930	0.4183	7.215	12.03	74.98
4.82	2.694	5.456	4.937	0.4179	7.102	11.97	75.18
4.84	2.716	5.493	4.945	0.4176	6.992	11.92	75.38
4.86	2.739	5.531	4.952	0.4173	6.883	11.87	75.58
4.88	2.761	5.569	4.959	0.4170	6.776	11.83	75.77
4.90	2.784	5.607	4.966	0.4167	6.671	11.78	75.97
4.92	2.807	5.645	4.973	0.4164	6.568	11.73	76.16
4.94	2.830	5.684	4.980	0.4161	6.466	11.68	76.35
4.96	2.853	5.722	4.986	0.4158	6.367	11.63	76.54
4.98	2.876	5.761	4.993	0.4155	6.269	11.58	76.73
5.00	2.900	5.800	5.000	0.4152	6.173	11.54	76.92

ISENTROPIC FUNCTIONS FOR SUPERSONIC COMPRESSIBLE FLOW

$\gamma = 7/5$

M	p/p°	T/T°	ρ/ρ°	A/A^*	q/p°	u/u_m	u/a^*	$F(\gamma;M)$
5.00	1.890 −3	1.134 −2	1.667 −1	2.500 +1	3.308 −2	0.9129	2.236	2.100
5.05	1.784	1.088	1.639	2.601	3.184	0.9144	2.240	2.102
5.10	1.684	1.044	1.612	2.707	3.065	0.9158	2.243	2.104
5.15	1.590	1.002	1.586	2.815	2.951	0.9173	2.247	2.106
5.20	1.502	9.622 −3	1.561	2.928	2.842	0.9187	2.250	2.108
5.25	1.419	9.241	1.536	3.044	2.738	0.9200	2.254	2.111
5.30	1.341	8.877	1.511	3.164	2.637	0.9214	2.257	2.113
5.35	1.269	8.529	1.487	3.289	2.541	0.9226	2.260	2.115
5.40	1.200	8.198	1.464	3.417	2.449	0.9239	2.263	2.117
5.45	1.136	7.881	1.441	3.549	2.361	0.9252	2.266	2.118
5.50	1.075	7.579	1.419	3.686	2.276	0.9264	2.269	2.120
5.55	1.018	7.290	1.397	3.828	2.195	0.9275	2.272	2.122
5.60	9.645 −4	7.014	1.375	3.973	2.117	0.9287	2.275	2.124
5.65	9.141	6.749	1.354	4.124	2.042	0.9298	2.278	2.126
5.70	8.665	6.497	1.334	4.279	1.971	0.9309	2.280	2.127
5.75	8.218	6.255	1.314	4.439	1.902	0.9320	2.283	2.129
5.80	7.796	6.024	1.294	4.604	1.836	0.9331	2.285	2.131
5.85	7.398	5.803	1.275	4.775	1.772	0.9341	2.288	2.132
5.90	7.023	5.591	1.256	4.950	1.711	0.9351	2.291	2.134
5.95	6.669	5.389	1.238	5.131	1.653	0.9361	2.293	2.135
6.00	6.335	5.194	1.220	5.317	1.596	0.9370	2.295	2.137
6.05	6.020	5.008	1.202	5.509	1.542	0.9380	2.298	2.138
6.10	5.722	4.830	1.185	5.707	1.490	0.9389	2.300	2.140
6.15	5.440	4.659	1.168	5.910	1.440	0.9398	2.302	2.141
6.20	5.175	4.495	1.151	6.120	1.392	0.9407	2.304	2.143
6.25	4.923	4.338	1.135	6.336	1.346	0.9416	2.306	2.144
6.30	4.685	4.188	1.119	6.558	1.302	0.9424	2.308	2.145
6.35	4.461	4.043	1.103	6.787	1.259	0.9432	2.310	2.146
6.40	4.248	3.904	1.088	7.022	1.218	0.9440	2.312	2.148
6.45	4.046	3.771	1.073	7.264	1.178	0.9448	2.314	2.149
6.50	3.856	3.643	1.058	7.512	1.140	0.9456	2.316	2.150
6.55	3.675	3.520	1.044	7.768	1.104	0.9464	2.318	2.151
6.60	3.504	3.402	1.030	8.031	1.068	0.9471	2.320	2.153
6.65	3.341	3.289	1.016	8.301	1.034	0.9478	2.322	2.154
6.70	3.187	3.180	1.002	8.579	1.002	0.9486	2.324	2.155
6.75	3.042	3.076	9.889 −2	8.865	9.700 −3	0.9493	2.325	2.156
6.80	2.903	2.975	9.759	9.158	9.396	0.9500	2.327	2.157
6.85	2.772	2.878	9.630	9.459	9.103	0.9506	2.329	2.158
6.90	2.647	2.785	9.504	9.769	8.821	0.9513	2.330	2.159
6.95	2.529	2.695	9.381	1.009 +2	8.549	0.9519	2.332	2.160
7.00	2.416	2.609	9.260	1.041	8.287	0.9526	2.333	2.161
7.05	2.309	2.526	9.141	1.075	8.034	0.9532	2.335	2.162
7.10	2.208	2.446	9.024	1.109	7.790	0.9538	2.336	2.163
7.15	2.111	2.369	8.910	1.144	7.554	0.9544	2.338	2.164
7.20	2.019	2.295	8.797	1.181	7.327	0.9550	2.339	2.165
7.25	1.932	2.224	8.687	1.218	7.108	0.9556	2.341	2.166
7.30	1.849	2.155	8.578	1.256	6.896	0.9561	2.342	2.167
7.35	1.770	2.089	8.472	1.295	6.692	0.9567	2.343	2.168
7.40	1.695	2.025	8.367	1.335	6.495	0.9572	2.345	2.168
7.45	1.623	1.964	8.265	1.376	6.305	0.9578	2.346	2.169
7.50	1.555	1.904	8.164	1.418	6.121	0.9583	2.347	2.170

NORMAL SHOCK FUNCTIONS FOR SUPERSONIC COMPRESSIBLE FLOW

$$\gamma = 7/5$$

M	p_{21}	T_{21}	ρ_{21}	M_2	p^o_{21}	μ	ν
5.00	2.900 +1	5.800 +0	5.000	0.4152	6.173 -2	11.54	76.92
5.05	2.958	5.897	5.016	0.4145	5.939	11.42	77.38
5.10	3.018	5.996	5.033	0.4138	5.716	11.31	77.84
5.15	3.077	6.096	5.048	0.4132	5.502	11.20	78.29
5.20	3.138	6.197	5.064	0.4125	5.297	11.09	78.73
5.25	3.199	6.298	5.079	0.4119	5.101	10.98	79.17
5.30	3.260	6.401	5.093	0.4113	4.913	10.88	79.59
5.35	3.322	6.505	5.108	0.4107	4.733	10.77	80.02
5.40	3.385	6.609	5.122	0.4101	4.561	10.67	80.43
5.45	3.448	6.715	5.135	0.4095	4.395	10.57	80.84
5.50	3.512	6.821	5.149	0.4090	4.237	10.48	81.24
5.55	3.577	6.929	5.162	0.4084	4.085	10.38	81.64
5.60	3.642	7.037	5.175	0.4079	3.939	10.29	82.03
5.65	3.707	7.147	5.187	0.4074	3.799	10.19	82.41
5.70	3.774	7.257	5.200	0.4069	3.665	10.10	82.79
5.75	3.840	7.369	5.212	0.4064	3.536	10.02	83.17
5.80	3.908	7.481	5.224	0.4059	3.413	9.929	83.54
5.85	3.976	7.594	5.235	0.4055	3.294	9.843	83.90
5.90	4.044	7.709	5.246	0.4050	3.180	9.759	84.25
5.95	4.113	7.824	5.257	0.4046	3.071	9.676	84.61
6.00	4.183	7.940	5.268	0.4042	2.966	9.594	84.95
6.05	4.253	8.057	5.279	0.4037	2.865	9.514	85.30
6.10	4.324	8.175	5.289	0.4033	2.768	9.436	85.63
6.15	4.396	8.295	5.299	0.4029	2.674	9.358	85.97
6.20	4.468	8.415	5.309	0.4025	2.585	9.282	86.29
6.25	4.540	8.536	5.319	0.4022	2.499	9.207	86.62
6.30	4.614	8.658	5.329	0.4018	2.416	9.134	86.94
6.35	4.687	8.781	5.338	0.4014	2.336	9.061	87.25
6.40	4.762	8.905	5.347	0.4011	2.260	8.990	87.56
6.45	4.837	9.030	5.356	0.4007	2.186	8.919	87.87
6.50	4.912	9.156	5.365	0.4004	2.115	8.850	88.17
6.55	4.988	9.283	5.374	0.4001	2.047	8.782	88.46
6.60	5.065	9.411	5.382	0.3997	1.981	8.715	88.76
6.65	5.142	9.540	5.390	0.3994	1.918	8.649	89.05
6.70	5.220	9.669	5.399	0.3991	1.857	8.584	89.33
6.75	5.299	9.800	5.407	0.3988	1.798	8.520	89.62
6.80	5.378	9.932	5.414	0.3985	1.742	8.457	89.89
6.85	5.457	1.006 +1	5.422	0.3982	1.687	8.395	90.17
6.90	5.537	1.020	5.430	0.3979	1.635	8.333	90.44
6.95	5.618	1.033	5.437	0.3976	1.584	8.273	90.71
7.00	5.700	1.047	5.444	0.3974	1.535	8.213	90.97
7.05	5.782	1.061	5.452	0.3971	1.488	8.155	91.23
7.10	5.864	1.074	5.459	0.3968	1.443	8.097	91.49
7.15	5.947	1.088	5.465	0.3966	1.399	8.040	91.74
7.20	6.031	1.102	5.472	0.3963	1.357	7.984	92.00
7.25	6.115	1.116	5.479	0.3961	1.316	7.928	92.24
7.30	6.200	1.130	5.485	0.3958	1.277	7.874	92.49
7.35	6.286	1.145	5.492	0.3956	1.239	7.820	92.73
7.40	6.372	1.159	5.498	0.3954	1.203	7.767	92.97
7.45	6.458	1.173	5.504	0.3951	1.167	7.714	93.21
7.50	6.545	1.188	5.510	0.3949	1.133	7.662	93.44

ISENTROPIC FUNCTIONS FOR SUPERSONIC COMPRESSIBLE FLOW

$\gamma = 7/5$

M	$p/p°$	$T/T°$	$\rho/\rho°$	$A/A*$	$q/p°$	u/u_m	$u/a*$	$F(\gamma;M)$
7.50	1.555 −4	1.904 −3	8.164 −2	1.418 +2	6.121 −3	0.9583	2.347	2.170
7.55	1.490	1.847	8.065	1.461	5.943	0.9588	2.349	2.171
7.60	1.428	1.792	7.967	1.506	5.772	0.9593	2.350	2.172
7.65	1.368	1.738	7.872	1.551	5.606	0.9598	2.351	2.172
7.70	1.312	1.687	7.778	1.597	5.445	0.9603	2.352	2.173
7.75	1.258	1.637	7.685	1.645	5.291	0.9608	2.354	2.174
7.80	1.207	1.590	7.595	1.694	5.141	0.9613	2.355	2.175
7.85	1.158	1.543	7.505	1.744	4.996	0.9617	2.356	2.175
7.90	1.112	1.499	7.418	1.795	4.856	0.9622	2.357	2.176
7.95	1.067	1.455	7.332	1.847	4.720	0.9626	2.358	2.177
8.00	1.024	1.414	7.247	1.901	4.589	0.9631	2.359	2.178
8.05	9.839 −5	1.373	7.163	1.956	4.463	0.9635	2.360	2.178
8.10	9.450	1.335	7.082	2.012	4.340	0.9639	2.361	2.179
8.15	9.079	1.297	7.001	2.069	4.221	0.9644	2.362	2.180
8.20	8.725	1.260	6.922	2.128	4.106	0.9648	2.363	2.180
8.25	8.386	1.225	6.844	2.188	3.995	0.9652	2.364	2.181
8.30	8.062	1.191	6.767	2.250	3.887	0.9656	2.365	2.181
8.35	7.752	1.158	6.692	2.313	3.783	0.9660	2.366	2.182
8.40	7.455	1.127	6.618	2.377	3.682	0.9663	2.367	2.183
8.45	7.171	1.096	6.545	2.443	3.584	0.9667	2.368	2.183
8.50	6.900	1.066	6.473	2.511	3.489	0.9671	2.369	2.184
8.55	6.640	1.037	6.402	2.579	3.397	0.9675	2.370	2.184
8.60	6.391	1.009	6.333	2.650	3.308	0.9678	2.371	2.185
8.65	6.152	9.821 −4	6.264	2.722	3.222	0.9682	2.372	2.186
8.70	5.924	9.560	6.197	2.795	3.139	0.9685	2.372	2.186
8.75	5.705	9.306	6.131	2.870	3.057	0.9689	2.373	2.187
8.80	5.495	9.060	6.065	2.947	2.979	0.9692	2.374	2.187
8.85	5.294	8.822	6.001	3.026	2.902	0.9695	2.375	2.188
8.90	5.102	8.592	5.938	3.106	2.829	0.9699	2.376	2.188
8.95	4.917	8.368	5.876	3.188	2.757	0.9702	2.376	2.189
9.00	4.740	8.152	5.814	3.271	2.687	0.9705	2.377	2.189
9.05	4.569	7.942	5.754	3.357	2.620	0.9708	2.378	2.190
9.10	4.406	7.738	5.694	3.444	2.554	0.9711	2.379	2.190
9.15	4.250	7.540	5.636	3.533	2.490	0.9714	2.380	2.191
9.20	4.099	7.349	5.578	3.624	2.429	0.9717	2.380	2.191
9.25	3.955	7.163	5.521	3.717	2.369	0.9720	2.381	2.192
9.30	3.817	6.983	5.465	3.812	2.311	0.9723	2.382	2.192
9.35	3.683	6.808	5.410	3.908	2.254	0.9726	2.382	2.193
9.40	3.556	6.639	5.356	4.007	2.199	0.9729	2.383	2.193
9.45	3.433	6.474	5.302	4.108	2.146	0.9731	2.384	2.193
9.50	3.315	6.314	5.250	4.211	2.094	0.9734	2.384	2.194
9.55	3.201	6.159	5.198	4.316	2.044	0.9737	2.385	2.194
9.60	3.092	6.008	5.146	4.423	1.995	0.9739	2.386	2.195
9.65	2.987	5.862	5.096	4.532	1.947	0.9742	2.386	2.195
9.70	2.887	5.720	5.046	4.643	1.901	0.9744	2.387	2.195
9.75	2.790	5.582	4.997	4.757	1.856	0.9747	2.388	2.196
9.80	2.696	5.448	4.949	4.872	1.812	0.9749	2.388	2.196
9.85	2.606	5.318	4.901	4.990	1.770	0.9752	2.389	2.197
9.90	2.520	5.191	4.854	5.111	1.729	0.9754	2.389	2.197
9.95	2.437	5.068	4.808	5.234	1.689	0.9757	2.390	2.197
10.00	2.357	4.949	4.762	5.359	1.650	0.9759	2.391	2.198

NORMAL SHOCK FUNCTIONS FOR SUPERSONIC COMPRESSIBLE FLOW

$$\gamma = 7/5$$

M	P_{21}	T_{21}	ρ_{21}	M_2	P^o_{21}	μ	ν
7.50	6.545 +1	1.188 +1	5.510	0.3949	1.133 -2	7.662	93.44
7.55	6.633	1.203	5.516	0.3947	1.100	7.611	93.67
7.60	6.722	1.217	5.522	0.3945	1.068	7.561	93.90
7.65	6.811	1.232	5.528	0.3943	1.037	7.511	94.12
7.70	6.900	1.247	5.533	0.3941	1.008	7.462	94.34
7.75	6.990	1.262	5.539	0.3939	9.789 -3	7.414	94.56
7.80	7.081	1.277	5.544	0.3937	9.512	7.366	94.78
7.85	7.172	1.292	5.550	0.3935	9.243	7.319	94.99
7.90	7.264	1.308	5.555	0.3933	8.983	7.272	95.21
7.95	7.356	1.323	5.560	0.3931	8.732	7.226	95.42
8.00	7.450	1.339	5.565	0.3929	8.489	7.181	95.62
8.05	7.543	1.354	5.570	0.3927	8.254	7.136	95.83
8.10	7.637	1.370	5.575	0.3925	8.026	7.092	96.03
8.15	7.732	1.386	5.580	0.3924	7.806	7.048	96.23
8.20	7.828	1.402	5.585	0.3922	7.593	7.005	96.43
8.25	7.923	1.418	5.589	0.3920	7.387	6.962	96.62
8.30	8.020	1.434	5.594	0.3918	7.188	6.920	96.82
8.35	8.117	1.450	5.598	0.3917	6.994	6.878	97.01
8.40	8.215	1.466	5.603	0.3915	6.807	6.837	97.20
8.45	8.313	1.483	5.607	0.3914	6.626	6.797	97.39
8.50	8.412	1.499	5.612	0.3912	6.450	6.757	97.57
8.55	8.511	1.516	5.616	0.3911	6.280	6.717	97.75
8.60	8.611	1.532	5.620	0.3909	6.115	6.678	97.94
8.65	8.712	1.549	5.624	0.3908	5.955	6.639	98.11
8.70	8.813	1.566	5.628	0.3906	5.800	6.601	98.29
8.75	8.915	1.583	5.632	0.3905	5.650	6.563	98.47
8.80	9.017	1.600	5.636	0.3903	5.504	6.525	98.64
8.85	9.120	1.617	5.640	0.3902	5.363	6.488	98.81
8.90	9.224	1.634	5.644	0.3901	5.226	6.452	98.98
8.95	9.328	1.652	5.647	0.3899	5.093	6.415	99.15
9.00	9.433	1.669	5.651	0.3898	4.965	6.380	99.32
9.05	9.538	1.687	5.655	0.3897	4.839	6.344	99.48
9.10	9.644	1.704	5.658	0.3895	4.718	6.309	99.64
9.15	9.750	1.722	5.662	0.3894	4.600	6.275	99.81
9.20	9.857	1.740	5.665	0.3893	4.486	6.240	99.97
9.25	9.965	1.758	5.669	0.3892	4.375	6.206	100.1
9.30	1.007 +2	1.776	5.672	0.3891	4.267	6.173	100.38
9.35	1.018	1.794	5.675	0.3889	4.163	6.140	100.4
9.40	1.029	1.812	5.679	0.3888	4.061	6.107	100.7
9.45	1.040	1.831	5.682	0.3887	3.963	6.075	100.7
9.50	1.051	1.849	5.685	0.3886	3.867	6.043	100.9
9.55	1.062	1.868	5.688	0.3885	3.774	6.011	101.0
9.60	1.073	1.886	5.691	0.3884	3.683	5.979	101.2
9.65	1.085	1.905	5.694	0.3883	3.595	5.948	101.3
9.70	1.096	1.924	5.697	0.3882	3.510	5.917	101.5
9.75	1.107	1.943	5.700	0.3881	3.427	5.887	101.6
9.80	1.119	1.962	5.703	0.3880	3.346	5.857	101.8
9.85	1.130	1.981	5.706	0.3879	3.268	5.827	101.9
9.90	1.142	2.000	5.709	0.3878	3.192	5.797	102.0
9.95	1.153	2.019	5.712	0.3877	3.117	5.768	102.2
10.00	1.165	2.039	5.714	0.3876	3.045	5.739	102.3

ISENTROPIC FUNCTIONS FOR SUPERSONIC COMPRESSIBLE FLOW

$$\gamma = 7/5$$

M	$p/p°$	$T/T°$	$\rho/\rho°$	A/A^*	$q/p°$	u/u_m	u/a^*	$F(\gamma;M)$
10.0	2.357 −5	4.949 −4	4.762 −2	5.359 +2	1.650 −3	0.9759	2.391	2.198
10.1	2.205	4.720	4.673	5.616	1.575	0.9764	2.392	2.198
10.2	2.065	4.503	4.586	5.884	1.504	0.9768	2.393	2.199
10.3	1.935	4.298	4.501	6.161	1.437	0.9772	2.394	2.200
10.4	1.814	4.104	4.419	6.450	1.373	0.9777	2.395	2.201
10.5	1.701	3.921	4.339	6.749	1.313	0.9781	2.396	2.201
10.6	1.596	3.747	4.261	7.059	1.256	0.9785	2.397	2.202
10.7	1.499	3.582	4.185	7.381	1.201	0.9789	2.398	2.202
10.8	1.408	3.426	4.111	7.714	1.150	0.9792	2.399	2.203
10.9	1.324	3.278	4.039	8.060	1.101	0.9796	2.400	2.204
11.0	1.245	3.137	3.968	8.418	1.054	0.9800	2.400	2.204
11.1	1.172	3.004	3.900	8.789	1.010	0.9803	2.401	2.205
11.2	1.103	2.877	3.833	9.173	9.684 −4	0.9806	2.402	2.205
11.3	1.039	2.757	3.768	9.570	9.285	0.9810	2.403	2.206
11.4	9.789 −6	2.642	3.705	9.982	8.905	0.9813	2.404	2.206
11.5	9.229	2.533	3.643	1.041 +3	8.544	0.9816	2.405	2.207
11.6	8.706	2.430	3.583	1.085	8.200	0.9819	2.405	2.207
11.7	8.216	2.331	3.524	1.130	7.872	0.9822	2.406	2.208
11.8	7.757	2.237	3.467	1.177	7.560	0.9825	2.407	2.208
11.9	7.326	2.148	3.411	1.226	7.262	0.9828	2.407	2.209
12.0	6.923	2.063	3.356	1.276	6.978	0.9831	2.408	2.209
12.1	6.545	1.982	3.302	1.328	6.708	0.9833	2.409	2.210
12.2	6.190	1.905	3.250	1.381	6.449	0.9836	2.409	2.210
12.3	5.857	1.831	3.199	1.437	6.203	0.9839	2.410	2.210
12.4	5.545	1.760	3.150	1.494	5.967	0.9841	2.411	2.211
12.5	5.251	1.693	3.101	1.553	5.743	0.9844	2.411	2.211
12.6	4.974	1.629	3.053	1.613	5.528	0.9846	2.412	2.212
12.7	4.714	1.568	3.007	1.676	5.323	0.9849	2.412	2.212
12.8	4.470	1.509	2.962	1.741	5.126	0.9851	2.413	2.212
12.9	4.240	1.453	2.917	1.807	4.938	0.9853	2.414	2.213
13.0	4.023	1.400	2.874	1.876	4.759	0.9855	2.414	2.213
13.1	3.819	1.349	2.831	1.947	4.587	0.9857	2.415	2.213
13.2	3.626	1.300	2.790	2.019	4.422	0.9860	2.415	2.214
13.3	3.445	1.253	2.749	2.094	4.265	0.9862	2.416	2.214
13.4	3.273	1.208	2.709	2.172	4.114	0.9864	2.416	2.214
13.5	3.112	1.165	2.670	2.251	3.969	0.9866	2.417	2.215
13.6	2.959	1.124	2.632	2.333	3.831	0.9868	2.417	2.215
13.7	2.815	1.085	2.595	2.417	3.696	0.9869	2.418	2.215
13.8	2.679	1.047	2.558	2.504	3.571	0.9871	2.418	2.216
13.9	2.550	1.011	2.523	2.593	3.448	0.9873	2.418	2.216
14.0	2.428	9.761 −5	2.488	2.685	3.331	0.9875	2.419	2.216
14.1	2.313	9.428	2.453	2.779	3.219	0.9877	2.419	2.216
14.2	2.204	9.108	2.420	2.876	3.111	0.9878	2.420	2.217
14.3	2.101	8.802	2.387	2.976	3.007	0.9880	2.420	2.217
14.4	2.003	8.507	2.355	3.079	2.907	0.9882	2.421	2.217
14.5	1.911	8.225	2.323	3.184	2.812	0.9883	2.421	2.217
14.6	1.823	7.953	2.292	3.292	2.720	0.9885	2.421	2.218
14.7	1.740	7.692	2.262	3.403	2.631	0.9886	2.422	2.218
14.8	1.661	7.441	2.232	3.517	2.546	0.9888	2.422	2.218
14.9	1.586	7.200	2.203	3.635	2.465	0.9889	2.422	2.218
15.0	1.515	6.969	2.174	3.755	2.386	0.9891	2.423	2.219

NORMAL SHOCK FUNCTIONS FOR SUPERSONIC COMPRESSIBLE FLOW

$$\gamma = 7/5$$

M	P_{21}	T_{21}	ρ_{21}	M_2	P^o_{21}	μ	ν
10.00	1.165 +2	2.039 +1	5.714	0.3876	3.045 −3	5.739	102.3
10.10	1.188	2.078	5.720	0.3874	2.907	5.682	102.6
10.20	1.212	2.117	5.725	0.3872	2.776	5.626	102.8
10.30	1.236	2.157	5.730	0.3870	2.652	5.572	103.1
10.40	1.260	2.197	5.735	0.3869	2.534	5.518	103.4
10.50	1.285	2.238	5.740	0.3867	2.422	5.465	103.6
10.60	1.309	2.279	5.744	0.3865	2.317	5.413	103.9
10.70	1.334	2.320	5.749	0.3864	2.217	5.363	104.1
10.80	1.359	2.362	5.753	0.3862	2.121	5.313	104.3
10.90	1.384	2.404	5.758	0.3861	2.031	5.264	104.6
11.00	1.410	2.447	5.762	0.3859	1.945	5.216	104.8
11.10	1.436	2.490	5.766	0.3858	1.864	5.169	105.0
11.20	1.462	2.533	5.770	0.3856	1.786	5.123	105.2
11.30	1.488	2.577	5.774	0.3855	1.713	5.077	105.5
11.40	1.514	2.621	5.778	0.3854	1.642	5.033	105.7
11.50	1.541	2.666	5.781	0.3853	1.576	4.989	105.9
11.60	1.568	2.711	5.785	0.3851	1.512	4.946	106.1
11.70	1.595	2.756	5.789	0.3850	1.452	4.903	106.3
11.80	1.623	2.802	5.792	0.3849	1.394	4.862	106.5
11.90	1.650	2.848	5.795	0.3848	1.339	4.821	106.7
12.00	1.678	2.894	5.799	0.3847	1.287	4.780	106.9
12.10	1.706	2.941	5.802	0.3846	1.237	4.741	107.1
12.20	1.735	2.988	5.805	0.3844	1.189	4.702	107.3
12.30	1.763	3.036	5.808	0.3843	1.144	4.663	107.4
12.40	1.792	3.084	5.811	0.3842	1.100	4.626	107.6
12.50	1.821	3.132	5.814	0.3841	1.059	4.589	107.8
12.60	1.850	3.181	5.817	0.3840	1.019	4.552	108.0
12.70	1.880	3.230	5.820	0.3839	9.812 −4	4.516	108.2
12.80	1.910	3.280	5.822	0.3839	9.450	4.481	108.3
12.90	1.940	3.330	5.825	0.3838	9.103	4.446	108.5
13.00	1.970	3.380	5.828	0.3837	8.772	4.412	108.7
13.10	2.000	3.431	5.830	0.3836	8.455	4.378	108.8
13.20	2.031	3.482	5.833	0.3835	8.151	4.345	109.0
13.30	2.062	3.534	5.835	0.3834	7.861	4.312	109.1
13.40	2.093	3.586	5.837	0.3833	7.582	4.280	109.4
13.50	2.124	3.638	5.840	0.3833	7.316	4.248	109.4
13.60	2.156	3.691	5.842	0.3832	7.060	4.217	109.6
13.70	2.188	3.744	5.844	0.3831	6.815	4.186	109.7
13.80	2.220	3.797	5.846	0.3830	6.580	4.156	109.9
13.90	2.252	3.851	5.849	0.3830	6.355	4.126	110.0
14.00	2.285	3.905	5.851	0.3829	6.139	4.096	110.2
14.10	2.318	3.960	5.853	0.3828	5.931	4.067	110.3
14.20	2.351	4.015	5.855	0.3828	5.732	4.038	110.5
14.30	2.384	4.070	5.857	0.3827	5.541	4.010	110.6
14.40	2.417	4.126	5.859	0.3826	5.357	3.982	110.7
14.50	2.451	4.182	5.861	0.3826	5.181	3.955	110.9
14.60	2.485	4.239	5.862	0.3825	5.011	3.928	111.0
14.70	2.519	4.296	5.864	0.3824	4.848	3.901	111.1
14.80	2.554	4.353	5.866	0.3824	4.691	3.874	111.3
14.90	2.588	4.411	5.868	0.3823	4.541	3.848	111.4
15.00	2.623	4.469	5.870	0.3823	4.396	3.823	111.5

ISENTROPIC FUNCTIONS FOR SUPERSONIC COMPRESSIBLE FLOW

$$\gamma = 7/5$$

M	$p/p°$	$T/T°$	$\rho/\rho°$	A/A^*	$q/p°$	u/u_m	u/a^*	$F(\gamma;M)$
15.0	1.515 −6	6.969 −5	2.174 −2	3.755 +3	2.386 −4	0.9891	2.423	2.219
15.1	1.448	6.746	2.146	3.878	2.310	0.9892	2.423	2.219
15.2	1.384	6.531	2.118	4.005	2.238	0.9894	2.423	2.219
15.3	1.323	6.325	2.091	4.135	2.167	0.9895	2.424	2.219
15.4	1.265	6.126	2.065	4.269	2.100	0.9896	2.424	2.220
15.5	1.210	5.935	2.039	4.406	2.035	0.9898	2.424	2.220
15.6	1.158	5.751	2.013	4.546	1.972	0.9899	2.425	2.220
15.7	1.108	5.574	1.988	4.690	1.912	0.9900	2.425	2.220
15.8	1.061	5.403	1.964	4.838	1.854	0.9901	2.425	2.220
15.9	1.016	5.239	1.939	4.989	1.798	0.9903	2.426	2.221
16.0	9.732 −7	5.080	1.916	5.144	1.744	0.9904	2.426	2.221
16.1	9.325	4.927	1.893	5.303	1.692	0.9905	2.426	2.221
16.2	8.936	4.780	1.870	5.466	1.642	0.9906	2.427	2.221
16.3	8.567	4.638	1.847	5.633	1.593	0.9907	2.427	2.221
16.4	8.214	4.500	1.825	5.804	1.546	0.9908	2.427	2.221
16.5	7.878	4.368	1.804	5.979	1.501	0.9909	2.427	2.222
16.6	7.557	4.240	1.782	6.158	1.458	0.9910	2.428	2.222
16.7	7.252	4.117	1.761	6.342	1.416	0.9912	2.428	2.222
16.8	6.960	3.998	1.741	6.530	1.375	0.9913	2.428	2.222
16.9	6.682	3.883	1.721	6.723	1.336	0.9914	2.428	2.222
17.0	6.416	3.772	1.701	6.920	1.298	0.9915	2.429	2.222
17.1	6.162	3.665	1.681	7.121	1.261	0.9916	2.429	2.223
17.2	5.920	3.562	1.662	7.328	1.226	0.9917	2.429	2.223
17.3	5.688	3.461	1.643	7.539	1.192	0.9917	2.429	2.223
17.4	5.467	3.365	1.625	7.755	1.159	0.9918	2.430	2.223
17.5	5.255	3.271	1.607	7.976	1.127	0.9919	2.430	2.223
17.6	5.053	3.181	1.589	8.202	1.096	0.9920	2.430	2.223
17.7	4.860	3.093	1.571	8.433	1.066	0.9921	2.430	2.223
17.8	4.675	3.009	1.554	8.669	1.037	0.9922	2.430	2.224
17.9	4.497	2.927	1.537	8.911	1.009	0.9923	2.431	2.224
18.0	4.328	2.848	1.520	9.158	9.815 −5	0.9924	2.431	2.224
18.1	4.166	2.771	1.503	9.411	9.553	0.9925	2.431	2.224
18.2	4.011	2.697	1.487	9.669	9.299	0.9925	2.431	2.224
18.3	3.862	2.625	1.471	9.932	9.053	0.9926	2.431	2.224
18.4	3.719	2.556	1.455	1.020 +4	8.814	0.9927	2.432	2.224
18.5	3.583	2.488	1.440	1.048	8.583	0.9928	2.432	2.225
18.6	3.452	2.423	1.425	1.076	8.359	0.9929	2.432	2.225
18.7	3.327	2.360	1.410	1.105	8.143	0.9929	2.432	2.225
18.8	3.206	2.299	1.395	1.134	7.933	0.9930	2.432	2.225
18.9	3.091	2.239	1.381	1.164	7.729	0.9931	2.433	2.225
19.0	2.981	2.182	1.366	1.194	7.531	0.9931	2.433	2.225
19.1	2.875	2.126	1.352	1.226	7.340	0.9932	2.433	2.225
19.2	2.773	2.072	1.338	1.258	7.154	0.9933	2.433	2.225
19.3	2.675	2.019	1.325	1.290	6.974	0.9934	2.433	2.225
19.4	2.581	1.969	1.311	1.323	6.800	0.9934	2.433	2.226
19.5	2.491	1.919	1.298	1.357	6.630	0.9935	2.434	2.226
19.6	2.405	1.871	1.285	1.392	6.466	0.9936	2.434	2.226
19.7	2.322	1.825	1.272	1.427	6.306	0.9936	2.434	2.226
19.8	2.242	1.780	1.259	1.463	6.151	0.9937	2.434	2.226
19.9	2.165	1.736	1.247	1.500	6.001	0.9937	2.434	2.226
20.0	2.091	1.694	1.235	1.537	5.855	0.9938	2.434	2.226

NORMAL SHOCK FUNCTIONS FOR SUPERSONIC COMPRESSIBLE FLOW

$$\gamma = 7/5$$

M	p_{21}	T_{21}	ρ_{21}	M_2	p°_{21}	μ	ν
15.0	2.623 +2	4.469 +1	5.870	0.3823	4.396 -4	3.823	111.5
15.1	2.658	4.528	5.871	0.3822	4.256	3.797	111.6
15.2	2.694	4.587	5.873	0.3822	4.122	3.772	111.8
15.3	2.729	4.646	5.875	0.3821	3.993	3.748	111.9
15.4	2.765	4.706	5.876	0.3820	3.868	3.723	112.0
15.5	2.801	4.766	5.878	0.3820	3.749	3.699	112.1
15.6	2.837	4.826	5.879	0.3819	3.633	3.675	112.2
15.7	2.874	4.887	5.881	0.3819	3.522	3.652	112.3
15.8	2.911	4.948	5.882	0.3818	3.415	3.629	112.5
15.9	2.948	5.010	5.884	0.3818	3.312	3.606	112.6
16.0	2.985	5.072	5.885	0.3817	3.212	3.583	112.7
16.1	3.022	5.134	5.886	0.3817	3.116	3.561	112.8
16.2	3.060	5.197	5.888	0.3817	3.024	3.539	112.9
16.3	3.098	5.260	5.889	0.3816	2.934	3.517	113.0
16.4	3.136	5.324	5.890	0.3816	2.848	3.496	113.1
16.5	3.174	5.388	5.892	0.3815	2.765	3.475	113.2
16.6	3.213	5.452	5.893	0.3815	2.685	3.454	113.3
16.7	3.252	5.517	5.894	0.3814	2.607	3.433	113.4
16.8	3.291	5.582	5.896	0.3814	2.532	3.413	113.5
16.9	3.330	5.648	5.897	0.3814	2.460	3.392	113.6
17.0	3.370	5.714	5.898	0.3813	2.390	3.372	113.7
17.1	3.410	5.780	5.899	0.3813	2.323	3.353	113.8
17.2	3.450	5.847	5.900	0.3812	2.257	3.333	113.9
17.3	3.490	5.914	5.901	0.3812	2.194	3.314	114.0
17.4	3.530	5.981	5.903	0.3812	2.133	3.295	114.1
17.5	3.571	6.049	5.904	0.3811	2.075	3.276	114.2
17.6	3.612	6.117	5.905	0.3811	2.018	3.257	114.3
17.7	3.653	6.186	5.906	0.3811	1.962	3.239	114.4
17.8	3.695	6.255	5.907	0.3810	1.909	3.221	114.5
17.9	3.736	6.324	5.908	0.3810	1.857	3.203	114.5
18.0	3.778	6.394	5.909	0.3810	1.807	3.185	114.6
18.1	3.820	6.464	5.910	0.3809	1.759	3.167	114.7
18.2	3.863	6.535	5.911	0.3809	1.712	3.150	114.8
18.3	3.905	6.606	5.912	0.3809	1.667	3.133	114.9
18.4	3.948	6.677	5.913	0.3808	1.623	3.116	115.0
18.5	3.991	6.749	5.914	0.3808	1.580	3.099	115.1
18.6	4.034	6.821	5.915	0.3808	1.539	3.082	115.1
18.7	4.078	6.894	5.915	0.3807	1.499	3.066	115.2
18.8	4.122	6.966	5.916	0.3807	1.461	3.049	115.3
18.9	4.166	7.040	5.917	0.3807	1.423	3.033	115.4
19.0	4.210	7.113	5.918	0.3806	1.387	3.017	115.5
19.1	4.254	7.187	5.919	0.3806	1.351	3.001	115.5
19.2	4.299	7.262	5.920	0.3806	1.317	2.986	115.6
19.3	4.344	7.337	5.921	0.3806	1.284	2.970	115.7
19.4	4.389	7.412	5.921	0.3805	1.252	2.955	115.8
19.5	4.434	7.488	5.922	0.3805	1.221	2.940	115.8
19.6	4.480	7.564	5.923	0.3805	1.190	2.925	115.9
19.7	4.526	7.640	5.924	0.3805	1.161	2.910	116.0
19.8	4.572	7.717	5.924	0.3804	1.133	2.895	116.1
19.9	4.618	7.794	5.925	0.3804	1.105	2.880	116.1
20.0	4.665	7.872	5.926	0.3804	1.078	2.866	116.2

ISENTROPIC FUNCTIONS FOR SUPERSONIC COMPRESSIBLE FLOW

$$\gamma = 7/5$$

M	$p/p°$	$T/T°$	$\rho/\rho°$	A/A^*	$q/p°$	u/u_m	u/a^*	$F(\gamma;M)$
20.0	2.091 −7	1.694 −5	1.235 −2	1.537 +4	5.855 −5	0.9938	2.434	2.226
21.0	1.492	1.331	1.121	1.956	4.606	0.9944	2.436	2.227
22.0	1.081	1.057	1.023	2.460	3.663	0.9949	2.437	2.228
23.0	7.945 −8	8.485 −6	9.364 −3	3.065	2.942	0.9953	2.438	2.229
24.0	5.914	6.871	8.606	3.783	2.384	0.9957	2.439	2.229
25.0	4.454	5.612	7.937	4.630	1.949	0.9960	2.440	2.230
26.0	3.392	4.620	7.343	5.623	1.605	0.9963	2.441	2.230
27.0	2.609	3.830	6.812	6.780	1.332	0.9966	2.441	2.231
28.0	2.026	3.197	6.337	8.120	1.112	0.9968	2.442	2.231
29.0	1.587	2.686	5.910	9.665	9.344 −6	0.9970	2.442	2.231
30.0	1.254	2.269	5.525	1.144 +5	7.898	0.9972	2.443	2.232
31.0	9.978 −9	1.928	5.176	1.346	6.712	0.9974	2.443	2.232
32.0	7.998	1.646	4.859	1.576	5.733	0.9976	2.444	2.232
33.0	6.455	1.412	4.571	1.837	4.921	0.9977	2.444	2.232
34.0	5.243	1.217	4.307	2.131	4.242	0.9978	2.444	2.233
35.0	4.283	1.054	4.065	2.461	3.673	0.9980	2.445	2.233
36.0	3.520	9.157 −7	3.843	2.832	3.193	0.9981	2.445	2.233
37.0	2.907	7.989	3.639	3.245	2.786	0.9982	2.445	2.233
38.0	2.414	6.995	3.451	3.706	2.440	0.9983	2.445	2.233
39.0	2.014	6.146	3.277	4.218	2.144	0.9984	2.446	2.233
40.0	1.688	5.417	3.115	4.785	1.890	0.9984	2.446	2.234
41.0	1.421	4.790	2.966	5.411	1.671	0.9985	2.446	2.234
42.0	1.201	4.248	2.827	6.102	1.482	0.9986	2.446	2.234
43.0	1.019	3.777	2.697	6.861	1.318	0.9987	2.446	2.234
44.0	8.677−10	3.368	2.576	7.694	1.176	0.9987	2.446	2.234
45.0	7.417	3.011	2.463	8.606	1.051	0.9988	2.447	2.234
46.0	6.361	2.698	2.357	9.602	9.422 −7	0.9988	2.447	2.234
47.0	5.474	2.424	2.258	1.069 +6	8.464	0.9989	2.447	2.234
48.0	4.726	2.182	2.166	1.187	7.621	0.9989	2.447	2.234
49.0	4.092	1.969	2.078	1.316	6.877	0.9990	2.447	2.234
50.0	3.553	1.780	1.996	1.455	6.218	0.9990	2.447	2.234
51.0	3.094	1.613	1.919	1.606	5.633	0.9990	2.447	2.235
52.0	2.702	1.464	1.846	1.770	5.113	0.9991	2.447	2.235
53.0	2.365	1.331	1.777	1.946	4.650	0.9991	2.447	2.235
54.0	2.075	1.212	1.712	2.137	4.236	0.9991	2.447	2.235
55.0	1.826	1.106	1.650	2.341	3.865	0.9992	2.448	2.235
56.0	1.609	1.011	1.592	2.562	3.533	0.9992	2.448	2.235
57.0	1.422	9.256 −8	1.537	2.798	3.234	0.9992	2.448	2.235
58.0	1.259	8.486	1.484	3.052	2.966	0.9993	2.448	2.235
59.0	1.118	7.792	1.434	3.324	2.723	0.9993	2.448	2.235
60.0	9.937−11	7.165	1.387	3.615	2.504	0.9993	2.448	2.235
65.0	5.678	4.804	1.182	5.391	1.679	0.9994	2.448	2.235
70.0	3.382	3.318	1.019	7.805	1.160	0.9995	2.448	2.235
75.0	2.088	2.351	8.881 −4	1.102 +7	8.220 −8	0.9996	2.448	2.235
80.0	1.329	1.703	7.807	1.521	5.955	0.9996	2.449	2.235
85.0	8.698−12	1.258	6.916	2.058	4.399	0.9997	2.449	2.236
90.0	5.831	9.453 −9	6.169	2.739	3.306	0.9997	2.449	2.236
95.0	3.995	7.215	5.537	3.588	2.524	0.9997	2.449	2.236
100.0	2.790	5.583	4.998	4.636	1.953	0.9997	2.449	2.236
∞	0.000	0.000	0.000	∞	0.000	1.000	2.450	2.236

NORMAL SHOCK FUNCTIONS FOR SUPERSONIC COMPRESSIBLE FLOW

$$\gamma = 7/5$$

M	P_{21}	T_{21}	ρ_{21}	M_2	P^0_{21}	μ	ν
20.0	4.665 +2	7.872 +1	5.926	0.3804	1.078 -4	2.866	116.2
21.0	5.143	8.669	5.933	0.3802	8.479 -5	2.729	116.9
22.0	5.645	9.505	5.939	0.3800	6.742	2.605	117.5
23.0	6.170	1.038 +2	5.944	0.3798	5.415	2.492	118.0
24.0	6.718	1.129	5.948	0.3796	4.388	2.388	118.6
25.0	7.290	1.225	5.952	0.3795	3.586	2.293	119.0
26.0	7.885	1.324	5.956	0.3794	2.954	2.204	119.5
27.0	8.503	1.427	5.959	0.3793	2.450	2.123	119.9
28.0	9.145	1.534	5.962	0.3792	2.046	2.047	120.3
29.0	9.810	1.645	5.965	0.3791	1.719	1.976	120.6
30.0	1.050 +3	1.759	5.967	0.3790	1.453	1.910	120.9
31.0	1.121	1.878	5.969	0.3790	1.235	1.849	121.2
32.0	1.194	2.000	5.971	0.3789	1.055	1.791	121.5
33.0	1.270	2.127	5.973	0.3789	9.053 -6	1.737	121.8
34.0	1.348	2.257	5.974	0.3788	7.805	1.685	122.0
35.0	1.429	2.391	5.976	0.3788	6.758	1.637	122.3
36.0	1.512	2.529	5.977	0.3787	5.874	1.592	122.5
37.0	1.597	2.671	5.978	0.3787	5.126	1.549	122.7
38.0	1.684	2.817	5.979	0.3786	4.489	1.508	122.9
39.0	1.774	2.967	5.980	0.3786	3.944	1.469	123.1
40.0	1.866	3.120	5.981	0.3786	3.477	1.433	123.3
41.0	1.961	3.278	5.982	0.3785	3.075	1.398	123.5
42.0	2.058	3.439	5.983	0.3785	2.727	1.364	123.6
43.0	2.157	3.605	5.984	0.3785	2.426	1.333	123.8
44.0	2.258	3.774	5.985	0.3785	2.163	1.302	124.0
45.0	2.362	3.947	5.985	0.3784	1.934	1.273	124.1
46.0	2.468	4.124	5.986	0.3784	1.733	1.246	124.2
47.0	2.577	4.305	5.986	0.3784	1.557	1.219	124.4
48.0	2.688	4.489	5.987	0.3784	1.402	1.194	124.5
49.0	2.801	4.678	5.988	0.3784	1.265	1.169	124.6
50.0	2.916	4.870	5.988	0.3784	1.144	1.146	124.7
51.0	3.034	5.067	5.988	0.3783	1.036	1.124	124.8
52.0	3.154	5.267	5.989	0.3783	9.406 -7	1.102	125.0
53.0	3.277	5.471	5.989	0.3783	8.554	1.081	125.1
54.0	3.402	5.679	5.990	0.3783	7.792	1.061	125.2
55.0	3.529	5.891	5.990	0.3783	7.111	1.042	125.3
56.0	3.658	6.107	5.990	0.3783	6.499	1.023	125.3
57.0	3.790	6.327	5.991	0.3783	5.950	1.005	125.4
58.0	3.924	6.550	5.991	0.3783	5.455	0.988	125.5
59.0	4.061	6.778	5.991	0.3782	5.009	0.971	125.6
60.0	4.200	7.009	5.992	0.3782	4.606	0.955	125.7
65.0	4.929	8.225	5.993	0.3782	3.089	0.882	126.1
70.0	5.716	9.537	5.994	0.3782	2.134	0.819	126.4
75.0	6.562	1.095 +3	5.995	0.3781	1.512	0.764	126.6
80.0	7.466	1.245	5.995	0.3781	1.095	0.716	126.9
85.0	8.429	1.406	5.996	0.3781	8.092 -8	0.674	127.1
90.0	9.450	1.576	5.996	0.3781	6.082	0.637	127.3
95.0	1.053 +4	1.756	5.997	0.3781	4.642	0.603	127.4
100.0	1.167	1.945	5.997	0.3781	3.593	0.573	127.6
∞	∞	∞	6.000	0.3780	0.000	0.000	130.5

TABLE A.3

ISENTROPIC FLOW FUNCTIONS

AND

NORMAL SHOCK FUNCTIONS

$\gamma = 4/3$

ISENTROPIC FUNCTIONS FOR SUBSONIC COMPRESSIBLE FLOW

$$\gamma = 4/3$$

M	p/p°	T/T°	ρ/ρ°	A/A^*	q/p°	u/u_m	u/a^*	$F(\gamma;M)$
0.00	1.000 +0	1.000 +0	1.000 +0	∞	0.000 +0	0.0000	0.0000	∞
0.01	1.000	1.000	1.000	5.831	6.666	0.0041	0.0108	75.01
0.02	9.997 -1	9.998 -1	1.000	2.916	2.666 -4	0.0082	0.0216	37.52
0.03	9.994	9.996	9.999 -1	1.944	5.996	0.0122	0.0324	25.03
0.04	9.989	9.992	9.997	1.459	1.066 -3	0.0163	0.0432	18.79
0.05	9.983	9.988	9.996	1.168	1.664	0.0204	0.0540	15.05
0.06	9.976	9.982	9.994	9.737 +0	2.394	0.0245	0.0648	12.56
0.07	9.967	9.976	9.992	8.353	3.256	0.0286	0.0756	10.78
0.08	9.957	9.968	9.989	7.315	4.249	0.0326	0.0864	9.450
0.09	9.946	9.960	9.987	6.509	5.371	0.0367	0.0971	8.418
0.10	9.934	9.950	9.983	5.864	6.622	0.0408	0.1079	7.594
0.11	9.920	9.940	9.980	5.338	8.002	0.0449	0.1187	6.921
0.12	9.905	9.928	9.976	4.899	9.508	0.0489	0.1295	6.362
0.13	9.888	9.916	9.972	4.529	1.114 -2	0.0530	0.1402	5.891
0.14	9.870	9.903	9.967	4.212	1.290	0.0571	0.1510	5.488
0.15	9.851	9.888	9.963	3.938	1.478	0.0611	0.1617	5.140
0.16	9.831	9.873	9.958	3.699	1.678	0.0652	0.1725	4.837
0.17	9.810	9.857	9.952	3.488	1.890	0.0692	0.1832	4.571
0.18	9.787	9.840	9.946	3.301	2.114	0.0733	0.1939	4.335
0.19	9.763	9.822	9.940	3.134	2.350	0.0773	0.2046	4.125
0.20	9.738	9.803	9.934	2.984	2.597	0.0814	0.2153	3.937
0.21	9.711	9.783	9.927	2.848	2.855	0.0854	0.2260	3.768
0.22	9.684	9.762	9.920	2.726	3.125	0.0895	0.2367	3.615
0.23	9.655	9.740	9.913	2.614	3.405	0.0935	0.2473	3.476
0.24	9.625	9.717	9.905	2.512	3.696	0.0975	0.2580	3.349
0.25	9.594	9.694	9.897	2.418	3.997	0.1015	0.2686	3.233
0.26	9.562	9.670	9.889	2.332	4.309	0.1056	0.2793	3.127
0.27	9.528	9.644	9.880	2.253	4.631	0.1096	0.2899	3.029
0.28	9.494	9.618	9.871	2.179	4.962	0.1136	0.3005	2.939
0.29	9.458	9.591	9.862	2.111	5.303	0.1176	0.3111	2.856
0.30	9.422	9.563	9.852	2.047	5.653	0.1216	0.3216	2.779
0.31	9.384	9.535	9.842	1.988	6.012	0.1256	0.3322	2.708
0.32	9.346	9.505	9.832	1.933	6.380	0.1295	0.3427	2.641
0.33	9.306	9.475	9.822	1.882	6.756	0.1335	0.3532	2.579
0.34	9.265	9.444	9.811	1.833	7.140	0.1375	0.3638	2.522
0.35	9.223	9.412	9.800	1.788	7.532	0.1415	0.3742	2.468
0.36	9.181	9.379	9.789	1.745	7.932	0.1454	0.3847	2.417
0.37	9.137	9.346	9.777	1.705	8.339	0.1494	0.3952	2.370
0.38	9.093	9.311	9.765	1.667	8.753	0.1533	0.4056	2.326
0.39	9.047	9.277	9.753	1.632	9.174	0.1572	0.4160	2.284
0.40	9.001	9.241	9.740	1.598	9.601	0.1612	0.4264	2.245
0.41	8.954	9.205	9.727	1.566	1.003 -1	0.1651	0.4368	2.209
0.42	8.906	9.167	9.714	1.536	1.047	0.1690	0.4471	2.174
0.43	8.857	9.130	9.701	1.508	1.092	0.1729	0.4575	2.141
0.44	8.807	9.091	9.687	1.481	1.137	0.1768	0.4678	2.111
0.45	8.757	9.052	9.674	1.455	1.182	0.1807	0.4781	2.082
0.46	8.705	9.012	9.659	1.431	1.228	0.1846	0.4883	2.055
0.47	8.654	8.972	9.645	1.408	1.274	0.1884	0.4986	2.029
0.48	8.601	8.931	9.630	1.386	1.321	0.1923	0.5088	2.004
0.49	8.548	8.890	9.615	1.365	1.368	0.1962	0.5190	1.981
0.50	8.493	8.847	9.600	1.345	1.416	0.2000	0.5291	1.960

ISENTROPIC FUNCTIONS FOR SUBSONIC COMPRESSIBLE FLOW

$$\gamma = 4/3$$

M	$p/p°$	$T/T°$	$\rho/\rho°$	A/A^*	$q/p°$	u/u_m	u/a^*	$F(\gamma;M)$
0.50	8.493 −1	8.847 −1	9.600 −1	1.345 +0	1.416 −1	0.2000	0.5291	1.960
0.51	8.439	8.805	9.585	1.326	1.463	0.2038	0.5393	1.939
0.52	8.384	8.761	9.569	1.308	1.511	0.2077	0.5494	1.920
0.53	8.328	8.717	9.553	1.291	1.559	0.2115	0.5595	1.901
0.54	8.271	8.673	9.537	1.275	1.608	0.2153	0.5696	1.884
0.55	8.214	8.628	9.520	1.259	1.656	0.2191	0.5796	1.867
0.56	8.156	8.583	9.503	1.244	1.705	0.2229	0.5897	1.852
0.57	8.098	8.537	9.486	1.230	1.754	0.2266	0.5996	1.837
0.58	8.040	8.490	9.469	1.217	1.803	0.2304	0.6096	1.823
0.59	7.981	8.444	9.452	1.204	1.852	0.2342	0.6196	1.809
0.60	7.921	8.396	9.434	1.192	1.901	0.2379	0.6295	1.797
0.61	7.861	8.348	9.416	1.180	1.950	0.2417	0.6393	1.785
0.62	7.801	8.300	9.398	1.169	1.999	0.2454	0.6492	1.774
0.63	7.740	8.252	9.380	1.158	2.048	0.2491	0.6590	1.763
0.64	7.679	8.203	9.361	1.148	2.097	0.2528	0.6688	1.753
0.65	7.617	8.153	9.342	1.138	2.145	0.2565	0.6786	1.744
0.66	7.555	8.104	9.323	1.129	2.194	0.2602	0.6883	1.735
0.67	7.493	8.054	9.304	1.120	2.242	0.2638	0.6980	1.726
0.68	7.431	8.003	9.284	1.112	2.291	0.2675	0.7077	1.718
0.69	7.368	7.953	9.265	1.104	2.339	0.2711	0.7174	1.710
0.70	7.305	7.902	9.245	1.096	2.386	0.2748	0.7270	1.703
0.71	7.242	7.850	9.225	1.089	2.434	0.2784	0.7366	1.697
0.72	7.179	7.799	9.205	1.082	2.481	0.2820	0.7461	1.690
0.73	7.115	7.747	9.184	1.076	2.528	0.2856	0.7556	1.684
0.74	7.051	7.695	9.164	1.070	2.574	0.2892	0.7651	1.679
0.75	6.988	7.643	9.143	1.064	2.620	0.2928	0.7746	1.673
0.76	6.924	7.590	9.122	1.058	2.666	0.2963	0.7840	1.668
0.77	6.860	7.537	9.101	1.053	2.711	0.2999	0.7934	1.664
0.78	6.795	7.485	9.079	1.048	2.756	0.3034	0.8028	1.659
0.79	6.731	7.431	9.058	1.043	2.801	0.3069	0.8121	1.655
0.80	6.667	7.378	9.036	1.039	2.845	0.3105	0.8214	1.652
0.81	6.603	7.325	9.014	1.035	2.888	0.3140	0.8307	1.648
0.82	6.538	7.271	8.992	1.031	2.931	0.3174	0.8399	1.645
0.83	6.474	7.218	8.970	1.028	2.973	0.3209	0.8491	1.642
0.84	6.410	7.164	8.948	1.024	3.015	0.3244	0.8582	1.639
0.85	6.346	7.110	8.925	1.021	3.057	0.3278	0.8674	1.637
0.86	6.282	7.056	8.903	1.018	3.097	0.3313	0.8765	1.634
0.87	6.218	7.002	8.880	1.015	3.137	0.3347	0.8855	1.632
0.88	6.154	6.948	8.857	1.013	3.177	0.3381	0.8945	1.630
0.89	6.090	6.894	8.834	1.011	3.216	0.3415	0.9035	1.629
0.90	6.026	6.839	8.811	1.009	3.254	0.3449	0.9125	1.627
0.91	5.962	6.785	8.787	1.007	3.292	0.3482	0.9214	1.626
0.92	5.899	6.731	8.764	1.006	3.328	0.3516	0.9303	1.624
0.93	5.835	6.677	8.740	1.004	3.365	0.3549	0.9391	1.623
0.94	5.772	6.622	8.716	1.003	3.400	0.3583	0.9479	1.623
0.95	5.709	6.568	8.693	1.002	3.435	0.3616	0.9567	1.622
0.96	5.647	6.514	8.669	1.001	3.469	0.3649	0.9654	1.621
0.97	5.584	6.460	8.644	1.001	3.503	0.3682	0.9741	1.621
0.98	5.522	6.405	8.620	1.000	3.535	0.3715	0.9828	1.620
0.99	5.460	6.351	8.596	1.000	3.567	0.3747	0.9914	1.620
1.00	5.398	6.297	8.571	1.000	3.598	0.3780	1.000	1.620

ISENTROPIC FUNCTIONS FOR SUPERSONIC COMPRESSIBLE FLOW

$\gamma = 4/3$

M	p/p°	T/T°	ρ/ρ°	A/A^*	q/p°	u/u_m	u/a^*	$F(\gamma;M)$
1.00	5.398 −1	6.297 −1	8.571 −1	1.000 +0	3.598 −1	0.3780	1.000	1.620
1.01	5.336	6.243	8.547	1.000	3.629	0.3812	1.009	1.620
1.02	5.275	6.190	8.522	1.000	3.659	0.3844	1.017	1.620
1.03	5.214	6.136	8.498	1.001	3.688	0.3876	1.026	1.621
1.04	5.153	6.082	8.473	1.001	3.716	0.3908	1.034	1.621
1.05	5.093	6.029	8.448	1.002	3.743	0.3940	1.042	1.622
1.06	5.033	5.975	8.423	1.003	3.770	0.3971	1.051	1.622
1.07	4.973	5.922	8.398	1.004	3.796	0.4003	1.059	1.623
1.08	4.914	5.869	8.372	1.005	3.821	0.4034	1.067	1.624
1.09	4.855	5.816	8.347	1.007	3.845	0.4066	1.076	1.624
1.10	4.796	5.763	8.322	1.008	3.869	0.4097	1.084	1.625
1.11	4.738	5.710	8.296	1.010	3.891	0.4128	1.092	1.626
1.12	4.680	5.658	8.271	1.012	3.913	0.4158	1.100	1.628
1.13	4.622	5.606	8.245	1.014	3.934	0.4189	1.108	1.629
1.14	4.565	5.553	8.220	1.016	3.955	0.4219	1.116	1.630
1.15	4.508	5.502	8.194	1.018	3.974	0.4250	1.124	1.631
1.16	4.451	5.450	8.168	1.020	3.993	0.4280	1.132	1.633
1.17	4.395	5.398	8.142	1.023	4.011	0.4310	1.140	1.634
1.18	4.340	5.347	8.116	1.026	4.028	0.4340	1.148	1.636
1.19	4.285	5.296	8.091	1.029	4.045	0.4370	1.156	1.637
1.20	4.230	5.245	8.065	1.032	4.061	0.4399	1.164	1.639
1.21	4.175	5.194	8.039	1.035	4.075	0.4429	1.172	1.641
1.22	4.122	5.144	8.012	1.038	4.090	0.4458	1.180	1.642
1.23	4.068	5.094	7.986	1.041	4.103	0.4487	1.187	1.644
1.24	4.015	5.044	7.960	1.045	4.116	0.4516	1.195	1.646
1.25	3.962	4.994	7.934	1.049	4.127	0.4545	1.203	1.648
1.26	3.910	4.945	7.908	1.052	4.138	0.4574	1.210	1.650
1.27	3.858	4.896	7.881	1.056	4.149	0.4603	1.218	1.652
1.28	3.807	4.847	7.855	1.060	4.158	0.4631	1.225	1.654
1.29	3.756	4.798	7.829	1.065	4.167	0.4660	1.233	1.656
1.30	3.706	4.750	7.802	1.069	4.175	0.4688	1.240	1.658
1.31	3.656	4.702	7.776	1.073	4.183	0.4716	1.248	1.660
1.32	3.607	4.654	7.750	1.078	4.189	0.4744	1.255	1.662
1.33	3.558	4.607	7.723	1.083	4.195	0.4772	1.263	1.664
1.34	3.509	4.559	7.697	1.088	4.201	0.4799	1.270	1.667
1.35	3.461	4.513	7.670	1.093	4.205	0.4827	1.277	1.669
1.36	3.414	4.466	7.644	1.098	4.209	0.4854	1.284	1.671
1.37	3.367	4.420	7.617	1.103	4.212	0.4881	1.292	1.673
1.38	3.320	4.374	7.591	1.109	4.215	0.4908	1.299	1.676
1.39	3.274	4.328	7.564	1.114	4.217	0.4935	1.306	1.678
1.40	3.228	4.283	7.538	1.120	4.218	0.4962	1.313	1.681
1.41	3.183	4.238	7.511	1.126	4.219	0.4989	1.320	1.683
1.42	3.138	4.193	7.485	1.132	4.219	0.5015	1.327	1.685
1.43	3.094	4.149	7.458	1.138	4.218	0.5042	1.334	1.688
1.44	3.050	4.105	7.432	1.144	4.217	0.5068	1.341	1.690
1.45	3.007	4.061	7.405	1.151	4.215	0.5094	1.348	1.693
1.46	2.964	4.017	7.379	1.157	4.212	0.5120	1.355	1.695
1.47	2.922	3.974	7.352	1.164	4.209	0.5146	1.361	1.698
1.48	2.880	3.931	7.326	1.171	4.206	0.5171	1.368	1.700
1.49	2.839	3.889	7.299	1.178	4.201	0.5197	1.375	1.703
1.50	2.798	3.847	7.273	1.185	4.196	0.5222	1.382	1.706

NORMAL SHOCK FUNCTIONS FOR SUPERSONIC COMPRESSIBLE FLOW

$$\gamma = 4/3$$

M	p_{21}	T_{21}	ρ_{21}	M_2	p^o_{21}	μ	ν
1.00	1.000 +0	1.000 +0	1.000	1.000	1.000 +0	90.00	0.000
1.01	1.023	1.006	1.017	0.9901	1.000	81.93	0.046
1.02	1.046	1.011	1.034	0.9805	1.000	78.64	0.129
1.03	1.070	1.017	1.052	0.9711	1.000	76.14	0.236
1.04	1.093	1.023	1.069	0.9620	1.000	74.06	0.361
1.05	1.117	1.028	1.087	0.9530	9.999 -1	72.25	0.502
1.06	1.141	1.034	1.104	0.9443	9.997	70.63	0.656
1.07	1.166	1.039	1.122	0.9358	9.996	69.16	0.822
1.08	1.190	1.045	1.139	0.9275	9.994	67.81	0.998
1.09	1.215	1.050	1.157	0.9193	9.992	66.55	1.184
1.10	1.240	1.056	1.175	0.9114	9.989	65.38	1.378
1.11	1.265	1.061	1.193	0.9036	9.986	64.28	1.581
1.12	1.291	1.066	1.210	0.8961	9.982	63.24	1.791
1.13	1.316	1.072	1.228	0.8886	9.977	62.25	2.007
1.14	1.342	1.077	1.246	0.8814	9.972	61.31	2.230
1.15	1.369	1.083	1.264	0.8743	9.966	60.41	2.459
1.16	1.395	1.088	1.282	0.8673	9.960	59.55	2.694
1.17	1.422	1.093	1.300	0.8605	9.953	58.73	2.934
1.18	1.448	1.099	1.318	0.8539	9.945	57.94	3.178
1.19	1.476	1.104	1.337	0.8474	9.936	57.18	3.427
1.20	1.503	1.109	1.355	0.8410	9.927	56.44	3.680
1.21	1.530	1.115	1.373	0.8347	9.916	55.74	3.938
1.22	1.558	1.120	1.391	0.8286	9.905	55.05	4.199
1.23	1.586	1.125	1.410	0.8226	9.894	54.39	4.464
1.24	1.614	1.131	1.428	0.8167	9.881	53.75	4.732
1.25	1.643	1.136	1.446	0.8109	9.868	53.13	5.003
1.26	1.671	1.141	1.465	0.8053	9.853	52.53	5.277
1.27	1.700	1.147	1.483	0.7997	9.838	51.94	5.554
1.28	1.730	1.152	1.501	0.7943	9.823	51.38	5.834
1.29	1.759	1.157	1.520	0.7890	9.806	50.82	6.116
1.30	1.789	1.163	1.538	0.7837	9.789	50.29	6.400
1.31	1.818	1.168	1.557	0.7786	9.770	49.76	6.687
1.32	1.848	1.173	1.575	0.7736	9.751	49.25	6.976
1.33	1.879	1.179	1.594	0.7686	9.731	48.75	7.267
1.34	1.909	1.184	1.612	0.7637	9.710	48.27	7.559
1.35	1.940	1.190	1.631	0.7590	9.689	47.80	7.854
1.36	1.971	1.195	1.649	0.7543	9.667	47.33	8.150
1.37	2.002	1.200	1.668	0.7497	9.643	46.88	8.447
1.38	2.034	1.206	1.686	0.7452	9.620	46.44	8.746
1.39	2.065	1.211	1.705	0.7407	9.595	46.01	9.046
1.40	2.097	1.217	1.724	0.7364	9.569	45.59	9.348
1.41	2.129	1.222	1.742	0.7321	9.543	45.17	9.651
1.42	2.161	1.228	1.761	0.7279	9.516	44.77	9.955
1.43	2.194	1.233	1.779	0.7237	9.488	44.37	10.26
1.44	2.227	1.239	1.798	0.7197	9.460	43.98	10.57
1.45	2.260	1.244	1.816	0.7157	9.431	43.60	10.87
1.46	2.293	1.250	1.835	0.7117	9.401	43.23	11.18
1.47	2.327	1.255	1.853	0.7079	9.370	42.87	11.49
1.48	2.360	1.261	1.872	0.7041	9.339	42.51	11.80
1.49	2.394	1.266	1.891	0.7003	9.307	42.16	12.11
1.50	2.428	1.272	1.909	0.6966	9.274	41.81	12.42

ISENTROPIC FUNCTIONS FOR SUPERSONIC COMPRESSIBLE FLOW

$$\gamma = 4/3$$

M	$p/p°$	$T/T°$	$\rho/\rho°$	$A/A*$	$q/p°$	u/u_m	$u/a*$	$F(\gamma;M)$
1.50	2.798 −1	3.847 −1	7.273 −1	1.185 +0	4.196 −1	0.5222	1.382	1.706
1.51	2.757	3.805	7.246	1.192	4.191	0.5247	1.388	1.708
1.52	2.717	3.764	7.220	1.199	4.185	0.5273	1.395	1.711
1.53	2.678	3.723	7.194	1.207	4.179	0.5298	1.402	1.713
1.54	2.639	3.682	7.167	1.215	4.172	0.5322	1.408	1.716
1.55	2.600	3.641	7.141	1.222	4.164	0.5347	1.415	1.719
1.56	2.562	3.601	7.115	1.230	4.156	0.5372	1.421	1.721
1.57	2.524	3.561	7.088	1.239	4.148	0.5396	1.428	1.724
1.58	2.487	3.522	7.062	1.247	4.139	0.5420	1.434	1.727
1.59	2.450	3.483	7.036	1.255	4.130	0.5445	1.441	1.729
1.60	2.414	3.444	7.009	1.264	4.120	0.5469	1.447	1.732
1.61	2.378	3.405	6.983	1.273	4.109	0.5492	1.453	1.735
1.62	2.343	3.367	6.957	1.281	4.099	0.5516	1.460	1.737
1.63	2.308	3.330	6.931	1.290	4.087	0.5540	1.466	1.740
1.64	2.273	3.292	6.905	1.300	4.076	0.5563	1.472	1.743
1.65	2.239	3.255	6.879	1.309	4.064	0.5587	1.478	1.745
1.66	2.205	3.218	6.853	1.318	4.051	0.5610	1.484	1.748
1.67	2.172	3.182	6.827	1.328	4.038	0.5633	1.490	1.751
1.68	2.139	3.146	6.801	1.338	4.025	0.5656	1.496	1.754
1.69	2.107	3.110	6.775	1.348	4.012	0.5679	1.503	1.756
1.70	2.075	3.075	6.749	1.358	3.998	0.5701	1.509	1.759
1.71	2.044	3.039	6.724	1.368	3.983	0.5724	1.514	1.762
1.72	2.012	3.005	6.698	1.379	3.969	0.5747	1.520	1.764
1.73	1.982	2.970	6.672	1.389	3.954	0.5769	1.526	1.767
1.74	1.951	2.936	6.646	1.400	3.939	0.5791	1.532	1.770
1.75	1.922	2.902	6.621	1.411	3.923	0.5813	1.538	1.773
1.76	1.892	2.869	6.595	1.422	3.907	0.5835	1.544	1.775
1.77	1.863	2.836	6.570	1.433	3.891	0.5857	1.550	1.778
1.78	1.834	2.803	6.544	1.445	3.874	0.5878	1.555	1.781
1.79	1.806	2.770	6.519	1.456	3.858	0.5900	1.561	1.784
1.80	1.778	2.738	6.494	1.468	3.840	0.5921	1.567	1.786
1.81	1.751	2.706	6.468	1.480	3.823	0.5943	1.572	1.789
1.82	1.723	2.675	6.443	1.492	3.806	0.5964	1.578	1.792
1.83	1.697	2.644	6.418	1.504	3.788	0.5985	1.584	1.794
1.84	1.670	2.613	6.393	1.517	3.770	0.6006	1.589	1.797
1.85	1.644	2.582	6.368	1.529	3.752	0.6027	1.595	1.800
1.86	1.619	2.552	6.343	1.542	3.733	0.6047	1.600	1.802
1.87	1.593	2.522	6.318	1.555	3.714	0.6068	1.605	1.805
1.88	1.568	2.492	6.293	1.569	3.695	0.6088	1.611	1.808
1.89	1.544	2.463	6.268	1.582	3.676	0.6109	1.616	1.811
1.90	1.520	2.434	6.244	1.596	3.657	0.6129	1.622	1.813
1.91	1.496	2.405	6.219	1.609	3.638	0.6149	1.627	1.816
1.92	1.472	2.377	6.194	1.623	3.618	0.6169	1.632	1.819
1.93	1.449	2.349	6.170	1.637	3.598	0.6189	1.637	1.821
1.94	1.426	2.321	6.145	1.652	3.578	0.6209	1.643	1.824
1.95	1.404	2.293	6.121	1.666	3.558	0.6228	1.648	1.826
1.96	1.382	2.266	6.097	1.681	3.538	0.6248	1.653	1.829
1.97	1.360	2.239	6.073	1.696	3.518	0.6267	1.658	1.832
1.98	1.338	2.213	6.048	1.711	3.497	0.6286	1.663	1.834
1.99	1.317	2.186	6.024	1.727	3.477	0.6305	1.668	1.837
2.00	1.296	2.160	6.000	1.742	3.456	0.6324	1.673	1.840

NORMAL SHOCK FUNCTIONS FOR SUPERSONIC COMPRESSIBLE FLOW

$$\gamma = 4/3$$

M	P_{21}	T_{21}	ρ_{21}	M_2	P°_{21}	μ	ν
1.50	2.428 +0	1.272 +0	1.909	0.6966	9.274 -1	41.81	12.42
1.51	2.463	1.278	1.928	0.6930	9.240	41.47	12.73
1.52	2.497	1.283	1.946	0.6895	9.206	41.14	13.04
1.53	2.532	1.289	1.964	0.6860	9.172	40.81	13.35
1.54	2.567	1.295	1.983	0.6825	9.136	40.49	13.66
1.55	2.603	1.300	2.001	0.6791	9.101	40.18	13.98
1.56	2.638	1.306	2.020	0.6758	9.064	39.87	14.29
1.57	2.674	1.312	2.038	0.6725	9.027	39.57	14.60
1.58	2.710	1.318	2.057	0.6692	8.989	39.27	14.91
1.59	2.746	1.323	2.075	0.6660	8.951	38.97	15.23
1.60	2.783	1.329	2.093	0.6629	8.912	38.68	15.54
1.61	2.819	1.335	2.112	0.6598	8.873	38.40	15.85
1.62	2.856	1.341	2.130	0.6568	8.833	38.12	16.17
1.63	2.893	1.347	2.148	0.6538	8.793	37.84	16.48
1.64	2.931	1.353	2.167	0.6508	8.752	37.57	16.80
1.65	2.968	1.359	2.185	0.6479	8.711	37.31	17.11
1.66	3.006	1.365	2.203	0.6450	8.669	37.04	17.42
1.67	3.044	1.371	2.221	0.6422	8.627	36.79	17.74
1.68	3.083	1.377	2.239	0.6394	8.585	36.53	18.05
1.69	3.121	1.383	2.257	0.6367	8.542	36.28	18.36
1.70	3.160	1.389	2.275	0.6340	8.498	36.03	18.67
1.71	3.199	1.395	2.294	0.6313	8.454	35.79	18.99
1.72	3.238	1.401	2.312	0.6287	8.410	35.55	19.30
1.73	3.277	1.407	2.330	0.6261	8.366	35.31	19.61
1.74	3.317	1.413	2.347	0.6235	8.321	35.08	19.92
1.75	3.357	1.419	2.365	0.6210	8.276	34.85	20.24
1.76	3.397	1.425	2.383	0.6185	8.230	34.62	20.55
1.77	3.437	1.432	2.401	0.6161	8.184	34.40	20.86
1.78	3.478	1.438	2.419	0.6137	8.138	34.18	21.17
1.79	3.519	1.444	2.437	0.6113	8.092	33.96	21.48
1.80	3.560	1.450	2.454	0.6089	8.045	33.75	21.79
1.81	3.601	1.457	2.472	0.6066	7.999	33.54	22.10
1.82	3.642	1.463	2.490	0.6043	7.951	33.33	22.40
1.83	3.684	1.469	2.507	0.6021	7.904	33.12	22.71
1.84	3.726	1.476	2.525	0.5999	7.857	32.92	23.02
1.85	3.768	1.482	2.542	0.5977	7.809	32.72	23.33
1.86	3.811	1.489	2.560	0.5955	7.761	32.52	23.63
1.87	3.853	1.495	2.577	0.5934	7.713	32.33	23.94
1.88	3.896	1.502	2.595	0.5913	7.665	32.14	24.25
1.89	3.939	1.508	2.612	0.5892	7.616	31.95	24.55
1.90	3.983	1.515	2.629	0.5871	7.568	31.76	24.86
1.91	4.026	1.521	2.647	0.5851	7.519	31.57	25.16
1.92	4.070	1.528	2.664	0.5831	7.470	31.39	25.46
1.93	4.114	1.534	2.681	0.5811	7.421	31.21	25.77
1.94	4.158	1.541	2.698	0.5792	7.372	31.03	26.07
1.95	4.203	1.548	2.715	0.5772	7.323	30.85	26.37
1.96	4.247	1.554	2.732	0.5753	7.274	30.68	26.67
1.97	4.292	1.561	2.749	0.5735	7.225	30.51	26.97
1.98	4.337	1.568	2.766	0.5716	7.176	30.34	27.27
1.99	4.383	1.575	2.783	0.5698	7.127	30.17	27.57
2.00	4.428	1.582	2.800	0.5680	7.077	30.00	27.87

ISENTROPIC FUNCTIONS FOR SUPERSONIC COMPRESSIBLE FLOW

$$\gamma = 4/3$$

M	$p/p°$	$T/T°$	$\rho/\rho°$	$A/A*$	$q/p°$	u/u_m	$u/a*$	$F(\gamma;M)$
2.00	1.296 −1	2.160 −1	6.000 −1	1.742 +0	3.456 −1	0.6324	1.673	1.840
2.01	1.276	2.134	5.976	1.758	3.435	0.6343	1.678	1.842
2.02	1.255	2.109	5.952	1.774	3.414	0.6362	1.683	1.845
2.03	1.235	2.084	5.929	1.790	3.394	0.6381	1.688	1.847
2.04	1.216	2.059	5.905	1.807	3.372	0.6399	1.693	1.850
2.05	1.196	2.034	5.881	1.823	3.351	0.6418	1.698	1.853
2.06	1.177	2.010	5.858	1.840	3.330	0.6436	1.703	1.855
2.07	1.158	1.986	5.834	1.857	3.309	0.6454	1.708	1.858
2.08	1.140	1.962	5.811	1.874	3.288	0.6473	1.713	1.860
2.09	1.122	1.938	5.787	1.892	3.266	0.6491	1.717	1.863
2.10	1.104	1.915	5.764	1.910	3.245	0.6509	1.722	1.865
2.11	1.086	1.892	5.741	1.928	3.223	0.6526	1.727	1.868
2.12	1.069	1.869	5.718	1.946	3.202	0.6544	1.731	1.870
2.13	1.052	1.847	5.694	1.964	3.180	0.6562	1.736	1.873
2.14	1.035	1.824	5.671	1.983	3.159	0.6579	1.741	1.875
2.15	1.018	1.802	5.649	2.002	3.137	0.6597	1.745	1.878
2.16	1.002	1.781	5.626	2.021	3.115	0.6614	1.750	1.880
2.17	9.856 −2	1.759	5.603	2.041	3.094	0.6631	1.754	1.883
2.18	9.697	1.738	5.580	2.060	3.072	0.6648	1.759	1.885
2.19	9.541	1.717	5.558	2.080	3.050	0.6665	1.763	1.888
2.20	9.388	1.696	5.535	2.100	3.029	0.6682	1.768	1.890
2.21	9.236	1.675	5.513	2.121	3.007	0.6699	1.772	1.893
2.22	9.087	1.655	5.490	2.141	2.986	0.6715	1.777	1.895
2.23	8.941	1.635	5.468	2.162	2.964	0.6732	1.781	1.898
2.24	8.797	1.615	5.446	2.184	2.942	0.6748	1.785	1.900
2.25	8.655	1.596	5.424	2.205	2.921	0.6765	1.790	1.902
2.26	8.515	1.576	5.402	2.227	2.899	0.6781	1.794	1.905
2.27	8.378	1.557	5.380	2.249	2.878	0.6797	1.798	1.907
2.28	8.242	1.538	5.358	2.271	2.856	0.6813	1.803	1.910
2.29	8.109	1.520	5.336	2.294	2.835	0.6829	1.807	1.912
2.30	7.978	1.501	5.315	2.316	2.813	0.6845	1.811	1.914
2.31	7.849	1.483	5.293	2.340	2.792	0.6861	1.815	1.917
2.32	7.722	1.465	5.272	2.363	2.771	0.6876	1.819	1.919
2.33	7.597	1.447	5.250	2.387	2.749	0.6892	1.823	1.921
2.34	7.474	1.429	5.229	2.411	2.728	0.6907	1.828	1.924
2.35	7.353	1.412	5.207	2.435	2.707	0.6923	1.832	1.926
2.36	7.234	1.395	5.186	2.459	2.686	0.6938	1.836	1.928
2.37	7.117	1.378	5.165	2.484	2.665	0.6953	1.840	1.931
2.38	7.002	1.361	5.144	2.509	2.644	0.6968	1.844	1.933
2.39	6.889	1.345	5.123	2.535	2.623	0.6983	1.848	1.935
2.40	6.777	1.328	5.102	2.561	2.602	0.6998	1.852	1.937
2.41	6.667	1.312	5.081	2.587	2.581	0.7013	1.856	1.940
2.42	6.559	1.296	5.061	2.613	2.561	0.7028	1.859	1.942
2.43	6.453	1.280	5.040	2.640	2.540	0.7043	1.863	1.944
2.44	6.349	1.265	5.020	2.667	2.520	0.7057	1.867	1.946
2.45	6.246	1.249	4.999	2.694	2.499	0.7072	1.871	1.949
2.46	6.145	1.234	4.979	2.722	2.479	0.7086	1.875	1.951
2.47	6.045	1.219	4.959	2.750	2.459	0.7100	1.879	1.953
2.48	5.947	1.204	4.938	2.778	2.438	0.7115	1.882	1.955
2.49	5.851	1.190	4.918	2.807	2.418	0.7129	1.886	1.957
2.50	5.756	1.175	4.898	2.836	2.398	0.7143	1.890	1.960

NORMAL SHOCK FUNCTIONS FOR SUPERSONIC COMPRESSIBLE FLOW

$$\gamma = 4/3$$

M	p_{21}	T_{21}	ρ_{21}	M_2	p_{21}^0	μ	ν
2.00	4.428 +0	1.582 +0	2.800	0.5680	7.077 -1	30.00	27.87
2.01	4.474	1.588	2.817	0.5662	7.028	29.84	28.16
2.02	4.520	1.595	2.833	0.5644	6.979	29.67	28.46
2.03	4.566	1.602	2.850	0.5627	6.929	29.51	28.76
2.04	4.613	1.609	2.867	0.5610	6.880	29.35	29.05
2.05	4.660	1.616	2.883	0.5593	6.831	29.20	29.35
2.06	4.707	1.623	2.900	0.5576	6.781	29.04	29.64
2.07	4.754	1.630	2.916	0.5559	6.732	28.89	29.93
2.08	4.801	1.637	2.933	0.5543	6.683	28.74	30.22
2.09	4.849	1.644	2.949	0.5527	6.634	28.59	30.52
2.10	4.897	1.651	2.965	0.5511	6.585	28.44	30.81
2.11	4.945	1.658	2.982	0.5495	6.536	28.29	31.10
2.12	4.993	1.666	2.998	0.5479	6.487	28.15	31.39
2.13	5.042	1.673	3.014	0.5464	6.438	28.00	31.67
2.14	5.091	1.680	3.030	0.5449	6.389	27.86	31.96
2.15	5.140	1.687	3.046	0.5434	6.341	27.72	32.25
2.16	5.189	1.695	3.062	0.5419	6.292	27.58	32.53
2.17	5.238	1.702	3.078	0.5404	6.244	27.44	32.82
2.18	5.288	1.709	3.094	0.5390	6.195	27.31	33.10
2.19	5.338	1.717	3.110	0.5375	6.147	27.17	33.39
2.20	5.388	1.724	3.125	0.5361	6.099	27.04	33.67
2.21	5.439	1.731	3.141	0.5347	6.051	26.90	33.95
2.22	5.489	1.739	3.157	0.5333	6.003	26.77	34.23
2.23	5.540	1.746	3.172	0.5319	5.956	26.64	34.51
2.24	5.591	1.754	3.188	0.5306	5.908	26.52	34.79
2.25	5.642	1.761	3.203	0.5292	5.861	26.39	35.07
2.26	5.694	1.769	3.219	0.5279	5.814	26.26	35.35
2.27	5.746	1.777	3.234	0.5266	5.767	26.14	35.63
2.28	5.798	1.784	3.249	0.5253	5.720	26.02	35.90
2.29	5.850	1.792	3.265	0.5240	5.674	25.89	36.18
2.30	5.902	1.800	3.280	0.5227	5.627	25.77	36.45
2.31	5.955	1.807	3.295	0.5215	5.581	25.65	36.73
2.32	6.008	1.815	3.310	0.5202	5.535	25.53	37.00
2.33	6.061	1.823	3.325	0.5190	5.489	25.42	37.27
2.34	6.114	1.831	3.340	0.5178	5.443	25.30	37.54
2.35	6.168	1.839	3.355	0.5166	5.398	25.19	37.82
2.36	6.222	1.846	3.370	0.5154	5.353	25.07	38.08
2.37	6.276	1.854	3.384	0.5142	5.308	24.96	38.35
2.38	6.330	1.862	3.399	0.5131	5.263	24.85	38.62
2.39	6.385	1.870	3.414	0.5119	5.218	24.74	38.89
2.40	6.439	1.878	3.428	0.5108	5.174	24.63	39.15
2.41	6.494	1.886	3.443	0.5096	5.130	24.52	39.42
2.42	6.550	1.894	3.457	0.5085	5.086	24.41	39.68
2.43	6.605	1.902	3.472	0.5074	5.042	24.30	39.95
2.44	6.661	1.911	3.486	0.5063	4.999	24.20	40.21
2.45	6.717	1.919	3.501	0.5052	4.956	24.09	40.47
2.46	6.773	1.927	3.515	0.5042	4.913	23.99	40.73
2.47	6.829	1.935	3.529	0.5031	4.870	23.88	40.99
2.48	6.886	1.943	3.543	0.5021	4.828	23.78	41.25
2.49	6.942	1.952	3.557	0.5010	4.786	23.68	41.51
2.50	6.999	1.960	3.571	0.5000	4.744	23.58	41.77

ISENTROPIC FUNCTIONS FOR SUPERSONIC COMPRESSIBLE FLOW

$$\gamma = 4/3$$

M	$p/p°$	$T/T°$	$\rho/\rho°$	A/A^*	$q/p°$	u/u_m	u/a^*	$F(\gamma;M)$
2.50	5.756 -2	1.175 -1	4.898 -1	2.836 +0	2.398 -1	0.7143	1.890	1.960
2.51	5.663	1.161	4.878	2.865	2.378	0.7157	1.894	1.962
2.52	5.571	1.147	4.858	2.895	2.358	0.7171	1.897	1.964
2.53	5.481	1.133	4.839	2.925	2.339	0.7184	1.901	1.966
2.54	5.392	1.119	4.819	2.955	2.319	0.7198	1.904	1.968
2.55	5.305	1.105	4.799	2.986	2.300	0.7212	1.908	1.970
2.56	5.219	1.092	4.780	3.017	2.280	0.7225	1.912	1.972
2.57	5.135	1.079	4.760	3.048	2.261	0.7239	1.915	1.974
2.58	5.052	1.066	4.741	3.080	2.242	0.7252	1.919	1.977
2.59	4.970	1.053	4.722	3.112	2.222	0.7265	1.922	1.979
2.60	4.890	1.040	4.702	3.145	2.203	0.7278	1.926	1.981
2.61	4.811	1.027	4.683	3.178	2.185	0.7292	1.929	1.983
2.62	4.733	1.015	4.664	3.211	2.166	0.7305	1.933	1.985
2.63	4.656	1.002	4.645	3.245	2.147	0.7318	1.936	1.987
2.64	4.581	9.902 -2	4.626	3.279	2.128	0.7330	1.940	1.989
2.65	4.507	9.782	4.608	3.313	2.110	0.7343	1.943	1.991
2.66	4.435	9.664	4.589	3.348	2.092	0.7356	1.946	1.993
2.67	4.363	9.546	4.570	3.384	2.073	0.7369	1.950	1.995
2.68	4.293	9.431	4.552	3.419	2.055	0.7381	1.953	1.997
2.69	4.223	9.316	4.533	3.455	2.037	0.7394	1.956	1.999
2.70	4.155	9.203	4.515	3.492	2.019	0.7406	1.960	2.001
2.71	4.088	9.092	4.497	3.529	2.001	0.7418	1.963	2.003
2.72	4.022	8.982	4.478	3.566	1.984	0.7431	1.966	2.005
2.73	3.958	8.873	4.460	3.604	1.966	0.7443	1.969	2.007
2.74	3.894	8.766	4.442	3.642	1.949	0.7455	1.972	2.009
2.75	3.831	8.660	4.424	3.681	1.931	0.7467	1.976	2.010
2.76	3.770	8.555	4.406	3.720	1.914	0.7479	1.979	2.012
2.77	3.709	8.452	4.388	3.759	1.897	0.7491	1.982	2.014
2.78	3.649	8.350	4.371	3.799	1.880	0.7503	1.985	2.016
2.79	3.591	8.249	4.353	3.840	1.863	0.7515	1.988	2.018
2.80	3.533	8.149	4.335	3.881	1.846	0.7526	1.991	2.020
2.81	3.476	8.051	4.318	3.922	1.830	0.7538	1.994	2.022
2.82	3.421	7.954	4.301	3.964	1.813	0.7549	1.997	2.024
2.83	3.366	7.858	4.283	4.006	1.797	0.7561	2.000	2.025
2.84	3.312	7.763	4.266	4.049	1.781	0.7572	2.004	2.027
2.85	3.259	7.670	4.249	4.092	1.764	0.7584	2.007	2.029
2.86	3.207	7.578	4.232	4.136	1.748	0.7595	2.009	2.031
2.87	3.155	7.487	4.215	4.180	1.732	0.7606	2.012	2.033
2.88	3.105	7.397	4.198	4.225	1.717	0.7617	2.015	2.035
2.89	3.055	7.308	4.181	4.270	1.701	0.7628	2.018	2.036
2.90	3.006	7.220	4.164	4.315	1.685	0.7639	2.021	2.038
2.91	2.958	7.133	4.147	4.362	1.670	0.7650	2.024	2.040
2.92	2.911	7.048	4.131	4.408	1.655	0.7661	2.027	2.042
2.93	2.865	6.963	4.114	4.455	1.639	0.7672	2.030	2.043
2.94	2.819	6.880	4.098	4.503	1.624	0.7683	2.033	2.045
2.95	2.774	6.798	4.081	4.551	1.609	0.7693	2.036	2.047
2.96	2.730	6.716	4.065	4.600	1.594	0.7704	2.038	2.049
2.97	2.687	6.636	4.049	4.649	1.580	0.7715	2.041	2.050
2.98	2.644	6.557	4.032	4.699	1.565	0.7725	2.044	2.052
2.99	2.602	6.478	4.016	4.750	1.551	0.7735	2.047	2.054
3.00	2.561	6.401	4.000	4.800	1.536	0.7746	2.049	2.055

NORMAL SHOCK FUNCTIONS FOR SUPERSONIC COMPRESSIBLE FLOW

$$\gamma = 4/3$$

M	p_{21}	T_{21}	ρ_{21}	M_2	p^o_{21}	μ	ν
2.50	6.999 +0	1.960 +0	3.571	0.5000	4.744 −1	23.58	41.77
2.51	7.057	1.968	3.585	0.4990	4.702	23.48	42.03
2.52	7.114	1.977	3.599	0.4980	4.661	23.38	42.28
2.53	7.172	1.985	3.613	0.4970	4.619	23.28	42.54
2.54	7.230	1.993	3.627	0.4960	4.578	23.19	42.79
2.55	7.288	2.002	3.641	0.4950	4.538	23.09	43.04
2.56	7.346	2.010	3.654	0.4941	4.497	22.99	43.30
2.57	7.405	2.019	3.668	0.4931	4.457	22.90	43.55
2.58	7.464	2.027	3.681	0.4922	4.417	22.81	43.80
2.59	7.523	2.036	3.695	0.4912	4.378	22.71	44.05
2.60	7.582	2.045	3.708	0.4903	4.338	22.62	44.30
2.61	7.642	2.053	3.722	0.4894	4.299	22.53	44.55
2.62	7.701	2.062	3.735	0.4885	4.260	22.44	44.79
2.63	7.761	2.071	3.748	0.4876	4.222	22.35	45.04
2.64	7.822	2.079	3.761	0.4867	4.184	22.26	45.29
2.65	7.882	2.088	3.775	0.4858	4.146	22.17	45.53
2.66	7.943	2.097	3.788	0.4849	4.108	22.08	45.78
2.67	8.004	2.106	3.801	0.4841	4.070	22.00	46.02
2.68	8.065	2.115	3.814	0.4832	4.033	21.91	46.26
2.69	8.126	2.124	3.827	0.4824	3.996	21.82	46.50
2.70	8.188	2.132	3.840	0.4815	3.959	21.74	46.74
2.71	8.250	2.141	3.852	0.4807	3.923	21.66	46.98
2.72	8.312	2.150	3.865	0.4799	3.887	21.57	47.22
2.73	8.374	2.159	3.878	0.4791	3.851	21.49	47.46
2.74	8.436	2.168	3.890	0.4782	3.815	21.41	47.70
2.75	8.499	2.178	3.903	0.4774	3.780	21.32	47.93
2.76	8.562	2.187	3.916	0.4766	3.745	21.24	48.17
2.77	8.625	2.196	3.928	0.4759	3.710	21.16	48.40
2.78	8.689	2.205	3.940	0.4751	3.676	21.08	48.64
2.79	8.752	2.214	3.953	0.4743	3.641	21.00	48.87
2.80	8.816	2.223	3.965	0.4735	3.607	20.93	49.10
2.81	8.880	2.233	3.977	0.4728	3.574	20.85	49.34
2.82	8.945	2.242	3.990	0.4720	3.540	20.77	49.57
2.83	9.009	2.251	4.002	0.4713	3.507	20.69	49.80
2.84	9.074	2.261	4.014	0.4706	3.474	20.62	50.03
2.85	9.139	2.270	4.026	0.4698	3.441	20.54	50.25
2.86	9.204	2.280	4.038	0.4691	3.409	20.47	50.48
2.87	9.270	2.289	4.050	0.4684	3.377	20.39	50.71
2.88	9.336	2.298	4.062	0.4677	3.345	20.32	50.93
2.89	9.401	2.308	4.073	0.4670	3.313	20.25	51.16
2.90	9.468	2.318	4.085	0.4663	3.282	20.17	51.38
2.91	9.534	2.327	4.097	0.4656	3.251	20.10	51.61
2.92	9.601	2.337	4.109	0.4649	3.220	20.03	51.83
2.93	9.667	2.346	4.120	0.4642	3.189	19.96	52.05
2.94	9.735	2.356	4.132	0.4636	3.159	19.89	52.27
2.95	9.802	2.366	4.143	0.4629	3.129	19.82	52.49
2.96	9.869	2.376	4.155	0.4622	3.099	19.75	52.71
2.97	9.937	2.385	4.166	0.4616	3.070	19.68	52.93
2.98	1.001 +1	2.395	4.177	0.4609	3.040	19.61	53.15
2.99	1.007	2.405	4.189	0.4603	3.011	19.54	53.37
3.00	1.014	2.415	4.200	0.4596	2.982	19.47	53.58

ISENTROPIC FUNCTIONS FOR SUPERSONIC COMPRESSIBLE FLOW

$$\gamma = 4/3$$

M	p/p°	T/T°	ρ/ρ°	A/A^*	q/p°	u/u_m	u/a^*	$F(\gamma;M)$
3.00	2.561 −2	6.401 −2	4.000 −1	4.800 +0	1.536 −1	0.7746	2.049	2.055
3.02	2.480	6.249	3.968	4.904	1.508	0.7766	2.055	2.059
3.04	2.402	6.102	3.937	5.010	1.480	0.7787	2.060	2.062
3.06	2.327	5.957	3.906	5.118	1.452	0.7807	2.066	2.065
3.08	2.254	5.817	3.875	5.228	1.425	0.7826	2.071	2.069
3.10	2.183	5.680	3.844	5.341	1.399	0.7846	2.076	2.072
3.12	2.115	5.546	3.814	5.456	1.372	0.7865	2.081	2.075
3.14	2.049	5.416	3.783	5.574	1.347	0.7885	2.086	2.078
3.16	1.985	5.289	3.754	5.694	1.321	0.7903	2.091	2.081
3.18	1.923	5.165	3.724	5.817	1.297	0.7922	2.096	2.084
3.20	1.864	5.044	3.695	5.943	1.272	0.7941	2.101	2.087
3.22	1.806	4.926	3.666	6.071	1.248	0.7959	2.106	2.091
3.24	1.750	4.811	3.637	6.202	1.225	0.7977	2.111	2.094
3.26	1.696	4.699	3.609	6.335	1.201	0.7995	2.115	2.096
3.28	1.644	4.590	3.581	6.472	1.179	0.8012	2.120	2.099
3.30	1.593	4.484	3.553	6.611	1.156	0.8030	2.124	2.102
3.32	1.544	4.380	3.525	6.753	1.134	0.8047	2.129	2.105
3.34	1.497	4.279	3.498	6.899	1.113	0.8064	2.134	2.108
3.36	1.451	4.180	3.471	7.047	1.092	0.8081	2.138	2.111
3.38	1.406	4.084	3.444	7.198	1.071	0.8097	2.142	2.114
3.40	1.363	3.990	3.417	7.353	1.051	0.8114	2.147	2.116
3.42	1.322	3.898	3.391	7.510	1.031	0.8130	2.151	2.119
3.44	1.282	3.809	3.365	7.671	1.011	0.8146	2.155	2.122
3.46	1.243	3.722	3.339	7.835	9.917 −2	0.8162	2.159	2.124
3.48	1.205	3.637	3.313	8.003	9.728	0.8177	2.164	2.127
3.50	1.169	3.554	3.288	8.174	9.543	0.8193	2.168	2.130
3.52	1.133	3.473	3.263	8.348	9.361	0.8208	2.172	2.132
3.54	1.099	3.395	3.238	8.527	9.182	0.8223	2.176	2.135
3.56	1.066	3.318	3.213	8.708	9.007	0.8238	2.180	2.137
3.58	1.034	3.243	3.189	8.894	8.835	0.8253	2.184	2.140
3.60	1.003	3.170	3.165	9.083	8.666	0.8268	2.187	2.142
3.62	9.732 −3	3.098	3.141	9.276	8.501	0.8282	2.191	2.145
3.64	9.441	3.029	3.117	9.472	8.339	0.8296	2.195	2.147
3.66	9.160	2.961	3.094	9.673	8.180	0.8310	2.199	2.150
3.68	8.888	2.895	3.070	9.878	8.024	0.8324	2.202	2.152
3.70	8.624	2.830	3.047	1.009 +1	7.871	0.8338	2.206	2.154
3.72	8.369	2.767	3.025	1.030	7.720	0.8352	2.210	2.157
3.74	8.122	2.706	3.002	1.052	7.573	0.8365	2.213	2.159
3.76	7.883	2.646	2.980	1.074	7.429	0.8379	2.217	2.161
3.78	7.651	2.587	2.958	1.096	7.287	0.8392	2.220	2.164
3.80	7.427	2.530	2.936	1.119	7.149	0.8405	2.224	2.166
3.82	7.209	2.474	2.914	1.143	7.013	0.8418	2.227	2.168
3.84	6.999	2.420	2.892	1.167	6.879	0.8431	2.231	2.170
3.86	6.795	2.367	2.871	1.191	6.749	0.8443	2.234	2.172
3.88	6.597	2.315	2.850	1.216	6.620	0.8456	2.237	2.174
3.90	6.406	2.264	2.829	1.241	6.495	0.8468	2.241	2.177
3.92	6.220	2.215	2.808	1.267	6.371	0.8480	2.244	2.179
3.94	6.040	2.167	2.788	1.294	6.251	0.8492	2.247	2.181
3.96	5.866	2.120	2.768	1.320	6.132	0.8504	2.250	2.183
3.98	5.697	2.074	2.747	1.348	6.016	0.8516	2.253	2.185
4.00	5.534	2.029	2.727	1.376	5.902	0.8528	2.256	2.187

NORMAL SHOCK FUNCTIONS FOR SUPERSONIC COMPRESSIBLE FLOW

$$\gamma = 4/3$$

M	p_{21}	T_{21}	ρ_{21}	M_2	p^o_{21}	μ	ν
3.00	1.014 +1	2.415 +0	4.200	0.4596	2.982 -1	19.47	53.58
3.02	1.028	2.435	4.222	0.4584	2.926	19.34	54.01
3.04	1.042	2.455	4.244	0.4572	2.870	19.21	54.44
3.06	1.056	2.475	4.266	0.4559	2.815	19.08	54.87
3.08	1.070	2.495	4.288	0.4547	2.761	18.95	55.29
3.10	1.084	2.515	4.309	0.4536	2.708	18.82	55.71
3.12	1.098	2.536	4.331	0.4524	2.656	18.69	56.12
3.14	1.112	2.556	4.352	0.4513	2.605	18.57	56.53
3.16	1.127	2.577	4.372	0.4502	2.555	18.45	56.94
3.18	1.141	2.598	4.393	0.4491	2.506	18.33	57.35
3.20	1.156	2.619	4.414	0.4480	2.457	18.21	57.75
3.22	1.171	2.640	4.434	0.4469	2.410	18.09	58.15
3.24	1.185	2.661	4.454	0.4459	2.363	17.98	58.55
3.26	1.200	2.683	4.474	0.4449	2.317	17.86	58.95
3.28	1.215	2.704	4.494	0.4439	2.273	17.75	59.34
3.30	1.230	2.726	4.513	0.4429	2.229	17.64	59.73
3.32	1.245	2.747	4.533	0.4419	2.185	17.53	60.12
3.34	1.261	2.769	4.552	0.4409	2.143	17.42	60.50
3.36	1.276	2.791	4.571	0.4400	2.101	17.32	60.88
3.38	1.291	2.813	4.589	0.4390	2.061	17.21	61.26
3.40	1.307	2.836	4.608	0.4381	2.021	17.11	61.63
3.42	1.322	2.858	4.626	0.4372	1.982	17.00	62.01
3.44	1.338	2.881	4.645	0.4363	1.943	16.90	62.38
3.46	1.354	2.903	4.663	0.4355	1.905	16.80	62.74
3.48	1.370	2.926	4.681	0.4346	1.868	16.70	63.11
3.50	1.386	2.949	4.698	0.4338	1.832	16.60	63.47
3.52	1.402	2.972	4.716	0.4329	1.796	16.51	63.83
3.54	1.418	2.995	4.733	0.4321	1.762	16.41	64.19
3.56	1.434	3.018	4.751	0.4313	1.727	16.31	64.54
3.58	1.450	3.042	4.768	0.4305	1.694	16.22	64.90
3.60	1.467	3.065	4.785	0.4297	1.661	16.13	65.25
3.62	1.483	3.089	4.801	0.4289	1.629	16.04	65.59
3.64	1.500	3.113	4.818	0.4282	1.597	15.95	65.94
3.66	1.516	3.137	4.834	0.4274	1.566	15.86	66.28
3.68	1.533	3.161	4.851	0.4267	1.536	15.77	66.62
3.70	1.550	3.185	4.867	0.4260	1.506	15.68	66.96
3.72	1.567	3.209	4.883	0.4252	1.477	15.59	67.29
3.74	1.584	3.234	4.899	0.4245	1.448	15.51	67.62
3.76	1.601	3.258	4.914	0.4238	1.420	15.42	67.95
3.78	1.619	3.283	4.930	0.4232	1.393	15.34	68.28
3.80	1.636	3.308	4.945	0.4225	1.366	15.26	68.61
3.82	1.653	3.333	4.960	0.4218	1.340	15.18	68.93
3.84	1.671	3.358	4.975	0.4212	1.314	15.10	69.25
3.86	1.688	3.383	4.990	0.4205	1.289	15.02	69.57
3.88	1.706	3.409	5.005	0.4199	1.264	14.94	69.89
3.90	1.724	3.434	5.020	0.4192	1.239	14.86	70.20
3.92	1.742	3.460	5.034	0.4186	1.216	14.78	70.51
3.94	1.760	3.486	5.049	0.4180	1.192	14.70	70.82
3.96	1.778	3.511	5.063	0.4174	1.169	14.63	71.13
3.98	1.796	3.537	5.077	0.4168	1.147	14.55	71.44
4.00	1.814	3.564	5.091	0.4162	1.125	14.48	71.74

ISENTROPIC FUNCTIONS FOR SUPERSONIC COMPRESSIBLE FLOW

$\gamma = 4/3$

M	p/p°	T/T°	ρ/ρ°	A/A^*	q/p°	u/u_m	u/a^*	$F(\gamma;M)$
4.00	5.534 -3	2.029 -2	2.727 -1	1.376 +1	5.902 -2	0.8528	2.256	2.187
4.02	5.375	1.985	2.708	1.404	5.791	0.8539	2.259	2.189
4.04	5.222	1.943	2.688	1.433	5.681	0.8551	2.262	2.191
4.06	5.073	1.901	2.669	1.462	5.574	0.8562	2.265	2.193
4.08	4.929	1.860	2.650	1.492	5.469	0.8573	2.268	2.195
4.10	4.789	1.820	2.631	1.523	5.366	0.8585	2.271	2.197
4.12	4.653	1.782	2.612	1.554	5.265	0.8595	2.274	2.198
4.14	4.522	1.744	2.593	1.586	5.166	0.8606	2.277	2.200
4.16	4.394	1.707	2.575	1.618	5.069	0.8617	2.280	2.202
4.18	4.271	1.671	2.556	1.651	4.974	0.8628	2.283	2.204
4.20	4.151	1.635	2.538	1.685	4.881	0.8638	2.285	2.206
4.22	4.035	1.601	2.520	1.719	4.790	0.8649	2.288	2.208
4.24	3.922	1.567	2.502	1.754	4.700	0.8659	2.291	2.209
4.26	3.813	1.534	2.485	1.789	4.612	0.8669	2.294	2.211
4.28	3.707	1.502	2.467	1.826	4.526	0.8679	2.296	2.213
4.30	3.604	1.471	2.450	1.862	4.442	0.8689	2.299	2.215
4.32	3.504	1.440	2.433	1.900	4.359	0.8699	2.302	2.216
4.34	3.407	1.410	2.416	1.938	4.278	0.8709	2.304	2.218
4.36	3.314	1.381	2.399	1.977	4.199	0.8718	2.307	2.220
4.38	3.222	1.353	2.383	2.016	4.121	0.8728	2.309	2.221
4.40	3.134	1.325	2.366	2.057	4.045	0.8737	2.312	2.223
4.42	3.048	1.297	2.350	2.098	3.970	0.8747	2.314	2.225
4.44	2.965	1.271	2.334	2.139	3.897	0.8756	2.317	2.226
4.46	2.885	1.245	2.318	2.182	3.825	0.8765	2.319	2.228
4.48	2.806	1.219	2.302	2.225	3.755	0.8774	2.321	2.230
4.50	2.730	1.194	2.286	2.269	3.686	0.8783	2.324	2.231
4.52	2.656	1.170	2.270	2.314	3.618	0.8792	2.326	2.233
4.54	2.585	1.146	2.255	2.359	3.552	0.8801	2.328	2.234
4.56	2.515	1.123	2.239	2.406	3.487	0.8809	2.331	2.236
4.58	2.448	1.100	2.224	2.453	3.423	0.8818	2.333	2.237
4.60	2.382	1.078	2.209	2.501	3.360	0.8827	2.335	2.239
4.62	2.319	1.057	2.194	2.550	3.299	0.8835	2.338	2.240
4.64	2.257	1.035	2.180	2.599	3.239	0.8843	2.340	2.242
4.66	2.197	1.015	2.165	2.650	3.180	0.8852	2.342	2.243
4.68	2.139	9.945 -3	2.150	2.701	3.123	0.8860	2.344	2.245
4.70	2.082	9.747	2.136	2.754	3.066	0.8868	2.346	2.246
4.72	2.027	9.554	2.122	2.807	3.011	0.8876	2.348	2.247
4.74	1.974	9.365	2.108	2.861	2.956	0.8884	2.350	2.249
4.76	1.922	9.179	2.094	2.916	2.903	0.8892	2.353	2.250
4.78	1.872	8.999	2.080	2.972	2.851	0.8899	2.355	2.251
4.80	1.823	8.822	2.066	3.029	2.800	0.8907	2.357	2.253
4.82	1.775	8.649	2.053	3.087	2.749	0.8915	2.359	2.254
4.84	1.729	8.479	2.039	3.146	2.700	0.8922	2.361	2.255
4.86	1.684	8.314	2.026	3.206	2.652	0.8930	2.363	2.257
4.88	1.641	8.152	2.013	3.267	2.605	0.8937	2.365	2.258
4.90	1.598	7.994	1.999	3.329	2.558	0.8945	2.367	2.259
4.92	1.557	7.839	1.986	3.392	2.513	0.8952	2.368	2.261
4.94	1.517	7.687	1.974	3.456	2.468	0.8959	2.370	2.262
4.96	1.478	7.539	1.961	3.521	2.424	0.8966	2.372	2.263
4.98	1.440	7.394	1.948	3.588	2.381	0.8973	2.374	2.264
5.00	1.404	7.252	1.936	3.655	2.339	0.8980	2.376	2.266

NORMAL SHOCK FUNCTIONS FOR SUPERSONIC COMPRESSIBLE FLOW

$$\gamma = 4/3$$

M	P_{21}	T_{21}	ρ_{21}	M_2	P^o_{21}	μ	ν
4.00	1.814 +1	3.564 +0	5.091	0.4162	1.125 -1	14.48	71.74
4.02	1.832	3.590	5.105	0.4156	1.103	14.40	72.04
4.04	1.851	3.616	5.118	0.4151	1.082	14.33	72.34
4.06	1.869	3.643	5.132	0.4145	1.062	14.26	72.64
4.08	1.888	3.669	5.145	0.4139	1.041	14.19	72.94
4.10	1.907	3.696	5.159	0.4134	1.022	14.12	73.23
4.12	1.925	3.723	5.172	0.4128	1.002	14.05	73.52
4.14	1.944	3.750	5.185	0.4123	9.831 -2	13.98	73.81
4.16	1.963	3.777	5.198	0.4118	9.645	13.91	74.10
4.18	1.982	3.805	5.211	0.4113	9.462	13.84	74.38
4.20	2.002	3.832	5.223	0.4108	9.283	13.77	74.67
4.22	2.021	3.860	5.236	0.4102	9.107	13.71	74.95
4.24	2.040	3.887	5.248	0.4097	8.935	13.64	75.23
4.26	2.060	3.915	5.261	0.4092	8.766	13.58	75.51
4.28	2.079	3.943	5.273	0.4088	8.601	13.51	75.78
4.30	2.099	3.971	5.285	0.4083	8.439	13.45	76.05
4.32	2.118	3.999	5.297	0.4078	8.281	13.38	76.33
4.34	2.138	4.028	5.309	0.4073	8.125	13.32	76.60
4.36	2.158	4.056	5.321	0.4069	7.973	13.26	76.87
4.38	2.178	4.085	5.332	0.4064	7.824	13.20	77.13
4.40	2.198	4.113	5.344	0.4060	7.678	13.14	77.40
4.42	2.218	4.142	5.355	0.4055	7.535	13.08	77.66
4.44	2.239	4.171	5.367	0.4051	7.394	13.02	77.92
4.46	2.259	4.200	5.378	0.4046	7.257	12.96	78.18
4.48	2.279	4.230	5.389	0.4042	7.122	12.90	78.44
4.50	2.300	4.259	5.400	0.4038	6.990	12.84	78.70
4.52	2.320	4.288	5.411	0.4034	6.860	12.78	78.95
4.54	2.341	4.318	5.422	0.4030	6.734	12.73	79.20
4.56	2.362	4.348	5.432	0.4025	6.609	12.67	79.45
4.58	2.383	4.378	5.443	0.4021	6.487	12.61	79.70
4.60	2.404	4.408	5.454	0.4017	6.368	12.56	79.95
4.62	2.425	4.438	5.464	0.4014	6.251	12.50	80.20
4.64	2.446	4.468	5.474	0.4010	6.136	12.45	80.44
4.66	2.467	4.499	5.485	0.4006	6.024	12.39	80.69
4.68	2.489	4.529	5.495	0.4002	5.914	12.34	80.93
4.70	2.510	4.560	5.505	0.3998	5.806	12.29	81.17
4.72	2.532	4.591	5.515	0.3995	5.700	12.23	81.41
4.74	2.553	4.622	5.525	0.3991	5.596	12.18	81.64
4.76	2.575	4.653	5.534	0.3987	5.495	12.13	81.88
4.78	2.597	4.684	5.544	0.3984	5.395	12.08	82.11
4.80	2.619	4.715	5.554	0.3980	5.298	12.03	82.34
4.82	2.641	4.747	5.563	0.3977	5.202	11.97	82.58
4.84	2.663	4.778	5.573	0.3973	5.108	11.92	82.80
4.86	2.685	4.810	5.582	0.3970	5.016	11.87	83.03
4.88	2.707	4.842	5.591	0.3966	4.926	11.83	83.26
4.90	2.729	4.874	5.600	0.3963	4.838	11.78	83.48
4.92	2.752	4.906	5.609	0.3960	4.751	11.73	83.71
4.94	2.774	4.938	5.618	0.3956	4.666	11.68	83.93
4.96	2.797	4.970	5.627	0.3953	4.583	11.63	84.15
4.98	2.820	5.003	5.636	0.3950	4.501	11.58	84.37
5.00	2.843	5.036	5.645	0.3947	4.421	11.54	84.59

ISENTROPIC FUNCTIONS FOR SUPERSONIC COMPRESSIBLE FLOW

$\gamma = 4/3$

M	$p/p°$	$T/T°$	$\rho/\rho°$	A/A^*	$q/p°$	u/u_m	u/a^*	$F(\gamma;M)$
5.00	1.404 -3	7.252 -3	1.936 -1	3.655 +1	2.339 -2	0.8980	2.376	2.266
5.05	1.316	6.910	1.905	3.828	2.238	0.8997	2.381	2.269
5.10	1.235	6.587	1.875	4.009	2.141	0.9014	2.385	2.272
5.15	1.159	6.280	1.845	4.197	2.049	0.9031	2.389	2.275
5.20	1.088	5.990	1.816	4.393	1.961	0.9046	2.394	2.277
5.25	1.022	5.714	1.788	4.596	1.877	0.9062	2.398	2.280
5.30	9.599 -4	5.453	1.760	4.808	1.797	0.9077	2.402	2.283
5.35	9.022	5.205	1.733	5.029	1.721	0.9092	2.406	2.286
5.40	8.482	4.970	1.707	5.258	1.649	0.9107	2.409	2.288
5.45	7.978	4.747	1.681	5.497	1.580	0.9121	2.413	2.291
5.50	7.507	4.535	1.655	5.745	1.514	0.9135	2.417	2.293
5.55	7.067	4.334	1.630	6.003	1.451	0.9149	2.421	2.295
5.60	6.654	4.143	1.606	6.271	1.391	0.9162	2.424	2.298
5.65	6.268	3.961	1.582	6.549	1.334	0.9175	2.427	2.300
5.70	5.906	3.789	1.559	6.838	1.279	0.9188	2.431	2.302
5.75	5.568	3.625	1.536	7.138	1.227	0.9200	2.434	2.305
5.80	5.250	3.468	1.514	7.449	1.177	0.9212	2.437	2.307
5.85	4.953	3.320	1.492	7.773	1.130	0.9224	2.440	2.309
5.90	4.674	3.179	1.470	8.108	1.084	0.9236	2.444	2.311
5.95	4.412	3.044	1.449	8.456	1.041	0.9247	2.447	2.313
6.00	4.166	2.916	1.429	8.817	9.998 -3	0.9258	2.450	2.315
6.05	3.935	2.794	1.408	9.191	9.602	0.9269	2.452	2.317
6.10	3.719	2.678	1.389	9.578	9.224	0.9280	2.455	2.319
6.15	3.515	2.567	1.369	9.980	8.862	0.9290	2.458	2.321
6.20	3.324	2.462	1.350	1.040 +2	8.517	0.9300	2.461	2.323
6.25	3.144	2.361	1.332	1.083	8.186	0.9310	2.463	2.324
6.30	2.975	2.265	1.313	1.127	7.870	0.9320	2.466	2.326
6.35	2.815	2.173	1.295	1.174	7.568	0.9330	2.469	2.328
6.40	2.666	2.086	1.278	1.222	7.278	0.9339	2.471	2.330
6.45	2.525	2.003	1.261	1.271	7.001	0.9349	2.473	2.331
6.50	2.392	1.923	1.244	1.323	6.736	0.9358	2.476	2.333
6.55	2.267	1.847	1.227	1.376	6.483	0.9366	2.478	2.334
6.60	2.149	1.775	1.211	1.431	6.240	0.9375	2.480	2.336
6.65	2.038	1.705	1.195	1.487	6.007	0.9384	2.483	2.337
6.70	1.933	1.639	1.179	1.546	5.784	0.9392	2.485	2.339
6.75	1.834	1.576	1.164	1.607	5.570	0.9400	2.487	2.340
6.80	1.741	1.515	1.149	1.669	5.365	0.9408	2.489	2.342
6.85	1.653	1.458	1.134	1.734	5.169	0.9416	2.491	2.343
6.90	1.569	1.402	1.119	1.801	4.981	0.9424	2.493	2.345
6.95	1.491	1.349	1.105	1.871	4.800	0.9431	2.495	2.346
7.00	1.417	1.299	1.091	1.942	4.627	0.9439	2.497	2.347
7.05	1.347	1.250	1.077	2.016	4.461	0.9446	2.499	2.349
7.10	1.280	1.204	1.064	2.092	4.302	0.9453	2.501	2.350
7.15	1.218	1.159	1.050	2.171	4.149	0.9460	2.503	2.351
7.20	1.158	1.116	1.037	2.252	4.003	0.9467	2.505	2.353
7.25	1.102	1.076	1.025	2.336	3.862	0.9474	2.507	2.354
7.30	1.049	1.037	1.012	2.422	3.727	0.9480	2.508	2.355
7.35	9.987 -5	9.990 -4	9.997 -2	2.511	3.597	0.9487	2.510	2.356
7.40	9.511	9.631	9.876	2.603	3.472	0.9493	2.512	2.357
7.45	9.060	9.286	9.756	2.698	3.352	0.9500	2.513	2.358
7.50	8.633	8.956	9.639	2.796	3.237	0.9506	2.515	2.359

NORMAL SHOCK FUNCTIONS FOR SUPERSONIC COMPRESSIBLE FLOW

$$\gamma = 4/3$$

M	P_{21}	T_{21}	ρ_{21}	M_2	P^o_{21}	μ	ν
5.00	2.843 +1	5.036 +0	5.645	0.3947	4.421 -2	11.54	84.59
5.05	2.900	5.118	5.667	0.3939	4.228	11.42	85.13
5.10	2.958	5.201	5.688	0.3932	4.044	11.31	85.66
5.15	3.017	5.284	5.709	0.3924	3.868	11.20	86.18
5.20	3.076	5.369	5.729	0.3917	3.701	11.09	86.69
5.25	3.135	5.454	5.749	0.3910	3.543	10.98	87.20
5.30	3.196	5.541	5.768	0.3904	3.391	10.88	87.70
5.35	3.257	5.628	5.787	0.3897	3.247	10.77	88.19
5.40	3.318	5.715	5.805	0.3891	3.109	10.67	88.68
5.45	3.380	5.804	5.824	0.3884	2.978	10.57	89.15
5.50	3.443	5.894	5.841	0.3878	2.854	10.48	89.62
5.55	3.506	5.984	5.859	0.3872	2.734	10.38	90.08
5.60	3.569	6.075	5.876	0.3867	2.621	10.29	90.54
5.65	3.634	6.167	5.892	0.3861	2.513	10.19	90.99
5.70	3.699	6.260	5.909	0.3856	2.409	10.10	91.43
5.75	3.764	6.353	5.925	0.3850	2.311	10.02	91.87
5.80	3.830	6.447	5.940	0.3845	2.216	9.929	92.30
5.85	3.897	6.543	5.956	0.3840	2.127	9.843	92.72
5.90	3.964	6.639	5.971	0.3835	2.041	9.759	93.14
5.95	4.031	6.735	5.986	0.3830	1.959	9.676	93.55
6.00	4.100	6.833	6.000	0.3825	1.881	9.594	93.96
6.05	4.169	6.931	6.014	0.3821	1.806	9.514	94.36
6.10	4.238	7.031	6.028	0.3816	1.735	9.436	94.76
6.15	4.308	7.131	6.042	0.3812	1.666	9.358	95.15
6.20	4.379	7.231	6.055	0.3808	1.601	9.282	95.53
6.25	4.450	7.333	6.068	0.3803	1.539	9.207	95.91
6.30	4.521	7.436	6.081	0.3799	1.479	9.134	96.28
6.35	4.594	7.539	6.093	0.3795	1.422	9.061	96.65
6.40	4.667	7.643	6.106	0.3791	1.367	8.990	97.02
6.45	4.740	7.748	6.118	0.3788	1.315	8.919	97.38
6.50	4.814	7.854	6.129	0.3784	1.265	8.850	97.73
6.55	4.889	7.960	6.141	0.3780	1.217	8.782	98.08
6.60	4.964	8.068	6.152	0.3777	1.172	8.715	98.43
6.65	5.039	8.176	6.164	0.3773	1.128	8.649	98.77
6.70	5.116	8.285	6.175	0.3770	1.086	8.584	99.10
6.75	5.193	8.395	6.185	0.3766	1.046	8.520	99.43
6.80	5.270	8.505	6.196	0.3763	1.007	8.457	99.76
6.85	5.348	8.617	6.206	0.3760	9.700 -3	8.395	100.1
6.90	5.426	8.729	6.217	0.3757	9.346	8.333	100.4
6.95	5.506	8.842	6.227	0.3754	9.007	8.273	100.7
7.00	5.585	8.956	6.236	0.3751	8.681	8.213	101.0
7.05	5.666	9.071	6.246	0.3748	8.369	8.155	101.3
7.10	5.746	9.186	6.255	0.3745	8.069	8.097	101.6
7.15	5.828	9.303	6.265	0.3742	7.781	8.040	101.9
7.20	5.910	9.420	6.274	0.3739	7.506	7.984	102.2
7.25	5.992	9.538	6.283	0.3736	7.241	7.928	102.5
7.30	6.076	9.657	6.292	0.3734	6.987	7.874	102.8
7.35	6.159	9.776	6.300	0.3731	6.743	7.820	103.1
7.40	6.244	9.897	6.309	0.3728	6.508	7.767	103.4
7.45	6.328	1.002 +1	6.317	0.3726	6.283	7.714	103.7
7.50	6.414	1.014	6.325	0.3723	6.067	7.662	103.9

ISENTROPIC FUNCTIONS FOR SUPERSONIC COMPRESSIBLE FLOW

$$\gamma = 4/3$$

M	$p/p°$	$T/T°$	$\rho/\rho°$	A/A^*	$q/p°$	u/u_m	u/a^*	$F(\gamma;M)$
7.50	8.633 −5	8.956 −4	9.639 −2	2.796 +2	3.237 −3	0.9506	2.515	2.359
7.55	8.228	8.639	9.524	2.897	3.126	0.9512	2.517	2.361
7.60	7.844	8.335	9.411	3.000	3.020	0.9518	2.518	2.362
7.65	7.479	8.043	9.300	3.107	2.918	0.9524	2.520	2.363
7.70	7.134	7.762	9.190	3.218	2.820	0.9529	2.521	2.364
7.75	6.806	7.493	9.083	3.331	2.725	0.9535	2.523	2.365
7.80	6.495	7.235	8.977	3.448	2.634	0.9541	2.524	2.366
7.85	6.199	6.986	8.873	3.569	2.547	0.9546	2.526	2.367
7.90	5.919	6.748	8.771	3.693	2.462	0.9551	2.527	2.368
7.95	5.652	6.519	8.671	3.821	2.381	0.9557	2.528	2.369
8.00	5.399	6.298	8.572	3.952	2.303	0.9562	2.530	2.370
8.05	5.158	6.087	8.475	4.088	2.228	0.9567	2.531	2.371
8.10	4.930	5.883	8.379	4.227	2.156	0.9572	2.533	2.371
8.15	4.712	5.687	8.285	4.370	2.086	0.9577	2.534	2.372
8.20	4.505	5.499	8.193	4.517	2.019	0.9582	2.535	2.373
8.25	4.308	5.318	8.102	4.669	1.955	0.9586	2.536	2.374
8.30	4.121	5.143	8.012	4.825	1.893	0.9591	2.538	2.375
8.35	3.943	4.976	7.924	4.985	1.833	0.9596	2.539	2.376
8.40	3.773	4.814	7.837	5.150	1.775	0.9600	2.540	2.377
8.45	3.611	4.659	7.752	5.320	1.719	0.9605	2.541	2.377
8.50	3.458	4.509	7.668	5.494	1.665	0.9609	2.542	2.378
8.55	3.311	4.365	7.586	5.673	1.613	0.9613	2.543	2.379
8.60	3.171	4.226	7.504	5.857	1.563	0.9617	2.545	2.380
8.65	3.038	4.092	7.424	6.046	1.515	0.9622	2.546	2.380
8.70	2.911	3.963	7.345	6.240	1.469	0.9626	2.547	2.381
8.75	2.790	3.839	7.268	6.439	1.424	0.9630	2.548	2.382
8.80	2.674	3.719	7.191	6.644	1.381	0.9634	2.549	2.383
8.85	2.564	3.603	7.116	6.854	1.339	0.9638	2.550	2.383
8.90	2.459	3.492	7.042	7.070	1.298	0.9641	2.551	2.384
8.95	2.359	3.384	6.969	7.292	1.259	0.9645	2.552	2.385
9.00	2.263	3.281	6.897	7.519	1.222	0.9649	2.553	2.385
9.05	2.171	3.181	6.826	7.753	1.185	0.9653	2.554	2.386
9.10	2.084	3.084	6.756	7.992	1.150	0.9656	2.555	2.387
9.15	2.000	2.991	6.688	8.238	1.116	0.9660	2.556	2.387
9.20	1.921	2.901	6.620	8.490	1.084	0.9663	2.557	2.388
9.25	1.844	2.814	6.553	8.749	1.052	0.9667	2.558	2.389
9.30	1.771	2.730	6.488	9.014	1.021	0.9670	2.559	2.389
9.35	1.702	2.650	6.423	9.287	9.917 −4	0.9674	2.559	2.390
9.40	1.635	2.571	6.359	9.566	9.631	0.9677	2.560	2.390
9.45	1.571	2.496	6.296	9.852	9.355	0.9680	2.561	2.391
9.50	1.510	2.423	6.234	1.015 +3	9.087	0.9683	2.562	2.392
9.55	1.452	2.352	6.173	1.045	8.828	0.9686	2.563	2.392
9.60	1.396	2.284	6.113	1.075	8.578	0.9690	2.564	2.393
9.65	1.343	2.218	6.053	1.107	8.336	0.9693	2.564	2.393
9.70	1.292	2.155	5.995	1.139	8.101	0.9696	2.565	2.394
9.75	1.243	2.093	5.937	1.173	7.875	0.9699	2.566	2.394
9.80	1.196	2.033	5.880	1.207	7.655	0.9702	2.567	2.395
9.85	1.151	1.976	5.824	1.241	7.443	0.9704	2.568	2.395
9.90	1.108	1.920	5.769	1.277	7.237	0.9707	2.568	2.396
9.95	1.066	1.866	5.714	1.314	7.038	0.9710	2.569	2.396
10.00	1.027	1.814	5.661	1.351	6.845	0.9713	2.570	2.397

NORMAL SHOCK FUNCTIONS FOR SUPERSONIC COMPRESSIBLE FLOW

$$\gamma = 4/3$$

M	P_{21}	T_{21}	ρ_{21}	M_2	P^o_{21}	μ	ν
7.50	6.414 +1	1.014 +1	6.325	0.3723	6.067 -3	7.662	103.9
7.55	6.500	1.026	6.333	0.3721	5.859	7.611	104.2
7.60	6.586	1.039	6.341	0.3719	5.659	7.561	104.5
7.65	6.674	1.051	6.349	0.3716	5.467	7.511	104.8
7.70	6.761	1.064	6.357	0.3714	5.282	7.462	105.0
7.75	6.850	1.076	6.364	0.3712	5.105	7.414	105.3
7.80	6.938	1.089	6.372	0.3710	4.934	7.366	105.5
7.85	7.028	1.102	6.379	0.3707	4.770	7.319	105.8
7.90	7.118	1.115	6.386	0.3705	4.612	7.272	106.0
7.95	7.208	1.128	6.393	0.3703	4.460	7.226	106.3
8.00	7.300	1.141	6.400	0.3701	4.313	7.181	106.5
8.05	7.391	1.154	6.407	0.3699	4.173	7.136	106.8
8.10	7.484	1.167	6.413	0.3697	4.037	7.092	107.0
8.15	7.576	1.180	6.420	0.3695	3.906	7.048	107.2
8.20	7.670	1.193	6.427	0.3693	3.780	7.005	107.5
8.25	7.764	1.207	6.433	0.3691	3.659	6.962	107.7
8.30	7.858	1.220	6.439	0.3690	3.543	6.920	107.9
8.35	7.954	1.234	6.445	0.3688	3.430	6.878	108.2
8.40	8.049	1.248	6.451	0.3686	3.322	6.837	108.4
8.45	8.146	1.261	6.457	0.3684	3.217	6.797	108.6
8.50	8.242	1.275	6.463	0.3683	3.116	6.757	108.8
8.55	8.340	1.289	6.469	0.3681	3.019	6.717	109.1
8.60	8.438	1.303	6.475	0.3679	2.925	6.678	109.3
8.65	8.536	1.317	6.480	0.3678	2.835	6.639	109.5
8.70	8.635	1.331	6.486	0.3676	2.748	6.601	109.7
8.75	8.735	1.346	6.491	0.3674	2.664	6.563	109.9
8.80	8.835	1.360	6.497	0.3673	2.583	6.525	110.1
8.85	8.936	1.374	6.502	0.3671	2.504	6.488	110.3
8.90	9.038	1.389	6.507	0.3670	2.429	6.452	110.5
8.95	9.140	1.403	6.512	0.3668	2.356	6.415	110.7
9.00	9.242	1.418	6.517	0.3667	2.285	6.380	110.9
9.05	9.345	1.433	6.522	0.3666	2.217	6.344	111.1
9.10	9.449	1.448	6.527	0.3664	2.151	6.309	111.3
9.15	9.553	1.463	6.532	0.3663	2.088	6.275	111.5
9.20	9.658	1.478	6.537	0.3661	2.026	6.240	111.7
9.25	9.764	1.493	6.541	0.3660	1.967	6.206	111.9
9.30	9.870	1.508	6.546	0.3659	1.910	6.173	112.0
9.35	9.976	1.523	6.550	0.3657	1.854	6.140	112.2
9.40	1.008 +2	1.538	6.555	0.3656	1.801	6.107	112.4
9.45	1.019	1.554	6.559	0.3655	1.749	6.075	112.6
9.50	1.030	1.569	6.564	0.3654	1.699	6.043	112.8
9.55	1.041	1.585	6.568	0.3652	1.650	6.011	112.9
9.60	1.052	1.600	6.572	0.3651	1.603	5.979	113.1
9.65	1.063	1.616	6.576	0.3650	1.558	5.948	113.3
9.70	1.074	1.632	6.580	0.3649	1.514	5.917	113.5
9.75	1.085	1.648	6.584	0.3648	1.472	5.887	113.6
9.80	1.096	1.664	6.588	0.3647	1.431	5.857	113.8
9.85	1.107	1.680	6.592	0.3646	1.391	5.827	114.0
9.90	1.119	1.696	6.596	0.3644	1.352	5.797	114.1
9.95	1.130	1.712	6.600	0.3643	1.315	5.768	114.3
10.00	1.141	1.728	6.604	0.3642	1.279	5.739	114.5

ISENTROPIC FUNCTIONS FOR SUPERSONIC COMPRESSIBLE FLOW

$\gamma = 4/3$

M	$p/p°$	$T/T°$	$\rho/\rho°$	$A/A*$	$q/p°$	u/u_m	$u/a*$	$F(\gamma;M)$
10.0	1.027 −5	1.814 −4	5.661 −2	1.351 +3	6.845 −4	0.9713	2.570	2.397
10.1	9.524 −6	1.714	5.555	1.429	6.477	0.9718	2.571	2.398
10.2	8.841	1.621	5.453	1.510	6.132	0.9724	2.573	2.399
10.3	8.212	1.534	5.353	1.595	5.807	0.9729	2.574	2.400
10.4	7.632	1.452	5.256	1.684	5.503	0.9734	2.575	2.401
10.5	7.098	1.375	5.162	1.777	5.217	0.9739	2.577	2.402
10.6	6.605	1.303	5.070	1.875	4.947	0.9743	2.578	2.403
10.7	6.150	1.235	4.980	1.977	4.694	0.9748	2.579	2.403
10.8	5.730	1.171	4.893	2.084	4.455	0.9752	2.580	2.404
10.9	5.342	1.111	4.808	2.196	4.231	0.9757	2.581	2.405
11.0	4.983	1.055	4.725	2.312	4.019	0.9761	2.583	2.406
11.1	4.651	1.001	4.644	2.434	3.820	0.9765	2.584	2.407
11.2	4.343	9.513 −5	4.565	2.561	3.632	0.9769	2.585	2.407
11.3	4.058	9.041	4.488	2.694	3.454	0.9773	2.586	2.408
11.4	3.794	8.596	4.413	2.832	3.287	0.9777	2.587	2.409
11.5	3.548	8.176	4.340	2.977	3.128	0.9781	2.588	2.409
11.6	3.321	7.779	4.269	3.127	2.979	0.9784	2.589	2.410
11.7	3.109	7.405	4.199	3.284	2.838	0.9788	2.590	2.411
11.8	2.913	7.051	4.131	3.448	2.704	0.9791	2.591	2.411
11.9	2.730	6.717	4.065	3.618	2.578	0.9795	2.591	2.412
12.0	2.560	6.401	4.000	3.795	2.458	0.9798	2.592	2.412
12.1	2.402	6.102	3.937	3.980	2.345	0.9801	2.593	2.413
12.2	2.255	5.819	3.875	4.172	2.237	0.9804	2.594	2.414
12.3	2.118	5.551	3.815	4.372	2.136	0.9807	2.595	2.414
12.4	1.990	5.298	3.756	4.579	2.040	0.9810	2.596	2.415
12.5	1.870	5.058	3.698	4.796	1.948	0.9813	2.596	2.415
12.6	1.759	4.830	3.642	5.020	1.862	0.9816	2.597	2.416
12.7	1.655	4.614	3.587	5.253	1.780	0.9819	2.598	2.416
12.8	1.558	4.409	3.533	5.496	1.701	0.9822	2.599	2.417
12.9	1.467	4.215	3.480	5.748	1.627	0.9824	2.599	2.417
13.0	1.382	4.031	3.429	6.009	1.557	0.9827	2.600	2.418
13.1	1.303	3.856	3.378	6.280	1.490	0.9830	2.601	2.418
13.2	1.228	3.689	3.329	6.562	1.427	0.9832	2.601	2.419
13.3	1.159	3.531	3.281	6.854	1.366	0.9835	2.602	2.419
13.4	1.093	3.381	3.234	7.156	1.309	0.9837	2.603	2.420
13.5	1.032	3.238	3.187	7.470	1.254	0.9839	2.603	2.420
13.6	9.748 −7	3.102	3.142	7.796	1.202	0.9842	2.604	2.420
13.7	9.210	2.973	3.098	8.133	1.152	0.9844	2.604	2.421
13.8	8.705	2.850	3.055	8.483	1.105	0.9846	2.605	2.421
13.9	8.231	2.733	3.012	8.845	1.060	0.9848	2.606	2.422
14.0	7.785	2.621	2.970	9.219	1.017	0.9850	2.606	2.422
14.1	7.367	2.515	2.930	9.608	9.763 −5	0.9852	2.607	2.422
14.2	6.973	2.413	2.890	1.001 +4	9.373	0.9854	2.607	2.423
14.3	6.603	2.316	2.851	1.043	9.001	0.9856	2.608	2.423
14.4	6.255	2.224	2.812	1.086	8.646	0.9858	2.608	2.424
14.5	5.927	2.136	2.775	1.130	8.308	0.9860	2.609	2.424
14.6	5.619	2.052	2.738	1.176	7.984	0.9862	2.609	2.424
14.7	5.328	1.972	2.702	1.224	7.675	0.9864	2.610	2.425
14.8	5.054	1.896	2.666	1.273	7.380	0.9866	2.610	2.425
14.9	4.796	1.822	2.632	1.324	7.098	0.9868	2.611	2.425
15.0	4.552	1.753	2.598	1.376	6.828	0.9869	2.611	2.426

NORMAL SHOCK FUNCTIONS FOR SUPERSONIC COMPRESSIBLE FLOW

$$\gamma = 4/3$$

M	p_{21}	T_{21}	ρ_{21}	M_2	p°_{21}	μ	ν
10.0	1.141 +2	1.728 +1	6.604	0.3642	1.279 -3	5.739	114.6
10.1	1.164	1.761	6.611	0.3640	1.210	5.682	114.8
10.2	1.188	1.794	6.618	0.3638	1.146	5.626	115.1
10.3	1.211	1.828	6.625	0.3636	1.085	5.572	115.4
10.4	1.235	1.862	6.632	0.3634	1.028	5.518	115.7
10.5	1.259	1.896	6.639	0.3633	9.745 -4	5.465	116.0
10.6	1.283	1.930	6.645	0.3631	9.241	5.413	116.3
10.7	1.307	1.965	6.651	0.3629	8.767	5.363	116.6
10.8	1.332	2.000	6.658	0.3627	8.321	5.313	116.9
10.9	1.356	2.035	6.663	0.3626	7.901	5.264	117.1
11.0	1.381	2.071	6.669	0.3624	7.506	5.216	117.4
11.1	1.407	2.107	6.675	0.3622	7.133	5.169	117.7
11.2	1.432	2.144	6.680	0.3621	6.781	5.123	117.9
11.3	1.458	2.180	6.686	0.3619	6.449	5.077	118.2
11.4	1.484	2.218	6.691	0.3618	6.136	5.033	118.5
11.5	1.510	2.255	6.696	0.3617	5.840	4.989	118.7
11.6	1.536	2.293	6.701	0.3615	5.561	4.946	119.0
11.7	1.563	2.331	6.706	0.3614	5.297	4.903	119.2
11.8	1.590	2.369	6.711	0.3613	5.047	4.862	119.4
11.9	1.617	2.408	6.715	0.3611	4.811	4.821	119.7
12.0	1.644	2.447	6.720	0.3610	4.588	4.780	119.9
12.1	1.672	2.486	6.724	0.3609	4.376	4.741	120.1
12.2	1.700	2.526	6.729	0.3608	4.176	4.702	120.3
12.3	1.728	2.566	6.733	0.3606	3.986	4.663	120.6
12.4	1.756	2.606	6.737	0.3605	3.806	4.626	120.8
12.5	1.784	2.647	6.741	0.3604	3.636	4.589	121.0
12.6	1.813	2.688	6.745	0.3603	3.474	4.552	121.2
12.7	1.842	2.729	6.749	0.3602	3.321	4.516	121.4
12.8	1.871	2.771	6.753	0.3601	3.175	4.481	121.6
12.9	1.900	2.813	6.756	0.3600	3.037	4.446	121.8
13.0	1.930	2.855	6.760	0.3599	2.905	4.412	122.0
13.1	1.960	2.898	6.764	0.3598	2.780	4.378	122.2
13.2	1.990	2.940	6.767	0.3597	2.662	4.345	122.4
13.3	2.020	2.984	6.770	0.3596	2.549	4.312	122.6
13.4	2.051	3.027	6.774	0.3595	2.441	4.280	122.8
13.5	2.081	3.071	6.777	0.3594	2.339	4.248	123.0
13.6	2.112	3.115	6.780	0.3594	2.242	4.217	123.1
13.7	2.144	3.160	6.783	0.3593	2.150	4.186	123.3
13.8	2.175	3.205	6.786	0.3592	2.061	4.156	123.5
13.9	2.207	3.250	6.789	0.3591	1.977	4.126	123.7
14.0	2.238	3.296	6.792	0.3590	1.897	4.096	123.8
14.1	2.271	3.342	6.795	0.3590	1.821	4.067	124.0
14.2	2.303	3.388	6.798	0.3589	1.748	4.038	124.2
14.3	2.335	3.434	6.800	0.3588	1.679	4.010	124.3
14.4	2.368	3.481	6.803	0.3587	1.613	3.982	124.5
14.5	2.401	3.528	6.806	0.3587	1.549	3.955	124.7
14.6	2.435	3.576	6.808	0.3586	1.489	3.928	124.8
14.7	2.468	3.624	6.811	0.3585	1.431	3.901	125.0
14.8	2.502	3.672	6.813	0.3585	1.376	3.874	125.1
14.9	2.536	3.720	6.816	0.3584	1.324	3.848	125.3
15.0	2.570	3.769	6.818	0.3583	1.273	3.823	125.4

ISENTROPIC FUNCTIONS FOR SUPERSONIC COMPRESSIBLE FLOW

$\gamma = 4/3$

M	p/p°	T/T°	ρ/ρ°	A/A^*	q/p°	u/u_m	u/a^*	$F(\gamma;M)$
15.0	4.552 -7	1.753 -5	2.598 -2	1.376 $+4$	6.828 -5	0.9869	2.611	2.426
15.1	4.323	1.686	2.564	1.430	6.570	0.9871	2.612	2.426
15.2	4.106	1.622	2.531	1.486	6.324	0.9873	2.612	2.426
15.3	3.901	1.561	2.499	1.544	6.088	0.9874	2.613	2.426
15.4	3.708	1.503	2.468	1.604	5.862	0.9876	2.613	2.427
15.5	3.525	1.447	2.437	1.666	5.646	0.9877	2.613	2.427
15.6	3.352	1.393	2.406	1.729	5.439	0.9879	2.614	2.427
15.7	3.189	1.342	2.376	1.795	5.241	0.9880	2.614	2.428
15.8	3.035	1.293	2.347	1.863	5.051	0.9882	2.615	2.428
15.9	2.889	1.246	2.318	1.933	4.869	0.9883	2.615	2.428
16.0	2.751	1.201	2.290	2.005	4.695	0.9885	2.615	2.428
16.1	2.620	1.158	2.262	2.079	4.527	0.9886	2.616	2.429
16.2	2.496	1.117	2.235	2.156	4.367	0.9888	2.616	2.429
16.3	2.379	1.077	2.208	2.234	4.214	0.9889	2.616	2.429
16.4	2.268	1.039	2.182	2.316	4.066	0.9890	2.617	2.429
16.5	2.162	1.003	2.156	2.400	3.925	0.9892	2.617	2.430
16.6	2.063	9.678 -6	2.131	2.486	3.789	0.9893	2.617	2.430
16.7	1.968	9.343	2.106	2.575	3.658	0.9894	2.618	2.430
16.8	1.878	9.021	2.082	2.666	3.533	0.9895	2.618	2.430
16.9	1.793	8.712	2.058	2.761	3.413	0.9897	2.618	2.431
17.0	1.712	8.415	2.034	2.858	3.297	0.9898	2.619	2.431
17.1	1.635	8.130	2.011	2.958	3.186	0.9899	2.619	2.431
17.2	1.562	7.856	1.988	3.060	3.080	0.9900	2.619	2.431
17.3	1.492	7.592	1.965	3.166	2.977	0.9901	2.620	2.431
17.4	1.426	7.339	1.943	3.275	2.879	0.9902	2.620	2.432
17.5	1.364	7.096	1.922	3.387	2.784	0.9903	2.620	2.432
17.6	1.304	6.862	1.900	3.502	2.693	0.9905	2.621	2.432
17.7	1.247	6.637	1.879	3.620	2.605	0.9906	2.621	2.432
17.8	1.193	6.420	1.859	3.742	2.520	0.9907	2.621	2.432
17.9	1.142	6.212	1.838	3.867	2.439	0.9908	2.621	2.433
18.0	1.093	6.011	1.818	3.996	2.361	0.9909	2.622	2.433
18.1	1.047	5.818	1.799	4.128	2.286	0.9910	2.622	2.433
18.2	1.002	5.633	1.779	4.264	2.213	0.9911	2.622	2.433
18.3	9.600 -8	5.454	1.760	4.403	2.143	0.9912	2.622	2.433
18.4	9.197	5.281	1.741	4.547	2.076	0.9913	2.623	2.433
18.5	8.813	5.115	1.723	4.694	2.011	0.9913	2.623	2.434
18.6	8.448	4.955	1.705	4.845	1.948	0.9914	2.623	2.434
18.7	8.099	4.801	1.687	5.000	1.888	0.9915	2.623	2.434
18.8	7.766	4.652	1.669	5.160	1.830	0.9916	2.624	2.434
18.9	7.449	4.509	1.652	5.323	1.774	0.9917	2.624	2.434
19.0	7.146	4.371	1.635	5.491	1.720	0.9918	2.624	2.434
19.1	6.857	4.237	1.618	5.663	1.667	0.9919	2.624	2.435
19.2	6.580	4.109	1.602	5.840	1.617	0.9920	2.625	2.435
19.3	6.317	3.984	1.585	6.022	1.569	0.9920	2.625	2.435
19.4	6.065	3.865	1.569	6.208	1.522	0.9921	2.625	2.435
19.5	5.824	3.749	1.554	6.399	1.476	0.9922	2.625	2.435
19.6	5.594	3.638	1.538	6.594	1.433	0.9923	2.625	2.435
19.7	5.375	3.530	1.523	6.795	1.390	0.9924	2.626	2.435
19.8	5.164	3.426	1.507	7.001	1.350	0.9924	2.626	2.436
19.9	4.963	3.325	1.493	7.212	1.310	0.9925	2.626	2.436
20.0	4.771	3.228	1.478	7.428	1.272	0.9926	2.626	2.436

NORMAL SHOCK FUNCTIONS FOR SUPERSONIC COMPRESSIBLE FLOW

$$\gamma = 4/3$$

M	P_{21}	T_{21}	ρ_{21}	M_2	P^o_{21}	μ	ν
15.0	2.570 +2	3.769 +1	6.818	0.3583	1.273 -4	3.823	125.4
15.1	2.604	3.818	6.821	0.3583	1.225	3.797	125.6
15.2	2.639	3.868	6.823	0.3582	1.179	3.772	125.7
15.3	2.674	3.918	6.825	0.3582	1.135	3.748	125.9
15.4	2.709	3.968	6.827	0.3581	1.093	3.723	126.0
15.5	2.744	4.018	6.829	0.3580	1.053	3.699	126.1
15.6	2.780	4.069	6.832	0.3580	1.014	3.675	126.3
15.7	2.815	4.120	6.834	0.3579	9.771 -5	3.652	126.4
15.8	2.851	4.171	6.836	0.3579	9.417	3.629	126.6
15.9	2.888	4.223	6.838	0.3578	9.078	3.606	126.7
16.0	2.924	4.275	6.840	0.3578	8.753	3.583	126.8
16.1	2.961	4.328	6.842	0.3577	8.441	3.561	126.9
16.2	2.998	4.380	6.844	0.3577	8.142	3.539	127.1
16.3	3.035	4.433	6.845	0.3576	7.855	3.517	127.2
16.4	3.072	4.487	6.847	0.3576	7.580	3.496	127.3
16.5	3.110	4.541	6.849	0.3575	7.316	3.475	127.5
16.6	3.148	4.595	6.851	0.3575	7.063	3.454	127.6
16.7	3.186	4.649	6.853	0.3574	6.820	3.433	127.7
16.8	3.224	4.704	6.854	0.3574	6.586	3.413	127.8
16.9	3.263	4.759	6.856	0.3573	6.362	3.392	127.9
17.0	3.301	4.814	6.858	0.3573	6.147	3.372	128.0
17.1	3.340	4.870	6.859	0.3572	5.940	3.353	128.2
17.2	3.379	4.926	6.861	0.3572	5.741	3.337	128.3
17.3	3.419	4.982	6.862	0.3572	5.550	3.314	128.4
17.4	3.459	5.039	6.864	0.3571	5.366	3.295	128.5
17.5	3.498	5.096	6.865	0.3571	5.189	3.276	128.6
17.6	3.538	5.153	6.867	0.3570	5.019	3.257	128.7
17.7	3.579	5.211	6.868	0.3570	4.855	3.239	128.8
17.8	3.619	5.268	6.870	0.3570	4.698	3.221	128.9
17.9	3.660	5.327	6.871	0.3569	4.546	3.203	129.0
18.0	3.701	5.385	6.873	0.3569	4.400	3.185	129.1
18.1	3.742	5.444	6.874	0.3568	4.260	3.167	129.3
18.2	3.784	5.504	6.875	0.3568	4.124	3.150	129.4
18.3	3.826	5.563	6.877	0.3568	3.994	3.133	129.5
18.4	3.868	5.623	6.878	0.3567	3.869	3.116	129.6
18.5	3.910	5.683	6.879	0.3567	3.748	3.099	129.7
18.6	3.952	5.744	6.881	0.3567	3.631	3.082	129.8
18.7	3.995	5.805	6.882	0.3566	3.518	3.066	129.9
18.8	4.038	5.866	6.883	0.3566	3.410	3.049	129.9
18.9	4.081	5.928	6.884	0.3566	3.306	3.033	130.0
19.0	4.124	5.989	6.886	0.3565	3.205	3.017	130.1
19.1	4.168	6.052	6.887	0.3565	3.107	3.001	130.2
19.2	4.211	6.114	6.888	0.3565	3.014	2.986	130.3
19.3	4.255	6.177	6.889	0.3564	2.923	2.970	130.4
19.4	4.300	6.240	6.890	0.3564	2.836	2.955	130.5
19.5	4.344	6.304	6.891	0.3564	2.751	2.940	130.6
19.6	4.389	6.367	6.892	0.3564	2.670	2.925	130.7
19.7	4.434	6.432	6.893	0.3563	2.591	2.910	130.8
19.8	4.479	6.496	6.894	0.3563	2.515	2.895	130.9
19.9	4.524	6.561	6.896	0.3563	2.442	2.880	130.9
20.0	4.570	6.626	6.897	0.3563	2.371	2.866	131.0

ISENTROPIC FUNCTIONS FOR SUPERSONIC COMPRESSIBLE FLOW

$\gamma = 4/3$

M	$p/p°$	$T/T°$	$\rho/\rho°$	$A/A*$	$q/p°$	u/u_m	$u/a*$	$F(\gamma;M)$
20.0	4.771 −8	3.228 −6	1.478 −2	7.428 +4	1.272 −5	0.9926	2.626	2.436
21.0	3.247	2.419	1.342	9.907	9.546 −6	0.9933	2.628	2.437
22.0	2.249	1.836	1.225	1.304 +5	7.255	0.9939	2.630	2.438
23.0	1.582	1.411	1.122	1.697	5.580	0.9944	2.631	2.439
24.0	1.130	1.096	1.031	2.183	4.338	0.9948	2.632	2.440
25.0	8.177 −9	8.599 −7	9.509 −3	2.781	3.407	0.9952	2.633	2.441
26.0	5.992	6.810	8.798	3.510	2.700	0.9956	2.634	2.441
27.0	4.442	5.441	8.164	4.393	2.159	0.9959	2.635	2.442
28.0	3.328	4.382	7.595	5.453	1.739	0.9962	2.636	2.443
29.0	2.519	3.555	7.084	6.719	1.412	0.9965	2.636	2.443
30.0	1.924	2.905	6.623	8.221	1.154	0.9967	2.637	2.443
31.0	1.482	2.389	6.205	9.994	9.497 −7	0.9969	2.638	2.444
32.0	1.152	1.977	5.825	1.207 +6	7.862	0.9971	2.638	2.444
33.0	9.016−10	1.645	5.480	1.451	6.545	0.9973	2.639	2.444
34.0	7.110	1.377	5.164	1.733	5.479	0.9974	2.639	2.445
35.0	5.645	1.158	4.874	2.060	4.610	0.9976	2.639	2.445
36.0	4.511	9.787 −8	4.608	2.438	3.897	0.9977	2.640	2.445
37.0	3.626	8.310	4.364	2.871	3.309	0.9978	2.640	2.445
38.0	2.932	7.086	4.138	3.366	2.823	0.9979	2.640	2.446
39.0	2.384	6.067	3.929	3.931	2.417	0.9980	2.641	2.446
40.0	1.948	5.215	3.736	4.573	2.078	0.9981	2.641	2.446
41.0	1.600	4.499	3.557	5.299	1.793	0.9982	2.641	2.446
42.0	1.321	3.896	3.390	6.120	1.553	0.9983	2.641	2.446
43.0	1.095	3.384	3.235	7.045	1.349	0.9984	2.642	2.447
44.0	9.113−11	2.949	3.090	8.082	1.176	0.9985	2.642	2.447
45.0	7.618	2.578	2.954	9.245	1.028	0.9985	2.642	2.447
46.0	6.392	2.261	2.828	1.054 +7	9.017 −8	0.9986	2.642	2.447
47.0	5.385	1.988	2.709	1.199	7.930	0.9986	2.642	2.447
48.0	4.552	1.752	2.597	1.360	6.992	0.9987	2.642	2.447
49.0	3.861	1.549	2.493	1.538	6.181	0.9988	2.643	2.447
50.0	3.286	1.373	2.394	1.736	5.477	0.9988	2.643	2.447
51.0	2.806	1.219	2.302	1.955	4.865	0.9988	2.643	2.447
52.0	2.403	1.085	2.214	2.195	4.332	0.9989	2.643	2.447
53.0	2.064	9.684 −9	2.131	2.461	3.865	0.9989	2.643	2.448
54.0	1.778	8.658	2.053	2.752	3.456	0.9990	2.643	2.448
55.0	1.536	7.758	1.980	3.071	3.097	0.9990	2.643	2.448
56.0	1.330	6.964	1.910	3.421	2.780	0.9990	2.643	2.448
57.0	1.155	6.264	1.843	3.803	2.501	0.9991	2.643	2.448
58.0	1.005	5.644	1.780	4.221	2.254	0.9991	2.643	2.448
59.0	8.767−12	5.095	1.721	4.676	2.034	0.9991	2.644	2.448
60.0	7.665	4.607	1.664	5.171	1.840	0.9992	2.644	2.448
65.0	4.045	2.852	1.418	8.352	1.139	0.9993	2.644	2.448
70.0	2.237	1.829	1.223	1.302 +8	7.308 −9	0.9994	2.644	2.448
75.0	1.289	1.210	1.066	1.968	4.834	0.9995	2.644	2.449
80.0	7.696−13	8.217−10	9.366 −4	2.898	3.284	0.9995	2.645	2.449
85.0	4.741	5.713	8.298	4.168	2.283	0.9996	2.645	2.449
90.0	3.002	4.056	7.402	5.871	1.621	0.9996	2.645	2.449
95.0	1.948	2.933	6.644	8.119	1.172	0.9997	2.645	2.449
100.0	1.293	2.156	5.996	1.104 +9	8.620−10	0.9997	2.645	2.449
∞	0.000	0.000	0.000	∞	0.000	1.000	2.646	2.449

NORMAL SHOCK FUNCTIONS FOR SUPERSONIC COMPRESSIBLE FLOW

$$\gamma = 4/3$$

M	P_{21}	T_{21}	ρ_{21}	M_2	P_{21}^o	μ	ν
20.0	4.570 +2	6.626 +1	6.897	0.3563	2.371 −5	2.866	131.0
21.0	5.038	7.295	6.906	0.3560	1.779	2.729	131.8
22.0	5.530	7.997	6.914	0.3558	1.352	2.605	132.6
23.0	6.044	8.732	6.921	0.3556	1.040	2.492	133.2
24.0	6.581	9.499	6.928	0.3554	8.082 −6	2.388	133.9
25.0	7.141	1.030 +2	6.933	0.3553	6.346	2.293	134.4
26.0	7.724	1.113	6.938	0.3552	5.030	2.204	134.9
27.0	8.330	1.200	6.943	0.3550	4.021	2.123	135.4
28.0	8.958	1.290	6.947	0.3549	3.240	2.047	135.9
29.0	9.610	1.383	6.950	0.3548	2.630	1.976	136.3
30.0	1.028 +3	1.479	6.954	0.3548	2.150	1.910	136.7
31.0	1.098	1.578	6.957	0.3547	1.769	1.849	137.1
32.0	1.170	1.681	6.959	0.3546	1.464	1.791	137.4
33.0	1.244	1.787	6.962	0.3545	1.219	1.737	137.7
34.0	1.321	1.897	6.964	0.3545	1.020	1.685	138.0
35.0	1.400	2.010	6.966	0.3544	8.584 −7	1.637	138.3
36.0	1.481	2.125	6.968	0.3544	7.257	1.592	138.6
37.0	1.564	2.245	6.969	0.3543	6.163	1.549	138.8
38.0	1.650	2.367	6.971	0.3543	5.256	1.508	139.1
39.0	1.738	2.493	6.972	0.3543	4.501	1.469	139.3
40.0	1.828	2.622	6.974	0.3542	3.870	1.433	139.5
41.0	1.921	2.754	6.975	0.3542	3.339	1.398	139.7
42.0	2.016	2.889	6.976	0.3542	2.892	1.364	139.9
43.0	2.113	3.028	6.977	0.3541	2.512	1.333	140.1
44.0	2.212	3.170	6.978	0.3541	2.190	1.302	140.3
45.0	2.314	3.316	6.979	0.3541	1.915	1.273	140.5
46.0	2.418	3.464	6.980	0.3541	1.679	1.246	140.7
47.0	2.524	3.616	6.981	0.3540	1.476	1.219	140.8
48.0	2.633	3.771	6.982	0.3540	1.302	1.194	141.0
49.0	2.744	3.929	6.983	0.3540	1.151	1.169	141.1
50.0	2.857	4.091	6.983	0.3540	1.020	1.146	141.3
51.0	2.972	4.256	6.984	0.3540	9.059 −8	1.124	141.4
52.0	3.090	4.424	6.985	0.3540	8.066	1.102	141.5
53.0	3.210	4.596	6.985	0.3539	7.197	1.081	141.6
54.0	3.332	4.770	6.986	0.3539	6.435	1.061	141.8
55.0	3.457	4.948	6.986	0.3539	5.766	1.042	141.9
56.0	3.584	5.129	6.987	0.3539	5.177	1.023	142.0
57.0	3.713	5.314	6.987	0.3539	4.656	1.005	142.1
58.0	3.844	5.502	6.988	0.3539	4.196	0.988	142.2
59.0	3.978	5.693	6.988	0.3539	3.788	0.971	142.3
60.0	4.114	5.887	6.988	0.3539	3.425	0.955	142.4
65.0	4.828	6.907	6.990	0.3538	2.121	0.882	142.8
70.0	5.600	8.009	6.991	0.3538	1.361	0.819	143.2
75.0	6.428	9.193	6.993	0.3537	9.000 −9	0.764	143.5
80.0	7.314	1.046 +3	6.993	0.3537	6.114	0.716	143.8
85.0	8.257	1.181	6.994	0.3537	4.251	0.674	144.1
90.0	9.257	1.323	6.995	0.3537	3.018	0.637	144.3
95.0	1.031 +4	1.474	6.995	0.3537	2.182	0.603	144.5
100.0	1.143	1.634	6.996	0.3537	1.605	0.573	144.7
∞	∞	∞	7.000	0.3536	0.0000	0.000	148.1

TABLE B.1

FANNO FLOW FUNCTIONS

$\gamma = 5/3$

B.1-1

SUBSONIC FANNO FLOW FUNCTIONS

$$\gamma = 5/3$$

M	$G(M;\gamma)$	p/p^*	T/T^*	p°/p^*
0.00	∞	∞	1.333	∞
0.01	1498.1	115.5	1.333	115.5
0.02	373.3	57.73	1.333	57.75
0.03	165.2	38.48	1.333	38.51
0.04	92.37	28.86	1.333	28.90
0.05	58.71	23.08	1.332	23.13
0.06	40.45	19.23	1.332	19.29
0.07	29.46	16.48	1.331	16.55
0.08	22.33	14.42	1.330	14.50
0.09	17.46	12.81	1.330	12.90
0.10	13.99	11.53	1.329	11.62
0.11	11.42	10.48	1.328	10.58
0.12	9.475	9.599	1.327	9.715
0.13	7.966	8.857	1.326	8.983
0.14	6.773	8.221	1.325	8.356
0.15	5.814	7.669	1.323	7.814
0.16	5.032	7.186	1.322	7.341
0.17	4.387	6.760	1.321	6.924
0.18	3.849	6.381	1.319	6.554
0.19	3.396	6.041	1.317	6.225
0.20	3.011	5.735	1.316	5.928
0.21	2.682	5.459	1.314	5.661
0.22	2.398	5.207	1.312	5.419
0.23	2.152	4.977	1.310	5.199
0.24	1.937	4.766	1.308	4.998
0.25	1.749	4.571	1.306	4.813
0.26	1.583	4.392	1.304	4.644
0.27	1.437	4.226	1.302	4.487
0.28	1.306	4.071	1.299	4.342
0.29	1.190	3.927	1.297	4.208
0.30	1.087	3.793	1.294	4.083
0.31	0.9936	3.667	1.292	3.967
0.32	0.9099	3.548	1.289	3.859
0.33	0.8344	3.437	1.287	3.758
0.34	0.7660	3.333	1.284	3.663
0.35	0.7041	3.234	1.281	3.574
0.36	0.6478	3.140	1.278	3.491
0.37	0.5966	3.052	1.275	3.412
0.38	0.5499	2.968	1.272	3.338
0.39	0.5072	2.888	1.269	3.269
0.40	0.4681	2.813	1.266	3.203
0.41	0.4323	2.741	1.263	3.141
0.42	0.3994	2.672	1.259	3.082
0.43	0.3692	2.606	1.256	3.027
0.44	0.3414	2.544	1.253	2.974
0.45	0.3158	2.484	1.249	2.924
0.46	0.2922	2.426	1.245	2.877
0.47	0.2704	2.371	1.242	2.832
0.48	0.2502	2.318	1.238	2.789
0.49	0.2315	2.268	1.235	2.749
0.50	0.2143	2.219	1.231	2.710

SUBSONIC FANNO FLOW FUNCTIONS

$$\gamma = 5/3$$

M	$G(M;\gamma)$	p/p^*	T/T^*	$p°/p^*$
0.50	0.2143	2.219	1.231	2.710
0.51	0.1983	2.172	1.227	2.674
0.52	0.1834	2.127	1.223	2.639
0.53	0.1697	2.083	1.219	2.606
0.54	0.1569	2.041	1.215	2.574
0.55	0.1451	2.001	1.211	2.544
0.56	0.1340	1.962	1.207	2.516
0.57	0.1238	1.924	1.203	2.488
0.58	0.1143	1.888	1.199	2.462
0.59	0.1054	1.853	1.195	2.438
0.60	0.09721	1.818	1.190	2.414
0.61	0.08955	1.785	1.186	2.392
0.62	0.08243	1.753	1.182	2.370
0.63	0.07580	1.722	1.178	2.350
0.64	0.06964	1.692	1.173	2.331
0.65	0.06390	1.663	1.169	2.312
0.66	0.05857	1.635	1.164	2.294
0.67	0.05361	1.607	1.160	2.278
0.68	0.04900	1.581	1.155	2.262
0.69	0.04471	1.555	1.151	2.247
0.70	0.04073	1.529	1.146	2.232
0.71	0.03704	1.505	1.142	2.219
0.72	0.03361	1.481	1.137	2.206
0.73	0.03043	1.458	1.132	2.194
0.74	0.02749	1.435	1.128	2.182
0.75	0.02476	1.413	1.123	2.171
0.76	0.02224	1.391	1.118	2.161
0.77	0.01991	1.370	1.113	2.151
0.78	0.01777	1.350	1.109	2.142
0.79	0.01579	1.330	1.104	2.133
0.80	0.01398	1.310	1.099	2.125
0.81	0.01231	1.291	1.094	2.117
0.82	0.01079	1.273	1.089	2.110
0.83	0.009400	1.255	1.084	2.103
0.84	0.008134	1.237	1.079	2.097
0.85	0.006985	1.220	1.075	2.092
0.86	0.005946	1.203	1.070	2.086
0.87	0.005012	1.186	1.065	2.081
0.88	0.004176	1.170	1.060	2.077
0.89	0.003431	1.154	1.055	2.073
0.90	0.002774	1.138	1.050	2.069
0.91	0.002198	1.123	1.045	2.066
0.92	0.001700	1.108	1.040	2.063
0.93	0.001274	1.094	1.035	2.061
0.94	0.0009162	1.080	1.030	2.059
0.95	0.0006231	1.066	1.025	2.057
0.96	0.0003905	1.052	1.020	2.055
0.97	0.0002152	1.039	1.015	2.054
0.98	0.0000938	1.025	1.010	2.053
0.99	0.0000230	1.013	1.005	2.053
1.00	0.0000	1.000	1.000	2.053

B.1-3

SUPERSONIC FANNO FLOW FUNCTIONS

$\gamma = 5/3$

M	$G(M;\gamma)$	p/p^*	T/T^*	$p°/p^*$
1.00	0.0000	1.000	1.000	2.053
1.05	0.000510	0.9404	0.9750	2.057
1.10	0.001857	0.8861	0.9501	2.067
1.15	0.003818	0.8365	0.9254	2.084
1.20	0.006223	0.7910	0.9009	2.108
1.25	0.008942	0.7491	0.8767	2.137
1.30	0.01188	0.7104	0.8529	2.171
1.35	0.01495	0.6746	0.8294	2.210
1.40	0.01810	0.6414	0.8065	2.255
1.45	0.02128	0.6106	0.7839	2.304
1.50	0.02447	0.5819	0.7619	2.358
1.55	0.02762	0.5551	0.7404	2.416
1.60	0.03073	0.5301	0.7194	2.479
1.65	0.03378	0.5067	0.6990	2.546
1.70	0.03676	0.4848	0.6791	2.618
1.75	0.03966	0.4642	0.6598	2.695
1.80	0.04247	0.4448	0.6410	2.775
1.85	0.04520	0.4266	0.6228	2.861
1.90	0.04783	0.4094	0.6051	2.950
1.95	0.05038	0.3932	0.5880	3.045
2.00	0.05284	0.3780	0.5714	3.143
2.10	0.05748	0.3499	0.5398	3.355
2.20	0.06179	0.3247	0.5102	3.585
2.30	0.06577	0.3020	0.4825	3.834
2.40	0.06945	0.2816	0.4566	4.102
2.50	0.07285	0.2630	0.4324	4.391
2.60	0.07599	0.2462	0.4098	4.701
2.70	0.07890	0.2309	0.3887	5.031
2.80	0.08159	0.2169	0.3690	5.384
2.90	0.08408	0.2042	0.3506	5.760
3.00	0.08639	0.1925	0.3333	6.158
3.20	0.09052	0.1718	0.3021	7.028
3.40	0.09409	0.1542	0.2747	8.000
3.60	0.09719	0.1391	0.2506	9.078
3.80	0.09989	0.1260	0.2294	10.27
4.00	0.1023	0.1147	0.2105	11.58
4.20	0.1043	0.1048	0.1938	13.01
4.40	0.1062	0.0961	0.1789	14.58
4.60	0.1078	0.0885	0.1656	16.28
4.80	0.1093	0.0817	0.1536	18.12
5.00	0.1106	0.0756	0.1429	20.12
5.50	0.1133	0.0631	0.1203	25.79
6.00	0.1154	0.0534	0.1026	32.52
6.50	0.1171	0.0457	0.0884	40.42
7.00	0.1184	0.0396	0.0769	49.56
7.50	0.1195	0.0346	0.0675	60.05
8.00	0.1204	0.0305	0.0597	71.99
8.50	0.1212	0.0271	0.0532	85.47
9.00	0.1218	0.0242	0.0476	100.6
9.50	0.1224	0.0218	0.0429	117.4
10.00	0.1228	0.0197	0.0388	136.1
∞	0.1273	0.0000	0.0000	∞

TABLE B.2

FANNO FLOW FUNCTIONS

$\gamma = 7/5$

SUBSONIC FANNO FLOW FUNCTIONS

$$\gamma = 7/5$$

M	$G(M;\gamma)$	$p/p*$	$T/T*$	$p^{\circ}/p*$
0.00	∞	∞	1.200	∞
0.01	1783.6	109.5	1.200	109.5
0.02	444.6	54.77	1.200	54.79
0.03	196.8	36.51	1.200	36.53
0.04	110.1	27.38	1.200	27.41
0.05	70.01	21.90	1.199	21.94
0.06	48.26	18.25	1.199	18.30
0.07	35.16	15.64	1.199	15.70
0.08	26.68	13.68	1.198	13.75
0.09	20.87	12.16	1.198	12.23
0.10	16.73	10.94	1.198	11.02
0.11	13.67	9.947	1.197	10.03
0.12	11.35	9.116	1.197	9.208
0.13	9.552	8.412	1.196	8.512
0.14	8.128	7.809	1.195	7.917
0.15	6.983	7.287	1.195	7.402
0.16	6.049	6.829	1.194	6.952
0.17	5.279	6.425	1.193	6.556
0.18	4.636	6.066	1.192	6.205
0.19	4.094	5.745	1.191	5.891
0.20	3.633	5.455	1.190	5.610
0.21	3.239	5.194	1.190	5.356
0.22	2.899	4.955	1.188	5.125
0.23	2.604	4.738	1.187	4.916
0.24	2.347	4.538	1.186	4.724
0.25	2.121	4.355	1.185	4.548
0.26	1.922	4.185	1.184	4.386
0.27	1.746	4.028	1.183	4.237
0.28	1.589	3.882	1.181	4.099
0.29	1.450	3.746	1.180	3.971
0.30	1.325	3.619	1.179	3.852
0.31	1.213	3.500	1.177	3.741
0.32	1.112	3.389	1.176	3.638
0.33	1.021	3.284	1.174	3.541
0.34	0.9380	3.185	1.173	3.451
0.35	0.8631	3.092	1.171	3.366
0.36	0.7950	3.004	1.170	3.286
0.37	0.7330	2.921	1.168	3.211
0.38	0.6764	2.842	1.166	3.140
0.39	0.6246	2.767	1.165	3.073
0.40	0.5771	2.696	1.163	3.010
0.41	0.5336	2.628	1.161	2.950
0.42	0.4936	2.563	1.159	2.894
0.43	0.4568	2.502	1.157	2.841
0.44	0.4229	2.443	1.155	2.790
0.45	0.3916	2.386	1.153	2.742
0.46	0.3627	2.333	1.151	2.697
0.47	0.3360	2.281	1.149	2.653
0.48	0.3113	2.231	1.147	2.612
0.49	0.2885	2.184	1.145	2.573
0.50	0.2673	2.138	1.143	2.536

SUBSONIC FANNO FLOW FUNCTIONS

$$\gamma = 7/5$$

M	$G(M;\gamma)$	p/p^*	T/T^*	p°/p^*
0.50	0.2673	2.138	1.143	2.536
0.51	0.2476	2.094	1.141	2.501
0.52	0.2294	2.052	1.138	2.467
0.53	0.2124	2.011	1.136	2.435
0.54	0.1967	1.972	1.134	2.405
0.55	0.1820	1.934	1.132	2.376
0.56	0.1684	1.898	1.129	2.348
0.57	0.1557	1.862	1.127	2.321
0.58	0.1439	1.828	1.124	2.296
0.59	0.1329	1.795	1.122	2.272
0.60	0.1227	1.763	1.119	2.249
0.61	0.1132	1.733	1.117	2.227
0.62	0.1043	1.703	1.114	2.206
0.63	0.09603	1.674	1.112	2.187
0.64	0.08832	1.646	1.109	2.168
0.65	0.08115	1.618	1.107	2.150
0.66	0.07446	1.592	1.104	2.132
0.67	0.06824	1.566	1.101	2.116
0.68	0.06244	1.541	1.098	2.100
0.69	0.05705	1.517	1.096	2.086
0.70	0.05203	1.493	1.093	2.072
0.71	0.04737	1.471	1.090	2.058
0.72	0.04304	1.448	1.087	2.045
0.73	0.03901	1.427	1.084	2.033
0.74	0.03528	1.405	1.082	2.022
0.75	0.03182	1.385	1.079	2.011
0.76	0.02862	1.365	1.076	2.001
0.77	0.02565	1.345	1.073	1.991
0.78	0.02292	1.326	1.070	1.982
0.79	0.02039	1.307	1.067	1.973
0.80	0.01807	1.289	1.064	1.965
0.81	0.01594	1.272	1.061	1.958
0.82	0.01398	1.254	1.058	1.951
0.83	0.01219	1.237	1.055	1.944
0.84	0.01056	1.221	1.052	1.938
0.85	0.009082	1.205	1.048	1.932
0.86	0.007741	1.189	1.045	1.927
0.87	0.006532	1.173	1.042	1.922
0.88	0.005449	1.158	1.039	1.917
0.89	0.004483	1.144	1.036	1.913
0.90	0.003628	1.129	1.033	1.910
0.91	0.002879	1.115	1.029	1.906
0.92	0.002228	1.101	1.026	1.904
0.93	0.001672	1.088	1.023	1.901
0.94	0.001204	1.074	1.020	1.899
0.95	0.0008197	1.061	1.017	1.897
0.96	0.0005143	1.049	1.013	1.896
0.97	0.0002838	1.036	1.010	1.894
0.98	0.0001237	1.024	1.007	1.894
0.99	0.0000304	1.012	1.003	1.893
1.00	0.0000000	1.000	1.000	1.893

SUPERSONIC FANNO FLOW FUNCTIONS

$$\gamma = 7/5$$

M	$G(M;\gamma)$	$p/p*$	$T/T*$	$p°/p*$
1.00	0.0000	1.000	1.000	1.893
1.05	0.000678	0.9443	0.9832	1.897
1.10	0.002484	0.8936	0.9662	1.908
1.15	0.005133	0.8471	0.9490	1.926
1.20	0.008409	0.8044	0.9317	1.951
1.25	0.01214	0.7649	0.9143	1.981
1.30	0.01621	0.7285	0.8969	2.018
1.35	0.02050	0.6947	0.8794	2.061
1.40	0.02493	0.6632	0.8621	2.110
1.45	0.02946	0.6339	0.8448	2.165
1.50	0.03401	0.6065	0.8276	2.226
1.55	0.03857	0.5808	0.8105	2.293
1.60	0.04309	0.5568	0.7937	2.367
1.65	0.04756	0.5342	0.7770	2.446
1.70	0.05195	0.5130	0.7605	2.532
1.75	0.05626	0.4929	0.7442	2.625
1.80	0.06047	0.4741	0.7282	2.724
1.85	0.06458	0.4562	0.7124	2.830
1.90	0.06858	0.4394	0.6969	2.944
1.95	0.07247	0.4234	0.6816	3.065
2.00	0.07625	0.4082	0.6667	3.194
2.10	0.08346	0.3802	0.6376	3.477
2.20	0.09023	0.3549	0.6098	3.795
2.30	0.09656	0.3320	0.5831	4.151
2.40	0.1025	0.3111	0.5576	4.549
2.50	0.1080	0.2921	0.5333	4.991
2.60	0.1131	0.2747	0.5102	5.482
2.70	0.1180	0.2588	0.4882	6.025
2.80	0.1224	0.2441	0.4673	6.625
2.90	0.1266	0.2307	0.4474	7.287
3.00	0.1305	0.2182	0.4286	8.016
3.20	0.1376	0.1961	0.3937	9.694
3.40	0.1438	0.1770	0.3623	11.71
3.60	0.1492	0.1606	0.3341	14.10
3.80	0.1540	0.1462	0.3086	16.94
4.00	0.1583	0.1336	0.2857	20.29
4.20	0.1620	0.1226	0.2650	24.21
4.40	0.1654	0.1128	0.2463	28.79
4.60	0.1684	0.1041	0.2294	34.11
4.80	0.1710	0.0964	0.2140	40.25
5.00	0.1735	0.0894	0.2000	47.32
5.50	0.1785	0.0750	0.1702	69.79
6.00	0.1825	0.0638	0.1463	100.7
6.50	0.1856	0.0548	0.1270	142.2
7.00	0.1882	0.0476	0.1111	197.1
7.50	0.1903	0.0417	0.0980	268.5
8.00	0.1920	0.0369	0.0870	359.9
8.50	0.1935	0.0328	0.0777	475.3
9.00	0.1947	0.0293	0.0698	619.3
9.50	0.1958	0.0264	0.0630	797.2
10.00	0.1967	0.0239	0.0571	1014.5
∞	0.2053	0.0000	0.0000	∞

TABLE B.3

FANNO FLOW FUNCTIONS

$\gamma = 4/3$

SUBSONIC FANNO FLOW FUNCTIONS

$$\gamma = 4/3$$

M	$G(M;\gamma)$	p/p^*	T/T^*	p°/p^*
0.00	∞	∞	1.167	∞
0.01	1872.8	108.0	1.167	108.0
0.02	466.9	54.00	1.167	54.02
0.03	206.6	36.00	1.166	36.02
0.04	115.6	27.00	1.166	27.03
0.05	73.54	21.60	1.166	21.63
0.06	50.70	18.00	1.166	18.04
0.07	36.95	15.42	1.166	15.47
0.08	28.04	13.49	1.165	13.55
0.09	21.94	11.99	1.165	12.06
0.10	17.59	10.79	1.165	10.86
0.11	14.38	9.809	1.164	9.889
0.12	11.94	8.990	1.164	9.077
0.13	10.05	8.297	1.163	8.391
0.14	8.552	7.703	1.163	7.804
0.15	7.349	7.187	1.162	7.296
0.16	6.368	6.736	1.162	6.852
0.17	5.558	6.338	1.161	6.461
0.18	4.882	5.985	1.160	6.115
0.19	4.312	5.668	1.160	5.805
0.20	3.828	5.383	1.159	5.528
0.21	3.414	5.125	1.158	5.277
0.22	3.056	4.890	1.157	5.050
0.23	2.746	4.676	1.156	4.843
0.24	2.475	4.479	1.156	4.654
0.25	2.237	4.298	1.155	4.480
0.26	2.028	4.131	1.154	4.320
0.27	1.843	3.976	1.153	4.173
0.28	1.678	3.833	1.152	4.037
0.29	1.531	3.699	1.151	3.910
0.30	1.400	3.574	1.149	3.793
0.31	1.281	3.457	1.148	3.684
0.32	1.175	3.347	1.147	3.581
0.33	1.079	3.244	1.146	3.486
0.34	0.9920	3.147	1.145	3.396
0.35	0.9131	3.055	1.143	3.312
0.36	0.8413	2.968	1.142	3.233
0.37	0.7759	2.887	1.141	3.159
0.38	0.7162	2.809	1.139	3.089
0.39	0.6615	2.735	1.138	3.023
0.40	0.6115	2.665	1.136	2.961
0.41	0.5655	2.598	1.135	2.902
0.42	0.5233	2.535	1.133	2.846
0.43	0.4844	2.474	1.132	2.793
0.44	0.4486	2.416	1.130	2.743
0.45	0.4155	2.361	1.129	2.696
0.46	0.3850	2.308	1.127	2.651
0.47	0.3568	2.257	1.125	2.608
0.48	0.3307	2.208	1.124	2.567
0.49	0.3065	2.162	1.122	2.529
0.50	0.2840	2.117	1.120	2.492

B.3-2

SUBSONIC FANNO FLOW FUNCTIONS

$\gamma = 4/3$

M	$G(M;\gamma)$	p/p^*	T/T^*	$p°/p^*$
0.50	0.2840	2.117	1.120	2.492
0.51	0.2632	2.073	1.118	2.457
0.52	0.2439	2.032	1.116	2.424
0.53	0.2260	1.992	1.114	2.392
0.54	0.2093	1.953	1.113	2.362
0.55	0.1937	1.916	1.111	2.333
0.56	0.1793	1.880	1.109	2.305
0.57	0.1659	1.846	1.107	2.279
0.58	0.1533	1.812	1.105	2.254
0.59	0.1417	1.780	1.103	2.230
0.60	0.1308	1.749	1.101	2.207
0.61	0.1207	1.718	1.099	2.186
0.62	0.1113	1.689	1.096	2.165
0.63	0.1025	1.660	1.094	2.145
0.64	0.09429	1.633	1.092	2.127
0.65	0.08665	1.606	1.090	2.109
0.66	0.07954	1.580	1.088	2.092
0.67	0.07292	1.555	1.085	2.075
0.68	0.06675	1.531	1.083	2.060
0.69	0.06100	1.507	1.081	2.045
0.70	0.05566	1.484	1.079	2.031
0.71	0.05068	1.461	1.076	2.018
0.72	0.04606	1.439	1.074	2.005
0.73	0.04177	1.418	1.071	1.993
0.74	0.03778	1.397	1.069	1.982
0.75	0.03409	1.377	1.067	1.971
0.76	0.03067	1.357	1.064	1.960
0.77	0.02750	1.338	1.062	1.951
0.78	0.02458	1.319	1.059	1.942
0.79	0.02188	1.301	1.057	1.933
0.80	0.01939	1.283	1.054	1.925
0.81	0.01711	1.266	1.052	1.917
0.82	0.01501	1.249	1.049	1.910
0.83	0.01310	1.233	1.047	1.904
0.84	0.01135	1.216	1.044	1.898
0.85	0.009762	1.201	1.041	1.892
0.86	0.008323	1.185	1.039	1.887
0.87	0.007026	1.170	1.036	1.882
0.88	0.005862	1.155	1.033	1.877
0.89	0.004824	1.141	1.031	1.873
0.90	0.003906	1.127	1.028	1.869
0.91	0.003100	1.113	1.025	1.866
0.92	0.002401	1.099	1.022	1.863
0.93	0.001801	1.086	1.020	1.861
0.94	0.001298	1.073	1.017	1.859
0.95	0.0008836	1.060	1.014	1.857
0.96	0.0005547	1.048	1.011	1.855
0.97	0.0003061	1.035	1.009	1.854
0.98	0.0001336	1.023	1.006	1.853
0.99	0.0000328	1.012	1.003	1.853
1.00	0.0000	1.000	1.000	1.853

SUPERSONIC FANNO FLOW FUNCTIONS

$\gamma = 4/3$

M	$G(M;\gamma)$	p/p^*	T/T^*	p°/p^*
1.00	0.0000	1.000	1.000	1.853
1.05	0.000734	0.9455	0.9856	1.856
1.10	0.002691	0.8958	0.9709	1.868
1.15	0.005570	0.8502	0.9560	1.886
1.20	0.009139	0.8083	0.9409	1.911
1.25	0.01322	0.7697	0.9256	1.943
1.30	0.01767	0.7339	0.9103	1.980
1.35	0.02237	0.7007	0.8949	2.025
1.40	0.02726	0.6698	0.8794	2.075
1.45	0.03224	0.6410	0.8639	2.132
1.50	0.03728	0.6141	0.8485	2.195
1.55	0.04233	0.5889	0.8331	2.265
1.60	0.04736	0.5652	0.8178	2.341
1.65	0.05234	0.5429	0.8025	2.425
1.70	0.05724	0.5220	0.7874	2.516
1.75	0.06207	0.5022	0.7724	2.614
1.80	0.06679	0.4835	0.7576	2.720
1.85	0.07142	0.4659	0.7429	2.834
1.90	0.07593	0.4492	0.7284	2.956
1.95	0.08033	0.4334	0.7141	3.087
2.00	0.08460	0.4183	0.7000	3.228
2.10	0.09280	0.3905	0.6724	3.538
2.20	0.1005	0.3653	0.6458	3.892
2.30	0.1078	0.3424	0.6200	4.292
2.40	0.1146	0.3215	0.5952	4.744
2.50	0.1210	0.3024	0.5714	5.254
2.60	0.1269	0.2849	0.5486	5.827
2.70	0.1325	0.2688	0.5267	6.470
2.80	0.1378	0.2540	0.5058	7.191
2.90	0.1427	0.2403	0.4858	7.996
3.00	0.1473	0.2277	0.4667	8.895
3.20	0.1556	0.2052	0.4310	11.01
3.40	0.1629	0.1857	0.3986	13.62
3.60	0.1694	0.1688	0.3692	16.83
3.80	0.1751	0.1540	0.3425	20.74
4.00	0.1802	0.1410	0.3182	25.49
4.20	0.1848	0.1296	0.2961	31.22
4.40	0.1888	0.1194	0.2760	38.11
4.60	0.1924	0.1104	0.2577	46.34
4.80	0.1957	0.1023	0.2410	56.13
5.00	0.1986	0.0950	0.2258	67.72
5.50	0.2048	0.0799	0.1931	106.5
6.00	0.2097	0.0680	0.1667	163.4
6.50	0.2136	0.0586	0.1451	245.1
7.00	0.2167	0.0510	0.1273	359.8
7.50	0.2193	0.0447	0.1124	518.0
8.00	0.2215	0.0395	0.1000	732.3
8.50	0.2233	0.0352	0.0895	1017.9
9.00	0.2249	0.0315	0.0805	1393.2
9.50	0.2262	0.0284	0.0727	1879.8
10.00	0.2273	0.0257	0.0660	2503.3
∞	0.2381	0.0000	0.0000	∞

TABLE C.1

ISOTHERMAL FLOW FUNCTIONS

$\gamma = 5/3$

ISOTHERMAL FLOW FUNCTIONS

$$\gamma = 5/3$$

M	$H(M;\gamma)$	p/p_f	$T°/T_f^°$	M	$H(M;\gamma)$	p/p_f	$T°/T_f^°$
0.0010	149996.6	774.60	0.83333	0.0056	4780.5	138.321	0.83334
0.0011	123963.4	704.18	0.83333	0.0057	4614.1	135.894	0.83334
0.0012	104163.3	645.50	0.83333	0.0058	4456.3	133.551	0.83334
0.0013	88754.0	595.84	0.83333	0.0059	4306.4	131.288	0.83334
0.0014	76527.3	553.28	0.83333	0.0060	4164.0	129.099	0.83334
0.0015	66663.3	516.40	0.83333	0.0061	4028.5	126.983	0.83334
0.0016	58590.5	484.12	0.83333	0.0062	3899.5	124.935	0.83334
0.0017	51899.8	455.65	0.83333	0.0063	3776.6	122.952	0.83334
0.0018	46293.0	430.33	0.83333	0.0064	3659.5	121.031	0.83334
0.0019	41548.0	407.68	0.83333	0.0065	3547.7	119.169	0.83334
0.0020	37496.8	387.30	0.83333	0.0066	3440.9	117.363	0.83335
0.0021	34010.4	368.86	0.83333	0.0067	3338.9	115.611	0.83335
0.0022	30988.6	352.09	0.83333	0.0068	3241.3	113.911	0.83335
0.0023	28352.2	336.78	0.83333	0.0069	3148.0	112.260	0.83335
0.0024	26038.5	322.75	0.83333	0.0070	3058.6	110.657	0.83335
0.0025	23996.9	309.84	0.83333	0.0071	2973.0	109.098	0.83335
0.0026	22186.3	297.92	0.83333	0.0072	2890.9	107.583	0.83335
0.0027	20573.0	286.89	0.83333	0.0073	2812.2	106.109	0.83335
0.0028	19129.6	276.64	0.83333	0.0074	2736.7	104.675	0.83335
0.0029	17832.9	267.10	0.83333	0.0075	2664.1	103.280	0.83335
0.0030	16663.6	258.20	0.83334	0.0076	2594.4	101.921	0.83335
0.0031	15605.7	249.87	0.83334	0.0077	2527.4	100.597	0.83335
0.0032	14645.5	242.06	0.83334	0.0078	2462.9	99.307	0.83335
0.0033	13771.1	234.73	0.83334	0.0079	2400.9	98.050	0.83335
0.0034	12972.8	227.82	0.83334	0.0080	2341.2	96.825	0.83335
0.0035	12242.0	221.31	0.83334	0.0081	2283.7	95.629	0.83335
0.0036	11571.1	215.17	0.83334	0.0082	2228.3	94.463	0.83335
0.0037	10954.0	209.35	0.83334	0.0083	2174.9	93.325	0.83335
0.0038	10384.9	203.84	0.83334	0.0084	2123.3	92.214	0.83335
0.0039	9859.0	198.61	0.83334	0.0085	2073.6	91.129	0.83335
0.0040	9372.1	193.65	0.83334	0.0086	2025.6	90.069	0.83335
0.0041	8920.4	188.93	0.83334	0.0087	1979.3	89.034	0.83335
0.0042	8500.6	184.43	0.83334	0.0088	1934.5	88.022	0.83335
0.0043	8109.7	180.14	0.83334	0.0089	1891.2	87.033	0.83335
0.0044	7745.1	176.04	0.83334	0.0090	1849.4	86.066	0.83336
0.0045	7404.6	172.13	0.83334	0.0091	1808.9	85.121	0.83336
0.0046	7086.0	168.39	0.83334	0.0092	1769.7	84.195	0.83336
0.0047	6787.6	164.81	0.83334	0.0093	1731.8	83.290	0.83336
0.0048	6507.6	161.37	0.83334	0.0094	1695.1	82.404	0.83336
0.0049	6244.6	158.08	0.83334	0.0095	1659.6	81.536	0.83336
0.0050	5997.2	154.92	0.83334	0.0096	1625.2	80.687	0.83336
0.0051	5764.3	151.88	0.83334	0.0097	1591.8	79.855	0.83336
0.0052	5544.6	148.96	0.83334	0.0098	1559.4	79.040	0.83336
0.0053	5337.2	146.15	0.83334	0.0099	1528.0	78.242	0.83336
0.0054	5141.3	143.44	0.83334	0.0100	1497.6	77.460	0.83336
0.0055	4956.0	140.84	0.83334				

ISOTHERMAL FLOW FUNCTIONS

$$\gamma = 5/3$$

M	$H(M;\gamma)$	p/p_f	$T°/T_f^°$	M	$H(M;\gamma)$	p/p_f	$T°/T_f^°$
0.010	1497.575	77.460	0.83336	0.056	46.268	13.8321	0.83420
0.011	1237.292	70.418	0.83337	0.057	44.613	13.5894	0.83424
0.012	1039.333	64.550	0.83337	0.058	43.044	13.3551	0.83427
0.013	885.282	59.584	0.83338	0.059	41.554	13.1288	0.83430
0.014	763.050	55.328	0.83339	0.060	40.138	12.9099	0.83433
0.015	664.445	51.640	0.83340	0.061	38.791	12.6983	0.83437
0.016	583.748	48.412	0.83340	0.062	37.509	12.4935	0.83440
0.017	516.872	45.564	0.83341	0.063	36.288	12.2952	0.83444
0.018	460.832	43.033	0.83342	0.064	35.124	12.1031	0.83447
0.019	413.409	40.768	0.83343	0.065	34.014	11.9169	0.83451
0.020	372.922	38.730	0.83344	0.066	32.954	11.7363	0.83454
0.021	338.082	36.886	0.83346	0.067	31.941	11.5611	0.83458
0.022	307.887	35.209	0.83347	0.068	30.973	11.3911	0.83462
0.023	281.546	33.678	0.83348	0.069	30.047	11.2260	0.83466
0.024	258.430	32.275	0.83349	0.070	29.160	11.0657	0.83469
0.025	238.033	30.984	0.83351	0.071	28.311	10.9098	0.83473
0.026	219.946	29.792	0.83352	0.072	27.497	10.7583	0.83477
0.027	203.833	28.689	0.83354	0.073	26.717	10.6109	0.83481
0.028	189.417	27.664	0.83355	0.074	25.968	10.4675	0.83485
0.029	176.467	26.710	0.83357	0.075	25.249	10.3280	0.83490
0.030	164.791	25.820	0.83358	0.076	24.559	10.1921	0.83494
0.031	154.228	24.987	0.83360	0.077	23.895	10.0597	0.83498
0.032	144.641	24.206	0.83362	0.078	23.257	9.9307	0.83502
0.033	135.913	23.473	0.83364	0.079	22.643	9.8050	0.83507
0.034	127.945	22.782	0.83365	0.080	22.052	9.6825	0.83511
0.035	120.651	22.131	0.83367	0.081	21.483	9.5629	0.83516
0.036	113.956	21.517	0.83369	0.082	20.935	9.4463	0.83520
0.037	107.798	20.935	0.83371	0.083	20.407	9.3325	0.83525
0.038	102.121	20.384	0.83373	0.084	19.898	9.2214	0.83529
0.039	96.875	19.861	0.83376	0.085	19.406	9.1129	0.83534
0.040	92.018	19.365	0.83378	0.086	18.932	9.0069	0.83539
0.041	87.513	18.893	0.83380	0.087	18.474	8.9034	0.83544
0.042	83.327	18.443	0.83382	0.088	18.032	8.8022	0.83548
0.043	79.429	18.014	0.83385	0.089	17.605	8.7033	0.83553
0.044	75.795	17.604	0.83387	0.090	17.192	8.6066	0.83558
0.045	72.401	17.213	0.83390	0.091	16.793	8.5121	0.83563
0.046	69.227	16.839	0.83392	0.092	16.407	8.4195	0.83568
0.047	66.253	16.481	0.83395	0.093	16.033	8.3290	0.83574
0.048	63.464	16.137	0.83397	0.094	15.671	8.2404	0.83579
0.049	60.844	15.808	0.83400	0.095	15.321	8.1536	0.83584
0.050	58.380	15.492	0.83403	0.096	14.982	8.0687	0.83589
0.051	56.060	15.188	0.83406	0.097	14.653	7.9855	0.83595
0.052	53.873	14.896	0.83408	0.098	14.335	7.9040	0.83600
0.053	51.809	14.615	0.83411	0.099	14.026	7.8242	0.83606
0.054	49.859	14.344	0.83414	0.100	13.726	7.7460	0.83611
0.055	48.014	14.084	0.83417				

C.1-3

ISOTHERMAL FLOW FUNCTIONS

$$\gamma = 5/3$$

M	$H(M;\gamma)$	p/p_f	$T°/T_f^o$	M	$H(M;\gamma)$	p/p_f	$T°/T_f^o$
0.10	13.7264	7.7460	0.83611	0.61	0.03368	1.26983	0.93669
0.11	11.1708	7.0418	0.83669	0.62	0.02891	1.24935	0.94011
0.12	9.2343	6.4550	0.83733	0.63	0.02462	1.22952	0.94358
0.13	7.7333	5.9584	0.83803	0.64	0.02077	1.21031	0.94711
0.14	6.5477	5.5328	0.83878	0.65	0.01734	1.19169	0.95069
0.15	5.5958	5.1640	0.83958	0.66	0.01430	1.17363	0.95433
0.16	4.8208	4.8412	0.84044	0.67	0.01162	1.15611	0.95803
0.17	4.1820	4.5565	0.84136	0.68	0.00927	1.13911	0.96178
0.18	3.6499	4.3033	0.84233	0.69	0.00723	1.12260	0.96558
0.19	3.2025	4.0768	0.84336	0.70	0.00549	1.10657	0.96944
0.20	2.8230	3.8730	0.84444	0.71	0.00402	1.09098	0.97336
0.21	2.4987	3.6886	0.84558	0.72	0.00281	1.07583	0.97733
0.22	2.2198	3.5209	0.84678	0.73	0.00183	1.06109	0.98136
0.23	1.9784	3.3678	0.84803	0.74	0.00108	1.04675	0.98544
0.24	1.7683	3.2275	0.84933	0.75	0.00053	1.03279	0.98958
0.25	1.5846	3.0984	0.85069	0.76	0.00018	1.01921	0.99378
0.26	1.4231	2.9792	0.85211	0.77	0.00002	1.00597	0.99803
0.27	1.2807	2.8689	0.85358	M_f=0.7746	0.00000	1.00000	1.00000
0.28	1.1545	2.7664	0.85511	0.78	0.00002	0.99307	1.00233
0.29	1.0424	2.6710	0.85669	0.79	0.00019	0.98050	1.00669
0.30	0.9424	2.5820	0.85833	0.80	0.00051	0.96825	1.01111
0.31	0.8530	2.4987	0.86003	0.81	0.00097	0.95629	1.01558
0.32	0.7728	2.4206	0.86178	0.82	0.00156	0.94463	1.02011
0.33	0.7008	2.3473	0.86358	0.83	0.00228	0.93325	1.02469
0.34	0.6359	2.2782	0.86544	0.84	0.00311	0.92214	1.02933
0.35	0.5773	2.2131	0.86736	0.85	0.00406	0.91129	1.03403
0.36	0.5243	2.1517	0.86933	0.86	0.00511	0.90069	1.03878
0.37	0.4763	2.0935	0.87136	0.87	0.00625	0.89034	1.04358
0.38	0.4327	2.0384	0.87344	0.88	0.00749	0.88022	1.04844
0.39	0.3931	1.9861	0.87558	0.89	0.00881	0.87033	1.05336
0.40	0.3571	1.9365	0.87778	0.90	0.01021	0.86066	1.05833
0.41	0.3242	1.8893	0.88003	0.91	0.01169	0.85120	1.06336
0.42	0.2943	1.8443	0.88233	0.92	0.01324	0.84195	1.06844
0.43	0.2670	1.8014	0.88469	0.93	0.01485	0.83290	1.07358
0.44	0.2420	1.7604	0.88711	0.94	0.01653	0.82404	1.07878
0.45	0.2192	1.7213	0.88958	0.95	0.01826	0.81536	1.08403
0.46	0.1983	1.6839	0.89211	0.96	0.02006	0.80687	1.08933
0.47	0.1792	1.6481	0.89469	0.97	0.02190	0.79855	1.09469
0.48	0.1618	1.6137	0.89733	0.98	0.02379	0.79040	1.10011
0.49	0.1458	1.5808	0.90003	0.99	0.02573	0.78242	1.10558
0.50	0.1311	1.5492	0.90278	1.00	0.02771	0.77460	1.11111
0.51	0.1177	1.5492	0.90278	1.20	0.07303	0.64550	1.23333
0.52	0.1055	1.5188	0.90558	1.50	0.14711	0.51640	1.45833
0.53	0.0943	1.4896	0.90844	2.00	0.26178	0.38730	1.94444
0.54	0.0840	1.4615	0.91136	3.00	0.44368	0.25820	3.33333
0.55	0.0747	1.4344	0.91433	4.00	0.58023	0.19365	5.27777
0.56	0.0661	1.4084	0.91736	5.00	0.68843	0.15492	7.77777
0.57	0.0583	1.3832	0.92044	6.00	0.77775	0.12910	10.83332
0.58	0.0512	1.3589	0.92358	7.00	0.85372	0.11066	14.44443
0.59	0.0448	1.3355	0.92678	8.00	0.91977	0.09682	18.61108
0.60	0.0390	1.3129	0.93003	9.00	0.97817	0.08607	23.33330
				10.00	1.03050	0.07746	28.61107

TABLE C.2

ISOTHERMAL FLOW FUNCTIONS

$\gamma = 7/5$

ISOTHERMAL FLOW FUNCTIONS

$$\gamma = 7/5$$

M	$H(M;\gamma)$	p/p_f	$T°/T°_f$	M	$H(M;\gamma)$	p/p_f	$T°/T°_f$
0.0010	178568.1	845.15	0.87500	0.0056	5691.5	150.920	0.87501
0.0011	147576.1	768.32	0.87500	0.0057	5493.4	148.273	0.87501
0.0012	124004.5	704.30	0.87500	0.0058	5305.6	145.716	0.87501
0.0013	105660.1	650.12	0.87500	0.0059	5127.2	143.246	0.87501
0.0014	91104.5	603.68	0.87500	0.0060	4957.6	140.859	0.87501
0.0015	79361.7	563.44	0.87500	0.0061	4796.3	138.550	0.87501
0.0016	69751.1	528.22	0.87500	0.0062	4642.8	136.315	0.87501
0.0017	61786.1	497.15	0.87500	0.0063	4496.5	134.151	0.87501
0.0018	55111.3	469.53	0.87500	0.0064	4357.0	132.055	0.87501
0.0019	49462.5	444.82	0.87500	0.0065	4223.9	130.024	0.87501
0.0020	44639.6	422.58	0.87500	0.0066	4096.8	128.054	0.87501
0.0021	40489.2	402.45	0.87500	0.0067	3975.3	126.142	0.87501
0.0022	36891.7	384.16	0.87500	0.0068	3859.2	124.287	0.87501
0.0023	33753.2	367.46	0.87500	0.0069	3748.1	122.486	0.87501
0.0024	30998.8	352.15	0.87500	0.0070	3641.7	120.736	0.87501
0.0025	28568.3	338.06	0.87500	0.0071	3539.7	119.036	0.87501
0.0026	26412.8	325.06	0.87500	0.0072	3442.0	117.383	0.87501
0.0027	24492.3	313.02	0.87500	0.0073	3348.3	115.775	0.87501
0.0028	22773.9	301.84	0.87500	0.0074	3258.4	114.210	0.87501
0.0029	21230.1	291.43	0.87500	0.0075	3172.0	112.687	0.87501
0.0030	19838.2	281.72	0.87500	0.0076	3089.0	111.204	0.87501
0.0031	18578.8	272.63	0.87500	0.0077	3009.2	109.760	0.87501
0.0032	17435.6	264.11	0.87500	0.0078	2932.5	108.353	0.87501
0.0033	16394.7	256.11	0.87500	0.0079	2858.7	106.982	0.87501
0.0034	15444.4	248.57	0.87500	0.0080	2787.6	105.644	0.87501
0.0035	14574.3	241.47	0.87500	0.0081	2719.1	104.340	0.87501
0.0036	13775.7	234.77	0.87500	0.0082	2653.2	103.068	0.87501
0.0037	13041.0	228.42	0.87500	0.0083	2589.6	101.826	0.87501
0.0038	12363.5	222.41	0.87500	0.0084	2528.2	100.614	0.87501
0.0039	11737.5	216.71	0.87500	0.0085	2469.0	99.430	0.87501
0.0040	11157.8	211.29	0.87500	0.0086	2411.9	98.274	0.87501
0.0041	10620.0	206.14	0.87500	0.0087	2356.7	97.144	0.87501
0.0042	10120.2	201.23	0.87500	0.0088	2303.4	96.040	0.87501
0.0043	9654.9	196.55	0.87500	0.0089	2251.9	94.961	0.87501
0.0044	9220.9	192.08	0.87500	0.0090	2202.1	93.906	0.87501
0.0045	8815.5	187.81	0.87500	0.0091	2153.9	92.874	0.87501
0.0046	8436.3	183.73	0.87500	0.0092	2107.3	91.865	0.87501
0.0047	8081.0	179.82	0.87500	0.0093	2062.1	90.877	0.87502
0.0048	7747.7	176.07	0.87500	0.0094	2018.5	89.910	0.87502
0.0049	7434.6	172.48	0.87500	0.0095	1976.1	88.964	0.87502
0.0050	7140.0	169.03	0.87500	0.0096	1935.1	88.037	0.87502
0.0051	6862.7	165.72	0.87500	0.0097	1895.4	87.129	0.87502
0.0052	6601.2	162.53	0.87500	0.0098	1856.9	86.240	0.87502
0.0053	6354.3	159.46	0.87500	0.0099	1819.5	85.369	0.87502
0.0054	6121.1	156.51	0.87501	0.0100	1783.2	84.515	0.87502
0.0055	5900.4	153.66	0.87501				

ISOTHERMAL FLOW FUNCTIONS

$$\gamma = 7/5$$

M	$H(M;\gamma)$	p/p_f	$T°/T°_f$	M	$H(M;\gamma)$	p/p_f	$T°/T°_f$
0.010	1783.247	84.515	0.87502	0.056	55.335	15.0920	0.87555
0.011	1473.376	76.832	0.87502	0.057	53.364	14.8273	0.87557
0.012	1237.702	70.430	0.87503	0.058	51.494	14.5716	0.87559
0.013	1054.298	65.012	0.87503	0.059	49.718	14.3246	0.87561
0.014	908.779	60.368	0.87503	0.060	48.031	14.0859	0.87563
0.015	791.386	56.344	0.87504	0.061	46.426	13.8550	0.87565
0.016	695.312	52.822	0.87504	0.062	44.898	13.6315	0.87567
0.017	615.692	49.715	0.87505	0.063	43.443	13.4152	0.87569
0.018	548.972	46.953	0.87506	0.064	42.056	13.2055	0.87572
0.019	492.510	44.482	0.87506	0.065	40.733	13.0024	0.87574
0.020	444.307	42.258	0.87507	0.066	39.469	12.8054	0.87576
0.021	402.826	40.245	0.87508	0.067	38.262	12.6142	0.87579
0.022	366.875	38.416	0.87508	0.068	37.108	12.4287	0.87581
0.023	335.512	36.746	0.87509	0.069	36.004	12.2486	0.87583
0.024	307.989	35.215	0.87510	0.070	34.948	12.0736	0.87586
0.025	283.704	33.806	0.87511	0.071	33.935	11.9036	0.87588
0.026	262.168	32.506	0.87512	0.072	32.965	11.7383	0.87591
0.027	242.982	31.302	0.87513	0.073	32.035	11.5775	0.87593
0.028	225.816	30.184	0.87514	0.074	31.142	11.4210	0.87596
0.029	210.396	29.143	0.87515	0.075	30.285	11.2687	0.87598
0.030	196.494	28.172	0.87516	0.076	29.462	11.1205	0.87601
0.031	183.916	27.263	0.87517	0.077	28.670	10.9760	0.87604
0.032	172.499	26.411	0.87518	0.078	27.910	10.8353	0.87606
0.033	162.106	25.611	0.87519	0.079	27.178	10.6982	0.87609
0.034	152.617	24.857	0.87520	0.080	26.473	10.5644	0.87612
0.035	143.931	24.147	0.87521	0.081	25.795	10.4340	0.87615
0.036	135.959	23.477	0.87523	0.082	25.141	10.3068	0.87618
0.037	128.625	22.842	0.87524	0.083	24.511	10.1826	0.87621
0.038	121.863	22.241	0.87525	0.084	23.903	10.0614	0.87623
0.039	115.616	21.671	0.87527	0.085	23.317	9.9430	0.87626
0.040	109.832	21.129	0.87528	0.086	22.752	9.8274	0.87629
0.041	104.466	20.614	0.87529	0.087	22.206	9.7144	0.87632
0.042	99.480	20.123	0.87531	0.088	21.678	9.6040	0.87636
0.043	94.838	19.655	0.87532	0.089	21.169	9.4961	0.87639
0.044	90.510	19.208	0.87534	0.090	20.676	9.3906	0.87642
0.045	86.467	18.781	0.87535	0.091	20.200	9.2874	0.87645
0.046	82.686	18.373	0.87537	0.092	19.739	9.1865	0.87648
0.047	79.143	17.982	0.87539	0.093	19.293	9.0877	0.87651
0.048	75.821	17.607	0.87540	0.094	18.861	8.9910	0.87655
0.049	72.700	17.248	0.87542	0.095	18.443	8.8964	0.87658
0.050	69.765	16.903	0.87544	0.096	18.039	8.8037	0.87661
0.051	67.001	16.572	0.87545	0.097	17.646	8.7129	0.87665
0.052	64.396	16.253	0.87547	0.098	17.266	8.6240	0.87668
0.053	61.937	15.946	0.87549	0.099	16.898	8.5369	0.87671
0.054	59.613	15.651	0.87551	0.100	16.540	8.4515	0.87675
0.055	57.416	15.366	0.87553				

ISOTHERMAL FLOW FUNCTIONS

$$\gamma = 7/5$$

M	$H(M;\gamma)$	p/p_f	$T°/T^°_f$	M	$H(M;\gamma)$	p/p_f	$T°/T^°_f$
0.10	16.5400	8.4515	0.87675	0.61	0.06687	1.38550	0.94012
0.11	13.4885	7.6832	0.87712	0.62	0.05965	1.36315	0.94227
0.12	11.1748	7.0430	0.87752	0.63	0.05302	1.34151	0.94446
0.13	9.3804	6.5012	0.87796	0.64	0.04694	1.32055	0.94668
0.14	7.9618	6.0368	0.87843	0.65	0.04138	1.30024	0.94894
0.15	6.8221	5.6344	0.87894	0.66	0.03630	1.28054	0.95123
0.16	5.8933	5.2822	0.87948	0.67	0.03168	1.26142	0.95356
0.17	5.1271	4.9715	0.88006	0.68	0.02747	1.24287	0.95592
0.18	4.4882	4.6953	0.88067	0.69	0.02366	1.22486	0.95832
0.19	3.9503	4.4482	0.88132	0.70	0.02021	1.20736	0.96075
0.20	3.4937	4.2258	0.88200	0.71	0.01711	1.19036	0.96322
0.21	3.1030	4.0245	0.88272	0.72	0.01433	1.17382	0.96572
0.22	2.7666	3.8416	0.88347	0.73	0.01186	1.15775	0.96826
0.23	2.4749	3.6746	0.88426	0.74	0.00966	1.14210	0.97083
0.24	2.2208	3.5215	0.88508	0.75	0.00774	1.12687	0.97344
0.25	1.9981	3.3806	0.88594	0.76	0.00606	1.11204	0.97608
0.26	1.8022	3.2506	0.88683	0.77	0.00462	1.09760	0.97876
0.27	1.6290	3.1302	0.88776	0.78	0.00340	1.08353	0.98147
0.28	1.4753	3.0184	0.88872	0.79	0.00238	1.06981	0.98422
0.29	1.3385	2.9143	0.88972	0.80	0.00156	1.05644	0.98700
0.30	1.2163	2.8172	0.89075	0.81	0.00093	1.04340	0.98982
0.31	1.1067	2.7263	0.89182	0.82	0.00047	1.03067	0.99267
0.32	1.0083	2.6411	0.89292	0.83	0.00017	1.01826	0.99556
0.33	0.9196	2.5611	0.89406	0.84	0.00002	1.00613	0.99848
0.34	0.8394	2.4857	0.89523	M_f=0.8451	0.00000	1.00000	1.00000
0.35	0.7669	2.4147	0.89644	0.85	0.00002	0.99430	1.00144
0.36	0.7012	2.3477	0.89768	0.86	0.00015	0.98274	1.00443
0.37	0.6414	2.2842	0.89896	0.87	0.00041	0.97144	1.00746
0.38	0.5870	2.2241	0.90027	0.88	0.00079	0.96040	1.01052
0.39	0.5374	2.1671	0.90162	0.89	0.00129	0.94961	1.01362
0.40	0.4920	2.1129	0.90300	0.90	0.00190	0.93906	1.01675
0.41	0.4506	2.0614	0.90442	0.91	0.00260	0.92874	1.01992
0.42	0.4127	2.0123	0.90587	0.92	0.00340	0.91865	1.02312
0.43	0.3779	1.9655	0.90736	0.93	0.00430	0.90877	1.02636
0.44	0.3460	1.9208	0.90888	0.94	0.00528	0.89910	1.02963
0.45	0.3167	1.8781	0.91044	0.95	0.00633	0.88964	1.03294
0.46	0.2898	1.8373	0.91203	0.96	0.00747	0.88037	1.03628
0.47	0.2650	1.7982	0.91366	0.97	0.00868	0.87129	1.03966
0.48	0.2422	1.7607	0.91532	0.98	0.00995	0.86240	1.04307
0.49	0.2212	1.7248	0.91702	0.99	0.01129	0.85369	1.04652
0.50	0.2018	1.6903	0.91875	1.00	0.01269	0.84515	1.05000
0.51	0.2018	1.6903	0.91875	1.20	0.04929	0.70430	1.12700
0.52	0.1840	1.6572	0.92052	1.50	0.11622	0.56344	1.26875
0.53	0.1676	1.6253	0.92232	2.00	0.22533	0.42258	1.57500
0.54	0.1524	1.5946	0.92416	3.00	0.40327	0.28172	2.45000
0.55	0.1384	1.5651	0.92603	4.00	0.53843	0.21129	3.67500
0.56	0.1255	1.5366	0.92794	5.00	0.64598	0.16903	5.25000
0.57	0.1136	1.5092	0.92988	6.00	0.73496	0.14086	7.17499
0.58	0.1027	1.4827	0.93186	7.00	0.81072	0.12074	9.44999
0.59	0.0926	1.4572	0.93387	8.00	0.87663	0.10564	12.07499
0.60	0.0833	1.4325	0.93592	9.00	0.93493	0.09391	15.04998
				10.00	0.98720	0.08452	18.37497

TABLE C.3

ISOTHERMAL FLOW FUNCTIONS

$\gamma = 4/3$

ISOTHERMAL FLOW FUNCTIONS

$$\gamma = 4/3$$

M	$H(M;\gamma)$	p/p_f	$T°/T°_f$	M	$H(M;\gamma)$	p/p_f	$T°/T°_f$
0.0010	187496.5	866.03	0.88889	0.0056	5976.2	154.647	0.88889
0.0011	154955.1	787.30	0.88889	0.0057	5768.3	151.934	0.88889
0.0012	130204.9	721.69	0.88889	0.0058	5571.0	149.315	0.88889
0.0013	110943.3	666.17	0.88889	0.0059	5383.6	146.784	0.88889
0.0014	95659.9	618.59	0.88889	0.0060	5205.6	144.338	0.88889
0.0015	83329.9	577.35	0.88889	0.0061	5036.2	141.971	0.88889
0.0016	73238.8	541.27	0.88889	0.0062	4875.0	139.682	0.88889
0.0017	64875.6	509.43	0.88889	0.0063	4721.4	137.464	0.88889
0.0018	57867.0	481.13	0.88889	0.0064	4574.9	135.317	0.88889
0.0019	51935.8	455.80	0.88889	0.0065	4435.2	133.235	0.88889
0.0020	46871.7	433.01	0.88889	0.0066	4301.7	131.216	0.88889
0.0021	42513.8	412.39	0.88889	0.0067	4174.2	129.258	0.88889
0.0022	38736.4	393.65	0.88889	0.0068	4052.3	127.357	0.88890
0.0023	35441.0	376.53	0.88889	0.0069	3935.6	125.511	0.88890
0.0024	32548.9	360.84	0.88889	0.0070	3823.9	123.718	0.88890
0.0025	29996.8	346.41	0.88889	0.0071	3716.9	121.975	0.88890
0.0026	27733.5	333.09	0.88889	0.0072	3614.3	120.281	0.88890
0.0027	25717.0	320.75	0.88889	0.0073	3515.8	118.634	0.88890
0.0028	23912.7	309.29	0.88889	0.0074	3421.4	117.031	0.88890
0.0029	22291.8	298.63	0.88889	0.0075	3330.7	115.470	0.88890
0.0030	20830.2	288.68	0.88889	0.0076	3243.6	113.951	0.88890
0.0031	19507.8	279.36	0.88889	0.0077	3159.8	112.471	0.88890
0.0032	18307.5	270.63	0.88889	0.0078	3079.3	111.029	0.88890
0.0033	17214.6	262.43	0.88889	0.0079	3001.7	109.623	0.88890
0.0034	16216.7	254.71	0.88889	0.0080	2927.1	108.253	0.88890
0.0035	15303.1	247.44	0.88889	0.0081	2855.2	106.917	0.88890
0.0036	14464.6	240.56	0.88889	0.0082	2785.9	105.613	0.88890
0.0037	13693.2	234.06	0.88889	0.0083	2719.2	104.340	0.88890
0.0038	12981.8	227.90	0.88889	0.0084	2654.7	103.098	0.88890
0.0039	12324.5	222.06	0.88889	0.0085	2592.6	101.885	0.88890
0.0040	11715.8	216.51	0.88889	0.0086	2532.6	100.701	0.88890
0.0041	11151.2	211.23	0.88889	0.0087	2474.7	99.543	0.88890
0.0042	10626.4	206.20	0.88889	0.0088	2418.7	98.412	0.88890
0.0043	10137.7	201.40	0.88889	0.0089	2364.6	97.306	0.88890
0.0044	9682.0	196.82	0.88889	0.0090	2312.3	96.225	0.88890
0.0045	9256.4	192.45	0.88889	0.0091	2261.7	95.168	0.88890
0.0046	8858.2	188.27	0.88889	0.0092	2212.7	94.133	0.88890
0.0047	8485.2	184.26	0.88889	0.0093	2165.4	93.121	0.88890
0.0048	8135.2	180.42	0.88889	0.0094	2119.5	92.130	0.88890
0.0049	7806.4	176.74	0.88889	0.0095	2075.1	91.161	0.88890
0.0050	7497.2	173.21	0.88889	0.0096	2032.0	90.211	0.88890
0.0051	7206.0	169.81	0.88889	0.0097	1990.3	89.281	0.88890
0.0052	6931.4	166.54	0.88889	0.0098	1949.8	88.370	0.88890
0.0053	6672.2	163.40	0.88889	0.0099	1910.6	87.477	0.88890
0.0054	6427.3	160.38	0.88889	0.0100	1872.5	86.603	0.88890
0.0055	6195.6	157.46	0.88889				

ISOTHERMAL FLOW FUNCTIONS

$$\gamma = 4/3$$

M	$H(M;\gamma)$	p/p_f	$T°/T°_f$	M	$H(M;\gamma)$	p/p_f	$T°/T°_f$
0.010	1872.520	86.603	0.88890	0.056	58.170	15.4647	0.88935
0.011	1547.153	78.730	0.88891	0.057	56.100	15.1934	0.88937
0.012	1299.694	72.169	0.88891	0.058	54.135	14.9315	0.88939
0.013	1107.118	66.617	0.88891	0.059	52.271	14.6784	0.88940
0.014	954.321	61.859	0.88892	0.060	50.499	14.4338	0.88942
0.015	831.056	57.735	0.88892	0.061	48.813	14.1971	0.88944
0.016	730.177	54.127	0.88893	0.062	47.209	13.9682	0.88946
0.017	646.574	50.943	0.88893	0.063	45.681	13.7464	0.88948
0.018	576.517	48.113	0.88894	0.064	44.224	13.5317	0.88950
0.019	517.231	45.580	0.88894	0.065	42.834	13.3235	0.88951
0.020	466.616	43.301	0.88895	0.066	41.507	13.1216	0.88953
0.021	423.061	41.239	0.88895	0.067	40.239	12.9258	0.88955
0.022	385.310	39.365	0.88896	0.068	39.027	12.7357	0.88957
0.023	352.378	37.653	0.88897	0.069	37.868	12.5511	0.88959
0.024	323.478	36.084	0.88897	0.070	36.758	12.3718	0.88961
0.025	297.978	34.641	0.88898	0.071	35.694	12.1975	0.88963
0.026	275.364	33.309	0.88899	0.072	34.675	12.0281	0.88966
0.027	255.218	32.075	0.88900	0.073	33.698	11.8634	0.88968
0.028	237.192	30.929	0.88901	0.074	32.760	11.7031	0.88970
0.029	221.001	29.863	0.88901	0.075	31.860	11.5470	0.88972
0.030	206.402	28.868	0.88902	0.076	30.995	11.3951	0.88974
0.031	193.194	27.936	0.88903	0.077	30.164	11.2471	0.88977
0.032	181.206	27.063	0.88904	0.078	29.365	11.1029	0.88979
0.033	170.293	26.243	0.88905	0.079	28.596	10.9623	0.88981
0.034	160.328	25.471	0.88906	0.080	27.856	10.8253	0.88984
0.035	151.207	24.744	0.88907	0.081	27.143	10.6917	0.88986
0.036	142.836	24.056	0.88908	0.082	26.457	10.5613	0.88989
0.037	135.135	23.406	0.88909	0.083	25.795	10.4340	0.88991
0.038	128.034	22.790	0.88910	0.084	25.157	10.3098	0.88993
0.039	121.474	22.206	0.88911	0.085	24.541	10.1885	0.88996
0.040	115.400	21.651	0.88913	0.086	23.947	10.0701	0.88998
0.041	109.766	21.123	0.88914	0.087	23.373	9.9543	0.89001
0.042	104.529	20.620	0.88915	0.088	22.819	9.8412	0.89004
0.043	99.655	20.140	0.88916	0.089	22.284	9.7306	0.89006
0.044	95.109	19.682	0.88918	0.090	21.766	9.6225	0.89009
0.045	90.864	19.245	0.88919	0.091	21.266	9.5168	0.89012
0.046	86.893	18.827	0.88920	0.092	20.782	9.4133	0.89014
0.047	83.173	18.426	0.88922	0.093	20.313	9.3121	0.89017
0.048	79.684	18.042	0.88923	0.094	19.860	9.2130	0.89020
0.049	76.406	17.674	0.88924	0.095	19.421	9.1161	0.89023
0.050	73.324	17.321	0.88926	0.096	18.995	9.0211	0.89025
0.051	70.422	16.981	0.88927	0.097	18.583	8.9281	0.89028
0.052	67.685	16.654	0.88929	0.098	18.184	8.8370	0.89031
0.053	65.103	16.340	0.88930	0.099	17.796	8.7477	0.89034
0.054	62.663	16.038	0.88932	0.100	17.421	8.6603	0.89037
0.055	60.355	15.746	0.88934				

ISOTHERMAL FLOW FUNCTIONS

$$\gamma = 4/3$$

M	$H(M;\gamma)$	p/p_f	T°/T°_f	M	$H(M;\gamma)$	p/p_f	T°/T°_f
0.10	17.4206	8.6603	0.89037	0.61	0.07867	1.41971	0.94401
0.11	14.2142	7.8730	0.89068	0.62	0.07068	1.39682	0.94584
0.12	11.7826	7.2169	0.89102	0.63	0.06331	1.37464	0.94769
0.13	9.8965	6.6617	0.89139	0.64	0.05654	1.35316	0.94957
0.14	8.4052	6.1859	0.89179	0.65	0.05032	1.33235	0.95148
0.15	7.2067	5.7735	0.89222	0.66	0.04460	1.31216	0.95342
0.16	6.2299	5.4127	0.89268	0.67	0.03937	1.29257	0.95539
0.17	5.4238	5.0943	0.89317	0.68	0.03458	1.27357	0.95739
0.18	4.7516	4.8113	0.89369	0.69	0.03021	1.25511	0.95942
0.19	4.1855	4.5580	0.89424	0.70	0.02624	1.23718	0.96148
0.20	3.7047	4.3301	0.89481	0.71	0.02263	1.21975	0.96357
0.21	3.2933	4.1239	0.89542	0.72	0.01936	1.20281	0.96569
0.22	2.9388	3.9365	0.89606	0.73	0.01641	1.18634	0.96784
0.23	2.6315	3.7653	0.89673	0.74	0.01377	1.17030	0.97001
0.24	2.3636	3.6084	0.89742	0.75	0.01141	1.15470	0.97222
0.25	2.1288	3.4641	0.89815	0.76	0.00932	1.13951	0.97446
0.26	1.9221	3.3309	0.89890	0.77	0.00748	1.12471	0.97673
0.27	1.7393	3.2075	0.89969	0.78	0.00588	1.11029	0.97902
0.28	1.5770	3.0929	0.90050	0.79	0.00449	1.09623	0.98135
0.29	1.4325	2.9863	0.90135	0.80	0.00332	1.08253	0.98370
0.30	1.3033	2.8868	0.90222	0.81	0.00234	1.06917	0.98609
0.31	1.1874	2.7936	0.90313	0.82	0.00155	1.05613	0.98850
0.32	1.0833	2.7063	0.90406	0.83	0.00093	1.04340	0.99095
0.33	0.9894	2.6243	0.90502	0.84	0.00048	1.03098	0.99342
0.34	0.9045	2.5471	0.90601	0.85	0.00018	1.01885	0.99593
0.35	0.8276	2.4744	0.90704	0.86	0.00002	1.00701	0.99846
0.36	0.7579	2.4056	0.90809	M_f=0.8660	0.00000	1.00000	1.00000
0.37	0.6944	2.3406	0.90917	0.87	0.00001	0.99543	1.00102
0.38	0.6366	2.2790	0.91028	0.88	0.00013	0.98412	1.00361
0.39	0.5839	2.2206	0.91142	0.89	0.00037	0.97306	1.00624
0.40	0.5357	2.1651	0.91259	0.90	0.00072	0.96225	1.00889
0.41	0.4915	2.1123	0.91379	0.91	0.00119	0.95168	1.01157
0.42	0.4511	2.0620	0.91502	0.92	0.00176	0.94133	1.01428
0.43	0.4140	2.0140	0.91628	0.93	0.00242	0.93121	1.01702
0.44	0.3799	1.9682	0.91757	0.94	0.00318	0.92130	1.01979
0.45	0.3486	1.9245	0.91889	0.95	0.00403	0.91161	1.02259
0.46	0.3198	1.8827	0.92024	0.96	0.00496	0.90211	1.02542
0.47	0.2932	1.8426	0.92161	0.97	0.00597	0.89281	1.02828
0.48	0.2687	1.8042	0.92302	0.98	0.00705	0.88370	1.03117
0.49	0.2462	1.7674	0.92446	0.99	0.00820	0.87477	1.03409
0.50	0.2253	1.7321	0.92593	1.00	0.00942	0.86603	1.03704
0.51	0.2253	1.7321	0.92593	1.20	0.04329	0.72169	1.10222
0.52	0.2061	1.6981	0.92742	1.50	0.10799	0.57735	1.22222
0.53	0.1884	1.6654	0.92895	2.00	0.21537	0.43301	1.48148
0.54	0.1720	1.6340	0.93050	3.00	0.39206	0.28868	2.22222
0.55	0.1568	1.6038	0.93209	4.00	0.52679	0.21651	3.25926
0.56	0.1428	1.5746	0.93370	5.00	0.63414	0.17321	4.59259
0.57	0.1299	1.5465	0.93535	6.00	0.72301	0.14434	6.22222
0.58	0.1180	1.5193	0.93702	7.00	0.79870	0.12372	8.14814
0.59	0.1069	1.4931	0.93873	8.00	0.86457	0.10825	10.37036
0.60	0.0967	1.4678	0.94046	9.00	0.92285	0.09623	12.88888
				10.00	0.97509	0.08660	15.70368

TABLE D

PRANDTL'S LOGARITHMIC LAW

FOR

SKIN FRICTION COEFFICIENT

PRANDTL'S LOGARITHMIC LAW

FOR SKIN FRICTION COEFFICIENT

$10^{-3}Re$	$10^6 c_f$	$10^{-3}Re$	$10^6 c_f$	$10^{-4}Re$	$10^6 c_f$	$10^{-4}Re$	$10^6 c_f$
1.0	15651	5.5	9099	1.0	7722	5.5	5114
1.1	15131	5.6	9052	1.1	7530	5.6	5094
1.2	14677	5.7	9007	1.2	7361	5.7	5074
1.3	14277	5.8	8963	1.3	7211	5.8	5054
1.4	13920	5.9	8919	1.4	7075	5.9	5035
1.5	13598	6.0	8877	1.5	6952	6.0	5017
1.6	13307	6.1	8836	1.6	6840	6.1	4999
1.7	13041	6.2	8796	1.7	6737	6.2	4981
1.8	12798	6.3	8757	1.8	6642	6.3	4964
1.9	12573	6.4	8718	1.9	6554	6.4	4947
2.0	12366	6.5	8681	2.0	6472	6.5	4930
2.1	12172	6.6	8644	2.1	6395	6.6	4914
2.2	11992	6.7	8608	2.2	6323	6.7	4898
2.3	11823	6.8	8572	2.3	6255	6.8	4882
2.4	11665	6.9	8538	2.4	6191	6.9	4867
2.5	11516	7.0	8504	2.5	6131	7.0	4851
2.6	11375	7.1	8471	2.6	6074	7.1	4837
2.7	11242	7.2	8438	2.7	6019	7.2	4822
2.8	11116	7.3	8406	2.8	5968	7.3	4808
2.9	10996	7.4	8375	2.9	5918	7.4	4794
3.0	10882	7.5	8344	3.0	5871	7.5	4780
3.1	10773	7.6	8314	3.1	5826	7.6	4766
3.2	10670	7.7	8284	3.2	5783	7.7	4753
3.3	10570	7.8	8255	3.3	5742	7.8	4740
3.4	10475	7.9	8227	3.4	5702	7.9	4727
3.5	10384	8.0	8198	3.5	5664	8.0	4714
3.6	10297	8.1	8171	3.6	5627	8.1	4702
3.7	10213	8.2	8144	3.7	5592	8.2	4690
3.8	10132	8.3	8117	3.8	5558	8.3	4678
3.9	10054	8.4	8091	3.9	5525	8.4	4666
4.0	9979	8.5	8065	4.0	5493	8.5	4654
4.1	9906	8.6	8039	4.1	5462	8.6	4643
4.2	9836	8.7	8014	4.2	5432	8.7	4631
4.3	9769	8.8	7990	4.3	5403	8.8	4620
4.4	9703	8.9	7966	4.4	5375	8.9	4609
4.5	9640	9.0	7942	4.5	5348	9.0	4598
4.6	9578	9.1	7918	4.6	5322	9.1	4588
4.7	9519	9.2	7895	4.7	5296	9.2	4577
4.8	9461	9.3	7872	4.8	5271	9.3	4567
4.9	9405	9.4	7850	4.9	5247	9.4	4556
5.0	9350	9.5	7828	5.0	5223	9.5	4546
5.1	9297	9.6	7806	5.1	5200	9.6	4536
5.2	9245	9.7	7785	5.2	5178	9.7	4526
5.3	9195	9.8	7763	5.3	5156	9.8	4517
5.4	9146	9.9	7743	5.4	5135	9.9	4507
5.5	9099	10.0	7722	5.5	5114	10.0	4498

PRANDTL'S LOGARITHMIC LAW

FOR SKIN FRICTION COEFFICIENT

$10^{-5}Re$	$10^6 c_f$	$10^{-5}Re$	$10^6 c_f$	$10^{-6}Re$	$10^6 c_f$	$10^{-6}Re$	$10^6 c_f$
1.0	4498	5.5	3233	1.0	2911	5.5	2213
1.1	4410	5.6	3223	1.1	2864	5.6	2207
1.2	4331	5.7	3213	1.2	2822	5.7	2201
1.3	4261	5.8	3203	1.3	2784	5.8	2195
1.4	4197	5.9	3193	1.4	2750	5.9	2190
1.5	4139	6.0	3183	1.5	2719	6.0	2184
1.6	4086	6.1	3174	1.6	2690	6.1	2179
1.7	4037	6.2	3165	1.7	2663	6.2	2173
1.8	3991	6.3	3156	1.8	2638	6.3	2168
1.9	3949	6.4	3147	1.9	2615	6.4	2163
2.0	3909	6.5	3138	2.0	2593	6.5	2158
2.1	3872	6.6	3130	2.1	2573	6.6	2153
2.2	3837	6.7	3121	2.2	2553	6.7	2148
2.3	3804	6.8	3113	2.3	2535	6.8	2144
2.4	3773	6.9	3105	2.4	2518	6.9	2139
2.5	3744	7.0	3097	2.5	2501	7.0	2134
2.6	3716	7.1	3090	2.6	2486	7.1	2130
2.7	3689	7.2	3082	2.7	2471	7.2	2125
2.8	3663	7.3	3075	2.8	2457	7.3	2121
2.9	3639	7.4	3067	2.9	2443	7.4	2117
3.0	3616	7.5	3060	3.0	2430	7.5	2113
3.1	3593	7.6	3053	3.1	2417	7.6	2108
3.2	3572	7.7	3046	3.2	2405	7.7	2104
3.3	3551	7.8	3039	3.3	2394	7.8	2100
3.4	3532	7.9	3032	3.4	2383	7.9	2096
3.5	3513	8.0	3026	3.5	2372	8.0	2093
3.6	3494	8.1	3019	3.6	2362	8.1	2089
3.7	3476	8.2	3013	3.7	2351	8.2	2085
3.8	3459	8.3	3006	3.8	2342	8.3	2081
3.9	3443	8.4	3000	3.9	2332	8.4	2078
4.0	3427	8.5	2994	4.0	2323	8.5	2074
4.1	3411	8.6	2988	4.1	2314	8.6	2070
4.2	3396	8.7	2982	4.2	2306	8.7	2067
4.3	3381	8.8	2976	4.3	2297	8.8	2064
4.4	3367	8.9	2970	4.4	2289	8.9	2060
4.5	3353	9.0	2964	4.5	2282	9.0	2057
4.6	3340	9.1	2959	4.6	2274	9.1	2053
4.7	3327	9.2	2953	4.7	2266	9.2	2050
4.8	3314	9.3	2948	4.8	2259	9.3	2047
4.9	3302	9.4	2942	4.9	2252	9.4	2044
5.0	3289	9.5	2937	5.0	2245	9.5	2041
5.1	3278	9.6	2932	5.1	2238	9.6	2037
5.2	3266	9.7	2926	5.2	2232	9.7	2034
5.3	3255	9.8	2921	5.3	2225	9.8	2031
5.4	3244	9.9	2916	5.4	2219	9.9	2028
5.5	3233	10.0	2911	5.5	2213	10.0	2025

PRANDTL'S LOGARITHMIC LAW

FOR SKIN FRICTION COEFFICIENT

$10^{-7}Re$	$10^6 c_f$	$10^{-7}Re$	$10^6 c_f$	$10^{-8}Re$	$10^6 c_f$	$10^{-8}Re$	$10^6 c_f$
1.0	2025	5.5	1603	1.0	1485	5.5	1211
1.1	1998	5.6	1599	1.1	1467	5.6	1208
1.2	1973	5.7	1595	1.2	1451	5.7	1206
1.3	1950	5.8	1592	1.3	1437	5.8	1204
1.4	1930	5.9	1588	1.4	1424	5.9	1201
1.5	1911	6.0	1585	1.5	1412	6.0	1199
1.6	1894	6.1	1581	1.6	1401	6.1	1197
1.7	1878	6.2	1578	1.7	1390	6.2	1194
1.8	1863	6.3	1575	1.8	1381	6.3	1192
1.9	1849	6.4	1572	1.9	1372	6.4	1190
2.0	1836	6.5	1568	2.0	1363	6.5	1188
2.1	1823	6.6	1565	2.1	1355	6.6	1186
2.2	1812	6.7	1562	2.2	1348	6.7	1184
2.3	1801	6.8	1559	2.3	1341	6.8	1182
2.4	1790	6.9	1556	2.4	1334	6.9	1180
2.5	1780	7.0	1554	2.5	1327	7.0	1178
2.6	1771	7.1	1551	2.6	1321	7.1	1176
2.7	1762	7.2	1548	2.7	1315	7.2	1174
2.8	1753	7.3	1545	2.8	1310	7.3	1173
2.9	1745	7.4	1543	2.9	1304	7.4	1171
3.0	1737	7.5	1540	3.0	1299	7.5	1169
3.1	1729	7.6	1537	3.1	1294	7.6	1167
3.2	1722	7.7	1535	3.2	1289	7.7	1166
3.3	1715	7.8	1532	3.3	1285	7.8	1164
3.4	1708	7.9	1530	3.4	1280	7.9	1162
3.5	1701	8.0	1527	3.5	1276	8.0	1161
3.6	1695	8.1	1525	3.6	1272	8.1	1159
3.7	1689	8.2	1522	3.7	1267	8.2	1157
3.8	1683	8.3	1520	3.8	1264	8.3	1156
3.9	1677	8.4	1518	3.9	1260	8.4	1154
4.0	1671	8.5	1516	4.0	1256	8.5	1153
4.1	1666	8.6	1513	4.1	1252	8.6	1151
4.2	1660	8.7	1511	4.2	1249	8.7	1150
4.3	1655	8.8	1509	4.3	1246	8.8	1148
4.4	1650	8.9	1507	4.4	1242	8.9	1147
4.5	1645	9.0	1505	4.5	1239	9.0	1146
4.6	1641	9.1	1503	4.6	1236	9.1	1144
4.7	1636	9.2	1500	4.7	1233	9.2	1143
4.8	1631	9.3	1498	4.8	1230	9.3	1141
4.9	1627	9.4	1496	4.9	1227	9.4	1140
5.0	1623	9.5	1494	5.0	1224	9.5	1139
5.1	1619	9.6	1492	5.1	1221	9.6	1137
5.2	1614	9.7	1490	5.2	1219	9.7	1136
5.3	1610	9.8	1489	5.3	1216	9.8	1135
5.4	1607	9.9	1487	5.4	1213	9.9	1134
5.5	1603	10.0	1485	5.5	1211	10.0	1132

TABLE E.1

OBLIQUE SHOCKS IN A PERFECT GAS

$\gamma = 5/3$

OBLIQUE SHOCKS IN A PERFECT GAS

$\gamma = 5/3$ $M_1 = 1.10$

θ	β	M_1	p_{21}	ρ_{21}	T_{21}	p_{21}^o
0.0000	65.3800	1.1000	1.0000	1.0000	1.0000	1.0000
1.0000	70.5637	1.0295	1.0950	1.0560	1.0370	0.9999
* 1.2528	73.2087	1.0000	1.1363	1.0796	1.0525	0.9998
m 1.3491	76.2416	0.9715	1.1769	1.1025	1.0675	0.9997
1.0000	82.5345	0.9303	1.2370	1.1357	1.0891	0.9992
0.0000	90.0000	0.9131	1.2625	1.1496	1.0982	0.9990

$\gamma = 5/3$ $M_1 = 1.15$

θ	β	M_1	p_{21}	ρ_{21}	T_{21}	p_{21}^o
0.0000	60.4081	1.1500	1.0000	1.0000	1.0000	1.0000
1.0000	63.5275	1.0953	1.0746	1.0441	1.0292	1.0000
2.0000	68.3582	1.0234	1.1783	1.1033	1.0680	0.9996
* 2.2095	70.1877	1.0000	1.2132	1.1227	1.0806	0.9994
m 2.3663	73.7215	0.9605	1.2732	1.1555	1.1019	0.9989
2.0000	79.4348	0.9127	1.3475	1.1950	1.1277	0.9979
1.0000	85.3953	0.8843	1.3925	1.2183	1.1429	0.9971
0.0000	90.0000	0.8776	1.4031	1.2238	1.1465	0.9969

$\gamma = 5/3$ $M_1 = 1.20$

θ	β	M_1	p_{21}	ρ_{21}	T_{21}	p_{21}^o
0.0000	56.4427	1.2000	1.0000	1.0000	1.0000	1.0000
1.0000	58.8100	1.1515	1.0672	1.0398	1.0264	1.0000
2.0000	61.7352	1.0971	1.1464	1.0853	1.0562	0.9998
3.0000	66.0399	1.0273	1.2531	1.1446	1.0949	0.9991
* 3.2653	67.9489	1.0000	1.2963	1.1678	1.1100	0.9986
m 3.4778	71.8218	0.9514	1.3748	1.2092	1.1369	0.9974
3.0000	77.8744	0.8935	1.4706	1.2581	1.1689	0.9955
2.0000	82.8407	0.8628	1.5220	1.2836	1.1857	0.9941
1.0000	86.5903	0.8500	1.5436	1.2942	1.1927	0.9935
0.0000	90.0000	0.8462	1.5500	1.2973	1.1948	0.9933

* DENOTES $M_2 = 1$, m DENOTES θ_{max}

OBLIQUE SHOCKS IN A PERFECT GAS

$\gamma = 5/3 \qquad M_1 = 1.30$

	θ	β	M_1	p_{21}	ρ_{21}	T_{21}	p^o_{21}
	0.0000	50.2849	1.3000	1.0000	1.0000	1.0000	1.0000
	1.0000	51.9925	1.2567	1.0615	1.0365	1.0242	1.0000
	2.0000	53.8855	1.2117	1.1286	1.0752	1.0496	0.9999
	3.0000	56.0411	1.1638	1.2033	1.1172	1.0771	0.9995
	4.0000	58.6184	1.1109	1.2897	1.1643	1.1077	0.9987
	5.0000	62.0647	1.0468	1.3989	1.2216	1.1451	0.9970
*	5.5246	64.8678	1.0000	1.4814	1.2635	1.1725	0.9952
m	5.8229	69.1142	0.9378	1.5940	1.3186	1.2089	0.9920
	5.0000	76.3062	0.8568	1.7441	1.3886	1.2560	0.9865
	4.0000	79.9790	0.8277	1.7985	1.4132	1.2727	0.9842
	3.0000	82.8477	0.8110	1.8297	1.4270	1.2822	0.9828
	2.0000	85.3712	0.8008	1.8487	1.4354	1.2880	0.9819
	1.0000	87.7233	0.7952	1.8592	1.4399	1.2911	0.9814
	0.0000	90.0000	0.7934	1.8625	1.4414	1.2922	0.9813

$\gamma = 5/3 \qquad M_1 = 1.40$

	θ	β	M_1	p_{21}	ρ_{21}	T_{21}	p^o_{21}
	0.0000	45.5847	1.4000	1.0000	1.0000	1.0000	1.0000
	1.0000	46.9888	1.3586	1.0600	1.0356	1.0236	1.0000
	2.0000	48.4929	1.3165	1.1240	1.0726	1.0479	0.9999
	3.0000	50.1202	1.2734	1.1928	1.1114	1.0732	0.9996
	4.0000	51.9073	1.2286	1.2675	1.1524	1.0999	0.9989
	5.0000	53.9147	1.1811	1.3501	1.1963	1.1285	0.9978
	6.0000	56.2573	1.1293	1.4441	1.2447	1.1602	0.9961
	7.0000	59.2130	1.0688	1.5581	1.3013	1.1974	0.9931
*	7.8301	62.9297	1.0000	1.6926	1.3650	1.2400	0.9886
m	8.1762	67.2900	0.9293	1.8348	1.4292	1.2838	0.9826
	8.0000	70.3916	0.8855	1.9241	1.4680	1.3107	0.9783
	7.0000	75.3286	0.8277	2.0428	1.5177	1.3460	0.9719
	6.0000	78.2903	0.8002	2.0991	1.5406	1.3625	0.9687
	5.0000	80.6786	0.7823	2.1357	1.5553	1.3732	0.9665
	4.0000	82.7748	0.7698	2.1612	1.5654	1.3806	0.9650
	3.0000	84.6995	0.7610	2.1791	1.5725	1.3858	0.9639
	2.0000	86.5175	0.7552	2.1909	1.5771	1.3892	0.9632
	1.0000	88.2738	0.7519	2.1978	1.5798	1.3912	0.9628
	0.0000	90.0000	0.7508	2.2000	1.5807	1.3918	0.9626

* DENOTES $M_2 = 1$, m DENOTES θ_{max}

OBLIQUE SHOCKS IN A PERFECT GAS

$$\gamma = 5/3 \qquad M_1 = 1.50$$

	θ	β	M_1	p_{21}	ρ_{21}	T_{21}	P^o_{21}
	0.0000	41.8103	1.5000	1.0000	1.0000	1.0000	1.0000
	1.0000	43.0403	1.4590	1.0601	1.0356	1.0236	1.0000
	2.0000	44.3375	1.4179	1.1237	1.0724	1.0478	0.9999
	3.0000	45.7120	1.3763	1.1912	1.1105	1.0727	0.9996
	4.0000	47.1788	1.3341	1.2631	1.1500	1.0984	0.9990
	5.0000	48.7592	1.2907	1.3402	1.1911	1.1252	0.9980
	6.0000	50.4846	1.2456	1.4238	1.2344	1.1534	0.9965
	7.0000	52.4059	1.1980	1.5157	1.2805	1.1837	0.9943
	8.0000	54.6150	1.1463	1.6194	1.3307	1.2170	0.9912
	9.0000	57.3185	1.0870	1.7425	1.3879	1.2555	0.9866
	10.0000	61.2683	1.0080	1.9126	1.4631	1.3072	0.9788
*	10.0733	61.6922	1.0000	1.9300	1.4705	1.3125	0.9780
m	10.4345	66.0070	0.9244	2.0975	1.5400	1.3620	0.9688
	10.0000	70.6682	0.8544	2.2543	1.6017	1.4074	0.9592
	9.0000	74.4567	0.8070	2.3605	1.6417	1.4378	0.9522
	8.0000	77.0259	0.7800	2.4207	1.6638	1.4549	0.9481
	7.0000	79.1295	0.7612	2.4624	1.6789	1.4667	0.9452
	6.0000	80.9758	0.7472	2.4933	1.6900	1.4754	0.9430
	5.0000	82.6592	0.7367	2.5166	1.6982	1.4819	0.9413
	4.0000	84.2331	0.7287	2.5341	1.7044	1.4868	0.9401
	3.0000	85.7322	0.7228	2.5469	1.7089	1.4904	0.9391
	2.0000	87.1803	0.7188	2.5557	1.7119	1.4928	0.9385
	1.0000	88.5980	0.7164	2.5608	1.7137	1.4943	0.9381
	0.0000	90.0000	0.7157	2.5625	1.7143	1.4948	0.9380

* DENOTES $M_2 = 1$, m DENOTES θ_{max}

OBLIQUE SHOCKS IN A PERFECT GAS

$$\gamma = 5/3 \qquad M_1 = 1.60$$

	θ	β	M_1	p_{21}	ρ_{21}	T_{21}	p^o_{21}
	0.0000	38.6822	1.6000	1.0000	1.0000	1.0000	1.0000
	1.0000	39.7996	1.5587	1.0611	1.0362	1.0240	1.0000
	2.0000	40.9678	1.5175	1.1255	1.0735	1.0485	0.9999
	3.0000	42.1931	1.4762	1.1935	1.1118	1.0735	0.9996
	4.0000	43.4834	1.4347	1.2653	1.1512	1.0992	0.9990
	5.0000	44.8489	1.3926	1.3416	1.1918	1.1256	0.9980
	6.0000	46.3037	1.3498	1.4228	1.2339	1.1531	0.9965
	7.0000	47.8674	1.3057	1.5099	1.2776	1.1818	0.9945
	8.0000	49.5692	1.2598	1.6041	1.3234	1.2121	0.9917
	9.0000	51.4558	1.2113	1.7075	1.3719	1.2446	0.9880
	10.0000	53.6106	1.1586	1.8237	1.4243	1.2804	0.9831
	11.0000	56.2144	1.0987	1.9604	1.4834	1.3216	0.9764
	12.0000	59.8684	1.0209	2.1436	1.5585	1.3755	0.9661
*	12.1945	60.9110	1.0000	2.1936	1.5782	1.3900	0.9630
m	12.5476	65.0830	0.9222	2.3820	1.6497	1.4439	0.9507
	12.0000	70.1465	0.8407	2.5809	1.7207	1.4999	0.9367
	11.0000	73.5409	0.7945	2.6931	1.7589	1.5311	0.9284
	10.0000	75.9076	0.7667	2.7603	1.7812	1.5497	0.9233
	9.0000	77.8485	0.7466	2.8082	1.7968	1.5629	0.9196
	8.0000	79.5457	0.7313	2.8446	1.8085	1.5729	0.9168
	7.0000	81.0840	0.7193	2.8731	1.8176	1.5807	0.9145
	6.0000	82.5116	0.7097	2.8956	1.8248	1.5869	0.9128
	5.0000	83.8592	0.7021	2.9134	1.8303	1.5917	0.9114
	4.0000	85.1483	0.6963	2.9271	1.8346	1.5955	0.9103
	3.0000	86.3951	0.6919	2.9373	1.8378	1.5983	0.9095
	2.0000	87.6123	0.6888	2.9444	1.8400	1.6002	0.9089
	1.0000	88.8110	0.6870	2.9486	1.8413	1.6013	0.9086
	0.0000	90.0000	0.6864	2.9500	1.8418	1.6017	0.9085

* DENOTES $M_2 = 1$, m DENOTES θ_{max}

OBLIQUE SHOCKS IN A PERFECT GAS

$$\gamma = 5/3 \qquad M_1 = 1.80$$

θ	β	M_1	p_{21}	ρ_{21}	T_{21}	p^o_{21}
0.0000	33.7490	1.8000	1.0000	1.0000	1.0000	1.0000
1.0000	34.7304	1.7566	1.0645	1.0382	1.0253	1.0000
2.0000	35.7490	1.7137	1.1324	1.0774	1.0511	0.9998
3.0000	36.8072	1.6711	1.2037	1.1175	1.0772	0.9995
4.0000	37.9085	1.6287	1.2788	1.1585	1.1039	0.9988
5.0000	39.0568	1.5863	1.3579	1.2004	1.1312	0.9977
6.0000	40.2574	1.5436	1.4413	1.2433	1.1592	0.9961
7.0000	41.5164	1.5007	1.5294	1.2872	1.1881	0.9939
8.0000	42.8421	1.4571	1.6226	1.3322	1.2180	0.9911
9.0000	44.2453	1.4127	1.7216	1.3784	1.2490	0.9874
10.0000	45.7406	1.3672	1.8273	1.4259	1.2815	0.9829
11.0000	47.3492	1.3201	1.9409	1.4751	1.3157	0.9774
12.0000	49.1026	1.2707	2.0640	1.5264	1.3522	0.9707
13.0000	51.0533	1.2180	2.1997	1.5806	1.3917	0.9626
14.0000	53.2987	1.1601	2.3534	1.6391	1.4358	0.9527
15.0000	56.0674	1.0926	2.5380	1.7057	1.4879	0.9398
* 15.9708	60.1658	1.0000	2.7976	1.7934	1.5600	0.9204
m 16.2723	63.9111	0.9230	3.0167	1.8623	1.6199	0.9032
16.0000	67.3563	0.8589	3.1997	1.9166	1.6694	0.8885
15.0000	71.2345	0.7948	3.3808	1.9678	1.7181	0.8737
14.0000	73.6141	0.7602	3.4776	1.9941	1.7440	0.8657
13.0000	75.4884	0.7356	3.5457	2.0122	1.7621	0.8602
12.0000	77.0868	0.7166	3.5977	2.0258	1.7760	0.8559
11.0000	78.5075	0.7013	3.6392	2.0365	1.7870	0.8524
10.0000	79.8037	0.6887	3.6730	2.0452	1.7960	0.8497
9.0000	81.0077	0.6783	3.7010	2.0523	1.8034	0.8474
8.0000	82.1416	0.6695	3.7242	2.0581	1.8095	0.8454
7.0000	83.2209	0.6622	3.7435	2.0630	1.8146	0.8439
6.0000	84.2575	0.6562	3.7594	2.0669	1.8188	0.8425
5.0000	85.2603	0.6513	3.7723	2.0701	1.8222	0.8415
4.0000	86.2372	0.6474	3.7825	2.0727	1.8249	0.8406
3.0000	87.1944	0.6444	3.7903	2.0746	1.8270	0.8400
2.0000	88.1373	0.6424	3.7957	2.0759	1.8284	0.8396
1.0000	89.0711	0.6411	3.7989	2.0767	1.8293	0.8393
0.0000	90.0000	0.6407	3.8000	2.0770	1.8296	0.8392

* DENOTES $M_2 = 1$, m DENOTES θ_{max}

E.1-6

OBLIQUE SHOCKS IN A PERFECT GAS

$\gamma = 5/3 \qquad M_1 = 2.00$

	θ	β	M_1	p_{21}	ρ_{21}	T_{21}	p^o_{21}
	0.0000	30.0000	2.0000	1.0000	1.0000	1.0000	1.0000
	1.0000	30.9037	1.9534	1.0689	1.0408	1.0270	1.0000
	2.0000	31.8391	1.9075	1.1415	1.0825	1.0544	0.9998
	3.0000	32.8075	1.8622	1.2178	1.1252	1.0823	0.9994
	4.0000	33.8110	1.8173	1.2982	1.1689	1.1107	0.9986
	5.0000	34.8517	1.7726	1.3829	1.2134	1.1396	0.9973
	6.0000	35.9323	1.7281	1.4718	1.2587	1.1693	0.9954
	7.0000	37.0560	1.6835	1.5656	1.3049	1.1998	0.9929
	8.0000	38.2267	1.6388	1.6644	1.3519	1.2312	0.9896
	9.0000	39.4491	1.5937	1.7686	1.3997	1.2635	0.9855
	10.0000	40.7293	1.5482	1.8787	1.4484	1.2970	0.9805
	11.0000	42.0748	1.5019	1.9952	1.4980	1.3319	0.9745
	12.0000	43.4962	1.4546	2.1188	1.5486	1.3682	0.9675
	13.0000	45.0072	1.4061	2.2506	1.6003	1.4064	0.9594
	14.0000	46.6279	1.3559	2.3920	1.6533	1.4467	0.9501
	15.0000	48.3882	1.3032	2.5450	1.7082	1.4898	0.9393
	16.0000	50.3372	1.2471	2.7130	1.7656	1.5366	0.9268
	17.0000	52.5658	1.1857	2.9026	1.8269	1.5888	0.9122
	18.0000	55.2818	1.1145	3.1281	1.8957	1.6501	0.8943
	19.0000	59.3136	1.0159	3.4477	1.9860	1.7360	0.8682
*	19.1073	60.0003	1.0000	3.5000	2.0001	1.7499	0.8639
m	19.3457	63.2724	0.9275	3.7386	2.0617	1.8133	0.8443
	19.0000	67.0530	0.8509	3.9899	2.1227	1.8796	0.8236
	18.0000	70.5793	0.7869	4.1972	2.1702	1.9340	0.8068
	17.0000	72.8051	0.7506	4.3130	2.1957	1.9643	0.7975
	16.0000	74.5592	0.7244	4.3955	2.2134	1.9859	0.7909
	15.0000	76.0502	0.7039	4.4594	2.2269	2.0025	0.7858
	14.0000	77.3695	0.6872	4.5109	2.2376	2.0159	0.7817
	13.0000	78.5668	0.6732	4.5535	2.2464	2.0270	0.7783
	12.0000	79.6727	0.6614	4.5893	2.2537	2.0363	0.7755
	11.0000	80.7079	0.6513	4.6196	2.2599	2.0442	0.7731
	10.0000	81.6869	0.6426	4.6454	2.2651	2.0509	0.7711
	9.0000	82.6209	0.6352	4.6675	2.2695	2.0566	0.7694
	8.0000	83.5181	0.6288	4.6862	2.2732	2.0615	0.7679
	7.0000	84.3853	0.6234	4.7021	2.2764	2.0656	0.7667
	6.0000	85.2281	0.6189	4.7153	2.2790	2.0691	0.7657
	5.0000	86.0513	0.6152	4.7262	2.2811	2.0719	0.7648
	4.0000	86.8588	0.6122	4.7349	2.2828	2.0741	0.7641
	3.0000	87.6543	0.6099	4.7416	2.2841	2.0759	0.7636
	2.0000	88.4411	0.6083	4.7462	2.2851	2.0771	0.7633
	1.0000	89.2221	0.6073	4.7490	2.2856	2.0778	0.7630
	0.0000	90.0000	0.6070	4.7499	2.2858	2.0780	0.7630

* DENOTES $M_2 = 1$, m DENOTES θ_{max}

E.1-7

OBLIQUE SHOCKS IN A PERFECT GAS

$\gamma = 5/3$ $M_1 = 2.20$

θ	β	M_1	p_{21}	ρ_{21}	T_{21}	p_{21}^o
0.0000	27.0357	2.2000	1.0000	1.0000	1.0000	1.0000
1.0000	27.8899	2.1494	1.0738	1.0436	1.0289	1.0000
2.0000	28.7734	2.0997	1.1517	1.0884	1.0582	0.9998
3.0000	29.6872	2.0507	1.2340	1.1341	1.0881	0.9993
4.0000	30.6326	2.0024	1.3207	1.1808	1.1185	0.9983
5.0000	31.6110	1.9545	1.4121	1.2285	1.1495	0.9967
6.0000	32.6241	1.9070	1.5085	1.2769	1.1813	0.9945
7.0000	33.6739	1.8596	1.6100	1.3262	1.2140	0.9915
8.0000	34.7626	1.8123	1.7169	1.3762	1.2475	0.9876
9.0000	35.8930	1.7649	1.8295	1.4269	1.2821	0.9828
10.0000	37.0685	1.7172	1.9481	1.4782	1.3179	0.9770
11.0000	38.2930	1.6692	2.0732	1.5302	1.3549	0.9702
12.0000	39.5719	1.6207	2.2052	1.5827	1.3933	0.9623
13.0000	40.9114	1.5715	2.3447	1.6359	1.4333	0.9533
14.0000	42.3206	1.5214	2.4925	1.6897	1.4751	0.9430
15.0000	43.8107	1.4701	2.6494	1.7442	1.5190	0.9316
16.0000	45.3981	1.4172	2.8170	1.7997	1.5653	0.9189
17.0000	47.1064	1.3622	2.9972	1.8563	1.6146	0.9048
18.0000	48.9729	1.3042	3.1931	1.9147	1.6677	0.8890
19.0000	51.0617	1.2419	3.4103	1.9758	1.7260	0.8713
20.0000	53.5023	1.1721	3.6596	2.0417	1.7924	0.8508
21.0000	56.6532	1.0870	3.9718	2.1184	1.8749	0.8251
* 21.6735	60.0958	1.0000	4.2962	2.1920	1.9599	0.7988
m 21.8563	62.9263	0.9333	4.5467	2.2450	2.0252	0.7789
21.0000	68.6962	0.8112	5.0014	2.3337	2.1431	0.7437
20.0000	71.2721	0.7633	5.1762	2.3654	2.1883	0.7306
19.0000	73.1517	0.7311	5.2917	2.3857	2.2180	0.7221
18.0000	74.6940	0.7066	5.3784	2.4007	2.2404	0.7158
17.0000	76.0291	0.6869	5.4473	2.4123	2.2581	0.7108
16.0000	77.2219	0.6705	5.5040	2.4218	2.2727	0.7068
15.0000	78.3099	0.6565	5.5516	2.4297	2.2849	0.7034
14.0000	79.3176	0.6446	5.5920	2.4363	2.2953	0.7005
13.0000	80.2619	0.6342	5.6268	2.4420	2.3042	0.6981
12.0000	81.1548	0.6252	5.6569	2.4468	2.3119	0.6959
11.0000	82.0059	0.6174	5.6829	2.4510	2.3186	0.6941
10.0000	82.8222	0.6105	5.7055	2.4546	2.3244	0.6925
9.0000	83.6095	0.6045	5.7250	2.4577	2.3294	0.6912
8.0000	84.3726	0.5994	5.7418	2.4603	2.3337	0.6900
7.0000	85.1156	0.5950	5.7561	2.4626	2.3374	0.6890
6.0000	85.8420	0.5912	5.7681	2.4645	2.3405	0.6882
5.0000	86.5546	0.5881	5.7781	2.4661	2.3430	0.6875
4.0000	87.2563	0.5856	5.7861	2.4673	2.3451	0.6870
3.0000	87.9495	0.5837	5.7922	2.4683	2.3467	0.6865
2.0000	88.6365	0.5824	5.7965	2.4689	2.3478	0.6862
1.0000	89.3193	0.5816	5.7991	2.4693	2.3484	0.6861
0.0000	90.0000	0.5813	5.7999	2.4695	2.3486	0.6860

* DENOTES $M_2 = 1$, m DENOTES θ_{max}

OBLIQUE SHOCKS IN A PERFECT GAS

$\gamma = 5/3$ $M_1 = 2.40$

WEAK SHOCK SOLUTION

θ	β	M_1	P_{21}	ρ_{21}	T_{21}	P_{21}^o
0.0000	24.6243	2.4000	1.0000	1.0000	1.0000	1.0000
1.0000	25.4446	2.3446	1.0790	1.0467	1.0309	1.0000
2.0000	26.2935	2.2905	1.1628	1.0946	1.0623	0.9997
3.0000	27.1716	2.2373	1.2514	1.1436	1.0943	0.9991
4.0000	28.0798	2.1849	1.3452	1.1938	1.1269	0.9979
5.0000	29.0190	2.1331	1.4443	1.2448	1.1602	0.9961
6.0000	29.9906	2.0818	1.5490	1.2968	1.1944	0.9934
7.0000	30.9959	2.0308	1.6594	1.3496	1.2296	0.9898
8.0000	32.0362	1.9800	1.7759	1.4030	1.2658	0.9852
9.0000	33.1136	1.9293	1.8988	1.4571	1.3031	0.9795
10.0000	34.2301	1.8786	2.0282	1.5117	1.3417	0.9727
11.0000	35.3881	1.8277	2.1646	1.5668	1.3816	0.9648
12.0000	36.5910	1.7765	2.3084	1.6222	1.4230	0.9557
13.0000	37.8424	1.7250	2.4599	1.6780	1.4659	0.9453
14.0000	39.1475	1.6729	2.6196	1.7341	1.5107	0.9338
15.0000	40.5125	1.6201	2.7884	1.7904	1.5574	0.9211
16.0000	41.9456	1.5663	2.9669	1.8470	1.6063	0.9072
17.0000	43.4582	1.5114	3.1563	1.9040	1.6577	0.8920
18.0000	45.0658	1.4548	3.3582	1.9615	1.7121	0.8756
19.0000	46.7914	1.3962	3.5749	2.0198	1.7699	0.8577
20.0000	48.6709	1.3345	3.8100	2.0795	1.8322	0.8384
21.0000	50.7659	1.2684	4.0696	2.1413	1.9006	0.8171
22.0000	53.1995	1.1950	4.3663	2.2072	1.9782	0.7932
23.0000	56.3043	1.1063	4.7339	2.2826	2.0739	0.7642
* 23.7682	60.3006	1.0000	5.1825	2.3666	2.1899	0.7302
m 23.9071	62.7463	0.9394	5.4401	2.4111	2.2562	0.7114

* DENOTES $M_2 = 1$, m DENOTES θ_{max}

OBLIQUE SHOCKS IN A PERFECT GAS

$$\gamma = 5/3 \qquad M_1 = 2.40$$

STRONG SHOCK SOLUTION

θ	β	M_1	p_{21}	ρ_{21}	T_{21}	p^o_{21}
m 23.9071	62.7463	0.9394	5.4401	2.4111	2.2562	0.7114
23.0000	68.6011	0.8083	5.9914	2.4988	2.3977	0.6729
22.0000	71.0706	0.7594	6.1922	2.5284	2.4491	0.6596
21.0000	72.8813	0.7262	6.3261	2.5475	2.4833	0.6509
20.0000	74.3668	0.7008	6.4271	2.5615	2.5091	0.6444
19.0000	75.6508	0.6803	6.5077	2.5726	2.5296	0.6393
18.0000	76.7953	0.6631	6.5742	2.5816	2.5466	0.6352
17.0000	77.8369	0.6485	6.6303	2.5891	2.5609	0.6317
16.0000	78.7992	0.6358	6.6782	2.5954	2.5731	0.6287
15.0000	79.6984	0.6248	6.7197	2.6008	2.5837	0.6262
14.0000	80.5465	0.6151	6.7557	2.6055	2.5928	0.6240
13.0000	81.3523	0.6066	6.7871	2.6096	2.6009	0.6221
12.0000	82.1228	0.5991	6.8147	2.6131	2.6079	0.6204
11.0000	82.8638	0.5924	6.8388	2.6162	2.6140	0.6189
10.0000	83.5798	0.5866	6.8599	2.6189	2.6194	0.6177
9.0000	84.2744	0.5815	6.8782	2.6212	2.6241	0.6166
8.0000	84.9510	0.5771	6.8941	2.6232	2.6281	0.6156
7.0000	85.6125	0.5732	6.9078	2.6250	2.6316	0.6148
6.0000	86.2614	0.5700	6.9193	2.6264	2.6345	0.6141
5.0000	86.8997	0.5673	6.9288	2.6276	2.6369	0.6135
4.0000	87.5295	0.5651	6.9365	2.6286	2.6389	0.6131
3.0000	88.1529	0.5635	6.9424	2.6293	2.6404	0.6127
2.0000	88.7713	0.5623	6.9466	2.6298	2.6415	0.6125
1.0000	89.3866	0.5616	6.9491	2.6301	2.6421	0.6123
0.0000	90.0000	0.5613	6.9499	2.6302	2.6423	0.6123

* DENOTES $M_2 = 1$, m DENOTES θ_{max}

OBLIQUE SHOCKS IN A PERFECT GAS

$\gamma = 5/3$ $M_1 = 2.60$

WEAK SHOCK SOLUTION

θ	β	M_1	p_{21}	ρ_{21}	T_{21}	p^o_{21}
0.0000	22.6199	2.6000	1.0000	1.0000	1.0000	1.0000
1.0000	23.4159	2.5393	1.0845	1.0499	1.0330	1.0000
2.0000	24.2405	2.4801	1.1744	1.1011	1.0665	0.9997
3.0000	25.0940	2.4221	1.2698	1.1536	1.1007	0.9989
4.0000	25.9772	2.3651	1.3712	1.2073	1.1357	0.9975
5.0000	26.8908	2.3089	1.4786	1.2621	1.1715	0.9953
6.0000	27.8357	2.2532	1.5924	1.3178	1.2084	0.9921
7.0000	28.8128	2.1981	1.7127	1.3743	1.2462	0.9878
8.0000	29.8232	2.1432	1.8399	1.4315	1.2853	0.9823
9.0000	30.8681	2.0886	1.9743	1.4893	1.3257	0.9757
10.0000	31.9489	2.0340	2.1161	1.5475	1.3674	0.9677
11.0000	33.0674	1.9795	2.2656	1.6060	1.4107	0.9585
12.0000	34.2258	1.9249	2.4232	1.6647	1.4556	0.9479
13.0000	35.4266	1.8700	2.5892	1.7236	1.5022	0.9361
14.0000	36.6730	1.8149	2.7641	1.7825	1.5507	0.9230
15.0000	37.9690	1.7593	2.9484	1.8413	1.6013	0.9086
16.0000	39.3198	1.7031	3.1427	1.9000	1.6541	0.8931
17.0000	40.7322	1.6462	3.3479	1.9586	1.7093	0.8764
18.0000	42.2150	1.5883	3.5649	2.0172	1.7672	0.8586
19.0000	43.7806	1.5291	3.7952	2.0758	1.8283	0.8396
20.0000	45.4463	1.4682	4.0408	2.1346	1.8930	0.8195
21.0000	47.2379	1.4050	4.3047	2.1939	1.9621	0.7981
22.0000	49.1975	1.3384	4.5918	2.2542	2.0370	0.7753
23.0000	51.3994	1.2666	4.9109	2.3168	2.1197	0.7506
24.0000	54.0037	1.1856	5.2811	2.3839	2.2153	0.7229
25.0000	57.5279	1.0824	5.7642	2.4639	2.3395	0.6885
* 25.4847	60.5424	1.0000	6.1563	2.5232	2.4399	0.6620
m 25.5905	62.6622	0.9452	6.4178	2.5603	2.5067	0.6450

* DENOTES $M_2 = 1$, m DENOTES θ_{max}

OBLIQUE SHOCKS IN A PERFECT GAS

$\gamma = 5/3 \qquad M_1 = 2.60$

STRONG SHOCK SOLUTION

	θ	β	M_1	p_{21}	ρ_{21}	T_{21}	p^o_{21}
m	25.5905	62.6622	0.9452	6.4178	2.5603	2.5067	0.6450
	25.0000	67.3849	0.8327	6.9504	2.6303	2.6424	0.6123
	24.0000	70.2206	0.7719	7.2323	2.6647	2.7141	0.5959
	23.0000	72.1479	0.7337	7.4058	2.6850	2.7582	0.5861
	22.0000	73.6847	0.7052	7.5330	2.6995	2.7905	0.5790
	21.0000	74.9918	0.6825	7.6332	2.7107	2.8159	0.5736
	20.0000	76.1441	0.6636	7.7153	2.7197	2.8368	0.5691
	19.0000	77.1843	0.6475	7.7841	2.7272	2.8542	0.5655
	18.0000	78.1388	0.6337	7.8429	2.7335	2.8691	0.5624
	17.0000	79.0256	0.6216	7.8937	2.7390	2.8820	0.5597
	16.0000	79.8577	0.6109	7.9379	2.7436	2.8932	0.5574
	15.0000	80.6448	0.6015	7.9766	2.7477	2.9030	0.5554
	14.0000	81.3942	0.5932	8.0107	2.7512	2.9117	0.5537
	13.0000	82.1118	0.5857	8.0407	2.7544	2.9193	0.5521
	12.0000	82.8025	0.5792	8.0672	2.7571	2.9260	0.5508
	11.0000	83.4700	0.5733	8.0906	2.7595	2.9319	0.5496
	10.0000	84.1179	0.5681	8.1111	2.7616	2.9371	0.5485
	9.0000	84.7487	0.5636	8.1291	2.7634	2.9417	0.5476
	8.0000	85.3652	0.5596	8.1447	2.7650	2.9456	0.5468
	7.0000	85.9695	0.5562	8.1581	2.7664	2.9490	0.5462
	6.0000	86.5633	0.5533	8.1695	2.7675	2.9519	0.5456
	5.0000	87.1487	0.5509	8.1790	2.7685	2.9543	0.5451
	4.0000	87.7270	0.5489	8.1866	2.7693	2.9562	0.5447
	3.0000	88.3000	0.5474	8.1925	2.7699	2.9577	0.5444
	2.0000	88.8690	0.5463	8.1966	2.7703	2.9588	0.5442
	1.0000	89.4352	0.5457	8.1991	2.7705	2.9594	0.5441
	0.0000	90.0000	0.5455	8.1999	2.7706	2.9596	0.5441

* DENOTES $M_2 = 1$, m DENOTES θ_{max}

OBLIQUE SHOCKS IN A PERFECT GAS

$$\gamma = 5/3 \qquad M_1 = 2.80$$

WEAK SHOCK SOLUTION

θ	β	M_1	P_{21}	ρ_{21}	T_{21}	P^o_{21}
0.0000	20.9248	2.8000	1.0000	1.0000	1.0000	1.0000
1.0000	21.7029	2.7335	1.0901	1.0531	1.0351	0.9999
2.0000	22.5100	2.6687	1.1864	1.1078	1.0709	0.9996
3.0000	23.3460	2.6054	1.2890	1.1639	1.1074	0.9987
4.0000	24.2120	2.5433	1.3983	1.2214	1.1449	0.9970
5.0000	25.1084	2.4821	1.5146	1.2799	1.1833	0.9943
6.0000	26.0357	2.4216	1.6381	1.3395	1.2229	0.9905
7.0000	26.9947	2.3618	1.7691	1.4000	1.2637	0.9855
8.0000	27.9861	2.3024	1.9080	1.4611	1.3059	0.9791
9.0000	29.0107	2.2433	2.0549	1.5227	1.3495	0.9712
10.0000	30.0696	2.1844	2.2103	1.5847	1.3948	0.9620
11.0000	31.1640	2.1257	2.3744	1.6469	1.4418	0.9513
12.0000	32.2953	2.0670	2.5475	1.7091	1.4906	0.9391
13.0000	33.4655	2.0082	2.7300	1.7712	1.5413	0.9256
14.0000	34.6765	1.9493	2.9222	1.8331	1.5941	0.9107
15.0000	35.9315	1.8902	3.1246	1.8947	1.6492	0.8946
16.0000	37.2340	1.8308	3.3379	1.9559	1.7066	0.8772
17.0000	38.5884	1.7709	3.5624	2.0166	1.7666	0.8588
18.0000	40.0007	1.7104	3.7992	2.0768	1.8294	0.8393
19.0000	41.4787	1.6491	4.0492	2.1365	1.8952	0.8188
20.0000	43.0331	1.5867	4.3138	2.1958	1.9645	0.7974
21.0000	44.6790	1.5229	4.5950	2.2549	2.0378	0.7751
22.0000	46.4380	1.4571	4.8958	2.3139	2.1158	0.7518
23.0000	48.3444	1.3884	5.2207	2.3733	2.1997	0.7274
24.0000	50.4561	1.3154	5.5775	2.4339	2.2916	0.7015
25.0000	52.8879	1.2350	5.9821	2.4974	2.3953	0.6736
26.0000	55.9396	1.1397	6.4759	2.5683	2.5215	0.6413
* 26.9006	60.7871	1.0000	7.2155	2.6627	2.7099	0.5968
m 26.9819	62.6341	0.9505	7.4792	2.6934	2.7768	0.5820

* DENOTES $M_2 = 1$, m DENOTES θ_{max}

OBLIQUE SHOCKS IN A PERFECT GAS

$\gamma = 5/3 \qquad M_1 = 2.80$

STRONG SHOCK SOLUTION

	θ	β	M_1	p_{21}	ρ_{21}	T_{21}	p^0_{21}
m	26.9819	62.6341	0.9505	7.4792	2.6934	2.7768	0.5820
	26.0000	68.6072	0.8050	8.2460	2.7752	2.9713	0.5418
	25.0000	70.9344	0.7547	8.5042	2.8005	3.0366	0.5292
	24.0000	72.6523	0.7200	8.6786	2.8170	3.0807	0.5209
	23.0000	74.0612	0.6933	8.8109	2.8293	3.1142	0.5147
	22.0000	75.2766	0.6715	8.9169	2.8389	3.1410	0.5099
	21.0000	76.3570	0.6532	9.0046	2.8467	3.1632	0.5059
	20.0000	77.3372	0.6376	9.0789	2.8533	3.1820	0.5026
	19.0000	78.2397	0.6240	9.1428	2.8588	3.1981	0.4998
	18.0000	79.0802	0.6120	9.1982	2.8636	3.2121	0.4973
	17.0000	79.8700	0.6014	9.2467	2.8678	3.2243	0.4952
	16.0000	80.6177	0.5920	9.2894	2.8714	3.2351	0.4934
	15.0000	81.3301	0.5837	9.3272	2.8746	3.2447	0.4918
	14.0000	82.0123	0.5762	9.3606	2.8774	3.2531	0.4903
	13.0000	82.6688	0.5695	9.3903	2.8799	3.2606	0.4891
	12.0000	83.3031	0.5635	9.4166	2.8821	3.2672	0.4879
	11.0000	83.9183	0.5582	9.4399	2.8841	3.2731	0.4870
	10.0000	84.5171	0.5535	9.4604	2.8858	3.2783	0.4861
	9.0000	85.1018	0.5494	9.4784	2.8873	3.2828	0.4853
	8.0000	85.6742	0.5457	9.4941	2.8886	3.2868	0.4847
	7.0000	86.2362	0.5426	9.5076	2.8897	3.2902	0.4841
	6.0000	86.7894	0.5399	9.5191	2.8906	3.2931	0.4836
	5.0000	87.3352	0.5377	9.5287	2.8914	3.2955	0.4832
	4.0000	87.8751	0.5359	9.5364	2.8920	3.2975	0.4829
	3.0000	88.4104	0.5345	9.5423	2.8925	3.2990	0.4826
	2.0000	88.9423	0.5335	9.5465	2.8929	3.3000	0.4825
	1.0000	89.4718	0.5329	9.5490	2.8931	3.3007	0.4824
	0.0000	90.0000	0.5327	9.5499	2.8931	3.3009	0.4823

* DENOTES $M_2 = 1$, m DENOTES θ_{max}

OBLIQUE SHOCKS IN A PERFECT GAS

$$\gamma = 5/3 \qquad M_1 = 3.00$$

WEAK SHOCK SOLUTION

θ	β	M_1	p_{21}	ρ_{21}	T_{21}	p^o_{21}
0.0000	19.4712	3 0000	1.0000	1.0000	1.0000	1.0000
1.0000	20.2357	2.9271	1.0959	1.0565	1.0373	0.9999
2.0000	21.0294	2.8564	1.1987	1.1147	1.0754	0.9995
3.0000	21.8532	2.7873	1.3087	1.1745	1.1143	0.9984
4.0000	22.7073	2.7196	1.4264	1.2357	1.1543	0.9964
5.0000	23.5919	2.6530	1.5520	1.2983	1.1954	0.9933
6.0000	24.5078	2.5873	1.6858	1.3619	1.2379	0.9888
7.0000	25.4552	2.5223	1.8282	1.4263	1.2818	0.9829
8.0000	26.4347	2.4579	1.9795	1.4915	1.3272	0.9754
9.0000	27.4468	2.3939	2.1401	1.5571	1.3744	0.9663
10.0000	28.4923	2.3303	2.3101	1.6229	1.4234	0.9556
11.0000	29.5720	2.2669	2.4900	1.6888	1.4744	0.9432
12.0000	30.6868	2.2036	2.6801	1.7546	1.5275	0.9293
13.0000	31.8382	2.1405	2.8806	1.8200	1.5827	0.9140
14.0000	33.0279	2.0774	3.0921	1.8850	1.6403	0.8972
15.0000	34.2576	2.0142	3.3148	1.9494	1.7004	0.8791
16.0000	35.5302	1.9509	3.5493	2.0131	1.7631	0.8599
17.0000	36.8491	1.8875	3.7960	2.0760	1.8285	0.8395
18.0000	38.2183	1.8237	4.0558	2.1381	1.8969	0.8183
19.0000	39.6436	1.7594	4.3293	2.1992	1.9686	0.7961
20.0000	41.1322	1.6945	4.6178	2.2595	2.0437	0.7733
21.0000	42.6945	1.6287	4.9227	2.3190	2.1228	0.7497
22.0000	44.3441	1.5617	5.2462	2.3778	2.2063	0.7255
23.0000	46.1017	1.4928	5.5912	2.4362	2.2951	0.7006
24.0000	47.9986	1.4213	5.9626	2.4945	2.3903	0.6749
25.0000	50.0874	1.3458	6.3686	2.5534	2.4941	0.6482
26.0000	52.4695	1.2636	6.8250	2.6144	2.6105	0.6198
27.0000	55.3930	1.1681	7.3711	2.6810	2.7494	0.5880
28.0000	60.1956	1.0234	8.2206	2.7727	2.9648	0.5430
* 28.0779	61.0192	1.0000	8.3589	2.7864	2.9998	0.5362
m 28.1409	62.6382	0.9553	8.6234	2.8119	3.0668	0.5235

* DENOTES $M_2 = 1$, m DENOTES θ_{max}

OBLIQUE SHOCKS IN A PERFECT GAS

$$\gamma = 5/3 \qquad M_1 = 3.00$$

STRONG SHOCK SOLUTION

	θ	β	M_1	p_{21}	ρ_{21}	T_{21}	p^o_{21}
m	28.1409	62.6382	0.9553	8.6234	2.8119	3.0668	0.5235
	28.0000	64.9721	0.8939	8.9864	2.8451	3.1586	0.5067
	27.0000	69.0088	0.7959	9.5562	2.8936	3.3025	0.4821
	26.0000	71.1759	0.7479	9.8286	2.9154	3.3712	0.4709
	25.0000	72.8111	0.7140	10.0174	2.9300	3.4188	0.4634
	24.0000	74.1636	0.6875	10.1621	2.9410	3.4553	0.4578
	23.0000	75.3349	0.6658	10.2788	2.9496	3.4848	0.4533
	22.0000	76.3784	0.6476	10.3759	2.9567	3.5092	0.4497
	21.0000	77.3262	0.6318	10.4583	2.9627	3.5300	0.4466
	20.0000	78.1994	0.6181	10.5293	2.9678	3.5479	0.4440
	19.0000	79.0125	0.6060	10.5912	2.9721	3.5635	0.4417
	18.0000	79.7765	0.5953	10.6455	2.9759	3.5772	0.4398
	17.0000	80.4996	0.5858	10.6934	2.9793	3.5892	0.4381
	16.0000	81.1881	0.5772	10.7358	2.9822	3.5999	0.4365
	15.0000	81.8470	0.5696	10.7736	2.9848	3.6094	0.4352
	14.0000	82.4805	0.5627	10.8072	2.9871	3.6179	0.4340
	13.0000	83.0922	0.5565	10.8371	2.9892	3.6254	0.4329
	12.0000	83.6850	0.5510	10.8637	2.9910	3.6322	0.4320
	11.0000	84.2612	0.5461	10.8874	2.9926	3.6381	0.4312
	10.0000	84.8232	0.5417	10.9083	2.9940	3.6434	0.4304
	9.0000	85.3728	0.5378	10.9266	2.9952	3.6480	0.4298
	8.0000	85.9118	0.5344	10.9427	2.9963	3.6520	0.4292
	7.0000	86.4416	0.5315	10.9565	2.9973	3.6555	0.4288
	6.0000	86.9636	0.5290	10.9683	2.9980	3.6585	0.4284
	5.0000	87.4792	0.5269	10.9781	2.9987	3.6610	0.4280
	4.0000	87.9896	0.5252	10.9860	2.9992	3.6629	0.4277
	3.0000	88.4957	0.5239	10.9921	2.9996	3.6645	0.4275
	2.0000	88.9988	0.5230	10.9964	2.9999	3.6656	0.4274
	1.0000	89.5000	0.5224	10.9990	3.0001	3.6662	0.4273
	0.0000	90.0000	0.5222	10.9998	3.0001	3.6664	0.4273

$*$ DENOTES $M_2 = 1$, m DENOTES θ_{max}

OBLIQUE SHOCKS IN A PERFECT GAS

$\gamma = 5/3 \qquad M_1 = 3.50$

WEAK SHOCK SOLUTION

θ	β	M_1	p_{21}	ρ_{21}	T_{21}	p^o_{21}
0.0000	16.6015	3.5000	1.0000	1.0000	1.0000	1.0000
1.0000	17.3431	3.4092	1.1106	1.0650	1.0429	0.9999
2.0000	18.1166	3.3214	1.2306	1.1323	1.0868	0.9993
3.0000	18.9223	3.2360	1.3603	1.2016	1.1320	0.9977
4.0000	19.7604	3.1525	1.5003	1.2729	1.1786	0.9947
5.0000	20.6307	3.0707	1.6510	1.3456	1.2269	0.9901
6.0000	21.5337	2.9901	1.8130	1.4196	1.2771	0.9836
7.0000	22.4690	2.9106	1.9866	1.4944	1.3293	0.9750
8.0000	23.4369	2.8320	2.1724	1.5698	1.3838	0.9643
9.0000	24.4373	2.7541	2.3706	1.6455	1.4407	0.9515
10.0000	25.4704	2.6768	2.5818	1.7210	1.5002	0.9366
11.0000	26.5364	2.6001	2.8063	1.7962	1.5624	0.9197
12.0000	27.6354	2.5240	3.0445	1.8707	1.6274	0.9010
13.0000	28.7683	2.4483	3.2966	1.9443	1.6955	0.8806
14.0000	29.9354	2.3731	3.5631	2.0168	1.7668	0.8587
15.0000	31.1378	2.2983	3.8444	2.0879	1.8413	0.8356
16.0000	32.3769	2.2239	4.1407	2.1575	1.9192	0.8114
17.0000	33.6545	2.1499	4.4527	2.2255	2.0008	0.7863
18.0000	34.9728	2.0761	4.7808	2.2918	2.0860	0.7606
19.0000	36.3348	2.0027	5.1255	2.3564	2.1752	0.7344
20.0000	37.7446	1.9293	5.4878	2.4191	2.2685	0.7079
21.0000	39.2073	1.8560	5.8686	2.4801	2.3662	0.6813
22.0000	40.7299	1.7824	6.2692	2.5394	2.4687	0.6546
23.0000	42.3217	1.7084	6.6914	2.5971	2.5765	0.6279
24.0000	43.9960	1.6336	7.1379	2.6534	2.6901	0.6013
25.0000	45.7715	1.5574	7.6123	2.7084	2.8106	0.5747
26.0000	47.6773	1.4791	8.1206	2.7626	2.9395	0.5481
27.0000	49.7610	1.3972	8.6727	2.8165	3.0792	0.5212
28.0000	52.1118	1.3094	9.2873	2.8712	3.2346	0.4935
29.0000	54.9375	1.2096	10.0090	2.9294	3.4167	0.4638
30.0000	59.1084	1.0732	11.0260	3.0019	3.6730	0.4264
* 30.2675	61.5145	1.0000	11.5792	3.0373	3.8123	0.4080
m 30.3026	62.7112	0.9651	11.8436	3.0534	3.8788	0.3996

* DENOTES $M_2 = 1$, m DENOTES $\theta_{\max}$

OBLIQUE SHOCKS IN A PERFECT GAS

$\gamma = 5/3$ $M_1 = 3.50$

STRONG SHOCK SOLUTION

	θ	β	M_1	p_{21}	ρ_{21}	T_{21}	p^o_{21}
m	30.3026	62.7112	0.9651	11.8436	3.0534	3.8788	0.3996
	30.0000	66.0628	0.8725	12.5416	3.0934	4.0543	0.3788
	29.0000	69.4088	0.7877	13.1683	3.1265	4.2119	0.3616
	28.0000	71.4171	0.7408	13.5072	3.1434	4.2970	0.3528
	27.0000	72.9583	0.7069	13.7471	3.1550	4.3573	0.3468
	26.0000	74.2408	0.6801	13.9328	3.1637	4.4039	0.3422
	25.0000	75.3542	0.6581	14.0834	3.1707	4.4417	0.3386
	24.0000	76.3470	0.6394	14.2092	3.1764	4.4733	0.3357
	23.0000	77.2485	0.6233	14.3163	3.1812	4.5002	0.3332
	22.0000	78.0785	0.6092	14.4089	3.1854	4.5235	0.3310
	21.0000	78.8508	0.5967	14.4898	3.1889	4.5438	0.3292
	20.0000	79.5754	0.5855	14.5610	3.1920	4.5617	0.3276
	19.0000	80.2601	0.5756	14.6240	3.1948	4.5775	0.3261
	18.0000	80.9110	0.5666	14.6802	3.1972	4.5916	0.3249
	17.0000	81.5330	0.5586	14.7303	3.1993	4.6042	0.3238
	16.0000	82.1299	0.5513	14.7752	3.2013	4.6154	0.3228
	15.0000	82.7051	0.5447	14.8154	3.2030	4.6255	0.3219
	14.0000	83.2613	0.5387	14.8514	3.2045	4.6346	0.3211
	13.0000	83.8010	0.5334	14.8837	3.2059	4.6427	0.3204
	12.0000	84.3261	0.5286	14.9126	3.2071	4.6499	0.3198
	11.0000	84.8385	0.5242	14.9384	3.2081	4.6564	0.3192
	10.0000	85.3396	0.5204	14.9612	3.2091	4.6621	0.3187
	9.0000	85.8311	0.5170	14.9814	3.2099	4.6672	0.3183
	8.0000	86.3142	0.5140	14.9990	3.2107	4.6716	0.3179
	7.0000	86.7900	0.5114	15.0143	3.2113	4.6754	0.3176
	6.0000	87.2596	0.5091	15.0273	3.2118	4.6787	0.3173
	5.0000	87.7240	0.5073	15.0381	3.2123	4.6814	0.3171
	4.0000	88.1842	0.5057	15.0469	3.2127	4.6836	0.3169
	3.0000	88.6411	0.5046	15.0537	3.2129	4.6853	0.3167
	2.0000	89.0954	0.5038	15.0585	3.2131	4.6865	0.3166
	1.0000	89.5482	0.5033	15.0613	3.2133	4.6873	0.3166
	0.0000	90.0000	0.5031	15.0623	3.2133	4.6875	0.3165

* DENOTES $M_2 = 1$, m DENOTES θ_{max}

OBLIQUE SHOCKS IN A PERFECT GAS

$\gamma = 5/3 \qquad M_1 = 4.00$

WEAK SHOCK SOLUTION

θ	β	M_1	P_{21}	ρ_{21}	T_{21}	P_{21}^o
0.0000	14.4775	4.0000	1.0000	1.0000	1.0000	1.0000
1.0000	15.2058	3.8883	1.1259	1.0737	1.0486	0.9999
2.0000	15.9690	3.7808	1.2638	1.1503	1.0986	0.9990
3.0000	16.7672	3.6766	1.4145	1.2296	1.1503	0.9967
4.0000	17.6004	3.5751	1.5786	1.3112	1.2040	0.9925
5.0000	18.4684	3.4756	1.7570	1.3945	1.2600	0.9860
6.0000	19.3707	3.3779	1.9502	1.4791	1.3185	0.9769
7.0000	20.3071	3.2816	2.1589	1.5645	1.3799	0.9651
8.0000	21.2772	3.1866	2.3836	1.6503	1.4444	0.9506
9.0000	22.2806	3.0927	2.6250	1.7359	1.5122	0.9334
10.0000	23.3169	2.9998	2.8834	1.8209	1.5835	0.9137
11.0000	24.3858	2.9079	3.1593	1.9049	1.6585	0.8918
12.0000	25.4871	2.8170	3.4533	1.9875	1.7375	0.8678
13.0000	26.6207	2.7271	3.7655	2.0684	1.8205	0.8420
14.0000	27.7869	2.6381	4.0965	2.1474	1.9076	0.8149
15.0000	28.9857	2.5503	4.4465	2.2242	1.9992	0.7868
16.0000	30.2182	2.4634	4.8160	2.2986	2.0952	0.7579
17.0000	31.4852	2.3775	5.2054	2.3706	2.1958	0.7285
18.0000	32.7880	2.2927	5.6151	2.4401	2.3012	0.6989
19.0000	34.1288	2.2087	6.0456	2.5069	2.4116	0.6693
20.0000	35.5100	2.1257	6.4975	2.5712	2.5270	0.6400
21.0000	36.9350	2.0434	6.9717	2.6330	2.6479	0.6110
22.0000	38.4084	1.9617	7.4692	2.6923	2.7743	0.5825
23.0000	39.9363	1.8805	7.9915	2.7492	2.9068	0.5547
24.0000	41.5267	1.7994	8.5404	2.8040	3.0458	0.5274
25.0000	43.1908	1.7182	9.1188	2.8567	3.1920	0.5008
26.0000	44.9443	1.6363	9.7304	2.9077	3.3465	0.4749
27.0000	46.8110	1.5530	10.3816	2.9572	3.5107	0.4495
28.0000	48.8277	1.4673	11.0820	3.0056	3.6871	0.4244
29.0000	51.0601	1.3771	11.8494	3.0538	3.8803	0.3994
30.0000	53.6424	1.2785	12.7210	3.1031	4.0995	0.3738
31.0000	56.9664	1.1599	13.8064	3.1578	4.3722	0.3453
* 31.7433	61.8913	1.0000	15.3102	3.2234	4.7497	0.3113
m 31.7641	62.8076	0.9722	15.5732	3.2338	4.8157	0.3059

* DENOTES $M_2 = 1$, m DENOTES θ_{max}

OBLIQUE SHOCKS IN A PERFECT GAS

$\gamma = 5/3 \qquad M_1 = 4.00$

STRONG SHOCK SOLUTION

	θ	β	M_1	p_{21}	ρ_{21}	T_{21}	p^o_{21}
m	31.7641	62.8076	0.9722	15.5732	3.2338	4.8157	0.3059
	31.0000	67.9858	0.8264	16.9397	3.2839	5.1585	0.2802
	30.0000	70.4472	0.7640	17.5096	3.3028	5.3014	0.2705
	29.0000	72.1732	0.7230	17.8753	3.3145	5.3931	0.2646
	28.0000	73.5558	0.6920	18.1470	3.3229	5.4612	0.2603
	27.0000	74.7299	0.6669	18.3625	3.3294	5.5152	0.2570
	26.0000	75.7612	0.6459	18.5398	3.3347	5.5596	0.2543
	25.0000	76.6875	0.6280	18.6893	3.3391	5.5971	0.2521
	24.0000	77.5331	0.6123	18.8177	3.3428	5.6293	0.2502
	23.0000	78.3143	0.5986	18.9293	3.3460	5.6572	0.2486
	22.0000	79.0431	0.5863	19.0272	3.3488	5.6818	0.2472
	21.0000	79.7280	0.5754	19.1138	3.3512	5.7035	0.2460
	20.0000	80.3761	0.5655	19.1907	3.3534	5.7228	0.2449
	19.0000	80.9928	0.5566	19.2595	3.3553	5.7400	0.2439
	18.0000	81.5822	0.5486	19.3211	3.3570	5.7555	0.2431
	17.0000	82.1482	0.5413	19.3765	3.3585	5.7693	0.2423
	16.0000	82.6936	0.5347	19.4263	3.3599	5.7818	0.2416
	15.0000	83.2210	0.5287	19.4711	3.3611	5.7930	0.2410
	14.0000	83.7326	0.5233	19.5114	3.3622	5.8031	0.2404
	13.0000	84.2301	0.5184	19.5476	3.3632	5.8122	0.2400
	12.0000	84.7154	0.5140	19.5801	3.3641	5.8203	0.2395
	11.0000	85.1898	0.5100	19.6091	3.3649	5.8276	0.2391
	10.0000	85.6548	0.5064	19.6349	3.3656	5.8341	0.2388
	9.0000	86.1112	0.5033	19.6577	3.3662	5.8398	0.2385
	8.0000	86.5605	0.5005	19.6777	3.3667	5.8448	0.2382
	7.0000	87.0035	0.4981	19.6951	3.3672	5.8492	0.2380
	6.0000	87.4411	0.4960	19.7099	3.3676	5.8529	0.2378
	5.0000	87.8743	0.4943	19.7222	3.3679	5.8560	0.2376
	4.0000	88.3038	0.4929	19.7322	3.3682	5.8585	0.2375
	3.0000	88.7304	0.4918	19.7399	3.3684	5.8604	0.2374
	2.0000	89.1548	0.4910	19.7454	3.3685	5.8618	0.2373
	1.0000	89.5778	0.4906	19.7486	3.3686	5.8626	0.2372
	0.0000	90.0000	0.4904	19.7497	3.3686	5.8628	0.2372

* DENOTES $M_2 = 1$, m DENOTES θ_{max}

OBLIQUE SHOCKS IN A PERFECT GAS

$$\gamma = 5/3 \qquad M_1 = 4.50$$

WEAK SHOCK SOLUTION

θ	β	M_1	P_{21}	ρ_{21}	T_{21}	P^o_{21}
0.0000	12.8396	4.5000	1.0000	1.0000	1.0000	1.0000
1.0000	13.5596	4.3647	1.1414	1.0825	1.0544	0.9998
2.0000	14.3181	4.2349	1.2981	1.1688	1.1106	0.9986
3.0000	15.1146	4.1095	1.4710	1.2583	1.1691	0.9954
4.0000	15.9488	3.9875	1.6612	1.3504	1.2301	0.9897
5.0000	16.8203	3.8683	1.8695	1.4445	1.2943	0.9809
6.0000	17.7284	3.7513	2.0971	1.5398	1.3619	0.9688
7.0000	18.6724	3.6362	2.3445	1.6358	1.4333	0.9533
8.0000	19.6514	3.5228	2.6127	1.7317	1.5088	0.9343
9.0000	20.6647	3.4110	2.9023	1.8269	1.5887	0.9123
10.0000	21.7113	3.3007	3.2139	1.9207	1.6733	0.8873
11.0000	22.7906	3.1921	3.5481	2.0128	1.7628	0.8599
12.0000	23.9020	3.0850	3.9054	2.1026	1.8574	0.8306
13.0000	25.0447	2.9797	4.2860	2.1898	1.9573	0.7996
14.0000	26.2187	2.8761	4.6906	2.2741	2.0626	0.7676
15.0000	27.4239	2.7742	5.1194	2.3552	2.1736	0.7349
16.0000	28.6606	2.6742	5.5727	2.4331	2.2903	0.7019
17.0000	29.9292	2.5759	6.0510	2.5077	2.4129	0.6690
18.0000	31.2306	2.4794	6.5546	2.5789	2.5416	0.6364
19.0000	32.5662	2.3847	7.0839	2.6468	2.6764	0.6044
20.0000	33.9380	2.2916	7.6396	2.7114	2.8176	0.5732
21.0000	35.3484	2.2001	8.2224	2.7729	2.9653	0.5429
22.0000	36.8008	2.1100	8.8331	2.8313	3.1198	0.5137
23.0000	38.2998	2.0211	9.4730	2.8868	3.2815	0.4856
24.0000	39.8512	1.9333	10.1437	2.9396	3.4507	0.4585
25.0000	41.4633	1.8461	10.8476	2.9899	3.6281	0.4326
26.0000	43.1470	1.7592	11.5880	3.0379	3.8145	0.4077
27.0000	44.9178	1.6722	12.3698	3.0838	4.0111	0.3838
28.0000	46.7991	1.5841	13.2004	3.1281	4.2199	0.3608
29.0000	48.8274	1.4939	14.0919	3.1711	4.4439	0.3384
30.0000	51.0664	1.3996	15.0661	3.2134	4.6885	0.3165
31.0000	53.6474	1.2971	16.1686	3.2565	4.9651	0.2943
32.0000	56.9449	1.1751	17.5315	3.3035	5.3069	0.2702
* 32.7801	62.1753	1.0000	19.5473	3.3632	5.8121	0.2400
m 32.7932	62.8980	0.9775	19.8086	3.3702	5.8776	0.2364

* DENOTES $M_2 = 1$, m DENOTES θ_{max}

OBLIQUE SHOCKS IN A PERFECT GAS

$$\gamma = 5/3 \qquad M_1 = 4.50$$

STRONG SHOCK SOLUTION

	θ	β	M_1	p_{21}	ρ_{21}	T_{21}	p^o_{21}
m	32.7932	62.8980	0.9775	19.8086	3.3702	5.8776	0.2364
	32.0000	68.1409	0.8256	21.5532	3.4132	6.3147	0.2150
	31.0000	70.5476	0.7629	22.2549	3.4289	6.4904	0.2073
	30.0000	72.2427	0.7214	22.7077	3.4386	6.6038	0.2026
	29.0000	73.6013	0.6899	23.0447	3.4456	6.6882	0.1991
	28.0000	74.7550	0.6645	23.3120	3.4510	6.7551	0.1965
	27.0000	75.7678	0.6432	23.5322	3.4554	6.8103	0.1944
	26.0000	76.6771	0.6249	23.7180	3.4591	6.8568	0.1926
	25.0000	77.5062	0.6090	23.8775	3.4622	6.8967	0.1911
	24.0000	78.2718	0.5950	24.0163	3.4648	6.9315	0.1898
	23.0000	78.9852	0.5825	24.1381	3.4671	6.9620	0.1887
	22.0000	79.6552	0.5713	24.2459	3.4692	6.9890	0.1877
	21.0000	80.2885	0.5613	24.3419	3.4710	7.0130	0.1868
	20.0000	80.8906	0.5522	24.4277	3.4726	7.0345	0.1861
	19.0000	81.4656	0.5439	24.5047	3.4740	7.0538	0.1854
	18.0000	82.0172	0.5364	24.5740	3.4753	7.0711	0.1848
	17.0000	82.5482	0.5296	24.6364	3.4764	7.0867	0.1842
	16.0000	83.0612	0.5235	24.6927	3.4774	7.1008	0.1837
	15.0000	83.5582	0.5178	24.7435	3.4784	7.1135	0.1833
	14.0000	84.0412	0.5127	24.7893	3.4792	7.1250	0.1829
	13.0000	84.5117	0.5081	24.8306	3.4799	7.1353	0.1825
	12.0000	84.9713	0.5039	24.8676	3.4806	7.1446	0.1822
	11.0000	85.4211	0.5002	24.9008	3.4812	7.1529	0.1819
	10.0000	85.8623	0.4968	24.9304	3.4817	7.1603	0.1817
	9.0000	86.2960	0.4938	24.9565	3.4822	7.1669	0.1814
	8.0000	86.7231	0.4912	24.9794	3.4826	7.1726	0.1812
	7.0000	87.1446	0.4889	24.9993	3.4830	7.1776	0.1811
	6.0000	87.5613	0.4869	25.0163	3.4833	7.1818	0.1809
	5.0000	87.9738	0.4853	25.0305	3.4835	7.1854	0.1808
	4.0000	88.3830	0.4839	25.0420	3.4837	7.1883	0.1807
	3.0000	88.7895	0.4829	25.0508	3.4839	7.1905	0.1806
	2.0000	89.1942	0.4822	25.0571	3.4840	7.1921	0.1806
	1.0000	89.5974	0.4817	25.0609	3.4841	7.1930	0.1805
	0.0000	90.0000	0.4816	25.0621	3.4841	7.1933	0.1805

* DENOTES $M_2 = 1$, m DENOTES θ_{max}

OBLIQUE SHOCKS IN A PERFECT GAS

$\gamma = 5/3 \qquad M_1 = 5.00$

WEAK SHOCK SOLUTION

θ	β	M_1	p_{21}	ρ_{21}	T_{21}	p^o_{21}
0.0000	11.5370	5.0000	1.0000	1.0000	1.0000	1.0000
1.0000	12.2519	4.8382	1.1573	1.0915	1.0603	0.9998
2.0000	13.0086	4.6837	1.3334	1.1875	1.1228	0.9981
3.0000	13.8066	4.5348	1.5297	1.2874	1.1882	0.9939
4.0000	14.6453	4.3902	1.7477	1.3903	1.2571	0.9864
5.0000	15.5238	4.2490	1.9884	1.4952	1.3299	0.9749
6.0000	16.4413	4.1107	2.2534	1.6013	1.4072	0.9592
7.0000	17.3964	3.9748	2.5434	1.7076	1.4894	0.9394
8.0000	18.3879	3.8413	2.8596	1.8133	1.5770	0.9156
9.0000	19.4144	3.7100	3.2027	1.9175	1.6703	0.8882
10.0000	20.4747	3.5809	3.5736	2.0195	1.7695	0.8579
11.0000	21.5677	3.4543	3.9728	2.1187	1.8751	0.8250
12.0000	22.6924	3.3301	4.4009	2.2145	1.9873	0.7904
13.0000	23.8480	3.2085	4.8583	2.3068	2.1061	0.7546
14.0000	25.0337	3.0895	5.3455	2.3950	2.2319	0.7182
15.0000	26.2492	2.9733	5.8627	2.4792	2.3647	0.6817
16.0000	27.4946	2.8598	6.4104	2.5592	2.5048	0.6455
17.0000	28.7701	2.7490	6.9888	2.6351	2.6522	0.6100
18.0000	30.0759	2.6410	7.5983	2.7068	2.8071	0.5755
19.0000	31.4133	2.5355	8.2392	2.7746	2.9695	0.5421
20.0000	32.7840	2.4326	8.9122	2.8384	3.1398	0.5101
21.0000	34.1896	2.3321	9.6177	2.8986	3.3180	0.4795
22.0000	35.6333	2.2338	10.3566	2.9553	3.5044	0.4504
23.0000	37.1185	2.1375	11.1302	3.0088	3.6992	0.4228
24.0000	38.6501	2.0430	11.9398	3.0591	3.9030	0.3967
25.0000	40.2348	1.9499	12.7877	3.1067	4.1162	0.3719
26.0000	41.8809	1.8579	13.6770	3.1516	4.3397	0.3485
27.0000	43.6007	1.7666	14.6119	3.1942	4.5744	0.3264
28.0000	45.4113	1.6752	15.5991	3.2348	4.8222	0.3054
29.0000	47.3384	1.5829	16.6487	3.2738	5.0855	0.2854
30.0000	49.4236	1.4884	17.7778	3.3114	5.3686	0.2661
31.0000	51.7417	1.3892	19.0179	3.3485	5.6795	0.2473
32.0000	54.4554	1.2802	20.4387	3.3864	6.0355	0.2283
33.0000	58.1092	1.1446	22.2777	3.4294	6.4961	0.2071
* 33.5344	62.3914	1.0000	24.2882	3.4700	6.9995	0.1873
m 33.5430	62.9754	0.9814	24.5479	3.4748	7.0646	0.1850

* DENOTES $M_2 = 1$, m DENOTES θ_{max}

OBLIQUE SHOCKS IN A PERFECT GAS

$\gamma = 5/3 \qquad M_1 = 5.00$

STRONG SHOCK SOLUTION

	θ	β	M_1	p_{21}	ρ_{21}	T_{21}	p^o_{21}
m	33.5430	62.9754	0.9814	24.5479	3.4748	7.0646	0.1850
	33.0000	67.3443	0.8507	26.3629	3.5062	7.5189	0.1700
	32.0000	70.0857	0.7761	27.3740	3.5221	7.7720	0.1625
	31.0000	71.8911	0.7303	27.9805	3.5312	7.9238	0.1583
	30.0000	73.3053	0.6964	28.4207	3.5376	8.0340	0.1553
	29.0000	74.4912	0.6693	28.7653	3.5424	8.1202	0.1530
	28.0000	75.5243	0.6467	29.0469	3.5463	8.1907	0.1512
	27.0000	76.4463	0.6275	29.2832	3.5496	8.2498	0.1498
	26.0000	77.2835	0.6108	29.4853	3.5523	8.3004	0.1485
	25.0000	78.0536	0.5961	29.6606	3.5546	8.3443	0.1474
	24.0000	78.7692	0.5831	29.8142	3.5566	8.3827	0.1465
	23.0000	79.4396	0.5714	29.9499	3.5584	8.4167	0.1457
	22.0000	80.0719	0.5609	30.0706	3.5600	8.4469	0.1450
	21.0000	80.6716	0.5514	30.1785	3.5614	8.4739	0.1443
	20.0000	81.2433	0.5428	30.2753	3.5626	8.4981	0.1438
	19.0000	81.7907	0.5349	30.3624	3.5637	8.5199	0.1433
	18.0000	82.3169	0.5278	30.4410	3.5647	8.5395	0.1428
	17.0000	82.8244	0.5213	30.5120	3.5656	8.5573	0.1424
	16.0000	83.3154	0.5154	30.5761	3.5664	8.5734	0.1420
	15.0000	83.7918	0.5100	30.6341	3.5671	8.5879	0.1417
	14.0000	84.2553	0.5052	30.6864	3.5678	8.6010	0.1414
	13.0000	84.7073	0.5007	30.7336	3.5684	8.6128	0.1411
	12.0000	85.1491	0.4967	30.7761	3.5689	8.6234	0.1409
	11.0000	85.5820	0.4931	30.8141	3.5694	8.6329	0.1407
	10.0000	86.0068	0.4898	30.8480	3.5698	8.6414	0.1405
	9.0000	86.4248	0.4870	30.8780	3.5702	8.6489	0.1403
	8.0000	86.8366	0.4844	30.9044	3.5705	8.6555	0.1402
	7.0000	87.2430	0.4822	30.9272	3.5708	8.6612	0.1400
	6.0000	87.6451	0.4803	30.9468	3.5710	8.6661	0.1399
	5.0000	88.0432	0.4787	30.9631	3.5712	8.6702	0.1398
	4.0000	88.4382	0.4774	30.9763	3.5714	8.6735	0.1398
	3.0000	88.8309	0.4765	30.9865	3.5715	8.6761	0.1397
	2.0000	89.2217	0.4757	30.9938	3.5716	8.6779	0.1397
	1.0000	89.6111	0.4753	30.9981	3.5716	8.6790	0.1396
	0.0000	90.0000	0.4752	30.9995	3.5717	8.6793	0.1396

* DENOTES $M_2 = 1$, m DENOTES θ_{max}

E.1-24

OBLIQUE SHOCKS IN A PERFECT GAS

$\gamma = 5/3$ $M_1 = 6.00$

WEAK SHOCK SOLUTION

θ	β	M_1	p_{21}	ρ_{21}	T_{21}	p^o_{21}
0.0000	9.5941	6.0000	1.0000	1.0000	1.0000	1.0000
1.0000	10.3039	5.7772	1.1897	1.1097	1.0721	0.9996
2.0000	11.0625	5.5660	1.4068	1.2257	1.1477	0.9968
3.0000	11.8690	5.3633	1.6536	1.3468	1.2278	0.9900
4.0000	12.7220	5.1671	1.9324	1.4715	1.3132	0.9778
5.0000	13.6202	4.9759	2.2454	1.5983	1.4049	0.9598
6.0000	14.5613	4.7892	2.5944	1.7254	1.5037	0.9357
7.0000	15.5432	4.6065	2.9812	1.8514	1.6102	0.9060
8.0000	16.5638	4.4279	3.4072	1.9750	1.7252	0.8715
9.0000	17.6206	4.2535	3.8735	2.0949	1.8490	0.8332
10.0000	18.7117	4.0836	4.3812	2.2104	1.9821	0.7920
11.0000	19.8353	3.9183	4.9311	2.3206	2.1249	0.7491
12.0000	20.9895	3.7579	5.5237	2.4251	2.2777	0.7054
13.0000	22.1731	3.6026	6.1595	2.5237	2.4407	0.6618
14.0000	23.3850	3.4524	6.8390	2.6162	2.6141	0.6189
15.0000	24.6244	3.3073	7.5625	2.7028	2.7980	0.5774
16.0000	25.8909	3.1674	8.3301	2.7836	2.9926	0.5376
17.0000	27.1843	3.0324	9.1421	2.8588	3.1979	0.4998
18.0000	28.5049	2.9022	9.9987	2.9286	3.4141	0.4642
19.0000	29.8534	2.7767	10.9003	2.9935	3.6414	0.4307
20.0000	31.2308	2.6556	11.8471	3.0536	3.8797	0.3995
21.0000	32.6386	2.5386	12.8397	3.1094	4.1293	0.3705
22.0000	34.0792	2.4255	13.8788	3.1612	4.3904	0.3436
23.0000	35.5553	2.3158	14.9656	3.2093	4.6632	0.3186
24.0000	37.0707	2.2094	16.1013	3.2540	4.9482	0.2956
25.0000	38.6305	2.1057	17.2882	3.2956	5.2459	0.2742
26.0000	40.2410	2.0045	18.5292	3.3344	5.5570	0.2545
27.0000	41.9111	1.9051	19.8284	3.3707	5.8826	0.2362
28.0000	43.6530	1.8071	21.1922	3.4048	6.2242	0.2192
29.0000	45.4838	1.7097	22.6296	3.4369	6.5842	0.2034
30.0000	47.4293	1.6119	24.1553	3.4675	6.9663	0.1885
31.0000	49.5312	1.5122	25.7936	3.4968	7.3764	0.1745
32.0000	51.8642	1.4083	27.5893	3.5254	7.8259	0.1610
33.0000	54.5916	1.2946	29.6429	3.5544	8.3398	0.1475
34.0000	58.2613	1.1538	32.2969	3.5870	9.0039	0.1327
* 34.5330	62.6891	1.0000	35.2763	3.6183	9.7493	0.1188
m 34.5372	63.0930	0.9868	35.5337	3.6208	9.8137	0.1177

* DENOTES $M_2 = 1$, m DENOTES θ_{max}

OBLIQUE SHOCKS IN A PERFECT GAS

$$\gamma = 5/3 \quad M_1 = 6.00$$

STRONG SHOCK SOLUTION

	θ	β	M_1	P_{21}	ρ_{21}	T_{21}	P^0_{21}
m	34.5372	63.0930	0.9868	35.5337	3.6208	9.8137	0.1177
	34.0000	67.4190	0.8535	38.1143	3.6441	10.4593	0.1077
	33.0000	70.1494	0.7771	39.5606	3.6559	10.8210	0.1027
	32.0000	71.9406	0.7302	40.4248	3.6626	11.0372	0.09985
	31.0000	73.3407	0.6956	41.0510	3.6673	11.1938	0.09789
	30.0000	74.5128	0.6679	41.5408	3.6709	11.3163	0.09639
	29.0000	75.5323	0.6449	41.9406	3.6737	11.4163	0.09520
	28.0000	76.4412	0.6253	42.2760	3.6761	11.5002	0.09423
	27.0000	77.2655	0.6082	42.5628	3.6781	11.5719	0.09340
	26.0000	78.0229	0.5932	42.8114	3.6798	11.6341	0.09270
	25.0000	78.7258	0.5799	43.0294	3.6813	11.6886	0.09209
	24.0000	79.3837	0.5680	43.2220	3.6826	11.7368	0.09155
	23.0000	80.0035	0.5572	43.3934	3.6838	11.7797	0.09108
	22.0000	80.5909	0.5475	43.5467	3.6848	11.8180	0.09066
	21.0000	81.1503	0.5387	43.6843	3.6857	11.8524	0.09029
	20.0000	81.6853	0.5307	43.8083	3.6865	11.8834	0.08996
	19.0000	82.1990	0.5233	43.9203	3.6872	11.9115	0.08966
	18.0000	82.6941	0.5167	44.0216	3.6879	11.9368	0.08939
	17.0000	83.1725	0.5106	44.1134	3.6885	11.9597	0.08915
	16.0000	83.6364	0.5050	44.1965	3.6890	11.9805	0.08893
	15.0000	84.0871	0.4999	44.2718	3.6895	11.9994	0.08873
	14.0000	84.5263	0.4953	44.3399	3.6899	12.0164	0.08855
	13.0000	84.9551	0.4911	44.4014	3.6903	12.0318	0.08839
	12.0000	85.3746	0.4873	44.4567	3.6907	12.0456	0.08825
	11.0000	85.7861	0.4838	44.5064	3.6910	12.0580	0.08812
	10.0000	86.1904	0.4807	44.5507	3.6913	12.0691	0.08800
	9.0000	86.5884	0.4780	44.5900	3.6915	12.0789	0.08790
	8.0000	86.9807	0.4756	44.6245	3.6918	12.0876	0.08781
	7.0000	87.3682	0.4735	44.6545	3.6920	12.0951	0.08774
	6.0000	87.7517	0.4717	44.6801	3.6921	12.1015	0.08767
	5.0000	88.1316	0.4702	44.7015	3.6923	12.1068	0.08762
	4.0000	88.5087	0.4689	44.7188	3.6924	12.1112	0.08757
	3.0000	88.8835	0.4680	44.7323	3.6924	12.1145	0.08754
	2.0000	89.2566	0.4673	44.7418	3.6925	12.1169	0.08751
	1.0000	89.6286	0.4669	44.7474	3.6925	12.1183	0.08750
	0.0000	90.0000	0.4668	44.7493	3.6926	12.1188	0.08749

* DENOTES $M_2 = 1$, m DENOTES θ_{max}

OBLIQUE SHOCKS IN A PERFECT GAS

$\gamma = 5/3$ $M_1 = 8.00$

WEAK SHOCK SOLUTION

θ	β	M_1	p_{21}	ρ_{21}	T_{21}	p^o_{21}
0.0000	7.1808	8.0000	1.0000	1.0000	1.0000	1.0000
1.0000	7.8897	7.6231	1.2574	1.1469	1.0963	0.9990
2.0000	8.6614	7.2706	1.5643	1.3043	1.1994	0.9929
3.0000	9.4943	6.9346	1.9266	1.4691	1.3115	0.9781
4.0000	10.3849	6.6106	2.3495	1.6376	1.4347	0.9529
5.0000	11.3292	6.2970	2.8373	1.8062	1.5709	0.9173
6.0000	12.3231	5.9931	3.3939	1.9714	1.7216	0.8726
7.0000	13.3618	5.6994	4.0225	2.1303	1.8882	0.8210
8.0000	14.4411	5.4168	4.7253	2.2810	2.0717	0.7649
9.0000	15.5571	5.1458	5.5044	2.4219	2.2728	0.7067
10.0000	16.7063	4.8871	6.3609	2.5524	2.4922	0.6487
11.0000	17.8859	4.6411	7.2958	2.6722	2.7303	0.5922
12.0000	19.0932	4.4076	8.3097	2.7816	2.9874	0.5386
13.0000	20.3265	4.1866	9.4031	2.8810	3.2638	0.4885
14.0000	21.5842	3.9777	10.5760	2.9711	3.5597	0.4423
15.0000	22.8653	3.7802	11.8285	3.0525	3.8750	0.4001
16.0000	24.1690	3.5937	13.1604	3.1261	4.2099	0.3618
17.0000	25.4953	3.4174	14.5719	3.1925	4.5644	0.3273
18.0000	26.8442	3.2506	16.0626	3.2525	4.9385	0.2963
19.0000	28.2161	3.0927	17.6328	3.3068	5.3323	0.2685
20.0000	29.6119	2.9429	19.2823	3.3559	5.7457	0.2436
21.0000	31.0331	2.8005	21.0116	3.4005	6.1790	0.2213
22.0000	32.4815	2.6649	22.8215	3.4410	6.6323	0.2014
23.0000	33.9593	2.5355	24.7127	3.4778	7.1058	0.1836
24.0000	35.4698	2.4115	26.6870	3.5114	7.6000	0.1675
25.0000	37.0167	2.2925	28.7465	3.5422	8.1155	0.1532
26.0000	38.6053	2.1778	30.8948	3.5704	8.6531	0.1402
27.0000	40.2422	2.0667	33.1369	3.5963	9.2141	0.1285
28.0000	41.9364	1.9586	35.4800	3.6203	9.8003	0.1180
29.0000	43.6998	1.8529	37.9345	3.6425	10.4143	0.1083
30.0000	45.5499	1.7486	40.5171	3.6633	11.0603	0.09956
31.0000	47.5122	1.6446	43.2525	3.6828	11.7444	0.09147
32.0000	49.6282	1.5393	46.1834	3.7014	12.4774	0.08395
33.0000	51.9722	1.4301	49.3884	3.7193	13.2790	0.07683
34.0000	54.7052	1.3117	53.0425	3.7373	14.1927	0.06987
35.0000	58.3660	1.1660	57.7418	3.7573	15.3678	0.06234
* 35.5442	63.0037	1.0000	63.2647	3.7773	16.7488	0.05508
m 35.5455	63.2297	0.9924	63.5191	3.7781	16.8124	0.05478

* DENOTES $M_2 = 1$, m DENOTES θ_{max}

OBLIQUE SHOCKS IN A PERFECT GAS

$$\gamma = 5/3 \qquad M_1 = 8.00$$

STRONG SHOCK SOLUTION

	θ	β	M_1	p_{21}	ρ_{21}	T_{21}	p^0_{21}
m	35.5455	63.2297	0.9924	63.5191	3.7781	16.8124	0.05478
	35.0000	67.5655	0.8549	68.0979	3.7922	17.9573	0.04981
	34.0000	70.2606	0.7772	70.6233	3.7993	18.5887	0.04738
	33.0000	72.0296	0.7296	72.1339	3.8032	18.9664	0.04602
	32.0000	73.4114	0.6943	73.2281	3.8060	19.2400	0.04508
	31.0000	74.5670	0.6662	74.0837	3.8082	19.4539	0.04436
	30.0000	75.5714	0.6428	74.7819	3.8099	19.6284	0.04379
	29.0000	76.4656	0.6228	75.3673	3.8113	19.7748	0.04332
	28.0000	77.2760	0.6054	75.8679	3.8125	19.9000	0.04293
	27.0000	78.0198	0.5902	76.3019	3.8135	20.0085	0.04259
	26.0000	78.7095	0.5766	76.6824	3.8144	20.1036	0.04230
	25.0000	79.3544	0.5644	77.0187	3.8151	20.1877	0.04204
	24.0000	79.9614	0.5534	77.3181	3.8158	20.2626	0.04182
	23.0000	80.5361	0.5435	77.5860	3.8164	20.3295	0.04162
	22.0000	81.0830	0.5345	77.8267	3.8170	20.3897	0.04144
	21.0000	81.6055	0.5263	78.0438	3.8174	20.4440	0.04128
	20.0000	82.1068	0.5188	78.2401	3.8179	20.4931	0.04114
	19.0000	82.5893	0.5119	78.4179	3.8183	20.5375	0.04101
	18.0000	83.0553	0.5056	78.5793	3.8186	20.5779	0.04089
	17.0000	83.5066	0.4999	78.7257	3.8189	20.6145	0.04078
	16.0000	83.9448	0.4946	78.8586	3.8192	20.6477	0.04069
	15.0000	84.3713	0.4898	78.9792	3.8195	20.6779	0.04060
	14.0000	84.7873	0.4854	79.0885	3.8197	20.7052	0.04052
	13.0000	85.1940	0.4814	79.1873	3.8200	20.7299	0.04045
	12.0000	85.5924	0.4778	79.2763	3.8201	20.7521	0.04039
	11.0000	85.9834	0.4745	79.3563	3.8203	20.7721	0.04033
	10.0000	86.3679	0.4716	79.4277	3.8205	20.7900	0.04028
	9.0000	86.7466	0.4690	79.4911	3.8206	20.8059	0.04024
	8.0000	87.1203	0.4667	79.5469	3.8207	20.8198	0.04020
	7.0000	87.4894	0.4647	79.5953	3.8208	20.8319	0.04017
	6.0000	87.8549	0.4630	79.6367	3.8209	20.8423	0.04014
	5.0000	88.2172	0.4615	79.6714	3.8210	20.8509	0.04011
	4.0000	88.5769	0.4603	79.6995	3.8211	20.8579	0.04009
	3.0000	88.9344	0.4594	79.7211	3.8211	20.8634	0.04008
	2.0000	89.2905	0.4588	79.7365	3.8211	20.8672	0.04007
	1.0000	89.6455	0.4584	79.7457	3.8212	20.8695	0.04006
	0.0000	90.0000	0.4583	79.7488	3.8212	20.8703	0.04006

* DENOTES $M_2 = 1$, m DENOTES θ_{max}

OBLIQUE SHOCKS IN A PERFECT GAS

$$\gamma = 5/3 \qquad M_1 = 10.00$$

WEAK SHOCK SOLUTION

θ	β	M_1	p_{21}	ρ_{21}	T_{21}	p^{o}_{21}
0.0000	5.7392	10.0000	1.0000	1.0000	1.0000	1.0000
1.0000	6.4516	9.4276	1.3282	1.1848	1.1210	0.9982
2.0000	7.2415	8.8977	1.7361	1.3850	1.2535	0.9868
3.0000	8.1046	8.3952	2.2344	1.5940	1.4017	0.9604
4.0000	9.0350	7.9134	2.8326	1.8047	1.5696	0.9177
5.0000	10.0255	7.4510	3.5383	2.0102	1.7601	0.8608
6.0000	11.0691	7.0090	4.3576	2.2053	1.9760	0.7939
7.0000	12.1587	6.5893	5.2950	2.3863	2.2189	0.7219
8.0000	13.2883	6.1935	6.3538	2.5514	2.4904	0.6491
9.0000	14.4527	5.8224	7.5362	2.6999	2.7913	0.5788
10.0000	15.6477	5.4760	8.8436	2.8322	3.1225	0.5132
11.0000	16.8698	5.1538	10.2767	2.9495	3.4842	0.4534
12.0000	18.1163	4.8547	11.8358	3.0529	3.8769	0.3999
13.0000	19.3853	4.5772	13.5211	3.1441	4.3005	0.3525
14.0000	20.6753	4.3197	15.3322	3.2243	4.7552	0.3109
15.0000	21.9852	4.0808	17.2686	3.2949	5.2410	0.2745
16.0000	23.3148	3.8586	19.3301	3.3573	5.7577	0.2429
17.0000	24.6636	3.6517	21.5159	3.4123	6.3053	0.2155
18.0000	26.0320	3.4587	23.8258	3.4612	6.8838	0.1916
19.0000	27.4205	3.2781	26.2591	3.5045	7.4930	0.1708
20.0000	28.8302	3.1088	28.8160	3.5431	8.1329	0.1527
21.0000	30.2624	2.9495	31.4965	3.5777	8.8037	0.1369
22.0000	31.7187	2.7992	34.3010	3.6086	9.5053	0.1231
23.0000	33.2015	2.6571	37.2306	3.6364	10.2382	0.1110
24.0000	34.7137	2.5221	40.2873	3.6616	11.0028	0.1003
25.0000	36.2590	2.3935	43.4739	3.6843	11.7998	0.09086
26.0000	37.8418	2.2704	46.7946	3.7049	12.6303	0.08251
27.0000	39.4684	2.1521	50.2559	3.7238	13.4959	0.07508
28.0000	41.1463	2.0379	53.8670	3.7410	14.3989	0.06844
29.0000	42.8862	1.9269	57.6416	3.7569	15.3428	0.06249
30.0000	44.7027	1.8183	61.6004	3.7716	16.3327	0.05712
31.0000	46.6167	1.7109	65.7743	3.7853	17.3763	0.05224
32.0000	48.6612	1.6035	70.2146	3.7982	18.4865	0.04776
33.0000	50.8913	1.4940	75.0112	3.8104	19.6858	0.04361
34.0000	53.4149	1.3784	80.3444	3.8224	21.0192	0.03965
35.0000	56.5105	1.2474	86.6904	3.8349	22.6058	0.03566
36.0000	62.4258	1.0252	97.9644	3.8532	25.4244	0.03004
* 36.0185	63.1557	1.0000	99.2593	3.8550	25.7481	0.02949
m 36.0190	63.2998	0.9951	99.5122	3.8554	25.8113	0.02939

* DENOTES $M_2 = 1$, m DENOTES θ_{max}

OBLIQUE SHOCKS IN A PERFECT GAS

$$\gamma = 5/3 \qquad M_1 = 10.00$$

STRONG SHOCK SOLUTION

	θ	β	M_1	p_{21}	ρ_{21}	T_{21}	p_{21}^{o}
m	36.0190	63.2998	0.9951	99.5122	3.8554	25.8113	0.02939
	36.0000	64.1549	0.9663	100.9931	3.8574	26.1816	0.02878
	35.0000	69.0916	0.8124	108.8284	3.8673	28.1404	0.02589
	34.0000	71.2096	0.7526	111.7792	3.8707	28.8781	0.02493
	33.0000	72.7568	0.7114	113.7646	3.8729	29.3745	0.02431
	32.0000	74.0117	0.6797	115.2648	3.8745	29.7495	0.02387
	31.0000	75.0822	0.6538	116.4642	3.8758	30.0494	0.02352
	30.0000	76.0239	0.6319	117.4569	3.8768	30.2975	0.02323
	29.0000	76.8696	0.6131	118.2976	3.8776	30.5077	0.02300
	28.0000	77.6405	0.5967	119.0213	3.8784	30.6886	0.02280
	27.0000	78.3514	0.5821	119.6523	3.8790	30.8464	0.02263
	26.0000	79.0130	0.5691	120.2078	3.8795	30.9852	0.02248
	25.0000	79.6333	0.5574	120.7005	3.8800	31.1084	0.02235
	24.0000	80.2187	0.5469	121.1404	3.8804	31.2184	0.02223
	23.0000	80.7740	0.5373	121.5350	3.8808	31.3171	0.02213
	22.0000	81.3032	0.5285	121.8903	3.8811	31.4059	0.02204
	21.0000	81.8098	0.5206	122.2113	3.8814	31.4861	0.02196
	20.0000	82.2962	0.5133	122.5019	3.8817	31.5588	0.02188
	19.0000	82.7650	0.5066	122.7656	3.8820	31.6247	0.02182
	18.0000	83.2182	0.5005	123.0050	3.8822	31.6845	0.02176
	17.0000	83.6573	0.4949	123.2226	3.8824	31.7389	0.02170
	16.0000	84.0840	0.4898	123.4202	3.8826	31.7884	0.02165
	15.0000	84.4997	0.4851	123.5997	3.8827	31.8332	0.02161
	14.0000	84.9054	0.4808	123.7624	3.8829	31.8739	0.02157
	13.0000	85.3022	0.4769	123.9097	3.8830	31.9107	0.02153
	12.0000	85.6911	0.4734	124.0425	3.8831	31.9439	0.02150
	11.0000	86.0728	0.4702	124.1618	3.8832	31.9737	0.02147
	10.0000	86.4483	0.4674	124.2684	3.8833	32.0004	0.02144
	9.0000	86.8184	0.4648	124.3631	3.8834	32.0241	0.02142
	8.0000	87.1835	0.4626	124.4463	3.8835	32.0449	0.02140
	7.0000	87.5445	0.4606	124.5187	3.8836	32.0630	0.02138
	6.0000	87.9018	0.4589	124.5806	3.8836	32.0784	0.02136
	5.0000	88.2561	0.4575	124.6324	3.8837	32.0914	0.02135
	4.0000	88.6079	0.4563	124.6743	3.8837	32.1019	0.02134
	3.0000	88.9576	0.4554	124.7068	3.8837	32.1100	0.02133
	2.0000	89.3059	0.4548	124.7298	3.8838	32.1157	0.02133
	1.0000	89.6532	0.4544	124.7436	3.8838	32.1192	0.02132
	0.0000	90.0000	0.4543	124.7481	3.8838	32.1203	0.02132

* DENOTES $M_2 = 1$, m DENOTES θ_{max}

OBLIQUE SHOCKS IN A PERFECT GAS

$\gamma = 5/3$ $M_1 = 15.00$

WEAK SHOCK SOLUTION

θ	β	M_1	p_{21}	ρ_{21}	T_{21}	p^o_{21}
0.0000	3.8226	15.0000	1.0000	1.0000	1.0000	1.0000
1.0000	4.5501	13.7621	1.5200	1.2826	1.1851	0.9942
2.0000	5.3894	12.6399	2.2311	1.5927	1.4008	0.9607
3.0000	6.3277	11.5893	3.1664	1.9070	1.6604	0.8912
4.0000	7.3485	10.6059	4.3511	2.2039	1.9743	0.7944
5.0000	8.4358	9.6978	5.8028	2.4699	2.3494	0.6858
6.0000	9.5756	8.8718	7.5326	2.6995	2.7904	0.5790
7.0000	10.7569	8.1293	9.5473	2.8929	3.3002	0.4824
8.0000	11.9714	7.4669	11.8505	3.0538	3.8806	0.3994
9.0000	13.2132	6.8779	14.4442	3.1869	4.5323	0.3302
10.0000	14.4778	6.3545	17.3284	3.2969	5.2560	0.2735
11.0000	15.7619	5.8887	20.5026	3.3880	6.0515	0.2275
12.0000	17.0636	5.4729	23.9662	3.4639	6.9189	0.1903
13.0000	18.3815	5.1005	27.7172	3.5273	7.8579	0.1601
14.0000	19.7145	4.7655	31.7539	3.5807	8.8681	0.1355
15.0000	21.0623	4.4628	36.0746	3.6259	9.9490	0.1155
16.0000	22.4250	4.1880	40.6775	3.6645	11.1004	0.09905
17.0000	23.8027	3.9375	45.5603	3.6976	12.3216	0.08546
18.0000	25.1960	3.7080	50.7211	3.7261	13.6122	0.07416
19.0000	26.6058	3.4969	56.1593	3.7509	14.9721	0.06472
20.0000	28.0332	3.3017	61.8728	3.7726	16.4008	0.05677
21.0000	29.4798	3.1206	67.8619	3.7915	17.8983	0.05005
22.0000	30.9474	2.9518	74.1266	3.8083	19.4646	0.04433
23.0000	32.4382	2.7938	80.6684	3.8231	21.1002	0.03943
24.0000	33.9551	2.6452	87.4902	3.8363	22.8057	0.03521
25.0000	35.5014	2.5049	94.5967	3.8481	24.5824	0.03156
26.0000	37.0815	2.3718	101.9964	3.8588	26.4324	0.02838
27.0000	38.7007	2.2449	109.7003	3.8684	28.3584	0.02560
28.0000	40.3661	2.1233	117.7256	3.8771	30.3647	0.02316
29.0000	42.0870	2.0060	126.0983	3.8850	32.4579	0.002100
30.0000	43.8756	1.8922	134.8549	3.8923	34.6470	0.01908
31.0000	45.7498	1.7807	144.0528	3.8990	36.9464	0.01735
32.0000	47.7359	1.6702	153.7818	3.9052	39.3786	0.01579
33.0000	49.8762	1.5591	164.1932	3.9111	41.9813	0.01437
34.0000	52.2479	1.4443	175.5721	3.9168	44.8259	0.01304
35.0000	55.0199	1.3200	188.5604	3.9224	48.0728	0.01176
36.0000	58.7812	1.1661	205.4411	3.9287	52.2928	0.01038
* 36.4909	63.3097	1.0000	224.2539	3.9346	56.9957	0.009138
m 36.4910	63.3735	0.9978	224.5052	3.9346	57.0585	0.009123

* DENOTES $M_2 = 1$, m DENOTES θ_{max}

OBLIQUE SHOCKS IN A PERFECT GAS

$\gamma = 5/3$ $M_1 = 15.00$

STRONG SHOCK SOLUTION

	θ	β	M_1	p_{21}	ρ_{21}	T_{21}	p^o_{21}
m	36.4910	63.3735	0.9978	224.5052	3.9346	57.0585	0.009123
	36.0000	67.4792	0.8635	239.7358	3.9387	60.8660	0.008289
	35.0000	70.2502	0.7810	248.8816	3.9410	63.1523	0.007848
	34.0000	72.0325	0.7315	254.2328	3.9422	64.4900	0.007607
	33.0000	73.4148	0.6951	258.0808	3.9431	65.4520	0.007442
	32.0000	74.5665	0.6661	261.0780	3.9437	66.2012	0.007317
	31.0000	75.5646	0.6421	263.5178	3.9442	66.8111	0.007218
	30.0000	76.4513	0.6216	265.5601	3.9446	67.3217	0.007137
	29.0000	77.2534	0.6039	267.3041	3.9450	67.7576	0.007069
	28.0000	77.9885	0.5882	268.8152	3.9453	68.1354	0.007010
	27.0000	78.6690	0.5743	270.1388	3.9456	68.4663	0.006960
	26.0000	79.3045	0.5618	271.3087	3.9458	68.7587	0.006916
	25.0000	79.9020	0.5506	272.3497	3.9460	69.0190	0.006877
	24.0000	80.4669	0.5404	273.2814	3.9462	69.2519	0.006843
	23.0000	81.0038	0.5311	274.1190	3.9464	69.4612	0.006812
	22.0000	81.5165	0.5227	274.8748	3.9465	69.6502	0.006785
	21.0000	82.0077	0.5150	275.5586	3.9466	69.8211	0.006760
	20.0000	82.4800	0.5079	276.1786	3.9468	69.9761	0.006738
	19.0000	82.9356	0.5014	276.7418	3.9469	70.1169	0.006718
	18.0000	83.3764	0.4955	277.2538	3.9470	70.2449	0.006700
	17.0000	83.8039	0.4900	277.7195	3.9470	70.3613	0.006683
	16.0000	84.2197	0.4850	278.1429	3.9471	70.4672	0.006668
	15.0000	84.6247	0.4805	278.5276	3.9472	70.5633	0.006655
	14.0000	85.0205	0.4763	278.8767	3.9473	70.6506	0.006643
	13.0000	85.4076	0.4725	279.1927	3.9473	70.7296	0.006632
	12.0000	85.7872	0.4690	279.4780	3.9474	70.8009	0.006622
	11.0000	86.1601	0.4659	279.7344	3.9474	70.8650	0.006613
	10.0000	86.5269	0.4631	279.9636	3.9475	70.9223	0.006605
	9.0000	86.8884	0.4606	280.1671	3.9475	70.9732	0.006598
	8.0000	87.2453	0.4584	280.3461	3.9475	71.0179	0.006592
	7.0000	87.5983	0.4565	280.5019	3.9476	71.0569	0.006586
	6.0000	87.9476	0.4549	280.6350	3.9476	71.0901	0.006582
	5.0000	88.2941	0.4535	280.7465	3.9476	71.1180	0.006578
	4.0000	88.6381	0.4524	280.8369	3.9476	71.1406	0.006575
	3.0000	88.9803	0.4515	280.9067	3.9476	71.1581	0.006572
	2.0000	89.3210	0.4509	280.9562	3.9477	71.1705	0.006571
	1.0000	89.6607	0.4505	280.9858	3.9477	71.1779	0.006570
	0.0000	90.0000	0.4504	280.9957	3.9477	71.1803	0.006569

* DENOTES $M_2 = 1$, m DENOTES θ_{max}

OBLIQUE SHOCKS IN A PERFECT GAS

$\gamma = 5/3 \qquad M_1 = 20.00$

WEAK SHOCK SOLUTION

θ	β	M_1	P_{21}	ρ_{21}	T_{21}	p^o_{21}
0.0000	2.8660	20.0000	1.0000	1.0000	1.0000	1.0000
1.0000	3.6109	17.8528	1.7333	1.3837	1.2526	0.9870
2.0000	4.4988	15.9386	2.8263	1.8027	1.5678	0.9182
3.0000	5.5034	14.1841	4.3488	2.2034	1.9737	0.7946
4.0000	6.5955	12.6076	6.3462	2.5503	2.4884	0.6496
5.0000	7.7510	11.2274	8.8445	2.8323	3.1227	0.5132
6.0000	8.9522	10.0403	11.8571	3.0542	3.8822	0.3992
7.0000	10.1873	9.0271	15.3907	3.2266	4.7699	0.3097
8.0000	11.4483	8.1631	19.4476	3.3605	5.7871	0.2413
9.0000	12.7299	7.4240	24.0277	3.4650	6.9343	0.1897
10.0000	14.0285	6.7881	29.1293	3.5475	8.2113	0.1507
11.0000	15.3419	6.2371	34.7505	3.6131	9.6178	0.1211
12.0000	16.6689	5.7562	40.8879	3.6661	11.1530	0.09839
13.0000	18.0083	5.3333	47.5377	3.7092	12.8161	0.08081
14.0000	19.3602	4.9588	54.6971	3.7447	14.6065	0.06705
15.0000	20.7242	4.6250	62.3608	3.7742	16.5228	0.05617
16.0000	22.1008	4.3253	70.5262	3.7990	18.5644	0.04747
17.0000	23.4906	4.0546	79.1891	3.8200	20.7303	0.04045
18.0000	24.8942	3.8087	88.3457	3.8378	23.0196	0.03473
19.0000	26.3128	3.5841	97.9935	3.8532	25.4317	0.03003
20.0000	27.7477	3.3779	108.1302	3.8665	27.9659	0.02613
21.0000	29.2003	3.1875	118.7542	3.8781	30.6218	0.02287
22.0000	30.6728	3.0109	129.8664	3.8882	33.3999	0.02013
23.0000	32.1671	2.8463	141.4674	3.8972	36.3000	0.01781
24.0000	33.6863	2.6922	153.5634	3.9051	39.3240	0.01583
25.0000	35.2337	2.5472	166.1616	3.9121	42.4734	0.01413
26.0000	36.8134	2.4100	179.2743	3.9184	45.7514	0.01265
27.0000	38.4310	2.2797	192.9223	3.9241	49.1632	0.01138
28.0000	40.0928	2.1551	207.1324	3.9292	52.7156	0.01026
29.0000	41.8078	2.0354	221.9467	3.9339	56.4190	0.009277
30.0000	43.5879	1.9194	237.4283	3.9382	60.2891	0.008407
31.0000	45.4496	1.8062	253.6696	3.9421	64.3492	0.007632
32.0000	47.4177	1.6945	270.8194	3.9457	68.6364	0.006935
33.0000	49.5310	1.5825	289.1210	3.9491	73.2115	0.006300
34.0000	51.8585	1.4676	309.0266	3.9524	78.1876	0.005713
35.0000	54.5454	1.3447	331.5111	3.9556	83.8083	0.005152
36.0000	58.0411	1.1983	359.6592	3.9590	90.8449	0.004569
* 36.6571	63.3646	1.0000	399.2518	3.9631	100.7424	0.003917
m 36.6572	63.4004	0.9988	399.5026	3.9631	100.8051	0.003913

* DENOTES $M_2 = 1$, m DENOTES θ_{max}

OBLIQUE SHOCKS IN A PERFECT GAS

$\gamma = 5/3 \qquad M_1 = 20.00$

STRONG SHOCK SOLUTION

	θ	β	M_1	p_{21}	ρ_{21}	T_{21}	p^o_{21}
m	36.6572	63.4004	0.9988	399.5026	3.9631	100.8051	0.003913
	36.0000	68.1058	0.8453	430.2185	3.9657	108.4835	0.003507
	35.0000	70.6071	0.7715	444.6164	3.9669	112.0827	0.003341
	34.0000	72.2997	0.7247	453.5234	3.9675	114.3093	0.003244
	33.0000	73.6332	0.6897	460.0414	3.9680	115.9387	0.003176
	32.0000	74.7530	0.6616	465.1635	3.9683	117.2191	0.003125
	31.0000	75.7278	0.6382	469.3544	3.9686	118.2668	0.003083
	30.0000	76.5968	0.6181	472.8765	3.9688	119.1473	0.003049
	29.0000	77.3845	0.6007	475.8919	3.9690	119.9011	0.003021
	28.0000	78.1076	0.5853	478.5094	3.9692	120.5554	0.002996
	27.0000	78.7781	0.5716	480.8062	3.9694	121.1295	0.002975
	26.0000	79.4047	0.5593	482.8386	3.9695	121.6376	0.002957
	25.0000	79.9945	0.5482	484.6493	3.9696	122.0903	0.002940
	24.0000	80.5525	0.5382	486.2711	3.9697	122.4957	0.002926
	23.0000	81.0833	0.5290	487.7303	3.9698	122.8605	0.002913
	22.0000	81.5901	0.5207	489.0475	3.9699	123.1897	0.002901
	21.0000	82.0761	0.5130	490.2402	3.9699	123.4879	0.002891
	20.0000	82.5435	0.5060	491.3219	3.9700	123.7583	0.002881
	19.0000	82.9946	0.4996	492.3050	3.9701	124.0040	0.002873
	18.0000	83.4312	0.4937	493.1993	3.9701	124.2276	0.002865
	17.0000	83.8547	0.4883	494.0126	3.9702	124.4309	0.002858
	16.0000	84.2666	0.4834	494.7524	3.9702	124.6158	0.002852
	15.0000	84.6682	0.4788	495.4251	3.9703	124.7840	0.002846
	14.0000	85.0603	0.4747	496.0352	3.9703	124.9365	0.002841
	13.0000	85.4442	0.4709	496.5879	3.9703	125.0747	0.002836
	12.0000	85.8206	0.4675	497.0868	3.9704	125.1994	0.002832
	11.0000	86.1903	0.4644	497.5352	3.9704	125.3115	0.002828
	10.0000	86.5542	0.4616	497.9362	3.9704	125.4117	0.002825
	9.0000	86.9128	0.4592	498.2922	3.9704	125.5007	0.002822
	8.0000	87.2668	0.4570	498.6055	3.9705	125.5791	0.002819
	7.0000	87.6169	0.4551	498.8779	3.9705	125.6472	0.002817
	6.0000	87.9635	0.4534	499.1110	3.9705	125.7054	0.002815
	5.0000	88.3072	0.4521	499.3060	3.9705	125.7542	0.002813
	4.0000	88.6486	0.4510	499.4643	3.9705	125.7937	0.002812
	3.0000	88.9881	0.4501	499.5865	3.9705	125.8243	0.002811
	2.0000	89.3262	0.4495	499.6733	3.9705	125.8460	0.002810
	1.0000	89.6633	0.4491	499.7252	3.9705	125.8589	0.002810
	0.0000	90.0000	0.4490	499.7424	3.9705	125.8633	0.002810

* DENOTES $M_2 = 1$, m DENOTES θ_{max}

OBLIQUE SHOCKS IN A PERFECT GAS

$$\gamma = 5/3 \qquad M_1 = 25.00$$

WEAK SHOCK SOLUTION

θ	β	M_1	p_{21}	ρ_{21}	T_{21}	p^o_{21}
0.0000	2.2924	25.0000	1.0000	1.0000	1.0000	1.0000
1.0000	3.0552	21.7080	1.9693	1.4872	1.3242	0.9759
2.0000	3.9885	18.8212	3.5298	2.0079	1.7579	0.8615
3.0000	5.0485	16.2597	5.8000	2.4694	2.3487	0.6860
4.0000	6.1941	14.0743	8.8452	2.8322	3.1230	0.5132
5.0000	7.3963	12.2643	12.6966	3.1016	4.0935	0.3745
6.0000	8.6365	10.7823	17.3669	3.2980	5.2659	0.2729
7.0000	9.9039	9.5676	22.8615	3.4416	6.6427	0.2010
8.0000	11.1914	8.5649	29.1797	3.5479	8.2245	0.1504
9.0000	12.4951	7.7285	36.3202	3.6280	10.0111	0.1145
10.0000	13.8121	7.0232	44.2780	3.6893	12.0017	0.08873
11.0000	15.1410	6.4218	53.0493	3.7371	14.1954	0.06987
12.0000	16.4810	5.9036	62.6285	3.7749	16.5909	0.05586
13.0000	17.8317	5.4526	73.0091	3.8052	19.1865	0.04528
14.0000	19.1930	5.0567	84.1857	3.8299	21.9811	0.03716
15.0000	20.5652	4.7061	96.1508	3.8502	24.9728	0.03085
16.0000	21.9489	4.3933	108.8990	3.8671	28.1601	0.02588
17.0000	23.3446	4.1122	122.4239	3.8814	31.5415	0.02191
18.0000	24.7535	3.8579	136.7202	3.8934	35.1158	0.01871
19.0000	26.1765	3.6265	151.7826	3.9037	38.8816	0.01610
20.0000	27.6151	3.4146	167.6085	3.9126	42.8382	0.01395
21.0000	29.0708	3.2196	184.1941	3.9203	46.9847	0.01217
22.0000	30.5456	3.0392	201.5395	3.9270	51.3212	0.01068
23.0000	32.0418	2.8714	219.6486	3.9329	55.8486	0.009423
24.0000	33.5623	2.7145	238.5282	3.9382	60.5685	0.008354
25.0000	35.1105	2.5671	258.1893	3.9428	65.4838	0.007440
26.0000	36.6903	2.4280	278.6506	3.9469	70.5992	0.006654
27.0000	38.3072	2.2960	299.9430	3.9507	75.9224	0.005972
28.0000	39.9676	2.1700	322.1082	3.9540	81.4637	0.005378
29.0000	41.6802	2.0490	345.2094	3.9570	87.2391	0.004856
30.0000	43.4566	1.9321	369.3400	3.9598	93.2717	0.004396
31.0000	45.3132	1.8181	394.6446	3.9624	99.5980	0.003986
32.0000	47.2735	1.7057	421.3426	3.9647	106.2725	0.003619
33.0000	49.3752	1.5933	449.8005	3.9669	113.3870	0.003286
34.0000	51.6840	1.4783	480.6888	3.9691	121.1091	0.002978
35.0000	54.3358	1.3559	515.4312	3.9711	129.7947	0.002685
36.0000	57.7331	1.2121	558.3344	3.9733	140.5205	0.002385
* 36.7325	63.3893	1.0000	624.2507	3.9761	156.9996	0.002021
m 36.7326	63.4122	0.9992	624.5012	3.9761	157.0622	0.002020

* DENOTES $M_2 = 1$, m DENOTES θ_{max}

OBLIQUE SHOCKS IN A PERFECT GAS

$$\gamma = 5/3 \qquad M_1 = 25.00$$

STRONG SHOCK SOLUTION

	θ	β	M_1	p_{21}	ρ_{21}	T_{21}	p^o_{21}
m	36.7326	63.4122	0.9992	624.5012	3.9761	157.0622	0.002020
	36.0000	68.3610	0.8380	674.7633	3.9779	169.6278	0.001801
	35.0000	70.7616	0.7674	696.1797	3.9786	174.9819	0.001719
	34.0000	72.4173	0.7217	709.7074	3.9790	178.3638	0.001670
	33.0000	73.7299	0.6873	719.6771	3.9793	180.8563	0.001636
	32.0000	74.8358	0.6596	727.5406	3.9795	182.8221	0.001610
	31.0000	75.8005	0.6364	733.9905	3.9797	184.4346	0.001589
	30.0000	76.6617	0.6165	739.4193	3.9798	185.7918	0.001572
	29.0000	77.4430	0.5992	744.0717	3.9800	186.9549	0.001557
	28.0000	78.1609	0.5840	748.1150	3.9801	187.9657	0.001544
	27.0000	78.8269	0.5704	751.6649	3.9802	188.8532	0.001534
	26.0000	79.4497	0.5582	754.8080	3.9802	189.6390	0.001524
	25.0000	80.0358	0.5471	757.6089	3.9803	190.3392	0.001516
	24.0000	80.5909	0.5371	760.1194	3.9804	190.9668	0.001508
	23.0000	81.1187	0.5280	762.3782	3.9804	191.5316	0.001502
	22.0000	81.6231	0.5197	764.4185	3.9805	192.0416	0.001496
	21.0000	82.1067	0.5121	766.2659	3.9805	192.5034	0.001490
	20.0000	82.5720	0.5052	767.9421	3.9806	192.9225	0.001485
	19.0000	83.0211	0.4988	769.4659	3.9806	193.3035	0.001481
	18.0000	83.4557	0.4929	770.8516	3.9806	193.6499	0.001477
	17.0000	83.8775	0.4875	772.1127	3.9807	193.9651	0.001474
	16.0000	84.2877	0.4826	773.2598	3.9807	194.2519	0.001470
	15.0000	84.6876	0.4781	774.3025	3.9807	194.5126	0.001467
	14.0000	85.0783	0.4740	775.2491	3.9808	194.7493	0.001465
	13.0000	85.4605	0.4702	776.1060	3.9808	194.9635	0.001462
	12.0000	85.8356	0.4668	776.8796	3.9808	195.1569	0.001460
	11.0000	86.2039	0.4637	777.5752	3.9808	195.3308	0.001458
	10.0000	86.5664	0.4610	778.1974	3.9808	195.4863	0.001456
	9.0000	86.9237	0.4585	778.7496	3.9808	195.6244	0.001455
	8.0000	87.2764	0.4563	779.2357	3.9809	195.7459	0.001454
	7.0000	87.6252	0.4544	779.6583	3.9809	195.8516	0.001452
	6.0000	87.9707	0.4528	780.0200	3.9809	195.9420	0.001451
	5.0000	88.3132	0.4514	780.3226	3.9809	196.0176	0.001451
	4.0000	88.6533	0.4503	780.5681	3.9809	196.0790	0.001450
	3.0000	88.9916	0.4195	780.7577	3.9809	196.1264	0.001449
	2.0000	89.3285	0.4488	780.8923	3.9809	196.1601	0.001449
	1.0000	89.6645	0.4485	780.9730	3.9809	196.1802	0.001449
	0.0000	90.0000	0.4484	780.9996	3.9809	196.1869	0.001449

* DENOTES $M_2 = 1$, m DENOTES θ_{max}

OBLIQUE SHOCKS IN A PERFECT GAS

$\gamma = 5/3$ $M_1 = 35.00$

WEAK SHOCK SOLUTION

θ	β	M_1	P_{21}	ρ_{21}	T_{21}	P^o_{21}
0.0000	1.6372	35.0000	1.0000	1.0000	1.0000	1.0000
1.0000	2.4348	28.7421	2.5136	1.6971	1.4811	0.9415
2.0000	3.4465	23.4758	5.2840	2.3843	2.2162	0.7227
3.0000	4.5988	19.2087	9.5508	2.8931	3.3013	0.4823
4.0000	5.8036	15.9440	15.4071	3.2271	4.7743	0.3093
5.0000	7.0604	13.4798	22.8844	3.4421	6.6485	0.2008
6.0000	8.3434	11.5993	31.9915	3.5832	8.9281	0.1343
7.0000	9.6443	10.1359	42.7270	3.6790	11.6138	0.09295
8.0000	10.9585	8.9728	55.0853	3.7461	14.7046	0.06644
9.0000	12.2837	8.0297	69.0594	3.7947	18.1989	0.04888
10.0000	13.6183	7.2512	84.6389	3.8308	22.0944	0.03689
11.0000	14.9618	6.5981	101.8143	3.8582	26.3888	0.02846
12.0000	16.3140	6.0425	120.5744	3.8796	31.0791	0.02239
13.0000	17.6750	5.5640	140.9051	3.8965	36.1621	0.01792
14.0000	19.0450	5.1473	162.7947	3.9101	41.6347	0.01455
15.0000	20.4247	4.7808	186.2322	3.9212	47.4942	0.01198
16.0000	21.8147	4.4556	211.2021	3.9303	53.7369	0.009977
17.0000	23.2160	4.1647	237.6940	3.9379	60.3599	0.008397
18.0000	24.6294	3.9026	265.6952	3.9444	67.3603	0.007135
19.0000	26.0563	3.6649	295.1954	3.9499	74.7355	0.006113
20.0000	27.4982	3.4478	326.1910	3.9546	82.4844	0.005279
21.0000	28.9566	3.2486	358.6726	3.9586	90.6049	0.004590
22.0000	30.4337	3.0646	392.6427	3.9622	99.0975	0.004016
23.0000	31.9316	2.8938	428.1038	3.9653	107.9628	0.003535
24.0000	33.4532	2.7345	465.0695	3.9680	117.2043	0.003127
25.0000	35.0019	2.5850	503.5620	3.9705	126.8274	0.002780
26.0000	36.5818	2.4441	543.6196	3.9726	136.8418	0.002482
27.0000	38.1981	2.3106	585.2936	3.9745	147.2603	0.002224
28.0000	39.8572	2.1834	628.6689	3.9763	158.1042	0.002000
29.0000	41.5678	2.0613	673.8662	3.9779	169.4035	0.001804
30.0000	43.3411	1.9434	721.0636	3.9793	181.2029	0.001631
31.0000	45.1928	1.8287	770.5276	3.9806	193.5689	0.001478
32.0000	47.1465	1.7157	822.6852	3.9819	206.6083	0.001341
33.0000	49.2382	1.6029	878.2290	3.9830	220.4942	0.001216
34.0000	51.5308	1.4878	938.4042	3.9841	235.5380	0.001102
35.0000	54.1530	1.3658	1005.8502	3.9852	252.3995	0.0009938
36.0000	57.4711	1.2240	1088.2407	3.9863	272.9972	0.0008837
* 36.7998	63.4116	1.0000	1224.2493	3.9878	306.9993	0.0007413
m 36.7998	63.4233	0.9996	1224.4998	3.9878	307.0619	0.0007411

* DENOTES $M_2 = 1$, m DENOTES θ_{max}

OBLIQUE SHOCKS IN A PERFECT GAS

$\gamma = 5/3$ $M_1 = 35.00$

STRONG SHOCK SOLUTION

	θ	β	M_1	p_{21}	ρ_{21}	T_{21}	p^o_{21}
m	36.7998	63.4233	0.9996	1224.4998	3.9878	307.0619	0.0007411
	36.0000	68.5768	0.8318	1326.7153	3.9887	332.6158	0.0006575
	35.0000	70.8967	0.7638	1366.9907	3.9891	342.6846	0.0006288
	34.0000	72.5207	0.7191	1392.8552	3.9893	349.1508	0.0006115
	33.0000	73.8155	0.6852	1412.0339	3.9894	353.9454	0.0005991
	32.0000	74.9092	0.6578	1427.2087	3.9895	357.7392	0.0005896
	31.0000	75.8651	0.6348	1439.6809	3.9896	360.8572	0.0005820
	30.0000	76.7193	0.6151	1450.1912	3.9897	363.4847	0.0005757
	29.0000	77.4951	0.5980	1459.2107	3.9898	365.7396	0.0005704
	28.0000	78.2083	0.5828	1467.0532	3.9898	367.7002	0.0005659
	27.0000	78.8703	0.5693	1473.9434	3.9899	369.4228	0.0005619
	26.0000	79.4896	0.5572	1480.0471	3.9899	370.9487	0.0005585
	25.0000	80.0728	0.5462	1485.4895	3.9899	372.3093	0.0005554
	24.0000	80.6251	0.5362	1490.3687	3.9900	373.5291	0.0005527
	23.0000	81.1505	0.5272	1494.7598	3.9900	374.6269	0.0005503
	22.0000	81.6525	0.5189	1498.7266	3.9900	375.6186	0.0005481
	21.0000	82.1340	0.5113	1502.3196	3.9900	376.5168	0.0005461
	20.0000	82.5975	0.5044	1505.5811	3.9901	377.3322	0.0005444
	19.0000	83.0448	0.4980	1508.5457	3.9901	378.0734	0.0005428
	18.0000	83.4777	0.4922	1511.2422	3.9901	378.7475	0.0005413
	17.0000	83.8978	0.4868	1513.6958	3.9901	379.3609	0.0005400
	16.0000	84.3065	0.4819	1515.9290	3.9901	379.9192	0.0005388
	15.0000	84.7049	0.4774	1517.9585	3.9901	380.4266	0.0005377
	14.0000	85.0943	0.4733	1519.8013	3.9902	380.8872	0.0005368
	13.0000	85.4753	0.4696	1521.4695	3.9902	381.3043	0.0005359
	12.0000	85.8489	0.4662	1522.9758	3.9902	381.6808	0.0005351
	11.0000	86.2161	0.4631	1524.3303	3.9902	382.0195	0.0005344
	10.0000	86.5773	0.4604	1525.5415	3.9902	382.3223	0.0005338
	9.0000	86.9334	0.4579	1526.6172	3.9902	382.5912	0.0005332
	8.0000	87.2850	0.4557	1527.5635	3.9902	382.8278	0.0005327
	7.0000	87.6327	0.4539	1528.3870	3.9902	383.0337	0.0005323
	6.0000	87.9771	0.4522	1529.0913	3.9902	383.2098	0.0005319
	5.0000	88.3185	0.4509	1529.6809	3.9902	383.3572	0.0005316
	4.0000	88.6576	0.4498	1530.1589	3.9902	383.4767	0.0005314
	3.0000	88.9948	0.4489	1530.5283	3.9902	383.5690	0.0005312
	2.0000	89.3306	0.4483	1530.7905	3.9902	383.6346	0.0005310
	1.0000	89.6656	0.4479	1530.9475	3.9902	383.6738	0.0005309
	0.0000	90.0000	0.4478	1530.9993	3.9902	383.6867	0.0005309

* DENOTES $M_2 = 1$, m DENOTES θ_{max}

OBLIQUE SHOCKS IN A PERFECT GAS

$$\gamma = 5/3 \qquad M_1 = \infty$$

WEAK SHOCK SOLUTION

θ	β	M_1	P_{21}	ρ_{21}	T_{21}	P^o_{21}
0.0000	0.0000	∞	1.0000	1.0000	1.0000	1.0000
1.0000	1.3330	76.8806	∞	4.0000	∞	0.0000
2.0000	2.6668	38.4103	∞	4.0000	∞	0.0000
3.0000	4.0015	25.5770	∞	4.0000	∞	0.0000
4.0000	5.3376	19.1527	∞	4.0000	∞	0.0000
5.0000	6.6763	15.2895	∞	4.0000	∞	0.0000
6.0000	8.0165	12.7102	∞	4.0000	∞	0.0000
7.0000	9.3592	10.8633	∞	4.0000	∞	0.0000
8.0000	10.7058	9.4730	∞	4.0000	∞	0.0000
9.0000	12.0561	8.3878	∞	4.0000	∞	0.0000
10.0000	13.4107	7.5163	∞	4.0000	∞	0.0000
11.0000	14.7709	6.7995	∞	4.0000	∞	0.0000
12.0000	16.1366	6.1992	∞	4.0000	∞	0.0000
13.0000	17.5095	5.6881	∞	4.0000	∞	0.0000
14.0000	18.8883	5.2476	∞	4.0000	∞	0.0000
15.0000	20.2768	4.8627	∞	4.0000	∞	0.0000
16.0000	21.6730	4.5237	∞	4.0000	∞	0.0000
17.0000	23.0807	4.2218	∞	4.0000	∞	0.0000
18.0000	24.4993	3.9510	∞	4.0000	∞	0.0000
19.0000	25.9299	3.7064	∞	4.0000	∞	0.0000
20.0000	27.3758	3.4836	∞	4.0000	∞	0.0000
21.0000	28.8369	3.2797	∞	4.0000	∞	0.0000
22.0000	30.3158	3.0920	∞	4.0000	∞	0.0000
23.0000	31.8159	2.9179	∞	4.0000	∞	0.0000
24.0000	33.3386	2.7559	∞	4.0000	∞	0.0000
25.0000	34.8880	2.6042	∞	4.0000	∞	0.0000
26.0000	36.4679	2.4614	∞	4.0000	∞	0.0000
27.0000	38.0837	2.3262	∞	4.0000	∞	0.0000
28.0000	39.7412	2.1976	∞	4.0000	∞	0.0000
29.0000	41.4497	2.0744	∞	4.0000	∞	0.0000
30.0000	43.2196	1.9555	∞	4.0000	∞	0.0000
31.0000	45.0659	1.8400	∞	4.0000	∞	0.0000
32.0000	47.0131	1.7264	∞	4.0000	∞	0.0000
33.0000	49.0945	1.6131	∞	4.0000	∞	0.0000
34.0000	51.3700	1.4979	∞	4.0000	∞	0.0000
35.0000	53.9621	1.3762	∞	4.0000	∞	0.0000
36.0000	57.2028	1.2365	∞	4.0000	∞	0.0000
m*36.8716	63.4358	1.0000	∞	4.0000	∞	0.0000

* DENOTES $M_2 = 1$, m DENOTES θ_{max}

OBLIQUE SHOCKS IN A PERFECT GAS

$$\gamma = 5/3 \qquad M_1 = \infty$$

STRONG SHOCK SOLUTION

θ	β	M_1	P_{21}	ρ_{21}	T_{21}	P^o_{21}
m*36.8716	63.4358	1.0000	∞	4.0000	∞	0.0000
36.0000	68.7964	0.8256	∞	4.0000	∞	0.0000
35.0000	71.0381	0.7601	∞	4.0000	∞	0.0000
34.0000	72.6297	0.7163	∞	4.0000	∞	0.0000
33.0000	73.9061	0.6829	∞	4.0000	∞	0.0000
32.0000	74.9871	0.6559	∞	4.0000	∞	0.0000
31.0000	75.9337	0.6332	∞	4.0000	∞	0.0000
30.0000	76.7807	0.6137	∞	4.0000	∞	0.0000
29.0000	77.5504	0.5966	∞	4.0000	∞	0.0000
28.0000	78.2589	0.5816	∞	4.0000	∞	0.0000
27.0000	78.9166	0.5682	∞	4.0000	∞	0.0000
26.0000	79.5322	0.5561	∞	4.0000	∞	0.0000
25.0000	80.1122	0.5452	∞	4.0000	∞	0.0000
24.0000	80.6616	0.5353	∞	4.0000	∞	0.0000
23.0000	81.1844	0.5263	∞	4.0000	∞	0.0000
22.0000	81.6840	0.5180	∞	4.0000	∞	0.0000
21.0000	82.1634	0.5105	∞	4.0000	∞	0.0000
20.0000	82.6247	0.5036	∞	4.0000	∞	0.0000
19.0000	83.0700	0.4973	∞	4.0000	∞	0.0000
18.0000	83.5012	0.4914	∞	4.0000	∞	0.0000
17.0000	83.9196	0.4861	∞	4.0000	∞	0.0000
16.0000	84.3267	0.4812	∞	4.0000	∞	0.0000
15.0000	84.7236	0.4767	∞	4.0000	∞	0.0000
14.0000	85.1113	0.4727	∞	4.0000	∞	0.0000
13.0000	85.4910	0.4689	∞	4.0000	∞	0.0000
12.0000	85.8632	0.4655	∞	4.0000	∞	0.0000
11.0000	86.2290	0.4625	∞	4.0000	∞	0.0000
10.0000	86.5890	0.4597	∞	4.0000	∞	0.0000
9.0000	86.9439	0.4573	∞	4.0000	∞	0.0000
8.0000	87.2943	0.4551	∞	4.0000	∞	0.0000
7.0000	87.6407	0.4532	∞	4.0000	∞	0.0000
6.0000	87.9839	0.4516	∞	4.0000	∞	0.0000
5.0000	88.3241	0.4503	∞	4.0000	∞	0.0000
4.0000	88.6621	0.4491	∞	4.0000	∞	0.0000
3.0000	88.9981	0.4483	∞	4.0000	∞	0.0000
2.0000	89.3329	0.4477	∞	4.0000	∞	0.0000
1.0000	89.6667	0.4473	∞	4.0000	∞	0.0000
0.0000	90.0000	0.4472	∞	4.0000	∞	0.0000

* DENOTES $M_2 = 1$, m DENOTES θ_{max}

TABLE E.2

OBLIQUE SHOCKS IN A PERFECT GAS

$\gamma = 7/5$

E.2-1

OBLIQUE SHOCKS IN A PERFECT GAS

$$\gamma = 7/5 \quad M_1 = 1.10$$

	θ	β	M_1	P_{21}	ρ_{21}	T_{21}	P^o_{21}
	0.0000	65.3800	1.1000	1.0000	1.0000	1.0000	1.0000
	1.0000	69.8018	1.0393	1.0767	1.0542	1.0214	1.0000
*	1.4062	73.2502	1.0000	1.1278	1.0896	1.0350	0.9998
m	1.5152	76.2965	0.9711	1.1658	1.1157	1.0449	0.9996
	1.0000	83.5756	0.9249	1.2273	1.1573	1.0605	0.9991
	0.0000	90.0000	0.9118	1.2450	1.1691	1.0649	0.9989

$$\gamma = 7/5 \quad M_1 = 1.15$$

	θ	β	M_1	P_{21}	ρ_{21}	T_{21}	P^o_{21}
	0.0000	60.4081	1.1500	1.0000	1.0000	1.0000	1.0000
	1.0000	63.1605	1.1023	1.0617	1.0437	1.0173	1.0000
	2.0000	67.0021	1.0434	1.1407	1.0985	1.0384	0.9998
*	2.4923	70.2671	1.0000	1.2004	1.1391	1.0537	0.9994
m	2.6708	73.8223	0.9598	1.2565	1.1767	1.0678	0.9988
	2.0000	81.1747	0.9006	1.3399	1.2316	1.0880	0.9975
	1.0000	85.9859	0.8804	1.3687	1.2502	1.0948	0.9969
	0.0000	90.0000	0.8750	1.3762	1.2550	1.0966	0.9967

$$\gamma = 7/5 \quad M_1 = 1.20$$

	θ	β	M_1	P_{21}	ρ_{21}	T_{21}	P^o_{21}
	0.0000	56.4427	1.2000	1.0000	1.0000	1.0000	1.0000
	1.0000	58.5471	1.1579	1.0559	1.0396	1.0157	1.0000
	2.0000	61.0491	1.1113	1.1197	1.0841	1.0329	0.9999
	3.0000	64.3375	1.0558	1.1983	1.1377	1.0532	0.9994
*	3.7008	68.0757	1.0000	1.2791	1.1917	1.0733	0.9985
m	3.9442	71.9765	0.9502	1.3525	1.2397	1.0910	0.9972
	3.0000	80.0308	0.8760	1.4630	1.3102	1.1166	0.9944
	2.0000	83.8620	0.8551	1.4941	1.3297	1.1237	0.9934
	1.0000	87.0417	0.8452	1.5089	1.3388	1.1270	0.9929
	0.0000	90.0000	0.8422	1.5133	1.3416	1.1280	0.9928

* DENOTES $M_2 = 1$, m DENOTES θ_{max}

OBLIQUE SHOCKS IN A PERFECT GAS

$\gamma = 7/5$ $M_1 = 1.30$

θ	β	M_1	p_{21}	ρ_{21}	T_{21}	p^o_{21}
0.0000	50.2849	1.3000	1.0000	1.0000	1.0000	1.0000
1.0000	51.8115	1.2629	1.0514	1.0364	1.0144	1.0000
2.0000	53.4730	1.2244	1.1065	1.0749	1.0294	0.9999
3.0000	55.3163	1.1837	1.1665	1.1162	1.0451	0.9996
4.0000	57.4219	1.1398	1.2334	1.1613	1.0620	0.9991
5.0000	59.9603	1.0902	1.3109	1.2126	1.0810	0.9980
6.0000	63.4567	1.0275	1.4113	1.2775	1.1047	0.9958
* 6.3173	65.1147	1.0000	1.4559	1.3057	1.1150	0.9946
m 6.6621	69.3953	0.9359	1.5608	1.3709	1.1386	0.9911
6.0000	75.3739	0.8636	1.6793	1.4423	1.1643	0.9860
5.0000	78.9666	0.8308	1.7328	1.4738	1.1757	0.9833
4.0000	81.6497	0.8118	1.7634	1.4917	1.1822	0.9817
3.0000	83.9533	0.7996	1.7831	1.5031	1.1863	0.9806
2.0000	86.0583	0.7918	1.7957	1.5103	1.1889	0.9799
1.0000	88.0538	0.7874	1.8027	1.5144	1.1904	0.9795
0.0000	90.0000	0.7860	1.8050	1.5157	1.1909	0.9794

$\gamma = 7/5$ $M_1 = 1.40$

θ	β	M_1	p_{21}	ρ_{21}	T_{21}	p^o_{21}
0.0000	45.5847	1.4000	1.0000	1.0000	1.0000	1.0000
1.0000	46.8420	1.3651	1.0501	1.0356	1.0141	1.0000
2.0000	48.1728	1.3295	1.1030	1.0725	1.0284	0.9999
3.0000	49.5909	1.2930	1.1591	1.1111	1.0432	0.9997
4.0000	51.1168	1.2553	1.2189	1.1516	1.0584	0.9992
5.0000	52.7810	1.2158	1.2834	1.1946	1.0744	0.9984
6.0000	54.6324	1.1737	1.3539	1.2406	1.0913	0.9972
7.0000	56.7606	1.1277	1.4330	1.2912	1.1098	0.9953
8.0000	59.3662	1.0744	1.5263	1.3496	1.1309	0.9923
9.0000	63.1846	1.0025	1.6546	1.4276	1.1590	0.9871
* 9.0255	63.3250	1.0000	1.6592	1.4303	1.1600	0.9869
m 9.4272	67.7156	0.9266	1.7912	1.5077	1.1880	0.9802
9.0000	72.1899	0.8625	1.9061	1.5730	1.2117	0.9732
8.0000	75.8943	0.8183	1.9842	1.6163	1.2276	0.9681
7.0000	78.4134	0.7934	2.0278	1.6401	1.2363	0.9650
6.0000	80.4854	0.7762	2.0575	1.6562	1.2423	0.9629
5.0000	82.3130	0.7637	2.0791	1.6678	1.2466	0.9613
4.0000	83.9884	0.7545	2.0949	1.6763	1.2497	0.9601
3.0000	85.5646	0.7478	2.1063	1.6824	1.2520	0.9592
2.0000	87.0758	0.7432	2.1140	1.6865	1.2535	0.9586
1.0000	88.5475	0.7406	2.1185	1.6889	1.2544	0.9583
0.0000	90.0000	0.7397	2.1200	1.6897	1.2547	0.9582

* DENOTES $M_2 = 1$, m DENOTES θ_{max}

OBLIQUE SHOCKS IN A PERFECT GAS

$$\gamma = 7/5 \qquad M_1 = 1.50$$

	θ	β	M_1	P_{21}	ρ_{21}	T_{21}	P_{21}^o
	0.0000	41.8103	1.5000	1.0000	1.0000	1.0000	1.0000
	1.0000	42.9126	1.4660	1.0503	1.0357	1.0141	1.0000
	2.0000	44.0642	1.4316	1.1030	1.0725	1.0284	0.9999
	3.0000	45.2713	1.3969	1.1583	1.1105	1.0430	0.9997
	4.0000	46.5424	1.3615	1.2165	1.1500	1.0578	0.9992
	5.0000	47.8888	1.3253	1.2780	1.1910	1.0731	0.9985
	6.0000	49.3256	1.2879	1.3433	1.2337	1.0888	0.9974
	7.0000	50.8751	1.2490	1.4131	1.2786	1.1052	0.9958
	8.0000	52.5709	1.2079	1.4887	1.3263	1.1224	0.9936
	9.0000	54.4692	1.1637	1.5718	1.3776	1.1410	0.9907
	10.0000	56.6778	1.1144	1.6662	1.4345	1.1615	0.9866
	11.0000	59.4639	1.0555	1.7807	1.5017	1.1858	0.9807
*	11.6933	62.2568	1.0000	1.8895	1.5637	1.2083	0.9743
m	12.1127	66.5888	0.9213	2.0439	1.6489	1.2396	0.9638
	12.0000	68.7937	0.8849	2.1149	1.6869	1.2537	0.9586
	11.0000	73.4371	0.8171	2.2450	1.7550	1.2792	0.9483
	10.0000	75.9957	0.7854	2.3046	1.7855	1.2908	0.9433
	9.0000	77.9981	0.7638	2.3448	1.8058	1.2985	0.9398
	8.0000	79.7124	0.7476	2.3746	1.8207	1.3042	0.9372
	7.0000	81.2477	0.7350	2.3976	1.8321	1.3086	0.9352
	6.0000	82.6618	0.7250	2.4155	1.8410	1.3121	0.9336
	5.0000	83.9899	0.7172	2.4296	1.8479	1.3147	0.9324
	4.0000	85.2559	0.7112	2.4404	1.8533	1.3168	0.9314
	3.0000	86.4773	0.7067	2.4484	1.8572	1.3183	0.9307
	2.0000	87.6678	0.7035	2.4540	1.8599	1.3194	0.9302
	1.0000	88.8388	0.7017	2.4573	1.8615	1.3200	0.9299
	0.0000	90.0000	0.7011	2.4583	1.8621	1.3202	0.9298

* DENOTES $M_2 = 1$, m DENOTES θ_{max}

OBLIQUE SHOCKS IN A PERFECT GAS

$$\gamma = 7/5 \qquad M_1 = 1.60$$

	θ	β	M_1	p_{21}	ρ_{21}	T_{21}	p^o_{21}
	0.0000	38.6822	1.6000	1.0000	1.0000	1.0000	1.0000
	1.0000	39.6841	1.5662	1.0512	1.0363	1.0144	1.0000
	2.0000	40.7236	1.5323	1.1046	1.0736	1.0289	0.9999
	3.0000	41.8043	1.4982	1.1604	1.1120	1.0435	0.9997
	4.0000	42.9305	1.4638	1.2189	1.1516	1.0584	0.9992
	5.0000	44.1077	1.4289	1.2802	1.1924	1.0736	0.9985
	6.0000	45.3433	1.3934	1.3446	1.2346	1.0891	0.9974
	7.0000	46.6465	1.3570	1.4124	1.2782	1.1050	0.9958
	8.0000	48.0297	1.3196	1.4843	1.3235	1.1215	0.9938
	9.0000	49.5105	1.2806	1.5608	1.3709	1.1386	0.9911
	10.0000	51.1147	1.2397	1.6430	1.4206	1.1565	0.9877
	11.0000	52.8833	1.1961	1.7324	1.4736	1.1756	0.9833
	12.0000	54.8882	1.1483	1.8319	1.5311	1.1965	0.9778
	13.0000	57.2819	1.0936	1.9475	1.5961	1.2202	0.9705
	14.0000	60.5355	1.0232	2.0974	1.6776	1.2502	0.9599
*	14.2428	61.6581	1.0000	2.1469	1.7039	1.2600	0.9561
m	14.6515	65.8278	0.9188	2.3192	1.7929	1.2936	0.9420
	14.0000	70.8964	0.8320	2.5001	1.8824	1.3281	0.9260
	13.0000	73.8201	0.7886	2.5881	1.9246	1.3447	0.9178
	12.0000	75.9008	0.7611	2.6428	1.9504	1.3550	0.9126
	11.0000	77.6106	0.7409	2.6825	1.9689	1.3624	0.9087
	10.0000	79.1022	0.7250	2.7132	1.9831	1.3682	0.9057
	9.0000	80.4487	0.7122	2.7378	1.9944	1.3727	0.9033
	8.0000	81.6919	0.7018	2.7576	2.0035	1.3764	0.9014
	7.0000	82.8585	0.6932	2.7738	2.0109	1.3794	0.8998
	6.0000	83.9671	0.6862	2.7870	2.0168	1.3819	0.8985
	5.0000	85.0314	0.6805	2.7976	2.0216	1.3838	0.8974
	4.0000	86.0618	0.6761	2.8059	2.0254	1.3854	0.8966
	3.0000	87.0670	0.6727	2.8122	2.0282	1.3865	0.8960
	2.0000	88.0544	0.6703	2.8166	2.0302	1.3873	0.8955
	1.0000	89.0302	0.6689	2.8191	2.0314	1.3878	0.8953
	0.0000	90.0000	0.6684	2.8200	2.0317	1.3880	0.8952

* DENOTES $M_2 = 1$, m DENOTES θ_{max}

OBLIQUE SHOCKS IN A PERFECT GAS

$\gamma = 7/5 \quad M_1 = 1.80$

	θ	β	M_1	p_{21}	ρ_{21}	T_{21}	p^{o}_{21}
	0.0000	33.7490	1.8000	1.0000	1.0000	1.0000	1.0000
	1.0000	34.6297	1.7655	1.0540	1.0383	1.0151	1.0000
	2.0000	35.5379	1.7312	1.1104	1.0776	1.0304	0.9999
	3.0000	36.4750	1.6968	1.1692	1.1180	1.0458	0.9996
	4.0000	37.4428	1.6625	1.2305	1.1594	1.0613	0.9991
	5.0000	38.4436	1.6280	1.2946	1.2019	1.0771	0.9982
	6.0000	39.4801	1.5932	1.3614	1.2455	1.0931	0.9970
	7.0000	40.5553	1.5581	1.4313	1.2902	1.1094	0.9953
	8.0000	41.6730	1.5225	1.5043	1.3360	1.1260	0.9931
	9.0000	42.8381	1.4864	1.5808	1.3831	1.1430	0.9903
	10.0000	44.0563	1.4494	1.6611	1.4315	1.1604	0.9868
	11.0000	45.3353	1.4116	1.7455	1.4812	1.1784	0.9826
	12.0000	46.6855	1.3726	1.8345	1.5326	1.1970	0.9777
	13.0000	48.1206	1.3320	1.9288	1.5857	1.2164	0.9718
	14.0000	49.6606	1.2896	2.0295	1.6410	1.2367	0.9649
	15.0000	51.3359	1.2445	2.1379	1.6992	1.2582	0.9568
	16.0000	53.1969	1.1958	2.2568	1.7611	1.2815	0.9473
	17.0000	55.3391	1.1415	2.3907	1.8287	1.3073	0.9358
	18.0000	57.9936	1.0766	2.5515	1.9071	1.3379	0.9212
*	18.8388	61.2895	1.0000	2.7410	1.9959	1.3733	0.9030
m	19.1833	64.9872	0.9195	2.9376	2.0839	1.4096	0.8834
	19.0000	67.5826	0.8669	3.0636	2.1384	1.4327	0.8704
	18.0000	71.4248	0.7956	3.2298	2.2079	1.4628	0.8531
	17.0000	73.6237	0.7589	3.3129	2.2418	1.4778	0.8444
	16.0000	75.3243	0.7327	3.3707	2.2650	1.4882	0.8383
	15.0000	76.7578	0.7124	3.4150	2.2825	1.4961	0.8336
	14.0000	78.0201	0.6958	3.4505	2.2965	1.5025	0.8299
	13.0000	79.1621	0.6820	3.4797	2.3079	1.5077	0.8268
	12.0000	80.2149	0.6703	3.5042	2.3174	1.5121	0.8242
	11.0000	81.1989	0.6604	3.5248	2.3254	1.5158	0.8220
	10.0000	82.1284	0.6518	3.5424	2.3322	1.5189	0.8202
	9.0000	83.0143	0.6444	3.5574	2.3379	1.5216	0.8186
	8.0000	83.8647	0.6381	3.5702	2.3428	1.5239	0.8172
	7.0000	84.6865	0.6328	3.5809	2.3469	1.5258	0.8161
	6.0000	85.4846	0.6283	3.5899	2.3503	1.5274	0.8152
	5.0000	86.2639	0.6246	3.5973	2.3531	1.5287	0.8144
	4.0000	87.0281	0.6216	3.6032	2.3554	1.5298	0.8138
	3.0000	87.7809	0.6194	3.6077	2.3571	1.5306	0.8133
	2.0000	88.5253	0.6178	3.6108	2.3583	1.5311	0.8129
	1.0000	89.2641	0.6168	3.6127	2.3590	1.5315	0.8127
	0.0000	90.0000	0.6165	3.6133	2.3592	1.5316	0.8127

* DENOTES $M_2 = 1$, m DENOTES θ_{max}

OBLIQUE SHOCKS IN A PERFECT GAS

$$\gamma = 7/5 \quad M_1 = 2.00$$

WEAK SHOCK SOLUTION

θ	β	M_1	p_{21}	ρ_{21}	T_{21}	p^o_{21}
0.0000	30.0000	2.0000	1.0000	1.0000	1.0000	1.0000
1.0000	30.8111	1.9639	1.0577	1.0409	1.0162	1.0000
2.0000	31.6460	1.9281	1.1180	1.0829	1.0324	0.9999
3.0000	32.5052	1.8924	1.1809	1.1260	1.0488	0.9995
4.0000	33.3899	1.8568	1.2467	1.1702	1.0654	0.9989
5.0000	34.3013	1.8213	1.3154	1.2156	1.0821	0.9979
6.0000	35.2406	1.7856	1.3871	1.2620	1.0991	0.9964
7.0000	36.2095	1.7498	1.4619	1.3095	1.1164	0.9944
8.0000	37.2098	1.7138	1.5400	1.3581	1.1339	0.9919
9.0000	38.2436	1.6774	1.6215	1.4077	1.1518	0.9886
10.0000	39.3136	1.6405	1.7065	1.4584	1.1701	0.9846
11.0000	40.4227	1.6032	1.7954	1.5102	1.1889	0.9799
12.0000	41.5748	1.5652	1.8884	1.5631	1.2081	0.9744
13.0000	42.7746	1.5264	1.9856	1.6171	1.2279	0.9680
14.0000	44.0282	1.4866	2.0875	1.6723	1.2483	0.9606
15.0000	45.3431	1.4457	2.1946	1.7289	1.2694	0.9524
16.0000	46.7302	1.4034	2.3075	1.7869	1.2913	0.9430
17.0000	48.2036	1.3594	2.4270	1.8467	1.3143	0.9326
18.0000	49.7846	1.3131	2.5546	1.9086	1.3384	0.9209
19.0000	51.5056	1.2638	2.6920	1.9733	1.3642	0.9078
20.0000	53.4222	1.2102	2.8428	2.0420	1.3922	0.8929
21.0000	55.6434	1.1498	3.0137	2.1170	1.4236	0.8756
22.0000	58.4553	1.0761	3.2227	2.2050	1.4615	0.8539
* 22.7060	61.4853	1.0000	3.4365	2.2910	1.5000	0.8314
m 22.9735	64.6689	0.9243	3.6457	2.3715	1.5373	0.8093

* DENOTES $M_2 = 1$, m DENOTES θ_{max}

OBLIQUE SHOCKS IN A PERFECT GAS

$$\gamma = 7/5 \qquad M_1 = 2.00$$

STRONG SHOCK SOLUTION

	θ	β	M_1	p_{21}	ρ_{21}	T_{21}	p^o_{21}
m	22.9735	64.6689	0.9243	3.6457	2.3715	1.5373	0.8093
	22.0000	70.3327	0.8017	3.9714	2.4900	1.5950	0.7750
	21.0000	72.5912	0.7580	4.0823	2.5286	1.6145	0.7635
	20.0000	74.2708	0.7278	4.1570	2.5541	1.6276	0.7558
	19.0000	75.6578	0.7044	4.2136	2.5732	1.6375	0.7499
	18.0000	76.8619	0.6854	4.2589	2.5883	1.6454	0.7453
	17.0000	77.9389	0.6694	4.2962	2.6007	1.6520	0.7415
	16.0000	78.9217	0.6558	4.3277	2.6111	1.6575	0.7383
	15.0000	79.8320	0.6440	4.3546	2.6198	1.6621	0.7355
	14.0000	80.6846	0.6337	4.3777	2.6274	1.6662	0.7332
	13.0000	81.4902	0.6247	4.3978	2.6339	1.6697	0.7312
	12.0000	82.2573	0.6168	4.4153	2.6396	1.6727	0.7294
	11.0000	82.9923	0.6098	4.4305	2.6445	1.6754	0.7279
	10.0000	83.7004	0.6037	4.4438	2.6487	1.6777	0.7265
	9.0000	84.3857	0.5984	4.4553	2.6524	1.6797	0.7254
	8.0000	85.0520	0.5937	4.4653	2.6556	1.6815	0.7244
	7.0000	85.7024	0.5897	4.4738	2.6583	1.6829	0.7235
	6.0000	86.3393	0.5864	4.4810	2.6606	1.6842	0.7228
	5.0000	86.9654	0.5836	4.4869	2.6625	1.6852	0.7222
	4.0000	87.5825	0.5813	4.4917	2.6640	1.6861	0.7217
	3.0000	88.1928	0.5796	4.4954	2.6652	1.6867	0.7213
	2.0000	88.7980	0.5783	4.4979	2.6660	1.6871	0.7211
	1.0000	89.4000	0.5776	4.4995	2.6665	1.6874	0.7209
	0.0000	90.0000	0.5774	4.5000	2.6667	1.6875	0.7209

* DENOTES $M_2 = 1$, m DENOTES θ_{max}

OBLIQUE SHOCKS IN A PERFECT GAS

$$\gamma = 7/5 \qquad M_1 = 2.20$$

WEAK SHOCK SOLUTION

	θ	β	M_1	P_{21}	ρ_{21}	T_{21}	P^o_{21}
	0.0000	27.0357	2.2000	1.0000	1.0000	1.0000	1.0000
	1.0000	27.8025	2.1617	1.0618	1.0437	1.0173	1.0000
	2.0000	28.5915	2.1237	1.1265	1.0888	1.0347	0.9998
	3.0000	29.4031	2.0860	1.1944	1.1351	1.0522	0.9994
	4.0000	30.2379	2.0486	1.2654	1.1826	1.0700	0.9987
	5.0000	31.0968	2.0112	1.3396	1.2314	1.0879	0.9975
	6.0000	31.9806	1.9738	1.4173	1.2813	1.1061	0.9957
	7.0000	32.8902	1.9363	1.4984	1.3324	1.1246	0.9933
	8.0000	33.8267	1.8987	1.5832	1.3845	1.1435	0.9902
	9.0000	34.7911	1.8609	1.6717	1.4378	1.1627	0.9863
	10.0000	35.7852	1.8228	1.7641	1.4921	1.1823	0.9816
	11.0000	36.8103	1.7843	1.8605	1.5474	1.2024	0.9761
	12.0000	37.8686	1.7454	1.9611	1.6036	1.2229	0.9696
	13.0000	38.9624	1.7059	2.0660	1.6608	1.2440	0.9622
	14.0000	40.0945	1.6658	2.1756	1.7190	1.2656	0.9539
	15.0000	41.2684	1.6249	2.2899	1.7780	1.2879	0.9445
	16.0000	42.4887	1.5831	2.4095	1.8380	1.3109	0.9342
	17.0000	43.7607	1.5403	2.5346	1.8990	1.3347	0.9228
	18.0000	45.0920	1.4963	2.6657	1.9611	1.3593	0.9104
	19.0000	46.4923	1.4508	2.8037	2.0244	1.3849	0.8968
	20.0000	47.9750	1.4035	2.9493	2.0891	1.4118	0.8821
	21.0000	49.5598	1.3539	3.1041	2.1556	1.4400	0.8662
	22.0000	51.2766	1.3013	3.2703	2.2245	1.4701	0.8489
	23.0000	53.1760	1.2443	3.4515	2.2969	1.5027	0.8298
	24.0000	55.3549	1.1806	3.6551	2.3750	1.5390	0.8083
	25.0000	58.0547	1.1041	3.8992	2.4644	1.5822	0.7826
*	25.9015	61.9076	1.0000	4.2279	2.5780	1.6400	0.7485
m	26.1028	64.6203	0.9305	4.4426	2.6484	1.6775	0.7266

* DENOTES $M_2 = 1$, m DENOTES θ_{max}

OBLIQUE SHOCKS IN A PERFECT GAS

$$\gamma = 7/5 \qquad M_1 = 2.20$$

STRONG SHOCK SOLUTION

	θ	β	M_1	P_{21}	ρ_{21}	T_{21}	P^o_{21}
m	26.1028	64.6203	0.9305	4.4426	2.6484	1.6775	0.7266
	26.0000	66.4832	0.8848	4.5810	2.6922	1.7016	0.7128
	25.0000	70.4850	0.7928	4.8499	2.7742	1.7482	0.6865
	24.0000	72.5600	0.7439	4.9728	2.8103	1.7695	0.6747
	23.0000	74.1245	0.7179	5.0575	2.8347	1.7841	0.6667
	22.0000	75.4204	0.6936	5.1222	2.8531	1.7953	0.6607
	21.0000	76.5451	0.6736	5.1743	2.8678	1.8043	0.6558
	20.0000	77.5494	0.6567	5.2175	2.8799	1.8117	0.6519
	19.0000	78.4637	0.6422	5.2542	2.8900	1.8180	0.6485
	18.0000	79.3081	0.6296	5.2856	2.8987	1.8234	0.6456
	17.0000	80.0964	0.6185	5.3130	2.9062	1.8281	0.6431
	16.0000	80.8389	0.6086	5.3369	2.9127	1.8323	0.6410
	15.0000	81.5434	0.5999	5.3579	2.9184	1.8359	0.6391
	14.0000	82.2157	0.5921	5.3764	2.9235	1.8391	0.6374
	13.0000	82.8610	0.5851	5.3928	2.9279	1.8419	0.6359
	12.0000	83.4829	0.5789	5.4073	2.9318	1.8444	0.6346
	11.0000	84.0849	0.5734	5.4200	2.9352	1.8466	0.6335
	10.0000	84.6699	0.5686	5.4313	2.9382	1.8485	0.6325
	9.0000	85.2402	0.5643	5.4411	2.9409	1.8502	0.6316
	8.0000	85.7979	0.5605	5.4497	2.9431	1.8517	0.6308
	7.0000	86.3449	0.5573	5.4570	2.9451	1.8529	0.6302
	6.0000	86.8828	0.5545	5.4633	2.9468	1.8540	0.6296
	5.0000	87.4134	0.5522	5.4685	2.9482	1.8549	0.6292
	4.0000	87.9378	0.5503	5.4727	2.9493	1.8556	0.6288
	3.0000	88.4575	0.5489	5.4759	2.9501	1.8562	0.6285
	2.0000	88.9736	0.5479	5.4782	2.9507	1.8565	0.6283
	1.0000	89.4875	0.5473	5.4795	2.9511	1.8568	0.6282
	0.0000	90.0000	0.5471	5.4800	2.9512	1.8569	0.6281

* DENOTES $M_2 = 1$, m DENOTES θ_{max}

OBLIQUE SHOCKS IN A PERFECT GAS

$$\gamma = 7/5 \qquad M_1 = 2.40$$

WEAK SHOCK SOLUTION

θ	β	M_1	P_{21}	ρ_{21}	T_{21}	P^o_{21}
0.0000	24.6243	2.4000	1.0000	1.0000	1.0000	1.0000
1.0000	25.3607	2.3590	1.0661	1.0468	1.0185	1.0000
2.0000	26.1189	2.3184	1.1357	1.0951	1.0371	0.9998
3.0000	26.9992	2.2783	1.2088	1.1448	1.0559	0.9993
4.0000	27.7019	2.2383	1.2856	1.1960	1.0749	0.9984
5.0000	28.5276	2.1986	1.3661	1.2485	1.0942	0.9969
6.0000	29.3769	2.1589	1.4504	1.3023	1.1138	0.9948
7.0000	30.2504	2.1192	1.5388	1.3574	1.1337	0.9919
8.0000	31.1487	2.0795	1.6313	1.4136	1.1540	0.9882
9.0000	32.0725	2.0395	1.7281	1.4710	1.1747	0.9836
10.0000	33.0228	1.9994	1.8291	1.5295	1.1959	0.9780
11.0000	34.0006	1.9589	1.9347	1.5890	1.2176	0.9714
12.0000	35.0071	1.9181	2.0449	1.6494	1.2398	0.9638
13.0000	36.0434	1.8768	2.1599	1.7107	1.2626	0.9551
14.0000	37.1115	1.8350	2.2798	1.7728	1.2859	0.9454
15.0000	38.2133	1.7927	2.4048	1.8357	1.3100	0.9346
16.0000	39.3511	1.7497	2.5351	1.8993	1.3348	0.9227
17.0000	40.5280	1.7059	2.6710	1.9635	1.3603	0.9098
18.0000	41.7478	1.6613	2.8127	2.0285	1.3866	0.8959
19.0000	43.0149	1.6157	2.9607	2.0941	1.4139	0.8810
20.0000	44.3358	1.5689	3.1154	2.1604	1.4421	0.8651
21.0000	45.7182	1.5207	3.2776	2.2275	1.4714	0.8481
22.0000	47.1732	1.4709	3.4480	2.2955	1.5021	0.8302
23.0000	48.7163	1.4190	3.6280	2.3648	1.5342	0.8111
24.0000	50.3701	1.3644	3.8195	2.4357	1.5682	0.7909
25.0000	52.1712	1.3062	4.0257	2.5090	1.6045	0.7694
26.0000	54.1835	1.2426	4.2521	2.5861	1.6442	0.7460
27.0000	56.5414	1.1702	4.5107	2.6700	1.6894	0.7198
28.0000	59.6545	1.0779	4.8381	2.7707	1.7462	0.6876
* 28.5313	62.3995	1.0000	5.1109	2.8499	1.7933	0.6617
m 28.6814	64.7095	0.9370	5.3269	2.9100	1.8305	0.6419

* DENOTES $M_2 = 1$, m DENOTES θ_{max}

OBLIQUE SHOCKS IN A PERFECT GAS

$$\gamma = 7/5 \qquad M_1 = 2.40$$

STRONG SHOCK SOLUTION

θ	β	M_1	p_{21}	ρ_{21}	T_{21}	p^o_{21}
m 28.6814	64.7095	0.9370	5.3269	2.9100	1.8305	0.6419
28.0000	69.2918	0.8201	5.7131	3.0119	1.8968	0.6078
27.0000	71.7188	0.7631	5.8921	3.0569	1.9275	0.5926
26.0000	73.4000	0.7260	6.0049	3.0845	1.9468	0.5833
25.0000	74.7449	0.6979	6.0881	3.1046	1.9610	0.5765
24.0000	75.8887	0.6751	6.1539	3.1203	1.9722	0.5712
23.0000	76.8958	0.6560	6.2079	3.1330	1.9815	0.5669
22.0000	77.8027	0.6397	6.2534	3.1436	1.9892	0.5633
21.0000	78.6331	0.6254	6.2923	3.1527	1.9958	0.5602
20.0000	79.4024	0.6129	6.3260	3.1605	2.0016	0.5576
19.0000	80.1221	0.6018	6.3556	3.1673	2.0066	0.5553
18.0000	80.8008	0.5919	6.3816	3.1732	2.0111	0.5533
17.0000	81.4450	0.5831	6.4046	3.1785	2.0150	0.5515
16.0000	82.0598	0.5751	6.4251	3.1831	2.0185	0.5499
15.0000	82.6495	0.5679	6.4433	3.1872	2.0216	0.5485
14.0000	83.2174	0.5615	6.4596	3.1909	2.0244	0.5473
13.0000	83.7666	0.5557	6.4741	3.1942	2.0268	0.5461
12.0000	84.2994	0.5505	6.4870	3.1971	2.0290	0.5452
11.0000	84.8180	0.5458	6.4985	3.1997	2.0310	0.5443
10.0000	85.3242	0.5416	6.5087	3.2019	2.0327	0.5435
9.0000	85.8196	0.5380	6.5176	3.2039	2.0342	0.5428
8.0000	86.3058	0.5348	6.5254	3.2057	2.0356	0.5422
7.0000	86.7839	0.5320	6.5322	3.2072	2.0367	0.5417
6.0000	87.2553	0.5296	6.5379	3.2085	2.0377	0.5413
5.0000	87.7211	0.5276	6.5427	3.2095	2.0385	0.5409
4.0000	88.1823	0.5260	6.5466	3.2104	2.0392	0.5407
3.0000	88.6399	0.5247	6.5495	3.2111	2.0397	0.5404
2.0000	89.0948	0.5238	6.5517	3.2115	2.0400	0.5403
1.0000	89.5479	0.5233	6.5529	3.2118	2.0403	0.5402
0.0000	90.0000	0.5231	6.5533	3.2119	2.0403	0.5401

* DENOTES $M_2 = 1$, m DENOTES θ_{max}

OBLIQUE SHOCKS IN A PERFECT GAS

$\gamma = 7/5$ $M_1 = 2.60$

WEAK SHOCK SOLUTION

θ	β	M_1	P_{21}	ρ_{21}	T_{21}	p^o_{21}
0.0000	22.6199	2.6000	1.0000	1.0000	1.0000	1.0000
1.0000	23.3346	2.5559	1.0707	1.0500	1.0197	1.0000
2.0000	24.0712	2.5124	1.1453	1.1017	1.0396	0.9997
3.0000	24.8297	2.4693	1.2240	1.1551	1.0597	0.9992
4.0000	25.6108	2.4265	1.3069	1.2100	1.0801	0.9980
5.0000	26.4145	2.3840	1.3941	1.2665	1.1008	0.9963
6.0000	27.2415	2.3416	1.4858	1.3245	1.1218	0.9937
7.0000	28.0919	2.2992	1.5821	1.3839	1.1432	0.9902
8.0000	28.9663	2.2568	1.6831	1.4445	1.1651	0.9858
9.0000	29.8651	2.2143	1.7889	1.5064	1.1875	0.9803
10.0000	30.7889	2.1715	1.8998	1.5695	1.2104	0.9737
11.0000	31.7383	2.1285	2.0157	1.6336	1.2339	0.9659
12.0000	32.7140	2.0852	2.1369	1.6986	1.2580	0.9569
13.0000	33.7169	2.0415	2.2634	1.7645	1.2828	0.9467
14.0000	34.7481	1.9973	2.3954	1.8311	1.3082	0.9354
15.0000	35.8089	1.9527	2.5331	1.8983	1.3344	0.9229
16.0000	36.9005	1.9075	2.6766	1.9662	1.3613	0.9093
17.0000	38.0251	1.8617	2.8260	2.0345	1.3891	0.8946
18.0000	39.1846	1.8152	2.9817	2.1032	1.4177	0.8788
19.0000	40.3821	1.7680	3.1438	2.1723	1.4472	0.8621
20.0000	41.6208	1.7199	3.3126	2.2417	1.4777	0.8444
21.0000	42.9054	1.6708	3.4886	2.3114	1.5093	0.8259
22.0000	44.2416	1.6205	3.6723	2.3814	1.5421	0.8065
23.0000	45.6369	1.5689	3.8643	2.4519	1.5761	0.7862
24.0000	47.1018	1.5157	4.0657	2.5228	1.6116	0.7652
25.0000	48.6505	1.4604	4.2778	2.5946	1.6487	0.7434
26.0000	50.3045	1.4025	4.5027	2.6675	1.6880	0.7206
27.0000	52.0973	1.3410	4.7436	2.7423	1.7298	0.6968
28.0000	54.0873	1.2744	5.0066	2.8201	1.7753	0.6715
29.0000	56.3932	1.1993	5.3039	2.9037	1.8266	0.6440
30.0000	59.3512	1.1062	5.6705	3.0010	1.8895	0.6115
* 30.7012	62.8901	1.0000	6.0822	3.1032	1.9600	0.5770
m 30.8137	64.8656	0.9433	6.2972	3.1538	1.9967	0.5598

* DENOTES $M_2 = 1$, m DENOTES θ_{max}

OBLIQUE SHOCKS IN A PERFECT GAS

$\gamma = 7/5 \qquad M_1 = 2.60$

STRONG SHOCK SOLUTION

	θ	β	M_1	p_{21}	ρ_{21}	T_{21}	p_{21}^o
m	30.8137	64.8656	0.9433	6.2972	3.1538	1.9967	0.5598
	30.0000	69.7796	0.8111	6.7778	3.2609	2.0785	0.5235
	29.0000	72.0070	0.7560	6.9675	3.3009	2.1108	0.5100
	28.0000	73.5907	0.7189	7.0906	3.3263	2.1317	0.5014
	27.0000	74.8671	0.6905	7.1825	3.3450	2.1473	0.4951
	26.0000	75.9555	0.6673	7.2555	3.3596	2.1596	0.4901
	25.0000	76.9145	0.6479	7.3157	3.3715	2.1699	0.4861
	24.0000	77.7780	0.6311	7.3665	3.3815	2.1785	0.4828
	23.0000	78.5679	0.6165	7.4102	3.3900	2.1859	0.4799
	22.0000	79.2991	0.6035	7.4481	3.3974	2.1923	0.4774
	21.0000	79.9822	0.5920	7.4813	3.4038	2.1979	0.4753
	20.0000	80.6254	0.5817	7.5107	3.4095	2.2029	0.4734
	19.0000	81.2350	0.5724	7.5369	3.4145	2.2073	0.4717
	18.0000	81.8158	0.5641	7.5602	3.4189	2.2113	0.4702
	17.0000	82.3718	0.5565	7.5810	3.4229	2.2148	0.4689
	16.0000	82.9063	0.5497	7.5997	3.4264	2.2180	0.4677
	15.0000	83.4222	0.5435	7.6165	3.4296	2.2208	0.4666
	14.0000	83.9216	0.5378	7.6316	3.4324	2.2234	0.4657
	13.0000	84.4066	0.5328	7.6451	3.4350	2.2257	0.4648
	12.0000	84.8790	0.5282	7.6572	3.4372	2.2277	0.4641
	11.0000	85.3404	0.5241	7.6680	3.4393	2.2295	0.4634
	10.0000	85.7920	0.5204	7.6775	3.4411	2.2312	0.4628
	9.0000	86.2350	0.5171	7.6860	3.4426	2.2326	0.4622
	8.0000	86.6708	0.5143	7.6934	3.4440	2.2338	0.4618
	7.0000	87.1002	0.5118	7.6998	3.4452	2.2349	0.4614
	6.0000	87.5241	0.5096	7.7053	3.4462	2.2359	0.4610
	5.0000	87.9434	0.5079	7.7098	3.4471	2.2366	0.4608
	4.0000	88.3592	0.5064	7.7135	3.4478	2.2372	0.4605
	3.0000	88.7719	0.5053	7.7164	3.4483	2.2377	0.4603
	2.0000	89.1825	0.5045	7.7184	3.4487	2.2381	0.4602
	1.0000	89.5917	0.5040	7.7196	3.4489	2.2383	0.4601
	0.0000	90.0000	0.5039	7.7200	3.4490	2.2383	0.4601

* DENOTES $M_2 = 1$, m DENOTES θ_{max}

OBLIQUE SHOCKS IN A PERFECT GAS

$\gamma = 7/5$ $M_1 = 2.80$

WEAK SHOCK SOLUTION

θ	β	M_1	p_{21}	ρ_{21}	T_{21}	p_{21}^o
0.0000	20.9248	2.8000	1.0000	1.0000	1.0000	1.0000
1.0000	21.6233	2.7524	1.0754	1.0533	1.0210	1.0000
2.0000	22.3442	2.7056	1.1553	1.1085	1.0422	0.9997
3.0000	23.0875	2.6592	1.2398	1.1656	1.0637	0.9990
4.0000	23.8535	2.6133	1.3292	1.2246	1.0854	0.9977
5.0000	24.6424	2.5677	1.4235	1.2852	1.1076	0.9955
6.0000	25.4545	2.5222	1.5229	1.3476	1.1301	0.9925
7.0000	26.2900	2.4767	1.6277	1.4114	1.1532	0.9883
8.0000	27.1493	2.4313	1.7378	1.4768	1.1768	0.9830
9.0000	28.0325	2.3857	1.8536	1.5434	1.2009	0.9765
10.0000	28.9400	2.3399	1.9750	1.6113	1.2257	0.9687
11.0000	29.8720	2.2939	2.1023	1.6803	1.2512	0.9595
12.0000	30.8294	2.2476	2.2356	1.7502	1.2774	0.9490
13.0000	31.8122	2.2010	2.3749	1.8209	1.3043	0.9372
14.0000	32.8215	2.1540	2.5205	1.8923	1.3320	0.9241
15.0000	33.8578	2.1065	2.6724	1.9642	1.3606	0.9097
16.0000	34.9222	2.0585	2.8308	2.0366	1.3900	0.6941
17.0000	36.0160	2.0101	2.9959	2.1093	1.4203	0.8774
18.0000	37.1405	1.9610	3.1677	2.1822	1.4516	0.8596
19.0000	38.2975	1.9113	3.3464	2.2552	1.4838	0.8409
20.0000	39.4894	1.8610	3.5324	2.3283	1.5171	0.8212
21.0000	40.7190	1.8098	3.7258	2.4013	1.5516	0.8008
22.0000	41.9900	1.7578	3.9270	2.4743	1.5871	0.7797
23.0000	43.3069	1.7048	4.1365	2.5471	1.6240	0.7579
24.0000	44.6756	1.6507	4.3549	2.6199	1.6622	0.7355
25.0000	46.1043	1.5951	4.5829	2.6928	1.7019	0.7126
26.0000	47.6040	1.5379	4.8218	2.7658	1.7434	0.6892
27.0000	49.1900	1.4785	5.0732	2.8392	1.7868	0.6653
28.0000	50.8857	1.4163	5.3397	2.9135	1.8327	0.6407
29.0000	52.7285	1.3502	5.6255	2.9894	1.8818	0.6154
30.0000	54.7848	1.2783	5.9385	3.0683	1.9354	0.5888
31.0000	57.1977	1.1965	6.2956	3.1535	1.9964	0.5600
32.0000	60.4314	1.0910	6.7527	3.2555	2.0743	0.5254
* 32.5024	63.3494	1.0000	7.1397	3.3363	2.1400	0.4980
m 32.5874	65.0497	0.9490	7.3524	3.3788	2.1761	0.4837

* DENOTES $M_2 = 1$, m DENOTES θ_{max}

OBLIQUE SHOCKS IN A PERFECT GAS

$$\gamma = 7/5 \qquad M_1 = 2.80$$

STRONG SHOCK SOLUTION

	θ	β	M_1	p_{21}	ρ_{21}	T_{21}	p^o_{21}
m	32.5874	65.0497	0.9490	7.3524	3.3788	2.1761	0.4837
	32.0000	69.2129	0.8306	7.8280	3.4689	2.2566	0.4535
	31.0000	71.6774	0.7656	8.0761	3.5135	2.2986	0.4386
	30.0000	73.3288	0.7243	8.2272	3.5399	2.3241	0.4299
	29.0000	74.6310	0.6934	8.3375	3.5588	2.3428	0.4236
	28.0000	75.7277	0.6684	8.4241	3.5735	2.3574	0.4188
	27.0000	76.6859	0.6475	8.4949	3.5854	2.3693	0.4149
	26.0000	77.5433	0.6296	8.5544	3.5952	2.3794	0.4117
	25.0000	78.3235	0.6140	8.6054	3.6036	2.3880	0.4089
	24.0000	79.0426	0.6002	8.6495	3.6108	2.3954	0.4066
	23.0000	79.7118	0.5879	8.6882	3.6171	2.4020	0.4045
	22.0000	80.3398	0.5769	8.7224	3.6227	2.4077	0.4027
	21.0000	80.9330	0.5670	8.7528	3.6276	2.4129	0.4011
	20.0000	81.4964	0.5580	8.7800	3.6319	2.4174	0.3997
	19.0000	82.0343	0.5499	8.8043	3.6358	2.4215	0.3984
	18.0000	82.5500	0.5425	8.8262	3.6393	2.4252	0.3973
	17.0000	83.0462	0.5358	8.8459	3.6425	2.4286	0.3963
	16.0000	83.5254	0.5297	8.8637	3.6453	2.4316	0.3954
	15.0000	83.9897	0.5241	8.8797	3.6478	2.4343	0.3946
	14.0000	84.4407	0.5191	8.8942	3.6501	2.4367	0.3938
	13.0000	84.8800	0.5145	8.9072	3.6521	2.4389	0.3932
	12.0000	85.3089	0.5103	8.9188	3.6540	2.4409	0.3926
	11.0000	85.7287	0.5066	8.9293	3.6556	2.4426	0.3920
	10.0000	86.1404	0.5033	8.9386	3.6571	2.4442	0.3916
	9.0000	86.5451	0.5003	8.9468	3.6584	2.4456	0.3911
	8.0000	86.9437	0.4977	8.9540	3.6595	2.4468	0.3908
	7.0000	87.3368	0.4954	8.9602	3.6605	2.4478	0.3905
	6.0000	87.7254	0.4935	8.9656	3.6613	2.4487	0.3902
	5.0000	88.1102	0.4918	8.9700	3.6620	2.4495	0.3900
	4.0000	88.4919	0.4905	8.9737	3.6626	2.4501	0.3898
	3.0000	88.8710	0.4895	8.9764	3.6630	2.4506	0.3896
	2.0000	89.2485	0.4887	8.9784	3.6633	2.4509	0.3895
	1.0000	89.6246	0.4883	8.9796	3.6635	2.4511	0.3895
	0.0000	90.0000	0.4882	8.9800	3.6636	2.4512	0.3895

* DENOTES $M_2 = 1$, m DENOTES θ_{max}

OBLIQUE SHOCKS IN A PERFECT GAS

$\gamma = 7/5$ $M_1 = 3.00$

WEAK SHOCK SOLUTION

θ	β	M_1	p_{21}	ρ_{21}	T_{21}	p^{o}_{21}
0.0000	19.4712	3.0000	1.0000	1.0000	1.0000	1.0000
1.0000	20.1574	2.9487	1.0802	1.0566	1.0223	1.0000
2.0000	20.8666	2.8981	1.1655	1.1155	1.0448	0.9996
3.0000	21.5987	2.8483	1.2561	1.1765	1.0677	0.9988
4.0000	22.3542	2.7988	1.3522	1.2395	1.0909	0.9972
5.0000	23.1330	2.7497	1.4540	1.3045	1.1146	0.9947
6.0000	23.9353	2.7008	1.5616	1.3713	1.1387	0.9910
7.0000	24.7613	2.6520	1.6753	1.4399	1.1635	0.9862
8.0000	25.6111	2.6031	1.7952	1.5101	1.1888	0.9799
9.0000	26.4847	2.5542	1.9216	1.5817	1.2149	0.9723
10.0000	27.3824	2.5050	2.0544	1.6546	1.2417	0.9631
11.0000	28.3043	2.4556	2.1940	1.7286	1.2692	0.9524
12.0000	29.2507	2.4060	2.3403	1.8035	1.2976	0.9402
13.0000	30.2219	2.3560	2.4936	1.8793	1.3269	0.9266
14.0000	31.2182	2.3057	2.6540	1.9556	1.3571	0.9115
15.0000	32.2400	2.2549	2.8215	2.0324	1.3882	0.8950
16.0000	33.2882	2.2037	2.9963	2.1095	1.4204	0.8773
17.0000	34.3636	2.1521	3.1786	2.1868	1.4535	0.8585
18.0000	35.4670	2.1000	3.3684	2.2640	1.4878	0.8386
19.0000	36.5996	2.0474	3.5658	2.3411	1.5231	0.8177
20.0000	37.7633	1.9941	3.7712	2.4180	1.5596	0.7960
21.0000	38.9597	1.9403	3.9846	2.4946	1.5973	0.7736
22.0000	40.1916	1.8858	4.2063	2.5707	1.6362	0.7507
23.0000	41.4619	1.8306	4.4366	2.6464	1.6765	0.7272
24.0000	42.7749	1.7744	4.6760	2.7216	1.7181	0.7034
25.0000	44.1355	1.7173	4.9249	2.7963	1.7612	0.6793
26.0000	45.5507	1.6589	5.1842	2.8706	1.8060	0.6549
27.0000	47.0298	1.5991	5.4550	2.9446	1.8526	0.6304
28.0000	48.5856	1.5374	5.7387	3.0184	1.9012	0.6056
29.0000	50.2370	1.4733	6.0377	3.0925	1.9524	0.5806
30.0000	52.0132	1.4060	6.3558	3.1673	2.0067	0.5553
31.0000	53.9629	1.3338	6.6992	3.2439	2.0652	0.5293
32.0000	56.1812	1.2541	7.0808	3.3243	2.1300	0.5021
33.0000	58.9078	1.1594	7.5331	3.4138	2.2067	0.4719
34.0000	63.6690	1.0031	8.2675	3.5469	2.3309	0.4276
* 34.0083	63.7666	1.0000	8.2817	3.5493	2.3333	0.4268
m 34.0734	65.2408	0.9540	8.4916	3.5848	2.3688	0.4151

* DENOTES $M_2 = 1$, m DENOTES θ_{max}

OBLIQUE SHOCKS IN A PERFECT GAS

$\gamma = 7/5 \qquad M_1 = 3.00$

STRONG SHOCK SOLUTION

	θ	β	M_1	p_{21}	ρ_{21}	T_{21}	p^o_{21}
m	34.0734	65.2408	0.9540	8.4916	3.5848	2.3688	0.4151
	34.0000	66.7531	0.9081	8.6976	3.6187	2.4035	0.4040
	33.0000	70.7124	0.7944	9.1877	3.6955	2.4862	0.3792
	32.0000	72.6429	0.7428	9.3988	3.7271	2.5218	0.3691
	31.0000	74.0721	0.7063	9.5426	3.7481	2.5460	0.3624
	30.0000	75.2397	0.6778	9.6518	3.7638	2.5643	0.3574
	29.0000	76.2412	0.6544	9.7394	3.7763	2.5791	0.3535
	28.0000	77.1259	0.6345	9.8121	3.7865	2.5913	0.3503
	27.0000	77.9230	0.6173	9.8737	3.7951	2.6017	0.3476
	26.0000	78.6521	0.6022	9.9268	3.8024	2.6106	0.3453
	25.0000	79.3264	0.5888	9.9731	3.8088	2.6184	0.3433
	24.0000	79.9557	0.5768	10.0139	3.8144	2.6253	0.3416
	23.0000	80.5472	0.5660	10.0501	3.8193	2.6314	0.3400
	22.0000	81.1067	0.5563	10.0824	3.8237	2.6368	0.3387
	21.0000	81.6386	0.5474	10.1113	3.8276	2.6417	0.3375
	20.0000	82.1469	0.5394	10.1373	3.8311	2.6460	0.3364
	19.0000	82.6343	0.5320	10.1608	3.8343	2.6500	0.3354
	18.0000	83.1034	0.5253	10.1819	3.8371	2.6536	0.3345
	17.0000	83.5566	0.5192	10.2011	3.8397	2.6568	0.3337
	16.0000	83.9955	0.5136	10.2184	3.8420	2.6597	0.3330
	15.0000	84.4219	0.5085	10.2341	3.8440	2.6623	0.3324
	14.0000	84.8369	0.5038	10.2483	3.8459	2.6647	0.3318
	13.0000	85.2421	0.4996	10.2611	3.8476	2.6669	0.3313
	12.0000	85.6384	0.4958	10.2726	3.8491	2.6688	0.3308
	11.0000	86.0268	0.4923	10.2829	3.8505	2.6705	0.3304
	10.0000	86.4084	0.4892	10.2921	3.8517	2.6721	0.3300
	9.0000	86.7838	0.4865	10.3003	3.8528	2.6735	0.3297
	8.0000	87.1539	0.4841	10.3074	3.8537	2.6747	0.3294
	7.0000	87.5193	0.4819	10.3137	3.8546	2.6757	0.3291
	6.0000	87.8808	0.4801	10.3190	3.8553	2.6766	0.3289
	5.0000	88.2390	0.4786	10.3234	3.8558	2.6773	0.3287
	4.0000	88.5945	0.4774	10.3270	3.8563	2.6779	0.3286
	3.0000	88.9477	0.4764	10.3298	3.8567	2.6784	0.3285
	2.0000	89.2994	0.4757	10.3318	3.8569	2.6787	0.3284
	1.0000	89.6500	0.4753	10.3329	3.8571	2.6789	0.3284
	0.0000	90.0000	0.4752	10.3333	3.8571	2.6790	0.3283

* DENOTES $M_2 = 1$, m DENOTES θ_{max}

OBLIQUE SHOCKS IN A PERFECT GAS

$$\gamma = 7/5 \qquad M_1 = 3.50$$

WEAK SHOCK SOLUTION

θ	β	M_1	p_{21}	ρ_{21}	T_{21}	p^{o}_{21}
0.0000	16.6015	3.5000	1.0000	1.0000	1.0000	1.0000
1.0000	17.2672	3.4379	1.0925	1.0652	1.0256	0.9999
2.0000	17.9580	3.3769	1.1919	1.1334	1.0516	0.9994
3.0000	18.6738	3.3168	1.2984	1.2045	1.0780	0.9982
4.0000	19.4149	3.2574	1.4125	1.2782	1.1050	0.9958
5.0000	20.1810	3.1984	1.5343	1.3546	1.1327	0.9921
6.0000	20.9721	3.1396	1.6641	1.4333	1.1611	0.9867
7.0000	21.7882	3.0809	1.8023	1.5142	1.1903	0.9795
8.0000	22.6289	3.0222	1.9491	1.5970	1.2205	0.9704
9.0000	23.4942	2.9634	2.1047	1.6815	1.2517	0.9594
10.0000	24.3838	2.9044	2.2692	1.7674	1.2839	0.9463
11.0000	25.2975	2.8452	2.4430	1.8546	1.3173	0.9312
12.0000	26.2352	2.7857	2.6261	1.9426	1.3519	0.9142
13.0000	27.1967	2.7259	2.8187	2.0312	1.3877	0.8953
14.0000	28.1821	2.6657	3.0210	2.1202	1.4249	0.8748
15.0000	29.1912	2.6053	3.2330	2.2092	1.4634	0.8528
16.0000	30.2243	2.5446	3.4548	2.2982	1.5033	0.8294
17.0000	31.2815	2.4835	3.6866	2.3867	1.5446	0.8049
18.0000	32.3631	2.4222	3.9283	2.4747	1.5874	0.7795
19.0000	33.4697	2.3606	4.1801	2.5619	1.6316	0.7534
20.0000	34.6017	2.2986	4.4420	2.6482	1.6774	0.7267
21.0000	35.7604	2.2364	4.7142	2.7333	1.7247	0.6997
22.0000	36.9469	2.1739	4.9968	2.8173	1.7736	0.6725
23.0000	38.1627	2.1110	5.2898	2.8999	1.8242	0.6452
24.0000	39.4098	2.0478	5.5936	2.9811	1.8764	0.6181
25.0000	40.6906	1.9841	5.9083	3.0609	1.9303	0.5913
26.0000	42.0085	1.9199	6.2343	3.1392	1.9860	0.5648
27.0000	43.3674	1.8551	6.5722	3.2161	2.0435	0.5387
28.0000	44.7729	1.7894	6.9225	3.2916	2.1031	0.5131
29.0000	46.2322	1.7228	7.2864	3.3657	2.1649	0.4881
30.0000	47.7549	1.6549	7.6653	3.4388	2.2291	0.4635
31.0000	49.3547	1.5852	8.0612	3.5109	2.2961	0.4395
32.0000	51.0517	1.5131	8.4775	3.5825	2.3664	0.4159
33.0000	52.8775	1.4377	8.9194	3.6541	2.4410	0.3925
34.0000	54.8868	1.3570	9.3966	3.7268	2.5214	0.3692
35.0000	57.1894	1.2676	9.9287	3.8027	2.6110	0.3452
36.0000	60.0890	1.1594	10.5713	3.8879	2.7190	0.3189
* 36.8317	64.6230	1.0000	11.5000	4.0000	2.8750	0.2854
m 36.8670	65.6886	0.9643	11.7026	4.0229	2.9090	0.2787

* DENOTES $M_2 = 1$, m DENOTES θ_{max}

OBLIQUE SHOCKS IN A PERFECT GAS

$$\gamma = 7/5 \qquad M_1 = 3.50$$

STRONG SHOCK SOLUTION

	θ	β	M_1	P_{21}	ρ_{21}	T_{21}	P^o_{21}
m	36.8670	65.6886	0.9643	11.7026	4.0229	2.9090	0.2787
	36.0000	70.5456	0.8104	12.5397	4.1122	3.0494	0.2532
	35.0000	72.5932	0.7504	12.8460	4.1428	3.1008	0.2447
	34.0000	74.0491	0.7097	13.0457	4.1623	3.1342	0.2393
	33.0000	75.2168	0.6784	13.1945	4.1766	3.1592	0.2354
	32.0000	76.2070	0.6529	13.3126	4.1877	3.1790	0.2324
	31.0000	77.0744	0.6314	13.4099	4.1968	3.1953	0.2300
	30.0000	77.8508	0.6128	13.4920	4.2044	3.2090	0.2279
	29.0000	78.5569	0.5965	13.5625	4.2109	3.2208	0.2262
	28.0000	79.2068	0.5820	13.6238	4.2165	3.2311	0.2247
	27.0000	79.8104	0.5691	13.6777	4.2213	3.2401	0.2234
	26.0000	80.3756	0.5574	13.7255	4.2256	3.2481	0.2222
	25.0000	80.9082	0.5469	13.7681	4.2295	3.2553	0.2212
	24.0000	81.4126	0.5373	13.8064	4.2329	3.2617	0.2203
	23.0000	81.8929	0.5285	13.8408	4.2360	3.2674	0.2195
	22.0000	82.3520	0.5205	13.8719	4.2387	3.2727	0.2188
	21.0000	82.7924	0.5132	13.9000	4.2412	3.2774	0.2181
	20.0000	83.2164	0.5065	13.9256	4.2435	3.2817	0.2175
	19.0000	83.6258	0.5003	13.9488	4.2455	3.2855	0.2170
	18.0000	84.0221	0.4946	13.9700	4.2474	3.2891	0.2165
	17.0000	84.4068	0.4894	13.9892	4.2491	3.2923	0.2160
	16.0000	84.7811	0.4846	14.0068	4.2506	3.2952	0.2156
	15.0000	85.1461	0.4802	14.0227	4.2520	3.2979	0.2153
	14.0000	85.5027	0.4762	14.0371	4.2532	3.3003	0.2149
	13.0000	85.8518	0.4725	14.0502	4.2544	3.3025	0.2146
	12.0000	86.1942	0.4692	14.0620	4.2554	3.3045	0.2144
	11.0000	86.5307	0.4662	14.0727	4.2563	3.3063	0.2141
	10.0000	86.8618	0.4635	14.0822	4.2572	3.3079	0.2139
	9.0000	87.1882	0.4611	14.0906	4.2579	3.3093	0.2137
	8.0000	87.5106	0.4590	14.0980	4.2585	3.3105	0.2136
	7.0000	87.8294	0.4571	14.1045	4.2591	3.3116	0.2134
	6.0000	88.1450	0.4555	14.1100	4.2596	3.3125	0.2133
	5.0000	88.4581	0.4542	14.1147	4.2600	3.3133	0.2132
	4.0000	88.7690	0.4531	14.1184	4.2603	3.3139	0.2131
	3.0000	89.0783	0.4522	14.1213	4.2606	3.3144	0.2130
	2.0000	89.3862	0.4516	14.1234	4.2607	3.3148	0.2130
	1.0000	89.6933	0.4513	14.1246	4.2608	3.3150	0.2130
	0.0000	90.0000	0.4512	14.1250	4.2609	3.3150	0.2129

* DENOTES $M_2 = 1$, m DENOTES θ_{max}

OBLIQUE SHOCKS IN A PERFECT GAS

$\gamma = 7/5$ $\quad M_1 = 4.00$

WEAK SHOCK SOLUTION

θ	β	M_1	p_{21}	ρ_{21}	T_{21}	p^o_{21}
0.0000	14.4775	4.0000	1.0000	1.0000	1.0000	1.0000
1.0000	15.1310	3.9253	1.1052	1.0740	1.0290	0.9999
2.0000	15.8123	3.8521	1.2193	1.1519	1.0585	0.9992
3.0000	16.5213	3.7801	1.3428	1.2335	1.0887	0.9974
4.0000	17.2575	3.7089	1.4762	1.3185	1.1196	0.9940
5.0000	18.0211	3.6383	1.6199	1.4067	1.1515	0.9887
6.0000	18.8115	3.5679	1.7742	1.4980	1.1844	0.9811
7.0000	19.6285	3.4976	1.9397	1.5918	1.2186	0.9711
8.0000	20.4715	3.4273	2.1166	1.6879	1.2540	0.9585
9.0000	21.3401	3.3568	2.3053	1.7858	1.2909	0.9432
10.0000	22.2338	3.2861	2.5060	1.8852	1.3293	0.9254
11.0000	23.1522	3.2151	2.7189	1.9858	1.3692	0.9052
12.0000	24.0947	3.1439	2.9444	2.0869	1.4109	0.8827
13.0000	25.0608	3.0725	3.1825	2.1884	1.4543	0.8581
14.0000	26.0502	3.0009	3.4334	2.2898	1.4994	0.8317
15.0000	27.0625	2.9290	3.6972	2.3907	1.5465	0.8038
16.0000	28.0976	2.8571	3.9739	2.4908	1.5954	0.7748
17.0000	29.1551	2.7850	4.2637	2.5899	1.6463	0.7448
18.0000	30.2352	2.7129	4.5665	2.6877	1.6991	0.7142
19.0000	31.3380	2.6407	4.8825	2.7838	1.7539	0.6834
20.0000	32.4635	2.5686	5.2115	2.8782	1.8107	0.6524
21.0000	33.6124	2.4966	5.5536	2.9706	1.8695	0.6216
22.0000	34.7854	2.4246	5.9089	3.0610	1.9304	0.5912
23.0000	35.9831	2.3527	6.2773	3.1492	1.9933	0.5614
24.0000	37.2069	2.2809	6.6589	3.2352	2.0583	0.5323
25.0000	38.4585	2.2091	7.0539	3.3188	2.1254	0.5039
26.0000	39.7395	2.1374	7.4624	3.4002	2.1947	0.4765
27.0000	41.0529	2.0656	7.8848	3.4793	2.2662	0.4500
28.0000	42.4016	1.9936	8.3213	3.5561	2.3400	0.4245
29.0000	43.7900	1.9213	8.7726	3.6307	2.4162	0.4001
30.0000	45.2236	1.8485	9.2395	3.7033	2.4949	0.3767
31.0000	46.7095	1.7751	9.7233	3.7740	2.5764	0.3542
32.0000	48.2580	1.7006	10.2258	3.8429	2.6609	0.3327
33.0000	49.8827	1.6245	10.7497	3.9104	2.7490	0.3121
34.0000	51.6046	1.5463	11.2994	3.9768	2.8413	0.2922
35.0000	53.4563	1.4646	11.8818	4.0427	2.9391	0.2730
36.0000	55.4949	1.3776	12.5099	4.1091	3.0444	0.2541
37.0000	57.8380	1.2813	13.2106	4.1781	3.1619	0.2350
38.0000	60.8258	1.1637	14.0644	4.2556	3.3049	0.2143
* 38.7533	65.2565	1.0000	15.2298	4.3514	3.5000	0.1899
m 38.7738	66.0589	0.9717	15.4261	4.3665	3.5329	0.1861

* DENOTES $M_2 = 1$, m DENOTES θ_{max}

OBLIQUE SHOCKS IN A PERFECT GAS

$\gamma = 7/5 \qquad M_1 = 4.00$

STRONG SHOCK SOLUTION

	θ	β	M_1	P_{21}	ρ_{21}	T_{21}	P^o_{21}
m	38.7738	66.0589	0.9717	15.4261	4.3665	3.5329	0.1861
	38.0000	70.6022	0.8195	16.4409	4.4404	3.7026	0.1683
	37.0000	72.7022	0.7542	16.8497	4.4683	3.7710	0.1618
	36.0000	74.1615	0.7109	17.1095	4.4855	3.8144	0.1578
	35.0000	75.3205	0.6779	17.3013	4.4979	3.8465	0.1550
	34.0000	76.2970	0.6511	17.4525	4.5076	3.8718	0.1528
	33.0000	77.1484	0.6285	17.5765	4.5155	3.8925	0.1511
	32.0000	77.9078	0.6090	17.6808	4.5220	3.9099	0.1496
	31.0000	78.5961	0.5920	17.7702	4.5276	3.9249	0.1484
	30.0000	79.2277	0.5769	17.8479	4.5324	3.9379	0.1473
	29.0000	79.8129	0.5634	17.9161	4.5366	3.9493	0.1464
	28.0000	80.3596	0.5512	17.9765	4.5402	3.9594	0.1455
	27.0000	80.8734	0.5402	18.0304	4.5435	3.9684	0.1448
	26.0000	81.3592	0.5302	18.0787	4.5464	3.9764	0.1442
	25.0000	81.8206	0.5210	18.1222	4.5491	3.9837	0.1436
	24.0000	82.2607	0.5126	18.1615	4.5514	3.9903	0.1431
	23.0000	82.6821	0.5049	18.1971	4.5535	3.9963	0.1426
	22.0000	83.0871	0.4978	18.2296	4.5555	4.0017	0.1422
	21.0000	83.4773	0.4913	18.2591	4.5572	4.0066	0.1418
	20.0000	83.8543	0.4852	18.2861	4.5588	4.0111	0.1415
	19.0000	84.2196	0.4797	18.3106	4.5603	4.0152	0.1412
	18.0000	84.5743	0.4746	18.3331	4.5616	4.0190	0.1409
	17.0000	84.9195	0.4698	18.3536	4.5628	4.0224	0.1406
	16.0000	85.2562	0.4655	18.3723	4.5639	4.0255	0.1404
	15.0000	85.5852	0.4615	18.3894	4.5649	4.0284	0.1402
	14.0000	85.9073	0.4579	18.4049	4.5659	4.0310	0.1400
	13.0000	86.2231	0.4545	18.4190	4.5667	4.0333	0.1398
	12.0000	86.5333	0.4515	18.4317	4.5674	4.0355	0.1396
	11.0000	86.8385	0.4488	18.4432	4.5681	4.0374	0.1395
	10.0000	87.1393	0.4463	18.4535	4.5687	4.0391	0.1393
	9.0000	87.4361	0.4441	18.4626	4.5692	4.0406	0.1392
	8.0000	87.7293	0.4421	18.4707	4.5697	4.0420	0.1391
	7.0000	88.0195	0.4404	18.4777	4.5701	4.0431	0.1390
	6.0000	88.3072	0.4390	18.4837	4.5705	4.0441	0.1390
	5.0000	88.5926	0.4377	18.4887	4.5708	4.0450	0.1389
	4.0000	88.8763	0.4367	18.4928	4.5710	4.0457	0.1388
	3.0000	89.1584	0.4359	18.4960	4.5712	4.0462	0.1388
	2.0000	89.4396	0.4354	18.4982	4.5713	4.0466	0.1388
	1.0000	89.7200	0.4351	18.4995	4.5714	4.0468	0.1388
	0.0000	90.0000	0.4350	18.5000	4.5714	4.0469	0.1388

* DENOTES $M_2 = 1$, m DENOTES θ_{max}

OBLIQUE SHOCKS IN A PERFECT GAS

$\gamma = 7/5$ $M_1 = 4.50$

WEAK SHOCK SOLUTION

θ	β	M_1	P_{21}	ρ_{21}	T_{21}	P^o_{21}
0.0000	12.8396	4.5000	1.0000	1.0000	1.0000	1.0000
1.0000	13.4856	4.4109	1.1181	1.0830	1.0325	0.9999
2.0000	14.1622	4.3239	1.2476	1.1708	1.0656	0.9989
3.0000	14.8691	4.2383	1.3890	1.2632	1.0996	0.9964
4.0000	15.6057	4.1538	1.5431	1.3600	1.1346	0.9917
5.0000	16.3719	4.0698	1.7104	1.4607	1.1710	0.9845
6.0000	17.1670	3.9862	1.8915	1.5648	1.2087	0.9742
7.0000	17.9904	3.9025	2.0870	1.6721	1.2482	0.9607
8.0000	18.8413	3.8188	2.2973	1.7818	1.2893	0.9439
9.0000	19.7190	3.7350	2.5229	1.8934	1.3325	0.9239
10.0000	20.6227	3.6509	2.7641	2.0064	1.3776	0.9007
11.0000	21.5515	3.5665	3.0212	2.1203	1.4249	0.8748
12.0000	22.5046	3.4820	3.2945	2.2343	1.4745	0.8463
13.0000	23.4814	3.3974	3.5841	2.3481	1.5264	0.8158
14.0000	24.4812	3.3127	3.8903	2.4612	1.5807	0.7835
15.0000	25.5033	3.2281	4.2130	2.5730	1.6374	0.7500
16.0000	26.5472	3.1436	4.5525	2.6832	1.6966	0.7156
17.0000	27.6125	3.0594	4.9085	2.7915	1.7584	0.6809
18.0000	28.6991	2.9754	5.2813	2.8975	1.8227	0.6460
19.0000	29.8067	2.8918	5.6707	3.0010	1.8896	0.6114
20.0000	30.9353	2.8087	6.0767	3.1018	1.9590	0.5774
21.0000	32.0852	2.7261	6.4992	3.1998	2.0311	0.5442
22.0000	33.2566	2.6440	6.9381	3.2948	2.1058	0.5120
23.0000	34.4501	2.5625	7.3934	3.3868	2.1830	0.4810
24.0000	35.6667	2.4816	7.8651	3.4757	2.2629	0.4512
25.0000	36.9073	2.4012	8.3531	3.5615	2.3454	0.4228
26.0000	38.1733	2.3214	8.8575	3.6443	2.4305	0.3957
27.0000	39.4667	2.2420	9.3784	3.7241	2.5183	0.3700
28.0000	40.7897	2.1630	9.9160	3.8010	2.6088	0.3458
29.0000	42.1455	2.0844	10.4707	3.8750	2.7021	0.3229
30.0000	43.5377	2.0058	11.0431	3.9464	2.7983	0.3013
31.0000	44.9715	1.9273	11.6341	4.0152	2.8975	0.2810
32.0000	46.4533	1.8484	12.2448	4.0816	3.0000	0.2618
33.0000	47.9919	1.7690	12.8773	4.1459	3.1060	0.2438
34.0000	49.5999	1.6885	13.5344	4.2083	3.2161	0.2269
35.0000	51.2947	1.6063	14.2205	4.2691	3.3310	0.2108
36.0000	53.1040	1.5214	14.9430	4.3288	3.4520	0.1955
37.0000	55.0734	1.4321	15.7144	4.3882	3.5811	0.1808
38.0000	57.2913	1.3353	16.5599	4.4486	3.7225	0.1664
39.0000	59.9754	1.2231	17.5433	4.5134	3.8870	0.1515
40.0000	64.3371	1.0519	19.0273	4.6015	4.1350	0.1323
* 40.1138	65.7278	1.0000	19.4662	4.6256	4.2083	0.1273
m 40.1264	66.3528	0.9771	19.6574	4.6359	4.2403	0.1252

* DENOTES $M_2 = 1$, m DENOTES θ_{max}

OBLIQUE SHOCKS IN A PERFECT GAS

$$\gamma = 7/5 \qquad M_1 = 4.50$$

STRONG SHOCK SOLUTION

θ	β	M_1	P_{21}	ρ_{21}	T_{21}	P^o_{21}
m 40.1264	66.3528	0.9771	19.6574	4.6359	4.2403	0.1252
40.0000	68.2523	0.9093	20.2150	4.6649	4.3334	0.1193
39.0000	71.7002	0.7927	21.1292	4.7099	4.4861	0.1105
38.0000	73.4735	0.7364	21.5467	4.7294	4.5559	0.1068
37.0000	74.7837	0.6965	21.8309	4.7424	4.6033	0.1043
36.0000	75.8489	0.6653	22.0463	4.7521	4.6393	0.1025
35.0000	76.7576	0.6396	22.2186	4.7597	4.6681	0.1011
34.0000	77.5557	0.6177	22.3613	4.7659	4.6919	0.09995
33.0000	78.2711	0.5988	22.4821	4.7712	4.7121	0.09899
32.0000	78.9217	0.5821	22.5861	4.7756	4.7294	0.09818
31.0000	79.5202	0.5673	22.6767	4.7795	4.7446	0.09747
30.0000	80.0757	0.5540	22.7566	4.7829	4.7579	0.09686
29.0000	80.5951	0.5420	22.8275	4.7859	4.7698	0.09632
28.0000	81.0837	0.5311	22.8908	4.7885	4.7803	0.09584
27.0000	81.5460	0.5211	22.9477	4.7909	4.7898	0.09542
26.0000	81.9854	0.5120	22.9991	4.7931	4.7984	0.09503
25.0000	82.4044	0.5037	23.0456	4.7950	4.8062	0.09469
24.0000	82.8058	0.4959	23.0878	4.7967	4.8132	0.09437
23.0000	83.1914	0.4889	23.1263	4.7983	4.8196	0.09409
22.0000	83.5629	0.4823	23.1614	4.7998	4.8255	0.09383
21.0000	83.9219	0.4762	23.1935	4.8011	4.8309	0.09360
20.0000	84.2695	0.4707	23.2228	4.8023	4.8358	0.09339
19.0000	84.6069	0.4655	23.2496	4.8034	4.8402	0.09319
18.0000	84.9352	0.4607	23.2742	4.8044	4.8443	0.09301
17.0000	85.2553	0.4563	23.2967	4.8053	4.8481	0.09285
16.0000	85.5678	0.4523	23.3172	4.8062	4.8515	0.09270
15.0000	85.8736	0.4485	23.3360	4.8069	4.8547	0.09257
14.0000	86.1732	0.4451	23.3531	4.8076	4.8575	0.09245
13.0000	86.4675	0.4420	23.3686	4.8083	4.8601	0.09234
12.0000	86.7567	0.4391	23.3827	4.8088	4.8625	0.09224
11.0000	87.0415	0.4366	23.3954	4.8093	4.8646	0.09214
10.0000	87.3223	0.4342	23.4068	4.8098	4.8665	0.09206
9.0000	87.5997	0.4322	23.4169	4.8102	4.8682	0.09199
8.0000	87.8739	0.4303	23.4258	4.8106	4.8697	0.09193
7.0000	88.1454	0.4287	23.4336	4.8109	4.8710	0.09187
6.0000	88.4145	0.4273	23.4402	4.8112	4.8721	0.09183
5.0000	88.6817	0.4262	23.4458	4.8114	4.8730	0.09179
4.0000	88.9473	0.4252	23.4504	4.8116	4.8738	0.09175
3.0000	89.2116	0.4245	23.4539	4.8117	4.8743	0.09173
2.0000	89.4750	0.4240	23.4563	4.8118	4.8748	0.09171
1.0000	89.7377	0.4237	23.4578	4.8119	4.8750	0.09170
0.0000	90.0000	0.4236	23.4583	4.8119	4.8751	0.09170

* DENOTES $M_2 = 1$, m DENOTES θ_{max}

OBLIQUE SHOCKS IN A PERFECT GAS

$\gamma = 7/5$ $M_1 = 5.00$

WEAK SHOCK SOLUTION

θ	β	M_1	P_{21}	ρ_{21}	T_{21}	P_{21}^o
0.0000	11.5370	5.0000	1.0000	1.0000	1.0000	1.0000
1.0000	12.1784	4.8948	1.1313	1.0921	1.0359	0.9998
2.0000	12.8530	4.7923	1.2766	1.1901	1.0727	0.9985
3.0000	13.5606	4.6916	1.4369	1.2937	1.1106	0.9952
4.0000	14.3006	4.5921	1.6129	1.4025	1.1500	0.9890
5.0000	15.0724	4.4932	1.8056	1.5160	1.1910	0.9793
6.0000	15.8752	4.3946	2.0157	1.6336	1.2339	0.9659
7.0000	16.7080	4.2960	2.2441	1.7545	1.2790	0.9483
8.0000	17.5697	4.1973	2.4911	1.8780	1.3264	0.9268
9.0000	18.4594	4.0983	2.7575	2.0034	1.3764	0.9014
10.0000	19.3757	3.9992	3.0436	2.1298	1.4290	0.8725
11.0000	20.3176	3.8999	3.3498	2.2566	1.4844	0.8405
12.0000	21.2842	3.8007	3.6765	2.3830	1.5428	0.8060
13.0000	22.2743	3.7015	4.0238	2.5083	1.6042	0.7696
14.0000	23.2869	3.6026	4.3918	2.6320	1.6686	0.7318
15.0000	24.3214	3.5041	4.7807	2.7535	1.7363	0.6932
16.0000	25.3769	3.4061	5.1905	2.8723	1.8070	0.6543
17.0000	26.4530	3.3088	5.6211	2.9882	1.8811	0.6157
18.0000	27.5491	3.2123	6.0725	3.1008	1.9583	0.5778
19.0000	28.6650	3.1168	6.5446	3.2100	2.0388	0.5408
20.0000	29.8005	3.0222	7.0372	3.3154	2.1226	0.5051
21.0000	30.9557	2.9287	7.5503	3.4170	2.2096	0.4708
22.0000	32.1305	2.8363	8.0835	3.5148	2.2998	0.4382
23.0000	33.3256	2.7450	8.6369	3.6088	2.3933	0.4073
24.0000	34.5415	2.6549	9.2102	3.6989	2.4900	0.3781
25.0000	35.7790	2.5659	9.8033	3.7853	2.5899	0.3507
26.0000	37.0393	2.4780	10.4162	3.8680	2.6929	0.3250
27.0000	38.3237	2.3911	11.0487	3.9471	2.7992	0.3011
28.0000	39.6343	2.3051	11.7012	4.0227	2.9088	0.2788
29.0000	40.9733	2.2201	12.3736	4.0951	3.0216	0.2580
30.0000	42.3438	2.1357	13.0665	4.1643	3.1377	0.2388
31.0000	43.7497	2.0518	13.7804	4.2306	3.2573	0.2209
32.0000	45.1959	1.9683	14.5164	4.2940	3.3806	0.2044
33.0000	46.6890	1.8848	15.2759	4.3549	3.5077	0.1890
34.0000	48.2379	1.8010	16.0614	4.4135	3.6391	0.1747
35.0000	49.8548	1.7164	16.8762	4.4700	3.7754	0.1614
36.0000	51.5576	1.6303	17.7258	4.5248	3.9175	0.1490
37.0000	53.3734	1.5417	18.6187	4.5783	4.0667	0.1373
38.0000	55.3484	1.4489	19.5707	4.6312	4.2258	0.1262
39.0000	57.5703	1.3485	20.6122	4.6848	4.3998	0.1154
40.0000	60.2581	1.2325	21.8218	4.7420	4.6018	0.1044
41.0000	64.6520	1.0549	23.6542	4.8197	4.9078	0.09032
* 41.1094	66.0840	1.0000	24.2065	4.8413	5.0000	0.08660
m 41.1176	66.5842	0.9812	24.3938	4.8484	5.0313	0.08539

* DENOTES $M_2 = 1$, m DENOTES θ_{max}

OBLIQUE SHOCKS IN A PERFECT GAS

$$\gamma = 7/5 \qquad M_1 = 5.00$$

STRONG SHOCK SOLUTION

	θ	β	M_1	p_{21}	ρ_{21}	T_{21}	p^o_{21}
m	41.1176	66.5842	0.9812	24.3938	4.8484	5.0313	0.08539
	41.0000	68.4065	0.9142	25.0497	4.8728	5.1407	0.08132
	40.0000	71.8697	0.7937	26.1757	4.9122	5.3287	0.07494
	39.0000	73.6292	0.7361	26.6830	4.9291	5.4133	0.07229
	38.0000	74.9256	0.6955	27.0272	4.9403	5.4708	0.07057
	37.0000	75.9776	0.6637	27.2876	4.9486	5.5143	0.06930
	36.0000	76.8736	0.6375	27.4957	4.9551	5.5490	0.06831
	35.0000	77.6597	0.6153	27.6678	4.9604	5.5777	0.06751
	34.0000	78.3633	0.5960	27.8133	4.9649	5.6020	0.06684
	33.0000	79.0026	0.5791	27.9386	4.9687	5.6229	0.06627
	32.0000	79.5900	0.5640	28.0477	4.9720	5.6411	0.06578
	31.0000	80.1347	0.5505	28.1438	4.9749	5.6571	0.06536
	30.0000	80.6435	0.5383	28.2291	4.9775	5.6714	0.06498
	29.0000	81.1220	0.5272	28.3053	4.9798	5.6841	0.06465
	28.0000	81.5740	0.5170	28.3737	4.9818	5.6955	0.06435
	27.0000	82.0032	0.5077	28.4355	4.9836	5.7058	0.06408
	26.0000	82.4124	0.4992	28.4915	4.9853	5.7151	0.06384
	25.0000	82.8038	0.4914	28.5423	4.9868	5.7236	0.06363
	24.0000	83.1796	0.4841	28.5886	4.9881	5.7314	0.06343
	23.0000	83.5414	0.4774	28.6309	4.9893	5.7384	0.06325
	22.0000	83.8906	0.4712	28.6696	4.9905	5.7449	0.06309
	21.0000	84.2286	0.4655	28.7050	4.9915	5.7508	0.06294
	20.0000	84.5563	0.4602	28.7375	4.9924	5.7562	0.06280
	19.0000	84.8749	0.4553	28.7672	4.9933	5.7612	0.06268
	18.0000	85.1853	0.4507	28.7945	4.9941	5.7657	0.06256
	17.0000	85.4881	0.4466	28.8195	4.9948	5.7699	0.06246
	16.0000	85.7841	0.4427	28.8424	4.9955	5.7737	0.06236
	15.0000	86.0740	0.4391	28.8633	4.9961	5.7772	0.06228
	14.0000	86.3582	0.4359	28.8823	4.9966	5.7804	0.06220
	13.0000	86.6375	0.4329	28.8997	4.9971	5.7833	0.06213
	12.0000	86.9123	0.4302	28.9154	4.9976	5.7859	0.06206
	11.0000	87.1830	0.4277	28.9295	4.9980	5.7882	0.06200
	10.0000	87.4500	0.4255	28.9423	4.9983	5.7904	0.06195
	9.0000	87.7138	0.4235	28.9536	4.9987	5.7923	0.06191
	8.0000	87.9747	0.4217	28.9636	4.9990	5.7939	0.06187
	7.0000	88.2332	0.4202	28.9723	4.9992	5.7954	0.06183
	6.0000	88.4895	0.4188	28.9797	4.9994	5.7966	0.06180
	5.0000	88.7439	0.4177	28.9860	4.9996	5.7977	0.06177
	4.0000	88.9970	0.4168	28.9911	4.9997	5.7985	0.06175
	3.0000	89.2488	0.4161	28.9950	4.9999	5.7992	0.06174
	2.0000	89.4997	0.4156	28.9978	4.9999	5.7996	0.06173
	1.0000	89.7500	0.4153	28.9994	5.0000	5.7999	0.06172
	0.0000	90.0000	0.4152	29.0000	5.0000	5.8000	0.06172

* DENOTES $M_2 = 1$, m DENOTES θ_{max}

OBLIQUE SHOCKS IN A PERFECT GAS

$\gamma = 7/5$ $M_1 = 6.00$

WEAK SHOCK SOLUTION

θ	β	M_1	p_{21}	ρ_{21}	T_{21}	p^o_{21}
0.0000	9.5941	6.0000	1.0000	1.0000	1.0000	1.0000
1.0000	10.2305	5.8576	1.1582	1.1105	1.0430	0.9997
2.0000	10.9061	5.7193	1.3368	1.2295	1.0872	0.9975
3.0000	11.6202	5.5835	1.5373	1.3564	1.1333	0.9920
4.0000	12.3716	5.4493	1.7613	1.4904	1.1817	0.9818
5.0000	13.1595	5.3157	2.0102	1.6306	1.2328	0.9663
6.0000	13.9821	5.1823	2.2853	1.7756	1.2870	0.9449
7.0000	14.8380	5.0488	2.5877	1.9244	1.3447	0.9178
8.0000	15.7252	4.9152	2.9184	2.0755	1.4061	0.8853
9.0000	16.6419	4.7814	3.2781	2.2277	1.4715	0.8481
10.0000	17.5866	4.6478	3.6676	2.3797	1.5412	0.8069
11.0000	18.5573	4.5146	4.0873	2.5303	1.6153	0.7630
12.0000	19.5526	4.3821	4.5375	2.6785	1.6940	0.7171
13.0000	20.5708	4.2506	5.0185	2.8235	1.7774	0.6704
14.0000	21.6109	4.1204	5.5304	2.9646	1.8655	0.6237
15.0000	22.6715	3.9918	6.0733	3.1010	1.9585	0.5777
16.0000	23.7518	3.8651	6.6469	3.2325	2.0563	0.5331
17.0000	24.8507	3.7405	7.2511	3.3587	2.1589	0.4904
18.0000	25.9679	3.6180	7.8859	3.4795	2.2664	0.4499
19.0000	27.1027	3.4980	8.5508	3.5946	2.3788	0.4119
20.0000	28.2547	3.3803	9.2455	3.7042	2.4959	0.3764
21.0000	29.4240	3.2652	9.9698	3.8084	2.6179	0.3435
22.0000	30.6104	3.1526	10.7232	3.9071	2.7445	0.3131
23.0000	31.8143	3.0424	11.5053	4.0006	2.8759	0.2852
24.0000	33.0359	2.9347	12.3159	4.0891	3.0119	0.2597
25.0000	34.2759	2.8294	13.1544	4.1727	3.1525	0.2365
26.0000	35.5351	2.7264	14.0208	4.2518	3.2976	0.2153
27.0000	36.8148	2.6256	14.9145	4.3265	3.4472	0.1961
28.0000	38.1162	2.5268	15.8356	4.3971	3.6014	0.1786
29.0000	39.4412	2.4299	16.7840	4.4638	3.7600	0.1628
30.0000	40.7921	2.3347	17.7598	4.5269	3.9232	0.1485
31.0000	42.1716	2.2410	18.7634	4.5866	4.0909	0.1355
32.0000	43.5835	2.1487	19.7954	4.6432	4.2633	0.1237
33.0000	45.0322	2.0574	20.8569	4.6968	4.4407	0.1130
34.0000	46.5239	1.9668	21.9498	4.7478	4.6232	0.1033
35.0000	48.0665	1.8765	23.0769	4.7963	4.8114	0.09446
36.0000	49.6710	1.7861	24.2422	4.8427	5.0059	0.08637
37.0000	51.3529	1.6949	25.4522	4.8872	5.2079	0.07896
38.0000	53.1357	1.6018	26.7172	4.9302	5.4191	0.07212
39.0000	55.0574	1.5055	28.0552	4.9723	5.6424	0.06575
40.0000	57.1874	1.4031	29.5001	5.0141	5.8834	0.05972
41.0000	59.6810	1.2887	31.1302	5.0574	6.1554	0.05380
42.0000	63.1033	1.1406	33.2380	5.1080	6.5070	0.04729
* 42.4358	66.5725	1.0000	35.1941	5.1504	6.8333	0.04219
m 42.4398	66.9139	0.9867	35.3757	5.1541	6.8636	0.04176

* DENOTES $M_2 = 1$, m DENOTES θ_{max}

OBLIQUE SHOCKS IN A PERFECT GAS

$$\gamma = 7/5 \qquad M_1 = 6.00$$

STRONG SHOCK SOLUTION

	θ	β	M_1	p_{21}	ρ_{21}	T_{21}	p^o_{21}
m	42.4398	66.9139	0.9867	35.3757	5.1541	6.8636	0.04176
	42.0000	70.3053	0.8592	37.0632	5.1872	7.1451	0.03801
	41.0000	72.7757	0.7721	38.1506	5.2073	7.3264	0.03584
	40.0000	74.3192	0.7204	38.7652	5.2181	7.4289	0.03469
	39.0000	75.5010	0.6823	39.2007	5.2257	7.5016	0.03390
	38.0000	76.4765	0.6519	39.5366	5.2314	7.5576	0.03332
	37.0000	77.3153	0.6267	39.8082	5.2359	7.6029	0.03285
	36.0000	78.0554	0.6051	40.0342	5.2397	7.6406	0.03247
	35.0000	78.7205	0.5862	40.2265	5.2429	7.6726	0.03215
	34.0000	79.3264	0.5696	40.3926	5.2456	7.7003	0.03188
	33.0000	79.8842	0.5548	40.5377	5.2479	7.7245	0.03165
	32.0000	80.4022	0.5414	40.6658	5.2500	7.7459	0.03144
	31.0000	80.8866	0.5293	40.7797	5.2518	7.7649	0.03126
	30.0000	81.3422	0.5183	40.8816	5.2534	7.7819	0.03110
	29.0000	81.7730	0.5083	40.9733	5.2549	7.7972	0.03095
	28.0000	82.1821	0.4990	41.0562	5.2562	7.8110	0.03083
	27.0000	82.5721	0.4905	41.1314	5.2574	7.8235	0.03071
	26.0000	82.9453	0.4827	41.1998	5.2585	7.8349	0.03060
	25.0000	83.3035	0.4754	41.2622	5.2595	7.8453	0.03051
	24.0000	83.6483	0.4687	41.3193	5.2603	7.8549	0.03042
	23.0000	83.9810	0.4625	41.3715	5.2612	7.8636	0.03034
	22.0000	84.3029	0.4568	41.4194	5.2619	7.8716	0.03027
	21.0000	84.6151	0.4514	41.4634	5.2626	7.8789	0.03020
	20.0000	84.9183	0.4465	41.5038	5.2632	7.8856	0.03014
	19.0000	85.2135	0.4419	41.5409	5.2638	7.8918	0.03009
	18.0000	85.5015	0.4376	41.5749	5.2643	7.8975	0.03003
	17.0000	85.7828	0.4337	41.6062	5.2648	7.9027	0.02999
	16.0000	86.0582	0.4301	41.6348	5.2652	7.9075	0.02994
	15.0000	86.3280	0.4267	41.6611	5.2656	7.9119	0.02991
	14.0000	86.5930	0.4237	41.6850	5.2660	7.9158	0.02987
	13.0000	86.8534	0.4208	41.7068	5.2664	7.9195	0.02984
	12.0000	87.1099	0.4183	41.7266	5.2667	7.9228	0.02981
	11.0000	87.3626	0.4159	41.7444	5.2669	7.9258	0.02978
	10.0000	87.6122	0.4138	41.7604	5.2672	7.9284	0.02976
	9.0000	87.8589	0.4120	41.7747	5.2674	7.9308	0.02974
	8.0000	88.1030	0.4103	41.7873	5.2676	7.9329	0.02972
	7.0000	88.3448	0.4088	41.7983	5.2678	7.9347	0.02970
	6.0000	88.5849	0.4076	41.8077	5.2679	7.9363	0.02969
	5.0000	88.8232	0.4065	41.8156	5.2680	7.9376	0.02968
	4.0000	89.0601	0.4057	41.8220	5.2681	7.9387	0.02967
	3.0000	89.2960	0.4050	41.8270	5.2682	7.9395	0.02966
	2.0000	89.5311	0.4045	41.8305	5.2683	7.9401	0.02965
	1.0000	89.7657	0.4043	41.8326	5.2683	7.9405	0.02965
	0.0000	90.0000	0.4042	41.8333	5.2683	7.9406	0.02965

* DENOTES $M_2 = 1$, m DENOTES θ_{max}

OBLIQUE SHOCKS IN A PERFECT GAS

$\gamma = 7/5 \qquad M_1 = 8.00$

WEAK SHOCK SOLUTION

θ	β	M_1	P_{21}	ρ_{21}	T_{21}	p^o_{21}
0.0000	7.1808	8.0000	1.0000	1.0000	1.0000	1.0000
1.0000	7.8157	7.7631	1.2141	1.1484	1.0572	0.9993
2.0000	8.5015	7.5343	1.4652	1.3116	1.1171	0.9943
3.0000	9.2367	7.3098	1.7571	1.4880	1.1808	0.9820
4.0000	10.0190	7.0870	2.0933	1.6754	1.2494	0.9602
5.0000	10.8457	6.8647	2.4770	1.8712	1.3238	0.9281
6.0000	11.7135	6.6423	2.9108	2.0722	1.4047	0.8861
7.0000	12.6189	6.4202	3.3969	2.2754	1.4929	0.8355
8.0000	13.5585	6.1988	3.9371	2.4778	1.5889	0.7786
9.0000	14.5292	5.9793	4.5327	2.6770	1.6932	0.7176
10.0000	15.5281	5.7624	5.1846	2.8707	1.8060	0.6549
11.0000	16.5522	5.5492	5.8934	3.0572	1.9277	0.5925
12.0000	17.5994	5.3405	6.6595	3.2353	2.0584	0.5322
13.0000	18.6676	5.1371	7.4829	3.4041	2.1982	0.4752
14.0000	19.7550	4.9394	8.3634	3.5633	2.3471	0.4222
15.0000	20.8601	4.7478	9.3009	3.7126	2.5053	0.3737
16.0000	21.9818	4.5627	10.2948	3.8521	2.6725	0.3299
17.0000	23.1191	4.3840	11.3447	3.9821	2.8489	0.2907
18.0000	24.2713	4.2119	12.4496	4.1029	3.0343	0.2558
19.0000	25.4378	4.0461	13.6091	4.2151	3.2286	0.2250
20.0000	26.6183	3.8867	14.8223	4.3191	3.4318	0.1980
21.0000	27.8126	3.7333	16.0881	4.4154	3.6436	0.1742
22.0000	29.0207	3.5857	17.4059	4.5047	3.8640	0.1535
23.0000	30.2428	3.4437	18.7747	4.5873	4.0928	0.1354
24.0000	31.4791	3.3070	20.1934	4.6638	4.3298	0.1196
25.0000	32.7303	3.1752	21.6614	4.7347	4.5750	0.1058
26.0000	33.9970	3.0481	23.1777	4.8005	4.8282	0.09372
27.0000	35.2801	2.9254	24.7415	4.8615	5.0893	0.08320
28.0000	36.5808	2.8067	26.3522	4.9182	5.3581	0.07401
29.0000	37.9005	2.6918	28.0092	4.9709	5.6347	0.06596
30.0000	39.2410	2.5803	29.7121	5.0199	5.9188	0.05890
31.0000	40.6046	2.4718	31.4610	5.0657	6.2106	0.05270
32.0000	41.9938	2.3662	33.2561	5.1084	6.5101	0.04724
33.0000	43.4120	2.2630	35.0983	5.1484	6.8173	0.04243
34.0000	44.8636	2.1619	36.9888	5.1858	7.1327	0.03817
35.0000	46.3541	2.0624	38.9306	5.2210	7.4565	0.03439
36.0000	47.8906	1.9642	40.9272	5.2542	7.7895	0.03103
37.0000	49.4829	1.8666	42.9848	5.2855	8.1326	0.02802
38.0000	51.1446	1.7691	45.1129	5.3152	8.4875	0.02533
39.0000	52.8957	1.6706	47.3264	5.3437	8.8565	0.02289
40.0000	54.7677	1.5698	49.6505	5.3711	9.2440	0.02067
41.0000	56.8154	1.4644	52.1314	5.3979	9.6577	0.01862
42.0000	59.1509	1.3500	54.8670	5.4250	10.1138	0.01668
43.0000	62.0958	1.2135	58.1466	5.4544	10.6605	0.01470
* 43.7896	67.0870	1.0000	63.1820	5.4941	11.5000	0.01225
m 43.7908	67.2752	0.9924	63.3573	5.4954	11.5292	0.01218

* DENOTES $M_2 = 1$, m DENOTES θ_{max}

OBLIQUE SHOCKS IN A PERFECT GAS

$$\gamma = 7/5 \qquad M_1 = 8.00$$

STRONG SHOCK SOLUTION

	θ	β	M_1	p_{21}	ρ_{21}	T_{21}	p^o_{21}
m	43.7908	67.2752	0.9924	63.3573	5.4954	11.5292	0.01218
	43.0000	71.6844	0.8219	67.1264	5.5214	12.1575	0.01071
	42.0000	73.6565	0.7510	68.5876	5.5308	12.4011	0.01021
	41.0000	75.0201	0.7041	69.5114	5.5365	12.5551	0.009912
	40.0000	76.0970	0.6684	70.1892	5.5406	12.6681	0.009700
	39.0000	76.9994	0.6394	70.7213	5.5438	12.7568	0.009538
	38.0000	77.7821	0.6151	71.1558	5.5464	12.8293	0.009408
	37.0000	78.4767	0.5941	71.5203	5.5485	12.8900	0.009301
	36.0000	79.1034	0.5758	71.8318	5.5503	12.9419	0.009211
	35.0000	79.6758	0.5595	72.1018	5.5519	12.9869	0.009134
	34.0000	80.2039	0.5449	72.3385	5.5532	13.0264	0.009067
	33.0000	80.6949	0.5318	72.5479	5.5544	13.0613	0.009009
	32.0000	81.1545	0.5198	72.7345	5.5555	13.0924	0.008957
	31.0000	81.5872	0.5089	72.9017	5.5564	13.1203	0.008911
	30.0000	81.9964	0.4990	73.0525	5.5573	13.1454	0.008870
	29.0000	82.3853	0.4898	73.1889	5.5580	13.1682	0.008833
	28.0000	82.7561	0.4814	73.3128	5.5587	13.1888	0.008799
	27.0000	83.1108	0.4736	73.4257	5.5593	13.2076	0.008769
	26.0000	83.4514	0.4663	73.5288	5.5599	13.2248	0.008742
	25.0000	83.7791	0.4596	73.6232	5.5604	13.2406	0.008716
	24.0000	84.0953	0.4534	73.7098	5.5609	13.2550	0.008694
	23.0000	84.4012	0.4476	73.7893	5.5613	13.2682	0.008673
	22.0000	84.6976	0.4423	73.8623	5.5617	13.2804	0.008653
	21.0000	84.9855	0.4373	73.9295	5.5621	13.2916	0.008636
	20.0000	85.2658	0.4327	73.9913	5.5625	13.3019	0.008619
	19.0000	85.5390	0.4284	74.0482	5.5628	13.3114	0.008605
	18.0000	85.8057	0.4244	74.1006	5.5631	13.3201	0.008591
	17.0000	86.0667	0.4207	74.1486	5.5633	13.3281	0.008578
	16.0000	86.3223	0.4173	74.1928	5.5636	13.3355	0.008567
	15.0000	86.5731	0.4142	74.2332	5.5638	13.3422	0.008556
	14.0000	86.8196	0.4113	74.2701	5.5640	13.3484	0.008547
	13.0000	87.0620	0.4086	74.3038	5.5642	13.3540	0.008538
	12.0000	87.3009	0.4062	74.3344	5.5643	13.3591	0.008530
	11.0000	87.5365	0.4040	74.3620	5.5645	13.3637	0.008523
	10.0000	87.7693	0.4020	74.3868	5.5646	13.3679	0.008517
	9.0000	87.9994	0.4003	74.4090	5.5647	13.3715	0.008511
	8.0000	88.2273	0.3987	74.4285	5.5648	13.3748	0.008506
	7.0000	88.4531	0.3973	74.4456	5.5649	13.3776	0.008502
	6.0000	88.6773	0.3961	74.4602	5.5650	13.3801	0.008498
	5.0000	88.8999	0.3951	74.4725	5.5651	13.3821	0.008495
	4.0000	89.1214	0.3943	74.4824	5.5651	13.3838	0.008492
	3.0000	89.3419	0.3937	74.4901	5.5652	13.3851	0.008490
	2.0000	89.5616	0.3932	74.4956	5.5652	13.3860	0.008489
	1.0000	89.7809	0.3930	74.4989	5.5652	13.3865	0.008488
	0.0000	90.0000	0.3929	74.5000	5.5652	13.3867	0.008488

* DENOTES $M_2 = 1$, m DENOTES θ_{max}

OBLIQUE SHOCKS IN A PERFECT GAS

$\gamma = 7/5$ $\qquad M_1 = 10.00$

WEAK SHOCK SOLUTION

θ	β	M_1	P_{21}	ρ_{21}	T_{21}	P^0_{21}
0.0000	5.7392	10.0000	1.0000	1.0000	1.0000	1.0000
1.0000	6.3767	9.6423	1.2725	1.1873	1.0717	0.9986
2.0000	7.0771	9.2981	1.6043	1.3973	1.1481	0.9893
3.0000	7.8371	8.9597	2.0026	1.6264	1.2313	0.9668
4.0000	8.6528	8.6229	2.4740	1.8697	1.3232	0.9284
5.0000	9.5190	8.2862	3.0240	2.1215	1.4254	0.8745
6.0000	10.4303	7.9505	3.6571	2.3757	1.5394	0.8081
7.0000	11.3814	7.6174	4.3766	2.6270	1.6660	0.7333
8.0000	12.3673	7.2894	5.1851	2.8708	1.8061	0.6548
9.0000	13.3835	6.9687	6.0840	3.1036	1.9603	0.5768
10.0000	14.4262	6.6574	7.0745	3.3230	2.1289	0.5025
11.0000	15.4922	6.3569	8.1570	3.5277	2.3123	0.4339
12.0000	16.5788	6.0684	9.3318	3.7172	2.5105	0.3722
13.0000	17.6836	5.7925	10.5981	3.8913	2.7235	0.3179
14.0000	18.8049	5.5295	11.9558	4.0508	2.9515	0.2707
15.0000	19.9412	5.2794	13.4038	4.1962	3.1943	0.2301
16.0000	21.0914	5.0418	14.9413	4.3287	3.4517	0.1956
17.0000	22.2547	4.8164	16.5671	4.4491	3.7237	0.1663
18.0000	23.4303	4.6026	18.2798	4.5585	4.0101	0.1416
19.0000	24.6180	4.3998	20.0783	4.6579	4.3106	0.1207
20.0000	25.8174	4.2074	21.9608	4.7482	4.6250	0.1032
21.0000	27.0284	4.0247	23.9260	4.8304	4.9532	0.08847
22.0000	28.2512	3.8511	25.9723	4.9053	5.2947	0.07604
23.0000	29.4858	3.6859	28.0980	4.9735	5.6495	0.06556
24.0000	30.7328	3.5285	30.3016	5.0359	6.0172	0.05670
25.0000	31.9926	3.3783	32.5815	5.0928	6.3975	0.04920
26.0000	33.2662	3.2347	34.9365	5.1450	6.7903	0.04282
27.0000	34.5540	3.0972	37.3645	5.1929	7.1953	0.03739
28.0000	35.8575	2.9654	39.8648	5.2369	7.6123	0.03276
29.0000	37.1778	2.8386	42.4361	5.2774	8.0411	0.02878
30.0000	38.5167	2.7166	45.0776	5.3148	8.4816	0.02537
31.0000	39.8760	2.5987	47.7886	5.3493	8.9336	0.02242
32.0000	41.2584	2.4846	50.5695	5.3813	9.3973	0.01988
33.0000	42.6665	2.3739	53.4203	5.4110	9.8726	0.01767
34.0000	44.1042	2.2661	56.3429	5.4386	10.3598	0.01574
35.0000	45.5764	2.1608	59.3402	5.4643	10.8595	0.01406
36.0000	47.0888	2.0574	62.4161	5.4884	11.3723	0.01258
37.0000	48.6497	1.9555	65.5780	5.5110	11.8994	0.01128
38.0000	50.2697	1.8544	68.8365	5.5323	12.4426	0.01013
39.0000	51.9646	1.7533	72.2089	5.5525	13.0048	0.009104
40.0000	53.7575	1.6511	75.7224	5.5717	13.5905	0.008183
41.0000	55.6864	1.5461	79.4252	5.5903	14.2077	0.007347
42.0000	57.8199	1.4354	83.4082	5.6085	14.8716	0.006576
43.0000	60.3129	1.3127	87.8833	5.6272	15.6176	0.005838
44.0000	63.7339	1.1547	93.6516	5.6488	16.5791	0.005047
* 44.4285	67.3351	1.0000	99.1763	5.6672	17.5000	0.004424
m 44.4290	67.4543	0.9951	99.3487	5.6678	17.5287	0.004406

* DENOTES $M_2 = 1$, m DENOTES θ_{max}

OBLIQUE SHOCKS IN A PERFECT GAS

$\gamma = 7/5 \quad M_1 = 10.00$

STRONG SHOCK SOLUTION

θ	β	M_1	p_{21}	ρ_{21}	T_{21}	p_{21}^o
m 44.4290	67.4543	0.9951	99.3487	5.6678	17.5287	0.004406
44.0000	70.7528	0.8636	103.8223	5.6813	18.2744	0.003980
43.0000	73.1507	0.7724	106.7432	5.6896	18.7612	0.003732
42.0000	74.7014	0.7187	108.3780	5.6940	19.0337	0.003603
41.0000	75.8533	0.6792	109.5310	5.6971	19.2259	0.003515
40.0000	76.8011	0.6479	110.4175	5.6994	19.3737	0.003450
39.0000	77.6139	0.6218	111.1322	5.7012	19.4928	0.003398
38.0000	78.3295	0.5995	111.7261	5.7027	19.5918	0.003357
37.0000	78.9711	0.5801	112.2303	5.7040	19.6758	0.003322
36.0000	79.5544	0.5630	112.6651	5.7051	19.7483	0.003292
35.0000	80.0905	0.5477	113.0448	5.7060	19.8116	0.003266
34.0000	80.5873	0.5339	113.3795	5.7068	19.8674	0.003244
33.0000	81.0510	0.5214	113.6769	5.7075	19.9169	0.003224
32.0000	81.4864	0.5101	113.9430	5.7082	19.9613	0.003206
31.0000	81.8974	0.4997	114.1823	5.7088	20.0012	0.003191
30.0000	82.2871	0.4901	114.3986	5.7093	20.0372	0.003177
29.0000	82.6580	0.4814	114.5947	5.7098	20.0699	0.003164
28.0000	83.0124	0.4732	114.7733	5.7102	20.0997	0.003153
27.0000	83.3519	0.4657	114.9363	5.7106	20.1269	0.003142
26.0000	83.6782	0.4588	115.0854	5.7109	20.1517	0.003133
25.0000	83.9926	0.4523	115.2221	5.7113	20.1745	0.003124
24.0000	84.2963	0.4463	115.3476	5.7116	20.1954	0.003116
23.0000	84.5903	0.4407	115.4630	5.7118	20.2146	0.003109
22.0000	84.8755	0.4356	115.5692	5.7121	20.2323	0.003102
21.0000	85.1527	0.4307	115.6669	5.7123	20.2486	0.003096
20.0000	85.4226	0.4263	115.7569	5.7125	20.2636	0.003091
19.0000	85.6860	0.4221	115.8398	5.7127	20.2775	0.003085
18.0000	85.9433	0.4182	115.9161	5.7129	20.2902	0.003081
17.0000	86.1951	0.4147	115.9862	5.7131	20.3019	0.003076
16.0000	86.4419	0.4113	116.0506	5.7132	20.3126	0.003072
15.0000	86.6842	0.4083	116.1097	5.7134	20.3224	0.003069
14.0000	86.9223	0.4055	116.1637	5.7135	20.3314	0.003065
13.0000	87.1566	0.4029	116.2129	5.7136	20.3396	0.003062
12.0000	87.3875	0.4006	116.2576	5.7137	20.3471	0.003060
11.0000	87.6154	0.3984	116.2980	5.7138	20.3538	0.003057
10.0000	87.8405	0.3965	116.3343	5.7139	20.3599	0.003055
9.0000	88.0632	0.3948	116.3667	5.7140	20.3653	0.003053
8.0000	88.2837	0.3932	116.3953	5.7140	20.3700	0.003051
7.0000	88.5023	0.3919	116.4203	5.7141	20.3742	0.003050
6.0000	88.7192	0.3907	116.4417	5.7142	20.3778	0.003048
5.0000	88.9348	0.3898	116.4597	5.7142	20.3808	0.003047
4.0000	89.1492	0.3890	116.4743	5.7142	20.3832	0.003046
3.0000	89.3627	0.3884	116.4856	5.7143	20.3851	0.003046
2.0000	89.5755	0.3879	116.4936	5.7143	20.3864	0.003045
1.0000	89.7879	0.3877	116.4984	5.7143	20.3872	0.003045
0.0000	90.0000	0.3876	116.5000	5.7143	20.3875	0.003045

* DENOTES $M_2 = 1$, m DENOTES θ_{max}

OBLIQUE SHOCKS IN A PERFECT GAS

$\gamma = 7/5$ $\quad$ $M_1 = 15.00$

WEAK SHOCK SOLUTION

θ	β	M_1	p_{21}	ρ_{21}	T_{21}	p^o_{21}
0.0000	3.8226	15.0000	1.0000	1.0000	1.0000	1.0000
1.0000	4.4721	14.2271	1.4293	1.2889	1.1089	0.9954
2.0000	5.2129	13.4858	2.0002	1.6251	1.2308	0.9670
3.0000	6.0359	12.7513	2.7358	1.9935	1.3724	0.9035
4.0000	6.9295	12.0192	3.6543	2.3747	1.5389	0.8084
5.0000	7.8814	11.2970	4.7691	2.7500	1.7342	0.6943
6.0000	8.8806	10.5961	6.0892	3.1048	1.9612	0.5764
7.0000	9.9175	9.9266	7.6200	3.4302	2.2214	0.4664
8.0000	10.9850	9.2959	9.3647	3.7221	2.5160	0.3707
9.0000	12.0771	8.7079	11.3246	3.9798	2.8456	0.2914
10.0000	13.1895	8.1635	13.4999	4.2051	3.2103	0.2277
11.0000	14.3189	7.6619	15.8895	4.4011	3.6104	0.1777
12.0000	15.4628	7.2008	18.4924	4.5710	4.0456	0.1389
13.0000	16.6194	6.7773	21.3067	4.7183	4.5158	0.1089
14.0000	17.7873	6.3884	24.3299	4.8460	5.0206	0.08580
15.0000	18.9654	6.0307	27.5594	4.9571	5.5596	0.06802
16.0000	20.1530	5.7014	30.9920	5.0539	6.1323	0.05427
17.0000	21.3498	5.3975	34.6251	5.1385	6.7384	0.04360
18.0000	22.5554	5.1164	38.4551	5.2127	7.3772	0.03526
19.0000	23.7696	4.8557	42.4783	5.2780	8.0481	0.02872
20.0000	24.9925	4.6134	46.6911	5.3358	8.7506	0.02356
21.0000	26.2243	4.3876	51.0901	5.3869	9.4841	0.01945
22.0000	27.4653	4.1766	55.6711	5.4325	10.2478	0.01616
23.0000	28.7157	3.9788	60.4303	5.4731	11.0413	0.01351
24.0000	29.9762	3.7930	65.3639	5.5096	11.8637	0.01136
25.0000	31.2473	3.6179	70.4679	5.5423	12.7146	0.009615
26.0000	32.5298	3.4526	75.7385	5.5718	13.5932	0.008179
27.0000	33.8248	3.2959	81.1728	5.5985	14.4990	0.006994
28.0000	35.1331	3.1472	86.7666	5.6227	15.4315	0.006011
29.0000	36.4561	3.0055	92.5173	5.6447	16.3900	0.005190
30.0000	37.7953	2.8702	98.4218	5.6648	17.3742	0.004502
31.0000	39.1527	2.7406	104.4793	5.6832	18.3839	0.003922
32.0000	40.5303	2.6161	110.6884	5.7001	19.4188	0.003430
33.0000	41.9309	2.4962	117.0490	5.7156	20.4790	0.003012
34.0000	43.3576	2.3802	123.5627	5.7299	21.5647	0.002653
35.0000	44.8147	2.2677	130.2341	5.7431	22.6767	0.002345
36.0000	46.3071	2.1581	137.0695	5.7554	23.8159	0.002079
37.0000	47.8417	2.0508	144.0811	5.7668	24.9846	0.001848
38.0000	49.4273	1.9451	151.2863	5.7775	26.1855	0.001647
39.0000	51.0763	1.8404	158.7135	5.7875	27.4234	0.001470
40.0000	52.8063	1.7356	166.4070	5.7970	28.7057	0.001313
41.0000	54.6448	1.6295	174.4408	5.8060	30.0448	0.001173
42.0000	56.6376	1.5200	182.9466	5.8148	31.4624	0.001047
43.0000	58.8739	1.4032	192.1908	5.8234	33.0032	0.0009306
44.0000	61.5790	1.2698	202.8702	5.8324	34.7831	0.0008174
45.0000	66.2173	1.0585	219.6438	5.8449	37.5788	0.0006752
* 45.0676	67.5865	1.0000	224.1708	5.8479	38.3333	0.0006428
m 45.0677	67.6389	0.9978	224.3401	5.8481	38.3615	0.0006416

* DENOTES $M_2 = 1$, m DENOTES θ_{max}

OBLIQUE SHOCKS IN A PERFECT GAS

$\gamma = 7/5 \qquad M_1 = 15.00$

STRONG SHOCK SOLUTION

	θ	β	M_1	p_{21}	ρ_{21}	T_{21}	p_{21}^{o}
m	45.0677	67.6389	0.9978	224.3401	5.8481	38.3615	0.0006416
	45.0000	68.9932	0.9416	228.6000	5.8508	39.0715	0.0006131
	44.0000	72.6390	0.7984	238.9609	5.8571	40.7983	0.0005509
	43.0000	74.3517	0.7353	243.2351	5.8596	41.5107	0.0005278
	42.0000	75.5960	0.6913	246.0897	5.8612	41.9865	0.0005131
	41.0000	76.5972	0.6571	248.2293	5.8623	42.3431	0.0005025
	40.0000	77.4443	0.6291	249.9280	5.8632	42.6262	0.0004942
	39.0000	78.1833	0.6053	251.3253	5.8640	42.8591	0.0004876
	38.0000	78.8416	0.5848	252.5025	5.8646	43.0553	0.0004821
	37.0000	79.4371	0.5667	253.5121	5.8651	43.2236	0.0004775
	36.0000	79.9819	0.5507	254.3896	5.8656	43.3698	0.0004735
	35.0000	80.4851	0.5362	255.1604	5.8660	43.4983	0.0004701
	34.0000	80.9535	0.5232	255.8434	5.8663	43.6121	0.0004670
	33.0000	81.3921	0.5114	256.4528	5.8666	43.7137	0.0004643
	32.0000	81.8051	0.5006	256.9998	5.8669	43.8049	0.0004620
	31.0000	82.1960	0.4906	257.4933	5.8672	43.8871	0.0004598
	30.0000	82.5673	0.4815	257.9404	5.8674	43.9616	0.0004579
	29.0000	82.9213	0.4731	258.3469	5.8676	44.0294	0.0004561
	28.0000	83.2601	0.4653	258.7175	5.8678	44.0912	0.0004546
	27.0000	83.5851	0.4580	259.0565	5.8680	44.1477	0.0004531
	26.0000	83.8979	0.4513	259.3670	5.8681	44.1994	0.0004518
	25.0000	84.1996	0.4451	259.6521	5.8682	44.2469	0.0004506
	24.0000	84.4913	0.4393	259.9142	5.8684	44.2906	0.0004495
	23.0000	84.7739	0.4339	260.1553	5.8685	44.3308	0.0004485
	22.0000	85.0483	0.4289	260.3775	5.8686	44.3678	0.0004476
	21.0000	85.3152	0.4242	260.5822	5.8687	44.4019	0.0004467
	20.0000	85.5752	0.4199	260.7708	5.8688	44.4334	0.0004459
	19.0000	85.8291	0.4158	260.9446	5.8689	44.4624	0.0004452
	18.0000	86.0772	0.4121	261.1047	5.8690	44.4890	0.0004446
	17.0000	86.3202	0.4086	261.2520	5.8690	44.5136	0.0004439
	16.0000	86.5584	0.4054	261.3873	5.8691	44.5361	0.0004434
	15.0000	86.7924	0.4024	261.5114	5.8692	44.5568	0.0004429
	14.0000	87.0224	0.3997	261.6249	5.8692	44.5757	0.0004424
	13.0000	87.2488	0.3972	261.7285	5.8693	44.5930	0.0004420
	12.0000	87.4720	0.3949	261.8226	5.8693	44.6087	0.0004416
	11.0000	87.6924	0.3928	261.9077	5.8694	44.6229	0.0004413
	10.0000	87.9101	0.3909	261.9841	5.8694	44.6356	0.0004409
	9.0000	88.1254	0.3893	262.0524	5.8694	44.6470	0.0004407
	8.0000	88.3388	0.3878	262.1127	5.8695	44.6570	0.0004404
	7.0000	88.5503	0.3865	262.1652	5.8695	44.6658	0.0004402
	6.0000	88.7602	0.3853	262.2104	5.8695	44.6733	0.0004400
	5.0000	88.9689	0.3844	262.2483	5.8695	44.6796	0.0004399
	4.0000	89.1764	0.3836	262.2791	5.8695	44.6848	0.0004397
	3.0000	89.3831	0.3830	262.3029	5.8696	44.6887	0.0004396
	2.0000	89.5891	0.3826	262.3198	5.8696	44.6916	0.0004396
	1.0000	89.7947	0.3823	262.3300	5.8696	44.6932	0.0004395
	0.0000	90.0000	0.3823	262.3333	5.8696	44.6938	0.0004395

* DENOTES $M_2 = 1$, m DENOTES θ_{max}

OBLIQUE SHOCKS IN A PERFECT GAS

$\gamma = 7/5$ $M_1 = 20.00$

WEAK SHOCK SOLUTION

θ	β	M_1	p_{21}	ρ_{21}	T_{21}	p^o_{21}
0.0000	2.8660	20.0000	1.0000	1.0000	1.0000	1.0000
1.0000	3.5296	18.6521	1.6021	1.3960	1.1476	0.9894
2.0000	4.3108	17.3566	2.4700	1.8678	1.3224	0.9287
3.0000	5.1908	16.0690	3.6532	2.3743	1.5387	0.8085
4.0000	6.1483	14.8057	5.1864	2.8712	1.8064	0.6547
5.0000	7.1636	13.6014	7.0903	3.3263	2.1316	0.5014
6.0000	8.2217	12.4836	9.3767	3.7238	2.5180	0.3701
7.0000	9.3118	11.4664	12.0513	4.0611	2.9675	0.2677
8.0000	10.4260	10.5522	15.1159	4.3425	3.4809	0.1921
9.0000	11.5590	9.7358	18.5705	4.5755	4.0587	0.1379
10.0000	12.7072	9.0088	22.4136	4.7682	4.7006	0.09953
11.0000	13.8677	8.3612	26.6418	4.9278	5.4065	0.07250
12.0000	15.0388	7.7832	31.2523	5.0605	6.1758	0.05339
13.0000	16.2191	7.2659	36.2401	5.1714	7.0078	0.03978
14.0000	17.4078	6.8010	41.6013	5.2647	7.9019	0.02999
15.0000	18.6042	6.3817	47.3302	5.3437	8.8572	0.02289
16.0000	19.8079	6.0018	53.4213	5.4110	9.8727	0.01767
17.0000	21.0188	5.6564	59.8685	5.4686	10.9476	0.01379
18.0000	22.2369	5.3408	66.6665	5.5183	12.0809	0.01088
19.0000	23.4621	5.0514	73.8081	5.5615	13.2714	0.008668
20.0000	24.6947	4.7850	81.2867	5.5990	14.5180	0.006972
21.0000	25.9350	4.5388	89.0953	5.6320	15.8196	0.005658
22.0000	27.1834	4.3105	97.2277	5.6609	17.1752	0.004631
23.0000	28.4402	4.0980	105.6760	5.6866	18.5833	0.003820
24.0000	29.7063	3.8995	114.4341	5.7094	20.0431	0.003174
25.0000	30.9820	3.7136	123.4940	5.7297	21.5532	0.002657
26.0000	32.2684	3.5388	132.8492	5.7479	23.1125	0.002238
27.0000	33.5663	3.3740	142.4931	5.7643	24.7199	0.001897
28.0000	34.8768	3.2180	152.4198	5.7791	26.3745	0.001618
29.0000	36.2013	3.0701	162.6236	5.7924	28.0752	0.001387
30.0000	37.5412	2.9293	173.0997	5.8046	29.8212	0.001195
31.0000	38.8984	2.7949	183.8446	5.8156	31.6121	0.001035
32.0000	40.2748	2.6662	194.8547	5.8257	33.4472	0.0009004
33.0000	41.6733	2.5425	206.1316	5.8350	35.3267	0.0007867
34.0000	43.0967	2.4233	217.6759	5.8435	37.2508	0.0006900
35.0000	44.5492	2.3079	229.4950	5.8514	39.2207	0.0006074
36.0000	46.0355	2.1957	241.5986	5.8586	41.2380	0.0005365
37.0000	47.5619	2.0862	254.0054	5.8654	43.3058	0.0004753
38.0000	49.1369	1.9786	266.7437	5.8717	45.4289	0.0004221
39.0000	50.7716	1.8724	279.8578	5.8776	47.6146	0.0003757
40.0000	52.4825	1.7664	293.4197	5.8831	49.8749	0.0003349
41.0000	54.2941	1.6596	307.5444	5.8884	52.2291	0.0002987
42.0000	56.2467	1.5500	322.4348	5.8934	54.7108	0.0002662
43.0000	58.4159	1.4344	338.4866	5.8984	57.3862	0.0002364
44.0000	60.9774	1.3052	356.6581	5.9035	60.4148	0.0002081
45.0000	64.6776	1.1312	381.1292	5.9096	64.4933	0.0001769
* 45.2931	67.6760	1.0000	399.1688	5.9136	67.4999	0.0001580
m 45.2932	67.7054	0.9988	399.3367	5.9137	67.5279	0.0001578

* DENOTES $M_2 = 1$, m DENOTES θ_{max}

OBLIQUE SHOCKS IN A PERFECT GAS

$$\gamma = 7/5 \qquad M_1 = 20.00$$

STRONG SHOCK SOLUTION

	θ	β	M_1	p_{21}	ρ_{21}	T_{21}	p^o_{21}
m	45.2932	67.7054	0.9988	399.3367	5.9137	67.5279	0.0001578
	45.0000	70.4410	0.8862	414.1979	5.9167	70.0048	0.0001443
	44.0000	73.1455	0.7815	427.2687	5.9192	72.1833	0.0001337
	43.0000	74.7113	0.7240	434.0533	5.9205	73.3141	0.0001286
	42.0000	75.8850	0.6825	438.7462	5.9213	74.0962	0.0001253
	41.0000	76.8424	0.6498	442.3194	5.9219	74.6917	0.0001228
	40.0000	77.6589	0.6228	445.1820	5.9224	75.1688	0.0001209
	39.0000	78.3749	0.5998	447.5507	5.9228	75.5636	0.0001193
	38.0000	79.0149	0.5798	449.5549	5.9232	75.8977	0.0001180
	37.0000	79.5954	0.5621	451.2792	5.9235	76.1850	0.0001169
	36.0000	80.1276	0.5464	452.7816	5.9237	76.4354	0.0001160
	35.0000	80.6200	0.5323	454.1039	5.9239	76.6558	0.0001151
	34.0000	81.0789	0.5195	455.2774	5.9241	76.8514	0.0001144
	33.0000	81.5091	0.5079	456.3259	5.9243	77.0262	0.0001138
	32.0000	81.9147	0.4972	457.2684	5.9245	77.1832	0.0001132
	31.0000	82.2987	0.4875	458.1192	5.9246	77.3250	0.0001127
	30.0000	82.6638	0.4785	458.8908	5.9247	77.4536	0.0001122
	29.0000	83.0121	0.4702	459.5927	5.9248	77.5706	0.0001118
	28.0000	83.3456	0.4625	460.2333	5.9249	77.6774	0.0001114
	27.0000	83.6657	0.4553	460.8193	5.9250	77.7750	0.0001111
	26.0000	83.9738	0.4487	461.3566	5.9251	77.8646	0.0001108
	25.0000	84.2711	0.4425	461.8500	5.9252	77.9468	0.0001105
	24.0000	84.5587	0.4368	462.3036	5.9253	78.0225	0.0001102
	23.0000	84.8374	0.4315	462.7214	5.9253	78.0921	0.0001099
	22.0000	85.1081	0.4265	463.1063	5.9254	78.1562	0.0001097
	21.0000	85.3714	0.4219	463.4611	5.9254	78.2153	0.0001095
	20.0000	85.6281	0.4176	463.7881	5.9255	78.2699	0.0001093
	19.0000	85.8787	0.4136	464.0895	5.9255	78.3201	0.0001092
	18.0000	86.1237	0.4099	464.3671	5.9256	78.3664	0.0001090
	17.0000	86.3636	0.4065	464.6227	5.9256	78.4090	0.0001088
	16.0000	86.5989	0.4033	464.8574	5.9257	78.4481	0.0001087
	15.0000	86.8299	0.4004	465.0728	5.9257	78.4840	0.0001086
	14.0000	87.0571	0.3977	465.2699	5.9257	78.5168	0.0001085
	13.0000	87.2808	0.3952	465.4496	5.9258	78.5468	0.0001084
	12.0000	87.5014	0.3929	465.6130	5.9258	78.5740	0.0001083
	11.0000	87.7191	0.3908	465.7607	5.9258	78.5986	0.0001082
	10.0000	87.9342	0.3890	465.8936	5.9258	78.6208	0.0001081
	9.0000	88.1470	0.3873	466.0120	5.9259	78.6405	0.0001081
	8.0000	88.3579	0.3858	466.1166	5.9259	78.6580	0.0001080
	7.0000	88.5669	0.3845	466.2081	5.9259	78.6732	0.0001079
	6.0000	88.7744	0.3834	466.2864	5.9259	78.6862	0.0001079
	5.0000	88.9807	0.3825	466.3522	5.9259	78.6972	0.0001079
	4.0000	89.1858	0.3817	466.4056	5.9259	78.7061	0.0001078
	3.0000	89.3901	0.3811	466.4470	5.9259	78.7130	0.0001078
	2.0000	89.5938	0.3807	466.4765	5.9259	78.7179	0.0001078
	1.0000	89.7970	0.3805	466.4941	5.9259	78.7209	0.0001078
	0.0000	90.0000	0.3804	466.4999	5.9259	78.7218	0.0001078

* DENOTES $M_2 = 1$, m DENOTES θ_{max}

OBLIQUE SHOCKS IN A PERFECT GAS

$\gamma = 7/5$ $M_1 = 25.00$

WEAK SHOCK SOLUTION

θ	β	M_1	p_{21}	ρ_{21}	T_{21}	p^o_{21}
0.0000	2.2924	25.0000	1.0000	1.0000	1.0000	1.0000
1.0000	2.9705	22.9188	1.7916	1.5080	1.1881	0.9801
2.0000	3.7900	20.9104	3.0192	2.1194	1.4246	0.8750
3.0000	4.7192	18.9283	4.7688	2.7499	1.7342	0.6943
4.0000	5.7262	17.0443	7.0921	3.3266	2.1319	0.5013
5.0000	6.7865	15.3280	10.0157	3.8146	2.6256	0.3415
6.0000	7.8836	13.8101	13.5511	4.2098	3.2189	0.2265
7.0000	9.0065	12.4890	17.7030	4.5234	3.9137	0.1493
8.0000	10.1484	11.3469	22.4710	4.7707	4.7102	0.09908
9.0000	11.3048	10.3599	27.8532	4.9661	5.6086	0.06666
10.0000	12.4726	9.5045	33.8448	5.1216	6.6083	0.04562
11.0000	13.6499	8.7595	40.4411	5.2464	7.7084	0.03180
12.0000	14.8353	8.1069	47.6356	5.3475	8.9081	0.02258
13.0000	16.0280	7.5318	55.4212	5.4302	10.2062	0.01632
14.0000	17.2272	7.0217	63.7900	5.4985	11.6013	0.01199
15.0000	18.4330	6.5665	72.7348	5.5555	13.0925	0.008957
16.0000	19.6448	6.1581	82.2451	5.6034	14.6778	0.006789
17.0000	20.8628	5.7895	92.3130	5.6440	16.3560	0.005217
18.0000	22.0870	5.4551	102.9276	5.6787	18.1252	0.004060
19.0000	23.3177	5.1502	114.0800	5.7085	19.9841	0.003197
20.0000	24.5551	4.8710	125.7578	5.7344	21.9306	0.002546
21.0000	25.7996	4.6142	137.9521	5.7569	23.9631	0.002048
22.0000	27.0516	4.3768	150.6510	5.7766	26.0796	0.001663
23.0000	28.3116	4.1567	163.8437	5.7939	28.2785	0.001362
24.0000	29.5803	3.9518	177.5186	5.8093	30.5577	0.001125
25.0000	30.8583	3.7603	191.6653	5.8229	32.9156	0.0009368
26.0000	32.1467	3.5807	206.2735	5.8351	35.3503	0.0007853
27.0000	33.4461	3.4118	221.3309	5.8460	37.8600	0.0006628
28.0000	34.7577	3.2523	236.8290	5.8559	40.4430	0.0005630
29.0000	36.0830	3.1013	252.7587	5.8647	43.0980	0.0004810
30.0000	37.4233	2.9578	269.1124	5.8728	45.8237	0.0004132
31.0000	38.7804	2.8210	285.8833	5.8801	48.6189	0.0003568
32.0000	40.1564	2.6902	303.0688	5.8868	51.4831	0.0003095
33.0000	41.5540	2.5647	320.6665	5.8929	54.4161	0.0002698
34.0000	42.9761	2.4438	338.6805	5.8985	57.4185	0.0002361
35.0000	44.4265	2.3270	357.1178	5.9036	60.4914	0.0002074
36.0000	45.9101	2.2135	375.9965	5.9084	63.6379	0.0001829
37.0000	47.4329	2.1029	395.3416	5.9128	66.8621	0.0001617
38.0000	49.0030	1.9945	415.1947	5.9169	70.1709	0.0001434
39.0000	50.6315	1.8875	435.6243	5.9207	73.5759	0.0001275
40.0000	52.3339	1.7810	456.7343	5.9244	77.0942	0.0001135
41.0000	54.1338	1.6737	478.6965	5.9278	80.7546	0.0001012
42.0000	56.0692	1.5640	501.8073	5.9311	84.6064	0.00009008
43.0000	58.2100	1.4488	526.6370	5.9343	88.7447	0.00007999
44.0000	60.7147	1.3212	554.5269	5.9376	93.3930	0.00007044
45.0000	64.1819	1.1561	590.6954	5.9413	99.4212	0.00006028
* 45.3979	67.7177	1.0000	624.1676	5.9445	104.9999	0.00005262
m 45.3979	67.7365	0.9992	624.3352	5.9445	105.0278	0.00005258

* DENOTES $M_2 = 1$, m DENOTES θ_{max}

OBLIQUE SHOCKS IN A PERFECT GAS

$$\gamma = 7/5 \qquad M_1 = 25.00$$

STRONG SHOCK SOLUTION

	θ	β	M_1	p_{21}	ρ_{21}	T_{21}	p_{21}^o
m	45.3979	67.7365	0.9992	624.3352	5.9445	105.0278	0.00005258
	45.0000	70.8941	0.8695	650.8801	5.9467	109.4520	0.00004745
	44.0000	73.3640	0.7742	669.2357	5.9482	112.5112	0.00004430
	43.0000	74.8714	0.7190	679.3334	5.9489	114.1942	0.00004269
	42.0000	76.0152	0.6786	686.4150	5.9495	115.3745	0.00004161
	41.0000	76.9536	0.6465	691.8424	5.9498	116.2790	0.00004081
	40.0000	77.7565	0.6199	696.2076	5.9502	117.0066	0.00004018
	39.0000	78.4623	0.5972	699.8292	5.9504	117.6102	0.00003967
	38.0000	79.0941	0.5775	702.8993	5.9506	118.1218	0.00003924
	37.0000	79.6679	0.5600	705.5439	5.9508	118.5626	0.00003888
	36.0000	80.1944	0.5445	707.8512	5.9510	118.9472	0.00003857
	35.0000	80.6819	0.5305	709.8834	5.9511	119.2859	0.00003829
	34.0000	81.1365	0.5178	711.6886	5.9512	119.5867	0.00003805
	33.0000	81.5629	0.5063	713.3025	5.9513	119.8557	0.00003784
	32.0000	81.9650	0.4957	714.7534	5.9514	120.0975	0.00003765
	31.0000	82.3459	0.4860	716.0643	5.9515	120.3160	0.00003748
	30.0000	82.7081	0.4771	717.2532	5.9516	120.5141	0.00003733
	29.0000	83.0539	0.4688	718.3352	5.9517	120.6945	0.00003719
	28.0000	83.3849	0.4612	719.3230	5.9517	120.8591	0.00003706
	27.0000	83.7027	0.4541	720.2271	5.9518	121.0098	0.00003695
	26.0000	84.0088	0.4475	721.0560	5.9519	121.1480	0.00003684
	25.0000	84.3041	0.4414	721.8174	5.9519	121.2749	0.00003675
	24.0000	84.5898	0.4357	722.5177	5.9520	121.3916	0.00003666
	23.0000	84.8667	0.4304	723.1626	5.9520	121.4991	0.00003658
	22.0000	85.1357	0.4254	723.7567	5.9520	121.5981	0.00003650
	21.0000	85.3974	0.4208	724.3047	5.9521	121.6894	0.00003644
	20.0000	85.6525	0.4166	724.8097	5.9521	121.7736	0.00003637
	19.0000	85.9015	0.4126	725.2751	5.9521	121.8512	0.00003632
	18.0000	86.1451	0.4089	725.7040	5.9522	121.9226	0.00003626
	17.0000	86.3836	0.4055	726.0988	5.9522	121.9884	0.00003621
	16.0000	86.6175	0.4023	726.4614	5.9522	122.0489	0.00003617
	15.0000	86.8472	0.3994	726.7943	5.9522	122.1044	0.00003613
	14.0000	87.0731	0.3967	727.0988	5.9523	122.1551	0.00003609
	13.0000	87.2956	0.3942	727.3765	5.9523	122.2014	0.00003606
	12.0000	87.5149	0.3920	727.6289	5.9523	122.2435	0.00003603
	11.0000	87.7314	0.3899	727.8574	5.9523	122.2815	0.00003600
	10.0000	87.9453	0.3881	728.0625	5.9523	122.3157	0.00003597
	9.0000	88.1570	0.3864	728.2456	5.9523	122.3462	0.00003595
	8.0000	88.3667	0.3849	728.4075	5.9523	122.3732	0.00003593
	7.0000	88.5746	0.3837	728.5487	5.9524	122.3967	0.00003591
	6.0000	88.7810	0.3826	728.6698	5.9524	122.4169	0.00003590
	5.0000	88.9861	0.3816	728.7716	5.9524	122.4339	0.00003589
	4.0000	89.1902	0.3809	728.8542	5.9524	122.4477	0.00003588
	3.0000	89.3934	0.3803	728.9181	5.9524	122.4583	0.00003587
	2.0000	89.5960	0.3799	728.9636	5.9524	122.4659	0.00003586
	1.0000	89.7981	0.3796	728.9907	5.9524	122.4704	0.00003586
	0.0000	90.0000	0.3795	728.9998	5.9524	122.4719	0.00003586

* DENOTES $M_2 = 1$, m DENOTES θ_{max}

OBLIQUE SHOCKS IN A PERFECT GAS

$\gamma = 7/5$ $M_1 = 35.00$

WEAK SHOCK SOLUTION

θ	β	M_1	p_{21}	ρ_{21}	T_{21}	p^o_{21}
0.0000	1.6372	35.0000	1.0000	1.0000	1.0000	1.0000
1.0000	2.3440	30.9777	2.2241	1.7442	1.2751	0.9500
2.0000	3.2311	27.0837	4.3737	2.6261	1.6655	0.7336
3.0000	4.2359	23.4150	7.6304	3.4322	2.2232	0.4657
4.0000	5.3099	20.2232	12.0731	4.0634	2.9712	0.2670
5.0000	6.4253	17.5758	17.7311	4.5251	3.9183	0.1489
6.0000	7.5663	15.4180	24.6119	4.8567	5.0677	0.08401
7.0000	8.7246	13.6598	32.7159	5.0960	6.4199	0.04880
8.0000	9.8950	12.2165	42.0364	5.2714	7.9744	0.02935
9.0000	11.0746	11.0187	52.5662	5.4024	9.7302	0.01829
10.0000	12.2616	10.0130	64.2938	5.5021	11.6853	0.01179
11.0000	13.4551	9.1587	77.2102	5.5794	13.8385	0.007832
12.0000	14.6540	8.4254	91.2997	5.6403	16.1871	0.005350
13.0000	15.8582	7.7894	106.5484	5.6890	18.7288	0.003748
14.0000	17.0674	7.2327	122.9422	5.7286	21.4613	0.002685
15.0000	18.2816	6.7415	140.4637	5.7610	24.3817	0.001963
16.0000	19.5009	6.3046	159.0947	5.7880	27.4870	0.001461
17.0000	20.7253	5.9134	178.8175	5.8106	30.7742	0.001106
18.0000	21.9553	5.5608	199.6138	5.8298	34.2404	0.0008498
19.0000	23.1909	5.2411	221.4614	5.8461	37.8817	0.0006619
20.0000	24.4326	4.9497	244.3406	5.8602	41.6950	0.0005220
21.0000	25.6809	4.6828	268.2297	5.8724	45.6765	0.0004165
22.0000	26.9361	4.4371	293.1060	5.8830	49.8227	0.0003358
23.0000	28.1990	4.2099	318.9508	5.8923	54.1302	0.0002733
24.0000	29.4701	3.9989	345.7397	5.9005	58.5950	0.0002245
25.0000	30.7502	3.8023	373.4514	5.9078	63.2137	0.0001859
26.0000	32.0402	3.6184	402.0659	5.9142	67.9828	0.0001552
27.0000	33.3411	3.4457	431.5612	5.9200	72.8987	0.0001305
28.0000	34.6537	3.2829	461.9163	5.9252	77.9579	0.0001104
29.0000	35.9798	3.1291	493.1177	5.9299	83.1581	0.00009403
30.0000	37.3204	2.9831	525.1449	5.9341	88.4960	0.00008055
31.0000	38.6775	2.8442	557.9872	5.9379	93.9698	0.00006937
32.0000	40.0533	2.7115	591.6385	5.9414	99.5783	0.00006004
33.0000	41.4501	2.5843	626.0962	5.9446	105.3213	0.00005222
34.0000	42.8710	2.4620	661.3606	5.9476	111.1987	0.00004561
35.0000	44.3197	2.3439	697.4489	5.9502	117.2134	0.00004000
36.0000	45.8010	2.2293	734.3926	5.9527	123.3707	0.00003521
37.0000	47.3207	2.1178	772.2407	5.9550	129.6787	0.00003110
38.0000	48.8869	2.0085	811.0736	5.9572	136.1509	0.00002754
39.0000	50.5100	1.9008	851.0107	5.9592	142.8071	0.00002445
40.0000	52.2053	1.7937	892.2539	5.9610	149.6810	0.00002175
41.0000	53.9953	1.6862	935.1219	5.9628	156.8256	0.00001936
42.0000	55.9161	1.5763	980.1627	5.9645	164.3324	0.00001723
43.0000	58.0338	1.4614	1028.4255	5.9662	172.3763	0.00001529
44.0000	60.4927	1.3350	1082.2979	5.9678	181.3550	0.00001347
45.0000	63.7986	1.1760	1150.3872	5.9697	192.7032	0.00001158
* 45.4893	67.7542	1.0000	1224.1660	5.9715	204.9997	0.000009924
m 45.4893	67.7638	0.9996	1224.3333	5.9716	205.0276	0.000009921

* DENOTES $M_2 = 1$, m DENOTES θ_{max}

OBLIQUE SHOCKS IN A PERFECT GAS

$$\gamma = 7/5 \qquad M_1 = 35.00$$

STRONG SHOCK SOLUTION

	θ	β	M_1	p_{21}	ρ_{21}	T_{21}	p^o_{21}
m	45.4893	67.7638	0.9996	1224.3333	5.9716	205.0276	0.000009921
	45.0000	71.2401	0.8569	1281.1826	5.9728	214.5024	0.000008863
	44.0000	73.5473	0.7682	1314.3584	5.9735	220.0317	0.000008318
	43.0000	75.0077	0.7147	1333.3599	5.9739	223.1986	0.000008026
	42.0000	76.1268	0.6752	1346.8342	5.9741	225.4444	0.000007828
	41.0000	77.0492	0.6437	1357.2161	5.9743	227.1747	0.000007680
	40.0000	77.8408	0.6175	1365.5950	5.9745	228.5712	0.000007564
	39.0000	78.5377	0.5950	1372.5608	5.9746	229.7321	0.000007469
	38.0000	79.1626	0.5755	1378.4753	5.9747	230.7179	0.000007389
	37.0000	79.7306	0.5582	1383.5769	5.9748	231.5682	0.000007322
	36.0000	80.2522	0.5428	1388.0308	5.9749	232.3105	0.000007264
	35.0000	80.7355	0.5289	1391.9580	5.9750	232.9650	0.000007213
	34.0000	81.1864	0.5163	1395.4475	5.9750	233.5466	0.000007168
	33.0000	81.6095	0.5049	1398.5693	5.9751	234.0669	0.000007128
	32.0000	82.0087	0.4944	1401.3774	5.9751	234.5349	0.000007093
	31.0000	82.3869	0.4847	1403.9148	5.9752	234.9578	0.000007061
	30.0000	82.7467	0.4759	1406.2175	5.9752	235.3416	0.000007032
	29.0000	83.0901	0.4676	1408.3140	5.9753	235.6910	0.000007006
	28.0000	83.4190	0.4600	1410.2280	5.9753	236.0100	0.000006983
	27.0000	83.7349	0.4530	1411.9797	5.9753	236.3020	0.000006961
	26.0000	84.0391	0.4464	1413.5864	5.9753	236.5698	0.000006942
	25.0000	84.3327	0.4404	1415.0627	5.9754	236.8158	0.000006924
	24.0000	84.6168	0.4347	1416.4207	5.9754	237.0421	0.000006907
	23.0000	84.8922	0.4294	1417.6714	5.9754	237.2506	0.000006892
	22.0000	85.1596	0.4245	1418.8242	5.9754	237.4427	0.000006878
	21.0000	85.4199	0.4199	1419.8867	5.9755	237.6198	0.000006865
	20.0000	85.6737	0.4157	1420.8665	5.9755	237.7831	0.000006854
	19.0000	85.9214	0.4117	1421.7695	5.9755	237.9336	0.000006843
	18.0000	86.1637	0.4080	1422.6023	5.9755	238.0724	0.000006833
	17.0000	86.4010	0.4046	1423.3679	5.9755	238.2000	0.000006824
	16.0000	86.6337	0.4015	1424.0718	5.9755	238.3174	0.000006815
	15.0000	86.8623	0.3986	1424.7175	5.9755	238.4250	0.000006808
	14.0000	87.0871	0.3959	1425.3088	5.9755	238.5235	0.000006801
	13.0000	87.3084	0.3934	1425.8481	5.9756	238.6134	0.000006794
	12.0000	87.5267	0.3912	1426.3384	5.9756	238.6951	0.000006788
	11.0000	87.7421	0.3891	1426.7815	5.9756	238.7690	0.000006783
	10.0000	87.9550	0.3873	1427.1799	5.9756	238.8354	0.000006778
	9.0000	88.1657	0.3856	1427.5354	5.9756	238.8946	0.000006774
	8.0000	88.3744	0.3842	1427.8494	5.9756	238.9469	0.000006771
	7.0000	88.5813	0.3829	1428.1235	5.9756	238.9926	0.000006767
	6.0000	88.7867	0.3818	1428.3591	5.9756	239.0319	0.000006765
	5.0000	88.9908	0.3809	1428.5564	5.9756	239.0648	0.000006762
	4.0000	89.1940	0.3801	1428.7168	5.9756	239.0915	0.000006760
	3.0000	89.3962	0.3795	1428.8411	5.9756	239.1122	0.000006759
	2.0000	89.5978	0.3791	1428.9294	5.9756	239.1269	0.000006758
	1.0000	89.7991	0.3788	1428.9822	5.9756	239.1357	0.000006757
	0.0000	90.0000	0.3788	1428.9995	5.9756	239.1386	0.000006757

* DENOTES $M_2 = 1$, m DENOTES θ_{max}

OBLIQUE SHOCKS IN A PERFECT GAS

$\gamma = 7/5$ $M_1 = \infty$

WEAK SHOCK SOLUTION

θ	β	M_1	P_{21}	ρ_{21}	T_{21}	p^o_{21}
0.0000	0.0000	∞	1.0000	1.0000	1.0000	1.0000
1.0000	1.2000	108.2638	∞	6.0000	∞	0.0000
2.0000	2.4000	54.1084	∞	6.0000	∞	0.0000
3.0000	3.6012	36.0354	∞	6.0000	∞	0.0000
4.0000	4.8023	26.9960	∞	6.0000	∞	0.0000
5.0000	6.0040	21.5653	∞	6.0000	∞	0.0000
6.0000	7.2075	17.9365	∞	6.0000	∞	0.0000
7.0000	8.4122	15.3394	∞	6.0000	∞	0.0000
8.0000	9.6175	13.3885	∞	6.0000	∞	0.0000
9.0000	10.8251	11.8659	∞	6.0000	∞	0.0000
10.0000	12.0347	10.6442	∞	6.0000	∞	0.0000
11.0000	13.2473	9.6404	∞	6.0000	∞	0.0000
12.0000	14.4614	8.8015	∞	6.0000	∞	0.0000
13.0000	15.6781	8.0886	∞	6.0000	∞	0.0000
14.0000	16.8987	7.4744	∞	6.0000	∞	0.0000
15.0000	18.1226	6.9394	∞	6.0000	∞	0.0000
16.0000	19.3500	6.4688	∞	6.0000	∞	0.0000
17.0000	20.5815	6.0511	∞	6.0000	∞	0.0000
18.0000	21.8172	5.6775	∞	6.0000	∞	0.0000
19.0000	23.0576	5.3411	∞	6.0000	∞	0.0000
20.0000	24.3041	5.0358	∞	6.0000	∞	0.0000
21.0000	25.5564	4.7575	∞	6.0000	∞	0.0000
22.0000	26.8158	4.5023	∞	6.0000	∞	0.0000
23.0000	28.0811	4.2673	∞	6.0000	∞	0.0000
24.0000	29.3557	4.0496	∞	6.0000	∞	0.0000
25.0000	30.6377	3.8475	∞	6.0000	∞	0.0000
26.0000	31.9296	3.6587	∞	6.0000	∞	0.0000
27.0000	33.2315	3.4820	∞	6.0000	∞	0.0000
28.0000	34.5453	3.3157	∞	6.0000	∞	0.0000
29.0000	35.8724	3.1587	∞	6.0000	∞	0.0000
30.0000	37.2130	3.0101	∞	6.0000	∞	0.0000
31.0000	38.5708	2.8688	∞	6.0000	∞	0.0000
32.0000	39.9468	2.7340	∞	6.0000	∞	0.0000
33.0000	41.3421	2.6051	∞	6.0000	∞	0.0000
34.0000	42.7625	2.4812	∞	6.0000	∞	0.0000
35.0000	44.2092	2.3617	∞	6.0000	∞	0.0000
36.0000	45.6883	2.2460	∞	6.0000	∞	0.0000
37.0000	47.2050	2.1334	∞	6.0000	∞	0.0000
38.0000	48.7669	2.0232	∞	6.0000	∞	0.0000
39.0000	50.3848	1.9148	∞	6.0000	∞	0.0000
40.0000	52.0728	1.8071	∞	6.0000	∞	0.0000
41.0000	53.8524	1.6992	∞	6.0000	∞	0.0000
42.0000	55.7593	1.5891	∞	6.0000	∞	0.0000
43.0000	57.8529	1.4744	∞	6.0000	∞	0.0000
44.0000	60.2680	1.3492	∞	6.0000	∞	0.0000
45.0000	63.4336	1.1953	∞	6.0000	∞	0.0000
m*45.5847	67.7923	1.0000	∞	6.0000	∞	0.0000

* DENOTES $M_2 = 1$, m DENOTES θ_{max}

OBLIQUE SHOCKS IN A PERFECT GAS

$$\gamma = 7/5 \qquad M_1 = \infty$$

STRONG SHOCK SOLUTION

θ	β	M_1	p_{21}	ρ_{21}	T_{21}	p^o_{21}
m*45.5847	67.7923	1.0000	∞	6.0000	∞	0.0000
45.0000	71.5654	0.8451	∞	6.0000	∞	0.0000
44.0000	73.7320	0.7621	∞	6.0000	∞	0.0000
43.0000	75.1465	0.7103	∞	6.0000	∞	0.0000
42.0000	76.2412	0.6717	∞	6.0000	∞	0.0000
41.0000	77.1476	0.6408	∞	6.0000	∞	0.0000
40.0000	77.9276	0.6149	∞	6.0000	∞	0.0000
39.0000	78.6157	0.5928	∞	6.0000	∞	0.0000
38.0000	79.2334	0.5734	∞	6.0000	∞	0.0000
37.0000	79.7954	0.5563	∞	6.0000	∞	0.0000
36.0000	80.3121	0.5411	∞	6.0000	∞	0.0000
35.0000	80.7911	0.5273	∞	6.0000	∞	0.0000
34.0000	81.2381	0.5148	∞	6.0000	∞	0.0000
33.0000	81.6579	0.5034	∞	6.0000	∞	0.0000
32.0000	82.0540	0.4930	∞	6.0000	∞	0.0000
31.0000	82.4294	0.4834	∞	6.0000	∞	0.0000
30.0000	82.7867	0.4746	∞	6.0000	∞	0.0000
29.0000	83.1277	0.4664	∞	6.0000	∞	0.0000
28.0000	83.4544	0.4589	∞	6.0000	∞	0.0000
27.0000	83.7684	0.4519	∞	6.0000	∞	0.0000
26.0000	84.0706	0.4453	∞	6.0000	∞	0.0000
25.0000	84.3624	0.4393	∞	6.0000	∞	0.0000
24.0000	84.6448	0.4336	∞	6.0000	∞	0.0000
23.0000	84.9185	0.4284	∞	6.0000	∞	0.0000
22.0000	85.1845	0.4235	∞	6.0000	∞	0.0000
21.0000	85.4433	0.4190	∞	6.0000	∞	0.0000
20.0000	85.6957	0.4147	∞	6.0000	∞	0.0000
19.0000	85.9420	0.4108	∞	6.0000	∞	0.0000
18.0000	86.1830	0.4071	∞	6.0000	∞	0.0000
17.0000	86.4190	0.4037	∞	6.0000	∞	0.0000
16.0000	86.6505	0.4006	∞	6.0000	∞	0.0000
15.0000	86.8779	0.3977	∞	6.0000	∞	0.0000
14.0000	87.1015	0.3950	∞	6.0000	∞	0.0000
13.0000	87.3217	0.3926	∞	6.0000	∞	0.0000
12.0000	87.5388	0.3903	∞	6.0000	∞	0.0000
11.0000	87.7532	0.3883	∞	6.0000	∞	0.0000
10.0000	87.9650	0.3865	∞	6.0000	∞	0.0000
9.0000	88.1747	0.3848	∞	6.0000	∞	0.0000
8.0000	88.3823	0.3834	∞	6.0000	∞	0.0000
7.0000	88.5882	0.3821	∞	6.0000	∞	0.0000
6.0000	88.7926	0.3810	∞	6.0000	∞	0.0000
5.0000	88.9958	0.3801	∞	6.0000	∞	0.0000
4.0000	89.1979	0.3793	∞	6.0000	∞	0.0000
3.0000	89.3991	0.3787	∞	6.0000	∞	0.0000
2.0000	89.5998	0.3783	∞	6.0000	∞	0.0000
1.0000	89.8000	0.3780	∞	6.0000	∞	0.0000
0.0000	90.0000	0.3780	∞	6.0000	∞	0.0000

* DENOTES $M_2 = 1$, m DENOTES θ_{max}

TABLE E.3

OBLIQUE SHOCKS IN A PERFECT GAS

$\gamma = 4/3$

OBLIQUE SHOCKS IN A PERFECT GAS

$$\gamma = 4/3 \quad M_1 = 1.10$$

	θ	β	M_1	p_{21}	ρ_{21}	T_{21}	p^o_{21}
	0.0000	65.3800	1.1000	1.0000	1.0000	1.0000	1.0000
	1.0000	69.6308	1.0416	1.0725	1.0539	1.0177	1.0000
*	1.4507	73.2622	1.0000	1.1253	1.0925	1.0300	0.9998
m	1.5633	76.3124	0.9710	1.1626	1.1195	1.0385	0.9996
	1.0000	83.8177	0.9237	1.2240	1.1634	1.0521	0.9991
	0.0000	90.0000	0.9114	1.2400	1.1748	1.0555	0.9989

$$\gamma = 4/3 \quad M_1 = 1.15$$

	θ	β	M_1	p_{21}	ρ_{21}	T_{21}	p^o_{21}
	0.0000	60.4081	1.1500	1.0000	1.0000	1.0000	1.0000
	1.0000	63.0705	1.1041	1.0586	1.0436	1.0143	1.0000
	2.0000	66.7110	1.0479	1.1323	1.0976	1.0316	0.9998
*	2.5748	70.2902	1.0000	1.1967	1.1440	1.0461	0.9994
m	2.7597	73.8516	0.9596	1.2517	1.1830	1.0580	0.9988
	2.0000	81.5649	0.8980	1.3360	1.2419	1.0758	0.9973
	1.0000	86.1316	0.8793	1.3617	1.2595	1.0811	0.9968
	0.0000	90.0000	0.8743	1.3686	1.2643	1.0825	0.9966

$$\gamma = 4/3 \quad M_1 = 1.20$$

	θ	β	M_1	p_{21}	ρ_{21}	T_{21}	p^o_{21}
	0.0000	56.4427	1.2000	1.0000	1.0000	1.0000	1.0000
	1.0000	58.4830	1.1594	1.0531	1.0396	1.0130	1.0000
	2.0000	60.8863	1.1148	1.1133	1.0838	1.0272	0.9999
	3.0000	63.9791	1.0622	1.1861	1.1364	1.0437	0.9995
*	3.8286	68.1129	1.0000	1.2742	1.1988	1.0629	0.9984
m	4.0812	72.0218	0.9499	1.3461	1.2488	1.0779	0.9971
	4.0000	74.2798	0.9249	1.3820	1.2735	1.0853	0.9963
	3.0000	80.5055	0.8722	1.4581	1.3250	1.1005	0.9942
	2.0000	84.1105	0.8531	1.4855	1.3433	1.1059	0.9933
	1.0000	87.1518	0.8438	1.4988	1.3521	1.1085	0.9928
	0.0000	90.0000	0.8410	1.5029	1.3548	1.1093	0.9927

* DENOTES $M_2 = 1$, m DENOTES θ_{max}

E.3-2

OBLIQUE SHOCKS IN A PERFECT GAS

$\gamma = 4/3 \quad M_1 = 1.30$

	θ	β	M_1	p_{21}	ρ_{21}	T_{21}	p^o_{21}
	0.0000	50.2849	1.3000	1.0000	1.0000	1.0000	1.0000
	1.0000	51.7664	1.2645	1.0488	1.0364	1.0120	1.0000
	2.0000	53.3729	1.2275	1.1011	1.0749	1.0244	0.9999
	3.0000	55.1432	1.1887	1.1577	1.1160	1.0374	0.9997
	4.0000	57.1460	1.1469	1.2201	1.1607	1.0512	0.9991
	5.0000	59.5176	1.1001	1.2916	1.2110	1.0665	0.9982
	6.0000	62.6238	1.0432	1.3802	1.2722	1.0849	0.9964
*	6.5528	65.1879	1.0000	1.4484	1.3185	1.0986	0.9945
m	6.9115	69.4784	0.9353	1.5512	1.3868	1.1186	0.9908
	6.0000	76.3687	0.8530	1.6813	1.4709	1.1431	0.9848
	5.0000	79.5518	0.8249	1.7251	1.4986	1.1511	0.9825
	4.0000	82.0468	0.8078	1.7516	1.5153	1.1559	0.9810
	3.0000	84.2239	0.7965	1.7690	1.5262	1.1591	0.9800
	2.0000	86.2284	0.7892	1.7802	1.5332	1.1611	0.9793
	1.0000	88.1354	0.7851	1.7865	1.5371	1.1623	0.9790
	0.0000	90.0000	0.7837	1.7886	1.5384	1.1626	0.9788

$\gamma = 4/3 \quad M_1 = 1.40$

	θ	β	M_1	p_{21}	ρ_{21}	T_{21}	p^o_{21}
	0.0000	45.5847	1.4000	1.0000	1.0000	1.0000	1.0000
	1.0000	46.8055	1.3667	1.0477	1.0355	1.0117	1.0000
	2.0000	48.0946	1.3327	1.0979	1.0725	1.0236	0.9999
	3.0000	49.4627	1.2979	1.1509	1.1111	1.0358	0.9997
	4.0000	50.9280	1.2619	1.2072	1.1515	1.0484	0.9993
	5.0000	52.5156	1.2244	1.2676	1.1942	1.0615	0.9985
	6.0000	54.2652	1.1845	1.3331	1.2398	1.0752	0.9974
	7.0000	56.2465	1.1413	1.4056	1.2895	1.0900	0.9957
	8.0000	58.5997	1.0924	1.4891	1.3457	1.1066	0.9931
	9.0000	61.7217	1.0316	1.5944	1.4150	1.1268	0.9890
*	9.3847	63.4438	1.0000	1.6494	1.4505	1.1371	0.9864
m	9.8031	67.8428	0.9258	1.7785	1.5321	1.1608	0.9794
	9.0000	73.8430	0.8413	1.9237	1.6211	1.1867	0.9700
	8.0000	76.8386	0.8072	1.9810	1.6554	1.1967	0.9659
	7.0000	79.0900	0.7856	2.0169	1.6767	1.2029	0.9632
	6.0000	80.9984	0.7702	2.0423	1.6916	1.2073	0.9613
	5.0000	82.7065	0.7587	2.0610	1.7026	1.2105	0.9598
	4.0000	84.2851	0.7501	2.0749	1.7107	1.2129	0.9587
	3.0000	85.7784	0.7439	2.0850	1.7166	1.2146	0.9579
	2.0000	87.2141	0.7397	2.0918	1.7205	1.2158	0.9574
	1.0000	88.6150	0.7372	2.0958	1.7229	1.2165	0.9570
	0.0000	90.0000	0.7364	2.0971	1.7236	1.2167	0.9569

* DENOTES $M_2 = 1$, m DENOTES θ_{max}

OBLIQUE SHOCKS IN A PERFECT GAS

$\gamma = 4/3 \quad M_1 = 1.50$

	θ	β	M_1	p_{21}	ρ_{21}	T_{21}	p^o_{21}
	0.0000	41.8103	1.5000	1.0000	1.0000	1.0000	1.0000
	1.0000	42.8812	1.4677	1.0479	1.0357	1.0118	1.0000
	2.0000	43.9968	1.4351	1.0978	1.0725	1.0236	0.9999
	3.0000	45.1636	1.4021	1.1502	1.1106	1.0357	0.9997
	4.0000	46.3886	1.3684	1.2052	1.1500	1.0479	0.9993
	5.0000	47.6803	1.3340	1.2630	1.1910	1.0605	0.9986
	6.0000	49.0529	1.2985	1.3241	1.2336	1.0734	0.9976
	7.0000	50.5226	1.2617	1.3892	1.2783	1.0867	0.9961
	8.0000	52.1175	1.2229	1.4590	1.3256	1.1007	0.9941
	9.0000	53.8784	1.1815	1.5350	1.3761	1.1155	0.9915
	10.0000	55.8799	1.1362	1.6195	1.4312	1.1315	0.9879
	11.0000	58.2786	1.0842	1.7177	1.4940	1.1498	0.9829
	12.0000	61.5730	1.0167	1.8459	1.5737	1.1729	0.9753
*	12.1851	62.4283	1.0000	1.8777	1.5932	1.1786	0.9732
m	12.6220	66.7645	0.9203	2.0284	1.6834	1.2049	0.9624
	12.0000	71.7868	0.8399	2.1774	1.7697	1.2303	0.9503
	11.0000	74.8229	0.7985	2.2523	1.8121	1.2429	0.9437
	10.0000	76.9839	0.7727	2.2981	1.8377	1.2506	0.9396
	9.0000	78.7673	0.7539	2.3310	1.8559	1.2560	0.9366
	8.0000	80.3306	0.7394	2.3560	1.8696	1.2602	0.9343
	7.0000	81.7500	0.7280	2.3756	1.8803	1.2634	0.9324
	6.0000	83.0692	0.7189	2.3911	1.8888	1.2660	0.9310
	5.0000	84.3158	0.7117	2.4033	1.8954	1.2680	0.9298
	4.0000	85.5081	0.7061	2.4128	1.9006	1.2695	0.9289
	3.0000	86.6620	0.7019	2.4199	1.9044	1.2707	0.9282
	2.0000	87.7894	0.6989	2.4247	1.9070	1.2715	0.9277
	1.0000	88.8990	0.6972	2.4276	1.9086	1.2720	0.9275
	0.0000	90.0000	0.6966	2.4286	1.9091	1.2721	0.9274

* DENOTES $M_2 = 1$, m DENOTES θ_{max}

OBLIQUE SHOCKS IN A PERFECT GAS

$\gamma = 4/3 \quad M_1 = 1.60$

θ	β	M_1	p_{21}	ρ_{21}	T_{21}	p^0_{21}
0.0000	38.6822	1.6000	1.0000	1.0000	1.0000	1.0000
1.0000	39.6556	1.5681	1.0487	1.0363	1.0120	1.0000
2.0000	40.6637	1.5360	1.0994	1.0736	1.0240	0.9999
3.0000	41.7088	1.5038	1.1523	1.1121	1.0362	0.9997
4.0000	42.7960	1.4711	1.2076	1.1517	1.0485	0.9993
5.0000	43.9291	1.4381	1.2653	1.1926	1.0610	0.9986
6.0000	45.1141	1.4044	1.3258	1.2348	1.0737	0.9976
7.0000	46.3587	1.3700	1.3894	1.2785	1.0867	0.9961
8.0000	47.6725	1.3345	1.4563	1.3237	1.1001	0.9942
9.0000	49.0709	1.2978	1.5272	1.3709	1.1140	0.9918
10.0000	50.5715	1.2594	1.6027	1.4204	1.1284	0.9886
11.0000	52.2056	1.2187	1.6841	1.4726	1.1436	0.9847
12.0000	54.0201	1.1749	1.7730	1.5287	1.1598	0.9798
13.0000	56.1041	1.1262	1.8729	1.5903	1.1777	0.9735
14.0000	58.6679	1.0686	1.9917	1.6618	1.1986	0.9651
* 14.8702	61.8872	1.0000	2.1332	1.7445	1.2229	0.9540
m 15.2956	66.0547	0.9178	2.3009	1.8392	1.2510	0.9394
15.0000	69.4121	0.8576	2.4211	1.9050	1.2709	0.9281
14.0000	72.9811	0.8004	2.5322	1.9644	1.2890	0.9172
13.0000	75.2072	0.7688	2.5921	1.9959	1.2987	0.9111
12.0000	76.9702	0.7462	2.6341	2.0177	1.3055	0.9068
11.0000	78.4795	0.7288	2.6662	2.0342	1.3107	0.9035
10.0000	79.8255	0.7147	2.6916	2.0472	1.3147	0.9009
9.0000	81.0572	0.7032	2.7122	2.0577	1.3180	0.8987
8.0000	82.2048	0.6937	2.7290	2.0663	1.3207	0.8969
7.0000	83.2886	0.6859	2.7429	2.0733	1.3229	0.8955
6.0000	84.3235	0.6794	2.7542	2.0791	1.3247	0.8943
5.0000	85.3205	0.6742	2.7634	2.0837	1.3262	0.8933
4.0000	86.2884	0.6700	2.7706	2.0873	1.3274	0.8925
3.0000	87.2343	0.6668	2.7760	2.0900	1.3282	0.8919
2.0000	88.1647	0.6646	2.7799	2.0920	1.3288	0.8915
1.0000	89.0844	0.6633	2.7821	2.0931	1.3292	0.8913
0.0000	90.0000	0.6629	2.7829	2.0935	1.3293	0.8912

* DENOTES $M_2 = 1$, m DENOTES θ_{max}

OBLIQUE SHOCKS IN A PERFECT GAS

$$\gamma = 4/3 \quad M_1 = 1.80$$

θ	β	M_1	p_{21}	ρ_{21}	T_{21}	p^o_{21}
0.0000	33.7490	1.8000	1.0000	1.0000	1.0000	1.0000
1.0000	34.6051	1.7677	1.0514	1.0383	1.0126	1.0000
2.0000	35.4859	1.7355	1.1049	1.0777	1.0253	0.9999
3.0000	36.3929	1.7033	1.1606	1.1181	1.0380	0.9996
4.0000	37.3288	1.6710	1.2187	1.1597	1.0509	0.9992
5.0000	38.2940	1.6386	1.2791	1.2023	1.0639	0.9984
6.0000	39.2921	1.6058	1.3421	1.2461	1.0771	0.9972
7.0000	40.3239	1.5728	1.4077	1.2910	1.0904	0.9956
8.0000	41.3941	1.5393	1.4761	1.3370	1.1040	0.9936
9.0000	42.5060	1.5052	1.5476	1.3844	1.1179	0.9910
10.0000	43.6642	1.4704	1.6223	1.4330	1.1321	0.9877
11.0000	44.8747	1.4348	1.7005	1.4831	1.1466	0.9838
12.0000	46.1451	1.3981	1.7826	1.5346	1.1616	0.9792
13.0000	47.4856	1.3602	1.8690	1.5879	1.1770	0.9738
14.0000	48.9104	1.3207	1.9605	1.6432	1.1931	0.9674
15.0000	50.4395	1.2792	2.0580	1.7008	1.2100	0.9600
16.0000	52.1041	1.2349	2.1630	1.7615	1.2279	0.9515
17.0000	53.9561	1.1868	2.2780	1.8265	1.2472	0.9414
18.0000	56.0956	1.1327	2.4078	1.8979	1.2687	0.9294
19.0000	58.7731	1.0672	2.5648	1.9816	1.2943	0.9139
* 19.7305	61.6397	1.0000	2.7245	2.0640	1.3200	0.8974
m 20.0875	65.3202	0.9185	2.9144	2.1586	1.3502	0.8770
20.0000	67.0802	0.8816	2.9984	2.1992	1.3634	0.8678
19.0000	71.3394	0.7990	3.1809	2.2853	1.3919	0.8474
18.0000	73.5444	0.7604	3.2629	2.3229	1.4046	0.8382
17.0000	75.2239	0.7333	3.3191	2.3485	1.4133	0.8319
16.0000	76.6289	0.7122	3.3620	2.3677	1.4199	0.8270
15.0000	77.8600	0.6951	3.3962	2.3829	1.4252	0.8231
14.0000	78.9696	0.6808	3.4244	2.3954	1.4296	0.8199
13.0000	79.9890	0.6686	3.4481	2.4059	1.4332	0.8172
12.0000	80.9391	0.6582	3.4682	2.4147	1.4363	0.8149
11.0000	81.8343	0.6492	3.4853	2.4222	1.4389	0.8130
10.0000	82.6853	0.6414	3.5000	2.4286	1.4412	0.8113
9.0000	83.5002	0.6347	3.5125	2.4340	1.4431	0.8099
8.0000	84.2856	0.6290	3.5233	2.4387	1.4447	0.8087
7.0000	85.0467	0.6240	3.5324	2.4426	1.4461	0.8077
6.0000	85.7878	0.6199	3.5400	2.4459	1.4473	0.8068
5.0000	86.5128	0.6165	3.5463	2.4486	1.4483	0.8061
4.0000	87.2249	0.6137	3.5513	2.4508	1.4490	0.8055
3.0000	87.9271	0.6116	3.5552	2.4525	1.4496	0.8051
2.0000	88.6222	0.6101	3.5579	2.4536	1.4500	0.8048
1.0000	89.3123	0.6092	3.5595	2.4543	1.4503	0.8046
0.0000	90.0000	0.6089	3.5600	2.4545	1.4504	0.8045

* DENOTES $M_2 = 1$, m DENOTES θ_{max}

OBLIQUE SHOCKS IN A PERFECT GAS

$$\gamma = 4/3 \quad M_1 = 2.00$$

WEAK SHOCK SOLUTION

θ	β	M_1	p_{21}	ρ_{21}	T_{21}	p^o_{21}
0.0000	30.0000	2.0000	1.0000	1.0000	1.0000	1.0000
1.0000	30.7886	1.9665	1.0549	1.0409	1.0135	1.0000
2.0000	31.5982	1.9332	1.1122	1.0830	1.0270	0.9999
3.0000	32.4305	1.9001	1.1718	1.1262	1.0406	0.9996
4.0000	33.2863	1.8669	1.2341	1.1706	1.0543	0.9990
5.0000	34.1664	1.8337	1.2989	1.2161	1.0681	0.9980
6.0000	35.0719	1.8004	1.3665	1.2628	1.0821	0.9967
7.0000	36.0043	1.7669	1.4369	1.3107	1.0963	0.9948
8.0000	36.9647	1.7331	1.5101	1.3597	1.1107	0.9924
9.0000	37.9553	1.6990	1.5864	1.4098	1.1253	0.9893
10.0000	38.9778	1.6644	1.6659	1.4610	1.1402	0.9856
11.0000	40.0347	1.6294	1.7487	1.5135	1.1554	0.9812
12.0000	41.1290	1.5937	1.8349	1.5670	1.1710	0.9760
13.0000	42.2642	1.5573	1.9249	1.6218	1.1869	0.9700
14.0000	43.4448	1.5201	2.0188	1.6778	1.2033	0.9631
15.0000	44.6763	1.4819	2.1170	1.7351	1.2201	0.9553
16.0000	45.9663	1.4425	2.2199	1.7939	1.2375	0.9466
17.0000	47.3243	1.4017	2.3281	1.8543	1.2555	0.9369
18.0000	48.7639	1.3591	2.4423	1.9165	1.2744	0.9260
19.0000	50.3044	1.3142	2.5637	1.9810	1.2941	0.9140
20.0000	51.9757	1.2665	2.6939	2.0485	1.3151	0.9006
21.0000	53.8272	1.2146	2.8361	2.1200	1.3378	0.8855
22.0000	55.9547	1.1563	2.9957	2.1980	1.3630	0.8681
23.0000	58.5948	1.0860	3.1873	2.2882	1.3929	0.8467
* 23.8381	61.9539	1.0000	3.4180	2.3926	1.4286	0.8207
m 24.1137	65.1070	0.9233	3.6186	2.4796	1.4593	0.7979

* DENOTES $M_2 = 1$, m DENOTES θ_{max}

OBLIQUE SHOCKS IN A PERFECT GAS

$\gamma = 4/3$ $M_1 = 2.00$

STRONG SHOCK SOLUTION

θ	β	M_1	p_{21}	ρ_{21}	T_{21}	p^o_{21}
m 24.1137	65.1070	0.9233	3.6186	2.4796	1.4593	0.7979
24.0000	67.0503	0.8782	3.7335	2.5280	1.4769	0.7848
23.0000	70.9759	0.7929	3.9428	2.6136	1.5086	0.7612
22.0000	73.0497	0.7515	4.0400	2.6522	1.5233	0.7503
21.0000	74.6219	0.7221	4.1071	2.6784	1.5334	0.7429
20.0000	75.9294	0.6992	4.1584	2.6983	1.5411	0.7372
19.0000	77.0683	0.6803	4.1996	2.7141	1.5473	0.7327
18.0000	78.0883	0.6644	4.2338	2.7272	1.5524	0.7289
17.0000	79.0199	0.6508	4.2627	2.7382	1.5568	0.7257
16.0000	79.8827	0.6389	4.2875	2.7475	1.5605	0.7230
15.0000	80.6906	0.6285	4.3089	2.7556	1.5637	0.7207
14.0000	81.4536	0.6193	4.3276	2.7626	1.5665	0.7186
13.0000	82.1796	0.6113	4.3439	2.7687	1.5690	0.7169
12.0000	82.8745	0.6041	4.3582	2.7740	1.5711	0.7153
11.0000	83.5433	0.5978	4.3708	2.7786	1.5730	0.7139
10.0000	84.1898	0.5922	4.3817	2.7827	1.5746	0.7128
9.0000	84.8176	0.5873	4.3913	2.7862	1.5761	0.7117
8.0000	85.4294	0.5831	4.3995	2.7893	1.5773	0.7108
7.0000	86.0278	0.5794	4.4066	2.7919	1.5784	0.7101
6.0000	86.6148	0.5763	4.4126	2.7941	1.5792	0.7094
5.0000	87.1926	0.5737	4.4176	2.7960	1.5800	0.7089
4.0000	87.7628	0.5716	4.4216	2.7974	1.5806	0.7084
3.0000	88.3271	0.5700	4.4247	2.7986	1.5810	0.7081
2.0000	88.8872	0.5689	4.4268	2.7994	1.5814	0.7079
1.0000	89.4444	0.5682	4.4281	2.7998	1.5816	0.7077
0.0000	90.0000	0.5680	4.4286	2.8000	1.5816	0.7077

* DENOTES $M_2 = 1$, m DENOTES θ_{max}

OBLIQUE SHOCKS IN A PERFECT GAS

$\gamma = 4/3 \quad M_1 = 2.20$

WEAK SHOCK SOLUTION

θ	β	M_1	P_{21}	ρ_{21}	T_{21}	P^o_{21}
0.0000	27.0357	2.2000	1.0000	1.0000	1.0000	1.0000
1.0000	27.7809	2.1648	1.0588	1.0438	1.0144	1.0000
2.0000	28.5462	2.1298	1.1203	1.0889	1.0288	0.9998
3.0000	29.3327	2.0950	1.1846	1.1353	1.0434	0.9995
4.0000	30.1406	2.0603	1.2518	1.1831	1.0581	0.9988
5.0000	30.9705	2.0257	1.3219	1.2321	1.0729	0.9976
6.0000	31.8232	1.9910	1.3951	1.2824	1.0879	0.9960
7.0000	32.6995	1.9562	1.4715	1.3339	1.1031	0.9937
8.0000	33.5999	1.9212	1.5511	1.3867	1.1186	0.9908
9.0000	34.5259	1.8859	1.6340	1.4406	1.1343	0.9872
10.0000	35.4782	1.8504	1.7204	1.4957	1.1503	0.9828
11.0000	36.4583	1.8144	1.8104	1.5519	1.1666	0.9775
12.0000	37.4675	1.7780	1.9040	1.6092	1.1832	0.9714
13.0000	38.5079	1.7411	2.0014	1.6675	1.2002	0.9644
14.0000	39.5814	1.7036	2.1028	1.7269	1.2177	0.9565
15.0000	40.6906	1.6653	2.2084	1.7874	1.2356	0.9476
16.0000	41.8388	1.6263	2.3183	1.8488	1.2539	0.9378
17.0000	43.0299	1.5864	2.4328	1.9114	1.2728	0.9270
18.0000	44.2690	1.5454	2.5523	1.9750	1.2923	0.9152
19.0000	45.5624	1.5032	2.6771	2.0399	1.3124	0.9024
20.0000	46.9187	1.4595	2.8080	2.1060	1.3333	0.8885
21.0000	48.3498	1.4141	2.9455	2.1737	1.3551	0.8736
22.0000	49.8720	1.3664	3.0909	2.2433	1.3779	0.8575
23.0000	51.5106	1.3160	3.2460	2.3152	1.4020	0.8401
24.0000	53.3059	1.2616	3.4135	2.3906	1.4279	0.8212
25.0000	55.3313	1.2015	3.5988	2.4712	1.4563	0.8001
26.0000	57.7507	1.1314	3.8136	2.5611	1.4890	0.7758
27.0000	61.1532	1.0361	4.1010	2.6761	1.5325	0.7436
* 27.2445	62.4848	1.0000	4.2080	2.7173	1.5486	0.7317
m 27.4505	65.1575	0.9297	4.4122	2.7940	1.5792	0.7094

* DENOTES $M_2 = 1$, m DENOTES θ_{max}

OBLIQUE SHOCKS IN A PERFECT GAS

$\gamma = 4/3 \quad M_1 = 2.20$

STRONG SHOCK SOLUTION

	θ	β	M_1	p_{21}	ρ_{21}	T_{21}	p^o_{21}
m	27.4505	65.1575	0.9297	4.4122	2.7940	1.5792	0.7094
	27.0000	68.8684	0.8374	4.6697	2.8868	1.6176	0.6820
	26.0000	71.6267	0.7737	4.8390	2.9456	1.6428	0.6644
	25.0000	73.4113	0.7350	4.9377	2.9791	1.6574	0.6543
	24.0000	74.8117	0.7063	5.0089	3.0030	1.6680	0.6471
	23.0000	75.9919	0.6833	5.0645	3.0214	1.6762	0.6415
	22.0000	77.0260	0.6642	5.1098	3.0363	1.6829	0.6370
	21.0000	77.9548	0.6479	5.1477	3.0486	1.6885	0.6333
	20.0000	78.8034	0.6338	5.1800	3.0591	1.6933	0.6301
	19.0000	79.5890	0.6214	5.2079	3.0681	1.6974	0.6273
	18.0000	80.3237	0.6104	5.2323	3.0760	1.7010	0.6250
	17.0000	81.0162	0.6007	5.2537	3.0828	1.7042	0.6229
	16.0000	81.6737	0.5920	5.2726	3.0888	1.7070	0.6210
	15.0000	82.3013	0.5842	5.2893	3.0942	1.7094	0.6194
	14.0000	82.9036	0.5772	5.3041	3.0989	1.7116	0.6180
	13.0000	83.4840	0.5709	5.3173	3.1031	1.7136	0.6167
	12.0000	84.0456	0.5653	5.3290	3.1068	1.7153	0.6156
	11.0000	84.5909	0.5603	5.3394	3.1100	1.7168	0.6146
	10.0000	85.1221	0.5558	5.3486	3.1129	1.7182	0.6137
	9.0000	85.6412	0.5519	5.3566	3.1154	1.7194	0.6129
	8.0000	86.1498	0.5485	5.3636	3.1176	1.7204	0.6123
	7.0000	86.6494	0.5455	5.3697	3.1195	1.7213	0.6117
	6.0000	87.1415	0.5429	5.3748	3.1212	1.7221	0.6112
	5.0000	87.6270	0.5408	5.3791	3.1225	1.7227	0.6108
	4.0000	88.1075	0.5391	5.3825	3.1236	1.7232	0.6104
	3.0000	88.5843	0.5378	5.3852	3.1244	1.7236	0.6102
	2.0000	89.0578	0.5368	5.3871	3.1250	1.7239	0.6100
	1.0000	89.5295	0.5363	5.3882	3.1253	1.7240	0.6099
	0.0000	90.0000	0.5361	5.3886	3.1255	1.7241	0.6099

* DENOTES $M_2 = 1$, m DENOTES θ_{max}

OBLIQUE SHOCKS IN A PERFECT GAS

$\gamma = 4/3$ $M_1 = 2.40$

WEAK SHOCK SOLUTION

θ	β	M_1	p_{21}	ρ_{21}	T_{21}	p^o_{21}
0.0000	24.6243	2.4000	1.0000	1.0000	1.0000	1.0000
1.0000	25.3403	2.3626	1.0630	1.0469	1.0154	1.0000
2.0000	26.0756	2.3255	1.1290	1.0952	1.0309	0.9998
3.0000	26.8318	2.2887	1.1983	1.1451	1.0464	0.9994
4.0000	27.6087	2.2520	1.2709	1.1965	1.0622	0.9985
5.0000	28.4068	2.2155	1.3470	1.2494	1.0781	0.9971
6.0000	29.2267	2.1789	1.4265	1.3037	1.0942	0.9951
7.0000	30.0686	2.1422	1.5097	1.3594	1.1106	0.9924
8.0000	30.9331	2.1054	1.5966	1.4164	1.1272	0.9889
9.0000	31.8210	2.0684	1.6872	1.4747	1.1442	0.9845
10.0000	32.7326	2.0311	1.7818	1.5342	1.1614	0.9793
11.0000	33.6691	1.9935	1.8804	1.5948	1.1791	0.9730
12.0000	34.6311	1.9555	1.9831	1.6566	1.1971	0.9658
13.0000	35.6196	1.9171	2.0900	1.7195	1.2155	0.9575
14.0000	36.6361	1.8781	2.2012	1.7833	1.2343	0.9482
15.0000	37.6817	1.8386	2.3169	1.8481	1.2537	0.9379
16.0000	38.7585	1.7984	2.4371	1.9137	1.2735	0.9265
17.0000	39.8683	1.7575	2.5621	1.9802	1.2939	0.9142
18.0000	41.0139	1.7159	2.6921	2.0475	1.3148	0.9008
19.0000	42.1984	1.6733	2.8272	2.1156	1.3364	0.8865
20.0000	43.4260	1.6298	2.9678	2.1845	1.3586	0.8712
21.0000	44.7017	1.5850	3.1143	2.2542	1.3815	0.8549
22.0000	46.0323	1.5390	3.2671	2.3249	1.4053	0.8377
23.0000	47.4268	1.4913	3.4271	2.3966	1.4300	0.8196
24.0000	48.8976	1.4417	3.5950	2.4696	1.4557	0.8005
25.0000	50.4626	1.3897	3.7724	2.5442	1.4828	0.7804
26.0000	52.1491	1.3345	3.9614	2.6210	1.5114	0.7591
27.0000	54.0025	1.2748	4.1660	2.7012	1.5422	0.7364
28.0000	56.1085	1.2085	4.3931	2.7869	1.5763	0.7115
29.0000	58.6728	1.1297	4.6605	2.8835	1.6162	0.6830
30.0000	62.6739	1.0115	5.0528	3.0175	1.6745	0.6427
* 30.0565	63.0735	1.0000	5.0900	3.0298	1.6800	0.6390
m 30.2092	65.3379	0.9363	5.2938	3.0956	1.7101	0.6190

* DENOTES $M_2 = 1$, m DENOTES θ_{max}

OBLIQUE SHOCKS IN A PERFECT GAS

$\gamma = 4/3 \quad M_1 = 2.40$

STRONG SHOCK SOLUTION

	θ	β	M_1	P_{21}	ρ_{21}	T_{21}	P^{o}_{21}
m	30.2092	65.3379	0.9363	5.2938	3.0956	1.7101	0.6190
	30.0000	67.8539	0.8684	5.5045	3.1614	1.7412	0.5989
	29.0000	71.1531	0.7844	5.7530	3.2362	1.7777	0.5760
	28.0000	73.0238	0.7397	5.8788	3.2729	1.7962	0.5648
	27.0000	74.4447	0.7075	5.9666	3.2982	1.8091	0.5571
	26.0000	75.6220	0.6821	6.0341	3.3173	1.8189	0.5512
	25.0000	76.6417	0.6611	6.0886	3.3327	1.8269	0.5465
	24.0000	77.5496	0.6432	6.1340	3.3454	1.8336	0.5427
	23.0000	78.3734	0.6277	6.1726	3.3561	1.8392	0.5394
	22.0000	79.1312	0.6140	6.2059	3.3653	1.8441	0.5366
	21.0000	79.8360	0.6020	6.2350	3.3733	1.8484	0.5342
	20.0000	80.4972	0.5912	6.2606	3.3802	1.8521	0.5320
	19.0000	81.1218	0.5816	6.2832	3.3864	1.8554	0.5302
	18.0000	81.7154	0.5729	6.3033	3.3919	1.8584	0.5285
	17.0000	82.2825	0.5650	6.3213	3.3967	1.8610	0.5270
	16.0000	82.8268	0.5579	6.3374	3.4011	1.8633	0.5257
	15.0000	83.3512	0.5515	6.3517	3.4050	1.8654	0.5245
	14.0000	83.8583	0.5457	6.3646	3.4084	1.8673	0.5235
	13.0000	84.3502	0.5405	6.3762	3.4115	1.8690	0.5225
	12.0000	84.8288	0.5357	6.3865	3.4143	1.8705	0.5217
	11.0000	85.2959	0.5315	6.3957	3.4168	1.8719	0.5209
	10.0000	85.7527	0.5277	6.4039	3.4190	1.8731	0.5203
	9.0000	86.2006	0.5244	6.4111	3.4209	1.8741	0.5197
	8.0000	86.6409	0.5214	6.4174	3.4226	1.8750	0.5192
	7.0000	87.0745	0.5189	6.4228	3.4240	1.8758	0.5187
	6.0000	87.5025	0.5167	6.4275	3.4252	1.8765	0.5184
	5.0000	87.9258	0.5149	6.4314	3.4263	1.8771	0.5181
	4.0000	88.3452	0.5134	6.4345	3.4271	1.8775	0.5178
	3.0000	88.7615	0.5122	6.4369	3.4278	1.8779	0.5176
	2.0000	89.1757	0.5114	6.4386	3.4282	1.8781	0.5175
	1.0000	89.5882	0.5109	6.4397	3.4285	1.8783	0.5174
	0.0000	90.0000	0.5108	6.4400	3.4286	1.8783	0.5174

* DENOTES $M_2 = 1$, m DENOTES θ_{max}

OBLIQUE SHOCKS IN A PERFECT GAS

$\gamma = 4/3$ $M_1 = 2.60$

WEAK SHOCK SOLUTION

θ	β	M_1	P_{21}	ρ_{21}	T_{21}	P_{21}^o
0.0000	22.6199	2.6000	1.0000	1.0000	1.0000	1.0000
1.0000	23.3147	2.5600	1.0673	1.0501	1.0164	1.0000
2.0000	24.0290	2.5205	1.1382	1.1019	1.0329	0.9998
3.0000	24.7643	2.4813	1.2127	1.1554	1.0496	0.9992
4.0000	25.5203	2.4423	1.2912	1.2107	1.0665	0.9982
5.0000	26.2974	2.4034	1.3735	1.2676	1.0835	0.9965
6.0000	27.0958	2.3646	1.4599	1.3262	1.1008	0.9941
7.0000	27.9158	2.3256	1.5505	1.3863	1.1185	0.9908
8.0000	28.7578	2.2866	1.6454	1.4479	1.1364	0.9866
9.0000	29.6222	2.2474	1.7446	1.5109	1.1547	0.9814
10.0000	30.5092	2.2079	1.8483	1.5752	1.1734	0.9751
11.0000	31.4195	2.1681	1.9566	1.6408	1.1925	0.9677
12.0000	32.3534	2.1279	2.0696	1.7076	1.2120	0.9591
13.0000	33.3117	2.0874	2.1873	1.7754	1.2320	0.9494
14.0000	34.2952	2.0463	2.3099	1.8442	1.2525	0.9385
15.0000	35.3046	2.0048	2.4375	1.9139	1.2736	0.9265
16.0000	36.3409	1.9627	2.5701	1.9844	1.2952	0.9134
17.0000	37.4057	1.9199	2.7079	2.0556	1.3174	0.8991
18.0000	38.5003	1.8765	2.8511	2.1274	1.3401	0.8839
19.0000	39.6267	1.8324	2.9997	2.1999	1.3636	0.8676
20.0000	40.7872	1.7874	3.1540	2.2728	1.3877	0.8505
21.0000	41.9846	1.7416	3.3142	2.3462	1.4126	0.8324
22.0000	43.2229	1.6948	3.4805	2.4201	1.4382	0.8135
23.0000	44.5065	1.6468	3.6535	2.4944	1.4646	0.7939
24.0000	45.8414	1.5975	3.8334	2.5693	1.4920	0.7735
25.0000	47.2360	1.5467	4.0212	2.6448	1.5204	0.7524
26.0000	48.7013	1.4940	4.2177	2.7210	1.5500	0.7307
27.0000	50.2529	1.4390	4.4243	2.7984	1.5810	0.7081
28.0000	51.9146	1.3811	4.6433	2.8775	1.6137	0.6848
29.0000	53.7246	1.3190	4.8783	2.9590	1.6486	0.6604
30.0000	55.7521	1.2509	5.1360	3.0448	1.6868	0.6344
31.0000	58.1495	1.1722	5.4314	3.1388	1.7304	0.6058
32.0000	61.4435	1.0676	5.8175	3.2551	1.7872	0.5702
* 32.3833	63.6492	1.0000	6.0608	3.3249	1.8229	0.5489
m 32.4971	65.5763	0.9427	6.2620	3.3806	1.8523	0.5319

* DENOTES $M_2 = 1$, m DENOTES θ_{max}

OBLIQUE SHOCKS IN A PERFECT GAS

$\gamma = 4/3$ $M_1 = 2.60$

STRONG SHOCK SOLUTION

	θ	β	M_1	p_{21}	ρ_{21}	T_{21}	p_{21}^o
m	32.4971	65.5763	0.9427	6.2620	3.3806	1.8523	0.5319
	32.0000	69.3364	0.8361	6.6208	3.4760	1.9047	0.5030
	31.0000	71.8867	0.7684	6.8361	3.5308	1.9361	0.4865
	30.0000	73.5473	0.7267	6.9631	3.5624	1.9546	0.4770
	29.0000	74.8460	0.6955	7.0549	3.5848	1.9680	0.4703
	28.0000	75.9353	0.6705	7.1266	3.6022	1.9784	0.4652
	27.0000	76.8847	0.6496	7.1851	3.6162	1.9869	0.4610
	26.0000	77.7328	0.6317	7.2341	3.6278	1.9941	0.4575
	25.0000	78.5039	0.6160	7.2760	3.6377	2.0002	0.4546
	24.0000	79.2137	0.6023	7.3123	3.6462	2.0054	0.4521
	23.0000	79.8740	0.5900	7.3440	3.6537	2.0100	0.4499
	22.0000	80.4931	0.5790	7.3721	3.6602	2.0141	0.4480
	21.0000	81.0777	0.5691	7.3970	3.6660	2.0177	0.4463
	20.0000	81.6330	0.5601	7.4193	3.6711	2.0210	0.4447
	19.0000	82.1628	0.5520	7.4392	3.6757	2.0239	0.4434
	18.0000	82.6706	0.5446	7.4571	3.6798	2.0265	0.4422
	17.0000	83.1592	0.5379	7.4732	3.6835	2.0288	0.4411
	16.0000	83.6310	0.5318	7.4878	3.6869	2.0309	0.4401
	15.0000	84.0879	0.5262	7.5009	3.6899	2.0328	0.4392
	14.0000	84.5318	0.5212	7.5127	3.6926	2.0346	0.4384
	13.0000	84.9640	0.5166	7.5233	3.6950	2.0361	0.4377
	12.0000	85.3860	0.5125	7.5329	3.6971	2.0375	0.4371
	11.0000	85.7991	0.5087	7.5414	3.6991	2.0387	0.4365
	10.0000	86.2041	0.5054	7.5490	3.7008	2.0398	0.4360
	9.0000	86.6022	0.5024	7.5557	3.7023	2.0408	0.4356
	8.0000	86.9941	0.4998	7.5616	3.7037	2.0417	0.4352
	7.0000	87.3808	0.4975	7.5667	3.7048	2.0424	0.4349
	6.0000	87.7631	0.4956	7.5711	3.7058	2.0430	0.4346
	5.0000	88.1416	0.4940	7.5747	3.7066	2.0436	0.4343
	4.0000	88.5169	0.4926	7.5777	3.7073	2.0440	0.4341
	3.0000	88.8898	0.4916	7.5800	3.7078	2.0443	0.4340
	2.0000	89.2609	0.4909	7.5816	3.7082	2.0446	0.4339
	1.0000	89.6308	0.4904	7.5825	3.7084	2.0447	0.4338
	0.0000	90.0000	0.4903	7.5829	3.7085	2.0447	0.4338

* DENOTES $M_2 = 1$, m DENOTES θ_{max}

OBLIQUE SHOCKS IN A PERFECT GAS

$\gamma = 4/3$ $M_1 = 2.80$

WEAK SHOCK SOLUTION

θ	β	M_1	p_{21}	ρ_{21}	T_{21}	p^o_{21}
0.0000	20.9248	2.8000	1.0000	1.0000	1.0000	1.0000
1.0000	21.6039	2.7572	1.0718	1.0534	1.0175	1.0000
2.0000	22.3035	2.7149	1.1476	1.1087	1.0351	0.9997
3.0000	23.0237	2.6730	1.2277	1.1661	1.0529	0.9991
4.0000	23.7650	2.6313	1.3122	1.2254	1.0709	0.9978
5.0000	24.5278	2.5898	1.4013	1.2866	1.0891	0.9958
6.0000	25.3119	2.5484	1.4950	1.3496	1.1077	0.9929
7.0000	26.1178	2.5068	1.5935	1.4144	1.1266	0.9890
8.0000	26.9455	2.4652	1.6970	1.4808	1.1460	0.9840
9.0000	27.7952	2.4234	1.8055	1.5488	1.1657	0.9778
10.0000	28.6671	2.3813	1.9191	1.6183	1.1859	0.9704
11.0000	29.5614	2.3389	2.0380	1.6891	1.2066	0.9616
12.0000	30.4786	2.2962	2.1623	1.7611	1.2278	0.9515
13.0000	31.4189	2.2530	2.2920	1.8342	1.2495	0.9402
14.0000	32.3827	2.2094	2.4272	1.9084	1.2719	0.9275
15.0000	33.3707	2.1653	2.5681	1.9833	1.2948	0.9136
16.0000	34.3833	2.1207	2.7146	2.0590	1.3184	0.8984
17.0000	35.4216	2.0755	2.8670	2.1353	1.3427	0.8822
18.0000	36.4865	2.0297	3.0253	2.2121	1.3676	0.8648
19.0000	37.5791	1.9833	3.1896	2.2893	1.3933	0.8465
20.0000	38.7010	1.9362	3.3600	2.3668	1.4196	0.8272
21.0000	39.8540	1.8884	3.5367	2.4445	1.4468	0.8072
22.0000	41.0406	1.8397	3.7199	2.5224	1.4748	0.7864
23.0000	42.2634	1.7901	3.9098	2.6003	1.5036	0.7649
24.0000	43.5261	1.7396	4.1067	2.6783	1.5333	0.7429
25.0000	44.8336	1.6879	4.3111	2.7564	1.5640	0.7204
26.0000	46.1920	1.6348	4.5235	2.8346	1.5958	0.6975
27.0000	47.6093	1.5803	4.7446	2.9130	1.6288	0.6742
28.0000	49.0970	1.5238	4.9757	2.9919	1.6631	0.6505
29.0000	50.6712	1.4650	5.2182	3.0714	1.6989	0.6263
30.0000	52.3560	1.4031	5.4749	3.1523	1.7368	0.6017
31.0000	54.1914	1.3369	5.7500	3.2353	1.7773	0.5763
32.0000	56.2501	1.2642	6.0516	3.3223	1.8215	0.5497
33.0000	58.6964	1.1799	6.3983	3.4175	1.8722	0.5207
34.0000	62.1476	1.0652	6.8614	3.5372	1.9398	0.4846
* 34.3198	64.1828	1.0000	7.1178	3.6000	1.9771	0.4658
m 34.4054	65.8342	0.9485	7.3155	3.6470	2.0059	0.4519

* DENOTES $M_2 = 1$, m DENOTES θ_{max}

OBLIQUE SHOCKS IN A PERFECT GAS

$$\gamma = 4/3 \quad M_1 = 2.80$$

STRONG SHOCK SOLUTION

	θ	β	M_1	P_{21}	ρ_{21}	T_{21}	P^o_{21}
m	34.4054	65.8342	0.9485	7.3155	3.6470	2.0059	0.4519
	34.0000	69.2025	0.8477	7.6875	3.7319	2.0599	0.4269
	33.0000	71.8733	0.7725	7.9498	3.7893	2.0980	0.4103
	32.0000	73.5458	0.7279	8.0983	3.8208	2.1195	0.4013
	31.0000	74.8375	0.6949	8.2042	3.8430	2.1348	0.3950
	30.0000	75.9131	0.6685	8.2863	3.8599	2.1468	0.3902
	29.0000	76.8458	0.6466	8.3531	3.8736	2.1564	0.3863
	28.0000	77.6756	0.6278	8.4089	3.8849	2.1645	0.3831
	27.0000	78.4272	0.6114	8.4565	3.8945	2.1714	0.3804
	26.0000	79.1172	0.5969	8.4978	3.9028	2.1774	0.3781
	25.0000	79.7572	0.5841	8.5338	3.9100	2.1826	0.3761
	24.0000	80.3557	0.5725	8.5657	3.9163	2.1872	0.3743
	23.0000	80.9195	0.5621	8.5940	3.9219	2.1913	0.3727
	22.0000	81.4535	0.5526	8.6193	3.9269	2.1949	0.3713
	21.0000	81.9621	0.5440	8.6420	3.9313	2.1982	0.3701
	20.0000	82.4483	0.5362	8.6624	3.9353	2.2012	0.3690
	19.0000	82.9152	0.5291	8.6808	3.9389	2.2038	0.3680
	18.0000	83.3650	0.5225	8.6975	3.9422	2.2063	0.3671
	17.0000	83.7997	0.5166	8.7126	3.9451	2.2084	0.3663
	16.0000	84.2210	0.5111	8.7263	3.9478	2.2104	0.3655
	15.0000	84.6305	0.5061	8.7387	3.9502	2.2122	0.3649
	14.0000	85.0293	0.5016	8.7499	3.9524	2.2138	0.3643
	13.0000	85.4188	0.4975	8.7600	3.9543	2.2153	0.3637
	12.0000	85.7999	0.4937	8.7691	3.9561	2.2166	0.3632
	11.0000	86.1736	0.4903	8.7772	3.9576	2.2178	0.3628
	10.0000	86.5407	0.4873	8.7845	3.9590	2.2188	0.3624
	9.0000	86.9021	0.4846	8.7910	3.9603	2.2198	0.3621
	8.0000	87.2584	0.4822	8.7966	3.9614	2.2206	0.3618
	7.0000	87.6103	0.4802	8.8016	3.9623	2.2213	0.3615
	6.0000	87.9584	0.4784	8.8058	3.9631	2.2219	0.3613
	5.0000	88.3034	0.4769	8.8093	3.9638	2.2224	0.3611
	4.0000	88.6459	0.4757	8.8121	3.9644	2.2228	0.3609
	3.0000	88.9861	0.4747	8.8143	3.9648	2.2232	0.3608
	2.0000	89.3250	0.4741	8.8159	3.9651	2.2234	0.3608
	1.0000	89.6628	0.4737	8.8168	3.9653	2.2235	0.3607
	0.0000	90.0000	0.4735	8.8171	3.9653	2.2236	0.3607

* DENOTES $M_2 = 1$, m DENOTES θ_{max}

OBLIQUE SHOCKS IN A PERFECT GAS

$\gamma = 4/3 \quad M_1 = 3.00$

WEAK SHOCK SOLUTION

θ	β	M_1	p_{21}	ρ_{21}	T_{21}	p_{21}^o
0.0000	19.4712	3.0000	1.0000	1.0000	1.0000	1.0000
1.0000	20.1379	2.9541	1.0763	1.0567	1.0186	1.0000
2.0000	20.8260	2.9088	1.1573	1.1157	1.0373	0.9997
3.0000	21.5357	2.8639	1.2431	1.1770	1.0562	0.9989
4.0000	22.2670	2.8193	1.3340	1.2405	1.0754	0.9974
5.0000	23.0201	2.7748	1.4301	1.3061	1.0949	0.9950
6.0000	23.7949	2.7305	1.5315	1.3738	1.1148	0.9916
7.0000	24.5917	2.6860	1.6384	1.4434	1.1351	0.9870
8.0000	25.4103	2.6415	1.7510	1.5149	1.1558	0.9811
9.0000	26.2511	2.5968	1.8694	1.5881	1.1771	0.9737
10.0000	27.1140	2.5518	1.9937	1.6629	1.1989	0.9650
11.0000	27.9989	2.5064	2.1240	1.7391	1.2213	0.9547
12.0000	28.9062	2.4607	2.2604	1.8167	1.2443	0.9430
13.0000	29.8360	2.4146	2.4031	1.8953	1.2679	0.9298
14.0000	30.7885	2.3680	2.5521	1.9749	1.2923	0.9152
15.0000	31.7640	2.3210	2.7075	2.0554	1.3173	0.8992
16.0000	32.7628	2.2735	2.8694	2.1365	1.3430	0.8819
17.0000	33.7854	2.2254	3.0378	2.2181	1.3696	0.8634
18.0000	34.8326	2.1768	3.2128	2.3000	1.3969	0.8439
19.0000	35.9051	2.1276	3.3946	2.3822	1.4250	0.8233
20.0000	37.0039	2.0778	3.5831	2.4645	1.4539	0.8019
21.0000	38.1301	2.0273	3.7785	2.5467	1.4837	0.7797
22.0000	39.2857	1.9761	3.9810	2.6288	1.5144	0.7569
23.0000	40.4722	1.9243	4.1905	2.7107	1.5459	0.7337
24.0000	41.6924	1.8716	4.4075	2.7923	1.5785	0.7100
25.0000	42.9491	1.8180	4.6321	2.8735	1.6120	0.6860
26.0000	44.2466	1.7634	4.8648	2.9544	1.6466	0.6618
27.0000	45.5898	1.7077	5.1059	3.0350	1.6823	0.6374
28.0000	46.9856	1.6506	5.3562	3.1153	1.7193	0.6130
29.0000	48.4431	1.5919	5.6166	3.1955	1.7577	0.5885
30.0000	49.9751	1.5311	5.8886	3.2758	1.7976	0.5639
31.0000	51.6003	1.4678	6.1744	3.3566	1.8395	0.5393
32.0000	53.3477	1.4009	6.4775	3.4385	1.8838	0.5144
33.0000	55.2663	1.3290	6.8038	3.5227	1.9314	0.4889
34.0000	57.4527	1.2490	7.1658	3.6116	1.9841	0.4624
35.0000	60.1567	1.1529	7.5957	3.7114	2.0466	0.4329
* 35.9427	64.6648	1.0000	8.2594	3.8544	2.1429	0.3917
m 36.0079	66.0909	0.9537	8.4533	3.8939	2.1709	0.3806

* DENOTES $M_2 = 1$, m DENOTES θ_{max}

OBLIQUE SHOCKS IN A PERFECT GAS

$\gamma = 4/3$ $M_1 = 3.00$

STRONG SHOCK SOLUTION

	θ	β	M_1	P_{21}	ρ_{21}	T_{21}	P^o_{21}
m	36.0079	66.0909	0.9537	8.4533	3.8939	2.1709	0.3806
	36.0000	66.5863	0.9378	8.5187	3.9070	2.1804	0.3769
	35.0000	71.2068	0.7966	9.0754	4.0141	2.2609	0.3473
	34.0000	73.1049	0.7425	9.2741	4.0505	2.2896	0.3375
	33.0000	74.4914	0.7048	9.4075	4.0745	2.3089	0.3310
	32.0000	75.6163	0.6753	9.5081	4.0923	2.3234	0.3263
	31.0000	76.5767	0.6511	9.5886	4.1064	2.3350	0.3226
	30.0000	77.4218	0.6306	9.6551	4.1180	2.3446	0.3195
	29.0000	78.1812	0.6128	9.7114	4.1277	2.3527	0.3170
	28.0000	78.8736	0.5972	9.7598	4.1360	2.3597	0.3148
	27.0000	79.5123	0.5834	9.8021	4.1432	2.3658	0.3129
	26.0000	80.1070	0.5709	9.8392	4.1495	2.3712	0.3112
	25.0000	80.6648	0.5597	9.8722	4.1551	2.3759	0.3098
	24.0000	81.1912	0.5496	9.9016	4.1600	2.3802	0.3085
	23.0000	81.6906	0.5404	9.9280	4.1645	2.3840	0.3074
	22.0000	82.1668	0.5319	9.9518	4.1684	2.3874	0.3063
	21.0000	82.6225	0.5242	9.9733	4.1720	2.3905	0.3054
	20.0000	83.0604	0.5172	9.9927	4.1753	2.3933	0.3046
	19.0000	83.4824	0.5107	10.0103	4.1782	2.3959	0.3038
	18.0000	83.8903	0.5048	10.0263	4.1808	2.3982	0.3031
	17.0000	84.2857	0.4993	10.0409	4.1832	2.4003	0.3025
	16.0000	84.6700	0.4943	10.0541	4.1854	2.4022	0.3019
	15.0000	85.0443	0.4898	10.0661	4.1874	2.4039	0.3014
	14.0000	85.4097	0.4856	10.0770	4.1892	2.4055	0.3010
	13.0000	85.7671	0.4818	10.0868	4.1908	2.4069	0.3006
	12.0000	86.1174	0.4783	10.0957	4.1923	2.4082	0.3002
	11.0000	86.4613	0.4752	10.1037	4.1936	2.4093	0.2998
	10.0000	86.7996	0.4724	10.1108	4.1948	2.4103	0.2995
	9.0000	87.1330	0.4699	10.1171	4.1958	2.4113	0.2993
	8.0000	87.4621	0.4677	10.1227	4.1967	2.4121	0.2990
	7.0000	87.7872	0.4658	10.1275	4.1975	2.4128	0.2988
	6.0000	88.1092	0.4641	10.1317	4.1982	2.4133	0.2987
	5.0000	88.4284	0.4628	10.1351	4.1987	2.4138	0.2985
	4.0000	88.7454	0.4616	10.1379	4.1992	2.4143	0.2984
	3.0000	89.0606	0.4608	10.1401	4.1996	2.4146	0.2983
	2.0000	89.3745	0.4601	10.1416	4.1998	2.4148	0.2983
	1.0000	89.6875	0.4598	10.1425	4.2000	2.4149	0.2982
	0.0000	90.0000	0.4596	10.1429	4.2000	2.4150	0.2982

* DENOTES $M_2 = 1$, m DENOTES θ_{max}

OBLIQUE SHOCKS IN A PERFECT GAS

$\gamma = 4/3$ $M_1 = 3.50$

WEAK SHOCK SOLUTION

θ	β	M_1	p_{21}	ρ_{21}	T_{21}	p^o_{21}
0.0000	16.6015	3.5000	1.0000	1.0000	1.0000	1.0000
1.0000	17.2482	3.4451	1.0880	1.0653	1.0213	0.9999
2.0000	17.9185	3.3911	1.1823	1.1337	1.0429	0.9995
3.0000	18.6123	3.3376	1.2832	1.2052	1.0648	0.9983
4.0000	19.3295	3.2845	1.3910	1.2796	1.0871	0.9961
5.0000	20.0703	3.2317	1.5059	1.3569	1.1098	0.9925
6.0000	20.8344	3.1789	1.6281	1.4368	1.1332	0.9875
7.0000	21.6216	3.1260	1.7580	1.5193	1.1571	0.9807
8.0000	22.4318	3.0730	1.8956	1.6041	1.1817	0.9720
9.0000	23.2648	3.0197	2.0413	1.6910	1.2071	0.9614
10.0000	24.1204	2.9660	2.1951	1.7798	1.2333	0.9487
11.0000	24.9982	2.9120	2.3573	1.8703	1.2604	0.9341
12.0000	25.8981	2.8576	2.5279	1.9622	1.2883	0.9176
13.0000	26.8200	2.8027	2.7071	2.0552	1.3172	0.8992
14.0000	27.7635	2.7473	2.8950	2.1491	1.3471	0.8791
15.0000	28.7288	2.6914	3.0917	2.2436	1.3780	0.8574
16.0000	29.7155	2.6351	3.2971	2.3385	1.4099	0.8343
17.0000	30.7238	2.5784	3.5114	2.4335	1.4429	0.8100
18.0000	31.7539	2.5211	3.7346	2.5285	1.4770	0.7847
19.0000	32.8059	2.4634	3.9667	2.6231	1.5122	0.7585
20.0000	33.8801	2.4053	4.2078	2.7173	1.5485	0.7318
21.0000	34.9772	2.3467	4.4578	2.8107	1.5860	0.7045
22.0000	36.0977	2.2877	4.7167	2.9033	1.6246	0.6771
23.0000	37.2426	2.2282	4.9847	2.9949	1.6644	0.6495
24.0000	38.4129	2.1683	5.2618	3.0854	1.7054	0.6221
25.0000	39.6104	2.1079	5.5480	3.1747	1.7476	0.5948
26.0000	40.8367	2.0469	5.8434	3.2627	1.7910	0.5679
27.0000	42.0942	1.9854	6.1483	3.3494	1.8357	0.5415
28.0000	43.3861	1.9232	6.4630	3.4347	1.8817	0.5155
29.0000	44.7161	1.8602	6.7878	3.5187	1.9291	0.4901
30.0000	46.0892	1.7963	7.1232	3.6013	1.9779	0.4654
31.0000	47.5120	1.7313	7.4701	3.6828	2.0284	0.4413
32.0000	48.9935	1.6648	7.8298	3.7633	2.0806	0.4178
33.0000	50.5461	1.5965	8.2038	3.8429	2.1348	0.3950
34.0000	52.1878	1.5257	8.5951	3.9221	2.1914	0.3727
35.0000	53.9470	1.4515	9.0079	4.0015	2.2511	0.3508
36.0000	55.8709	1.3723	9.4501	4.0821	2.3150	0.3290
37.0000	58.0528	1.2849	9.9373	4.1660	2.3853	0.3070
38.0000	60.7310	1.1810	10.5107	4.2588	2.4680	0.2833
* 38.9964	65.6500	1.0000	11.4771	4.4022	2.6071	0.2484
m 39.0313	66.6721	0.9641	11.6618	4.4279	2.6337	0.2424

* DENOTES $M_2 = 1$, m DENOTES θ_{max}

OBLIQUE SHOCKS IN A PERFECT GAS

$\gamma = 4/3$ $M_1 = 3.50$

STRONG SHOCK SOLUTION

	θ	β	M_1	P_{21}	ρ_{21}	T_{21}	P°_{21}
m	39.0313	66.6721	0.9641	11.6618	4.4279	2.6337	0.2424
	39.0000	67.6179	0.9313	11.8272	4.4505	2.6575	0.2371
	38.0000	71.7228	0.7956	12.4802	4.5360	2.7514	0.2178
	37.0000	73.5421	0.7392	12.7334	4.5676	2.7878	0.2108
	36.0000	74.8696	0.6997	12.9033	4.5883	2.8122	0.2063
	35.0000	75.9440	0.6689	13.0313	4.6038	2.8306	0.2030
	34.0000	76.8585	0.6436	13.1335	4.6159	2.8453	0.2004
	33.0000	77.6611	0.6221	13.2178	4.6259	2.8574	0.1983
	32.0000	78.3803	0.6034	13.2892	4.6342	2.8676	0.1965
	31.0000	79.0341	0.5871	13.3505	4.6413	2.8764	0.1950
	30.0000	79.6358	0.5725	13.4040	4.6475	2.8841	0.1937
	29.0000	80.1944	0.5594	13.4511	4.6529	2.8909	0.1926
	28.0000	80.7169	0.5476	13.4928	4.6577	2.8969	0.1916
	27.0000	81.2088	0.5369	13.5301	4.6620	2.9022	0.1907
	26.0000	81.6742	0.5271	13.5636	4.6658	2.9070	0.1899
	25.0000	82.1169	0.5182	13.5938	4.6692	2.9114	0.1892
	24.0000	82.5395	0.5100	13.6211	4.6723	2.9153	0.1886
	23.0000	82.9444	0.5024	13.6459	4.6751	2.9189	0.1880
	22.0000	83.3336	0.4955	13.6685	4.6776	2.9221	0.1875
	21.0000	83.7089	0.4891	13.6890	4.6799	2.9250	0.1870
	20.0000	84.0717	0.4832	13.7078	4.6820	2.9277	0.1866
	19.0000	84.4234	0.4777	13.7249	4.6840	2.9302	0.1862
	18.0000	84.7651	0.4727	13.7406	4.6857	2.9325	0.1858
	17.0000	85.0976	0.4681	13.7549	4.6873	2.9345	0.1855
	16.0000	85.4221	0.4638	13.7680	4.6888	2.9364	0.1852
	15.0000	85.7392	0.4599	13.7799	4.6901	2.9381	0.1849
	14.0000	86.0498	0.4563	13.7907	4.6913	2.9396	0.1847
	13.0000	86.3543	0.4530	13.8005	4.6924	2.9411	0.1844
	12.0000	86.6536	0.4501	13.8094	4.6934	2.9423	0.1842
	11.0000	86.9480	0.4474	13.8175	4.6942	2.9435	0.1841
	10.0000	87.2382	0.4449	13.8246	4.6950	2.9445	0.1839
	9.0000	87.5246	0.4427	13.8310	4.6957	2.9454	0.1838
	8.0000	87.8077	0.4408	13.8367	4.6964	2.9462	0.1836
	7.0000	88.0879	0.4391	13.8416	4.6969	2.9469	0.1835
	6.0000	88.3656	0.4377	13.8458	4.6974	2.9475	0.1834
	5.0000	88.6411	0.4365	13.8493	4.6978	2.9481	0.1834
	4.0000	88.9150	0.4355	13.8521	4.6981	2.9485	0.1833
	3.0000	89.1874	0.4347	13.8543	4.6983	2.9488	0.1832
	2.0000	89.4589	0.4342	13.8559	4.6985	2.9490	0.1832
	1.0000	89.7296	0.4339	13.8568	4.6986	2.9491	0.1832
	0.0000	90.0000	0.4338	13.8571	4.6986	2.9492	0.1832

* DENOTES $M_2 = 1$, m DENOTES θ_{max}

OBLIQUE SHOCKS IN A PERFECT GAS

$\gamma = 4/3 \quad M_1 = 4.00$

WEAK SHOCK SOLUTION

θ	β	M_1	P_{21}	ρ_{21}	T_{21}	p^o_{21}
0.0000	14.4775	4.0000	1.0000	1.0000	1.0000	1.0000
1.0000	15.1128	3.9346	1.1001	1.0742	1.0242	0.9999
2.0000	15.7734	3.8703	1.2083	1.1523	1.0486	0.9993
3.0000	16.4604	3.8068	1.3253	1.2344	1.0736	0.9976
4.0000	17.1729	3.7438	1.4512	1.3203	1.0991	0.9944
5.0000	17.9110	3.6810	1.5866	1.4099	1.1253	0.9893
6.0000	18.6742	3.6182	1.7318	1.5028	1.1523	0.9821
7.0000	19.4623	3.5553	1.8871	1.5989	1.1802	0.9726
8.0000	20.2748	3.4921	2.0529	1.6978	1.2091	0.9605
9.0000	21.1112	3.4286	2.2293	1.7992	1.2391	0.9458
10.0000	21.9710	3.3646	2.4168	1.9027	1.2702	0.9285
11.0000	22.8537	3.3001	2.6153	2.0080	1.3025	0.9088
12.0000	23.7587	3.2352	2.8252	2.1146	1.3360	0.8867
13.0000	24.6858	3.1697	3.0466	2.2223	1.3709	0.8625
14.0000	25.6341	3.1038	3.2795	2.3305	1.4072	0.8363
15.0000	26.6036	3.0375	3.5241	2.4391	1.4449	0.8086
16.0000	27.5938	2.9707	3.7804	2.5475	1.4840	0.7795
17.0000	28.6043	2.9035	4.0484	2.6555	1.5245	0.7494
18.0000	29.6351	2.8361	4.3281	2.7627	1.5666	0.7186
19.0000	30.6859	2.7683	4.6194	2.8690	1.6101	0.6873
20.0000	31.7570	2.7003	4.9225	2.9740	1.6552	0.6559
21.0000	32.8483	2.6321	5.2371	3.0775	1.7017	0.6245
22.0000	33.9602	2.5637	5.5632	3.1793	1.7498	0.5934
23.0000	35.0930	2.4952	5.9009	3.2793	1.7994	0.5628
24.0000	36.2476	2.4266	6.2499	3.3773	1.8505	0.5329
25.0000	37.4247	2.3578	6.6104	3.4733	1.9032	0.5038
26.0000	38.6253	2.2890	6.9823	3.5671	1.9574	0.4756
27.0000	39.8510	2.2199	7.3655	3.6587	2.0132	0.4484
28.0000	41.1036	2.1507	7.7603	3.7480	2.0705	0.4223
29.0000	42.3853	2.0812	8.1667	3.8352	2.1294	0.3972
30.0000	43.6993	2.0114	8.5850	3.9201	2.1900	0.3732
31.0000	45.0492	1.9411	9.0157	4.0029	2.2523	0.3504
32.0000	46.4397	1.8702	9.4593	4.0837	2.3163	0.3286
33.0000	47.8774	1.7985	9.9168	4.1626	2.3824	0.3078
34.0000	49.3708	1.7256	10.3895	4.2397	2.4505	0.2881
35.0000	50.9320	1.6511	10.8797	4.3154	2.5211	0.2693
36.0000	52.5783	1.5745	11.3905	4.3900	2.5947	0.2513
37.0000	54.3364	1.4946	11.9272	4.4640	2.6719	0.2340
38.0000	56.2515	1.4100	12.4992	4.5384	2.7541	0.2172
39.0000	58.4101	1.3173	13.1252	4.6149	2.8441	0.2006
40.0000	61.0286	1.2088	13.8527	4.6981	2.9485	0.1833
41.0000	65.3734	1.0379	14.9677	4.8150	3.1086	0.1603
* 41.0841	66.3771	1.0000	15.2067	4.8385	3.1429	0.1559
m 41.1043	67.1418	0.9716	15.3836	4.8556	3.1682	0.1527

* DENOTES $M_2 = 1$, m DENOTES θ_{max}

OBLIQUE SHOCKS IN A PERFECT GAS

$\gamma = 4/3 \quad M_1 = 4.00$

STRONG SHOCK SOLUTION

	θ	β	M_1	p_{21}	ρ_{21}	T_{21}	p^o_{21}
m	41.1043	67.1418	0.9716	15.3836	4.8556	3.1682	0.1527
	41.0000	68.8165	0.9105	15.7551	4.8906	3.2215	0.1463
	40.0000	72.2650	0.7908	16.4461	4.9528	3.3206	0.1353
	39.0000	73.9905	0.7341	16.7520	4.9791	3.3644	0.1307
	38.0000	75.2596	0.6941	16.9590	4.9966	3.3941	0.1278
	37.0000	76.2890	0.6627	17.1155	5.0096	3.4166	0.1256
	36.0000	77.1655	0.6369	17.2406	5.0199	3.4345	0.1239
	35.0000	77.9341	0.6149	17.3438	5.0283	3.4493	0.1225
	34.0000	78.6223	0.5959	17.4312	5.0353	3.4618	0.1214
	33.0000	79.2473	0.5791	17.5064	5.0413	3.4726	0.1204
	32.0000	79.8217	0.5642	17.5718	5.0465	3.4820	0.1195
	31.0000	80.3543	0.5508	17.6295	5.0511	3.4902	0.1188
	30.0000	80.8518	0.5386	17.6806	5.0552	3.4975	0.1182
	29.0000	81.3196	0.5276	17.7263	5.0588	3.5041	0.1176
	28.0000	81.7615	0.5175	17.7674	5.0620	3.5100	0.1171
	27.0000	82.1811	0.5083	17.8044	5.0649	3.5153	0.1166
	26.0000	82.5812	0.4998	17.8380	5.0675	3.5201	0.1162
	25.0000	82.9639	0.4920	17.8685	5.0699	3.5245	0.1158
	24.0000	83.3314	0.4848	17.8963	5.0720	3.5284	0.1155
	23.0000	83.6850	0.4782	17.9216	5.0740	3.5321	0.1151
	22.0000	84.0266	0.4720	17.9448	5.0758	3.5354	0.1149
	21.0000	84.3570	0.4663	17.9660	5.0774	3.5384	0.1146
	20.0000	84.6774	0.4610	17.9855	5.0789	3.5412	0.1144
	19.0000	84.9889	0.4561	18.0033	5.0803	3.5438	0.1142
	18.0000	85.2923	0.4516	18.0197	5.0815	3.5461	0.1140
	17.0000	85.5885	0.4474	18.0347	5.0827	3.5483	0.1138
	16.0000	85.8779	0.4436	18.0484	5.0837	3.5502	0.1136
	15.0000	86.1613	0.4400	18.0609	5.0847	3.5520	0.1135
	14.0000	86.4392	0.4368	18.0723	5.0855	3.5537	0.1133
	13.0000	86.7123	0.4338	18.0827	5.0863	3.5552	0.1132
	12.0000	86.9809	0.4311	18.0921	5.0871	3.5565	0.1131
	11.0000	87.2455	0.4286	18.1006	5.0877	3.5577	0.1130
	10.0000	87.5067	0.4264	18.1082	5.0883	3.5588	0.1129
	9.0000	87.7646	0.4244	18.1150	5.0888	3.5598	0.1128
	8.0000	88.0197	0.4227	18.1210	5.0893	3.5606	0.1127
	7.0000	88.2724	0.4211	18.1262	5.0897	3.5614	0.1127
	6.0000	88.5231	0.4198	18.1307	5.0900	3.5620	0.1126
	5.0000	88.7719	0.4187	18.1345	5.0903	3.5626	0.1126
	4.0000	89.0192	0.4178	18.1375	5.0905	3.5630	0.1125
	3.0000	89.2655	0.4171	18.1398	5.0907	3.5633	0.1125
	2.0000	89.5108	0.4166	18.1415	5.0908	3.5636	0.1125
	1.0000	89.7556	0.4163	18.1425	5.0909	3.5637	0.1125
	0.0000	90.0000	0.4162	18.1428	5.0909	3.5638	0.1125

* DENOTES $M_2 = 1$, m DENOTES θ_{max}

OBLIQUE SHOCKS IN A PERFECT GAS

$\gamma = 4/3$ $M_1 = 4.50$

WEAK SHOCK SOLUTION

θ	β	M_1	p_{21}	ρ_{21}	T_{21}	p^o_{21}
0.0000	12.8396	4.5000	1.0000	1.0000	1.0000	1.0000
1.0000	13.4676	4.4226	1.1124	1.0831	1.0270	0.9999
2.0000	14.1238	4.3467	1.2352	1.1713	1.0545	0.9990
3.0000	14.8084	4.2717	1.3689	1.2645	1.0826	0.9966
4.0000	15.5211	4.1973	1.5143	1.3624	1.1115	0.9922
5.0000	16.2617	4.1231	1.6719	1.4649	1.1413	0.9853
6.0000	17.0298	4.0488	1.8422	1.5715	1.1723	0.9755
7.0000	17.8240	3.9743	2.0255	1.6817	1.2044	0.9626
8.0000	18.6436	3.8994	2.2222	1.7952	1.2379	0.9464
9.0000	19.4887	3.8240	2.4330	1.9115	1.2728	0.9269
10.0000	20.3582	3.7481	2.6580	2.0300	1.3093	0.9044
11.0000	21.2512	3.6716	2.8976	2.1503	1.3475	0.8788
12.0000	22.1670	3.5945	3.1518	2.2718	1.3874	0.8507
13.0000	23.1049	3.5170	3.4209	2.3939	1.4290	0.8203
14.0000	24.0639	3.4390	3.7050	2.5161	1.4725	0.7881
15.0000	25.0436	3.3607	4.0041	2.6380	1.5179	0.7544
16.0000	26.0433	3.2821	4.3183	2.7591	1.5651	0.7196
17.0000	27.0627	3.2032	4.6476	2.8790	1.6143	0.6843
18.0000	28.1011	3.1243	4.9918	2.9973	1.6655	0.6488
19.0000	29.1585	3.0454	5.3510	3.1137	1.7185	0.6135
20.0000	30.2344	2.9665	5.7251	3.2279	1.7736	0.5786
21.0000	31.3291	2.8878	6.1138	3.3397	1.8306	0.5444
22.0000	32.4423	2.8092	6.5171	3.4490	1.8896	0.5112
23.0000	33.5744	2.7309	6.9349	3.5554	1.9505	0.4791
24.0000	34.7257	2.6529	7.3670	3.6590	2.0134	0.4483
25.0000	35.8967	2.5751	7.8131	3.7596	2.0782	0.4189
26.0000	37.0883	2.4977	8.2733	3.8573	2.1449	0.3909
27.0000	38.3014	2.4206	8.7475	3.9519	2.2135	0.3644
28.0000	39.5374	2.3437	9.2354	4.0435	2.2840	0.3394
29.0000	40.7978	2.2671	9.7373	4.1322	2.3565	0.3158
30.0000	42.0847	2.1906	10.2531	4.2179	2.4309	0.2936
31.0000	43.4008	2.1142	10.7830	4.3008	2.5072	0.2729
32.0000	44.7495	2.0378	11.3274	4.3810	2.5856	0.2534
33.0000	46.1348	1.9611	11.8868	4.4585	2.6661	0.2353
34.0000	47.5624	1.8840	12.4622	4.5337	2.7488	0.2183
35.0000	49.0399	1.8063	13.0550	4.6066	2.8340	0.2024
36.0000	50.5772	1.7274	13.6671	4.6775	2.9219	0.1875
37.0000	52.1887	1.6469	14.3018	4.7467	3.0130	0.1735
38.0000	53.8957	1.5639	14.9642	4.8146	3.1081	0.1604
39.0000	55.7327	1.4771	15.6630	4.8820	3.2083	0.1478
40.0000	57.7610	1.3842	16.4142	4.9500	3.3160	0.1358
41.0000	60.1131	1.2802	17.2538	5.0209	3.4364	0.1237
42.0000	63.2204	1.1485	18.3019	5.1029	3.5866	0.1106
* 42.5676	66.9179	1.0000	19.4428	5.1848	3.7500	0.09832
m 42.5800	67.5106	0.9770	19.6138	5.1964	3.7745	0.09663

* DENOTES $M_2 = 1$, m DENOTES θ_{max}

OBLIQUE SHOCKS IN A PERFECT GAS

$\gamma = 4/3 \quad M_1 = 4.50$

STRONG SHOCK SOLUTION

	θ	β	M_1	p_{21}	ρ_{21}	T_{21}	P^o_{21}
m	42.5800	67.5106	0.9770	19.6138	5.1964	3.7745	0.09663
	42.0000	71.2672	0.8370	20.6130	5.2617	3.9176	0.08751
	41.0000	73.4568	0.7602	21.1237	5.2933	3.9907	0.08329
	40.0000	74.8939	0.7119	21.4282	5.3115	4.0343	0.08089
	39.0000	76.0101	0.6757	21.6475	5.3245	4.0657	0.07923
	38.0000	76.9380	0.6465	21.8179	5.3344	4.0901	0.07796
	37.0000	77.7393	0.6221	21.9563	5.3423	4.1099	0.07696
	36.0000	78.4484	0.6012	22.0720	5.3489	4.1264	0.07613
	35.0000	79.0870	0.5829	22.1705	5.3545	4.1405	0.07543
	34.0000	79.6697	0.5667	22.2558	5.3593	4.1527	0.07483
	33.0000	80.2068	0.5522	22.3304	5.3635	4.1634	0.07432
	32.0000	80.7061	0.5391	22.3964	5.3672	4.1729	0.07386
	31.0000	81.1732	0.5273	22.4551	5.3704	4.1813	0.07347
	30.0000	81.6131	0.5165	22.5076	5.3733	4.1888	0.07311
	29.0000	82.0291	0.5066	22.5550	5.3759	4.1956	0.07279
	28.0000	82.4245	0.4975	22.5978	5.3783	4.2017	0.07250
	27.0000	82.8015	0.4891	22.6366	5.3804	4.2072	0.07225
	26.0000	83.1625	0.4814	22.6720	5.3823	4.2123	0.07201
	25.0000	83.5090	0.4742	22.7042	5.3841	4.2169	0.07180
	24.0000	83.8427	0.4676	22.7337	5.3857	4.2211	0.07161
	23.0000	84.1649	0.4615	22.7608	5.3871	4.2250	0.07143
	22.0000	84.4765	0.4558	22.7856	5.3885	4.2286	0.07127
	21.0000	84.7788	0.4505	22.8083	5.3897	4.2318	0.07112
	20.0000	85.0726	0.4457	22.8292	5.3908	4.2348	0.07098
	19.0000	85.3586	0.4411	22.8485	5.3919	4.2376	0.07086
	18.0000	85.6376	0.4369	22.8661	5.3928	4.2401	0.07074
	17.0000	85.9103	0.4330	22.8823	5.3937	4.2424	0.07064
	16.0000	86.1771	0.4294	22.8971	5.3945	4.2445	0.07054
	15.0000	86.4387	0.4261	22.9107	5.3952	4.2465	0.07046
	14.0000	86.6956	0.4231	22.9231	5.3959	4.2482	0.07038
	13.0000	86.9480	0.4203	22.9344	5.3965	4.2499	0.07030
	12.0000	87.1968	0.4178	22.9446	5.3970	4.2513	0.07024
	11.0000	87.4419	0.4155	22.9539	5.3975	4.2527	0.07018
	10.0000	87.6839	0.4134	22.9622	5.3980	4.2538	0.07013
	9.0000	87.9231	0.4115	22.9696	5.3984	4.2549	0.07008
	8.0000	88.1599	0.4099	22.9761	5.3987	4.2558	0.07004
	7.0000	88.3945	0.4084	22.9818	5.3990	4.2567	0.07000
	6.0000	88.6272	0.4072	22.9867	5.3993	4.2574	0.06997
	5.0000	88.8584	0.4061	22.9908	5.3995	4.2579	0.06994
	4.0000	89.0883	0.4053	22.9941	5.3997	4.2584	0.06992
	3.0000	89.3171	0.4046	22.9967	5.3998	4.2588	0.06991
	2.0000	89.5452	0.4042	22.9985	5.3999	4.2590	0.06990
	1.0000	89.7727	0.4039	22.9996	5.4000	4.2592	0.06989
	0.0000	90.0000	0.4038	23.0000	5.4000	4.2593	0.06989

* DENOTES $M_2 = 1$, m DENOTES θ_{max}

OBLIQUE SHOCKS IN A PERFECT GAS

$\gamma = 4/3 \quad M_1 = 5.00$

WEAK SHOCK SOLUTION

θ	β	M_1	p_{21}	ρ_{21}	T_{21}	p^o_{21}
0.0000	11.5370	5.0000	1.0000	1.0000	1.0000	1.0000
1.0000	12.1600	4.9092	1.1249	1.0922	1.0299	0.9998
2.0000	12.8144	4.8202	1.2626	1.1907	1.0604	0.9986
3.0000	13.4997	4.7324	1.4141	1.2953	1.0917	0.9955
4.0000	14.2156	4.6451	1.5802	1.4057	1.1241	0.9896
5.0000	14.9614	4.5581	1.7615	1.5215	1.1577	0.9804
6.0000	15.7364	4.4709	1.9587	1.6421	1.1928	0.9675
7.0000	16.5396	4.3833	2.1726	1.7670	1.2295	0.9507
8.0000	17.3701	4.2951	2.4036	1.8956	1.2680	0.9298
9.0000	18.2267	4.2063	2.6523	2.0271	1.3084	0.9050
10.0000	19.1086	4.1168	2.9190	2.1608	1.3509	0.8765
11.0000	20.0145	4.0267	3.2040	2.2960	1.3955	0.8449
12.0000	20.9435	3.9361	3.5077	2.4319	1.4424	0.8105
13.0000	21.8945	3.8450	3.8301	2.5679	1.4915	0.7739
14.0000	22.8666	3.7536	4.1714	2.7033	1.5431	0.7358
15.0000	23.8589	3.6621	4.5317	2.8376	1.5970	0.6966
16.0000	24.8707	3.5705	4.9109	2.9701	1.6535	0.6570
17.0000	25.9013	3.4791	5.3089	3.1004	1.7123	0.6175
18.0000	26.9502	3.3878	5.7258	3.2281	1.7737	0.5785
19.0000	28.0168	3.2970	6.1613	3.3530	1.8376	0.5404
20.0000	29.1010	3.2066	6.6153	3.4746	1.9039	0.5034
21.0000	30.2024	3.1168	7.0876	3.5927	1.9727	0.4680
22.0000	31.3209	3.0276	7.5778	3.7073	2.0440	0.4341
23.0000	32.4567	2.9391	8.0859	3.8182	2.1177	0.4020
24.0000	33.6100	2.8514	8.6115	3.9253	2.1938	0.3718
25.0000	34.7811	2.7645	9.1544	4.0287	2.2723	0.3434
26.0000	35.9706	2.6784	9.7144	4.1282	2.3532	0.3168
27.0000	37.1793	2.5931	10.2912	4.2240	2.4363	0.2921
28.0000	38.4082	2.5085	10.8847	4.3161	2.5219	0.2691
29.0000	39.6586	2.4247	11.4947	4.4047	2.6097	0.2478
30.0000	40.9320	2.3415	12.1211	4.4897	2.6998	0.2282
31.0000	42.2307	2.2589	12.7640	4.5713	2.7922	0.2100
32.0000	43.5571	2.1767	13.4236	4.6498	2.8869	0.1932
33.0000	44.9145	2.0949	14.1002	4.7251	2.9841	0.1778
34.0000	46.3071	2.0132	14.7944	4.7976	3.0837	0.1636
35.0000	47.7404	1.9314	15.5073	4.8674	3.1860	0.1505
36.0000	49.2215	1.8491	16.2403	4.9346	3.2911	0.1384
37.0000	50.7602	1.7661	16.9959	4.9997	3.3994	0.1273
38.0000	52.3703	1.6817	17.7778	5.0628	3.5115	0.1169
39.0000	54.0723	1.5951	18.5916	5.1244	3.6281	0.1073
40.0000	55.8988	1.5051	19.4476	5.1851	3.7507	0.09827
41.0000	57.9075	1.4092	20.3638	5.2459	3.8819	0.08968
42.0000	60.2203	1.3027	21.3806	5.3087	4.0275	0.08126
43.0000	63.2125	1.1709	22.6253	5.3798	4.2056	0.07232
* 43.6564	67.3268	1.0000	24.1831	5.4607	4.4286	0.06287
m 43.6644	67.7994	0.9811	24.3494	5.4689	4.4524	0.06196

* DENOTES $M_2 = 1$, m DENOTES θ_{max}

OBLIQUE SHOCKS IN A PERFECT GAS

$\gamma = 4/3$ $M_1 = 5.00$

STRONG SHOCK SOLUTION

	θ	β	M_1	P_{21}	ρ_{21}	T_{21}	P_{21}^o
m	43.6644	67.7994	0.9811	24.3494	5.4689	4.4524	0.06196
	43.0000	71.7646	0.8290	25.6308	5.5290	4.6357	0.05550
	42.0000	73.8230	0.7547	26.2108	5.5547	4.7187	0.05287
	41.0000	75.2043	0.7070	26.5653	5.5700	4.7694	0.05134
	40.0000	76.2839	0.6709	26.8222	5.5808	4.8061	0.05027
	39.0000	77.1837	0.6417	27.0227	5.5892	4.8348	0.04945
	38.0000	77.9617	0.6173	27.1857	5.5959	4.8581	0.04880
	37.0000	78.6506	0.5963	27.3221	5.6015	4.8776	0.04827
	36.0000	79.2710	0.5779	27.4384	5.6062	4.8943	0.04782
	35.0000	79.8372	0.5615	27.5390	5.6103	4.9087	0.04743
	34.0000	80.3589	0.5470	27.6272	5.6138	4.9213	0.04710
	33.0000	80.8437	0.5338	27.7051	5.6169	4.9324	0.04681
	32.0000	81.2972	0.5218	27.7744	5.6197	4.9424	0.04655
	31.0000	81.7239	0.5109	27.8366	5.6221	4.9512	0.04632
	30.0000	82.1274	0.5009	27.8925	5.6244	4.9592	0.04611
	29.0000	82.5106	0.4918	27.9432	5.6263	4.9665	0.04593
	28.0000	82.8759	0.4833	27.9891	5.6282	4.9731	0.04576
	27.0000	83.2252	0.4755	28.0310	5.6298	4.9790	0.04561
	26.0000	83.5605	0.4683	28.0692	5.6313	4.9845	0.04547
	25.0000	83.8831	0.4615	28.1042	5.6326	4.9895	0.04535
	24.0000	84.1944	0.4553	28.1362	5.6339	4.9941	0.04523
	23.0000	84.4953	0.4495	28.1656	5.6350	4.9983	0.04513
	22.0000	84.7870	0.4442	28.1927	5.6361	5.0022	0.04503
	21.0000	85.0703	0.4392	28.2176	5.6371	5.0057	0.04494
	20.0000	85.3460	0.4346	28.2405	5.6379	5.0090	0.04486
	19.0000	85.6146	0.4303	28.2615	5.6387	5.0120	0.04479
	18.0000	85.8770	0.4263	28.2809	5.6395	5.0148	0.04472
	17.0000	86.1336	0.4226	28.2987	5.6402	5.0173	0.04466
	16.0000	86.3850	0.4192	28.3150	5.6408	5.0197	0.04460
	15.0000	86.6316	0.4160	28.3299	5.6414	5.0218	0.04455
	14.0000	86.8739	0.4131	28.3436	5.6419	5.0238	0.04450
	13.0000	87.1123	0.4105	28.3560	5.6424	5.0255	0.04445
	12.0000	87.3470	0.4080	28.3674	5.6428	5.0272	0.04441
	11.0000	87.5787	0.4058	28.3776	5.6432	5.0286	0.04438
	10.0000	87.8075	0.4039	28.3867	5.6436	5.0299	0.04435
	9.0000	88.0337	0.4021	28.3949	5.6439	5.0311	0.04432
	8.0000	88.2576	0.4005	28.4021	5.6442	5.0321	0.04429
	7.0000	88.4797	0.3991	28.4084	5.6444	5.0330	0.04427
	6.0000	88.7000	0.3979	28.4139	5.6446	5.0338	0.04425
	5.0000	88.9188	0.3969	28.4184	5.6448	5.0345	0.04424
	4.0000	89.1365	0.3961	28.4221	5.6449	5.0350	0.04422
	3.0000	89.3531	0.3955	28.4249	5.6450	5.0354	0.04421
	2.0000	89.5692	0.3950	28.4269	5.6451	5.0357	0.04421
	1.0000	89.7847	0.3948	28.4282	5.6452	5.0359	0.04420
	0.0000	90.0000	0.3947	28.4286	5.6452	5.0359	0.04420

* DENOTES $M_2 = 1$, m DENOTES θ_{max}

OBLIQUE SHOCKS IN A PERFECT GAS

$\gamma = 4/3 \quad M_1 = 6.00$

WEAK SHOCK SOLUTION

θ	β	M_1	p_{21}	ρ_{21}	T_{21}	p^o_{21}
0.0000	9.5941	6.0000	1.0000	1.0000	1.0000	1.0000
1.0000	10.2123	5.8780	1.1504	1.1107	1.0357	0.9997
2.0000	10.8678	5.7587	1.3197	1.2306	1.0724	0.9977
3.0000	11.5586	5.6412	1.5090	1.3589	1.1104	0.9924
4.0000	12.2854	5.5242	1.7199	1.4954	1.1502	0.9828
5.0000	13.0463	5.4071	1.9537	1.6391	1.1919	0.9679
6.0000	13.8402	5.2895	2.2115	1.7891	1.2361	0.9473
7.0000	14.6655	5.1710	2.4943	1.9443	1.2829	0.9210
8.0000	15.5204	5.0517	2.8029	2.1035	1.3325	0.8891
9.0000	16.4034	4.9314	3.1382	2.2655	1.3853	0.8522
10.0000	17.3127	4.8105	3.5007	2.4289	1.4413	0.8113
11.0000	18.2469	4.6890	3.8907	2.5926	1.5007	0.7671
12.0000	19.2041	4.5672	4.3087	2.7555	1.5637	0.7207
13.0000	20.1832	4.4455	4.7548	2.9166	1.6303	0.6731
14.0000	21.1826	4.3241	5.2291	3.0749	1.7005	0.6253
15.0000	22.2014	4.2033	5.7315	3.2298	1.7746	0.5780
16.0000	23.2383	4.0834	6.2620	3.3807	1.8523	0.5319
17.0000	24.2926	3.9646	6.8204	3.5269	1.9338	0.4877
18.0000	25.3634	3.8472	7.4065	3.6682	2.0191	0.4456
19.0000	26.4502	3.7313	8.0198	3.8042	2.1081	0.4060
20.0000	27.5525	3.6171	8.6602	3.9349	2.2009	0.3691
21.0000	28.6699	3.5047	9.3271	4.0601	2.2973	0.3349
22.0000	29.8022	3.3941	10.0201	4.1798	2.3973	0.3034
23.0000	30.9495	3.2855	10.7389	4.2941	2.5009	0.2745
24.0000	32.1117	3.1788	11.4828	4.4030	2.6080	0.2482
25.0000	33.2890	3.0740	12.2514	4.5067	2.7185	0.2243
26.0000	34.4820	2.9712	13.0444	4.6053	2.8325	0.2027
27.0000	35.6911	2.8703	13.8611	4.6991	2.9497	0.1831
28.0000	36.9171	2.7712	14.7011	4.7881	3.0703	0.1654
29.0000	38.1608	2.6738	15.5640	4.8727	3.1941	0.1495
30.0000	39.4236	2.5781	16.4495	4.9531	3.3211	0.1352
31.0000	40.7070	2.4839	17.3574	5.0294	3.4512	0.1223
32.0000	42.0130	2.3910	18.2876	5.1018	3.5845	0.1108
33.0000	43.3438	2.2995	19.2400	5.1707	3.7209	0.1004
34.0000	44.7025	2.2089	20.2149	5.2363	3.8606	0.09101
35.0000	46.0927	2.1192	21.2130	5.2987	4.0035	0.08258
36.0000	47.5193	2.0300	22.2353	5.3582	4.1498	0.07498
37.0000	48.9887	1.9411	23.2835	5.4150	4.2998	0.06811
38.0000	50.5092	1.8520	24.3602	5.4694	4.4539	0.06190
39.0000	52.0924	1.7622	25.4696	5.5217	4.6126	0.05626
40.0000	53.7553	1.6710	26.6181	5.5722	4.7769	0.05112
41.0000	55.5237	1.5772	27.8166	5.6214	4.9484	0.04639
42.0000	57.4411	1.4791	29.0841	5.6698	5.1297	0.04200
43.0000	59.5909	1.3731	30.4588	5.7186	5.3263	0.03785
44.0000	62.1807	1.2508	32.0393	5.7705	5.5523	0.03371
45.0000	66.3934	1.0633	34.4021	5.8406	5.8901	0.02858
* 45.1116	67.8883	1.0000	35.1706	5.8618	6.0000	0.02714
m 45.1154	68.2090	0.9866	35.3303	5.8661	6.0228	0.02685

* DENOTES $M_2 = 1$, m DENOTES θ_{max}

OBLIQUE SHOCKS IN A PERFECT GAS

$\gamma = 4/3 \quad M_1 = 6.00$

STRONG SHOCK SOLUTION

	θ	β	M_1	p_{21}	ρ_{21}	T_{21}	p^o_{21}
m	45.1154	68.2090	0.9866	35.3303	5.8661	6.0228	0.02685
	45.0000	69.9142	0.9169	36.1475	5.8875	6.1397	0.02544
	44.0000	73.1715	0.7901	37.5516	5.9226	6.3404	0.02324
	43.0000	74.8074	0.7297	38.1744	5.9375	6.4234	0.02234
	42.0000	76.0050	0.6871	38.5937	5.9472	6.4894	0.02176
	41.0000	76.9717	0.6538	38.9091	5.9545	6.5344	0.02134
	40.0000	77.7912	0.6264	39.1600	5.9601	6.5703	0.02101
	39.0000	78.5071	0.6031	39.3667	5.9648	6.5998	0.02075
	38.0000	79.1454	0.5830	39.5409	5.9687	6.6248	0.02053
	37.0000	79.7231	0.5652	39.6905	5.9720	6.6461	0.02034
	36.0000	80.2520	0.5494	39.8205	5.9748	6.6647	0.02018
	35.0000	80.7408	0.5352	39.9348	5.9773	6.6811	0.02004
	34.0000	81.1957	0.5223	40.0361	5.9795	6.6955	0.01992
	33.0000	81.6219	0.5106	40.1265	5.9815	6.7085	0.01981
	32.0000	82.0235	0.4999	40.2077	5.9832	6.7201	0.01972
	31.0000	82.4034	0.4901	40.2810	5.9848	6.7305	0.01963
	30.0000	82.7644	0.4811	40.3473	5.9862	6.7400	0.01955
	29.0000	83.1088	0.4727	40.4077	5.9875	6.7487	0.01948
	28.0000	83.4382	0.4650	40.4627	5.9887	6.7565	0.01942
	27.0000	83.7545	0.4578	40.5131	5.9898	6.7637	0.01936
	26.0000	84.0588	0.4512	40.5592	5.9907	6.7703	0.01930
	25.0000	84.3524	0.4450	40.6015	5.9916	6.7764	0.01926
	24.0000	84.6362	0.4392	40.6405	5.9925	6.7819	0.01921
	23.0000	84.9113	0.4339	40.6763	5.9932	6.7871	0.01917
	22.0000	85.1783	0.4289	40.7093	5.9939	6.7918	0.01913
	21.0000	85.4381	0.4243	40.7397	5.9946	6.7961	0.01910
	20.0000	85.6913	0.4199	40.7678	5.9951	6.8001	0.01907
	19.0000	85.9383	0.4159	40.7936	5.9957	6.8038	0.01904
	18.0000	86.1800	0.4122	40.8174	5.9962	6.8072	0.01901
	17.0000	86.4165	0.4087	40.8393	5.9966	6.8103	0.01898
	16.0000	86.6485	0.4056	40.8594	5.9971	6.8132	0.01896
	15.0000	86.8762	0.4026	40.8778	5.9975	6.8159	0.01894
	14.0000	87.1002	0.3999	40.8947	5.9978	6.8183	0.01892
	13.0000	87.3207	0.3974	40.9101	5.9981	6.8205	0.01890
	12.0000	87.5381	0.3951	40.9241	5.9984	6.8225	0.01889
	11.0000	87.7526	0.3931	40.9367	5.9987	6.8243	0.01887
	10.0000	87.9646	0.3912	40.9481	5.9989	6.8259	0.01886
	9.0000	88.1744	0.3895	40.9582	5.9991	6.8274	0.01885
	8.0000	88.3821	0.3880	40.9672	5.9993	6.8286	0.01884
	7.0000	88.5881	0.3867	40.9750	5.9995	6.8298	0.01883
	6.0000	88.7925	0.3856	40.9817	5.9996	6.8307	0.01882
	5.0000	88.9957	0.3847	40.9874	5.9997	6.8315	0.01882
	4.0000	89.1978	0.3839	40.9919	5.9998	6.8322	0.01881
	3.0000	89.3992	0.3833	40.9955	5.9999	6.8327	0.01881
	2.0000	89.5997	0.3829	40.9980	6.0000	6.8330	0.01881
	1.0000	89.8000	0.3826	40.9995	6.0000	6.8333	0.01880
	0.0000	90.0000	0.3825	41.0000	6.0000	6.8333	0.01880

* DENOTES $M_2 = 1$, m DENOTES θ_{max}

OBLIQUE SHOCKS IN A PERFECT GAS

$\gamma = 4/3 \quad M_1 = 8.00$

WEAK SHOCK SOLUTION

θ	β	M_1	p_{21}	ρ_{21}	T_{21}	p^o_{21}
0.0000	7.1808	8.0000	1.0000	1.0000	1.0000	1.0000
1.0000	7.7977	7.7987	1.2035	1.1489	1.0476	0.9993
2.0000	8.4620	7.6030	1.4410	1.3135	1.0971	0.9947
3.0000	9.1732	7.4093	1.7161	1.4929	1.1495	0.9830
4.0000	9.9291	7.2158	2.0318	1.6855	1.2055	0.9621
5.0000	10.7271	7.0210	2.3912	1.8888	1.2660	0.9310
6.0000	11.5641	6.8245	2.7965	2.1003	1.3315	0.8898
7.0000	12.4370	6.6263	3.2497	2.3169	1.4026	0.8397
8.0000	13.3428	6.4269	3.7525	2.5359	1.4797	0.7827
9.0000	14.2782	6.2271	4.3062	2.7545	1.5633	0.7210
10.0000	15.2406	6.0277	4.9115	2.9703	1.6535	0.6570
11.0000	16.2272	5.8297	5.5689	3.1811	1.7507	0.5929
12.0000	17.2359	5.6339	6.2789	3.3852	1.8548	0.5305
13.0000	18.2645	5.4410	7.0414	3.5815	1.9660	0.4713
14.0000	19.3113	5.2519	7.8563	3.7690	2.0844	0.4162
15.0000	20.3749	5.0670	8.7232	3.9472	2.2100	0.3657
16.0000	21.4539	4.8867	9.6418	4.1157	2.3427	0.3201
17.0000	22.5473	4.7114	10.6114	4.2745	2.4825	0.2794
18.0000	23.6544	4.5412	11.6314	4.4237	2.6293	0.2434
19.0000	24.7745	4.3762	12.7011	4.5636	2.7831	0.2117
20.0000	25.9068	4.2165	13.8193	4.6945	2.9438	0.1840
21.0000	27.0513	4.0620	14.9855	4.8167	3.1111	0.1600
22.0000	28.2077	3.9126	16.1984	4.9309	3.2851	0.1391
23.0000	29.3759	3.7681	17.4572	5.0374	3.4655	0.1210
24.0000	30.5559	3.6285	18.7608	5.1367	3.6523	0.1054
25.0000	31.7479	3.4935	20.1081	5.2293	3.8453	0.09197
26.0000	32.9523	3.3628	21.4980	5.3157	4.0443	0.08036
27.0000	34.1696	3.2363	22.9296	5.3962	4.2492	0.07034
28.0000	35.4002	3.1138	24.4017	5.4714	4.4598	0.06168
29.0000	36.6452	2.9949	25.9134	5.5416	4.6761	0.05420
30.0000	37.9052	2.8795	27.4637	5.6072	4.8979	0.04772
31.0000	39.1818	2.7674	29.0521	5.6686	5.1251	0.04211
32.0000	40.4762	2.6581	30.6775	5.7260	5.3575	0.03724
33.0000	41.7903	2.5516	32.3396	5.7799	5.5952	0.03300
34.0000	43.1262	2.4475	34.0381	5.8304	5.8381	0.02930
35.0000	44.4867	2.3455	35.7732	5.8778	6.0862	0.02607
36.0000	45.8752	2.2454	37.5456	5.9225	6.3395	0.02324
37.0000	47.2959	2.1468	39.3563	5.9645	6.5984	0.02076
38.0000	48.7543	2.0493	41.2075	6.0043	6.8630	0.01857
39.0000	50.2576	1.9525	43.1026	6.0420	7.1339	0.01664
40.0000	51.8157	1.8559	45.0475	6.0778	7.4118	0.01493
41.0000	53.4425	1.7587	47.0507	6.1120	7.6981	0.01340
42.0000	55.1584	1.6600	49.1265	6.1448	7.9948	0.01202
43.0000	56.9966	1.5583	51.2995	6.1767	8.3054	0.01078
44.0000	59.0154	1.4510	53.6151	6.2081	8.6363	0.009638
45.0000	61.3406	1.3326	56.1758	6.2402	9.0022	0.008554
46.0000	64.3750	1.1859	59.3194	6.2762	9.4514	0.007434
* 46.6026	68.4806	1.0000	63.1583	6.3158	10.0000	0.006316
m 46.6038	68.6564	0.9924	63.3109	6.3173	10.0218	0.006276

* DENOTES $M_2 = 1$, m DENOTES θ_{max}

OBLIQUE SHOCKS IN A PERFECT GAS

$\gamma = 4/3$ $M_1 = 8.00$

STRONG SHOCK SOLUTION

	θ	β	M_1	P_{21}	ρ_{21}	T_{21}	P^o_{21}
m	46.6038	58.6564	0.9924	63.3109	6.3173	10.0218	0.006276
	46.0000	72.3487	0.8376	66.2749	6.3449	10.4453	0.005567
	45.0000	74.4081	0.7563	67.7158	6.3576	10.6512	0.005261
	44.0000	75.7593	0.7051	68.5738	6.3649	10.7738	0.005090
	43.0000	76.8051	0.6668	69.1889	6.3700	10.8617	0.004971
	42.0000	77.6711	0.6360	69.6652	6.3739	10.9297	0.004882
	41.0000	78.4159	0.6103	70.0506	6.3770	10.9848	0.004811
	40.0000	79.0726	0.5882	70.3716	6.3796	11.0307	0.004753
	39.0000	79.6619	0.5688	70.6444	6.3818	11.0697	0.004705
	38.0000	80.1976	0.5517	70.8799	6.3837	11.1033	0.004663
	37.0000	80.6898	0.5364	71.0856	6.3853	11.1327	0.004628
	36.0000	81.1458	0.5226	71.2671	6.3867	11.1586	0.004597
	35.0000	81.5710	0.5101	71.4284	6.3880	11.1817	0.004569
	34.0000	81.9700	0.4987	71.5727	6.3891	11.2023	0.004545
	33.0000	82.3463	0.4882	71.7025	6.3901	11.2208	0.004523
	32.0000	82.7027	0.4786	71.8199	6.3910	11.2376	0.004503
	31.0000	83.0417	0.4697	71.9265	6.3918	11.2528	0.004486
	30.0000	83.3651	0.4615	72.0235	6.3926	11.2667	0.004470
	29.0000	83.6746	0.4539	72.1121	6.3933	11.2794	0.004455
	28.0000	83.9717	0.4468	72.1933	6.3939	11.2910	0.004442
	27.0000	84.2577	0.4402	72.2677	6.3945	11.3016	0.004430
	26.0000	84.5336	0.4341	72.3362	6.3950	11.3114	0.004419
	25.0000	84.8003	0.4284	72.3992	6.3955	11.3204	0.004408
	24.0000	85.0588	0.4231	72.4573	6.3959	11.3287	0.004399
	23.0000	85.3097	0.4181	72.5109	6.3963	11.3364	0.004390
	22.0000	85.5537	0.4135	72.5604	6.3967	11.3434	0.004382
	21.0000	85.7914	0.4092	72.6060	6.3970	11.3499	0.004375
	20.0000	86.0233	0.4052	72.6482	6.3974	11.3560	0.004368
	19.0000	86.2500	0.4014	72.6871	6.3977	11.3615	0.004362
	18.0000	86.4719	0.3979	72.7230	6.3979	11.3667	0.004356
	17.0000	86.6893	0.3947	72.7560	6.3982	11.3714	0.004351
	16.0000	86.9027	0.3917	72.7864	6.3984	11.3757	0.004346
	15.0000	87.1125	0.3890	72.8144	6.3986	11.3797	0.004342
	14.0000	87.3189	0.3864	72.8399	6.3988	11.3834	0.004338
	13.0000	87.5222	0.3841	72.8633	6.3990	11.3867	0.004334
	12.0000	87.7228	0.3819	72.8845	6.3991	11.3897	0.004331
	11.0000	87.9209	0.3800	72.9037	6.3993	11.3925	0.004328
	10.0000	88.1167	0.3782	72.9210	6.3994	11.3949	0.004325
	9.0000	88.3106	0.3767	72.9364	6.3995	11.3971	0.004323
	8.0000	88.5026	0.3753	72.9500	6.3996	11.3991	0.004321
	7.0000	88.6931	0.3741	72.9619	6.3997	11.4008	0.004319
	6.0000	88.8823	0.3730	72.9721	6.3998	11.4023	0.004317
	5.0000	89.0703	0.3721	72.9807	6.3999	11.4035	0.004316
	4.0000	89.2574	0.3714	72.9877	6.3999	11.4045	0.004315
	3.0000	89.4437	0.3708	72.9931	6.4000	11.4052	0.004314
	2.0000	89.6294	0.3704	72.9969	6.4000	11.4058	0.004313
	1.0000	89.8148	0.3702	72.9992	6.4000	11.4061	0.004313
	0.0000	90.0000	0.3701	73.0000	6.4000	11.4062	0.004313

* DENOTES $M_2 = 1$, m DENOTES θ_{max}

OBLIQUE SHOCKS IN A PERFECT GAS

$\gamma = 4/3 \quad M_1 = 10.00$
WEAK SHOCK SOLUTION

θ	β	M_1	p_{21}	ρ_{21}	T_{21}	p^o_{21}
0.0000	5.7392	10.0000	1.0000	1.0000	1.0000	1.0000
1.0000	6.3582	9.6974	1.2588	1.1880	1.0596	0.9987
2.0000	7.0369	9.4032	1.5724	1.4006	1.1226	0.9899
3.0000	7.7713	9.1116	1.9468	1.6349	1.1907	0.9684
4.0000	8.5591	8.8181	2.3886	1.8874	1.2655	0.9312
5.0000	9.3947	8.5217	2.9024	2.1527	1.3483	0.8783
6.0000	10.2738	8.2225	3.4925	2.4253	1.4400	0.8122
7.0000	11.1910	7.9220	4.1619	2.6997	1.5416	0.7368
8.0000	12.1417	7.6221	4.9130	2.9708	1.6538	0.6568
9.0000	13.1218	7.3250	5.7473	3.2345	1.7769	0.5765
10.0000	14.1275	7.0328	6.6657	3.4876	1.9113	0.4995
11.0000	15.1556	6.7473	7.6687	3.7277	2.0572	0.4282
12.0000	16.2034	6.4700	8.7563	3.9536	2.2148	0.3639
13.0000	17.2687	6.2019	9.9281	4.1645	2.3840	0.3074
14.0000	18.3497	5.9436	11.1838	4.3603	2.5649	0.2584
15.0000	19.4448	5.6957	12.5225	4.5413	2.7575	0.2166
16.0000	20.5530	5.4582	13.9431	4.7081	2.9615	0.1813
17.0000	21.6730	5.2312	15.4444	4.8614	3.1770	0.1516
18.0000	22.8043	5.0143	17.0254	5.0021	3.4036	0.1269
19.0000	23.9464	4.8073	18.6845	5.1312	3.6414	0.1063
20.0000	25.0986	4.6097	20.4201	5.2495	3.8899	0.08918
21.0000	26.2609	4.4212	22.2309	5.3579	4.1492	0.07501
22.0000	27.4331	4.2413	24.1149	5.4573	4.4188	0.06325
23.0000	28.6152	4.0694	26.0708	5.5486	4.6986	0.05349
24.0000	29.8074	3.9050	28.0964	5.6323	4.9884	0.04537
25.0000	31.0098	3.7478	30.1903	5.7093	5.2879	0.03861
26.0000	32.2228	3.5972	32.3503	5.7802	5.5967	0.03297
27.0000	33.4469	3.4527	34.5750	5.8455	5.9148	0.02825
28.0000	34.6827	3.3139	36.8625	5.9057	6.2419	0.02428
29.0000	35.9310	3.1805	39.2110	5.9613	6.5776	0.02095
30.0000	37.1927	3.0519	41.6191	6.0127	6.9218	0.01813
31.0000	38.4688	2.9278	44.0850	6.0604	7.2743	0.01574
32.0000	39.7609	2.8078	46.6078	6.1046	7.6348	0.01372
33.0000	41.0703	2.6916	49.1860	6.1457	8.0033	0.01199
34.0000	42.3990	2.5787	51.8190	6.1839	8.3796	0.01051
35.0000	43.7495	2.4689	54.5065	6.2196	8.7637	0.009241
36.0000	45.1248	2.3617	57.2489	6.2529	9.1556	0.008147
37.0000	46.5283	2.2568	60.0468	6.2841	9.5554	0.007203
38.0000	47.9647	2.1538	62.9029	6.3133	9.9635	0.006383
39.0000	49.4398	2.0522	65.8205	6.3409	10.3804	0.005669
40.0000	50.9618	1.9515	68.8060	6.3668	10.8070	0.005044
41.0000	52.5411	1.8511	71.8688	6.3914	11.2446	0.004495
42.0000	54.1934	1.7502	75.0245	6.4148	11.6955	0.004010
43.0000	55.9418	1.6476	78.2984	6.4373	12.1633	0.003577
44.0000	57.8246	1.5416	81.7349	6.4591	12.6543	0.003188
45.0000	59.9132	1.4289	85.4212	6.4806	13.1810	0.002830
46.0000	62.3729	1.3023	89.5678	6.5029	13.7734	0.002489
47.0000	65.8930	1.1313	95.0771	6.5298	14.5605	0.002115
* 47.3087	68.7668	1.0000	99.1526	6.5478	15.1428	0.001886
m 47.3091	68.8778	0.9951	99.3019	6.5485	15.1641	0.001878

* DENOTES $M_2 = 1$, m DENOTES θ_{max}

OBLIQUE SHOCKS IN A PERFECT GAS

$\gamma = 4/3$ $M_1 = 10.00$

STRONG SHOCK SOLUTION

θ	β	M_1	p_{21}	ρ_{21}	T_{21}	p^o_{21}
m 47.3091	68.8778	0.9951	99.3019	6.5485	15.1641	0.001878
47.0000	71.5580	0.8792	102.7058	6.5625	15.6505	0.001712
46.0000	74.0939	0.7753	105.5588	6.5736	16.0581	0.001588
45.0000	75.5699	0.7175	107.0457	6.5791	16.2705	0.001527
44.0000	76.6754	0.6757	108.0725	6.5829	16.4172	0.001488
43.0000	77.5757	0.6427	108.8526	6.5857	16.5287	0.001458
42.0000	78.3423	0.6154	109.4765	6.5879	16.6178	0.001436
41.0000	79.0135	0.5921	109.9919	6.5897	16.6914	0.001417
40.0000	79.6124	0.5718	110.4273	6.5912	16.7536	0.001402
39.0000	80.1547	0.5539	110.8014	6.5925	16.8071	0.001389
38.0000	80.6511	0.5380	111.1270	6.5937	16.8536	0.001377
37.0000	81.1096	0.5237	111.4131	6.5946	16.8945	0.001368
36.0000	81.5362	0.5107	111.6670	6.5955	16.9307	0.001359
35.0000	81.9357	0.4988	111.8937	6.5963	16.9631	0.001351
34.0000	82.3114	0.4880	112.0972	6.5970	16.9922	0.001345
33.0000	82.6669	0.4780	112.2809	6.5976	17.0185	0.001338
32.0000	83.0043	0.4689	112.4475	6.5982	17.0423	0.001333
31.0000	83.3257	0.4604	112.5990	6.5987	17.0639	0.001328
30.0000	83.6329	0.4525	112.7373	6.5991	17.0837	0.001324
29.0000	83.9275	0.4452	112.8639	6.5996	17.1017	0.001319
28.0000	84.2106	0.4384	112.9800	6.5999	17.1183	0.001316
27.0000	84.4834	0.4321	113.0866	6.6003	17.1336	0.001312
26.0000	84.7468	0.4262	113.1848	6.6006	17.1476	0.001309
25.0000	85.0018	0.4207	113.2753	6.6009	17.1605	0.001306
24.0000	85.2490	0.4155	113.3588	6.6012	17.1725	0.001304
23.0000	85.4892	0.4107	113.4359	6.6015	17.1835	0.001301
22.0000	85.7229	0.4063	113.5071	6.6017	17.1936	0.001299
21.0000	85.9508	0.4021	113.5730	6.6019	17.2030	0.001297
20.0000	86.1733	0.3982	113.6338	6.6021	17.2117	0.001295
19.0000	86.3909	0.3946	113.6899	6.6023	17.2198	0.001293
18.0000	86.6039	0.3912	113.7418	6.6025	17.2272	0.001291
17.0000	86.8128	0.3881	113.7895	6.6026	17.2340	0.001290
16.0000	87.0178	0.3852	113.8335	6.6028	17.2403	0.001289
15.0000	87.2195	0.3825	113.8739	6.6029	17.2460	0.001287
14.0000	87.4180	0.3801	113.9109	6.6030	17.2513	0.001286
13.0000	87.6136	0.3778	113.9447	6.6031	17.2561	0.001285
12.0000	87.8065	0.3757	113.9754	6.6032	17.2605	0.001284
11.0000	87.9972	0.3738	114.0032	6.6033	17.2645	0.001283
10.0000	88.1857	0.3721	114.0283	6.6034	17.2681	0.001282
9.0000	88.3724	0.3706	114.0506	6.6035	17.2713	0.001282
8.0000	88.5573	0.3692	114.0704	6.6035	17.2741	0.001281
7.0000	88.7408	0.3681	114.0876	6.6036	17.2766	0.001281
6.0000	88.9230	0.3670	114.1025	6.6037	17.2787	0.001280
5.0000	89.1041	0.3662	114.1149	6.6037	17.2805	0.001280
4.0000	89.2844	0.3655	114.1250	6.6037	17.2819	0.001279
3.0000	89.4639	0.3649	114.1328	6.6037	17.2830	0.001279
2.0000	89.6428	0.3645	114.1384	6.6038	17.2838	0.001279
1.0000	89.8215	0.3643	114.1417	6.6038	17.2843	0.001279
0.0000	90.0000	0.3642	114.1428	6.6038	17.2845	0.001279

* DENOTES $M_2 = 1$, m DENOTES θ_{max}

OBLIQUE SHOCKS IN A PERFECT GAS
$\gamma = 4/3$ $M_1 = 15.00$
WEAK SHOCK SOLUTION

θ	β	M_1	p_{21}	ρ_{21}	T_{21}	p^o_{21}
0.0000	3.8226	15.0000	1.0000	1.0000	1.0000	1.0000
1.0000	4.4528	14.3480	1.4071	1.2905	1.0903	0.9957
2.0000	5.1697	13.7132	1.9449	1.6338	1.1904	0.9685
3.0000	5.9642	13.0751	2.6334	2.0174	1.3054	0.9069
4.0000	6.8266	12.4279	3.4903	2.4243	1.4397	0.8124
5.0000	7.7458	11.7767	4.5282	2.8363	1.5965	0.6970
6.0000	8.7101	11.1324	5.7540	3.2365	1.7779	0.5759
7.0000	9.7113	10.5050	7.1740	3.6135	1.9853	0.4618
8.0000	10.7426	9.9027	8.7913	3.9603	2.2198	0.3621
9.0000	11.7976	9.3317	10.6062	4.2737	2.4817	0.2796
10.0000	12.8734	8.7945	12.6213	4.5537	2.7717	0.2139
11.0000	13.9644	8.2934	14.8318	4.8014	3.0891	0.1629
12.0000	15.0701	7.8269	17.2400	5.0198	3.4344	0.1239
13.0000	16.1877	7.3942	19.8424	5.2118	3.8072	0.09444
14.0000	17.3159	6.9931	22.6374	5.3804	4.2074	0.07224
15.0000	18.4537	6.6215	25.6219	5.5286	4.6344	0.05554
16.0000	19.6003	6.2769	28.7935	5.6590	5.0881	0.04296
17.0000	20.7548	5.9572	32.1484	5.7739	5.5679	0.03345
18.0000	21.9171	5.6599	35.6839	5.8755	6.0734	0.02623
19.0000	23.0869	5.3832	39.3961	5.9654	6.6041	0.02071
20.0000	24.2639	5.1250	43.2812	6.0454	7.1594	0.01647
21.0000	25.4482	4.8836	47.3355	6.1166	7.7389	0.01320
22.0000	26.6400	4.6574	51.5551	6.1803	8.3419	0.01065
23.0000	27.8394	4.4451	55.9358	6.2373	8.9679	0.008648
24.0000	29.0466	4.2452	60.4736	6.2886	9.6164	0.007072
25.0000	30.2620	4.0567	65.1637	6.3349	10.2865	0.005820
26.0000	31.4860	3.8784	70.0024	6.3767	10.9779	0.004820
27.0000	32.7192	3.7095	74.9849	6.4145	11.6898	0.004015
28.0000	33.9622	3.5491	80.1074	6.4490	12.4217	0.003365
29.0000	35.2158	3.3964	85.3657	6.4803	13.1730	0.002835
30.0000	36.4809	3.2507	90.7558	6.5090	13.9431	0.002401
31.0000	37.7586	3.1114	96.2740	6.5352	14.7315	0.002044
32.0000	39.0500	2.9778	101.9166	6.5593	15.5377	0.001749
33.0000	40.3567	2.8493	107.6807	6.5815	16.3612	0.001503
34.0000	41.6805	2.7256	113.5637	6.6019	17.2017	0.001297
35.0000	43.0234	2.6060	119.5645	6.6208	18.0590	0.001124
36.0000	44.3879	2.4902	125.6817	6.6382	18.9330	0.0009781
37.0000	45.7773	2.3775	131.9165	6.6545	19.8237	0.0008542
38.0000	47.1954	2.2677	138.2720	6.6696	20.7317	0.0007485
39.0000	48.6471	2.1601	144.7523	6.6837	21.6575	0.0006579
40.0000	50.1389	2.0543	151.3684	6.6969	22.6027	0.0005799
41.0000	51.6794	1.9496	158.1345	6.7093	23.5693	0.0005124
42.0000	53.2803	1.8453	165.0751	6.7211	24.5609	0.0004536
43.0000	54.9590	1.7405	172.2293	6.7322	25.5829	0.0004021
44.0000	56.7418	1.6339	179.6623	6.7429	26.6448	0.0003564
45.0000	58.6736	1.5233	187.4918	6.7532	27.7633	0.0003156
46.0000	60.8425	1.4046	195.9605	6.7635	28.9732	0.0002781
47.0000	63.4753	1.2677	205.7161	6.7744	30.3669	0.0002419
48.0000	68.4403	1.0282	222.2766	6.7907	32.7327	0.0001936
* 48.0164	69.0571	1.0000	224.1470	6.7923	32.9999	0.0001890
m 48.0165	69.1058	0.9978	224.2928	6.7925	33.0207	0.0001887

* DENOTES $M_2 = 1$, m DENOTES θ_{max}

OBLIQUE SHOCKS IN A PERFECT GAS
$\gamma = 4/3$ $M_1 = 15.00$
STRONG SHOCK SOLUTION

	θ	β	M_1	p_{21}	ρ_{21}	T_{21}	p^o_{21}
m	48.0165	69.1058	0.9978	224.2928	6.7925	33.0207	0.0001887
	48.0000	69.7550	0.9685	226.2094	6.7942	33.2946	0.0001841
	47.0000	73.7266	0.7977	236.8081	6.8031	34.8087	0.0001613
	46.0000	75.3666	0.7314	240.5880	6.8061	35.3487	0.0001541
	45.0000	76.5428	0.6855	243.0735	6.8081	35.7038	0.0001496
	44.0000	77.4824	0.6500	244.9203	6.8095	35.9676	0.0001463
	43.0000	78.2731	0.6209	246.3774	6.8106	36.1758	0.0001439
	42.0000	78.9599	0.5963	247.5703	6.8115	36.3462	0.0001419
	41.0000	79.5694	0.5750	248.5715	6.8122	36.4892	0.0001402
	40.0000	80.1187	0.5563	249.4274	6.8128	36.6115	0.0001388
	39.0000	80.6198	0.5397	250.1691	6.8134	36.7174	0.0001376
	38.0000	81.0812	0.5248	250.8192	6.8138	36.8103	0.0001366
	37.0000	81.5093	0.5113	251.3942	6.8142	36.8924	0.0001357
	36.0000	81.9093	0.4990	251.9064	6.8146	36.9656	0.0001349
	35.0000	82.2849	0.4878	252.3656	6.8149	37.0312	0.0001342
	34.0000	82.6394	0.4775	252.7793	6.8152	37.0903	0.0001336
	33.0000	82.9753	0.4680	253.1539	6.8155	37.1438	0.0001330
	32.0000	83.2949	0.4593	253.4943	6.8157	37.1925	0.0001325
	31.0000	83.5999	0.4511	253.8047	6.8160	37.2368	0.0001320
	30.0000	83.8918	0.4436	254.0885	6.8162	37.2773	0.0001316
	29.0000	84.1722	0.4366	254.3487	6.8163	37.3145	0.0001312
	28.0000	84.4419	0.4301	254.5877	6.8165	37.3487	0.0001308
	27.0000	84.7021	0.4240	254.8076	6.8167	37.3801	0.0001305
	26.0000	84.9536	0.4183	255.0103	6.8168	37.4090	0.0001302
	25.0000	85.1973	0.4130	255.1974	6.8169	37.4357	0.0001299
	24.0000	85.4338	0.4080	255.3702	6.8171	37.4604	0.0001297
	23.0000	85.6637	0.4034	255.5298	6.8172	37.4832	0.0001294
	22.0000	85.8876	0.3991	255.6775	6.8173	37.5043	0.0001292
	21.0000	86.1060	0.3951	255.8140	6.8174	37.5238	0.0001290
	20.0000	86.3193	0.3913	255.9402	6.8175	37.5419	0.0001288
	19.0000	86.5280	0.3878	256.0568	6.8175	37.5585	0.0001287
	18.0000	86.7325	0.3845	256.1645	6.8176	37.5739	0.0001285
	17.0000	86.9331	0.3815	256.2639	6.8177	37.5881	0.0001284
	16.0000	87.1301	0.3787	256.3553	6.8177	37.6012	0.0001282
	15.0000	87.3239	0.3761	256.4393	6.8178	37.6132	0.0001281
	14.0000	87.5147	0.3737	256.5164	6.8179	37.6242	0.0001280
	13.0000	87.7027	0.3715	256.5868	6.8179	37.6342	0.0001279
	12.0000	87.8884	0.3695	256.6508	6.8180	37.6434	0.0001278
	11.0000	88.0718	0.3677	256.7087	6.8180	37.6517	0.0001277
	10.0000	88.2532	0.3660	256.7609	6.8180	37.6591	0.0001277
	9.0000	88.4328	0.3645	256.8076	6.8181	37.6658	0.0001276
	8.0000	88.6108	0.3632	256.8488	6.8181	37.6717	0.0001275
	7.0000	88.7874	0.3621	256.8848	6.8181	37.6768	0.0001275
	6.0000	88.9628	0.3611	256.9156	6.8181	37.6812	0.0001274
	5.0000	89.1373	0.3602	256.9416	6.8182	37.6849	0.0001274
	4.0000	89.3108	0.3595	256.9627	6.8182	37.6880	0.0001274
	3.0000	89.4837	0.3590	256.9791	6.8182	37.6903	0.0001273
	2.0000	89.6561	0.3586	256.9907	6.8182	37.6919	0.0001273
	1.0000	89.8281	0.3584	256.9976	6.8182	37.6929	0.0001273
	0.0000	90.0000	0.3583	256.9999	6.8182	37.6933	0.0001273

* DENOTES $M_2 = 1$, m DENOTES θ_{max}

OBLIQUE SHOCKS IN A PERFECT GAS
$\gamma = 4/3$ $M_1 = 20.00$
WEAK SHOCK SOLUTION

θ	β	M_1	P_{21}	ρ_{21}	T_{21}	P^o_{21}
0.0000	2.8660	20.0000	1.0000	1.0000	1.0000	1.0000
1.0000	3.5095	18.8623	1.5702	1.3992	1.1222	0.9900
2.0000	4.2645	17.7473	2.3849	1.8854	1.2649	0.9315
3.0000	5.1141	16.6141	3.4895	2.4240	1.4396	0.8125
4.0000	6.0383	15.4738	4.9158	2.9717	1.6542	0.6565
5.0000	7.0189	14.3575	6.6832	3.4921	1.9138	0.4982
6.0000	8.0415	13.2950	8.8032	3.9626	2.2215	0.3614
7.0000	9.0956	12.3056	11.2812	4.3744	2.5789	0.2550
8.0000	10.1736	11.3987	14.1195	4.7272	2.9868	0.1774
9.0000	11.2702	10.5751	17.3177	5.0261	3.4455	0.1229
10.0000	12.3813	9.8314	20.8740	5.2780	3.9549	0.08532
11.0000	13.5045	9.1611	24.7863	5.4899	4.5149	0.05965
12.0000	14.6378	8.5569	29.0508	5.6686	5.1249	0.04211
13.0000	15.7797	8.0117	33.6631	5.8196	5.7845	0.03007
14.0000	16.9292	7.5186	38.6190	5.9478	6.4930	0.02173
15.0000	18.0858	7.0712	43.9136	6.0572	7.2498	0.01590
16.0000	19.2488	6.6641	49.5406	6.1511	8.0540	0.01177
17.0000	20.4181	6.2923	55.4954	6.2320	8.9050	0.008825
18.0000	21.5933	5.9516	61.7705	6.3020	9.8017	0.006692
19.0000	22.7746	5.6383	68.3607	6.3631	10.7433	0.005131
20.0000	23.9619	5.3493	75.2584	6.4165	11.7289	0.003977
21.0000	25.1553	5.0818	82.4570	6.4634	12.7575	0.003113
22.0000	26.3550	4.8333	89.9485	6.5049	13.8278	0.002460
23.0000	27.5614	4.6018	97.7269	6.5417	14.9391	0.001962
24.0000	28.7747	4.3855	105.7833	6.5744	16.0901	0.001578
25.0000	29.9953	4.1828	114.1105	6.6037	17.2799	0.001280
26.0000	31.2239	3.9922	122.7010	6.6299	18.5071	0.001046
27.0000	32.4608	3.8125	131.5469	6.6536	19.7709	0.0008609
28.0000	33.7069	3.6427	140.6401	6.6749	21.0700	0.0007136
29.0000	34.9628	3.4817	149.9735	6.6942	22.4034	0.0005953
30.0000	36.2296	3.3287	159.5406	6.7118	23.7702	0.0004997
31.0000	37.5083	3.1829	169.3333	6.7278	25.1692	0.0004219
32.0000	38.7999	3.0436	179.3454	6.7424	26.5995	0.0003582
33.0000	40.1061	2.9101	189.5716	6.7558	28.0605	0.0003058
34.0000	41.4287	2.7819	200.0078	6.7681	29.5514	0.0002623
35.0000	42.7694	2.6583	210.6490	6.7795	31.0716	0.0002260
36.0000	44.1308	2.5389	221.4941	6.7899	32.6209	0.0001956
37.0000	45.5158	2.4231	232.5437	6.7996	34.1995	0.0001700
38.0000	46.9282	2.3105	243.8015	6.8086	35.8077	0.0001483
39.0000	48.3726	2.2005	255.2753	6.8170	37.4469	0.0001298
40.0000	49.8551	2.0925	266.9805	6.8248	39.1191	0.0001140
41.0000	51.3835	1.9860	278.9396	6.8321	40.8275	0.0001004
42.0000	52.9687	1.8803	291.1909	6.8390	42.5778	0.00008860
43.0000	54.6262	1.7744	303.7953	6.8456	44.3784	0.00007832
44.0000	56.3797	1.6672	316.8541	6.8518	46.2439	0.00006928
45.0000	58.2680	1.5567	330.5449	6.8578	48.1998	0.00006124
46.0000	60.3639	1.4395	345.2189	6.8637	50.2961	0.00005394
47.0000	62.8387	1.3078	361.7361	6.8698	52.6557	0.00004705
48.0000	66.4455	1.1275	383.9955	6.8772	55.8356	0.00003951
* 48.2667	69.1606	1.0000	399.1450	6.8818	57.9999	0.00003527
m 48.2667	69.1879	0.9988	399.2894	6.8819	58.0205	0.00003523

* DENOTES $M_2 = 1$, m DENOTES θ_{max}

OBLIQUE SHOCKS IN A PERFECT GAS
$\gamma = 4/3 \quad M_1 = 20.00$
STRONG SHOCK SOLUTION

	θ	β	M_1	p_{21}	ρ_{21}	T_{21}	p^o_{21}
m	48.2667	69.1879	0.9988	399.2894	6.8819	58.0205	0.00003523
	48.0000	71.6642	0.8883	411.7590	6.8854	59.8019	0.00003219
	47.0000	74.2751	0.7778	423.4220	6.8885	61.4680	0.00002966
	46.0000	75.7539	0.7182	429.3157	6.8900	62.3100	0.00002848
	45.0000	76.8539	0.6753	433.3535	6.8910	62.8868	0.00002771
	44.0000	77.7466	0.6415	436.4081	6.8918	63.3232	0.00002714
	43.0000	78.5045	0.6136	438.8434	6.8923	63.6711	0.00002670
	42.0000	79.1668	0.5898	440.8511	6.8928	63.9579	0.00002635
	41.0000	79.7567	0.5692	442.5437	6.8932	64.1997	0.00002605
	40.0000	80.2902	0.5510	443.9962	6.8936	64.4072	0.00002580
	39.0000	80.7779	0.5348	445.2585	6.8939	64.5875	0.00002559
	38.0000	81.2278	0.5202	446.3674	6.8941	64.7460	0.00002540
	37.0000	81.6459	0.5070	447.3499	6.8944	64.8863	0.00002524
	36.0000	82.0370	0.4950	448.2265	6.8946	65.0115	0.00002509
	35.0000	82.4046	0.4840	449.0132	6.8948	65.1239	0.00002496
	34.0000	82.7520	0.4739	449.7233	6.8949	65.2254	0.00002485
	33.0000	83.0814	0.4646	450.3664	6.8951	65.3172	0.00002474
	32.0000	83.3949	0.4559	450.9513	6.8952	65.4008	0.00002465
	31.0000	83.6943	0.4479	451.4851	6.8953	65.4771	0.00002456
	30.0000	83.9811	0.4405	451.9734	6.8954	65.5468	0.00002448
	29.0000	84.2566	0.4336	452.4216	6.8955	65.6108	0.00002441
	28.0000	84.5218	0.4271	452.8335	6.8956	65.6697	0.00002435
	27.0000	84.7776	0.4211	453.2124	6.8957	65.7238	0.00002429
	26.0000	85.0251	0.4155	453.5620	6.8958	65.7737	0.00002423
	25.0000	85.2649	0.4103	453.8847	6.8959	65.8198	0.00002418
	24.0000	85.4977	0.4054	454.1828	6.8959	65.8624	0.00002414
	23.0000	85.7240	0.4008	454.4585	6.8960	65.9018	0.00002409
	22.0000	85.9446	0.3966	454.7134	6.8960	65.9382	0.00002405
	21.0000	86.1596	0.3926	454.9492	6.8961	65.9719	0.00002402
	20.0000	86.3698	0.3889	455.1672	6.8962	66.0031	0.00002398
	19.0000	86.5755	0.3854	455.3688	6.8962	66.0319	0.00002395
	18.0000	86.7770	0.3822	455.5548	6.8962	66.0584	0.00002392
	17.0000	86.9747	0.3792	455.7266	6.8963	66.0830	0.00002390
	16.0000	87.1690	0.3764	455.8848	6.8963	66.1056	0.00002387
	15.0000	87.3601	0.3738	456.0301	6.8963	66.1263	0.00002385
	14.0000	87.5481	0.3715	456.1633	6.8964	66.1454	0.00002383
	13.0000	87.7336	0.3693	456.2850	6.8964	66.1628	0.00002381
	12.0000	87.9167	0.3673	456.3958	6.8964	66.1786	0.00002379
	11.0000	88.0976	0.3655	456.4961	6.8965	66.1929	0.00002378
	10.0000	88.2765	0.3638	456.5862	6.8965	66.2058	0.00002376
	9.0000	88.4537	0.3624	456.6670	6.8965	66.2173	0.00002375
	8.0000	88.6293	0.3611	456.7382	6.8965	66.2275	0.00002374
	7.0000	88.8036	0.3599	456.8006	6.8965	66.2364	0.00002373
	6.0000	88.9766	0.3589	456.8540	6.8965	66.2440	0.00002372
	5.0000	89.1487	0.3581	456.8990	6.8965	66.2505	0.00002372
	4.0000	89.3199	0.3574	456.9355	6.8965	66.2557	0.00002371
	3.0000	89.4906	0.3569	456.9637	6.8966	66.2597	0.00002371
	2.0000	89.6607	0.3566	456.9838	6.8966	66.2626	0.00002370
	1.0000	89.8304	0.3563	456.9959	6.8966	66.2643	0.00002370
	0.0000	90.0000	0.3563	456.9998	6.8966	66.2649	0.00002370

* DENOTES $M_2 = 1$, m DENOTES θ_{max}

OBLIQUE SHOCKS IN A PERFECT GAS
$\gamma = 4/3$ $M_1 = 25.00$
WEAK SHOCK SOLUTION

θ	β	M_1	p_{21}	ρ_{21}	T_{21}	p_{21}^o
0.0000	2.2924	25.0000	1.0000	1.0000	1.0000	1.0000
1.0000	2.9496	23.2408	1.7485	1.5134	1.1554	0.9812
2.0000	3.7411	?1.4998	2.8982	2.1506	1.3476	0.8788
3.0000	4.6382	19.7295	4.5277	2.8361	1.5964	0.6970
4.0000	5.6109	17.9904	6.6852	3.4926	1.9141	0.4980
5.0000	6.6361	16.3551	9.3962	4.0725	2.3072	0.3316
6.0000	7.6974	14.8684	12.6717	4.5599	2.7789	0.2125
7.0000	8.7846	13.5443	16.5166	4.9589	3.3307	0.1342
8.0000	9.8904	12.3782	20.9307	5.2815	3.9631	0.08485
9.0000	11.0106	11.3554	25.9123	5.5416	4.6760	0.05420
10.0000	12.1417	10.4585	31.4563	5.7518	5.4689	0.03516
11.0000	13.2820	9.6698	37.5589	5.9228	6.3414	0.02323
12.0000	14.4299	8.9737	44.2130	6.0627	7.2926	0.01563
13.0000	15.5844	8.3563	51.4123	6.1783	8.3215	0.01072
14.0000	16.7450	7.8061	59.1490	6.2744	9.4271	0.007489
15.0000	17.9112	7.3131	67.4158	6.3550	10.6083	0.005323
16.0000	19.0825	6.8693	76.2020	6.4231	11.8637	0.003847
17.0000	20.2592	6.4677	85.5005	6.4811	13.1923	0.002823
18.0000	21.4410	6.1026	95.3007	6.5308	14.5925	0.002102
19.0000	22.6278	5.7692	105.5911	6.5737	16.0627	0.001586
20.0000	23.8202	5.4634	116.3632	6.6109	17.6017	0.001212
21.0000	25.0180	5.1818	127.6050	6.6434	19.2078	0.0009375
22.0000	26.2217	4.9215	139.3048	6.6719	20.8792	0.0007330
23.0000	27.4314	4.6800	151.4516	6.6971	22.6146	0.0005790
24.0000	28.6477	4.4551	164.0331	6.7194	24.4120	0.0004619
25.0000	29.8710	4.2450	177.0370	6.7392	26.2698	0.0003717
26.0000	31.1017	4.0481	190.4517	6.7569	28.1862	0.0003017
27.0000	32.3405	3.8629	204.2647	6.7728	30.1595	0.0002469
28.0000	33.5881	3.6883	218.4645	6.7871	32.1881	0.0002035
29.0000	34.8453	3.5231	233.0387	6.8000	34.2702	0.0001689
30.0000	36.1129	3.3665	247.9762	6.8118	36.4041	0.0001412
31.0000	37.3921	3.2175	263.2652	6.8224	38.5883	0.0001187
32.0000	38.6840	3.0753	278.8967	6.8321	40.8214	0.0001004
33.0000	39.9901	2.9393	294.8605	6.8410	43.1020	0.00008543
34.0000	41.3120	2.8089	311.1492	6.8491	45.4290	0.00007305
35.0000	42.6518	2.6834	327.7577	6.8566	47.8016	0.00006277
36.0000	44.0118	2.5622	344.6826	6.8635	50.2195	0.00005419
37.0000	45.3949	2.4449	361.9230	6.8699	52.6824	0.00004698
38.0000	46.8049	2.3309	379.4855	6.8758	55.1914	0.00004090
39.0000	48.2460	2.2196	397.3802	6.8813	57.7477	0.00003573
40.0000	49.7244	2.1106	415.6301	6.8864	60.3549	0.00003132
41.0000	51.2475	2.0033	434.2687	6.8912	63.0176	0.00002754
42.0000	52.8257	1.8968	453.3510	6.8957	65.7436	0.00002427
43.0000	54.4740	1.7904	472.9675	6.9000	68.5460	0.00002142
44.0000	56.2148	1.6828	493.2669	6.9041	71.4459	0.00001893
45.0000	58.0844	1.5723	514.5055	6.9080	74.4800	0.00001672
46.0000	60.1498	1.4556	537.1862	6.9118	77.7201	0.00001472
47.0000	62.5633	1.3259	562.4943	6.9157	81.3356	0.00001285
48.0000	65.9099	1.1564	595.1394	6.9203	85.9991	0.00001088
* 48.3830	69.2089	1.0000	624.1440	6.9240	90.1427	0.000009453
m 48.3830	69.2263	0.9992	624.2877	6.9240	90.1632	0.000009446

* DENOTES $M_2 = 1$, m DENOTES θ_{max}

OBLIQUE SHOCKS IN A PERFECT GAS
$\gamma = 4/3$ $M_1 = 25.00$
STRONG SHOCK SOLUTION

θ	β	M_1	p_{21}	ρ_{21}	T_{21}	p^o_{21}
m 48.3830	69.2263	0.9992	624.2877	6.9240	90.1632	0.000009446
48.0000	72.1604	0.8685	647.1053	6.9266	93.4229	0.000008495
47.0000	74.5096	0.7695	663.1921	6.9284	95.7210	0.000007899
46.0000	75.9257	0.7124	671.9021	6.9293	96.9653	0.000007600
45.0000	76.9938	0.6707	677.9635	6.9299	97.8312	0.000007401
44.0000	77.8661	0.6377	682.5835	6.9304	98.4912	0.000007254
43.0000	78.6098	0.6103	686.2836	6.9308	99.0197	0.000007138
42.0000	79.2611	0.5869	689.3424	6.9311	99.4567	0.000007045
41.0000	79.8424	0.5666	691.9271	6.9313	99.8260	0.000006967
40.0000	80.3687	0.5486	694.1484	6.9316	100.1433	0.000006902
39.0000	80.8504	0.5326	696.0817	6.9317	100.4195	0.000006845
38.0000	81.2951	0.5181	697.7814	6.9319	100.6623	0.000006796
37.0000	81.7086	0.5051	699.2886	6.9320	100.8776	0.000006753
36.0000	82.0957	0.4931	700.6344	6.9322	101.0699	0.000006714
35.0000	82.4597	0.4822	701.8430	6.9323	101.2425	0.000006680
34.0000	82.8037	0.4722	702.9338	6.9324	101.3984	0.000006649
33.0000	83.1301	0.4629	703.9227	6.9325	101.5396	0.000006622
32.0000	83.4409	0.4544	704.8228	6.9326	101.6682	0.000006597
31.0000	83.7378	0.4464	705.6438	6.9327	101.7855	0.000006574
30.0000	84.0222	0.4391	706.3956	6.9327	101.8929	0.000006553
29.0000	84.2955	0.4322	707.0853	6.9328	101.9914	0.000006534
28.0000	84.5586	0.4258	707.7194	6.9329	102.0820	0.000006517
27.0000	84.8125	0.4198	708.3033	6.9329	102.1654	0.000006501
26.0000	85.0581	0.4142	708.8418	6.9330	102.2424	0.000006487
25.0000	85.2961	0.4090	709.3390	6.9330	102.3134	0.000006473
24.0000	85.5272	0.4042	709.7983	6.9330	102.3790	0.000006461
23.0000	85.7519	0.3996	710.2233	6.9331	102.4397	0.000006449
22.0000	85.9709	0.3954	710.6163	6.9331	102.4959	0.000006439
21.0000	86.1845	0.3914	710.9797	6.9332	102.5478	0.000006429
20.0000	86.3932	0.3877	711.3158	6.9332	102.5958	0.000006420
19.0000	86.5975	0.3843	711.6266	6.9332	102.64C2	0.000006412
18.0000	86.7977	0.3811	711.9136	6.9332	102.6812	0.000006404
17.0000	86.9941	0.3781	712.1785	6.9333	102.7190	0.000006397
16.0000	87.1870	0.3753	712.4222	6.9333	102.7538	0.000006390
15.0000	87.3768	0.3728	712.6464	6.9333	102.7859	0.000006385
14.0000	87.5637	0.3704	712.8519	6.9333	102.8152	0.000006379
13.0000	87.7479	0.3683	713.0397	6.9333	102.8421	0.000006374
12.0000	87.9298	0.3663	713.2104	6.9334	102.8664	0.000006370
11.0000	88.1096	0.3645	713.3654	6.9334	102.8886	0.000006365
10.0000	88.2873	0.3628	713.5045	6.9334	102.9084	0.000006362
9.0000	88.4634	0.3614	713.6290	6.9334	102.9263	0.000006359
8.0000	88.6379	0.3601	713.7390	6.9334	102.9420	0.000006356
7.0000	88.8111	0.3589	713.8350	6.9334	102.9557	0.000006353
6.0000	88.9830	0.3580	713.9176	6.9334	102.9675	0.000006351
5.0000	89.1541	0.3571	713.9869	6.9334	102.9774	0.000006349
4.0000	89.3242	0.3565	714.0433	6.9334	102.9854	0.000006348
3.0000	89.4937	0.3559	714.0869	6.9334	102.9917	0.000006346
2.0000	89.6628	0.3556	714.1179	6.9334	102.9961	0.000006346
1.0000	89.8315	0.3554	714.1365	6.9334	102.9987	0.000006345
0.0000	90.0000	0.3553	714.1426	6.9334	102.9996	0.000006345

* DENOTES $M_2 = 1$, m DENOTES θ_{max}

OBLIQUE SHOCKS IN A PERFECT GAS
$\gamma = 4/3$ $M_1 = 35.00$
WEAK SHOCK SOLUTION

θ	β	M_1	p_{21}	ρ_{21}	T_{21}	p^o_{21}
0.0000	1.6372	35.0000	1.0000	1.0000	1.0000	1.0000
1.0000	2.3217	31.5858	2.1547	1.7568	1.2265	0.9522
2.0000	3.1782	28.1535	4.1604	2.6991	1.5414	0.7370
3.0000	4.1488	24.7701	7.1849	3.6161	1.9869	0.4610
4.0000	5.1879	21.6963	11.3036	4.3776	2.5822	0.2543
5.0000	6.2678	19.0566	16.5442	4.9613	3.3347	0.1338
6.0000	7.3736	16.8474	22.9161	5.3955	4.2472	0.07042
7.0000	8.4964	15.0123	30.4183	5.7172	5.3205	0.03796
8.0000	9.6312	13.4843	39.0451	5.9576	6.5539	0.02116
9.0000	10.7750	12.2028	48.7892	6.1396	7.9466	0.01223
10.0000	11.9260	11.1181	59.6419	6.2797	9.4975	0.007330
11.0000	13.0827	10.1914	71.5893	6.3892	11.2047	0.004542
12.0000	14.2445	9.3920	84.6215	6.4761	13.0667	0.002903
13.0000	15.4110	8.6962	98.7222	6.5460	15.0813	0.001908
14.0000	16.5817	8.0856	113.8774	6.6029	17.2465	0.001287
15.0000	17.7567	7.5455	130.0713	6.6498	19.5601	0.0008886
16.0000	18.9358	7.0644	147.2847	6.6889	22.0193	0.0006265
17.0000	20.1191	6.6330	165.5011	6.7218	24.6217	0.0004503
18.0000	21.3068	6.2438	184.7013	6.7496	27.3647	0.0003294
19.0000	22.4989	5.8907	204.8629	6.7734	30.2450	0.0002448
20.0000	23.6958	5.5688	225.9676	6.7940	33.2600	0.0001846
21.0000	24.8976	5.2738	247.9920	6.8118	36.4064	0.0001412
22.0000	26.1048	5.0022	270.9146	6.8273	39.6811	0.0001093
23.0000	27.3176	4.7512	294.7104	6.8409	43.0805	0.00008556
24.0000	28.5366	4.5183	319.3590	6.8529	46.6018	0.00006771
25.0000	29.7622	4.3014	344.8364	6.8636	50.2415	0.00005412
26.0000	30.9949	4.0985	371.1158	6.8731	53.9957	0.00004366
27.0000	32.2354	3.9083	398.1771	6.8815	57.8616	0.00003552
28.0000	33.4843	3.7293	425.9927	6.8892	61.8353	0.00002914
29.0000	34.7426	3.5603	454.5430	6.8960	65.9139	0.00002408
30.0000	36.0110	3.4003	483.8012	6.9022	70.0936	0.00002004
31.0000	37.2907	3.2484	513.7489	6.9078	74.3719	0.00001679
32.0000	38.5828	3.1036	544.3636	6.9130	78.7455	0.00001416
33.0000	39.8889	2.9654	575.6292	6.9176	83.2119	0.00001201
34.0000	41.2103	2.8329	607.5271	6.9219	87.7688	0.00001024
35.0000	42.5494	2.7056	640.0510	6.9258	92.4151	0.000008775
36.0000	43.9082	2.5829	673.1864	6.9294	97.1487	0.000007557
37.0000	45.2898	2.4642	706.9381	6.9328	101.9704	0.000006538
38.0000	46.6976	2.3489	741.3132	6.9359	106.8811	0.000005680
39.0000	48.1361	2.2366	776.3320	6.9387	111.8838	0.000004954
40.0000	49.6109	2.1267	812.0370	6.9414	116.9845	0.000004336
41.0000	51.1294	2.0185	848.4869	6.9439	122.1917	0.000003806
42.0000	52.7018	1.9114	885.7874	6.9462	127.5203	0.000003350
43.0000	54.3423	1.8044	924.1083	6.9485	132.9947	0.000002954
44.0000	56.0723	1.6965	963.7190	6.9506	138.6534	0.000002607
45.0000	57.9264	1.5859	1005.0980	6.9526	144.5646	0.000002301
46.0000	59.9668	1.4697	1049.1538	6.9546	150.8583	0.000002026
47.0000	62.3317	1.3414	1097.9839	6.9566	157.8340	0.000001769
48.0000	65.5042	1.1790	1159.1743	6.9588	166.5755	0.000001506
* 48.4844	69.2511	1.0000	1224.1426	6.9610	175.8567	0.000001280
m 48.4844	69.2599	0.9996	1224.2861	6.9610	175.8772	0.000001279

* DENOTES $M_2 = 1$, m DENOTES θ_{max}

OBLIQUE SHOCKS IN A PERFECT GAS
$\gamma = 4/3$ $M_1 = 35.00$
STRONG SHOCK SOLUTION

	θ	β	M_1	p_{21}	ρ_{21}	T_{21}	P^o_{21}
m	48.4844	69.2599	0.9996	1224.2861	6.9610	175.8772	0.000001279
	48.0000	72.5316	0.8539	1273.7056	6.9625	182.9372	0.000001137
	47.0000	74.7054	0.7625	1302.4434	6.9634	187.0425	0.000001064
	46.0000	76.0715	0.7074	1318.7385	6.9638	189.3704	0.000001025
	45.0000	77.1134	0.6668	1330.2217	6.9641	191.0109	0.0000009993
	44.0000	77.9688	0.6344	1339.0283	6.9643	192.2690	0.0000009798
	43.0000	78.7003	0.6074	1346.1069	6.9645	193.2802	0.0000009645
	42.0000	79.3424	0.5844	1351.9729	6.9647	194.1182	0.0000009521
	41.0000	79.9164	0.5643	1356.9395	6.9648	194.8277	0.0000009418
	40.0000	80.4365	0.5465	1361.2131	6.9649	195.4382	0.0000009330
	39.0000	80.9130	0.5306	1364.9363	6.9650	195.9701	0.0000009254
	38.0000	81.3533	0.5163	1368.2129	6.9651	196.4382	0.0000009188
	37.0000	81.7630	0.5034	1371.1204	6.9652	196.8535	0.0000009130
	36.0000	82.1465	0.4915	1373.7175	6.9652	197.2245	0.0000009079
	35.0000	82.5074	0.4807	1376.0518	6.9653	197.5580	0.0000009033
	34.0000	82.8486	0.4707	1378.1594	6.9654	197.8591	0.0000008992
	33.0000	83.1725	0.4615	1380.0706	6.9654	198.1321	0.0000008955
	32.0000	83.4809	0.4530	1381.8105	6.9654	198.3807	0.0000008921
	31.0000	83.7755	0.4451	1383.3984	6.9655	198.6075	0.0000008891
	30.0000	84.0579	0.4378	1384.8528	6.9655	198.8153	0.0000008863
	29.0000	84.3292	0.4310	1386.1870	6.9656	199.0059	0.0000008838
	28.0000	84.5905	0.4246	1387.4143	6.9656	199.1812	0.0000008815
	27.0000	84.8427	0.4187	1388.5444	6.9656	199.3427	0.0000008793
	26.0000	85.0867	0.4131	1389.5869	6.9656	199.4916	0.0000008774
	25.0000	85.3232	0.4080	1390.5496	6.9657	199.6291	0.0000008755
	24.0000	85.5528	0.4031	1391.4392	6.9657	199.7562	0.0000008739
	23.0000	85.7761	0.3986	1392.2617	6.9657	199.8737	0.0000008723
	22.0000	85.9937	0.3944	1393.0229	6.9657	199.9825	0.0000008709
	21.0000	86.2060	0.3904	1393.7271	6.9657	200.0831	0.0000008696
	20.0000	86.4135	0.3868	1394.3782	6.9658	200.1761	0.0000008684
	19.0000	86.6166	0.3833	1394.9802	6.9658	200.2621	0.0000008673
	18.0000	86.8155	0.3801	1395.5364	6.9658	200.3415	0.0000008663
	17.0000	87.0108	0.3772	1396.0496	6.9658	200.4149	0.0000008653
	16.0000	87.2026	0.3744	1396.5220	6.9658	200.4823	0.0000008644
	15.0000	87.3913	0.3719	1396.9565	6.9658	200.5444	0.0000008636
	14.0000	87.5771	0.3695	1397.3545	6.9658	200.6013	0.0000008629
	13.0000	87.7603	0.3674	1397.7183	6.9658	200.6532	0.0000008622
	12.0000	87.9412	0.3654	1398.0496	6.9658	200.7005	0.0000008616
	11.0000	88.1199	0.3636	1398.3499	6.9659	200.7435	0.0000008611
	10.0000	88.2967	0.3620	1398.6196	6.9659	200.7820	0.0000008606
	9.0000	88.4718	0.3605	1398.8611	6.9659	200.8165	0.0000008601
	8.0000	88.6453	0.3592	1399.0742	6.9659	200.8469	0.0000008597
	7.0000	88.8176	0.3581	1399.2605	6.9659	200.8735	0.0000008594
	6.0000	88.9886	0.3571	1399.4204	6.9659	200.8964	0.0000008591
	5.0000	89.1586	0.3563	1399.5549	6.9659	200.9156	0.0000008589
	4.0000	89.3279	0.3556	1399.6641	6.9659	200.9312	0.0000008587
	3.0000	89.4965	0.3551	1399.7488	6.9659	200.9433	0.0000008585
	2.0000	89.6646	0.3547	1399.8088	6.9659	200.9519	0.0000008584
	1.0000	89.8324	0.3545	1399.8450	6.9659	200.9571	0.0000008583
	0.0000	90.0000	0.3544	1399.8567	6.9659	200.9587	0.0000008583

* DENOTES $M_2 = 1$, m DENOTES θ_{max}

OBLIQUE SHOCKS IN A PERFECT GAS
$\gamma = 4/3$ $M_1 = \infty$

WEAK SHOCK SOLUTION

θ	β	M_1	p_{21}	ρ_{21}	T_{21}	p^o_{21}
0.0000	0.0000	∞	1.0000	1.0000	1.0000	1.0000
1.0000	1.1670	121.4920	∞	7.0000	∞	0.0000
2.0000	2.3334	60.7384	∞	7.0000	∞	0.0000
3.0000	3.5011	40.4522	∞	7.0000	∞	0.0000
4.0000	4.6688	30.3066	∞	7.0000	∞	0.0000
5.0000	5.8370	24.2119	∞	7.0000	∞	0.0000
6.0000	7.0058	20.1426	∞	7.0000	∞	0.0000
7.0000	8.1754	17.2304	∞	7.0000	∞	0.0000
8.0000	9.3469	15.0401	∞	7.0000	∞	0.0000
9.0000	10.5197	13.3323	∞	7.0000	∞	0.0000
10.0000	11.6940	11.9622	∞	7.0000	∞	0.0000
11.0000	12.8690	10.8386	∞	7.0000	∞	0.0000
12.0000	14.0466	9.8982	∞	7.0000	∞	0.0000
13.0000	15.2272	9.0989	∞	7.0000	∞	0.0000
14.0000	16.4092	8.4114	∞	7.0000	∞	0.0000
15.0000	17.5936	7.8128	∞	7.0000	∞	0.0000
16.0000	18.7817	7.2861	∞	7.0000	∞	0.0000
17.0000	19.9718	6.8193	∞	7.0000	∞	0.0000
18.0000	21.1661	6.4016	∞	7.0000	∞	0.0000
19.0000	22.3639	6.0256	∞	7.0000	∞	0.0000
20.0000	23.5656	5.6851	∞	7.0000	∞	0.0000
21.0000	24.7715	5.3748	∞	7.0000	∞	0.0000
22.0000	25.9830	5.0904	∞	7.0000	∞	0.0000
23.0000	27.1985	4.8289	∞	7.0000	∞	0.0000
24.0000	28.4204	4.5869	∞	7.0000	∞	0.0000
25.0000	29.6484	4.3624	∞	7.0000	∞	0.0000
26.0000	30.8837	4.1530	∞	7.0000	∞	0.0000
27.0000	32.1262	3.9571	∞	7.0000	∞	0.0000
28.0000	33.3764	3.7733	∞	7.0000	∞	0.0000
29.0000	34.6362	3.6001	∞	7.0000	∞	0.0000
30.0000	35.9054	3.4365	∞	7.0000	∞	0.0000
31.0000	37.1850	3.2814	∞	7.0000	∞	0.0000
32.0000	38.4781	3.1338	∞	7.0000	∞	0.0000
33.0000	39.7840	2.9931	∞	7.0000	∞	0.0000
34.0000	41.1053	2.8584	∞	7.0000	∞	0.0000
35.0000	42.4436	2.7292	∞	7.0000	∞	0.0000
36.0000	43.8012	2.6048	∞	7.0000	∞	0.0000
37.0000	45.1805	2.4846	∞	7.0000	∞	0.0000
38.0000	46.5867	2.3680	∞	7.0000	∞	0.0000
39.0000	48.0225	2.2545	∞	7.0000	∞	0.0000
40.0000	49.4929	2.1436	∞	7.0000	∞	0.0000
41.0000	51.0071	2.0346	∞	7.0000	∞	0.0000
42.0000	52.5733	1.9267	∞	7.0000	∞	0.0000
43.0000	54.2061	1.8192	∞	7.0000	∞	0.0000
44.0000	55.9258	1.7109	∞	7.0000	∞	0.0000
45.0000	57.7638	1.6002	∞	7.0000	∞	0.0000
46.0000	59.7799	1.4843	∞	7.0000	∞	0.0000
47.0000	62.0982	1.3573	∞	7.0000	∞	0.0000
48.0000	65.1227	1.2008	∞	7.0000	∞	0.0000
m*48.5904	69.2952	1.0000	∞	7.0000	∞	0.0000

* DENOTES $M_2 = 1$, m DENOTES θ_{max}

OBLIQUE SHOCKS IN A PERFECT GAS
$\gamma = 4/3$ $M_1 = \infty$
STRONG SHOCK SOLUTION

θ	β	M_1	p_{21}	ρ_{21}	T_{21}	p^o_{21}
m*48.5904	69.2952	1.0000	∞	7.0000	∞	0.0000
48.0000	72.8774	0.8404	∞	7.0000	∞	0.0000
47.0000	74.9021	0.7555	∞	7.0000	∞	0.0000
46.0000	76.2198	0.7024	∞	7.0000	∞	0.0000
45.0000	77.2360	0.6628	∞	7.0000	∞	0.0000
44.0000	78.0744	0.6310	∞	7.0000	∞	0.0000
43.0000	78.7937	0.6045	∞	7.0000	∞	0.0000
42.0000	79.4264	0.5817	∞	7.0000	∞	0.0000
41.0000	79.9928	0.5619	∞	7.0000	∞	0.0000
40.0000	80.5067	0.5443	∞	7.0000	∞	0.0000
39.0000	80.9779	0.5286	∞	7.0000	∞	0.0000
38.0000	81.4135	0.5144	∞	7.0000	∞	0.0000
37.0000	81.8192	0.5016	∞	7.0000	∞	0.0000
36.0000	82.1992	0.4899	∞	7.0000	∞	0.0000
35.0000	82.5569	0.4791	∞	7.0000	∞	0.0000
34.0000	82.8952	0.4692	∞	7.0000	∞	0.0000
33.0000	83.2163	0.4601	∞	7.0000	∞	0.0000
32.0000	83.5223	0.4516	∞	7.0000	∞	0.0000
31.0000	83.8147	0.4438	∞	7.0000	∞	0.0000
30.0000	84.0949	0.4365	∞	7.0000	∞	0.0000
29.0000	84.3642	0.4297	∞	7.0000	∞	0.0000
28.0000	84.6236	0.4234	∞	7.0000	∞	0.0000
27.0000	84.8742	0.4175	∞	7.0000	∞	0.0000
26.0000	85.1165	0.4120	∞	7.0000	∞	0.0000
25.0000	85.3513	0.4068	∞	7.0000	∞	0.0000
24.0000	85.5794	0.4020	∞	7.0000	∞	0.0000
23.0000	85.8012	0.3975	∞	7.0000	∞	0.0000
22.0000	86.0174	0.3933	∞	7.0000	∞	0.0000
21.0000	86.2284	0.3894	∞	7.0000	∞	0.0000
20.0000	86.4346	0.3857	∞	7.0000	∞	0.0000
19.0000	86.6364	0.3823	∞	7.0000	∞	0.0000
18.0000	86.8341	0.3791	∞	7.0000	∞	0.0000
17.0000	87.0281	0.3762	∞	7.0000	∞	0.0000
16.0000	87.2188	0.3734	∞	7.0000	∞	0.0000
15.0000	87.4064	0.3709	∞	7.0000	∞	0.0000
14.0000	87.5911	0.3686	∞	7.0000	∞	0.0000
13.0000	87.7732	0.3664	∞	7.0000	∞	0.0000
12.0000	87.9530	0.3645	∞	7.0000	∞	0.0000
11.0000	88.1307	0.3627	∞	7.0000	∞	0.0000
10.0000	88.3065	0.3610	∞	7.0000	∞	0.0000
9.0000	88.4805	0.3596	∞	7.0000	∞	0.0000
8.0000	88.6530	0.3583	∞	7.0000	∞	0.0000
7.0000	88.8242	0.3572	∞	7.0000	∞	0.0000
6.0000	88.9944	0.3562	∞	7.0000	∞	0.0000
5.0000	89.1634	0.3554	∞	7.0000	∞	0.0000
4.0000	89.3317	0.3547	∞	7.0000	∞	0.0000
3.0000	89.4993	0.3542	∞	7.0000	∞	0.0000
2.0000	89.6665	0.3538	∞	7.0000	∞	0.0000
1.0000	89.8334	0.3536	∞	7.0000	∞	0.0000
0.0000	90.0000	0.3536	∞	7.0000	∞	0.0000

* DENOTES $M_2 = 1$, m DENOTES θ_{max}

TABLE F

AXISYMMETRIC

SUPERSONIC FLOW ABOUT A CONE

$\gamma = 7/5$

F.1a Cone Surface Conditions at the Minimum
     Free-Stream Mach Number.

F.1b Conditions Immediately behind the Shock at the
     Minimum Free-Stream Mach Number.

F.2 Solutions as a Function of the Cone Half-Angle.

SYMBOLS FOR TABLE F

$a*$ critical sound speed

C_{pc} pressure coefficient on cone surface

M_1 free-stream Mach number

M_2 Mach number immediately behind shock

M_c Mach number on cone surface

ω_c cone surface-speed

p_{21} static-pressure ratio across shock

p_c/p_1 total-pressure ratio across shock

P_c/P_1 ratio of cone static-pressure to
free-stream static pressure

T_c/T_1 ratio of cone static-temperature to
free-stream static temperature

T_{21} static-temperature ratio across shock

β shock angle in degrees

θ_c cone half-angle in degrees

ρ_c/ρ_1 ratio of cone static-density to
free-stream static density

ψ_2 flow deflection across shock in degrees

TABLE F.1a

CONDITIONS ON THE CONE SURFACE
FOR THE CASE OF THE MINIMUM FREE-STREAM MACH NUMBER*

θ_c	M_1	P_c/P_1	ρ_c/ρ_1	T_c/T_1	P^o_{12}	C_{pc}
2.5	1.0121844	1.0143689	1.0102426	1.0040845	1.0000000	0.0200358
5.0	1.0383341	1.0463206	1.0328714	1.0040845	1.0000000	0.0613765
7.5	1.0735583	1.0918310	1.0647652	1.0254195	1.0000000	0.1138256
10.0	1.1159051	1.1505082	1.1053308	1.0408723	1.0000029	0.1726663
12.5	1.1643198	1.2232501	1.1548030	1.0592716	1.0000238	0.2352605
15.0	1.2182190	1.3118250	1.2138947	1.0806744	1.0001297	0.3001664
17.5	1.2773745	1.4188223	1.2836779	1.1052790	1.0005150	0.3666868
20.0	1.3419094	1.5477651	1.3655327	1.1334515	1.0016251	0.4345602
22.5	1.4123337	1.7033574	1.4611321	1.1657791	1.0042715	0.5037365
25.0	1.4895953	1.8919026	1.5724618	1.2031469	1.0097561	0.5742259
27.5	1.5751393	1.1219533	1.7018625	1.2468418	1.0200090	0.6460093
30.0	1.6710795	1.4055423	1.8521996	1.2987489	1.0378275	0.7190379

*   In all cases the converged asymptotic values used on the cone surface were w_c/a^* =1.0000042 and M_c = 1.0000050.

TABLE F.1b

CONDITIONS IMMEDIATELY BEHIND THE SHOCK
FOR THE CASE OF THE MINIMUM FREE-STREAM MACH NUMBER#

θ_c	M_1	β	M_2	ψ_2	P_{21}	T_{21}
2.5	1.0121844	1.4174491	1.0115757	0.0044250	1.0007160	1.0002045
5.0	1.0383341	1.3047785	1.0348001	0.0438714	1.0042298	1.0012068
7.5	1.0735583	1.2132282	1.0627586	0.1770606	1.0132358	1.0037640
10.0	1.1159051	1.1389087	1.0908819	0.4870256	1.0315698	1.0089209
12.5	1.1643198	1.0801535	1.1154959	1.0640227	1.0637671	1.0178258
15.0	1.2182190	1.0355562	1.1341844	1.9773346	1.1143125	1.0314463
17.5	1.2773745	1.0034478	1.1460602	3.2556551	1.1872042	1.0504054
20.0	1.3419094	0.9819651	1.1515059	4.8877253	1.2861542	1.0750485
22.5	1.4123337	0.9692859	1.1516153	6.8364121	1.4152502	1.1056665
25.0	1.4895953	0.9637981	1.1476910	9.0541083	1.5797773	1.1427373
27.5	1.5751393	0.9641647	1.1409427	11.492777	1.7871198	1.1871407
30.0	1.6710795	0.9693061	1.1323727	14.110196	1.0480351	1.2403971

\#   For reasons of space the columns for w_2/a^* and ρ_{21} of the original report have been omitted.

SUPERSONIC FLOW ABOUT A CONE

$$\gamma = 7/5$$

CONE HALF-ANGLE, $\theta_c = 2.5°$

M_1	C_{pc}	ρ_c/ρ_1	T_c/T_1	M_c	β	ψ_2	P^o_{i2}	M_2
1.0122	0.0200	1.0102	1.0041	1.0000	81.2139	0.0044	1.0000	1.0116
1.5000	0.0123	1.0138	1.0055	1.4867	41.8139	0.0034	1.0000	1.4999
1.7500	0.0114	1.0175	1.0069	1.7340	34.8542	0.0050	1.0000	1.7498
2.0000	0.0108	1.0215	1.0085	1.9809	30.0052	0.0072	1.0000	1.9997
2.5000	0.0098	1.0305	1.0121	2.4729	23.5875	0.0136	1.0000	2.4994
3.0000	0.0091	1.0408	1.0161	2.9628	19.4869	0.0232	1.0000	2.9988
3.5000	0.0086	1.0521	1.0205	3.4501	16.6255	0.0368	1.0000	3.4977
4.0000	0.0082	1.0644	1.0253	3.9348	14.5124	0.0549	1.0000	3.9958
4.5000	0.0078	1.0776	1.0303	4.4166	12.8887	0.0779	1.0000	4.4930
5.0000	0.0075	1.0916	1.0357	4.8955	11.6030	0.1060	1.0000	4.9837
6.0000	0.0069	1.1221	1.0472	5.8441	9.7036	0.1774	1.0000	5.9743
7.0000	0.0065	1.1556	1.0596	6.7797	8.3778	0.2666	1.0000	6.9495
8.0000	0.0062	1.1919	1.0728	7.7019	7.4089	0.3686	1.0000	7.9114
10.0000	0.0057	1.2725	1.1013	9.5046	6.1077	0.5901	1.0003	9.7866
12.0000	0.0054	1.3630	1.1325	11.2503	5.2924	0.8076	1.0013	11.5902
15.0000	0.0050	1.5152	1.1836	13.7591	4.5361	1.0917	1.0060	14.1581
20.0000	0.0047	1.8035	1.2813	17.6378	3.8583	1.4382	1.0303	18.0821

CONE HALF-ANGLE, $\theta_c = 5.0°$

M_1	C_{pc}	ρ_c/ρ_1	T_c/T_1	M_c	β	ψ_2	P^o_{i2}	M_2
1.0383	0.0614	1.0329	1.0130	1.0000	74.7584	0.0439	1.0000	1.0348
1.5000	0.0397	1.0442	1.0175	1.4579	41.8708	0.0559	1.0000	1.4981
1.7500	0.0363	1.0550	1.0217	1.7005	34.9233	0.0823	1.0000	1.7472
2.0000	0.0340	1.0670	1.0263	1.9415	30.0945	0.1180	1.0000	1.9957
2.5000	0.0306	1.0939	1.0365	2.4194	23.7350	0.2189	1.0000	2.4906
3.0000	0.0282	1.1241	1.0479	2.8914	19.7154	0.3597	1.0000	2.9814
3.5000	0.0265	1.1573	1.0602	3.3572	16.9549	0.5354	1.0000	3.4666
4.0000	0.0251	1.1935	1.0733	3.8165	14.9565	0.7369	1.0000	3.9447
4.5000	0.0240	1.2323	1.0872	4.2691	13.4552	0.9535	1.0001	4.4150
5.0000	0.0230	1.2737	1.1018	4.7148	12.2945	1.1755	1.0003	4.8766
6.0000	0.0216	1.3638	1.1327	5.5852	10.6367	1.6077	1.0013	5.7731
7.0000	0.0206	1.4629	1.1662	6.4269	9.5271	1.9986	1.0039	6.6334
8.0000	0.0198	1.5700	1.2020	7.2391	8.7444	2.3374	1.0089	7.4581
10.0000	0.0187	1.8033	1.2812	8.7725	7.7341	2.8685	1.0303	9.0040
12.0000	0.0180	2.0542	1.3711	10.1821	7.1275	3.2464	1.0715	10.4159
15.0000	0.0174	2.4444	1.5280	12.0635	6.5873	3.6251	1.1806	12.2928
20.0000	0.0168	3.0749	1.8531	14.6135	6.1290	3.9802	1.5202	14.8298

SUPERSONIC FLOW ABOUT A CONE

$\gamma = 7/5$

CONE HALF-ANGLE, $\theta_c = 7.5°$

M_1	C_{pc}	ρ_c/ρ_1	T_c/T_1	M_c	β	ψ_2	$p_{i_2}^o$	M_2
1.0736	0.1138	1.0648	1.0254	1.0000	69.5130	0.1770	1.0000	1.0628
1.5000	0.0774	1.0856	1.0334	1.4197	42.1061	0.2722	1.0000	1.4907
1.7500	0.0704	1.1057	1.0410	1.6568	35.2048	0.3965	1.0000	1.7364
2.0000	0.0656	1.1280	1.0493	1.8912	30.4470	0.5541	1.0000	1.9799
2.5000	0.0590	1.1781	1.0678	2.3529	24.2688	0.9534	1.0000	2.4594
3.0000	0.0545	1.2347	1.0881	2.8048	20.4552	1.4232	1.0001	2.9271
3.5000	0.0513	1.2973	1.1099	3.2467	17.8980	1.9148	1.0005	3.3820
4.0000	0.0489	1.3653	1.1333	3.6784	16.0875	2.3927	1.0013	3.8237
4.5000	0.0470	1.4385	1.1580	4.0995	14.7519	2.8367	1.0031	4.2522
5.0000	0.0454	1.5162	1.1840	4.5097	13.7349	3.2395	1.0061	4.6676
6.0000	0.0432	1.6838	1.2403	5.2969	12.3077	3.9162	1.0175	5.4604
7.0000	0.0416	1.8640	1.3024	6.0385	11.3709	4.4433	1.0384	6.2034
8.0000	0.0405	2.0529	1.3706	6.7337	10.7206	4.8518	1.0713	6.8975
10.0000	0.0390	2.4422	1.5270	7.9852	9.9001	5.4236	1.1798	8.1435
12.0000	0.0381	2.8266	1.7116	9.0583	9.4211	5.7886	1.3559	9.2101
15.0000	0.0373	3.3598	2.0444	10.3683	9.0069	6.1232	1.7787	10.5113
20.0000	0.0366	4.0824	2.7537	11.9195	8.6671	6.4114	3.0823	12.0521

CONE HALF-ANGLE, $\theta_c = 10.0°$

M_1	C_{pc}	ρ_c/ρ_1	T_c/T_1	M_c	β	ψ_2	$p_{i_2}^o$	M_2
1.1159	0.1727	1.1053	1.0409	1.0000	65.2547	0.4870	1.0000	1.0909
1.5000	0.1238	1.1357	1.0522	1.3749	42.6659	0.7792	1.0000	1.4735
1.7500	0.1123	1.1665	1.0636	1.6065	35.8459	1.0995	1.0000	1.7124
2.0000	0.1045	1.2011	1.0761	1.8341	31.2056	1.4759	1.0001	1.9468
2.5000	0.0941	1.2792	1.1037	2.2789	25.2881	2.3130	1.0004	2.4024
3.0000	0.0875	1.3677	1.1341	2.7101	21.7145	3.1547	1.0014	2.8405
3.5000	0.0828	1.4654	1.1670	3.1275	19.3602	3.9271	1.0040	3.2617
4.0000	0.0794	1.5712	1.2024	3.5306	17.7147	4.6026	1.0090	3.6663
4.5000	0.0768	1.6838	1.2403	3.9189	16.5138	5.1807	1.0175	4.0546
5.0000	0.0748	1.8022	1.2808	4.2922	15.6079	5.6711	1.0302	4.4270
6.0000	0.0718	2.0512	1.3700	4.9928	14.3520	6.4366	1.0709	5.1240
7.0000	0.0699	2.3092	1.4707	5.6318	13.5401	6.9884	1.1359	5.7586
8.0000	0.0685	2.5684	1.5837	6.2104	12.9849	7.3934	1.2289	6.3327
10.0000	0.0667	3.0672	1.8483	7.1978	12.2968	7.9292	1.5143	7.3122
12.0000	0.0657	3.5167	2.1665	7.9858	11.9032	8.2523	1.9646	8.0940
15.0000	0.0648	4.0773	2.7470	8.8728	11.5692	8.5371	3.0675	8.9743
20.0000	0.0640	4.7386	3.9945	9.8177	11.3004	8.7726	6.7300	9.9127

SUPERSONIC FLOW ABOUT A CONE

$$\gamma = 7/5$$

CONE HALF-ANGLE, $\theta_c = 12.5°$

M_1	C_{pc}	ρ_c/ρ_1	T_c/T_1	M_c	β	ψ_2	$p^o_{i_2}$	M_2
1.1643	0.2353	1.1548	1.0593	1.0000	61.8880	1.0640	1.0000	1.1155
1.5000	0.1781	1.1932	1.0732	1.3249	43.6353	1.6324	1.0001	1.4443
1.7500	0.1612	1.2361	1.0885	1.5513	36.8973	2.2196	1.0002	1.6742
2.0000	0.1501	1.2846	1.1056	1.7722	32.3824	2.8585	1.0004	1.8974
2.5000	0.1359	1.3947	1.1432	2.2001	26.7170	4.1345	1.0020	2.3271
3.0000	0.1270	1.5192	1.1850	2.6104	23.3537	5.2775	1.0063	2.7361
3.5000	0.1210	1.6556	1.2308	3.0026	21.1645	6.2384	1.0151	3.1256
4.0000	0.1167	1.8016	1.2806	3.3761	19.6496	7.0256	1.0301	3.4958
4.5000	0.1135	1.9547	1.3347	3.7308	18.5541	7.6656	1.0528	3.8468
5.0000	0.1110	2.1127	1.3931	4.0663	17.7342	8.1870	1.0842	4.1787
6.0000	0.1076	2.4352	1.5239	4.6802	16.6100	8.9651	1.1773	4.7860
7.0000	0.1053	2.7548	1.6742	5.2205	15.8933	9.5014	1.3166	5.3206
8.0000	0.1037	3.0614	1.8448	5.6923	15.4091	9.8818	1.5100	5.7877
10.0000	0.1018	3.6132	2.2488	6.4568	14.8167	10.3682	2.0990	6.5451
12.0000	0.1007	4.0708	2.7385	7.0291	14.4832	10.6530	3.0487	7.1125
15.0000	0.0997	4.5945	3.6361	7.6324	14.2030	10.8977	5.4873	7.7111
20.0000	0.0990	5.1535	3.5707	8.2281	13.9802	11.0959	14.2125	8.3026

CONE HALF-ANGLE, $\theta_c = 15.0°$

M_1	C_{pc}	ρ_c/ρ_1	T_c/T_1	M_c	β	ψ_2	$p^o_{i_2}$	M_2
1.2182	0.3002	1.2139	1.0807	1.0000	59.3332	1.9773	1.0001	1.1342
1.5000	0.2400	1.2573	1.0960	1.2707	45.0299	2.8035	1.0003	1.4037
1.7500	0.2169	1.3133	1.1155	1.4925	38.3309	3.6830	1.0007	1.6241
2.0000	0.2022	1.3772	1.1373	1.7069	33.9145	4.5785	1.0016	1.8362
2.5000	0.1840	1.5220	1.1860	2.1179	28.4542	6.2286	1.0064	2.2411
3.0000	0.1731	1.6849	1.2407	2.5068	25.2588	7.5883	1.0176	2.6232
3.5000	0.1658	1.8612	1.3014	2.8731	23.2013	8.6643	1.0381	2.9831
4.0000	0.1608	2.0467	1.3683	3.2167	21.7907	9.5071	1.0701	3.3209
4.5000	0.1570	2.2376	1.4417	3.5374	20.7789	10.1694	1.1154	3.6365
5.0000	0.1542	2.4305	1.5219	3.8356	20.0272	10.6943	1.1757	3.9302
6.0000	0.1503	2.8111	1.7034	4.3668	19.0067	11.4546	1.3472	4.4542
7.0000	0.1478	3.1716	1.9141	4.8178	18.3639	11.9622	1.5983	4.8997
8.0000	0.1461	3.5024	2.1549	5.1981	17.9336	12.3146	1.9462	5.2758
10.0000	0.1441	4.0628	2.7282	5.7867	17.4122	12.7558	3.0260	5.8586
12.0000	0.1429	4.4961	3.4258	6.2043	17.1217	13.0084	4.8314	6.2724
15.0000	0.1419	4.9604	4.7071	6.6229	16.8793	13.2227	9.6908	6.6876
20.0000	0.1411	5.4219	7.4718	7.0145	16.6874	13.3946	28.1455	7.0763

SUPERSONIC FLOW ABOUT A CONE

$$\gamma = 7/5$$

CONE HALF-ANGLE, $\theta_c = 17.5°$

M_1	C_{pc}	ρ_c/ρ_1	T_c/T_1	M_c	β	ψ_2	$p^o_{i_2}$	M_2
1.2774	0.3667	1.2837	1.1053	1.0000	57.4935	3.2555	1.0005	1.1461
1.5000	0.3095	1.3276	1.1204	1.2127	46.8318	4.2198	1.0009	1.3536
1.7500	0.2793	1.3973	1.1441	1.4306	40.0944	5.3875	1.0021	1.5654
2.0000	0.2608	1.4774	1.1711	1.6387	35.7354	6.5140	1.0045	1.7672
2.5000	0.2385	1.6588	1.2319	2.0328	30.4246	8.4643	1.0154	2.1494
3.0000	0.2256	1.8607	1.3012	2.3998	27.3599	9.9746	1.0381	2.5062
3.5000	0.2172	2.0756	1.3791	2.7401	25.4072	11.1177	1.0761	2.8381
4.0000	0.2114	2.2974	1.4659	3.0538	24.0803	11.9846	1.1325	3.1450
4.5000	0.2072	2.5209	1.5620	3.3414	23.1360	12.6486	1.2096	3.4271
5.0000	0.2041	2.7417	1.6677	3.6041	22.4393	13.1643	1.3099	3.6852
6.0000	0.1999	3.1631	1.9086	4.0597	21.5014	13.8954	1.5911	4.1339
7.0000	0.1972	3.5450	2.1901	4.4335	20.9164	14.3732	2.0024	4.5027
8.0000	0.1954	3.8815	2.5130	4.7389	20.5273	14.6998	2.5791	4.8045
10.0000	0.1932	4.4228	3.2842	5.1934	20.0592	15.1026	4.4196	5.2542
12.0000	0.1920	4.8182	4.2243	5.5020	19.8003	15.3306	7.6121	5.5599
15.0000	0.1910	5.2212	5.9528	5.7998	19.5848	15.5226	16.5588	5.8552
20.0000	0.1902	5.6019	9.6845	6.0679	19.4158	15.6744	52.1018	6.1212

CONE HALF-ANGLE, $\theta_c = 20.0°$

M_1	C_{pc}	ρ_c/ρ_1	T_c/T_1	M_c	β	ψ_2	$p^o_{i_2}$	M_2
1.3419	0.4346	1.3655	1.1334	1.0000	56.2627	4.8879	1.0016	1.1515
1.5000	0.3870	1.4039	1.1464	1.1506	49.0286	5.7989	1.0024	1.2956
1.7500	0.3482	1.4872	1.1744	1.3657	42.1468	7.2405	1.0050	1.5004
2.0000	0.3255	1.5840	1.2068	1.5677	37.7957	8.5703	1.0100	1.6930
2.5000	0.2992	1.8024	1.2810	1.9448	32.5807	10.7613	1.0304	2.0542
3.0000	0.2843	2.0421	1.3667	2.2900	29.6145	12.3770	1.0692	2.3871
3.5000	0.2748	2.2927	1.4640	2.6043	27.7443	13.5585	1.1312	2.6923
4.0000	0.2684	2.5460	1.5735	2.8890	26.4850	14.4317	1.2198	2.9698
4.5000	0.2638	2.7957	1.6954	3.1453	25.5946	15.0883	1.3387	3.2206
5.0000	0.2605	3.0370	1.8302	3.3752	24.9426	15.5902	1.4920	3.4461
6.0000	0.2559	3.4827	2.1390	3.7641	24.0699	16.2909	1.9213	3.8486
7.0000	0.2531	3.8706	2.5011	4.0731	23.5296	16.7412	2.5559	4.1333
8.0000	0.2512	4.2003	2.9173	4.3188	23.1721	17.0461	3.4608	4.3758
10.0000	0.2489	4.7083	3.9134	4.6724	22.7453	17.4179	6.4345	4.7255
12.0000	0.2477	5.0627	5.1288	4.9043	22.5098	17.6265	11.7664	4.9551
15.0000	0.2466	5.4104	7.3644	5.1216	22.3150	17.8006	27.2023	5.1705
20.0000	0.2458	5.7271	12.1925	5.3120	22.1620	17.9387	90.6340	5.3592

SUPERSONIC FLOW ABOUT A CONE

$$\gamma = 7/5$$

CONE HALF-ANGLE, $\theta_c = 22.5°$

M_1	C_{pc}	ρ_c/ρ_1	T_c/T_1	M_c	β	ψ_2	p^o_{i2}	M_2
1.4123	0.5037	1.4611	1.1658	1.0000	55.5362	6.8365	1.0043	1.1516
1.5000	0.4732	1.4863	1.1742	1.0837	51.6418	7.4639	1.0052	1.2302
1.7500	0.4238	1.5822	1.2063	1.2977	44.4644	9.1702	1.0100	1.4305
2.0000	0.3964	1.6955	1.2444	1.4941	40.0658	10.6817	1.0188	1.6151
2.5000	0.3657	1.9502	1.3331	1.8544	34.8925	13.0709	1.0522	1.9568
3.0000	0.3489	2.2255	1.4370	2.1777	31.9962	14.7634	1.1123	2.2669
3.5000	0.3384	2.5076	1.5561	2.4669	30.1903	15.9666	1.2045	2.5465
4.0000	0.3314	2.7868	1.6908	2.7240	28.9831	16.8387	1.3339	2.7966
4.5000	0.3265	3.0560	1.8416	2.9514	28.1362	17.4838	1.5061	3.0187
5.0000	0.3229	3.3106	2.0089	3.1518	27.5186	17.9720	1.7277	3.2150
6.0000	0.3181	3.7670	2.3933	3.4830	26.6975	18.6446	2.3523	3.5403
7.0000	0.3151	4.1500	2.8453	3.7388	26.1922	19.0715	3.2905	3.7923
8.0000	0.3131	4.4656	3.3654	3.9373	25.8593	19.3579	4.6530	3.9882
10.0000	0.3108	4.9346	4.6115	4.2154	25.4634	19.7057	9.2545	4.2629
12.0000	0.3095	5.2501	6.1327	4.3927	25.2457	19.8988	17.7407	4.4384
15.0000	0.3084	5.5507	8.9319	4.5553	25.0663	20.0598	42.9544	4.5993
20.0000	0.3076	5.8170	14.9777	4.6947	24.9259	20.1864	149.2495	4.7375

CONE HALF-ANGLE, $\theta_c = 25.0°$

M_1	C_{pc}	ρ_c/ρ_1	T_c/T_1	M_c	β	ψ_2	p^o_{i2}	M_2
1.4896	0.5742	1.5725	1.2031	1.0000	55.2217	9.0539	1.0098	1.1477
1.5000	0.5700	1.5758	1.2043	1.0101	54.7621	9.1421	1.0099	1.1570
1.7500	0.5062	1.6818	1.2398	1.2259	47.0479	11.1194	1.0178	1.3563
2.0000	0.4733	1.8109	1.2840	1.4175	42.5318	12.8004	1.0317	1.5341
2.5000	0.4379	2.1001	1.3885	1.7614	37.3437	15.3621	1.0817	1.8576
3.0000	0.4191	2.4075	1.5121	2.0636	34.4898	17.1154	1.1679	2.1461
3.5000	0.4077	2.7164	1.6550	2.3287	32.7291	18.3329	1.2972	2.4017
4.0000	0.4001	3.0157	1.8175	2.5601	31.5625	19.2004	1.4768	2.6262
4.5000	0.3948	3.2982	2.0001	2.7612	30.7484	19.8352	1.7153	2.8223
5.0000	0.3910	3.5601	2.2029	2.9357	30.1576	20.3102	2.0232	2.9930
6.0000	0.3859	4.0167	2.6702	3.2178	29.3761	20.9588	2.9007	3.2698
7.0000	0.3828	4.3877	3.2206	3.4303	28.8977	21.3667	4.2422	3.4789
8.0000	0.3808	4.6854	3.8544	3.5919	28.5837	21.6395	6.2251	3.6382
10.0000	0.3784	5.1148	5.3737	3.8131	28.2113	21.9672	13.0870	3.8566
12.0000	0.3770	5.3955	7.2292	3.9510	28.0073	22.1591	26.0433	3.9929
15.0000	0.3759	5.6567	10.6439	4.0753	27.8394	22.2995	65.3416	4.1159
20.0000	0.3751	5.8835	18.0201	4.1803	27.7082	22.4181	234.2901	4.2197

SUPERSONIC FLOW ABOUT A CONE

$$\gamma = 7/5$$

CONE HALF-ANGLE, $\theta_c = 27.5°$

M_1	C_{pc}	ρ_c/ρ_1	T_c/T_1	M_c	β	ψ_2	p^o_{i2}	M_2
1.5751	0.6460	1.7019	1.2468	1.0000	55.2423	11.4930	1.0200	1.1409
1.7500	0.5959	1.7856	1.2754	1.1496	49.9247	13.0377	1.0288	1.2776
2.0000	0.5562	1.9290	1.3257	1.3375	45.1960	14.8900	1.0490	1.4502
2.5000	0.5155	2.2499	1.4469	1.6660	39.9271	17.6144	1.1192	1.7570
3.0000	0.4946	2.5857	1.5919	1.9479	37.0864	19.4215	1.2366	2.0250
3.5000	0.4822	2.9164	1.7606	2.1904	35.3538	20.6517	1.4103	2.2581
4.0000	0.4740	3.2302	1.9531	2.3983	34.2147	21.5157	1.6505	2.4594
4.5000	0.4684	3.5206	2.1699	2.5761	33.4241	22.1408	1.9702	2.6324
5.0000	0.4643	3.7848	2.4113	2.7279	32.8528	22.6060	2.3854	2.7807
6.0000	0.4590	4.2343	2.9681	2.9688	32.1011	23.2357	3.5843	3.0168
7.0000	0.4558	4.5893	3.6244	3.1462	31.6427	23.6288	5.4495	3.1912
8.0000	0.4537	4.8677	4.3808	3.2788	31.3431	23.8900	8.2522	3.3217
10.0000	0.4512	5.2596	6.1946	3.4569	30.9878	24.2029	18.1583	3.4973
12.0000	0.4498	5.5098	8.4103	3.5660	30.7942	24.3759	37.2302	3.6050
15.0000	0.4487	5.7385	12.4883	3.6629	30.6349	24.5186	95.7538	3.7008
20.0000	0.4478	5.9339	21.2978	3.7438	30.5106	24.6309	352.7710	3.7807

CONE HALF-ANGLE, $\theta_c = 30.0°$

M_1	C_{pc}	ρ_c/ρ_1	T_c/T_1	M_c	β	ψ_2	p^o_{i2}	M_2
1.6711	0.7190	1.8522	1.2987	1.0000	55.5374	14.1102	1.0378	1.1324
1.7500	0.6941	1.8941	1.3135	1.0669	53.1716	14.8734	1.0438	1.1930
2.0000	0.6451	2.0491	1.3695	1.2536	48.0786	16.9171	1.0712	1.3631
2.5000	0.5981	2.3982	1.5083	1.5681	42.6435	19.8094	1.1650	1.6549
3.0000	0.5750	2.7580	1.6761	1.8310	39.7839	21.6721	1.3188	1.9038
3.5000	0.5615	3.1056	1.8723	2.0527	38.0610	22.9172	1.5445	2.1162
4.0000	0.5527	3.4291	2.0969	2.2394	36.9369	23.7806	1.8568	2.2966
4.5000	0.5467	3.7229	2.3503	2.3965	36.1617	24.4005	2.2747	2.4493
5.0000	0.5425	3.9857	2.6326	2.5289	35.6036	24.8589	2.8215	2.5783
6.0000	0.5369	4.4230	3.2848	2.7353	34.8719	25.4748	4.4214	2.7802
7.0000	0.5335	4.7600	4.0542	2.8843	34.4273	25.8570	6.9527	2.9265
8.0000	0.5313	5.0192	4.9412	2.9941	34.1380	26.1097	10.8127	3.0344
10.0000	0.5287	5.3769	7.0686	3.1392	33.7953	26.4116	24.7054	3.1773
12.0000	0.5273	5.6008	9.6678	3.2268	33.6086	26.5778	51.8876	3.2636
15.0000	0.5261	5.8027	14.4520	3.3038	33.4556	26.7153	136.8327	3.3396
20.0000	0.5252	5.9729	24.7875	3.3673	33.3364	26.8230	512.1409	3.4024

TABLE G.1

SELECTED VALUES

THERMOPHYSICAL PROPERTIES OF ARGON

G.1-1

THERMOPHYSICAL PROPERTIES OF ARGON

ISOBAR: 0.101325 MPa

T (K)	ρ (Pa·s)	a (m/s)	$10^6 \mu$ (Pa·s)	λ (J/s·m·K)
87.282[b]	5.781	170.7	7.25	0.00586
90.0	5.584	173.6	7.46	0.00601
95.0	5.265	178.9	7.85	0.00630
100.0	4.982	184.0	8.24	0.00658
105.0	4.730	188.9	8.63	0.00688
110.0	4.504	193.6	9.02	0.00717
115.0	4.299	198.2	9.42	0.00747
120.0	4.112	202.7	9.81	0.00777
130.0	3.785	211.3	10.6	0.00837
140.0	3.508	219.6	11.4	0.00898
150.0	3.269	227.5	12.2	0.00958
160.0	3.061	235.1	13.0	0.0102
170.0	2.878	242.4	13.7	0.0108
180.0	2.716	249.6	14.5	0.0114
190.0	2.571	256.5	15.2	0.0120
200.0	2.441	263.2	16.0	0.0125
210.0	2.324	269.8	16.7	0.0131
220.0	2.218	276.2	17.5	0.0137
230.0	2.121	282.4	18.2	0.0142
240.0	2.032	288.5	18.9	0.0148
250.0	1.950	294.5	19.6	0.0153
260.0	1.875	300.3	20.3	0.0158
270.0	1.805	306.1	20.9	0.0164
280.0	1.740	311.7	21.6	0.0169
300.0	1.624	322.7	22.9	0.0179
310.0	1.522	333.3	24.2	0.0189
340.0	1.432	343.5	25.4	0.0199
360.0	1.353	353.5	26.6	0.0208
380.0	1.281	363.2	27.8	0.0217
400.0	1.217	372.7	28.9	0.0217

[b]Liquid-vapor boundary

G.1-2

THERMOPHYSICAL PROPERTIES OF ARGON

ISOBAR: 0.60 MPa

T (K)	ρ (Pa·s)	a (m/s)	$10^6 \mu$ (Pa·s)	k (J/s·m·K)
108.385[b]	30.02	182.2	9.17	0.00804
110.0	29.53	184.2	9.29	0.00812
115.0	27.77	190.1	9.67	0.00834
120.0	26.25	195.7	10.1	0.00857
125.0	24.93	201.0	10.4	0.00882
130.0	23.75	206.0	10.8	0.00907
135.0	22.69	210.8	11.2	0.00933
140.0	21.74	215.5	11.6	0.00959
150.0	20.08	224.3	12.4	0.0101
160.0	18.67	232.6	13.1	0.0107
170.0	17.47	240.5	13.9	0.0112
180.0	16.42	248.1	14.6	0.0118
190.0	15.49	255.3	15.4	0.0123
200.0	14.67	262.4	16.1	0.0129
210.0	13.94	269.2	16.9	0.0134
220.0	13.27	275.8	17.6	0.0140
230.0	12.67	282.2	18.3	0.0145
240.0	12.12	288.4	19.0	0.0150
250.0	11.62	294.5	19.7	0.0156
260.0	11.16	300.5	20.4	0.0161
270.0	10.74	306.3	21.0	0.0166
280.0	10.35	312.0	21.7	0.0171
290.0	9.983	317.6	22.3	0.0176
300.0	9.644	323.1	23.0	0.0181
320.0	9.031	333.8	24.3	0.0191
340.0	8.493	344.2	25.5	0.0200
360.0	8.016	354.2	26.7	0.0210
380.0	7.590	364.0	27.8	0.0219
400.0	7.207	373.4	29.0	0.0228

[b]Liquid-vapor boundary

THERMOPHYSICAL PROPERTIES OF ARGON

ISOBAR: 1.00 MPa

T (K)	ρ (Pa·s)	a (m/s)	$10^6 \mu$ (Pa·s)	k (J/s·m·K)
116.550[b]	49.24	184.4	10.1	0.00923
120.0	47.18	189.1	10.3	0.00934
125.0	44.26	195.4	10.7	0.00951
130.0	41.78	201.2	11.0	0.00971
135.0	39.63	206.7	11.4	0.00992
140.0	37.75	211.9	11.8	0.0101
145.0	36.05	216.9	12.2	0.0104
150.0	34.55	221.6	12.5	0.0106
160.0	31.93	230.5	13.3	0.0111
170.0	29.73	238.9	14.0	0.0116
180.0	27.84	246.9	14.8	0.0121
190.0	26.19	254.4	15.5	0.0126
200.0	24.75	261.7	16.3	0.0132
210.0	23.46	268.7	17.0	0.0137
220.0	22.31	275.5	17.7	0.0142
230.0	21.27	282.0	18.4	0.0147
240.0	20.33	288.4	19.1	0.0153
250.0	19.48	294.6	19.8	0.0158
260.0	18.69	300.6	20.4	0.0163
270.0	17.97	306.5	21.1	0.0168
280.0	17.30	312.3	21.8	0.0173
290.0	16.68	318.0	22.4	0.0178
300.0	16.11	323.5	23.1	0.0183
320.0	15.08	334.3	24.3	0.0192
340.0	14.17	344.7	25.5	0.0202
360.0	13.37	354.8	26.7	0.0211
380.0	12.65	364.6	27.9	0.0220
400.0	12.01	374.1	29.0	0.0229

[b]Liquid-vapor boundary

THERMOPHYSICAL PROPERTIES OF ARGON

ISOBAR: 2.00 MPa

T (K)	ρ (Pa·s)	a (m/s)	$10^6\mu$ (Pa·s)	k (J/s·m·K)
129.714[b]	101.7	185.4	11.9	0.0120
130.0	101.4	185.8	11.9	0.0120
132.0	97.53	189.4	12.0	0.0119
134.0	94.12	192.7	12.1	0.0119
136.0	91.06	195.8	12.2	0.0119
138.0	88.33	198.8	12.3	0.0119
140.0	85.83	201.6	12.5	0.0119
145.0	80.41	208.2	12.8	0.0120
150.0	75.88	214.3	13.1	0.0121
155.0	72.00	219.9	13.5	0.0122
160.0	68.61	225.2	13.8	0.0123
165.0	65.61	230.2	14.1	0.0125
170.0	62.92	235.0	14.5	0.0127
175.0	60.49	239.5	14.8	0.0129
180.0	58.28	243.9	15.2	0.0131
190.0	54.39	252.3	15.9	0.0135
200.0	51.06	260.2	16.6	0.0139
210.0	48.16	267.7	17.3	0.0144
220.0	45.61	274.9	18.0	0.0149
230.0	43.34	281.8	18.7	0.0153
240.0	41.31	288.4	19.4	0.0158
250.0	39.47	294.9	20.0	0.0163
260.0	37.80	301.1	20.7	0.0168
270.0	36.28	307.2	21.4	0.0173
280.0	34.88	313.2	22.0	0.0177
290.0	33.60	318.9	22.6	0.0182
300.0	32.40	324.6	23.3	0.0187
320.0	30.27	335.6	24.5	0.0196
340.0	28.41	346.1	25.7	0.0205
360.0	26.77	356.3	26.9	0.0214
380.0	25.32	366.1	28.1	0.0223
400.0	24.02	375.7	29.2	0.0232

[b]Liquid-vapor boundary

THERMOPHYSICAL PROPERTIES OF ARGON

ISOBAR: 4.00 MPa

T (K)	ρ (Pa·s)	a (m/s)	$10^6 \mu$ (Pa·s)	k (J/s·m·K)
145.672[b]	261.4	6.543	16.5	0.0240
150.0	209.7	5.250	15.6	0.0190
152.0	197.7	4.949	15.5	0.0181
154.0	188.1	4.708	15.5	0.0175
156.0	180.0	4.506	15.5	0.0171
158.0	173.1	4.333	15.5	0.0168
160.0	167.1	4.182	15.5	0.0165
162.0	161.7	4.047	15.6	0.0162
164.0	156.8	3.926	15.7	0.0161
166.0	152.4	3.815	15.7	0.0159
170.0	144.7	3.621	15.9	0.0157
175.0	136.5	3.416	16.1	0.0155
180.0	129.5	3.242	16.4	0.0154
185.0	123.5	3.090	16.7	0.0154
190.0	118.1	2.957	16.9	0.0155
195.0	113.4	2.838	17.2	0.0155
200.0	109.1	2.731	17.5	0.0156
205.0	105.2	2.634	17.8	0.0158
210.0	101.7	2.545	18.2	0.0159
220.0	95.36	2.387	18.8	0.0162
230.0	89.95	2.252	19.4	0.0166
240.0	85.21	2.133	20.0	0.0170
250.0	81.01	2.028	20.7	0.0174
260.0	77.27	1.934	21.3	0.0178
270.0	73.90	1.850	21.9	0.0182
280.0	70.84	1.773	22.5	0.0187
290.0	68.05	1.704	23.2	0.0191
300.0	65.50	1.640	23.8	0.0195
310.0	63.14	1.581	24.4	0.0200
320.0	60.97	1.526	25.0	0.0204
340.0	57.07	1.429	26.1	0.0212
360.0	53.67	1.343	27.3	0.0221
380.0	50.68	1.269	28.4	0.0229
400.0	48.01	1.202	29.5	0.0238

[b]Liquid-vapor boundary

G.1-6

THERMOPHYSICAL PROPERTIES OF ARGON

ISOBAR: 4.80 MPa

T (K)	ρ (Pa·s)	a (m/s)	$10^6 \mu$ (Pa·s)	k (J/s·m·K)
150.304[b]	450.8	183.5	23.3	0.0596
155.0	264.2	198.2	17.4	0.0230
156.0	254.1	200.5	17.2	0.0221
157.0	245.6	202.7	17.1	0.0214
158.0	238.1	204.8	17.0	0.0208
160.0	225.6	208.7	16.8	0.0199
162.0	215.2	212.3	16.8	0.0192
164.0	206.5	215.6	16.7	0.0186
166.0	198.9	218.8	16.7	0.0182
168.0	192.1	221.8	16.7	0.0179
170.0	186.1	224.6	16.8	0.0176
172.0	180.7	227.3	16.8	0.0173
174.0	175.8	229.9	16.8	0.0171
180.0	163.1	237.2	17.1	0.0167
185.0	154.4	242.8	17.3	0.0165
190.0	146.9	248.1	17.5	0.0164
195.0	140.4	253.0	17.8	0.0164
200.0	134.6	257.8	18.0	0.0164
205.0	129.4	262.3	18.3	0.0165
210.0	124.7	266.6	18.6	0.0166
220.0	116.5	274.9	19.2	0.0168
230.0	109.5	282.6	19.8	0.0172
240.0	103.5	289.9	20.4	0.0175
250.0	98.20	296.9	21.0	0.0179
260.0	93.50	303.7	21.6	0.0183
270.0	89.29	310.1	22.2	0.0187
280.0	85.50	316.4	22.8	0.0191
290.0	82.05	322.5	23.4	0.0195
300.0	78.90	328.4	24.0	0.0199
310.0	76.01	334.1	24.6	0.0203
320.0	73.35	339.7	25.2	0.0207
330.0	70.88	345.2	25.7	0.0211
340.0	68.58	350.5	26.3	0.0215
360.0	64.45	360.9	27.5	0.0224
380.0	60.81	370.9	28.6	0.0232
400.0	57.59	380.6	29.7	0.0240

[b]Liquid-vapor boundary

TABLE G.2

SELECTED VALUES

THERMOPHYSICAL PROPERTIES OF PARAHYDROGEN

THERMOPHYSICAL PROPERTIES OF PARAHYDROGEN

ISOBAR: 0.01 MPa

T (K)	ρ (Pa·s)	a (m/s)	$10^6 \mu$ (Pa·s)	k (J/s·m·K)
14.430[b]	0.1714	312	7.74	12.77
15.0	0.1646	318	8.07	13.01
16.0	0.1539	329	8.64	13.45
17.0	0.1445	340	9.21	13.89
18.0	0.1363	350	9.77	14.40
19.0	0.1289	360	10.32	14.90
20.0	0.1223	369	10.87	15.53
21.0	0.1164	379	11.41	16.32
22.0	0.1110	388	11.95	17.11
23.0	0.1061	396	12.48	17.91
24.0	0.1016	405	13.01	18.70
25.0	0.0975	414	13.52	19.40
26.0	0.0937	422	14.04	20.28
27.0	0.0902	430	14.54	21.07
28.0	0.0869	438	15.04	21.85
29.0	0.0839	446	15.54	22.64
30.0	0.0811	453	16.03	23.42
31.0	0.0784	461	16.51	24.12
32.0	0.0760	468	16.99	24.82
33.0	0.0737	476	17.47	25.51
34.0	0.0715	483	17.93	26.22
35.0	0.0694	490	18.40	26.92
36.0	0.0675	497	18.86	27.63
37.0	0.0657	504	19.31	28.33
38.0	0.0639	511	19.76	29.04
39.0	0.0623	517	20.21	29.74
40.0	0.0607	524	20.65	30.45
42.0	0.0578	537	21.53	31.85
44.0	0.0552	549	22.38	33.25
46.0	0.0528	561	23.22	34.65
48.0	0.0506	573	24.05	36.05
50.0	0.0485	585	24.86	37.45
52.0	0.0467	596	25.66	38.85
54.0	0.0449	606	26.45	40.25
56.0	0.0433	617	27.22	41.65
58.0	0.0418	627	27.99	43.05
60.0	0.0404	636	28.74	44.51
65.0	0.0373	658	30.58	48.38
70.0	0.0346	679	32.36	52.26
75.0	0.0323	697	34.09	56.13
80.0	0.0303	714	35.77	60.04
85.0	0.0285	729	37.41	65.00
90.0	0.0269	744	39.01	69.96
100.0	0.0242	772	45.74	89.54
120.0	0.0202	827	54.09	117.1
140.0	0.0173	883	59.20	136.1

[b]Liquid-vapor boundary

THERMOPHYSICAL PROPERTIES OF PARAHYDROGEN

ISOBAR: 0.01 MPa

T (K)	ρ (Pa·s)	a (m/s)	$10^6 \mu$ (Pa·s)	k (J/s·m·K)
140.0	0.0173	883	59.20	136.1
160.0	0.0152	940	63.00	148.3
180.0	0.0135	997	66.17	155.6
200.0	0.0121	1053	69.01	160.2
220.0	0.0110	1109	71.65	163.4
240.0	0.0101	1162	74.18	166.2
260.0	0.0093	1213	76.63	169.1
280.0	0.0087	1262	79.03	172.3
300.0	0.0081	1309	81.39	175.9
350.0	0.0069	1418	87.21	186.1
400.0	0.0061	1517	92.95	197.4
450.0	0.0054	1610	98.65	209.3
500.0	0.0048	1697	104.2	221.2
550.0	0.0044	1780	109.8	233.2
600.0	0.0040	1858	115.3	245.2
700.0	0.0035	2006	126.1	269.3
800.0	0.0030	2142	136.6	293.7
900.0	0.0027	2268	146.9	318.5
1000.0	0.0024	2386	157.0	440.6
1200.0	0.0020	2601	176.4	511.2
1400.0	0.0017	2795	195.1	584.6
1600.0	0.0015	2970	213.1	668.0
1800.0	0.0013	3121	230.6	800.6
2000.0	0.0012	3235	247.7	1099.2
2500.0	0.0009	3476	292.9	4536.7
3000.0	0.0006	3904	346.3	15311.6

THERMOPHYSICAL PROPERTIES OF PARAHYDROGEN

ISOBAR: 0.101325 MPa

T (K)	ρ (Pa·s)	a (m/s)	$10^6 \mu$ (Pa·s)	k (J/s·m·K)
20.268[b]	1.338	355	11.28	16.94
21.0	1.282	363	11.67	17.40
22.0	1.209	374	12.19	18.08
23.0	1.146	385	12.71	18.78
24.0	1.090	395	13.23	19.50
25.0	1.039	404	13.74	20.24
26.0	0.9936	413	14.24	20.98
27.0	0.9522	422	14.74	21.72
28.0	0.9143	431	15.24	22.47
29.0	0.8795	439	15.73	23.21
30.0	0.8474	448	16.21	23.96
31.0	0.8177	456	16.69	24.65
32.0	0.7901	463	17.16	25.34
33.0	0.7644	471	17.63	26.02
34.0	0.7404	479	18.10	26.72
35.0	0.7179	486	18.56	27.41
36.0	0.6968	493	19.02	28.10
37.0	0.6770	501	19.47	28.80
38.0	0.5682	508	19.91	29.49
39.0	0.6406	514	20.36	30.19
40.0	0.6238	521	20.80	30.88
42.0	0.5929	535	21.66	32.27
44.0	0.5650	547	22.51	33.65
46.0	0.5397	560	23.35	35.03
48.0	0.5166	572	24.17	36.42
50.0	0.4954	583	24.98	37.81
52.0	0.4759	595	25.78	39.19
54.0	0.4579	606	26.56	40.58
56.0	0.4412	616	27.33	41.97
58.0	0.4257	626	28.09	43.36
60.0	0.4113	636	28.84	44.81
65.0	0.3793	658	30.68	48.67
70.0	0.3519	679	32.45	52.53
75.0	0.3282	697	34.17	56.37
80.0	0.3075	714	35.85	60.27
85.0	0.2893	730	37.48	65.22
90.0	0.2732	744	39.08	70.17
95.0	0.2587	759	40.64	75.12
100.0	0.2457	773	45.74	89.54
120.0	0.2047	827	54.08	117.1
140.0	0.1754	883	59.19	136.1
160.0	0.1535	940	63.00	148.3
180.0	0.1364	998	66.17	155.7
200.0	0.1228	1054	69.01	160.2
220.0	0.1116	1110	71.66	163.4
240.0	0.1023	1163	74.19	166.2
260.0	0.0944	1214	76.64	169.1

[b]Liquid-vapor boundary

THERMOPHYSICAL PROPERTIES OF PARAHYDROGEN

ISOBAR: 0.101325 MPa

T (K)	ρ (Pa·s)	a (m/s)	$10^6 \mu$ (Pa·s)	k (J/s·m·K)
260.0	0.0944	1214	76.64	169.1
280.0	0.0877	1263	79.04	172.4
300.0	0.0818	1310	81.41	175.9
350.0	0.0702	1419	87.23	186.1
400.0	0.0614	1518	92.97	197.5
450.0	0.0546	1611	98.65	209.3
500.0	0.0491	1698	104.3	221.3
550.0	0.0447	1780	109.8	233.3
600.0	0.0409	1859	115.3	245.3
700.0	0.0351	2006	126.1	269.4
800.0	0.0307	2142	136.7	293.8
900.0	0.0273	2268	147.0	318.6
1000.0	0.0246	2386	157.0	440.6
1200.0	0.0205	2601	176.5	511.2
1400.0	0.0175	2795	195.2	583.7
1600.0	0.0154	2974	213.2	659.2
1800.0	0.0136	3136	230.7	749.3
2000.0	0.0123	3277	247.7	891.5
2500.0	0.0097	3539	289.8	2124.6
3000.0	0.0076	3839	338.7	6725.3

G.2-5

THERMOPHYSICAL PROPERTIES OF PARAHYDROGEN

ISOBAR: 0.50 MPa

T (K)	ρ (Pa·s)	a (m/s)	$10^6\mu$ (Pa·s)	k (J/s·m·K)
27.116[b]	6.139	378	16.10	27.63
28.0	5.701	391	16.44	27.33
29.0	5.308	405	16.84	27.35
30.0	4.987	417	17.25	27.58
31.0	4.715	428	17.67	27.99
32.0	4.480	439	18.10	28.46
33.0	4.273	449	18.53	28.97
34.0	4.090	458	18.96	29.50
36.0	3.775	477	19.81	30.63
37.0	3.639	485	20.23	31.22
38.0	3.514	494	20.66	31.83
39.0	3.398	502	21.08	32.44
40.0	3.291	510	21.49	33.06
42.0	3.098	525	22.32	34.31
44.0	2.930	539	23.14	35.58
46.0	2.780	553	23.95	36.87
48.0	2.647	566	24.74	38.17
50.0	2.527	579	25.53	39.48
52.0	2.418	591	26.30	40.80
54.0	2.318	603	37.07	42.12
56.0	2.228	614	27.82	43.46
58.0	2.144	624	28.56	44.79
60.0	2.067	634	29.30	46.20
65.0	1.897	658	31.10	49.96
70.0	1.754	679	32.84	53.73
75.0	1.632	698	34.54	57.44
80.0	1.526	715	36.19	61.28
85.0	1.434	731	37.81	66.19
90.0	1.352	746	39.39	71.11
95.0	1.279	761	40.93	76.03
100.0	1.214	775	45.75	89.72
120.0	1.009	830	54.04	117.1
140.0	0.8639	886	59.17	136.1
160.0	0.7555	944	63.00	148.3
180.0	0.6714	1001	66.19	155.7
200.0	0.6042	1058	69.05	160.3
220.0	0.5493	1113	71.70	163.5
240.0	0.5035	1166	74.24	166.4
260.0	0.4648	1217	76.69	169.3
280.0	0.4317	1266	79.10	172.5
300.0	0.4030	1313	81.48	176.1
350.0	0.3455	1422	87.32	186.4
400.0	0.3024	1521	93.07	197.7
450.0	0.2688	1613	98.76	209.6
500.0	0.2420	1701	104.4	221.6
550.0	0.2200	1783	110.0	233.6
600.0	0.2017	1862	115.5	245.7

[b]Liquid-vapor boundary

G.2-6

THERMOPHYSICAL PROPERTIES OF PARAHYDROGEN

ISOBAR: 0.50 MPa

T (K)	ρ (Pa·s)	a (m/s)	$10^6\mu$ (Pa·s)	k (J/s·m·K)
600.0	0.2017	1862	115.5	245.7
700.0	0.1729	2009	126.3	269.8
800.0	0.1513	2144	136.9	294.3
900.0	0.1345	2270	147.2	319.1
1000.0	0.1211	2388	157.3	440.6
1200.0	0.1009	2603	176.8	511.1
1400.0	0.0865	2797	195.6	583.5
1600.0	0.0757	2976	213.6	657.0
1800.0	0.0673	3142	231.1	736.4
2000.0	0.0606	3292	248.1	839.1
2500.0	0.0482	3590	289.4	1493.0
3000.0	0.0390	3853	333.4	3780.4

THERMOPHYSICAL PROPERTIES OF PARAHYDROGEN

ISOBAR: 1.00 MPa

T (K)	ρ (Pa·s)	a (m/s)	$10^6\mu$ (Pa·s)	k (J/s·m·K)
31.248[b]	14.36	374	21.45	49.56
32.0	12.43	393	20.84	40.71
33.0	11.05	411	20.72	37.54
34.0	10.11	427	20.83	36.34
35.0	9.406	442	21.05	35.87
36.0	8.834	454	21.32	35.75
37.0	8.355	466	21.62	35.84
38.0	7.944	477	21.95	36.05
39.0	7.586	487	22.29	36.36
40.0	7.268	497	22.64	36.72
42.0	6.725	515	23.36	37.58
44.0	6.276	531	24.09	38.56
46.0	5.894	547	24.83	39.62
48.0	5.565	561	25.57	40.73
50.0	5.276	575	26.31	41.88
52.0	5.020	588	27.04	43.06
54.0	4.791	600	27.77	44.27
56.0	4.584	612	28.49	45.50
58.0	4.397	623	29.21	46.75
60.0	4.226	634	29.91	48.08
65.0	3.856	659	31.66	51.68
70.0	3.551	681	33.36	55.32
75.0	3.293	700	35.02	58.83
80.0	3.072	718	36.64	62.59
85.0	2.880	734	38.23	67.45
90.0	2.712	749	39.78	72.32
95.0	2.563	764	41.30	77.19
100.0	2.430	778	45.80	90.00
120.0	2.015	834	54.01	117.1
140.0	1.723	891	59.17	136.1
160.0	1.506	948	63.01	148.3
180.0	1.338	1005	66.22	155.7
200.0	1.204	1062	69.09	160.4
220.0	1.095	1117	71.76	163.6
240.0	1.004	1171	74.30	166.5
260.0	0.9268	1222	76.77	169.4
280.0	0.8607	1271	79.18	172.7
300.0	0.8036	1317	81.56	176.3
350.0	0.6891	1426	87.42	186.6
400.0	0.6033	1525	93.20	198.0
450.0	0.5365	1617	98.91	210.0
500.0	0.4830	1704	104.6	222.0

[b]Liquid-vapor boundary

THERMOPHYSICAL PROPERTIES OF PARAHYDROGEN

ISOBAR: 1.00 MPa

T (K)	ρ (Pa·s)	a (m/s)	$10^6 \mu$ (Pa·s)	k (J/s·m·K)
500.0	0.4830	1704	104.6	222.0
550.0	0.4392	1786	110.2	234.1
600.0	0.4027	1865	115.7	246.1
700.0	0.3454	2012	126.5	270.4
800.0	0.3023	2147	137.1	294.9
900.0	0.2688	2273	147.5	319.8
1000.0	0.2420	2391	157.6	440.6
1200.0	0.2017	2605	177.2	511.1
1400.0	0.1729	2799	196.0	583.4
1600.0	0.1513	2978	214.2	656.4
1800.0	0.1345	3145	231.7	733.3
2000.0	0.1211	3297	248.7	826.5
2500.0	0.0965	3609	290.0	1340.5
3000.0	0.0788	3869	332.5	3028.1

G.2-9

THERMOPHYSICAL PROPERTIES OF PARAHYDROGEN

ISOBAR: 2.00 MPa

T (K)	ρ (Pa·s)	a (m/s)	$10^6\mu$ (Pa·s)	k (J/s·m·K)
14.457[b]	78.12	1319	267.2	80.58
15.0	77.71	1311	249.2	85.23
16.0	76.95	1288	221.8	91.81
17.0	76.13	1264	199.6	96.32
18.0	75.27	1240	181.3	99.64
19.0	74.37	1218	166.0	101.7
20.0	73.42	1196	153.0	103.7
21.0	72.43	1173	141.8	105.9
22.0	71.38	1150	132.0	107.4
23.0	70.28	1125	123.3	108.2
24.0	69.10	1098	115.4	108.5
25.0	67.84	1070	108.3	108.2
26.0	66.49	1039	101.7	107.4
27.0	65.04	1006	95.46	106.1
28.0	53.47	969	89.61	104.4
29.0	61.75	929	83.98	102.2
30.0	59.84	884	78.48	99.54
31.0	57.69	835	73.03	95.97
32.0	55.20	780	67.50	91.79
33.0	52.23	719	61.76	87.82
34.0	48.50	649	55.59	84.92
35.0	43.43	568	48.63	82.44
36.0	36.34	494	40.96	80.21
37.0	28.91	466	34.76	70.15
38.0	24.17	468	31.40	60.34
39.0	21.23	477	29.55	54.70
40.0	19.21	488	28.51	51.56
42.0	16.52	508	27.62	48.56
44.0	14.76	527	27.48	47.51
46.0	13.43	544	27.66	47.26
48.0	12.39	560	18.01	47.48
50.0	11.54	575	28.47	47.96
52.0	10.83	589	29.00	48.62
54.0	10.22	602	29.56	49.42
56.0	9.696	615	30.15	50.30
58.0	9.230	627	30.76	51.25
60.0	8.815	638	31.37	52.34
70.0	7.255	687	34.50	58.75
75.0	6.686	707	36.06	61.79
80.0	6.209	725	37.60	65.35
85.0	5.801	742	39.12	70.09
90.0	5.447	757	40.61	74.84
95.0	5.136	772	42.08	79.60
100.0	4.861	787	45.98	90.72

[b]Liquid-vapor boundary

THERMOPHYSICAL PROPERTIES OF PARAHYDROGEN

ISOBAR: 2.00 MPa

T (K)	ρ (Pa·s)	a (m/s)	$10^6 \mu$ (Pa·s)	k (J/s·m·K)
100.0	4.861	787	45.98	90.72
120.0	4.015	842	54.04	117.3
140.0	3.428	899	59.21	136.1
160.0	2.994	957	63.09	148.4
180.0	2.659	1014	66.32	155.9
200.0	2.393	1071	69.20	160.6
220.0	2.175	1126	71.88	163.9
240.0	1.995	1180	74.44	166.9
260.0	1.842	1231	76.92	169.9
280.0	1.711	1279	79.35	173.2
300.0	1.598	1326	81.75	176.8
350.0	1.371	1434	87.64	187.2
400.0	1.201	1533	93.45	198.6
450.0	1.068	1624	99.20	210.7
500.0	0.9622	1711	104.9	222.7
550.0	0.8753	1793	110.5	234.9
600.0	0.8028	1871	116.1	247.1
700.0	0.6887	2018	127.0	271.4
800.0	0.6030	2153	137.7	296.1
900.0	0.5363	2278	148.1	321.2
1000.0	0.4829	2396	158.3	440.6
1200.0	0.4027	2610	178.0	511.1
1400.0	0.3453	2804	196.9	583.4
1600.0	0.3023	2982	215.2	656.1
1800.0	0.2688	3149	232.8	731.1
2000.0	0.2419	3303	250.0	817.6
2500.0	0.1931	3627	291.3	1232.3
3000.0	0.1586	3890	332.9	2485.6

G.2-11

THERMOPHYSICAL PROPERTIES OF PARAHYDROGEN

ISOBAR: 5.00 MPa

T (K)	ρ (Pa·s)	a (m/s)	$10^6 \mu$ (Pa·s)	k (J/s·m·K)
15.382[b]	79.64	1390	285.8	91.17
16.0	79.23	1384	265.4	95.52
17.0	78.54	1370	237.9	100.9
18.0	77.82	1354	215.4	105.0
19.0	77.06	1337	196.8	107.8
20.0	76.28	1322	181.1	110.4
21.0	75.46	1302	167.8	113.4
22.0	74.61	1284	156.4	115.7
23.0	73.74	1264	146.4	117.3
24.0	72.82	1245	137.6	118.4
25.0	71.87	1225	129.7	119.0
26.0	70.88	1204	122.6	119.1
27.0	69.85	1181	116.2	118.8
28.0	68.77	1158	110.3	118.3
29.0	67.65	1134	104.8	117.4
30.0	66.48	1109	99.78	116.3
31.0	65.25	1084	95.05	114.5
32.0	63.97	1059	90.59	112.6
33.0	62.63	1033	86.38	110.4
34.0	61.23	1005	82.39	108.2
35.0	59.77	977	78.58	105.9
36.0	58.24	949	74.94	103.4
37.0	56.63	921	71.46	100.9
38.0	54.95	893	68.14	98.22
39.0	53.21	865	64.97	95.62
40.0	51.41	839	61.98	93.45
42.0	47.65	791	56.55	89.06
44.0	43.84	751	51.94	84.77
46.0	40.14	720	48.24	80.84
48.0	36.71	701	45.39	77.38
50.0	33.66	689	43.31	74.53
52.0	31.04	684	41.84	72.32
54.0	26.79	684	40.81	70.71
56.0	26.88	687	40.11	69.61
58.0	25.23	692	39.64	68.90
60.0	23.80	698	39.37	68.64
65.0	20.94	715	39.24	69.25
70.0	18.78	733	39.67	70.78
75.0	17.09	749	40.43	72.04
80.0	15.71	765	41.39	74.67
85.0	14.56	780	42.47	78.78
90.0	13.60	794	43.63	83.01
95.0	12.77	808	44.83	87.34
100.0	12.04	821	47.31	94.20
120.0	9.869	873	54.78	119.0
140.0	8.398	929	59.80	137.2

[b]Liquid-vapor boundary

THERMOPHYSICAL PROPERTIES OF PARAHYDROGEN

ISOBAR: 5.00 MPa

T (K)	ρ (Pa·s)	a (m/s)	$10^6 \mu$ (Pa·s)	k (J/s·m·K)
140.0	8.398	929	59.80	137.2
160.0	7.327	986	63.63	149.3
180.0	6.508	1043	66.84	156.9
200.0	5.858	1099	69.72	161.6
220.0	5.329	1154	72.40	165.1
240.0	4.890	1207	74.97	168.1
260.0	4.519	1257	77.47	171.2
280.0	4.201	1305	79.92	174.6
300.0	3.925	1352	82.34	178.3
350.0	3.374	1458	88.32	188.8
400.0	2.960	1556	94.22	200.5
450.0	2.637	1646	100.1	212.7
500.0	2.377	1732	105.9	225.0
550.0	2.164	1813	111.6	237.4
600.0	1.987	1891	117.3	249.8
700.0	1.707	2036	128.4	274.6
800.0	1.496	2170	139.3	299.7
900.0	1.332	2294	149.9	325.2
1000.0	1.200	2411	160.3	440.6
1200.0	1.001	2624	180.4	511.1
1400.0	0.8594	2816	199.7	583.4
1600.0	0.7527	2994	218.3	655.7
1800.0	0.6695	3161	226.3	729.2
2000.0	0.6029	3316	253.7	809.8
2500.0	0.4820	3651	295.6	1136.0
3000.0	0.3982	3924	337.0	1996.9

THERMOPHYSICAL PROPERTIES OF PARAHYDROGEN

ISOBAR: 10.00 MPa

T (K)	ρ (Pa·s)	a (m/s)	$10^6\mu$ (Pa·s)	k (J/s·m·K)
16.808[b]	81.87	1506	317.0	106.0
17.0	81.76	1503	310.0	107.1
18.0	81.16	1491	278.1	112.2
19.0	80.56	1476	252.4	116.0
20.0	79.92	1464	231.0	119.5
21.0	79.27	1450	213.0	123.5
22.0	78.59	1436	197.7	126.6
23.0	77.89	1424	184.6	129.0
24.0	77.18	1410	173.2	130.9
25.0	76.44	1399	163.2	132.3
26.0	75.69	1385	154.4	133.3
27.0	74.92	1370	146.6	133.9
28.0	74.14	1354	139.6	134.1
29.0	73.33	1338	133.2	134.1
30.0	72.50	1322	127.4	133.9
31.0	71.65	1307	122.1	132.9
32.0	70.79	1290	117.3	131.8
33.0	69.90	1273	112.8	130.6
34.0	69.00	1256	108.6	129.3
35.0	68.08	1239	104.7	127.9
36.0	67.14	1221	101.1	126.4
37.0	66.18	1204	97.62	124.9
38.0	65.20	1187	94.40	123.4
39.0	64.21	1170	91.36	121.8
40.0	63.21	1153	88.48	120.2
42.0	61.15	1121	83.17	116.9
44.0	59.05	1090	78.42	113.6
46.0	56.93	1059	74.16	110.4
48.0	54.79	1031	70.37	107.4
50.0	52.67	1005	67.02	104.6
52.0	50.58	982	64.09	102.3
54.0	48.53	962	61.53	100.1
56.0	46.55	945	59.34	98.15
58.0	44.65	931	57.47	96.42
60.0	42.85	920	55.91	95.05
65.0	38.75	901	53.06	92.92
70.0	35.28	891	51.41	91.90
75.0	32.36	886	50.57	90.89
80.0	29.90	886	50.28	91.69
85.0	27.80	888	50.34	94.55
90.0	25.99	892	50.65	97.71
95.0	24.41	898	51.15	101.1
100.0	23.03	905	51.83	103.7

[b]Liquid-vapor boundary

THERMOPHYSICAL PROPERTIES OF PARAHYDROGEN

ISOBAR: 10.00 MPa

T (K)	ρ (Pa·s)	a (m/s)	$10^6 \mu$ (Pa·s)	k (J/s·m·K)
100.0	23.03	905	51.83	103.7
120.0	18.88	942	57.97	125.4
140.0	16.08	990	62.22	141.8
160.0	14.08	1039	65.56	152.8
180.0	12.52	1094	68.46	159.9
200.0	11.29	1148	71.14	164.4
220.0	10.29	1202	73.69	167.7
240.0	9.461	1253	76.18	170.7
260.0	8.757	1302	78.63	173.8
280.0	8.153	1349	81.07	177.2
300.0	7.623	1395	83.49	181.0
350.0	6.574	1499	89.53	191.7
400.0	5.782	1594	95.56	203.6
450.0	5.162	1683	101.6	216.2
500.0	4.663	1767	107.5	228.8
550.0	4.252	1846	113.4	241.5
600.0	3.908	1922	119.3	254.3
700.0	3.365	2065	130.8	279.8
800.0	2.954	2197	142.0	305.7
900.0	2.633	2320	153.0	332.0
1000.0	2.375	2435	163.7	440.6
1200.0	1.986	2646	184.4	511.1
1400.0	1.706	2837	204.3	583.3
1600.0	1.496	3014	223.5	655.6
1800.0	1.331	3179	242.1	728.2
2000.0	1.200	3334	260.1	805.8
2500.0	0.9605	3674	303.2	1087.5
3000.0	0.7962	3955	345.2	1748.1

THERMOPHYSICAL PROPERTIES OF PARAHYDROGEN

ISOBAR: 20.00 MPa

T (K)	ρ (Pa·s)	a (m/s)	$10^6\mu$ (Pa·s)	k (J/s·m·K)
19.360[b]	85.54	1680	377.6	130.5
20.0	85.23	1675	353.3	133.8
21.0	84.73	1667	321.0	139.1
22.0	84.21	1659	294.2	143.5
23.0	83.69	1651	271.7	147.2
24.0	83.16	1642	252.5	150.2
25.0	82.62	1634	236.0	152.6
26.0	82.07	1625	221.8	154.5
27.0	81.51	1615	209.3	155.9
28.0	80.93	1606	198.3	157.1
29.0	80.35	1597	188.5	157.8
30.0	79.76	1588	179.9	158.4
31.0	79.16	1577	172.1	158.1
32.0	78.56	1566	165.0	157.6
33.0	77.94	1556	158.7	157.0
34.0	77.32	1546	152.9	156.3
36.0	76.06	1523	142.7	154.6
38.0	74.77	1503	134.0	152.7
39.0	74.12	1493	130.1	151.6
40.0	73.47	1482	126.5	150.5
42.0	72.14	1462	120.0	148.2
44.0	70.80	1441	114.2	145.8
46.0	69.45	1419	109.1	143.5
48.0	68.10	1398	104.5	141.1
50.0	66.74	1378	100.4	138.8
52.0	65.38	1358	96.59	136.6
54.0	64.03	1338	93.17	134.4
56.0	62.68	1319	90.06	132.4
58.0	61.35	1302	87.22	130.5
60.0	60.03	1286	84.63	128.9
65.0	56.82	1249	79.15	125.9
70.0	53.77	1217	74.87	123.6
75.0	50.91	1190	71.57	121.9
80.0	48.25	1167	69.08	121.0
85.0	45.79	1148	67.25	122.6
90.0	43.54	1134	65.95	124.4
100.0	39.57	1115	64.04	126.8
120.0	33.45	1109	68.24	144.7
140.0	28.98	1132	70.44	157.0
160.0	25.60	1169	72.14	165.0
180.0	22.96	1212	73.83	169.8
200.0	20.85	1259	75.64	172.9
220.0	19.11	1305	77.58	175.3
240.0	17.65	1350	79.64	177.7
260.0	16.42	1394	81.79	180.4
280.0	15.35	1436	84.03	183.5

[b]Liquid-vapor boundary

THERMOPHYSICAL PROPERTIES OF PARAHYDROGEN

ISOBAR: 20.00 MPa

T (K)	ρ (Pa·s)	a (m/s)	$10^6 \mu$ (Pa·s)	k (J/s·m·K)
280.0	15.35	1436	84.03	183.5
300.0	14.39	1484	86.33	187.1
350.0	12.49	1581	92.27	197.8
400.0	11.05	1671	98.38	210.0
450.0	9.906	1755	104.6	223.0
500.0	8.981	1835	110.8	236.1
550.0	8.216	1911	117.0	249.5
600.0	7.571	1984	123.2	263.0
700.0	6.546	2122	135.3	290.0
800.0	5.766	2250	147.3	317.4
900.0	5.153	2370	158.0	345.3
1000.0	4.657	2482	170.4	440.6
1200.0	3.907	2688	192.5	511.1
1400.0	3.364	2876	213.8	583.3
1600.0	2.954	3050	234.2	655.4
1800.0	2.633	3213	254.1	727.5
2000.0	2.375	3367	273.3	803.0
2500.0	1.905	3709	319.2	1053.1
3000.0	1.585	3996	363.5	1571.1

THERMOPHYSICAL PROPERTIES OF PARAHYDROGEN

ISOBAR: 100.00 MPa

T (K)	ρ (Pa·s)	a (m/s)	$10^6\mu$ (Pa·s)	k (J/s·m·K)
34.1691[b]	102.4	2527	937.8	276.3
35.0	102.1	2525	871.0	276.3
36.0	101.8	2522	800.9	276.3
37.0	101.6	2520	740.3	276.0
38.0	101.3	2517	687.5	275.5
39.0	101.0	2514	641.2	274.9
40.0	100.7	2511	600.5	274.2
42.0	100.1	2505	532.4	272.4
44.0	99.49	2498	478.2	270.3
46.0	98.90	2490	434.3	267.9
48.0	98.31	2481	398.2	265.4
50.0	97.72	2472	368.2	262.8
52.0	97.14	2462	343.0	260.2
54.0	96.55	2453	321.6	257.5
56.0	95.96	2443	303.3	254.9
58.0	95.38	2433	287.5	252.3
60.0	94.79	2423	273.7	250.1
65.0	93.34	2396	246.2	245.8
70.0	91.89	2367	225.8	241.7
75.0	90.45	2336	210.1	237.9
80.0	89.02	2304	197.8	234.6
85.0	87.62	2272	187.9	235.5
90.0	86.24	2240	179.8	236.5
95.0	84.89	2210	173.1	237.5
100.0	83.58	2181	133.0	245.1
120.0	78.37	2088	137.4	263.8
140.0	73.80	2039	135.5	269.1
160.0	69.59	2031	131.4	267.5
180.0	65.94	2005	127.2	262.9
200.0	62.60	1971	123.6	257.6
220.0	59.38	1981	120.9	252.8
240.0	56.56	2026	119.1	249.2
260.0	54.01	2044	117.9	246.9
280.0	51.71	2057	117.5	246.0
300.0	49.42	2112	117.6	246.3
350.0	44.76	2159	119.8	251.5
400.0	40.94	2208	124.0	261.3
450.0	37.74	2257	129.7	274.2
500.0	35.01	2306	136.3	289.2
550.0	32.66	2356	143.7	305.7
600.0	30.62	2406	151.6	323.3
700.0	27.22	2504	168.6	361.3
800.0	24.50	2599	186.6	402.3
900.0	22.28	2692	205.2	445.9
1000.0	20.43	2781	224.1	440.6
1200.0	17.51	2951	262.6	511.1

[b]Liquid-vapor boundary

THERMOPHYSICAL PROPERTIES OF PARAHYDROGEN

ISOBAR: 100.00 MPa

T (K)	ρ (Pa·s)	a (m/s)	$10^6\mu$ (Pa·s)	k (J/s·m·K)
1200.0	17.51	2951	262.6	511.1
1400.0	15.33	3110	301.3	583.3
1600.0	13.62	3262	339.7	655.3
1800.0	12.26	3408	377.8	726.6
2000.0	11.14	3548	415.3	799.2
2500.0	9.064	3871	506.7	1007.2
3000.0	7.625	4156	595.0	1333.5

TABLE G.3

SELECTED VALUES

THERMOPHYSICAL PROPERTIES OF NITROGEN

THERMOPHYSICAL PROPERTIES OF NITROGEN

ISOBAR: 0.01 MPa

T (K)	ρ (Pa·s)	a (m/s)	$10^6\mu$ (Pa·s)	k (J/s·m·K)
65.0	0.5220	163.7	4.35	0.00599
70.0	0.4840	170.0	4.61	0.00658
75.0	0.4513	176.1	5.06	0.00711
80.0	1.4228	181.9	5.40	0.00761
85.0	0.3977	187.6	5.74	0.00810
90.0	0.3754	193.1	6.08	0.00857
95.0	0.3555	198.4	6.42	0.00904
100.0	0.3376	203.6	6.75	0.00951
105.0	0.3215	208.7	7.09	0.00997
110.0	0.3068	213.6	7.42	0.0104
120.0	0.2811	223.2	8.08	0.0114
130.0	0.2594	232.3	8.72	0.0123
135.0	0.2498	236.7	9.04	0.0128
140.0	0.2409	241.1	9.36	0.0132
145.0	0.2325	245.4	9.67	0.0137
150.0	0.2248	249.6	9.98	0.0141
155.0	0.2175	253.7	10.3	0.0146
160.0	0.2107	257.8	10.6	0.0150
165.0	0.2043	261.8	10.9	0.0154
170.0	0.1983	265.7	11.2	0.0159
175.0	0.1926	269.6	11.5	0.0163
180.0	0.1873	273.4	11.8	0.0167
190.0	0.1774	280.9	12.3	0.0176
200.0	0.1685	288.3	12.9	0.0184
210.0	0.1605	295.4	13.4	0.0192
220.0	0.1532	302.3	14.0	0.0200
230.0	0.1465	309.1	14.5	0.0208
240.0	0.1404	315.8	15.0	0.0215
250.0	0.1348	322.3	15.5	0.0223
260.0	0.1296	328.7	16.0	0.0230
270.0	0.1248	334.9	16.5	0.0237
280.0	0.1204	341.1	17.0	0.0244
290.0	0.1162	347.1	17.5	0.0251
300.0	0.1123	353.0	17.9	0.0258
310.0	0.1087	358.9	18.4	0.0265
320.0	0.1053	364.6	18.8	0.0272
330.0	0.1021	370.2	19.3	0.0278
340.0	0.09911	375.8	19.7	0.0285
350.0	0.09628	381.2	20.2	0.0291
360.0	0.09360	386.6	20.6	0.0298
370.0	0.09107	391.9	21.0	0.0304
380.0	0.08868	397.1	21.4	0.0311
400.0	0.08424	407.3	22.2	0.0323
420.0	0.08023	417.3	23.0	0.0336
440.0	0.07658	426.9	23.8	0.0348
440.0	0.07658	426.9	23.8	0.0348
460.0	0.07325	436.3	24.6	0.0361
480.0	0.07020	445.5	25.3	0.0373
500.0	0.06739	454.4	26.1	0.0385

THERMOPHYSICAL PROPERTIES OF NITROGEN

ISOBAR: 0.01 MPa

T (K)	ρ (Pa·s)	a (m/s)	$10^6\mu$ (Pa·s)	k (J/s·m·K)
500.0	0.06739	454.4	26.1	0.0385
520.0	0.06480	463.2	26.8	0.0398
540.0	0.06240	471.7	27.5	0.0410
560.0	0.06017	480.0	28.2	0.0422
580.0	0.05810	488.1	28.9	0.0434
600.0	0.05616	496.1	29.5	0.0446
620.0	0.05435	503.9	30.2	0.0458
640.0	0.05265	511.6	30.9	0.0470
660.0	0.05105	519.1	31.5	0.0482
680.0	0.04955	526.5	32.1	0.0493
700.0	0.04814	533.7	32.8	0.0505
720.0	0.04680	540.9	33.4	0.0517
740.0	0.04554	547.9	34.0	0.0529
760.0	0.04434	554.8	34.6	0.0541
780.0	0.04320	561.6	35.2	0.0552
800.0	0.04212	568.3	35.8	0.0564
820.0	0.04109	575.0	36.4	0.0576
840.0	0.04011	581.5	37.0	0.0587
860.0	0.03918	588.0	37.6	0.0599
880.0	0.03829	594.3	38.2	0.0610
900.0	0.03744	600.6	38.7	0.0621
920.0	0.03663	606.8	39.3	0.0633
940.0	0.03585	613.0	39.8	0.0644
1000.0	0.03370	631.1	41.5	0.0678
1050.0	0.03209	645.7	42.8	0.0705
1100.0	0.03063	660.1	44.2	0.0732
1150.0	0.02930	674.1	45.5	0.0759
1200.0	0.02808	687.8	46.7	0.0785
1250.0	0.02696	701.2	48.0	0.0811
1300.0	0.02592	714.4	49.2	0.0836
1350.0	0.02496	727.4	50.4	0.0861
1400.0	0.02407	740.1	51.6	0.0886
1450.0	0.02324	752.7	52.8	0.0910
1500.0	0.02247	765.0	54.0	0.0934
1550.0	0.02174	777.1	55.2	0.0958
1600.0	0.02106	789.1	56.3	0.0981
1650.0	0.02042	800.9	57.4	0.100
1700.0	0.01982	812.5	58.6	0.103
1750.0	0.01926	824.0	59.7	0.105
1800.0	0.01872	835.3	60.8	0.107
1850.0	0.01822	846.4	61.8	0.109
1900.0	0.01774	857.5	62.9	0.112

THERMOPHYSICAL PROPERTIES OF NITROGEN

ISOBAR: 0.1015 MPa

T (K)	ρ (Pa·s)	a (m/s)	$10^6 \mu$ (Pa·s)	k (J/s·m·K)
77.363[b]	4.611	174.8	5.28	0.00783
80.0	4.440	178.3	5.46	0.00808
85.0	4.151	184.5	5.80	0.00853
90.0	3.900	190.4	6.14	0.00898
95.0	3.679	196.1	6.47	0.00943
100.0	3.484	201.6	6.80	0.00988
105.0	3.309	206.9	7.13	0.0103
110.0	3.151	212.1	7.46	0.0108
120.0	2.879	221.9	8.12	0.0117
130.0	2.651	231.4	8.76	0.0126
135.0	2.550	235.9	9.08	0.0131
140.0	2.457	240.3	9.39	0.0135
145.0	2.370	244.7	9.70	0.0140
150.0	2.290	249.0	10.0	0.0144
155.0	2.214	253.2	10.3	0.0148
160.0	2.144	257.3	10.6	0.0153
165.0	2.078	261.4	10.9	0.0157
170.0	2.016	265.4	11.2	0.0162
175.0	1.958	269.3	11.5	0.0166
180.0	1.903	273.2	11.8	0.0170
190.0	1.802	280.7	12.4	0.0178
200.0	1.711	288.1	12.9	0.0187
210.0	1.629	295.3	13.5	0.0195
220.0	1.554	302.3	14.0	0.0202
230.0	1.486	309.1	14.5	0.0210
240.0	1.424	315.8	15.0	0.0218
250.0	1.367	322.3	15.6	0.0225
260.0	1.314	328.8	16.1	0.0232
270.0	1.265	335.0	16.5	0.0240
280.0	1.220	341.2	17.0	0.0247
290.0	1.178	347.2	17.5	0.0254
300.0	1.138	353.2	18.0	0.0260
310.0	1.102	359.0	18.4	0.0267
320.0	1.067	364.8	18.9	0.0274
330.0	1.035	370.4	19.3	0.0281
340.0	1.004	376.0	19.7	0.0287
350.0	0.9754	381.4	20.2	0.0294
360.0	0.9483	386.8	20.6	0.0300
370.0	0.9226	392.1	21.0	0.0307
380.0	0.8983	397.3	21.4	0.0313
400.0	0.8533	407.6	22.2	0.0326
420.0	0.8127	417.5	23.0	0.0338
440.0	0.7757	427.2	23.8	0.0351
440.0	0.7757	427.2	23.8	0.0351
460.0	0.7420	436.6	24.6	0.0363
480.0	0.7110	445.7	25.3	0.0375

[b]Liquid-vapor boundary

THERMOPHYSICAL PROPERTIES OF NITROGEN

ISOBAR: 0.1015 MPa

T (K)	ρ (Pa·s)	a (m/s)	$10^6\mu$ (Pa·s)	k (J/s·m·K)
480.0	0.7110	445.7	25.3	0.0375
500.0	0.6826	454.7	26.1	0.0388
520.0	0.6563	463.4	26.8	0.0400
540.0	0.6320	471.9	27.5	0.0412
560.0	0.6094	480.2	28.2	0.0424
580.0	0.5884	488.4	28.9	0.0436
600.0	0.5688	496.4	29.5	0.0448
620.0	0.5505	504.2	30.2	0.0460
640.0	0.5333	511.8	30.9	0.0472
660.0	0.5171	519.3	31.5	0.0484
680.0	0.5019	526.7	32.2	0.0496
700.0	0.4876	534.0	32.8	0.0508
720.0	0.4740	541.1	33.4	0.0520
740.0	0.4612	548.1	34.0	0.0531
760.0	0.4491	555.0	34.6	0.0543
780.0	0.4376	561.9	35.2	0.0555
800.0	0.4266	568.6	35.8	0.0567
820.0	0.4162	575.2	36.4	0.0578
840.0	0.4063	581.7	37.0	0.0590
860.0	0.3969	588.2	37.6	0.0601
880.0	0.3878	594.5	38.2	0.0613
900.0	0.3792	600.8	38.7	0.0624
920.0	0.3710	607.1	39.3	0.0635
940.0	0.3631	613.2	39.9	0.0647
1000.0	0.3413	631.3	41.5	0.0680
1050.0	0.3251	645.9	42.8	0.0708
1100.0	0.3103	660.3	44.2	0.0735
1150.0	0.2968	674.3	25.5	0.0762
1200.0	0.2844	688.0	46.7	0.0788
1250.0	0.2731	701.4	48.0	0.0814
1300.0	0.2626	714.6	49.2	0.0839
1350.0	0.2528	727.6	50.5	0.0864
1400.0	0.2438	740.3	51.7	0.0889
1450.0	0.2354	752.8	52.8	0.0913
1500.0	0.2276	765.2	54.0	0.0937
1550.0	0.2202	777.3	55.2	0.0961
1600.0	0.2133	789.3	56.3	0.0984
1650.0	0.2069	801.0	57.4	0.101
1700.0	0.2008	812.7	58.6	0.103
1750.0	0.1951	824.1	59.7	0.105
1800.0	0.1896	835.4	60.8	0.107
1850.0	0.1845	846.6	61.8	0.110
1900.0	0.1797	857.6	62.9	0.112

G.3-5

THERMOPHYSICAL PROPERTIES OF NITROGEN

ISOBAR: 0.60 MPa

T (K)	ρ (Pa·s)	a (m/s)	$10^6 \mu$ (Pa·s)	k (J/s·m·K)
96.399[b]	24.63	183.3	6.86	0.0116
100.0	23.25	189.0	7.08	0.0118
105.0	21.65	196.2	7.40	0.0121
110.0	20.31	202.9	7.71	0.0124
115.0	19.15	209.1	8.03	0.0127
120.0	18.15	215.0	8.34	0.0131
125.0	16.47	226.0	8.96	0.0139
135.0	15.75	231.2	9.27	0.0143
140.0	15.10	236.2	9.58	0.0147
145.0	14.51	241.0	9.88	0.0151
150.0	13.97	245.8	10.2	0.0155
155.0	13.47	250.3	10.5	0.0159
160.0	13.01	254.8	10.8	0.0163
165.0	12.58	259.2	11.1	0.0167
170.0	12.17	263.4	11.4	0.0171
175.0	11.80	267.6	11.7	0.0175
180.0	11.45	271.7	11.9	0.0179
185.0	11.12	275.7	12.2	0.0183
190.0	10.81	279.7	12.5	0.0187
195.0	10.52	283.6	12.8	0.0191
200.0	10.24	287.4	13.0	0.0195
210.0	9.731	294.8	13.6	0.0203
220.0	9.271	302.1	14.1	0.0210
230.0	8.853	309.1	14.6	0.0218
240.0	8.473	316.0	15.1	0.0225
250.0	8.125	322.7	15.7	0.0233
260.0	7.805	329.2	15.1	0.0240
270.0	7.510	335.6	16.6	0.0247
280.0	7.236	341.8	17.1	0.0254
290.0	6.982	348.0	17.6	0.0261
300.0	6.746	354.0	18.0	0.0267
310.0	6.525	359.9	18.5	0.0274
320.0	6.319	365.7	18.9	0.0281
330.0	6.125	371.4	19.4	0.0287
340.0	5.943	377.0	19.8	0.0294
350.0	5.772	382.5	20.2	0.0300
360.0	5.610	387.9	20.7	0.0307
370.0	5.458	393.3	21.1	0.0313
380.0	5.313	398.5	21.5	0.0320
400.0	5.046	408.8	22.3	0.0332
420.0	4.805	418.7	23.1	0.0345
440.0	4.585	428.4	23.9	0.0357
460.0	4.385	437.9	24.6	0.0369
480.0	4.202	447.0	25.4	0.0382
480.0	4.202	447.0	25.4	0.0382
500.0	4.034	456.0	26.1	0.0394
520.0	3.878	464.7	26.8	0.0406
540.0	3.735	473.2	27.5	0.0418

[b]Liquid-vapor boundary

THERMOPHYSICAL PROPERTIES OF NITROGEN

ISOBAR: 0.60 MPa

T (K)	ρ (Pa·s)	a (m/s)	$10^6 \mu$ (Pa·s)	k (J/s·m·K)
540.0	3.735	473.2	27.5	0.0418
560.0	3.601	481.6	28.2	0.0430
580.0	3.477	489.7	28.9	0.0442
600.0	3.361	497.7	29.6	0.0454
620.0	3.253	505.5	30.2	0.0466
640.0	3.151	513.1	30.9	0.0478
660.0	3.056	520.6	31.6	0.0490
680.0	2.966	528.0	32.2	0.0502
700.0	2.881	535.3	32.8	0.0514
720.0	2.801	542.4	33.4	0.0526
740.0	2.726	549.4	34.1	0.0538
760.0	2.654	556.3	34.7	0.0549
780.0	2.586	563.1	35.3	0.0561
800.0	2.521	569.8	35.9	0.0573
820.0	2.460	576.4	36.5	0.0584
840.0	2.401	582.9	37.0	0.0596
860.0	2.346	589.4	37.6	0.0607
880.0	2.292	595.7	38.2	0.0619
900.0	2.242	602.0	38.8	0.0630
920.0	2.193	608.2	39.3	0.0642
940.0	2.146	614.4	39.9	0.0653
1000.0	2.018	632.4	41.5	0.0687
1050.0	1.922	647.1	42.9	0.0714
1100.0	1.834	661.4	44.2	0.0741
1150.0	1.755	675.3	45.5	0.0768
1200.0	1.682	689.0	46.8	0.0794
1250.0	1.615	702.5	48.0	0.0820
1300.0	1.553	715.6	49.3	0.0846
1350.0	1.495	728.6	50.5	0.0871
1400.0	1.442	741.3	51.7	0.0896
1450.0	1.392	753.8	52.9	0.0920
1500.0	1.346	766.1	54.0	0.0944
1550.0	1.303	778.2	55.2	0.0968
1600.0	1.262	790.2	56.3	0.0991
1650.0	1.224	801.9	57.5	0.101
1700.0	1.188	813.6	58.6	0.104
1750.0	1.154	825.0	59.7	0.106
1800.0	1.122	836.3	60.8	0.108
1850.0	1.092	847.5	61.9	0.110
1900.0	1.063	858.5	62.9	0.113

THERMOPHYSICAL PROPERTIES OF NITROGEN

ISOBAR: 1.00 MPa

T (K)	ρ (Pa·s)	a (m/s)	$10^6\mu$ (Pa·s)	k (J/s·m·K)
103.748[b]	41.23	183.2	7.61	0.0139
105.0	40.24	185.5	7.68	0.0139
106.0	39.49	187.4	7.74	0.0139
108.0	38.13	190.9	7.85	0.0139
110.0	36.89	194.2	7.97	0.0139
115.0	34.26	201.9	8.26	0.0141
120.0	32.08	208.9	8.55	0.0144
125.0	30.24	215.4	8.85	0.0146
130.0	28.65	221.5	9.15	0.0149
135.0	27.25	227.3	9.45	0.0153
140.0	26.00	232.8	9.75	0.0156
145.0	24.89	238.1	10.0	0.0160
150.0	23.88	243.2	10.3	0.0163
155.0	22.96	248.1	10.6	0.0167
160.0	22.11	252.9	10.9	0.0171
165.0	21.34	257.5	11.2	0.0175
170.0	20.62	262.0	11.5	0.0178
175.0	19.95	266.4	11.8	0.0182
180.0	19.33	270.6	12.1	0.0186
185.0	18.75	274.8	12.3	0.0190
190.0	18.21	278.9	12.6	0.0194
195.0	17.70	282.9	12.9	0.0197
200.0	17.21	286.9	13.2	0.0201
210.0	16.33	294.6	13.7	0.0209
220.0	15.54	302.0	14.2	0.0216
230.0	14.82	309.2	14.7	0.0223
240.0	14.17	316.2	15.2	0.0230
250.0	13.58	323.0	15.7	0.0238
260.0	13.04	329.6	16.2	0.0245
270.0	12.54	336.1	16.7	0.0251
280.0	12.08	342.4	17.2	0.0258
290.0	11.65	348.6	17.6	0.0265
300.0	11.25	354.7	18.1	0.0272
310.0	10.88	360.6	18.6	0.0278
320.0	10.53	366.5	19.0	0.0285
330.0	10.21	372.2	19.4	0.0292
340.0	9.902	377.8	19.9	0.0298
350.0	9.614	383.4	20.3	0.0304
360.0	9.344	388.8	20.7	0.0311
370.0	9.088	394.2	21.1	0.0317
380.0	8.846	399.5	21.5	0.0324
390.0	8.617	404.6	21.9	0.0330
400.0	8.400	409.8	22.4	0.0336
420.0	7.997	419.8	23.1	0.0348
420.0	7.997	419.8	23.1	0.0348
440.0	7.631	429.5	23.9	0.0361
460.0	7.298	438.9	24.7	0.0373
480.0	6.993	448.1	25.4	0.0385

[b]Liquid-vapor boundary

G.3-8

THERMOPHYSICAL PROPERTIES OF NITROGEN

ISOBAR: 1.00 MPa

T (K)	ρ (Pa·s)	a (m/s)	$10^6\mu$ (Pa·s)	k (J/s·m·K)
480.0	6.993	448.1	25.4	0.0385
500.0	6.712	457.0	26.1	0.0398
520.0	6.453	465.8	26.9	0.0410
540.0	6.214	474.3	27.6	0.0422
560.0	5.992	482.6	28.3	0.0434
580.0	5.785	490.8	28.9	0.0446
600.0	5.592	498.7	29.6	0.0458
620.0	5.412	506.5	30.3	0.0470
640.0	5.243	514.2	30.9	0.0482
660.0	5.084	521.7	31.6	0.0494
680.0	4.935	529.1	32.2	0.0506
700.0	4.794	536.3	32.8	0.0517
720.0	4.661	543.4	33.5	0.0529
740.0	4.535	550.4	34.1	0.0541
760.0	4.416	557.3	34.7	0.0553
780.0	4.303	564.1	35.3	0.0564
800.0	4.196	570.8	35.9	0.0576
820.0	4.093	577.4	36.5	0.0588
840.0	3.996	583.9	37.1	0.0599
860.0	3.903	590.4	37.6	0.0611
880.0	3.815	596.7	38.2	0.0622
900.0	3.730	603.0	38.8	0.0634
920.0	3.649	609.2	39.3	0.0645
940.0	3.572	615.3	39.9	0.0657
1000.0	3.358	633.4	41.5	0.0690
1050.0	3.199	648.0	42.9	0.0718
1100.0	3.054	662.2	44.2	0.0745
1150.0	2.921	676.2	45.4	0.0772
1200.0	2.800	689.9	46.8	0.0798
1250.0	2.688	703.3	48.0	0.0824
1300.0	2.585	716.4	49.3	0.0849
1350.0	2.489	729.4	50.5	0.0875
1400.0	2.401	742.1	51.7	0.0899
1450.0	2.318	754.6	52.9	0.0924
1500.0	2.241	766.9	54.0	0.0948
1550.0	2.169	779.0	55.2	0.0971
1600.0	2.101	790.9	56.3	0.0995
1650.0	2.038	802.7	57.5	0.102
1700.0	1.978	814.3	58.6	0.104
1750.0	1.921	825.7	59.7	0.106
1900.0	1.868	837.0	60.8	0.109
1850.0	1.818	848.1	61.9	0.111
1900.0	1.770	859.1	62.9	0.113

THERMOPHYSICAL PROPERTIES OF NITROGEN

ISOBAR: 2.00 MPa

T (K)	ρ (Pa·s)	a (m/s)	$10^6\mu$ (Pa·s)	k (J/s·m·K)
115.571[b]	90.49	178.6	9.31	0.0202
120.0	79.36	190.2	9.39	0.0190
122.0	75.76	194.6	9.46	0.0187
124.0	72.68	198.7	9.53	0.0185
126.0	69.98	202.5	9.62	0.0184
128.0	67.58	206.0	9.70	0.0183
130.0	65.43	209.4	9.79	0.0182
132.0	63.47	212.6	9.89	0.0182
134.0	61.68	215.7	9.98	0.0182
136.0	60.03	218.7	10.1	0.0182
138.0	58.50	221.5	10.2	0.0182
140.0	57.08	224.3	10.3	0.0183
142.0	55.75	227.0	10.4	0.0183
144.0	54.50	229.6	10.5	0.0184
146.0	53.33	232.1	10.6	0.0185
148.0	52.22	234.6	10.7	0.0185
150.0	51.17	237.0	10.8	0.0186
155.0	48.76	242.8	11.1	0.0189
160.0	46.63	248.4	11.3	0.0191
165.0	44.71	253.6	11.6	0.0194
170.0	42.98	258.7	11.9	0.0197
175.0	41.40	263.6	12.1	0.0200
180.0	39.96	268.4	12.4	0.0203
185.0	38.62	272.9	12.7	0.0206
190.0	37.39	277.4	12.9	0.0209
195.0	36.24	281.8	13.2	0.0213
200.0	35.17	286.0	13.4	0.0216
205.0	34.17	290.1	13.7	0.0219
210.0	33.23	294.2	14.0	0.0223
215.0	32.35	298.1	14.2	0.0226
220.0	31.51	302.0	14.5	0.0229
230.0	29.98	309.6	15.0	0.0236
240.0	28.60	316.9	15.5	0.0243
250.0	27.36	323.9	15.9	0.0249
260.0	26.22	330.8	16.4	0.0256
270.0	25.18	337.5	16.9	0.0262
280.0	24.22	344.0	17.4	0.0269
290.0	23.34	350.3	17.8	0.0275
300.0	22.53	356.5	18.3	0.0282
310.0	21.77	362.6	18.7	0.0288
320.0	21.06	368.6	19.2	0.0295
330.0	20.40	374.4	19.6	0.0301
340.0	19.78	380.1	20.0	0.0307
350.0	19.20	385.7	20.4	0.0314
350.0	19.20	385.7	20.4	0.0314
360.0	18.65	391.2	20.9	0.0320
370.0	18.13	396.6	21.3	0.0326
380.0	17.65	401.9	21.7	0.0332

[b]Liquid-vapor boundary

G.3-10

THERMOPHYSICAL PROPERTIES OF NITROGEN

ISOBAR: 2.00 MPa

T (K)	ρ (Pa·s)	a (m/s)	$10^6 \mu$ (Pa·s)	k (J/s·m·K)
380.0	17.65	401.9	21.7	0.0332
390.0	17.19	407.2	22.1	0.0338
400.0	16.75	412.3	22.5	0.0344
420.0	15.94	422.3	23.3	0.0357
440.0	15.21	432.1	24.0	0.0369
460.0	14.54	441.6	24.8	0.0381
480.0	13.93	450.8	25.5	0.0393
500.0	13.37	459.7	26.2	0.0405
520.0	12.85	468.5	27.0	0.0417
540.0	12.37	477.0	27.7	0.0429
560.0	11.93	485.3	28.3	0.0441
580.0	11.52	493.5	29.0	0.0453
600.0	11.14	501.4	29.7	0.0465
620.0	10.78	509.2	30.4	0.0477
640.0	10.44	516.8	31.0	0.0489
660.0	10.13	524.3	31.6	0.0501
680.0	9.828	531.7	32.3	0.0513
700.0	9.548	538.9	32.9	0.0525
720.0	9.283	546.0	33.5	0.0536
740.0	9.033	553.0	34.1	0.0548
760.0	8.796	559.9	34.7	0.0560
780.0	8.572	566.6	35.3	0.0571
800.0	8.358	573.3	35.9	0.0583
820.0	8.155	579.9	36.5	0.0595
840.0	7.962	586.4	37.1	0.0606
860.0	7.778	592.8	37.7	0.0618
880.0	7.602	599.1	38.3	0.0629
900.0	7.433	605.4	38.8	0.0641
920.0	7.273	611.6	39.4	0.0652
940.0	7.119	617.7	39.9	0.0663
1000.0	6.693	635.7	41.6	0.0697
1050.0	6.376	650.2	42.9	0.0725
1100.0	6.088	664.4	44.2	0.0752
1150.0	5.824	678.4	45.5	0.0779
1200.0	5.583	692.0	46.8	0.0805
1250.0	5.361	705.4	48.1	0.0831
1300.0	5.155	718.5	49.3	0.0856
1350.0	4.965	731.4	50.5	0.0882
1400.0	4.789	744.1	51.7	0.0906
1450.0	4.624	756.5	52.9	0.0931
1500.0	4.471	768.8	54.1	0.0955
1550.0	4.327	780.9	55.2	0.0978
1600.0	4.193	792.8	56.4	0.100
1650.0	4.066	804.5	57.5	0.102
1700.0	3.947	816.0	58.6	0.105
1700.0	3.947	816.0	58.6	0.105
1750.0	3.835	827.5	59.7	0.107
1800.0	3.729	838.7	60.8	0.109
1850.0	3.628	849.8	63.0	0.114
1900.0	3.533	860.8	63.0	0.114

THERMOPHYSICAL PROPERTIES OF NITROGEN

ISOBAR: 3.00 MPa

T (K)	ρ (Pa·s)	a (m/s)	$10^6\mu$ (Pa·s)	k (J/s·m·K)
123.620[b]	171.1	173.3	11.9	0.0332
125.0	150.5	179.9	11.4	0.0284
125.5	145.8	181.9	11.3	0.0275
126.0	141.7	183.7	11.2	0.0268
127.0	134.9	187.0	11.1	0.0258
127.5	132.1	188.5	11.1	0.0254
128.0	129.4	190.0	11.0	0.0250
129.0	124.8	192.8	11.0	0.0245
130.0	120.8	195.4	11.0	0.0240
131.0	117.2	197.9	11.0	0.0236
132.0	114.1	200.2	11.0	0.0233
133.0	111.2	202.4	11.0	0.0230
134.0	108.6	204.5	11.0	0.0228
135.0	106.1	206.6	11.0	0.0225
136.0	103.9	208.6	11.0	0.0224
138.0	99.87	212.4	11.0	0.0221
140.0	96.32	215.9	11.1	0.0219
142.0	93.15	219.3	11.1	0.0217
144.0	90.28	222.5	11.2	0.0216
146.0	87.68	225.6	11.3	0.0215
148.0	85.29	228.6	11.3	0.0214
150.0	83.08	231.5	11.4	0.0214
152.0	81.03	234.3	11.5	0.0214
154.0	79.12	236.9	11.6	0.0214
156.0	77.34	239.6	11.7	0.0214
158.0	75.66	242.1	11.8	0.0214
160.0	74.08	244.6	11.9	0.0214
162.0	72.58	247.0	12.0	0.0214
165.0	70.49	250.6	12.1	0.0215
170.0	67.33	256.2	12.3	0.0217
175.0	64.51	261.6	12.6	0.0219
180.0	61.97	266.8	12.8	0.0221
185.0	59.68	271.7	13.1	0.0223
190.0	57.58	276.5	13.3	0.0226
195.0	55.65	281.2	13.5	0.0228
200.0	53.87	285.7	13.8	0.0231
205.0	52.22	290.1	14.0	0.0234
210.0	56.68	294.3	14.3	0.0236
215.0	49.24	298.5	14.5	0.0239
220.0	47.90	302.6	14.8	0.0242
225.0	46.64	306.5	15.0	0.0245
230.0	45.45	310.4	15.2	0.0248
240.0	43.26	318.0	15.7	0.0254
250.0	41.30	325.3	16.2	0.0260
260.0	39.52	332.4	16.7	0.0267
260.0	39.52	332.4	16.7	0.0267
270.0	37.90	339.2	17.1	0.0273
280.0	36.43	345.9	17.6	0.0279
290.0	35.07	352.3	18.0	0.0285

[b]Liquid-vapor boundary

THERMOPHYSICAL PROPERTIES OF NITROGEN

ISOBAR: 3.00 MPa

T (K)	ρ (Pa·s)	a (m/s)	$10^6 \mu$ (Pa·s)	k (J/s·m·K)
290.0	35.07	352.3	18.0	0.0285
300.0	33.81	358.6	18.5	0.0291
310.0	32.65	364.8	18.9	0.0297
320.0	31.57	370.8	19.3	0.0304
330.0	30.56	376.7	19.8	0.0310
340.0	29.62	382.5	20.2	0.0316
350.0	28.74	388.2	20.6	0.0322
360.0	27.91	393.7	21.0	0.0328
370.0	27.13	399.2	21.4	0.0334
380.0	26.39	404.5	21.8	0.0340
400.0	25.04	414.9	22.6	0.0352
420.0	23.82	425.0	23.4	0.0364
440.0	22.72	434.8	24.1	0.0376
460.0	21.72	444.3	24.9	0.0388
480.0	20.81	453.5	25.6	0.0400
500.0	19.97	462.5	26.3	0.0412
520.0	19.20	471.2	27.0	0.0424
540.0	18.48	479.7	27.7	0.0436
560.0	17.82	488.1	28.4	0.0448
580.0	17.21	496.2	29.1	0.0460
600.0	16.63	504.1	29.8	0.0471
620.0	16.10	511.9	30.4	0.0483
640.0	15.59	519.5	31.1	0.0495
660.0	15.12	527.0	31.7	0.0507
680.0	14.68	534.3	32.3	0.0519
700.0	14.26	541.5	33.0	0.0531
720.0	13.87	548.6	33.6	0.0542
740.0	13.49	555.6	34.2	0.0554
760.0	13.14	562.4	34.8	0.0566
780.0	12.81	569.2	35.4	0.0577
800.0	12.49	575.8	36.0	0.0589
820.0	12.19	582.4	36.6	0.0601
840.0	11.90	588.9	37.2	0.0612
860.0	11.62	595.3	37.7	0.0624
880.0	11.36	601.6	38.3	0.0635
900.0	11.11	607.8	38.9	0.0647
920.0	10.87	614.0	39.4	0.0658
950.0	10.53	623.1	40.3	0.0675
1000.0	10.01	638.0	41.6	0.0703
1050.0	9.533	652.5	43.0	0.0730
1100.0	9.103	666.7	44.3	0.0758
1150.0	8.710	680.5	45.6	0.0784
1200.0	8.349	694.1	46.8	0.0811
1250.0	8.018	707.5	48.1	0.0837
1300.0	7.712	720.5	49.3	0.0862
1350.0	7.428	733.4	50.5	0.0887
1400.0	7.164	746.0	51.7	0.0912
1450.0	6.919	758.5	52.9	0.0937
1500.0	6.690	770.7	54.1	0.0961
1550.0	6.476	782.7	55.3	0.0984
1600.0	6.275	794.6	56.4	0.101

G.3-13

THERMOPHYSICAL PROPERTIES OF NITROGEN

ISOBAR: 3.00 MPa

T (K)	ρ (Pa·s)	a (m/s)	$10^6\mu$ (Pa·s)	k (J/s·m·K)
1600.0	6.275	794.6	56.4	0.101
1650.0	6.086	806.3	57.5	0.103
1700.0	5.908	817.8	58.6	0.105
1750.0	5.740	829.2	59.7	0.108
1800.0	5.581	840.0	60.8	0.110
1850.0	5.432	851.5	61.9	0.112
1900.0	5.289	862.5	63.0	0.114

TABLE G.4

SELECTED VALUES

THERMOPHYSICAL PROPERTIES OF OXYGEN

G.4-1

THERMOPHYSICAL PROPERTIES OF OXYGEN

ISOBAR: 0.02 MPa

T (K)	ρ (Pa·s)	a (m/s)	$10^6\mu$ (Pa·s)	k (J/s·m·K)
77.109[b]	1.003	166.4	5.63	0.00687
80.0	0.9715	169.6	5.86	0.00712
85.0	0.9128	175.0	6.26	0.00758
90.0	0.8610	180.2	6.65	0.00804
95.0	0.8148	185.3	7.04	0.00852
100.0	0.7734	190.2	7.43	0.00901
105.0	0.7361	195.0	7.82	0.00950
110.0	0.7012	199.6	8.20	0.01000
115.0	0.6715	204.1	8.58	0.0105
120.0	0.6433	208.6	8.96	0.0110
130.0	0.5934	217.2	9.72	0.0120
140.0	0.5508	225.4	10.5	0.0130
150.0	0.5139	233.4	11.2	0.0139
160.0	0.4817	241.1	11.9	0.0148
170.0	0.4532	248.5	12.6	0.0158
180.0	0.4280	255.7	13.3	0.0166
190.0	0.4054	262.8	14.0	0.0175
200.0	0.3851	269.6	14.6	0.0184
210.0	0.3667	276.3	15.3	0.0192
220.0	0.3500	282.8	15.9	0.0200
230.0	0.3348	289.1	16.5	0.0208
240.0	0.3208	295.3	17.2	0.0216
250.0	0.3080	301.3	17.8	0.0224
260.0	0.2961	307.2	18.3	0.0232
280.0	0.2750	318.7	19.5	0.0247
300.0	0.2566	329.7	20.6	0.0262
320.0	0.2406	340.3	21.7	0.0278
340.0	0.2264	350.5	22.8	0.0292
360.0	0.2138	360.3	23.8	0.0307
380.0	0.2026	369.8	24.8	0.0322
400.0	0.1924	378.9	25.8	0.0337

[b]Liquid-vapor boundary

THERMOPHYSICAL PROPERTIES OF OXYGEN

ISOBAR: 0.101325 MPa

T (K)	ρ (Pa·s)	a (m/s)	$10^6 \mu$ (Pa·s)	k (J/s·m·K)
90.191[b]	4.483	177.5	6.70	0.00846
95.0	4.228	182.7	7.07	0.00887
100.0	3.999	188.0	7.46	0.00932
105.0	3.795	193.0	7.85	0.00978
110.0	3.612	197.9	8.23	0.0102
115.0	3.447	202.6	8.61	0.0107
120.0	3.297	207.2	8.99	0.0112
130.0	3.034	216.1	9.74	0.0122
140.0	2.811	224.5	10.5	0.0131
150.0	2.619	232.7	11.2	0.0140
160.0	2.452	240.5	11.9	0.0150
170.0	2.306	248.0	12.6	0.0159
180.0	2.176	255.3	13.3	0.0167
190.0	2.060	262.4	14.0	0.0176
100.0	1.956	269.3	14.6	0.0184
210.0	1.862	276.0	15.3	0.0193
220.0	1.777	282.6	15.9	0.0201
230.0	1.699	289.0	16.6	0.0209
240.0	1.628	295.2	17.2	0.0217
250.0	1.562	301.3	17.8	0.0225
260.0	1.502	307.2	18.4	0.0232
280.0	1.394	318.7	19.5	0.0248
300.0	1.301	329.7	20.6	0.0263
310.0	1.219	340.3	21.7	0.0278
340.0	1.147	350.5	22.8	0.0293
360.0	1.083	360.4	23.8	0.0308
380.0	1.026	369.9	24.8	0.0322
400.0	0.9749	379.1	25.8	0.0337

[b]Liquid-vapor boundary

THERMOPHYSICAL PROPERTIES OF OXYGEN
ISOBAR: 0.60 MPa

T (K)	ρ (Pa·s)	a (m/s)	$10^6\mu$ (Pa·s)	k (J/s·m·K)
111.457[b]	23.45	187.6	8.53	0.0120
115.0	22.43	192.1	8.80	0.0122
120.0	21.17	198.0	9.17	0.0126
125.0	20.07	203.6	9.53	0.0129
130.0	19.10	208.9	9.90	0.0133
135.0	18.24	214.0	10.3	0.0137
140.0	17.47	218.8	10.6	0.0141
150.0	16.12	228.1	11.3	0.0149
160.0	14.98	236.7	12.0	0.0157
170.0	14.01	245.0	12.7	0.0165
180.0	13.17	252.8	13.4	0.0173
190.0	12.42	260.4	14.1	0.0181
200.0	11.76	267.7	14.7	0.0189
210.0	11.17	274.7	15.4	0.0197
220.0	10.64	281.5	16.0	0.0205
230.0	10.15	288.1	16.6	0.0212
240.0	9.715	294.6	17.2	0.0220
250.0	9.313	300.8	17.8	0.0228
260.0	8.944	306.9	18.4	0.0235
280.0	8.289	318.7	19.6	0.0250
300.0	7.725	329.9	20.7	0.0265
320.0	7.234	340.7	21.8	0.0280
340.0	6.802	351.0	22.8	0.0295
360.0	6.420	360.9	23.8	0.0310
380.0	6.079	370.5	24.8	0.0324
400.0	5.772	379.8	25.8	0.0339

[b]Liquid-vapor boundary

THERMOPHYSICAL PROPERTIES OF OXYGEN

ISOBAR: 1.00 MPa

T (K)	ρ (Pa·s)	a (m/s)	$10^6 \mu$ (Pa·s)	k (J/s·m·K)
119.623[b]	38.54	188.6	9.33	0.0139
120.0	38.33	189.1	9.36	0.0139
125.0	35.84	196.1	9.71	0.0141
130.0	33.76	202.4	10.1	0.0143
135.0	31.97	208.3	10.4	0.0146
140.0	30.41	213.9	10.8	0.0149
145.0	29.03	219.1	11.1	0.0152
150.0	27.79	224.2	11.5	0.0156
155.0	26.68	229.0	11.8	0.0159
160.0	25.66	233.6	12.2	0.0163
170.0	23.87	242.5	12.8	0.0170
180.0	22.34	250.8	13.5	0.0178
190.0	21.01	258.8	14.2	0.0185
200.0	19.85	266.4	14.8	0.0193
210.0	18.81	273.7	15.5	0.0200
220.0	17.89	280.7	16.1	0.0208
230.0	17.05	287.5	16.7	0.0215
240.0	16.30	294.1	17.3	0.0223
250.0	15.61	300.5	17.9	0.0230
260.0	14.98	306.7	18.5	0.0238
270.0	14.40	312.8	19.1	0.0245
280.0	13.86	318.7	19.6	0.0252
300.0	12.91	330.1	20.7	0.0267
320.0	12.08	341.0	21.8	0.0282
340.0	11.35	351.4	22.8	0.0296
360.0	10.71	361.4	23.9	0.0311
380.0	10.13	371.0	24.9	0.0326
400.0	9.619	380.4	25.8	0.0340

[b]Liquid-vapor boundary

THERMOPHYSICAL PROPERTIES OF OXYGEN

ISOBAR: 2.00 MPa

T (K)	ρ (Pa·s)	a (m/s)	$10^6\mu$ (Pa·s)	k (J/s·m·K)
132.746[b]	79.26	186.6	10.9	0.0180
135.0	75.93	190.9	11.0	0.0178
136.0	74.61	192.7	11.1	0.0178
138.0	72.17	196.1	11.2	0.0177
140.0	69.98	199.3	11.3	0.0177
145.0	65.31	206.7	11.6	0.0177
150.0	61.47	213.5	11.9	0.0178
155.0	58.21	219.7	12.3	0.0179
160.0	55.39	225.4	12.6	0.0181
165.0	52.90	230.9	12.9	0.0183
170.0	50.69	236.1	13.2	0.0185
175.0	48.70	241.0	13.5	0.0188
180.0	46.89	245.8	13.8	0.0191
190.0	43.72	254.8	14.5	0.0196
200.0	41.01	263.2	15.1	0.0203
210.0	38.66	271.2	15.7	0.0209
220.0	36.60	278.8	16.3	0.0216
230.0	34.77	286.1	16.9	0.0223
240.0	33.13	293.1	17.5	0.0230
250.0	31.65	299.9	18.1	0.0237
260.0	30.30	306.4	18.7	0.0244
270.0	29.08	312.7	19.2	0.0251
280.0	27.95	318.8	19.8	0.0258
300.0	25.96	330.6	20.9	0.0272
320.0	24.24	341.8	21.9	0.0286
340.0	22.75	352.4	23.0	0.0301
360.0	21.44	362.6	24.0	0.0315
380.0	20.27	372.4	25.0	0.0329
400.0	19.23	381.8	25.9	0.0344

[b]Liquid-vapor boundary

THERMOPHYSICAL PROPERTIES OF OXYGEN

ISOBAR: 5.00 MPa

T (K)	ρ (Pa·s)	a (m/s)	$10^6 \mu$ (Pa·s)	k (J/s·m·K)
154.361[b]	367.8	175.7	21.1	0.0767
155.0	287.3	178.9	17.9	0.0450
155.2	278.8	180.0	17.6	0.0431
155.4	271.9	180.9	17.4	0.0417
155.6	265.9	181.8	17.2	0.0405
155.8	260.7	182.7	17.1	0.0395
156.0	256.1	183.5	16.9	0.0387
156.5	246.3	185.5	16.7	0.0370
157.0	238.3	187.3	16.5	0.0357
157.5	231.5	189.0	16.3	0.0347
158.0	225.6	190.6	16.2	0.0338
159.0	215.7	193.5	16.0	0.0324
160.0	207.5	196.2	15.8	0.0314
161.0	200.5	198.7	15.7	0.0305
162.0	194.5	201.1	15.6	0.0298
164.0	184.3	205.5	15.5	0.0286
166.0	175.9	209.6	15.4	0.0277
168.0	168.8	213.3	15.4	0.0271
170.0	162.6	216.8	15.4	0.0265
172.0	157.2	220.1	15.4	0.0260
174.0	152.3	223.3	15.5	0.0257
176.0	147.9	226.3	15.5	0.0253
178.0	143.9	229.2	15.6	0.0251
180.0	140.2	231.9	15.6	0.0249
185.0	132.1	238.5	15.8	0.0245
190.0	125.2	244.6	16.0	0.0242
195.0	119.3	250.3	16.2	0.0241
200.0	114.2	255.6	16.4	0.0241
205.0	109.5	260.8	16.6	0.0241
210.0	105.4	265.7	16.9	0.0242
215.0	101.6	270.4	17.1	0.0243
220.0	98.20	274.9	17.4	0.0245
230.0	92.14	283.5	17.9	0.0249
240.0	86.98	291.6	18.4	0.0253
250.0	82.38	299.2	18.9	0.0258
260.0	78.36	306.5	19.4	0.0264
270.0	74.77	313.5	19.9	0.0269
280.0	71.54	310.2	20.5	0.0275
290.0	68.61	326.7	21.0	0.0281
300.0	65.95	333.0	21.5	0.0288
310.0	63.50	339.0	22.0	0.0294
320.0	61.25	344.8	22.5	0.0301
340.0	57.23	356.1	23.4	0.0314
360.0	53.75	366.7	24.4	0.0327
380.0	50.70	376.8	25.4	0.0341
400.0	47.99	386.5	26.3	0.0355

[b]Liquid-vapor boundary

TABLE G.5

THERMOPHYSICAL PROPERTIES OF AIR

G.5-1

THERMOPHYSICAL PROPERTIES OF AIR

ISOBAR: 0.01 MPa

T (K)	ρ (kg/m^3)	a (m/s)	μ (Pa·s)	k (J/s·m·K)	$\bar{m}$
288.15	1.2090 −1	340.3	1.7979 −5	0.0249 +0	28.9644
300.0	1.1612	347.2	1.8554	0.0257	28.9644
310.0	1.1237	352.9	1.9030	0.0264	28.9644
320.0	1.0886	358.5	1.9500	0.0271	28.9644
330.0	1.0556	364.1	1.9962	0.0278	28.9644
340.0	1.0246	369.5	2.0419	0.0285	28.9644
350.0	9.9532 −2	374.8	2.0869	0.0292	28.9644
360.0	9.6767	380.1	2.1313	0.0298	28.9644
370.0	9.4152	385.2	2.1751	0.0305	28.9644
380.0	9.1674	390.3	2.2184	0.0312	28.9644
390.0	8.9324	395.3	2.2612	0.0318	28.9644
400.0	8.7091	400.3	2.3035	0.0325	28.9644
410.0	8.4966	405.1	2.3454	0.0331	28.9644
420.0	8.2943	409.9	2.3867	0.0338	28.9644
430.0	8.1015	414.7	2.4276	0.0344	28.9644
440.0	7.9173	419.3	2.4681	0.0350	28.9644
450.0	7.7414	423.9	2.5082	0.0357	28.9644
460.0	7.5731	428.5	2.5479	0.0363	28.9644
470.0	7.4120	433.0	2.5872	0.0370	28.9644
480.0	7.2576	437.4	2.6261	0.0376	28.9644
490.0	7.1094	441.8	2.6647	0.0382	28.9644
500.0	6.9672	446.1	2.7029	0.0388	28.9644
510.0	6.8306	450.4	2.7408	0.0395	28.9644
520.0	6.6993	454.6	2.7783	0.0401	28.9644
530.0	6.5729	458.8	2.8155	0.0407	28.9644
540.0	6.4512	462.9	2.8525	0.0413	28.9644
550.0	6.3339	467.0	2.8891	0.0420	28.9644
560.0	6.2208	471.0	2.9254	0.0426	28.9644
580.0	6.0063	479.0	2.9972	0.0438	28.9644
600.0	5.8060	486.8	3.0680	0.0450	28.9644
620.0	5.6187	494.4	3.1377	0.0463	28.9644
640.0	5.4432	501.9	3.2065	0.0475	28.9644
660.0	5.2782	509.3	3.2743	0.0487	28.9644
680.0	5.1230	516.5	3.3412	0.0499	28.9644
700.0	4.9766	523.6	3.4073	0.0511	28.9644
720.0	4.8384	530.6	3.4726	0.0523	28.9644
740.0	4.7076	537.5	3.5371	0.0535	28.9644
760.0	4.5837	544.3	3.6008	0.0547	28.9644
780.0	4.4662	551.0	3.6639	0.0559	28.9644
800.0	4.3545	557.5	3.7263	0.0571	28.9644
820.0	4.2483	564.1	3.7880	0.0583	28.9644
840.0	4.1472	570.5	3.8490	0.0594	28.9644
860.0	4.0507	576.8	3.9095	0.0606	28.9644
880.0	3.9587	583.1	3.9693	0.0618	28.9644
900.0	3.8707	589.3	4.0286	0.0630	28.9644
920.0	3.7866	595.4	4.0873	0.0641	28.9644
940.0	3.7060	601.4	4.1455	0.0653	28.9644
960.0	3.6288	607.4	4.2032	0.0665	28.9645
980.0	3.5547	613.3	4.2603	0.0676	28.9644
1000.0	3.4836	619.2	4.3170	0.0688	28.9644

THERMOPHYSICAL PROPERTIES OF AIR

ISOBAR: 0.01 MPa

T (K)	h (J/kg)	e (J/kg)	s (J/kg·K)	c_p (J/kg·K)	γ	Pr
288.15	-1.4106 +1	-9.6876 +1	7.4960 +0	1.0052 +0	1.4001	0.7254
300.0	-2.1924 +0	-8.8367	7.5365	1.0055	1.3999	0.7246
310.0	7.8646	-8.1182	7.5695	1.0059	1.3997	0.7240
320.0	1.7926 +1	-7.3994	7.6015	1.0064	1.3994	0.7233
330.0	2.7993	-6.6799	7.6324	1.0070	1.3991	0.7228
340.0	3.8067	-5.9598	7.6625	1.0077	1.3987	0.7222
350.0	4.8148	-5.2389	7.6917	1.0085	1.3982	0.7217
360.0	5.8238	-4.5171	7.7202	1.0095	1.3977	0.7212
370.0	6.8337	-3.7944	7.7478	1.0105	1.3972	0.7207
380.0	7.8447	-3.0707	7.7748	1.0116	1.3966	0.7203
390.0	8.8569	-2.3458	7.8011	1.0127	1.3959	0.7199
400.0	9.8702	-1.6197	7.8267	1.0140	1.3953	0.7195
410.0	1.0885 +2	-8.9229 +0	7.8518	1.0153	1.3945	0.7191
420.0	1.1901	-1.6350	7.8763	1.0168	1.3938	0.7188
430.0	1.2918	5.6674	7.9002	1.0182	1.3930	0.7185
440.0	1.3937	1.2985 +1	7.9236	1.0198	1.3921	0.7182
450.0	1.4958	2.0319	7.9466	1.0214	1.3913	0.7179
460.0	1.5980	2.7669	7.9690	1.0231	1.3904	0.7177
470.0	1.7004	3.5036	7.9911	1.0249	1.3894	0.7175
480.0	1.8030	4.2421	8.0127	1.0267	1.3885	0.7173
490.0	1.9058	4.9825	8.0339	1.0285	1.3875	0.7171
500.0	2.0087	5.7247	8.0547	1.0304	1.3865	0.7170
510.0	2.1119	6.4689	8.0751	1.0324	1.3855	0.7169
520.0	2.2152	7.2150	8.0951	1.0344	1.3845	0.7167
530.0	2.3187	7.9632	8.1149	1.0365	1.3834	0.7167
540.0	2.4225	8.7134	8.1343	1.0385	1.3823	0.7166
550.0	2.5264	9.4658	8.1533	1.0407	1.3813	0.7165
560.0	2.6306	1.0220 +2	8.1721	1.0428	1.3802	0.7165
580.0	2.8396	1.1736	8.2088	1.0473	1.3780	0.7165
600.0	3.0495	1.3260	8.2444	1.0518	1.3757	0.7165
620.0	3.2603	1.4794	8.2789	1.0564	1.3735	0.7165
640.0	3.4721	1.6337	8.3125	1.0611	1.3712	0.7166
660.0	3.6848	1.7890	8.3453	1.0659	1.3689	0.7168
680.0	3.8984	1.9452	8.3771	1.0707	1.3667	0.7169
700.0	4.1131	2.1023	8.4083	1.0755	1.3644	0.7171
720.0	4.3286	2.2605	8.4386	1.0803	1.3622	0.7172
740.0	4.5452	2.4196	8.4683	1.0851	1.3600	0.7174
760.0	4.7627	2.5796	8.4973	1.0900	1.3578	0.7176
780.0	4.9812	2.7406	8.5257	1.0947	1.3557	0.7177
800.0	5.2006	2.9026	8.5534	1.0995	1.3537	0.7178
820.0	5.4210	3.0655	8.5806	1.1042	1.3516	0.7179
840.0	5.6423	3.2294	8.6073	1.1088	1.3496	0.7179
860.0	5.8645	3.3941	8.6335	1.1134	1.3477	0.7179
880.0	6.0876	3.5598	8.6591	1.1178	1.3458	0.7179
900.0	6.3116	3.7264	8.6843	1.1223	1.3440	0.7178
920.0	6.5365	3.8938	8.7090	1.1266	1.3422	0.7176
940.0	6.7622	4.0621	8.7333	1.1308	1.3405	0.7174
960.0	6.9888	4.2312	8.7571	1.1349	1.3389	0.7172
980.0	7.2162	4.4012	8.7806	1.1389	1.3373	0.7169
1000.0	7.4444	4.5719	8.8036	1.1428	1.3357	0.7165

THERMOPHYSICAL PROPERTIES OF AIR

ISOBAR: 0.01 MPa

T (K)	ρ (kg/m^3)	a (m/s)	μ (Pa·s)	k (J/s·m·K)	$\bar{m}$
1020.0	3.4153 −2	625.0	4.3745 −5	0.0697 +0	28.9644
1040.0	3.3496	630.8	4.4309	0.0708	28.9644
1060.0	3.2864	636.5	4.4870	0.0719	28.9644
1080.0	3.2256	642.2	4.5426	0.0730	28.9644
1100.0	3.1669	647.8	4.5978	0.0741	28.9644
1120.0	3.1104	653.3	4.6526	0.0754	28.9644
1140.0	3.0558	658.8	4.7070	0.0765	28.9644
1160.0	3.0031	664.2	4.7610	0.0776	28.9644
1180.0	2.9522	669.6	4.8147	0.0787	28.9644
1200.0	2.9030	674.9	4.8680	0.0797	28.9644
1220.0	2.8554	680.2	4.9209	0.0808	28.9644
1240.0	2.8094	685.5	4.9735	0.0819	28.9644
1260.0	2.7648	690.6	5.0258	0.0830	28.9644
1280.0	2.7216	695.8	5.0778	0.0841	28.9644
1300.0	2.6797	700.9	5.1294	0.0851	28.9644
1320.0	2.6391	705.9	5.1807	0.0862	28.9644
1340.0	2.5997	710.9	5.2318	0.0873	28.9644
1360.0	2.5615	715.9	5.2825	0.0884	28.9644
1380.0	2.5244	720.8	5.3329	0.0894	28.9644
1400.0	2.4883	725.7	5.3831	0.0905	28.9644
1420.0	2.4533	730.6	5.4330	0.0916	28.9644
1440.0	2.4192	735.4	5.4827	0.0927	28.9644
1460.0	2.3860	740.1	5.5320	0.0937	28.9643
1480.0	2.3538	744.8	5.5811	0.0948	28.9643
1500.0	2.3224	749.5	5.6300	0.0959	28.9643
1520.0	2.2918	754.2	5.6786	0.0969	28.9643
1540.0	2.2621	758.8	5.7270	0.0980	28.9642
1560.0	2.2331	763.3	5.7751	0.0991	28.9642
1580.0	2.2048	767.9	5.8230	0.1002	28.9641
1600.0	2.1772	772.4	5.8707	0.1013	28.9641
1620.0	2.1504	776.8	5.9182	0.1023	28.9640
1640.0	2.1241	781.2	5.9654	0.1034	28.9639
1660.0	2.0985	785.6	6.0125	0.1045	28.9638
1680.0	2.0735	789.9	6.0593	0.1056	28.9636
1700.0	2.0491	794.2	6.1059	0.1067	28.9635
1720.0	2.0253	798.5	6.1523	0.1078	28.9632
1740.0	2.0020	802.7	6.1985	0.1089	28.9630
1760.0	1.9792	806.9	6.2446	0.1112	28.9627
1780.0	1.9570	811.0	6.2904	0.1125	28.9623
1800.0	1.9352	815.1	6.3360	0.1139	28.9618
1820.0	1.9139	819.1	6.3815	0.1153	28.9613
1840.0	1.8930	823.1	6.4267	0.1168	28.9607
1860.0	1.8726	827.0	6.4718	0.1183	28.9600
1880.0	1.8527	830.9	6.5167	0.1199	28.9591
1900.0	1.8331	834.8	6.5615	0.1216	28.9582
1920.0	1.8139	838.5	6.6061	0.1234	28.9570
1940.0	1.7951	842.3	6.6505	0.1253	28.9557
1960.0	1.7767	845.9	6.6947	0.1272	28.9542
1980.0	1.7587	849.6	6.7388	0.1293	28.9525
2000.0	1.7410	853.1	6.7827	0.1315	28.9505

THERMOPHYSICAL PROPERTIES OF AIR

ISOBAR: 0.01 MPa

T	h	e	s	c_p	γ	Pr
(K)	(J/kg)	(J/kg)	(J/kg·K)	(J/kg·K)		
1020.0	7.6733 +2	4.7434 +2	8.8263 +0	1.1465 +0	1.3343	0.7185
1040.0	7.9030	4.9156	8.8486	1.1501	1.3329	0.7185
1060.0	8.1333	5.0885	8.8705	1.1537	1.3315	0.7185
1080.0	8.3644	5.2622	8.8921	1.1572	1.3302	0.7185
1100.0	8.5962	5.4365	8.9134	1.1608	1.3288	0.7186
1120.0	8.8288	5.6116	8.9343	1.1643	1.3275	0.7187
1140.0	9.0620	5.7873	8.9550	1.1678	1.3262	0.7188
1160.0	9.2959	5.9638	8.9753	1.1712	1.3249	0.7190
1180.0	9.5305	6.1409	8.9953	1.1747	1.3237	0.7191
1200.0	9.7657	6.3188	9.0151	1.1781	1.3224	0.7192
1220.0	1.0002 +3	6.4973	9.0346	1.1816	1.3212	0.7194
1240.0	1.0238	6.6765	9.0539	1.1850	1.3200	0.7195
1260.0	1.0476	6.8564	9.0728	1.1884	1.3188	0.7197
1280.0	1.0714	7.0369	9.0916	1.1918	1.3176	0.7199
1300.0	1.0952	7.2182	9.1101	1.1952	1.3164	0.7201
1320.0	1.1192	7.4001	9.1284	1.1987	1.3152	0.7202
1340.0	1.1432	7.5828	9.1464	1.2021	1.3140	0.7204
1360.0	1.1673	7.7661	9.1643	1.2055	1.3128	0.7206
1380.0	1.1914	7.9501	9.1819	1.2090	1.3116	0.7208
1400.0	1.2156	8.1348	9.1993	1.2125	1.3105	0.7210
1420.0	1.2399	8.3202	9.2165	1.2159	1.3093	0.7212
1440.0	1.2643	8.5063	9.2336	1.2195	1.3082	0.7214
1460.0	1.2887	8.6930	9.2504	1.2230	1.3070	0.7216
1480.0	1.3132	8.8805	9.2671	1.2266	1.3058	0.7218
1500.0	1.3377	9.0688	9.2836	1.2302	1.3047	0.7220
1520.0	1.3624	9.2577	9.2999	1.2338	1.3035	0.7222
1540.0	1.3871	9.4474	9.3160	1.2375	1.3023	0.7224
1560.0	1.4119	9.6378	9.3320	1.2413	1.3012	0.7225
1580.0	1.4368	9.8290	9.3479	1.2452	1.3000	0.7227
1600.0	1.4617	1.0021 +3	9.3635	1.2491	1.2988	0.7229
1620.0	1.4867	1.0214	9.3791	1.2531	1.2976	0.7231
1640.0	1.5118	1.0407	9.3945	1.2572	1.2964	0.7232
1660.0	1.5370	1.0602	9.4097	1.2615	1.2951	0.7234
1680.0	1.5623	1.0797	9.4249	1.2659	1.2939	0.7236
1700.0	1.5876	1.0993	9.4399	1.2705	1.2926	0.7237
1720.0	1.6131	1.1190	9.4548	1.2752	1.2913	0.7239
1740.0	1.6387	1.1388	9.4696	1.2802	1.2899	0.7240
1760.0	1.6643	1.1587	9.4842	1.2854	1.2885	0.7220
1780.0	1.6901	1.1787	9.4988	1.2909	1.2871	0.7217
1800.0	1.7159	1.1989	9.5132	1.2967	1.2856	0.7214
1820.0	1.7419	1.2191	9.5276	1.3028	1.2841	0.7209
1840.0	1.7681	1.2395	9.5419	1.3094	1.2825	0.7204
1860.0	1.7943	1.2600	9.5560	1.3163	1.2809	0.7199
1880.0	1.8207	1.2806	9.5702	1.3238	1.2791	0.7192
1900.0	1.8473	1.3014	9.5842	1.3318	1.2773	0.7184
1920.0	1.8740	1.3223	9.5982	1.3403	1.2755	0.7175
1940.0	1.9009	1.3435	9.6121	1.3495	1.2735	0.7165
1960.0	1.9280	1.3648	9.6260	1.3595	1.2715	0.7153
1980.0	1.9553	1.3863	9.6399	1.3702	1.2694	0.7140
2000.0	1.9828	1.4080	9.6537	1.3818	1.2671	0.7126

THERMOPHYSICAL PROPERTIES OF AIR

ISOBAR: 0.01 MPa

T (K)	ρ (kg/m^3)	a (m/s)	μ (Pa·s)	k (J/s·m·K)	$\bar{m}$
2040.0	1.7066 −2	860.1	6.8702 −5	0.1363 +0	28.9456
2080.0	1.6734	866.8	6.9570	0.1418	28.9393
2120.0	1.6413	873.2	7.0433	0.1481	28.9313
2160.0	1.6104	879.5	7.1290	0.1553	28.9213
2200.0	1.5804	885.5	7.2143	0.1636	28.9087
2240.0	1.5514	891.3	7.2991	0.1731	28.8931
2280.0	1.5231	897.0	7.3835	0.1839	28.8739
2320.0	1.4957	902.6	7.4675	0.1963	28.8507
2360.0	1.4689	908.1	7.5512	0.2104	28.8227
2400.0	1.4427	913.6	7.6346	0.2264	28.7892
2440.0	1.4171	919.1	7.7177	0.2444	28.7494
2480.0	1.3920	924.7	7.8007	0.2645	28.7027
2520.0	1.3673	930.5	7.8835	0.2867	28.6482
2560.0	1.3430	936.6	7.9663	0.3113	28.5850
2600.0	1.3189	942.8	8.0491	0.3380	28.5125
2640.0	1.2952	949.4	8.1319	0.3671	28.4298
2680.0	1.2717	956.3	8.2148	0.3978	28.3363
2720.0	1.2483	963.5	8.2979	0.4302	28.2316
2760.0	1.2252	971.2	8.3813	0.4636	28.1151
2800.0	1.2022	979.2	8.4650	0.4977	27.9867
2840.0	1.1793	987.6	8.5490	0.5318	27.8464
2880.0	1.1566	996.5	8.6333	0.5649	27.6946
2920.0	1.1340	1005.8	8.7181	0.5963	27.5319
2960.0	1.1117	1015.5	8.8033	0.6250	27.3592
3000.0	1.0896	1025.6	8.8888	0.6498	27.1778
3040.0	1.0678	1036.1	8.9747	0.6700	26.9894
3080.0	1.0464	1046.9	9.0610	0.6855	26.7960
3120.0	1.0254	1058.1	9.1474	0.6941	26.5996
3160.0	1.0049	1069.5	9.2340	0.6960	26.4027
3200.0	9.8502 −3	1081.2	9.3206	0.6912	26.2076
3240.0	9.6578	1093.2	9.4071	0.6799	26.0168
3280.0	9.4724	1105.3	9.4933	0.6626	25.8325
3320.0	9.2945	1117.6	9.5793	0.6403	25.6566
3360.0	9.1245	1130.0	9.6648	0.6141	25.4907
3400.0	8.9624	1142.5	9.7498	0.5852	25.3360
3440.0	8.8083	1155.0	9.8342	0.5549	25.1931
3480.0	8.6619	1167.6	9.9179	0.5245	25.0625
3520.0	8.5230	1180.2	1.0001 −4	0.4948	24.9440
3560.0	8.3911	1192.7	1.0083	0.4668	24.8373
3600.0	8.2659	1205.1	1.0165	0.4410	24.7415
3640.0	8.1468	1217.4	1.0246	0.4179	24.6560
3680.0	8.0334	1229.4	1.0327	0.3977	24.5798
3720.0	7.9251	1241.2	1.0407	0.3806	24.5120
3760.0	7.8214	1252.6	1.0486	0.3664	24.4514
3800.0	7.7219	1263.5	1.0565	0.3552	24.3973
3840.0	7.6262	1273.9	1.0643	0.3467	24.3486
3880.0	7.5340	1283.7	1.0721	0.3410	24.3046
3920.0	7.4448	1292.9	1.0799	0.3377	24.2645
3960.0	7.3584	1301.4	1.0876	0.3369	24.2277
4000.0	7.2745	1309.3	1.0953	0.3384	24.1935

THERMOPHYSICAL PROPERTIES OF AIR

ISOBAR: 0.01 MPa

T (K)	h (J/kg)	e (J/kg)	s (J/kg·K)	c_p (J/kg·K)	γ	Pr
2040.0	2.0386 +3	1.4522 +3	9.6813 +0	1.4078 +0	1.2624	0.7092
2080.0	2.0955	1.4975	9.7090	1.4383	1.2572	0.7052
2120.0	2.1537	1.5440	9.7367	1.4739	1.2516	0.7006
2160.0	2.2135	1.5921	9.7646	1.5155	1.2456	0.6952
2200.0	2.2750	1.6419	9.7928	1.5639	1.2392	0.6893
2240.0	2.3387	1.6937	9.8215	1.6200	1.2325	0.6828
2280.0	2.4047	1.7478	9.8507	1.6847	1.2255	0.6758
2320.0	2.4736	1.8045	9.8807	1.7590	1.2184	0.6685
2360.0	2.5456	1.8644	9.9115	1.8438	1.2112	0.6610
2400.0	2.6212	1.9277	9.9432	1.9400	1.2041	0.6535
2440.0	2.7010	1.9948	9.9762	2.0483	1.1971	0.6462
2480.0	2.7853	2.0664	1.0010 +1	2.1695	1.1904	0.6392
2520.0	2.8747	2.1428	1.0046	2.3039	1.1840	0.6326
2560.0	2.9698	2.2247	1.0084	2.4518	1.1780	0.6267
2600.0	3.0710	2.3123	1.0123	2.6132	1.1724	0.6215
2640.0	3.1790	2.4064	1.0164	2.7876	1.1674	0.6174
2680.0	3.2942	2.5073	1.0207	2.9739	1.1629	0.6140
2720.0	3.4171	2.6155	1.0253	3.1709	1.1589	0.6115
2760.0	3.5480	2.7312	1.0301	3.3762	1.1555	0.6102
2800.0	3.6872	2.8548	1.0351	3.5872	1.1526	0.6099
2840.0	3.8350	2.9864	1.0403	3.8001	1.1503	0.6107
2880.0	3.9912	3.1260	1.0458	4.0105	1.1485	0.6126
2920.0	4.1557	3.2733	1.0514	4.2133	1.1472	0.6156
2960.0	4.3281	3.4279	1.0573	4.4027	1.1464	0.6197
3000.0	4.5077	3.5893	1.0633	4.5724	1.1461	0.6249
3040.0	4.6935	3.7564	1.0695	4.7159	1.1462	0.6310
3080.0	4.8845	3.9282	1.0757	4.8269	1.1468	0.6377
3120.0	5.0792	4.1033	1.0820	4.8999	1.1479	0.6455
3160.0	5.2759	4.2802	1.0883	4.9307	1.1495	0.6539
3200.0	5.4730	4.4571	1.0945	4.9167	1.1516	0.6628
3240.0	5.6687	4.6325	1.1006	4.8575	1.1542	0.6719
3280.0	5.8611	4.8047	1.1065	4.7550	1.1573	0.6811
3320.0	6.0486	4.9719	1.1121	4.6138	1.1609	0.6901
3360.0	6.2297	5.1330	1.1176	4.4401	1.1651	0.6986
3400.0	6.4034	5.2869	1.1227	4.2417	1.1698	0.7065
3440.0	6.5688	5.4328	1.1275	4.0273	1.1751	0.7135
3480.0	6.7255	5.5703	1.1321	3.8053	1.1808	0.7195
3520.0	6.8733	5.6992	1.1363	3.5837	1.1871	0.7243
3560.0	7.0123	5.8198	1.1402	3.3692	1.1936	0.7277
3600.0	7.1430	5.9324	1.1439	3.1670	1.2005	0.7299
3640.0	7.2659	6.0376	1.1473	2.9808	1.2074	0.7307
3680.0	7.3817	6.1361	1.1504	2.8131	1.2143	0.7303
3720.0	7.4912	6.2285	1.1534	2.6651	1.2209	0.7287
3760.0	7.5952	6.3158	1.1562	2.5370	1.2271	0.7260
3800.0	7.6944	6.3985	1.1588	2.4285	1.2328	0.7223
3840.0	7.7897	6.4776	1.1613	2.3387	1.2376	0.7178
3880.0	7.8817	6.5535	1.1637	2.2666	1.2415	0.7126
3920.0	7.9712	6.6271	1.1660	2.2110	1.2444	0.7069
3960.0	8.0588	6.6989	1.1682	2.1708	1.2463	0.7007
4000.0	8.1451	6.7695	1.1704	2.1449	1.2471	0.6942

THERMOPHYSICAL PROPERTIES OF AIR

ISOBAR: 0.01 MPa

T (K)	ρ (kg/m^3)	a (m/s)	μ (Pa·s)	k (J/s·m·K)	$\overline{m}$
4040.0	7.1929 −3	1316.6	1.1030 −4	0.3421 +0	24.1613
4080.0	7.1134	1323.2	1.1107	0.3479	24.1307
4120.0	7.0357	1329.4	1.1184	0.3559	24.1012
4160.0	6.9597	1335.0	1.1261	0.3659	24.0723
4200.0	6.8852	1340.2	1.1337	0.3780	24.0437
4240.0	6.8121	1345.0	1.1414	0.3923	24.0150
4280.0	6.7403	1349.6	1.1491	0.4087	23.9859
4320.0	6.6696	1353.9	1.1568	0.4273	23.9561
4360.0	6.5999	1358.0	1.1645	0.4482	23.9252
4400.0	6.5311	1362.1	1.1723	0.4715	23.8930
4440.0	6.4631	1366.1	1.1800	0.4973	23.8592
4480.0	6.3958	1370.0	1.1879	0.5256	23.8234
4520.0	6.3291	1374.1	1.1957	0.5565	23.7855
4560.0	6.2629	1378.2	1.2037	0.5903	23.7451
4600.0	6.1972	1382.4	1.2116	0.6269	23.7019
4640.0	6.1318	1386.8	1.2197	0.6666	23.6557
4680.0	6.0666	1391.4	1.2278	0.7094	23.6062
4720.0	6.0017	1396.1	1.2360	0.7554	23.5532
4760.0	5.9369	1401.1	1.2444	0.8048	23.4963
4800.0	5.8721	1406.3	1.2528	0.8576	23.4353
4840.0	5.8074	1411.8	1.2613	0.9140	23.3699
4880.0	5.7425	1417.6	1.2699	0.9740	23.2999
4920.0	5.6775	1423.6	1.2787	0.1038 +1	23.2250
4960.0	5.6123	1429.9	1.2877	0.1105	23.1449
5000.0	5.5468	1436.5	1.2967	0.1176	23.0594
5040.0	5.4811	1443.5	1.3060	0.1251	22.9683
5080.0	5.4150	1450.7	1.3154	0.1330	22.8713
5120.0	5.3485	1458.3	1.3251	0.1412	22.7683
5160.0	5.2815	1466.2	1.3349	0.1498	22.6591
5200.0	5.2142	1474.5	1.3449	0.1587	22.5435
5240.0	5.1463	1483.1	1.3552	0.1680	22.4214
5280.0	5.0780	1492.1	1.3657	0.1775	22.2927
5320.0	5.0092	1501.5	1.3765	0.1873	22.1573
5360.0	4.9400	1511.2	1.3875	0.1973	22.0152
5400.0	4.8703	1521.3	1.3988	0.2076	21.8665
5440.0	4.8001	1531.8	1.4104	0.2180	21.7111
5480.0	4.7295	1542.7	1.4223	0.2285	21.5492
5520.0	4.6586	1553.9	1.4345	0.2390	21.3809
5560.0	4.5874	1565.6	1.4470	0.2495	21.2066
5600.0	4.5159	1577.7	1.4598	0.2598	21.0263
5640.0	4.4442	1590.1	1.4730	0.2700	20.8405
5680.0	4.3725	1603.0	1.4864	0.2799	20.6495
5720.0	4.3008	1616.2	1.5002	0.2895	20.4538
5760.0	4.2292	1629.8	1.5143	0.2985	20.2539
5800.0	4.1578	1643.8	1.5287	0.3070	20.0504
5840.0	4.0868	1658.1	1.5434	0.3149	19.8438
5880.0	4.0162	1672.8	1.5585	0.3220	19.6348
5920.0	3.9463	1687.8	1.5738	0.3282	19.4242
5960.0	3.8771	1703.0	1.5893	0.3335	19.2126
6000.0	3.8088	1718.6	1.6051	0.3377	19.0008

THERMOPHYSICAL PROPERTIES OF AIR

ISOBAR: 0.01 MPa

T	h	e	s	c_p	γ	Pr
(K)	(J/kg)	(J/kg)	(J/kg·K)	(J/kg·K)		
4040.0	8.2306 +3	6.8394 +3	1.1725 +1	2.1323 +0	1.2468	0.6875
4080.0	8.3158	6.9091	1.1746	2.1323	1.2455	0.6807
4120.0	8.4013	6.9791	1.1767	2.1441	1.2433	0.6738
4160.0	8.4875	7.0497	1.1788	2.1672	1.2403	0.6670
4200.0	8.5749	7.1215	1.1808	2.2013	1.2367	0.6602
4240.0	8.6638	7.1948	1.1829	2.2461	1.2324	0.6536
4280.0	8.7547	7.2701	1.1851	2.3014	1.2276	0.6471
4320.0	8.8480	7.3477	1.1873	2.3671	1.2225	0.6408
4360.0	8.9442	7.4280	1.1895	2.4434	1.2172	0.6348
4400.0	9.0436	7.5115	1.1917	2.5304	1.2117	0.6291
4440.0	9.1468	7.5985	1.1941	2.6281	1.2061	0.6237
4480.0	9.2540	7.6895	1.1965	2.7370	1.2005	0.6186
4520.0	9.3659	7.7848	1.1990	2.8573	1.1950	0.6139
4560.0	9.4828	7.8850	1.2015	2.9892	1.1896	0.6095
4600.0	9.6052	7.9905	1.2042	3.1334	1.1843	0.6056
4640.0	9.7336	8.1017	1.2070	3.2900	1.1793	0.6020
4680.0	9.8686	8.2191	1.2099	3.4596	1.1744	0.5988
4720.0	1.0011 +4	8.3433	1.2129	3.6427	1.1698	0.5960
4760.0	1.0160	8.4747	1.2161	3.8396	1.1655	0.5936
4800.0	1.0318	8.6138	1.2194	4.0508	1.1614	0.5917
4840.0	1.0484	8.7613	1.2228	4.2769	1.1575	0.5902
4880.0	1.0660	8.9177	1.2264	4.5181	1.1540	0.5891
4920.0	1.0846	9.0836	1.2302	4.7750	1.1506	0.5884
4960.0	1.1042	9.2595	1.2342	5.0479	1.1475	0.5882
5000.0	1.1250	9.4461	1.2384	5.3371	1.1447	0.5884
5040.0	1.1470	9.6440	1.2427	5.6430	1.1420	0.5890
5080.0	1.1702	9.8538	1.2473	5.9657	1.1396	0.5901
5120.0	1.1947	1.0076 +4	1.2521	6.3054	1.1374	0.5917
5160.0	1.2206	1.0312	1.2572	6.6621	1.1354	0.5937
5200.0	1.2480	1.0561	1.2625	7.0358	1.1336	0.5962
5240.0	1.2770	1.0825	1.2680	7.4262	1.1320	0.5992
5280.0	1.3075	1.1104	1.2738	7.8329	1.1306	0.6026
5320.0	1.3396	1.1399	1.2799	8.2555	1.1293	0.6064
5360.0	1.3735	1.1710	1.2862	8.6930	1.1281	0.6109
5400.0	1.4092	1.2037	1.2929	9.1447	1.1271	0.6159
5440.0	1.4467	1.2382	1.2998	9.6091	1.1263	0.6215
5480.0	1.4861	1.2745	1.3070	1.0085 +1	1.1255	0.6276
5520.0	1.5274	1.3126	1.3145	1.0570	1.1249	0.6342
5560.0	1.5707	1.3525	1.3223	1.1062	1.1244	0.6414
5600.0	1.6159	1.3943	1.3304	1.1559	1.1240	0.6491
5640.0	1.6631	1.4380	1.3388	1.2058	1.1237	0.6575
5680.0	1.7124	1.4835	1.3475	1.2556	1.1235	0.6664
5720.0	1.7636	1.5309	1.3565	1.3048	1.1234	0.6759
5760.0	1.8167	1.5801	1.3658	1.3530	1.1234	0.6860
5800.0	1.8718	1.6311	1.3753	1.3999	1.1234	0.6966
5840.0	1.9287	1.6838	1.3851	1.4449	1.1236	0.7079
5880.0	1.9874	1.7382	1.3951	1.4876	1.1238	0.7197
5920.0	2.0477	1.7941	1.4053	1.5274	1.1241	0.7320
5960.0	2.1095	1.8514	1.4157	1.5637	1.1245	0.7448
6000.0	2.1727	1.9100	1.4263	1.5961	1.1250	0.7581

G.5-9

THERMOPHYSICAL PROPERTIES OF AIR

ISOBAR: 0.1013250 MPa

T (K)	ρ (kg/m^3)	a (m/s)	μ (Pa·s)	k (J/s·m·K)	$\bar{m}$
288.15	1.2250 +0	340.3	1.7979 −5	0.0249 +0	28.9644
300.0	1.1766	347.2	1.8554	0.0257	28.9644
310.0	1.1386	352.9	1.9030	0.0264	28.9644
320.0	1.1031	358.5	1.9500	0.0271	28.9644
330.0	1.0696	364.1	1.9962	0.0278	28.9644
340.0	1.0382	369.5	2.0419	0.0285	28.9644
350.0	1.0085	374.8	2.0869	0.0292	28.9644
360.0	9.8050 −1	380.1	2.1313	0.0298	28.9644
370.0	9.5400	385.2	2.1751	0.0305	28.9644
380.0	9.2889	390.3	2.2184	0.0312	28.9644
390.0	9.0508	395.3	2.2612	0.0318	28.9644
400.0	8.8245	400.3	2.3035	0.0325	28.9644
410.0	8.6093	405.1	2.3454	0.0331	28.9644
420.0	8.4043	409.9	2.3867	0.0338	28.9644
430.0	8.2088	414.7	2.4276	0.0344	28.9644
440.0	8.0223	419.3	2.4681	0.0350	28.9644
450.0	7.8440	423.9	2.5082	0.0357	28.9644
460.0	7.6735	428.5	2.5479	0.0363	28.9644
470.0	7.5102	433.0	2.5872	0.0370	28.9644
480.0	7.3537	437.4	2.6261	0.0376	28.9644
490.0	7.2037	441.8	2.6647	0.0382	28.9644
500.0	7.0596	446.1	2.7029	0.0388	28.9644
510.0	6.9212	450.4	2.7408	0.0395	28.9644
520.0	6.7881	454.6	2.7783	0.0401	28.9644
530.0	6.6600	458.8	2.8155	0.0407	28.9644
540.0	6.5367	462.9	2.8525	0.0413	28.9644
550.0	6.4178	467.0	2.8891	0.0420	28.9644
560.0	6.3032	471.0	2.9254	0.0426	28.9644
580.0	6.0859	479.0	2.9972	0.0438	28.9644
600.0	5.8830	486.8	3.0680	0.0450	28.9644
620.0	5.6932	494.4	3.1377	0.0463	28.9644
640.0	5.5153	501.9	3.2065	0.0475	28.9644
660.0	5.3482	509.3	3.2743	0.0487	28.9644
680.0	5.1909	516.5	3.3412	0.0499	28.9644
700.0	5.0426	523.6	3.4073	0.0511	28.9644
720.0	4.9025	530.6	3.4726	0.0523	28.9644
740.0	4.7700	537.5	3.5371	0.0535	28.9644
760.0	4.6445	544.3	3.6008	0.0547	28.9644
780.0	4.5254	551.0	3.6639	0.0559	28.9644
800.0	4.4122	557.5	3.7263	0.0571	28.9644
820.0	4.3046	564.1	3.7880	0.0583	28.9644
840.0	4.2021	570.5	3.8490	0.0594	28.9644
860.0	4.1044	576.8	3.9095	0.0606	28.9644
880.0	4.0111	583.1	3.9693	0.0618	28.9644
900.0	3.9220	589.3	4.0286	0.0630	28.9644
920.0	3.8367	595.4	4.0873	0.0641	28.9644
940.0	3.7551	601.4	4.1455	0.0653	28.9644
960.0	3.6769	607.4	4.2032	0.0665	28.9645
980.0	3.6018	613.3	4.2603	0.0676	28.9644
1000.0	3.5298	619.2	4.3170	0.0688	28.9644

THERMOPHYSICAL PROPERTIES OF AIR

ISOBAR: 0.1013250 MPa

T (K)	h (J/kg)	e (J/kg)	s (J/kg·K)	c_p (J/kg·K)	γ	Pr
288.15	-1.4106 +1	-9.6876 +1	6.8308 +0	1.0052 +0	1.4001	0.7254
300.0	-2.1924 +0	-8.8367	6.8713	1.0055	1.3999	0.7246
310.0	7.8646	-8.1182	6.9043	1.0059	1.3997	0.7240
320.0	1.7926 +1	-7.3994	6.9363	1.0064	1.3994	0.7233
330.0	2.7993	-6.6799	6.9672	1.0070	1.3991	0.7228
340.0	3.8067	-5.9598	6.9973	1.0077	1.3987	0.7222
350.0	4.8148	-5.2389	7.0265	1.0085	1.3982	0.7217
360.0	5.8238	-4.5171	7.0550	1.0095	1.3977	0.7212
370.0	6.8337	-3.7944	7.0826	1.0105	1.3972	0.7207
380.0	7.8447	-3.0707	7.1096	1.0116	1.3966	0.7203
390.0	8.8569	-2.3458	7.1359	1.0127	1.3959	0.7199
400.0	9.8702	-1.6197	7.1615	1.0140	1.3953	0.7195
410.0	1.0885 +2	-8.9229 +0	7.1866	1.0153	1.3945	0.7191
420.0	1.1901	-1.6350	7.2111	1.0168	1.3938	0.7188
430.0	1.2918	5.6674	7.2350	1.0182	1.3930	0.7185
440.0	1.3937	1.2985 +1	7.2584	1.0198	1.3921	0.7182
450.0	1.4958	2.0319	7.2814	1.0214	1.3913	0.7179
460.0	1.5980	2.7669	7.3039	1.0231	1.3904	0.7177
470.0	1.7004	3.5036	7.3259	1.0249	1.3894	0.7175
480.0	1.8030	4.2421	7.3475	1.0267	1.3885	0.7173
490.0	1.9058	4.9825	7.3687	1.0285	1.3875	0.7171
500.0	2.0087	5.7247	7.3895	1.0304	1.3865	0.7170
510.0	2.1119	6.4689	7.4099	1.0324	1.3855	0.7169
520.0	2.2152	7.2150	7.4299	1.0344	1.3845	0.7167
530.0	2.3187	7.9632	7.4497	1.0365	1.3834	0.7167
540.0	2.4225	8.7134	7.4691	1.0385	1.3823	0.7166
550.0	2.5264	9.4658	7.4881	1.0407	1.3813	0.7165
560.0	2.6306	1.0220 +2	7.5069	1.0428	1.3802	0.7165
580.0	2.8396	1.1736	7.5436	1.0473	1.3780	0.7165
600.0	3.0495	1.3260	7.5792	1.0518	1.3757	0.7165
620.0	3.2603	1.4794	7.6137	1.0564	1.3735	0.7165
640.0	3.4721	1.6337	7.6473	1.0611	1.3712	0.7166
660.0	3.6848	1.7890	7.6801	1.0659	1.3689	0.7168
680.0	3.8984	1.9452	7.7120	1.0707	1.3667	0.7169
700.0	4.1131	2.1023	7.7431	1.0755	1.3644	0.7171
720.0	4.3286	2.2605	7.7734	1.0803	1.3622	0.7172
740.0	4.5452	2.4196	7.8031	1.0851	1.3600	0.7174
760.0	4.7627	2.5796	7.8321	1.0900	1.3578	0.7176
780.0	4.9812	2.7406	7.8605	1.0947	1.3557	0.7177
800.0	5.2006	2.9026	7.8882	1.0995	1.3537	0.7178
820.0	5.4210	3.0655	7.9155	1.1042	1.3516	0.7179
840.0	5.6423	3.2294	7.9421	1.1088	1.3496	0.7179
860.0	5.8645	3.3941	7.9683	1.1134	1.3477	0.7179
880.0	6.0876	3.5598	7.9939	1.1178	1.3458	0.7179
900.0	6.3116	3.7264	8.0191	1.1223	1.3440	0.7178
920.0	6.5365	3.8938	8.0438	1.1266	1.3422	0.7176
940.0	6.7622	4.0621	8.0681	1.1308	1.3405	0.7174
960.0	6.9888	4.2312	8.0919	1.1349	1.3389	0.7172
980.0	7.2162	4.4012	8.1154	1.1389	1.3373	0.7169
1000.0	7.4444	4.5719	8.1384	1.1428	1.3357	0.7165

G.5-11

THERMOPHYSICAL PROPERTIES OF AIR

ISOBAR: 0.1013250 MPa

T (K)	ρ (kg/m^3)	a (m/s)	μ (Pa·s)	k (J/s·m·K)	$\overline{m}$
1020.0	3.4606 −1	625.0	4.3745 −5	0.0697 +0	28.9644
1040.0	3.3940	630.8	4.4309	0.0708	28.9644
1060.0	3.3300	636.5	4.4870	0.0719	28.9644
1080.0	3.2683	642.2	4.5426	0.0730	28.9644
1100.0	3.2089	647.8	4.5978	0.0741	28.9644
1120.0	3.1516	653.3	4.6526	0.0754	28.9644
1140.0	3.0963	658.8	4.7070	0.0765	28.9644
1160.0	3.0429	664.2	4.7610	0.0776	28.9644
1180.0	2.9914	669.6	4.8147	0.0787	28.9644
1200.0	2.9415	674.9	4.8680	0.0797	28.9644
1220.0	2.8933	680.2	4.9209	0.0808	28.9644
1240.0	2.8466	685.5	4.9735	0.0819	28.9644
1260.0	2.8014	690.6	5.0258	0.0830	28.9644
1280.0	2.7577	695.8	5.0778	0.0841	28.9644
1300.0	2.7152	700.9	5.1294	0.0851	28.9644
1320.0	2.6741	705.9	5.1807	0.0862	28.9644
1340.0	2.6342	710.9	5.2318	0.0873	28.9644
1360.0	2.5954	715.9	5.2825	0.0884	28.9644
1380.0	2.5578	720.8	5.3329	0.0894	28.9644
1400.0	2.5213	725.7	5.3831	0.0905	28.9644
1420.0	2.4858	730.6	5.4330	0.0916	28.9644
1440.0	2.4512	735.4	5.4827	0.0927	28.9644
1460.0	2.4177	740.1	5.5320	0.0937	28.9644
1480.0	2.3850	744.9	5.5811	0.0948	28.9644
1500.0	2.3532	749.6	5.6300	0.0959	28.9644
1520.0	2.3222	754.2	5.6786	0.0969	28.9644
1540.0	2.2921	758.8	5.7270	0.0980	28.9643
1560.0	2.2627	763.4	5.7751	0.0991	28.9643
1580.0	2.2340	768.0	5.8231	0.1002	28.9643
1600.0	2.2061	772.5	5.8707	0.1013	28.9643
1620.0	2.1789	776.9	5.9182	0.1023	28.9643
1640.0	2.1523	781.4	5.9654	0.1034	28.9642
1660.0	2.1264	785.8	6.0125	0.1045	28.9642
1680.0	2.1011	790.2	6.0593	0.1056	28.9642
1700.0	2.0763	794.5	6.1059	0.1067	28.9641
1720.0	2.0522	798.8	6.1523	0.1078	28.9640
1740.0	2.0286	803.1	6.1985	0.1089	28.9639
1760.0	2.0055	807.3	6.2445	0.1099	28.9638
1780.0	1.9830	811.5	6.2903	0.1110	28.9637
1800.0	1.9609	815.7	6.3359	0.1121	28.9636
1820.0	1.9394	819.9	6.3814	0.1132	28.9634
1840.0	1.9183	824.0	6.4266	0.1143	28.9632
1860.0	1.8976	828.1	6.4717	0.1154	28.9630
1880.0	1.8774	832.1	6.5167	0.1176	28.9627
1900.0	1.8577	836.1	6.5614	0.1189	28.9624
1920.0	1.8383	840.1	6.6060	0.1202	28.9621
1940.0	1.8193	844.0	6.6504	0.1215	28.9617
1960.0	1.8007	847.9	6.6946	0.1229	28.9612
1980.0	1.7825	851.8	6.7386	0.1243	28.9606
2000.0	1.7646	855.7	6.7825	0.1258	28.9600

THERMOPHYSICAL PROPERTIES OF AIR

ISOBAR: 0.1013250 MPa

T (K)	h (J/kg)	e (J/kg)	s (J/kg·K)	c_p (J/kg·K)	γ	Pr
1020.0	7.6733 +2	4.7434 +2	8.1611 +0	1.1465 +0	1.3343	0.7185
1040.0	7.9030	4.9156	8.1834	1.1501	1.3329	0.7185
1060.0	8.1333	5.0885	8.2053	1.1537	1.3315	0.7185
1080.0	8.3644	5.2622	8.2269	1.1572	1.3302	0.7185
1100.0	8.5962	5.4365	8.2482	1.1608	1.3288	0.7186
1120.0	8.8288	5.6116	8.2691	1.1643	1.3275	0.7187
1140.0	9.0620	5.7873	8.2898	1.1678	1.3262	0.7188
1160.0	9.2959	5.9638	8.3101	1.1712	1.3249	0.7190
1180.0	9.5305	6.1409	8.3302	1.1747	1.3237	0.7191
1200.0	9.7657	6.3188	8.3499	1.1781	1.3224	0.7192
1220.0	1.0002 +3	6.4973	8.3694	1.1816	1.3212	0.7194
1240.0	1.0238	6.6765	8.3887	1.1850	1.3200	0.7195
1260.0	1.0476	6.8564	8.4077	1.1884	1.3188	0.7197
1280.0	1.0714	7.0369	8.4264	1.1918	1.3176	0.7199
1300.0	1.0952	7.2182	8.4449	1.1952	1.3164	0.7201
1320.0	1.1192	7.4001	8.4632	1.1986	1.3152	0.7202
1340.0	1.1432	7.5827	8.4812	1.2020	1.3140	0.7204
1360.0	1.1673	7.7660	8.4991	1.2055	1.3128	0.7206
1380.0	1.1914	7.9500	8.5167	1.2089	1.3117	0.7208
1400.0	1.2156	8.1347	8.5341	1.2123	1.3105	0.7210
1420.0	1.2399	8.3201	8.5513	1.2158	1.3094	0.7212
1440.0	1.2642	8.5061	8.5683	1.2192	1.3082	0.7214
1460.0	1.2887	8.6928	8.5852	1.2227	1.3071	0.7216
1480.0	1.3132	8.8803	8.6018	1.2262	1.3059	0.7218
1500.0	1.3377	9.0684	8.6183	1.2297	1.3048	0.7220
1520.0	1.3623	9.2573	8.6346	1.2332	1.3037	0.7222
1540.0	1.3870	9.4468	8.6508	1.2368	1.3025	0.7224
1560.0	1.4118	9.6371	8.6668	1.2403	1.3014	0.7225
1580.0	1.4367	9.8280	8.6826	1.2439	1.3003	0.7227
1600.0	1.4616	1.0020 +3	8.6983	1.2476	1.2992	0.7229
1620.0	1.4866	1.0212	8.7138	1.2512	1.2980	0.7231
1640.0	1.5116	1.0405	8.7292	1.2549	1.2969	0.7232
1660.0	1.5368	1.0599	8.7444	1.2587	1.2958	0.7234
1680.0	1.5620	1.0794	8.7595	1.2625	1.2947	0.7236
1700.0	1.5873	1.0989	8.7745	1.2663	1.2935	0.7237
1720.0	1.6126	1.1185	8.7893	1.2702	1.2924	0.7239
1740.0	1.6381	1.1382	8.8040	1.2742	1.2912	0.7240
1760.0	1.6636	1.1580	8.8186	1.2783	1.2901	0.7242
1780.0	1.6892	1.1779	8.8330	1.2824	1.2889	0.7243
1800.0	1.7149	1.1978	8.8474	1.2867	1.2877	0.7245
1820.0	1.7407	1.2179	8.8616	1.2910	1.2866	0.7246
1840.0	1.7665	1.2380	8.8758	1.2955	1.2854	0.7247
1860.0	1.7925	1.2582	8.8898	1.3001	1.2842	0.7249
1880.0	1.8185	1.2785	8.9037	1.3049	1.2829	0.7231
1900.0	1.8447	1.2989	8.9176	1.3098	1.2817	0.7229
1920.0	1.8709	1.3194	8.9313	1.3149	1.2804	0.7227
1940.0	1.8973	1.3400	8.9450	1.3202	1.2791	0.7224
1960.0	1.9237	1.3607	8.9585	1.3257	1.2778	0.7221
1980.0	1.9503	1.3815	8.9720	1.3314	1.2764	0.7217
2000.0	1.9770	1.4024	8.9854	1.3374	1.2751	0.7213

THERMOPHYSICAL PROPERTIES OF AIR

ISOBAR: 0.1013250 MPa

T (K)	ρ (kg/m^3)	a (m/s)	μ (Pa·s)	k (J/s·m·K)	$\bar{m}$
2040.0	1.7299 −1	863.2	6.8699 −5	0.1288 +0	28.9585
2080.0	1.6966	870.6	6.9566	0.1320	28.9565
2120.0	1.6644	877.9	7.0427	0.1355	28.9540
2160.0	1.6334	885.0	7.1283	0.1393	28.9508
2200.0	1.6035	892.0	7.2133	0.1435	28.9468
2240.0	1.5746	898.8	7.2978	0.1480	28.9419
2280.0	1.5466	905.5	7.3818	0.1530	28.9358
2320.0	1.5196	912.0	7.4653	0.1585	28.9284
2360.0	1.4934	918.4	7.5483	0.1645	28.9194
2400.0	1.4679	924.6	7.6309	0.1712	28.9086
2440.0	1.4432	930.8	7.7132	0.1787	28.8958
2480.0	1.4192	936.8	7.7950	0.1869	28.8806
2520.0	1.3958	942.8	7.8764	0.1960	28.8627
2560.0	1.3730	948.7	7.9576	0.2060	28.8418
2600.0	1.3507	954.6	8.0384	0.2169	28.8175
2640.0	1.3290	960.4	8.1190	0.2290	28.7895
2680.0	1.3077	966.3	8.1994	0.2421	28.7574
2720.0	1.2868	972.3	8.2795	0.2563	28.7208
2760.0	1.2663	978.3	8.3596	0.2717	28.6792
2800.0	1.2462	984.4	8.4395	0.2883	28.6324
2840.0	1.2264	990.7	8.5193	0.3059	28.5799
2880.0	1.2069	997.1	8.5990	0.3250	28.5213
2920.0	1.1876	1003.7	8.6788	0.3447	28.4565
2960.0	1.1686	1010.4	8.7586	0.3654	28.3849
3000.0	1.1499	1017.4	8.8385	0.3869	28.3066
3040.0	1.1313	1024.6	8.9186	0.4089	28.2211
3080.0	1.1130	1032.1	8.9987	0.4312	28.1285
3120.0	1.0948	1039.8	9.0791	0.4537	28.0288
3160.0	1.0768	1047.8	9.1597	0.4760	27.9220
3200.0	1.0590	1056.0	9.2405	0.4978	27.8083
3240.0	1.0414	1064.5	9.3215	0.5187	27.6880
3280.0	1.0240	1073.2	9.4029	0.5395	27.5616
3320.0	1.0068	1082.2	9.4844	0.5578	27.4295
3360.0	9.8989 −2	1091.4	9.5663	0.5741	27.2924
3400.0	9.7318	1100.9	9.6483	0.5882	27.1512
3440.0	9.5674	1110.7	9.7306	0.5994	27.0065
3480.0	9.4060	1120.6	9.8130	0.6080	26.8595
3520.0	9.2477	1130.8	9.8956	0.6136	26.7112
3560.0	9.0929	1141.1	9.9783	0.6160	26.5625
3600.0	8.9418	1151.6	1.0061 −4	0.6152	26.4145
3640.0	8.7946	1162.2	1.0144	0.6114	26.2683
3680.0	8.6515	1173.0	1.0226	0.6046	26.1250
3720.0	8.5127	1183.9	1.0309	0.5953	25.9853
3760.0	8.3784	1194.8	1.0391	0.5837	25.8502
3800.0	8.2485	1205.9	1.0473	0.5703	25.7202
3840.0	8.1232	1216.9	1.0554	0.5555	25.5961
3880.0	8.0024	1228.0	1.0636	0.5397	25.4782
3920.0	7.8861	1239.0	1.0717	0.5235	25.3667
3960.0	7.7742	1250.0	1.0797	0.5073	25.2619
4000.0	7.6665	1261.0	1.0877	0.4913	25.1636

THERMOPHYSICAL PROPERTIES OF AIR

ISOBAR: 0.1013250 MPa

T (K)	h (J/kg)	e (J/kg)	s (J/kg·K)	c_p (J/kg·K)	γ	Pr
2040.0	2.0307 +3	1.4446 +3	9.0120 +0	1.3504 +0	1.2722	0.7202
2080.0	2.0850	1.4874	9.0384	1.3646	1.2692	0.7189
2120.0	2.1399	1.5308	9.0645	1.3805	1.2660	0.7172
2160.0	2.1955	1.5748	9.0905	1.3982	1.2627	0.7152
2200.0	2.2518	1.6195	9.1163	1.4180	1.2592	0.7128
2240.0	2.3090	1.6650	9.1421	1.4402	1.2554	0.7101
2280.0	2.3671	1.7115	9.1678	1.4652	1.2515	0.7069
2320.0	2.4262	1.7590	9.1935	1.4933	1.2474	0.7033
2360.0	2.4866	1.8076	9.2193	1.5248	1.2431	0.6993
2400.0	2.5483	1.8575	9.2452	1.5600	1.2386	0.6948
2440.0	2.6114	1.9089	9.2713	1.5995	1.2340	0.6901
2480.0	2.6763	1.9618	9.2977	1.6434	1.2292	0.6850
2520.0	2.7430	2.0166	9.3244	1.6922	1.2244	0.6797
2560.0	2.8117	2.0732	9.3514	1.7462	1.2196	0.6741
2600.0	2.8827	2.1321	9.3790	1.8057	1.2147	0.6685
2640.0	2.9563	2.1933	9.4070	1.8710	1.2099	0.6628
2680.0	3.0325	2.2571	9.4357	1.9423	1.2051	0.6571
2720.0	3.1117	2.3238	9.4650	2.0197	1.2005	0.6516
2760.0	3.1942	2.3935	9.4951	2.1034	1.1961	0.6463
2800.0	3.2801	2.4665	9.5260	2.1934	1.1919	0.6413
2840.0	3.3697	2.5430	9.5578	2.2895	1.1879	0.6367
2880.0	3.4633	2.6232	9.5905	2.3916	1.1841	0.6327
2920.0	3.5611	2.7074	9.6242	2.4992	1.1807	0.6290
2960.0	3.6633	2.7957	9.6590	2.6119	1.1776	0.6258
3000.0	3.7701	2.8884	9.6948	2.7291	1.1747	0.6232
3040.0	3.8817	2.9855	9.7318	2.8497	1.1722	0.6212
3080.0	3.9981	3.0871	9.7698	2.9728	1.1701	0.6199
3120.0	4.1195	3.1934	9.8090	3.0972	1.1682	0.6192
3160.0	4.2459	3.3043	9.8492	3.2213	1.1667	0.6192
3200.0	4.3772	3.4198	9.8905	3.3436	1.1655	0.6199
3240.0	4.5134	3.5398	9.9328	3.4621	1.1646	0.6213
3280.0	4.6541	3.6640	9.9760	3.5750	1.1640	0.6231
3320.0	4.7993	3.7922	1.0020 +1	3.6801	1.1637	0.6258
3360.0	4.9484	3.9241	1.0065	3.7754	1.1638	0.6291
3400.0	5.1011	4.0593	1.0110	3.8586	1.1641	0.6329
3440.0	5.2569	4.1971	1.0155	3.9278	1.1648	0.6373
3480.0	5.4151	4.3372	1.0201	3.9811	1.1657	0.6422
3520.0	5.5752	4.4787	1.0247	4.0172	1.1670	0.6476
3560.0	5.7363	4.6212	1.0292	4.0348	1.1685	0.6534
3600.0	5.8977	4.7638	1.0337	4.0334	1.1703	0.6594
3640.0	6.0587	4.9058	1.0382	4.0129	1.1724	0.6656
3680.0	6.2185	5.0465	1.0426	3.9738	1.1748	0.6719
3720.0	6.3763	5.1853	1.0468	3.9172	1.1775	0.6781
3760.0	6.5316	5.3215	1.0510	3.8449	1.1805	0.6843
3800.0	6.6838	5.4545	1.0550	3.7589	1.1837	0.6901
3840.0	6.8322	5.5840	1.0589	3.6617	1.1872	0.6956
3880.0	6.9766	5.7096	1.0626	3.5560	1.1909	0.7006
3920.0	7.1166	5.8309	1.0662	3.4448	1.1948	0.7050
3960.0	7.2521	5.9479	1.0697	3.3306	1.1989	0.7088
4000.0	7.3831	6.0605	1.0729	3.2163	1.2030	0.7119

THERMOPHYSICAL PROPERTIES OF AIR

ISOBAR: 0.1013250 MPa

T (K)	ρ (kg/m^3)	a (m/s)	μ (Pa·s)	k (J/s·m·K)	$\bar{m}$
4040.0	7.5629 -2	1271.8	1.0957 -4	0.4761 +0	25.0718
4080.0	7.4632	1282.5	1.1036	0.4619	24.9863
4120.0	7.3673	1293.0	1.1115	0.4489	24.9068
4160.0	7.2748	1303.3	1.1193	0.4373	24.8329
4200.0	7.1856	1313.4	1.1271	0.4272	24.7643
4240.0	7.0995	1323.1	1.1349	0.4188	24.7006
4280.0	7.0162	1332.5	1.1426	0.4120	24.6412
4320.0	6.9356	1341.7	1.1503	0.4069	24.5859
4360.0	6.8575	1350.5	1.1579	0.4035	24.5340
4400.0	6.7817	1358.9	1.1656	0.4019	24.4854
4440.0	6.7080	1366.9	1.1732	0.4019	24.4394
4480.0	6.6362	1374.5	1.1808	0.4036	24.3958
4520.0	6.5663	1381.7	1.1883	0.4070	24.3541
4560.0	6.4980	1388.5	1.1959	0.4121	24.3141
4600.0	6.4312	1395.0	1.2035	0.4188	24.2755
4640.0	6.3659	1401.1	1.2110	0.4271	24.2378
4680.0	6.3018	1407.0	1.2186	0.4370	24.2008
4720.0	6.2390	1412.6	1.2261	0.4486	24.1642
4760.0	6.1773	1418.0	1.2337	0.4618	24.1278
4800.0	6.1165	1423.2	1.2413	0.4767	24.0913
4840.0	6.0567	1428.2	1.2489	0.4933	24.0545
4880.0	5.9977	1433.1	1.2565	0.5116	24.0172
4920.0	5.9395	1437.9	1.2641	0.5316	23.9791
4960.0	5.8820	1442.7	1.2718	0.5534	23.9400
5000.0	5.8252	1447.4	1.2795	0.5770	23.8998
5040.0	5.7689	1452.2	1.2873	0.6024	23.8582
5080.0	5.7131	1456.9	1.2950	0.6297	23.8150
5120.0	5.6578	1461.8	1.3029	0.6589	23.7701
5160.0	5.6029	1466.7	1.3108	0.6900	23.7233
5200.0	5.5483	1471.7	1.3188	0.7232	23.6744
5240.0	5.4941	1476.8	1.3268	0.7584	23.6232
5280.0	5.4401	1482.0	1.3349	0.7956	23.5696
5320.0	5.3863	1487.4	1.3431	0.8350	23.5134
5360.0	5.3327	1492.9	1.3513	0.8765	23.4544
5400.0	5.2792	1498.6	1.3597	0.9201	23.3925
5440.0	5.2258	1504.5	1.3681	0.9659	23.3276
5480.0	5.1725	1510.5	1.3767	0.1014 +1	23.2595
5520.0	5.1193	1516.8	1.3854	0.1064	23.1880
5560.0	5.0660	1523.2	1.3942	0.1116	23.1130
5600.0	5.0128	1529.9	1.4031	0.1171	23.0345
5640.0	4.9594	1536.8	1.4122	0.1228	22.9523
5680.0	4.9060	1543.9	1.4214	0.1286	22.8663
5720.0	4.8526	1551.2	1.4307	0.1347	22.7763
5760.0	4.7990	1558.7	1.4402	0.1410	22.6824
5800.0	4.7453	1566.5	1.4499	0.1474	22.5845
5840.0	4.6915	1574.6	1.4597	0.1541	22.4824
5880.0	4.6376	1582.8	1.4697	0.1609	22.3762
5920.0	4.5835	1591.4	1.4799	0.1678	22.2658
5960.0	4.5293	1600.2	1.4903	0.1749	22.1512
6000.0	4.4750	1609.2	1.5009	0.1821	22.0324

THERMOPHYSICAL PROPERTIES OF AIR

ISOBAR: 0.1013250 MPa

T	h	e	s	c_p	γ	Pr
(K)	(J/kg)	(J/kg)	(J/kg·K)	(J/kg·K)		
4040.0	7.5095 +3	6.1688 +3	1.0761 +1	3.1040 +0	1.2073	0.7142
4080.0	7.6314	6.2729	1.0791	2.9958	1.2115	0.7157
4120.0	7.7492	6.3729	1.0820	2.8934	1.2156	0.7163
4160.0	7.8630	6.4692	1.0847	2.7982	1.2195	0.7161
4200.0	7.9732	6.5621	1.0873	2.7111	1.2233	0.7152
4240.0	8.0800	6.6518	1.0899	2.6329	1.2267	0.7135
4280.0	8.1839	6.7388	1.0923	2.5639	1.2297	0.7110
4320.0	8.2852	6.8233	1.0947	2.5045	1.2323	0.7080
4360.0	8.3844	6.9058	1.0970	2.4546	1.2344	0.7043
4400.0	8.4817	6.9866	1.0992	2.4142	1.2359	0.7002
4440.0	8.5777	7.0661	1.1014	2.3831	1.2369	0.6956
4480.0	8.6725	7.1446	1.1035	2.3611	1.2373	0.6907
4520.0	8.7667	7.2225	1.1056	2.3480	1.2371	0.6855
4560.0	8.8605	7.3001	1.1076	2.3435	1.2364	0.6801
4600.0	8.9542	7.3777	1.1097	2.3473	1.2351	0.6746
4640.0	9.0484	7.4556	1.1117	2.3593	1.2334	0.6690
4680.0	9.1431	7.5342	1.1138	2.3793	1.2312	0.6634
4720.0	9.2388	7.6137	1.1158	2.4070	1.2287	0.6578
4760.0	9.3358	7.6944	1.1178	2.4423	1.2258	0.6524
4800.0	9.4343	7.7766	1.1199	2.4851	1.2226	0.6470
4840.0	9.5347	7.8606	1.1220	2.5353	1.2192	0.6418
4880.0	9.6372	7.9467	1.1241	2.5929	1.2157	0.6368
4920.0	9.7422	8.0351	1.1262	2.6579	1.2120	0.6320
4960.0	9.8499	8.1262	1.1284	2.7302	1.2082	0.6274
5000.0	9.9607	8.2201	1.1306	2.8098	1.2044	0.6231
5040.0	1.0075 +4	8.3172	1.1329	2.8969	1.2006	0.6191
5080.0	1.0193	8.4178	1.1352	2.9915	1.1968	0.6153
5120.0	1.0314	8.5221	1.1376	3.0937	1.1931	0.6118
5160.0	1.0440	8.6305	1.1401	3.2036	1.1895	0.6086
5200.0	1.0571	8.7432	1.1426	3.3212	1.1859	0.6056
5240.0	1.0706	8.8605	1.1452	3.4468	1.1825	0.6030
5280.0	1.0846	8.9827	1.1479	3.5804	1.1792	0.6007
5320.0	1.0992	9.1101	1.1506	3.7221	1.1760	0.5987
5360.0	1.1144	9.2430	1.1535	3.8720	1.1730	0.5970
5400.0	1.1302	9.3818	1.1564	4.0304	1.1701	0.5956
5440.0	1.1467	9.5267	1.1594	4.1972	1.1674	0.5945
5480.0	1.1638	9.6781	1.1626	4.3726	1.1648	0.5937
5520.0	1.1817	9.8362	1.1658	4.5567	1.1624	0.5933
5560.0	1.2003	1.0001 +4	1.1692	4.7496	1.1601	0.5931
5600.0	1.2197	1.0174	1.1727	4.9513	1.1579	0.5933
5640.0	1.2399	1.0355	1.1762	5.1618	1.1559	0.5938
5680.0	1.2610	1.0543	1.1800	5.3811	1.1541	0.5946
5720.0	1.2830	1.0740	1.1838	5.6093	1.1524	0.5957
5760.0	1.3059	1.0946	1.1878	5.8462	1.1508	0.5972
5800.0	1.3298	1.1161	1.1919	6.0917	1.1493	0.5990
5840.0	1.3546	1.1385	1.1962	6.3457	1.1479	0.6012
5880.0	1.3805	1.1619	1.2006	6.6080	1.1467	0.6036
5920.0	1.4075	1.1863	1.2052	6.8783	1.1456	0.6064
5960.0	1.4356	1.2117	1.2099	7.1563	1.1446	0.6096
6000.0	1.4648	1.2382	1.2148	7.4415	1.1437	0.6131

THERMOPHYSICAL PROPERTIES OF AIR

ISOBAR: 0.60 MPa

T (K)	ρ (kg/m^3)	a (m/s)	μ (Pa·s)	k (J/s·m·K)	$\bar{m}$
288.15	7.2538 +0	340.3	1.7979 -5	0.0249 +0	28.9644
300.0	6.9673	347.2	1.8554	0.0257	28.9644
310.0	6.7425	352.9	1.9030	0.0264	28.9644
320.0	6.5318	358.5	1.9500	0.0271	28.9644
330.0	6.3339	364.1	1.9962	0.0278	28.9644
340.0	6.1476	369.5	2.0419	0.0285	28.9644
350.0	5.9720	374.8	2.0869	0.0292	28.9644
360.0	5.8061	380.1	2.1313	0.0298	28.9644
370.0	5.6491	385.2	2.1751	0.0305	28.9644
380.0	5.5005	390.3	2.2184	0.0312	28.9644
390.0	5.3594	395.3	2.2612	0.0318	28.9644
400.0	5.2255	400.3	2.3035	0.0325	28.9644
410.0	5.0980	405.1	2.3454	0.0331	28.9644
420.0	4.9766	409.9	2.3867	0.0338	28.9644
430.0	4.8609	414.7	2.4276	0.0344	28.9644
440.0	4.7504	419.3	2.4681	0.0350	28.9644
450.0	4.6448	423.9	2.5082	0.0357	28.9644
460.0	4.5439	428.5	2.5479	0.0363	28.9644
470.0	4.4472	433.0	2.5872	0.0370	28.9644
480.0	4.3545	437.4	2.6261	0.0376	28.9644
490.0	4.2657	441.8	2.6647	0.0382	28.9644
500.0	4.1804	446.1	2.7029	0.0388	28.9644
510.0	4.0984	450.4	2.7408	0.0395	28.9644
520.0	4.0196	454.6	2.7783	0.0401	28.9644
530.0	3.9437	458.8	2.8155	0.0407	28.9644
540.0	3.8707	462.9	2.8525	0.0413	28.9644
550.0	3.8003	467.0	2.8891	0.0420	28.9644
560.0	3.7325	471.0	2.9254	0.0426	28.9644
580.0	3.6038	479.0	2.9972	0.0438	28.9644
600.0	3.4836	486.8	3.0680	0.0450	28.9644
620.0	3.3713	494.4	3.1377	0.0463	28.9644
640.0	3.2659	501.9	3.2065	0.0475	28.9644
660.0	3.1669	509.3	3.2743	0.0487	28.9644
680.0	3.0738	516.5	3.3412	0.0499	28.9644
700.0	2.9860	523.6	3.4073	0.0511	28.9644
720.0	2.9030	530.6	3.4726	0.0523	28.9644
740.0	2.8246	537.5	3.5371	0.0535	28.9644
760.0	2.7502	544.3	3.6008	0.0547	28.9644
780.0	2.6797	551.0	3.6639	0.0559	28.9644
800.0	2.6127	557.5	3.7263	0.0571	28.9644
820.0	2.5490	564.1	3.7880	0.0583	28.9644
840.0	2.4883	570.5	3.8490	0.0594	28.9644
860.0	2.4304	576.8	3.9095	0.0606	28.9644
880.0	2.3752	583.1	3.9693	0.0618	28.9644
900.0	2.3224	589.3	4.0286	0.0630	28.9644
920.0	2.2719	595.4	4.0873	0.0641	28.9644
940.0	2.2236	601.4	4.1455	0.0653	28.9644
960.0	2.1773	607.4	4.2032	0.0665	28.9645
980.0	2.1328	613.3	4.2603	0.0676	28.9644
1000.0	2.0902	619.2	4.3170	0.0688	28.9644

THERMOPHYSICAL PROPERTIES OF AIR

ISOBAR: 0.60 MPa

T (K)	h (J/kg)	e (J/kg)	s (J/kg·K)	c_p (J/kg·K)	γ	Pr
288.15	-1.4106 +1	-9.6876 +1	6.3199 +0	1.0052 +0	1.4001	0.7254
300.0	-2.1924 +0	-8.8367	6.3604	1.0055	1.3999	0.7246
310.0	7.8646	-8.1182	6.3934	1.0059	1.3997	0.7240
320.0	1.7926 +1	-7.3994	6.4254	1.0064	1.3994	0.7233
330.0	2.7993	-6.6799	6.4563	1.0070	1.3991	0.7228
340.0	3.8067	-5.9598	6.4864	1.0077	1.3987	0.7222
350.0	4.8148	-5.2389	6.5156	1.0085	1.3982	0.7217
360.0	5.8238	-4.5171	6.5441	1.0095	1.3977	0.7212
370.0	6.8337	-3.7944	6.5717	1.0105	1.3972	0.7207
380.0	7.8447	-3.0707	6.5987	1.0116	1.3966	0.7203
390.0	8.8569	-2.3458	6.6250	1.0127	1.3959	0.7199
400.0	9.8702	-1.6197	6.6506	1.0140	1.3953	0.7195
410.0	1.0885 +2	-8.9229 +0	6.6757	1.0153	1.3945	0.7191
420.0	1.1901	-1.6350	6.7002	1.0168	1.3938	0.7188
430.0	1.2918	5.6674	6.7241	1.0182	1.3930	0.7185
440.0	1.3937	1.2985 +1	6.7475	1.0198	1.3921	0.7182
450.0	1.4958	2.0319	6.7705	1.0214	1.3913	0.7179
460.0	1.5980	2.7669	6.7930	1.0231	1.3904	0.7177
470.0	1.7004	3.5036	6.8150	1.0249	1.3894	0.7175
480.0	1.8030	4.2421	6.8366	1.0267	1.3885	0.7173
490.0	1.9058	4.9825	6.8578	1.0285	1.3875	0.7171
500.0	2.0087	5.7247	6.8786	1.0304	1.3865	0.7170
510.0	2.1119	6.4689	6.8990	1.0324	1.3855	0.7169
520.0	2.2152	7.2150	6.9191	1.0344	1.3845	0.7167
530.0	2.3187	7.9632	6.9388	1.0365	1.3834	0.7167
540.0	2.4225	8.7134	6.9582	1.0385	1.3823	0.7166
550.0	2.5264	9.4658	6.9772	1.0407	1.3813	0.7165
560.0	2.6306	1.0220 +2	6.9960	1.0428	1.3802	0.7165
580.0	2.8396	1.1736	7.0327	1.0473	1.3780	0.7165
600.0	3.0495	1.3260	7.0683	1.0518	1.3757	0.7165
620.0	3.2603	1.4794	7.1028	1.0564	1.3735	0.7165
640.0	3.4721	1.6337	7.1364	1.0611	1.3712	0.7166
660.0	3.6848	1.7890	7.1692	1.0659	1.3689	0.7168
680.0	3.8984	1.9452	7.2011	1.0707	1.3667	0.7169
700.0	4.1131	2.1023	7.2322	1.0755	1.3644	0.7171
720.0	4.3286	2.2605	7.2625	1.0803	1.3622	0.7172
740.0	4.5452	2.4196	7.2922	1.0851	1.3600	0.7174
760.0	4.7627	2.5796	7.3212	1.0900	1.3578	0.7176
780.0	4.9812	2.7406	7.3496	1.0947	1.3557	0.7177
800.0	5.2006	2.9026	7.3773	1.0995	1.3537	0.7178
820.0	5.4210	3.0655	7.4046	1.1042	1.3516	0.7179
840.0	5.6423	3.2294	7.4312	1.1088	1.3496	0.7179
860.0	5.8645	3.3941	7.4574	1.1134	1.3477	0.7179
880.0	6.0876	3.5598	7.4830	1.1178	1.3458	0.7179
900.0	6.3116	3.7264	7.5082	1.1223	1.3440	0.7178
920.0	6.5365	3.8938	7.5329	1.1266	1.3422	0.7176
940.0	6.7622	4.0621	7.5572	1.1308	1.3405	0.7174
960.0	6.9888	4.2312	7.5810	1.1349	1.3389	0.7172
980.0	7.2162	4.4012	7.6045	1.1389	1.3373	0.7169
1000.0	7.4444	4.5719	7.6275	1.1428	1.3357	0.7165

THERMOPHYSICAL PROPERTIES OF AIR

ISOBAR: 0.60 MPa

T (K)	ρ (kg/m^3)	a (m/s)	μ (Pa·s)	k (J/s·m·K)	$\bar{m}$
1020.0	2.0492 +0	625.0	4.3745 −5	0.0697 +0	28.9644
1040.0	2.0098	630.8	4.4309	0.0708	28.9644
1060.0	1.9719	636.5	4.4870	0.0719	28.9644
1080.0	1.9354	642.2	4.5426	0.0730	28.9644
1100.0	1.9002	647.8	4.5978	0.0741	28.9644
1120.0	1.8662	653.3	4.6526	0.0754	28.9644
1140.0	1.8335	658.8	4.7070	0.0765	28.9644
1160.0	1.8019	664.2	4.7610	0.0776	28.9644
1180.0	1.7713	669.6	4.8147	0.0787	28.9644
1200.0	1.7418	674.9	4.8680	0.0797	28.9644
1220.0	1.7133	680.2	4.9209	0.0808	28.9644
1240.0	1.6856	685.5	4.9735	0.0819	28.9644
1260.0	1.6589	690.6	5.0258	0.0830	28.9644
1280.0	1.6330	695.8	5.0778	0.0841	28.9644
1300.0	1.6078	700.9	5.1294	0.0851	28.9644
1320.0	1.5835	705.9	5.1807	0.0862	28.9644
1340.0	1.5598	710.9	5.2318	0.0873	28.9644
1360.0	1.5369	715.9	5.2825	0.0884	28.9644
1380.0	1.5146	720.8	5.3329	0.0894	28.9644
1400.0	1.4930	725.7	5.3831	0.0905	28.9644
1420.0	1.4720	730.6	5.4330	0.0916	28.9644
1440.0	1.4515	735.4	5.4827	0.0927	28.9644
1460.0	1.4316	740.1	5.5320	0.0937	28.9644
1480.0	1.4123	744.9	5.5811	0.0948	28.9644
1500.0	1.3935	749.6	5.6300	0.0959	28.9644
1520.0	1.3751	754.2	5.6786	0.0969	28.9644
1540.0	1.3573	758.8	5.7270	0.0980	28.9644
1560.0	1.3399	763.4	5.7751	0.0991	28.9644
1580.0	1.3229	768.0	5.8231	0.1002	28.9644
1600.0	1.3064	772.5	5.8707	0.1013	28.9643
1620.0	1.2902	777.0	5.9182	0.1023	28.9643
1640.0	1.2745	781.4	5.9654	0.1034	28.9643
1660.0	1.2591	785.8	6.0125	0.1045	28.9643
1680.0	1.2442	790.2	6.0593	0.1056	28.9643
1700.0	1.2295	794.6	6.1059	0.1067	28.9643
1720.0	1.2152	798.9	6.1523	0.1078	28.9642
1740.0	1.2012	803.2	6.1985	0.1089	28.9642
1760.0	1.1876	807.5	6.2445	0.1099	28.9642
1780.0	1.1742	811.7	6.2903	0.1110	28.9641
1800.0	1.1612	815.9	6.3360	0.1121	28.9641
1820.0	1.1484	820.1	6.3814	0.1132	28.9640
1840.0	1.1360	824.2	6.4267	0.1143	28.9639
1860.0	1.1237	828.3	6.4717	0.1154	28.9638
1880.0	1.1118	832.4	6.5166	0.1165	28.9637
1900.0	1.1001	836.5	6.5613	0.1176	28.9636
1920.0	1.0886	840.5	6.6059	0.1187	28.9634
1940.0	1.0774	844.5	6.6503	0.1198	28.9633
1960.0	1.0664	848.5	6.6945	0.1209	28.9631
1980.0	1.0556	852.4	6.7386	0.1230	28.9628
2000.0	1.0450	856.4	6.7825	0.1242	28.9626

THERMOPHYSICAL PROPERTIES OF AIR

ISOBAR: 0.60 MPa

T (K)	h (J/kg)	e (J/kg)	s (J/kg·K)	c_p (J/kg·K)	γ	Pr
1020.0	7.6733 +2	4.7434 +2	7.6502 +0	1.1465 +0	1.3343	0.7185
1040.0	7.9030	4.9156	7.6725	1.1501	1.3329	0.7185
1060.0	8.1333	5.0885	7.6944	1.1537	1.3315	0.7185
1080.0	8.3644	5.2622	7.7160	1.1572	1.3302	0.7185
1100.0	8.5962	5.4365	7.7373	1.1608	1.3288	0.7186
1120.0	8.8288	5.6116	7.7582	1.1643	1.3275	0.7187
1140.0	9.0620	5.7873	7.7789	1.1678	1.3262	0.7188
1160.0	9.2959	5.9638	7.7992	1.1712	1.3249	0.7190
1180.0	9.5305	6.1409	7.8193	1.1747	1.3237	0.7191
1200.0	9.7657	6.3187	7.8390	1.1781	1.3224	0.7192
1220.0	1.0002 +3	6.4973	7.8585	1.1816	1.3212	0.7194
1240.0	1.0238	6.6765	7.8778	1.1850	1.3200	0.7195
1260.0	1.0476	6.8564	7.8968	1.1884	1.3188	0.7197
1280.0	1.0714	7.0369	7.9155	1.1918	1.3176	0.7199
1300.0	1.0952	7.2182	7.9340	1.1952	1.3164	0.7201
1320.0	1.1192	7.4001	7.9523	1.1986	1.3152	0.7202
1340.0	1.1432	7.5827	7.9703	1.2020	1.3140	0.7204
1360.0	1.1673	7.7660	7.9882	1.2054	1.3128	0.7206
1380.0	1.1914	7.9500	8.0058	1.2089	1.3117	0.7208
1400.0	1.2156	8.1347	8.0232	1.2123	1.3105	0.7210
1420.0	1.2399	8.3200	8.0404	1.2157	1.3094	0.7212
1440.0	1.2642	8.5061	8.0574	1.2192	1.3082	0.7214
1460.0	1.2887	8.6928	8.0743	1.2226	1.3071	0.7216
1480.0	1.3131	8.8802	8.0909	1.2261	1.3060	0.7218
1500.0	1.3377	9.0683	8.1074	1.2296	1.3048	0.7220
1520.0	1.3623	9.2571	8.1237	1.2331	1.3037	0.7222
1540.0	1.3870	9.4467	8.1399	1.2366	1.3026	0.7224
1560.0	1.4118	9.6369	8.1559	1.2401	1.3015	0.7225
1580.0	1.4366	9.8278	8.1717	1.2436	1.3004	0.7227
1600.0	1.4615	1.0019 +3	8.1873	1.2472	1.2993	0.7229
1620.0	1.4865	1.0212	8.2029	1.2507	1.2982	0.7231
1640.0	1.5116	1.0405	8.2182	1.2543	1.2971	0.7232
1660.0	1.5367	1.0599	8.2334	1.2579	1.2960	0.7234
1680.0	1.5619	1.0793	8.2485	1.2616	1.2949	0.7236
1700.0	1.5872	1.0988	8.2635	1.2652	1.2938	0.7237
1720.0	1.6125	1.1184	8.2783	1.2689	1.2927	0.7239
1740.0	1.6379	1.1381	8.2930	1.2726	1.2916	0.7240
1760.0	1.6634	1.1578	8.3076	1.2764	1.2905	0.7242
1780.0	1.6890	1.1777	8.3220	1.2802	1.2894	0.7243
1800.0	1.7146	1.1976	8.3363	1.2840	1.2883	0.7245
1820.0	1.7403	1.2175	8.3505	1.2878	1.2872	0.7246
1840.0	1.7661	1.2376	8.3646	1.2918	1.2862	0.7247
1860.0	1.7920	1.2577	8.3786	1.2957	1.2851	0.7249
1880.0	1.8179	1.2779	8.3925	1.2998	1.2840	0.7250
1900.0	1.8440	1.2982	8.4063	1.3038	1.2829	0.7251
1920.0	1.8701	1.3186	8.4200	1.3080	1.2818	0.7252
1940.0	1.8963	1.3390	8.4335	1.3122	1.2807	0.7253
1960.0	1.9226	1.3596	8.4470	1.3165	1.2796	0.7254
1980.0	1.9490	1.3802	8.4604	1.3210	1.2784	0.7239
2000.0	1.9754	1.4009	8.4737	1.3255	1.2773	0.7238

THERMOPHYSICAL PROPERTIES OF AIR

ISOBAR: 0.60 MPa

T (K)	ρ (kg/m^3)	a (m/s)	μ (Pa·s)	k (J/s·m·K)	$\bar{m}$
2040.0	1.0245 +0	864.1	6.8698 −5	0.1267 +0	28.9620
2080.0	1.0048	871.8	6.9565	0.1294	28.9611
2120.0	9.8579 −1	879.3	7.0426	0.1321	28.9601
2160.0	9.6749	886.8	7.1281	0.1350	28.9588
2200.0	9.4985	894.1	7.2130	0.1380	28.9572
2240.0	9.3282	901.3	7.2974	0.1411	28.9551
2280.0	9.1637	908.4	7.3813	0.1445	28.9526
2320.0	9.0048	915.4	7.4647	0.1480	28.9496
2360.0	8.8510	922.3	7.5476	0.1518	28.9459
2400.0	8.7022	929.0	7.6299	0.1559	28.9414
2440.0	8.5580	935.7	7.7119	0.1603	28.9361
2480.0	8.4181	942.3	7.7934	0.1649	28.9298
2520.0	8.2823	948.8	7.8745	0.1700	28.9223
2560.0	8.1505	955.2	7.9551	0.1755	28.9136
2600.0	8.0223	961.5	8.0354	0.1813	28.9035
2640.0	7.8975	967.7	8.1154	0.1877	28.8917
2680.0	7.7760	973.9	8.1949	0.1946	28.8782
2720.0	7.6575	980.0	8.2742	0.2019	28.8627
2760.0	7.5419	986.1	8.3532	0.2099	28.8451
2800.0	7.4290	992.2	8.4319	0.2134	28.8250
2840.0	7.3186	998.3	8.5103	0.2276	28.8024
2880.0	7.2106	1004.3	8.5885	0.2374	28.7770
2920.0	7.1048	1010.4	8.6665	0.2478	28.7486
2960.0	7.0011	1016.5	8.7443	0.2589	28.7170
3000.0	6.8993	1022.7	8.8220	0.2706	28.6819
3040.0	6.7994	1029.0	8.8995	0.2830	28.6432
3080.0	6.7011	1035.3	8.9769	0.2959	28.6007
3120.0	6.6044	1041.7	9.0543	0.3097	28.5542
3160.0	6.5093	1048.2	9.1316	0.3239	28.5035
3200.0	6.4155	1054.9	9.2089	0.3384	28.4485
3240.0	6.3230	1061.7	9.2862	0.3535	28.3891
3280.0	6.2319	1068.6	9.3635	0.3688	28.3251
3320.0	6.1419	1075.6	9.4409	0.3844	28.2565
3360.0	6.0530	1082.8	9.5183	0.4001	28.1834
3400.0	5.9653	1090.2	9.5959	0.4158	28.1056
3440.0	5.8787	1097.7	9.6737	0.4323	28.0233
3480.0	5.7931	1105.4	9.7515	0.4478	27.9365
3520.0	5.7086	1113.3	9.8295	0.4629	27.8454
3560.0	5.6252	1121.3	9.9076	0.4775	27.7502
3600.0	5.5428	1129.5	9.9859	0.4913	27.6512
3640.0	5.4616	1137.9	1.0064 −4	0.5042	27.5485
3680.0	5.3814	1146.4	1.0143	0.5162	27.4427
3720.0	5.3025	1155.1	1.0222	0.5270	27.3339
3760.0	5.2247	1164.0	1.0301	0.5364	27.2227
3800.0	5.1482	1173.0	1.0380	0.5443	27.1095
3840.0	5.0730	1182.2	1.0459	0.5509	26.9948
3880.0	4.9992	1191.4	1.0538	0.5559	26.8791
3920.0	4.9268	1200.8	1.0617	0.5593	26.7629
3960.0	4.8559	1210.3	1.0697	0.5611	26.6467
4000.0	4.7865	1220.0	1.0776	0.5613	26.5311

G.5-22

THERMOPHYSICAL PROPERTIES OF AIR

ISOBAR: 0.60 MPa

T (K)	h (J/kg)	e (J/kg)	s (J/kg·K)	c_p (J/kg·K)	γ	Pr
2040.0	2.0286 +3	1.4426 +3	8.5000 +0	1.3348 +0	1.2750	0.7235
2080.0	2.0822	1.4847	8.5261	1.3447	1.2727	0.7230
2120.0	2.1362	1.5272	8.5518	1.3552	1.2704	0.7223
2160.0	2.1906	1.5701	8.5772	1.3664	1.2679	0.7215
2200.0	2.2455	1.6134	8.6024	1.3784	1.2655	0.7205
2240.0	2.3009	1.6573	8.6273	1.3913	1.2629	0.7193
2280.0	2.3569	1.7017	8.6521	1.4053	1.2603	0.7179
2320.0	2.4134	1.7466	8.6766	1.4205	1.2575	0.7162
2360.0	2.4705	1.7922	8.7011	1.4371	1.2547	0.7143
2400.0	2.5284	1.8384	8.7254	1.4552	1.2518	0.7121
2440.0	2.5870	1.8854	8.7496	1.4750	1.2488	0.7096
2480.0	2.6464	1.9332	8.7737	1.4966	1.2457	0.7069
2520.0	2.7067	1.9818	8.7979	1.5201	1.2426	0.7039
2560.0	2.7680	2.0314	8.8220	1.5459	1.2393	0.7006
2600.0	2.8304	2.0820	8.8462	1.5739	1.2360	0.6971
2640.0	2.8940	2.1337	8.8705	1.6045	1.2326	0.6933
2680.0	2.9588	2.1867	8.8948	1.6377	1.2292	0.6894
2720.0	3.0250	2.2410	8.9194	1.6737	1.2258	0.6853
2760.0	3.0927	2.2966	8.9441	1.7125	1.2223	0.6810
2800.0	3.1621	2.3539	8.9690	1.7544	1.2189	0.6766
2840.0	3.2331	2.4128	8.9942	1.7994	1.2155	0.6722
2880.0	3.3061	2.4734	9.0197	1.8476	1.2122	0.6678
2920.0	3.3810	2.5359	9.0455	1.8989	1.2089	0.6634
2960.0	3.4580	2.6004	9.0717	1.9536	1.2058	0.6591
3000.0	3.5373	2.6671	9.0983	2.0114	1.2027	0.6549
3040.0	3.6190	2.7359	9.1254	2.0723	1.1998	0.6509
3080.0	3.7031	2.8072	9.1529	2.1363	1.1971	0.6471
3120.0	3.7899	2.8808	9.1809	2.2032	1.1945	0.6437
3160.0	3.8794	2.9570	9.2094	2.2727	1.1921	0.6405
3200.0	3.9718	3.0359	9.2384	2.3447	1.1898	0.6375
3240.0	4.0670	3.1175	9.2680	2.4187	1.1878	0.6349
3280.0	4.1653	3.2018	9.2981	2.4945	1.1860	0.6327
3320.0	4.2666	3.2890	9.3289	2.5715	1.1843	0.6309
3360.0	4.3710	3.3791	9.3601	2.6494	1.1829	0.6295
3400.0	4.4785	3.4721	9.3919	2.7275	1.1816	0.6285
3440.0	4.5892	3.5679	9.4243	2.8051	1.1806	0.6277
3480.0	4.7029	3.6665	9.4572	2.8817	1.1798	0.6275
3520.0	4.8197	3.7680	9.4905	2.9566	1.1792	0.6278
3560.0	4.9394	3.8721	9.5243	3.0288	1.1788	0.6285
3600.0	5.0620	3.9788	9.5586	3.0977	1.1786	0.6296
3640.0	5.1872	4.0879	9.5932	3.1625	1.1786	0.6312
3680.0	5.3149	4.1992	9.6281	3.2223	1.1788	0.6332
3720.0	5.4449	4.3126	9.6632	3.2763	1.1792	0.6355
3760.0	5.5769	4.4278	9.6985	3.3238	1.1798	0.6383
3800.0	5.7107	4.5445	9.7339	3.3641	1.1806	0.6412
3840.0	5.8460	4.6624	9.7693	3.3967	1.1816	0.6446
3880.0	5.9823	4.7813	9.8046	3.4209	1.1827	0.6483
3920.0	6.1195	4.9009	9.8398	3.4366	1.1841	0.6522
3960.0	6.2571	5.0207	9.8747	3.4434	1.1856	0.6562
4000.0	6.3949	5.1405	9.9093	3.4413	1.1873	0.6604

THERMOPHYSICAL PROPERTIES OF AIR

ISOBAR: 0.60 MPa

T (K)	ρ (kg/m^3)	a (m/s)	μ (Pa·s)	k (J/s·m·K)	$\overline{m}$
4040.0	4.7186 −1	1229.6	1.0856 −4	0.5601 +0	26.4165
4080.0	4.6523	1239.4	1.0935	0.5574	26.3034
4120.0	4.5877	1249.2	1.1014	0.5534	26.1922
4160.0	4.5247	1259.1	1.1093	0.5483	26.0835
4200.0	4.4634	1269.0	1.1172	0.5422	25.9774
4240.0	4.4038	1278.9	1.1251	0.5353	25.8744
4280.0	4.3458	1288.8	1.1330	0.5278	25.7746
4320.0	4.2895	1298.7	1.1408	0.5199	25.6784
4360.0	4.2348	1308.5	1.1486	0.5118	25.5857
4400.0	4.1817	1318.2	1.1564	0.5037	25.4968
4440.0	4.1302	1327.9	1.1642	0.4957	25.4116
4480.0	4.0802	1337.5	1.1719	0.4881	25.3301
4520.0	4.0317	1346.9	1.1796	0.4809	25.2523
4560.0	3.9845	1356.3	1.1873	0.4744	25.1782
4600.0	3.9388	1365.4	1.1950	0.4685	25.1074
4640.0	3.8944	1374.4	1.2026	0.4635	25.0400
4680.0	3.8512	1383.2	1.2102	0.4594	24.9757
4720.0	3.8091	1391.7	1.2178	0.4563	24.9144
4760.0	3.7683	1400.1	1.2253	0.4542	24.8558
4800.0	3.7284	1408.2	1.2329	0.4532	24.7997
4840.0	3.6896	1416.0	1.2404	0.4533	24.7460
4880.0	3.6517	1423.7	1.2479	0.4545	24.6943
4920.0	3.6147	1431.0	1.2554	0.4568	24.6446
4960.0	3.5786	1438.2	1.2629	0.4604	24.5965
5000.0	3.5432	1445.1	1.2703	0.4651	24.5498
5040.0	3.5086	1451.7	1.2778	0.4710	24.5044
5080.0	3.4747	1458.2	1.2853	0.4782	24.4600
5120.0	3.4414	1464.4	1.2928	0.4865	24.4164
5160.0	3.4087	1470.5	1.3002	0.4961	24.3735
5200.0	3.3766	1476.4	1.3077	0.5068	24.3310
5240.0	3.3450	1482.1	1.3152	0.5189	24.2888
5280.0	3.3139	1487.7	1.3227	0.5321	24.2467
5320.0	3.2833	1493.2	1.3302	0.5466	24.2046
5360.0	3.2531	1498.6	1.3377	0.5624	24.1622
5400.0	3.2233	1504.0	1.3453	0.5795	24.1195
5440.0	3.1938	1509.3	1.3529	0.5979	24.0762
5480.0	3.1647	1514.5	1.3605	0.6175	24.0323
5520.0	3.1359	1519.8	1.3682	0.6385	23.9876
5560.0	3.1074	1525.0	1.3759	0.6609	23.9419
5600.0	3.0792	1530.3	1.3836	0.6845	23.8951
5640.0	3.0512	1535.6	1.3914	0.7096	23.8471
5680.0	3.0235	1541.0	1.3992	0.7360	23.7978
5720.0	2.9959	1546.4	1.4071	0.7638	23.7471
5760.0	2.9686	1551.8	1.4150	0.7930	23.6948
5800.0	2.9414	1557.4	1.4230	0.8236	23.6409
5840.0	2.9144	1563.0	1.4311	0.8557	23.5851
5880.0	2.8875	1568.8	1.4392	0.8892	23.5275
5920.0	2.8607	1574.7	1.4474	0.9240	23.4679
5960.0	2.8340	1580.7	1.4557	0.9604	23.4062
6000.0	2.8075	1586.8	1.4641	0.9981	23.3424

THERMOPHYSICAL PROPERTIES OF AIR

ISOBAR: 0.60 MPa

T (K)	h (J/kg)	e (J/kg)	s (J/kg·K)	c_p (J/kg·K)	γ	Pr
4040.0	6.5323 +3	5.2599 +3	9.9435 +0	3.4305 +0	1.1891	0.6647
4080.0	6.6692	5.3786	9.9772	3.4112	1.1911	0.6691
4120.0	6.8051	5.4964	1.0010 +1	3.3841	1.1933	0.6733
4160.0	6.9398	5.6129	1.0043	3.3496	1.1955	0.6776
4200.0	7.0730	5.7278	1.0075	3.3085	1.1979	0.6816
4240.0	7.2044	5.8410	1.0106	3.2619	1.2004	0.6855
4280.0	7.3339	5.9523	1.0136	3.2106	1.2030	0.6890
4320.0	7.4612	6.0615	1.0166	3.1556	1.2057	0.6923
4360.0	7.5863	6.1685	1.0195	3.0982	1.2084	0.6952
4400.0	7.7091	6.2733	1.0223	3.0392	1.2111	0.6977
4440.0	7.8294	6.3757	1.0250	2.9798	1.2138	0.6997
4480.0	7.9474	6.4759	1.0276	2.9208	1.2165	0.7012
4520.0	8.0631	6.5739	1.0302	2.8632	1.2191	0.7022
4560.0	8.1765	6.6697	1.0327	2.8079	1.2216	0.7027
4600.0	8.2878	6.7635	1.0351	2.7554	1.2239	0.7027
4640.0	8.3970	6.8553	1.0375	2.7064	1.2260	0.7021
4680.0	8.5044	6.9453	1.0398	2.6615	1.2280	0.7010
4720.0	8.6100	7.0338	1.0421	2.6209	1.2297	0.6994
4760.0	8.7141	7.1208	1.0443	2.5851	1.2311	0.6974
4800.0	8.8169	7.2065	1.0464	2.5544	1.2322	0.6949
4840.0	8.9185	7.2912	1.0485	2.5288	1.2330	0.6920
4880.0	9.0192	7.3751	1.0506	2.5084	1.2335	0.6887
4920.0	9.1193	7.4583	1.0526	2.4935	1.2337	0.6852
4960.0	9.2188	7.5410	1.0546	2.4840	1.2336	0.6813
5000.0	9.3181	7.6235	1.0566	2.4799	1.2331	0.6773
5040.0	9.4173	7.7060	1.0586	2.4813	1.2324	0.6731
5080.0	9.5166	7.7887	1.0606	2.4880	1.2313	0.6687
5120.0	9.6164	7.8717	1.0625	2.5000	1.2300	0.6643
5160.0	9.7167	7.9553	1.0645	2.5174	1.2284	0.6598
5200.0	9.8178	8.0397	1.0664	2.5400	1.2266	0.6553
5240.0	9.9200	8.1250	1.0684	2.5678	1.2246	0.6509
5280.0	1.0023 +4	8.2116	1.0704	2.6008	1.2225	0.6465
5320.0	1.0128	8.2994	1.0723	2.6389	1.2201	0.6421
5360.0	1.0234	8.3889	1.0743	2.6821	1.2177	0.6379
5400.0	1.0343	8.4800	1.0763	2.7303	1.2152	0.6338
5440.0	1.0453	8.5731	1.0784	2.7836	1.2125	0.6299
5480.0	1.0565	8.6683	1.0804	2.8420	1.2099	0.6261
5520.0	1.0680	8.7658	1.0825	2.9054	1.2072	0.6225
5560.0	1.0798	8.8659	1.0846	2.9738	1.2045	0.6191
5600.0	1.0918	8.9685	1.0868	3.0473	1.2018	0.6159
5640.0	1.1042	9.0741	1.0890	3.1258	1.1992	0.6129
5680.0	1.1169	9.1827	1.0912	3.2094	1.1966	0.6101
5720.0	1.1299	9.2946	1.0935	3.2981	1.1940	0.6075
5760.0	1.1432	9.4099	1.0959	3.3919	1.1915	0.6052
5800.0	1.1570	9.5289	1.0982	3.4908	1.1891	0.6031
5840.0	1.1712	9.6516	1.1007	3.5949	1.1867	0.6012
5880.0	1.1858	9.7784	1.1032	3.7042	1.1844	0.5996
5920.0	1.2008	9.9094	1.1057	3.8187	1.1822	0.5981
5960.0	1.2163	1.0045 +4	1.1083	3.9385	1.1801	0.5970
6000.0	1.2323	1.0185	1.1110	4.0635	1.1781	0.5960

THERMOPHYSICAL PROPERTIES OF AIR

ISOBAR: 1.0 MPa

T (K)	ρ (kg/m^3)	a (m/s)	μ (Pa·s)	k (J/s·m·K)	$\bar{m}$
288.15	1.2090 +1	340.3	1.7979 −5	0.0249 +0	28.9644
300.0	1.1612	347.2	1.8554	0.0257	28.9644
310.0	1.1238	352.9	1.9030	0.0264	28.9644
320.0	1.0886	358.5	1.9500	0.0271	28.9644
330.0	1.0556	364.1	1.9962	0.0278	28.9644
340.0	1.0246	369.5	2.0419	0.0285	28.9644
350.0	9.9533 +0	374.8	2.0869	0.0292	28.9644
360.0	9.6768	380.1	2.1313	0.0298	28.9644
370.0	9.4152	385.2	2.1751	0.0305	28.9644
380.0	9.1675	390.3	2.2184	0.0312	28.9644
390.0	8.9324	395.3	2.2612	0.0318	28.9644
400.0	8.7091	400.3	2.3035	0.0325	28.9644
410.0	8.4967	405.1	2.3454	0.0331	28.9644
420.0	8.2944	409.9	2.3867	0.0338	28.9644
430.0	8.1015	414.7	2.4276	0.0344	28.9644
440.0	7.9174	419.3	2.4681	0.0350	28.9644
450.0	7.7414	423.9	2.5082	0.0357	28.9644
460.0	7.5731	428.5	2.5479	0.0363	28.9644
470.0	7.4120	433.0	2.5872	0.0370	28.9644
480.0	7.2576	437.4	2.6261	0.0376	28.9644
490.0	7.1095	441.8	2.6647	0.0382	28.9644
500.0	6.9673	446.1	2.7029	0.0388	28.9644
510.0	6.8307	450.4	2.7408	0.0395	28.9644
520.0	6.6993	454.6	2.7783	0.0401	28.9644
530.0	6.5729	458.8	2.8155	0.0407	28.9644
540.0	6.4512	462.9	2.8525	0.0413	28.9644
550.0	6.3339	467.0	2.8891	0.0420	28.9644
560.0	6.2208	471.0	2.9254	0.0426	28.9644
580.0	6.0063	479.0	2.9972	0.0438	28.9644
600.0	5.8061	486.8	3.0680	0.0450	28.9644
620.0	5.6188	494.4	3.1377	0.0463	28.9644
640.0	5.4432	501.9	3.2065	0.0475	28.9644
660.0	5.2782	509.3	3.2743	0.0487	28.9644
680.0	5.1230	516.5	3.3412	0.0499	28.9644
700.0	4.9766	523.6	3.4073	0.0511	28.9644
720.0	4.8384	530.6	3.4726	0.0523	28.9644
740.0	4.7076	537.5	3.5371	0.0535	28.9644
760.0	4.5837	544.3	3.6008	0.0547	28.9644
780.0	4.4662	551.0	3.6639	0.0559	28.9644
800.0	4.3545	557.5	3.7263	0.0571	28.9644
820.0	4.2483	564.1	3.7880	0.0583	28.9644
840.0	4.1472	570.5	3.8490	0.0594	28.9644
860.0	4.0507	576.8	3.9095	0.0606	28.9644
880.0	3.9587	583.1	3.9693	0.0618	28.9644
900.0	3.8707	589.3	4.0286	0.0630	28.9644
920.0	3.7866	595.4	4.0873	0.0641	28.9644
940.0	3.7060	601.4	4.1455	0.0653	28.9644
960.0	3.6288	607.4	4.2032	0.0665	28.9645
980.0	3.5547	613.3	4.2603	0.0676	28.9644
1000.0	3.4836	619.2	4.3170	0.0688	28.9644

THERMOPHYSICAL PROPERTIES OF AIR

ISOBAR: 1.0 MPa

T (K)	h (J/kg)	e (J/kg)	s (J/kg·K)	c_p (J/kg·K)	γ	Pr
288.15	-1.4106 +1	-9.6876 +1	6.1732 +0	1.0052 +0	1.4001	0.7254
300.0	-2.1924 +0	-8.8367	6.2137	1.0055	1.3999	0.7246
310.0	7.8646	-8.1182	6.2467	1.0059	1.3997	0.7240
320.0	1.7926 +1	-7.3994	6.2786	1.0064	1.3994	0.7233
330.0	2.7993	-6.6799	6.3096	1.0070	1.3991	0.7228
340.0	3.8067	-5.9598	6.3397	1.0077	1.3987	0.7222
350.0	4.8148	-5.2389	6.3689	1.0085	1.3982	0.7217
360.0	5.8238	-4.5171	6.3973	1.0095	1.3977	0.7212
370.0	6.8337	-3.7944	6.4250	1.0105	1.3972	0.7207
380.0	7.8447	-3.0707	6.4520	1.0116	1.3966	0.7203
390.0	8.8569	-2.3458	6.4783	1.0127	1.3959	0.7199
400.0	9.8702	-1.6197	6.5039	1.0140	1.3953	0.7195
410.0	1.0885 +2	-8.9229 +0	6.5290	1.0153	1.3945	0.7191
420.0	1.1901	-1.6350	6.5534	1.0168	1.3938	0.7188
430.0	1.2918	5.6674	6.5774	1.0182	1.3930	0.7185
440.0	1.3937	1.2985 +1	6.6008	1.0198	1.3921	0.7182
450.0	1.4958	2.0319	6.6238	1.0214	1.3913	0.7179
460.0	1.5980	2.7669	6.6462	1.0231	1.3904	0.7177
470.0	1.7004	3.5036	6.6682	1.0249	1.3894	0.7175
480.0	1.8030	4.2421	6.6898	1.0267	1.3885	0.7173
490.0	1.9058	4.9825	6.7110	1.0285	1.3875	0.7171
500.0	2.0087	5.7247	6.7318	1.0304	1.3865	0.7170
510.0	2.1119	6.4689	6.7522	1.0324	1.3855	0.7169
520.0	2.2152	7.2150	6.7723	1.0344	1.3845	0.7167
530.0	2.3187	7.9632	6.7920	1.0365	1.3834	0.7167
540.0	2.4225	8.7134	6.8114	1.0385	1.3823	0.7166
550.0	2.5264	9.4658	6.8305	1.0407	1.3813	0.7165
560.0	2.6306	1.0220 +2	6.8493	1.0428	1.3802	0.7165
580.0	2.8396	1.1736	6.8859	1.0473	1.3780	0.7165
600.0	3.0495	1.3260	6.9215	1.0518	1.3757	0.7165
620.0	3.2603	1.4794	6.9561	1.0564	1.3735	0.7165
640.0	3.4721	1.6337	6.9897	1.0611	1.3712	0.7166
660.0	3.6848	1.7890	7.0224	1.0659	1.3689	0.7168
680.0	3.8984	1.9452	7.0543	1.0707	1.3667	0.7169
700.0	4.1131	2.1023	7.0854	1.0755	1.3644	0.7171
720.0	4.3286	2.2605	7.1158	1.0803	1.3622	0.7172
740.0	4.5452	2.4196	7.1455	1.0851	1.3600	0.7174
760.0	4.7627	2.5796	7.1745	1.0900	1.3578	0.7176
780.0	4.9812	2.7406	7.2028	1.0947	1.3557	0.7177
800.0	5.2006	2.9026	7.2306	1.0995	1.3537	0.7178
820.0	5.4210	3.0655	7.2578	1.1042	1.3516	0.7179
840.0	5.6423	3.2294	7.2845	1.1088	1.3496	0.7179
860.0	5.8645	3.3941	7.3106	1.1134	1.3477	0.7179
880.0	6.0876	3.5598	7.3363	1.1178	1.3458	0.7179
900.0	6.3116	3.7264	7.3614	1.1223	1.3440	0.7178
920.0	6.5365	3.8938	7.3862	1.1266	1.3422	0.7176
940.0	6.7622	4.0621	7.4104	1.1308	1.3405	0.7174
960.0	6.9888	4.2312	7.4343	1.1349	1.3389	0.7172
980.0	7.2162	4.4012	7.4577	1.1389	1.3373	0.7169
1000.0	7.4444	4.5719	7.4808	1.1428	1.3357	0.7165

THERMOPHYSICAL PROPERTIES OF AIR
ISOBAR: 1.0 MPa

T (K)	ρ (kg/m^3)	a (m/s)	μ (Pa·s)	k (J/s·m·K)	$\bar{m}$
1020.0	3.4153 +0	625.0	4.3745 −5	0.0697 +0	28.9644
1040.0	3.3497	630.8	4.4309	0.0708	28.9644
1060.0	3.2865	636.5	4.4870	0.0719	28.9644
1080.0	3.2256	642.2	4.5426	0.0730	28.9644
1100.0	3.1669	647.8	4.5978	0.0741	28.9644
1120.0	3.1104	653.3	4.6526	0.0754	28.9644
1140.0	3.0558	658.8	4.7070	0.0765	28.9644
1160.0	3.0031	664.2	4.7610	0.0776	28.9644
1180.0	2.9522	669.6	4.8147	0.0787	28.9644
1200.0	2.9030	674.9	4.8680	0.0797	28.9644
1220.0	2.8554	680.2	4.9209	0.0808	28.9644
1240.0	2.8094	685.5	4.9735	0.0819	28.9644
1260.0	2.7648	690.6	5.0258	0.0830	28.9644
1280.0	2.7216	695.8	5.0778	0.0841	28.9644
1300.0	2.6797	700.9	5.1294	0.0851	28.9644
1320.0	2.6391	705.9	5.1807	0.0862	28.9644
1340.0	2.5997	710.9	5.2318	0.0873	28.9644
1360.0	2.5615	715.9	5.2825	0.0884	28.9644
1380.0	2.5244	720.8	5.3329	0.0894	28.9644
1400.0	2.4883	725.7	5.3831	0.0905	28.9644
1420.0	2.4533	730.6	5.4330	0.0916	28.9644
1440.0	2.4192	735.4	5.4827	0.0927	28.9644
1460.0	2.3861	740.1	5.5320	0.0937	28.9644
1480.0	2.3538	744.9	5.5811	0.0948	28.9644
1500.0	2.3224	749.6	5.6300	0.0959	28.9644
1520.0	2.2919	754.2	5.6786	0.0969	28.9644
1540.0	2.2621	758.8	5.7270	0.0980	28.9644
1560.0	2.2331	763.4	5.7751	0.0991	28.9644
1580.0	2.2048	768.0	5.8231	0.1002	28.9644
1600.0	2.1773	772.5	5.8707	0.1013	28.9644
1620.0	2.1504	777.0	5.9182	0.1023	28.9643
1640.0	2.1242	781.4	5.9654	0.1034	28.9643
1660.0	2.0986	785.8	6.0125	0.1045	28.9643
1680.0	2.0736	790.2	6.0593	0.1056	28.9643
1700.0	2.0492	794.6	6.1059	0.1067	28.9643
1720.0	2.0254	798.9	6.1523	0.1078	28.9643
1740.0	2.0021	803.2	6.1985	0.1089	28.9642
1760.0	1.9793	807.5	6.2445	0.1099	28.9642
1780.0	1.9571	811.7	6.2903	0.1110	28.9642
1800.0	1.9353	815.9	6.3360	0.1121	28.9641
1820.0	1.9141	820.1	6.3814	0.1132	28.9641
1840.0	1.8933	824.3	6.4267	0.1143	28.9640
1860.0	1.8729	828.4	6.4717	0.1154	28.9639
1880.0	1.8530	832.5	6.5166	0.1165	28.9639
1900.0	1.8335	836.5	6.5613	0.1176	28.9638
1920.0	1.8143	840.6	6.6059	0.1187	28.9637
1940.0	1.7956	844.6	6.6503	0.1198	28.9635
1960.0	1.7773	848.6	6.6945	0.1209	28.9634
1980.0	1.7593	852.5	6.7385	0.1220	28.9632
2000.0	1.7417	856.5	6.7824	0.1231	28.9630

THERMOPHYSICAL PROPERTIES OF AIR

ISOBAR: 1.0 MPa

T (K)	h (J/kg)	e (J/kg)	s (J/kg·K)	c_p (J/kg·K)	γ	Pr
1020.0	7.6733 +2	4.7434 +2	7.5034 +0	1.1465 +0	1.3343	0.7185
1040.0	7.9030	4.9156	7.5257	1.1501	1.3329	0.7185
1060.0	8.1333	5.0885	7.5477	1.1537	1.3315	0.7185
1080.0	8.3644	5.2622	7.5693	1.1572	1.3302	0.7185
1100.0	8.5962	5.4365	7.5905	1.1608	1.3288	0.7186
1120.0	8.8288	5.6116	7.6115	1.1643	1.3275	0.7187
1140.0	9.0620	5.7873	7.6321	1.1678	1.3262	0.7188
1160.0	9.2959	5.9638	7.6525	1.1712	1.3249	0.7190
1180.0	9.5305	6.1409	7.6725	1.1747	1.3237	0.7191
1200.0	9.7657	6.3187	7.6923	1.1781	1.3224	0.7192
1220.0	1.0002 +3	6.4973	7.7118	1.1816	1.3212	0.7194
1240.0	1.0238	6.6765	7.7310	1.1850	1.3200	0.7195
1260.0	1.0476	6.8564	7.7500	1.1884	1.3188	0.7197
1280.0	1.0714	7.0369	7.7688	1.1918	1.3176	0.7199
1300.0	1.0952	7.2182	7.7873	1.1952	1.3164	0.7201
1320.0	1.1192	7.4001	7.8055	1.1986	1.3152	0.7202
1340.0	1.1432	7.5827	7.8236	1.2020	1.3140	0.7204
1360.0	1.1673	7.7660	7.8414	1.2054	1.3128	0.7206
1380.0	1.1914	7.9500	7.8590	1.2089	1.3117	0.7208
1400.0	1.2156	8.1347	7.8765	1.2123	1.3105	0.7210
1420.0	1.2399	8.3200	7.8937	1.2157	1.3094	0.7212
1440.0	1.2642	8.5061	7.9107	1.2192	1.3082	0.7214
1460.0	1.2887	8.6928	7.9276	1.2226	1.3071	0.7216
1480.0	1.3131	8.8802	7.9442	1.2261	1.3060	0.7218
1500.0	1.3377	9.0683	7.9607	1.2295	1.3048	0.7220
1520.0	1.3623	9.2571	7.9770	1.2330	1.3037	0.7222
1540.0	1.3870	9.4466	7.9931	1.2365	1.3026	0.7224
1560.0	1.4118	9.6368	8.0091	1.2400	1.3015	0.7225
1580.0	1.4366	9.8277	8.0249	1.2436	1.3004	0.7227
1600.0	1.4615	1.0019 +3	8.0406	1.2471	1.2993	0.7229
1620.0	1.4865	1.0212	8.0561	1.2506	1.2982	0.7231
1640.0	1.5116	1.0405	8.0715	1.2542	1.2971	0.7232
1660.0	1.5367	1.0598	8.0867	1.2578	1.2960	0.7234
1680.0	1.5619	1.0793	8.1018	1.2614	1.2949	0.7236
1700.0	1.5871	1.0988	8.1167	1.2650	1.2938	0.7237
1720.0	1.6125	1.1184	8.1316	1.2687	1.2927	0.7239
1740.0	1.6379	1.1381	8.1462	1.2724	1.2916	0.7240
1760.0	1.6634	1.1578	8.1608	1.2761	1.2906	0.7242
1780.0	1.6889	1.1776	8.1753	1.2798	1.2895	0.7243
1800.0	1.7146	1.1975	8.1896	1.2836	1.2884	0.7245
1820.0	1.7403	1.2175	8.2038	1.2873	1.2873	0.7246
1840.0	1.7661	1.2375	8.2179	1.2912	1.2863	0.7247
1860.0	1.7919	1.2576	8.2318	1.2950	1.2852	0.7249
1880.0	1.8179	1.2778	8.2457	1.2989	1.2841	0.7250
1900.0	1.8439	1.2981	8.2595	1.3029	1.2831	0.7251
1920.0	1.8700	1.3184	8.2732	1.3069	1.2820	0.7252
1940.0	1.8962	1.3389	8.2867	1.3110	1.2809	0.7253
1960.0	1.9224	1.3594	8.3002	1.3151	1.2798	0.7254
1980.0	1.9488	1.3800	8.3136	1.3193	1.2788	0.7255
2000.0	1.9752	1.4007	8.3268	1.3236	1.2777	0.7256

THERMOPHYSICAL PROPERTIES OF AIR

ISOBAR: 1.0 MPa

T (K)	ρ (kg/m^3)	a (m/s)	μ (Pa·s)	k (J/s·m·K)	$\bar{m}$
2040.0	1.7076 +0	864.3	6.8698 −5	0.1264 +0	28.9625
2080.0	1.6747	872.0	6.9565	0.1290	28.9619
2120.0	1.6430	879.6	7.0425	0.1316	28.9611
2160.0	1.6126	887.0	7.1281	0.1343	28.9601
2200.0	1.5832	894.4	7.2130	0.1371	28.9588
2240.0	1.5548	901.7	7.2974	0.1401	28.9572
2280.0	1.5274	908.9	7.3812	0.1431	28.9553
2320.0	1.5010	915.9	7.4646	0.1464	28.9529
2360.0	1.4754	922.9	7.5474	0.1498	28.9500
2400.0	1.4506	929.8	7.6298	0.1535	28.9466
2440.0	1.4266	936.6	7.7117	0.1573	28.9424
2480.0	1.4034	943.3	7.7931	0.1615	28.9376
2520.0	1.3808	949.9	7.8742	0.1659	28.9318
2560.0	1.3589	956.4	7.9548	0.1706	28.9250
2600.0	1.3377	962.8	8.0350	0.1756	28.9171
2640.0	1.3170	969.2	8.1148	0.1810	28.9080
2680.0	1.2969	975.5	8.1942	0.1869	28.8975
2720.0	1.2773	981.7	8.2733	0.1931	28.8854
2760.0	1.2581	987.9	8.3521	0.1998	28.8717
2800.0	1.2395	994.1	8.4306	0.2069	28.8561
2840.0	1.2213	1000.2	8.5088	0.2145	28.8384
2880.0	1.2035	1006.3	8.5868	0.2227	28.8186
2920.0	1.1861	1012.4	8.6645	0.2313	28.7964
2960.0	1.1691	1018.5	8.7419	0.2405	28.7716
3000.0	1.1524	1024.6	8.8192	0.2503	28.7441
3040.0	1.1360	1030.8	8.8963	0.2606	28.7136
3080.0	1.1200	1037.0	8.9733	0.2714	28.6801
3120.0	1.1042	1043.3	9.0501	0.2828	28.6434
3160.0	1.0887	1049.6	9.1268	0.2949	28.6032
3200.0	1.0734	1056.1	9.2034	0.3072	28.5595
3240.0	1.0584	1062.6	9.2800	0.3200	28.5121
3280.0	1.0436	1069.2	9.3566	0.3332	28.4609
3320.0	1.0291	1075.9	9.4331	0.3467	28.4059
3360.0	1.0147	1082.8	9.5097	0.3605	28.3469
3400.0	1.0005	1089.8	9.5862	0.3744	28.2838
3440.0	9.8655 −1	1096.9	9.6629	0.3885	28.2167
3480.0	9.7275	1104.1	9.7396	0.4026	28.1456
3520.0	9.5913	1111.5	9.8165	0.4177	28.0705
3560.0	9.4568	1119.0	9.8934	0.4317	27.9916
3600.0	9.3241	1126.7	9.9704	0.4454	27.9088
3640.0	9.1931	1134.5	1.0048 −4	0.4587	27.8223
3680.0	9.0637	1142.5	1.0125	0.4714	27.7324
3720.0	8.9362	1150.6	1.0202	0.4835	27.6392
3760.0	8.8103	1158.8	1.0280	0.4949	27.5430
3800.0	8.6863	1167.2	1.0358	0.5053	27.4441
3840.0	8.5641	1175.8	1.0436	0.5148	27.3428
3880.0	8.4438	1184.5	1.0513	0.5231	27.2395
3920.0	8.3254	1193.3	1.0592	0.5302	27.1345
3960.0	8.2090	1202.2	1.0670	0.5362	27.0282
4000.0	8.0947	1211.2	1.0748	0.5409	26.9210

THERMOPHYSICAL PROPERTIES OF AIR

ISOBAR: 1.0 MPa

T	h	e	s	c_p	γ	Pr
(K)	(J/kg)	(J/kg)	(J/kg·K)	(J/kg·K)		
2040.0	2.0283 +3	1.4423 +3	8.3531 +0	1.3324 +0	1.2755	0.7240
2080.0	2.0818	1.4843	8.3791	1.3416	1.2733	0.7236
2120.0	2.1356	1.5266	8.4047	1.3512	1.2711	0.7231
2160.0	2.1899	1.5693	8.4301	1.3613	1.2688	0.7225
2200.0	2.2445	1.6125	8.4552	1.3721	1.2665	0.7218
2240.0	2.2997	1.6561	8.4800	1.3836	1.2641	0.7208
2280.0	2.3552	1.7001	8.5046	1.3959	1.2617	0.7197
2320.0	2.4113	1.7447	8.5290	1.4091	1.2592	0.7184
2360.0	2.4680	1.7897	8.5532	1.4233	1.2567	0.7169
2400.0	2.5252	1.8354	8.5772	1.4386	1.2541	0.7152
2440.0	2.5831	1.8817	8.6011	1.4553	1.2514	0.7132
2480.0	2.6417	1.9286	8.6250	1.4733	1.2487	0.7110
2520.0	2.7010	1.9763	8.6487	1.4928	1.2459	0.7085
2560.0	2.7611	2.0248	8.6724	1.5140	1.2430	0.7058
2600.0	2.8221	2.0741	8.6960	1.5370	1.2401	0.7029
2640.0	2.8841	2.1243	8.7197	1.5618	1.2371	0.6997
2680.0	2.9471	2.1755	8.7433	1.5888	1.2340	0.6964
2720.0	3.0112	2.2278	8.7671	1.6178	1.2310	0.6928
2760.0	3.0766	2.2812	8.7909	1.6492	1.2279	0.6891
2800.0	3.1432	2.3359	8.8149	1.6829	1.2248	0.6852
2840.0	3.2112	2.3919	8.8390	1.7190	1.2217	0.6813
2880.0	3.2807	2.4493	8.8633	1.7577	1.2187	0.6772
2920.0	3.3519	2.5082	8.8879	1.7990	1.2157	0.6731
2960.0	3.4247	2.5687	8.9126	1.8428	1.2128	0.6690
3000.0	3.4993	2.6310	8.9377	1.8893	1.2099	0.6650
3040.0	3.5759	2.6950	8.9630	1.9385	1.2071	0.6610
3080.0	3.6544	2.7609	8.9887	1.9902	1.2044	0.6572
3120.0	3.7351	2.8289	9.0147	2.0444	1.2019	0.6534
3160.0	3.8180	2.8989	9.0411	2.1010	1.1995	0.6500
3200.0	3.9032	2.9710	9.0679	2.1599	1.1972	0.6467
3240.0	3.9908	3.0454	9.0951	2.2209	1.1950	0.6437
3280.0	4.0809	3.1221	9.1228	2.2838	1.1931	0.6409
3320.0	4.1736	3.2012	9.1508	2.3483	1.1913	0.6384
3360.0	4.2688	3.2826	9.1794	2.4142	1.1896	0.6362
3400.0	4.3667	3.3666	9.2083	2.4810	1.1882	0.6344
3440.0	4.4673	3.4530	9.2377	2.5484	1.1869	0.6329
3480.0	4.5706	3.5419	9.2676	2.6161	1.1858	0.6318
3520.0	4.6766	3.6333	9.2979	2.6834	1.1849	0.6307
3560.0	4.7853	3.7271	9.3286	2.7499	1.1841	0.6303
3600.0	4.8966	3.8234	9.3597	2.8152	1.1836	0.6302
3640.0	5.0104	3.9219	9.3911	2.8785	1.1832	0.6306
3680.0	5.1268	4.0228	9.4229	2.9393	1.1830	0.6313
3720.0	5.2456	4.1257	9.4550	2.9971	1.1830	0.6324
3760.0	5.3665	4.2307	9.4873	3.0511	1.1832	0.6338
3800.0	5.4896	4.3376	9.5199	3.1009	1.1835	0.6356
3840.0	5.6145	4.4461	9.5526	3.1459	1.1840	0.6378
3880.0	5.7412	4.5561	9.5854	3.1855	1.1846	0.6402
3920.0	5.8693	4.6673	9.6183	3.2192	1.1855	0.6429
3960.0	5.9986	4.7796	9.6511	3.2467	1.1864	0.6459
4000.0	6.1289	4.8927	9.6838	3.2676	1.1876	0.6491

THERMOPHYSICAL PROPERTIES OF AIR

ISOBAR: 1.0 MPa

T (K)	ρ (kg/m^3)	a (m/s)	μ (Pa·s)	k (J/s·m·K)	$\overline{m}$
4040.0	7.9825 −1	1220.4	1.0826 −4	0.5443 +0	26.8133
4080.0	7.8725	1229.6	1.0905	0.5465	26.7056
4120.0	7.7647	1238.9	1.0983	0.5473	26.5983
4160.0	7.6592	1248.3	1.1062	0.5470	26.4917
4200.0	7.5561	1257.8	1.1140	0.5455	26.3863
4240.0	7.4553	1267.3	1.1218	0.5430	26.2823
4280.0	7.3570	1276.8	1.1296	0.5395	26.1802
4320.0	7.2610	1286.4	1.1374	0.5352	26.0803
4360.0	7.1675	1296.0	1.1452	0.5303	25.9827
4400.0	7.0764	1305.5	1.1530	0.5248	25.8877
4440.0	6.9876	1315.1	1.1608	0.5189	25.7955
4480.0	6.9013	1324.6	1.1685	0.5127	25.7063
4520.0	6.8173	1334.0	1.1762	0.5065	25.6201
4560.0	6.7355	1343.4	1.1839	0.5004	25.5370
4600.0	6.6561	1352.7	1.1916	0.4944	25.4570
4640.0	6.5787	1361.9	1.1993	0.4888	25.3800
4680.0	6.5035	1370.9	1.2069	0.4836	25.3062
4720.0	6.4304	1379.9	1.2145	0.4790	25.2353
4760.0	6.3591	1388.6	1.2221	0.4750	25.1673
4800.0	6.2898	1397.2	1.2297	0.4717	25.1021
4840.0	6.2223	1405.7	1.2372	0.4692	25.0395
4880.0	6.1564	1413.9	1.2447	0.4675	24.9794
4920.0	6.0923	1421.9	1.2523	0.4667	24.9215
4960.0	6.0296	1429.8	1.2597	0.4669	24.8659
5000.0	5.9685	1437.4	1.2672	0.4681	24.8122
5040.0	5.9087	1444.8	1.2747	0.4702	24.7602
5080.0	5.8503	1452.0	1.2821	0.4734	24.7099
5120.0	5.7931	1459.0	1.2896	0.4776	24.6611
5160.0	5.7371	1465.8	1.2970	0.4828	24.6135
5200.0	5.6822	1472.4	1.3045	0.4892	24.5669
5240.0	5.6283	1478.8	1.3119	0.4966	24.5213
5280.0	5.5755	1485.1	1.3194	0.5051	24.4764
5320.0	5.5235	1491.2	1.3268	0.5147	24.4321
5360.0	5.4725	1497.1	1.3343	0.5254	24.3882
5400.0	5.4222	1502.9	1.3417	0.5372	24.3446
5440.0	5.3727	1508.6	1.3492	0.5502	24.3011
5480.0	5.3240	1514.3	1.3567	0.5643	24.2575
5520.0	5.2758	1519.8	1.3642	0.5796	24.2138
5560.0	5.2284	1525.3	1.3718	0.5960	24.1697
5600.0	5.1815	1530.7	1.3793	0.6136	24.1252
5640.0	5.1351	1536.1	1.3869	0.6323	24.0802
5680.0	5.0893	1541.5	1.3946	0.6523	24.0344
5720.0	5.0439	1546.8	1.4022	0.6735	23.9879
5760.0	4.9989	1552.2	1.4099	0.6958	23.9404
5800.0	4.9544	1557.7	1.4177	0.7194	23.8918
5840.0	4.9102	1563.1	1.4255	0.7443	23.8421
5880.0	4.8664	1568.6	1.4333	0.7704	23.7911
5920.0	4.8229	1574.2	1.4412	0.7977	23.7388
5960.0	4.7797	1579.8	1.4492	0.8263	23.6850
6000.0	4.7367	1585.5	1.4572	0.8561	23.6297

G.5-32

THERMOPHYSICAL PROPERTIES OF AIR

ISOBAR: 1.0 MPa

T (K)	h (J/kg)	e (J/kg)	s (J/kg·K)	c_p (J/kg·K)	γ	Pr
4040.0	6.2600 +3	5.0064 +3	9.7164 +0	3.2817 +0	1.1889	0.6525
4080.0	6.3914	5.1203	9.7488	3.2888	1.1903	0.6561
4120.0	6.5230	5.2342	9.7809	3.2890	1.1919	0.6598
4160.0	6.6544	5.3479	9.8126	3.2823	1.1936	0.6636
4200.0	6.7855	5.4611	9.8440	3.2690	1.1954	0.6674
4240.0	6.9158	5.5736	9.8749	3.2494	1.1974	0.6712
4280.0	7.0453	5.6852	9.9053	3.2240	1.1994	0.6749
4320.0	7.1737	5.7956	9.9351	3.1933	1.2016	0.6785
4360.0	7.3007	5.9046	9.9644	3.1580	1.2038	0.6819
4400.0	7.4263	6.0122	9.9931	3.1187	1.2061	0.6851
4440.0	7.5502	6.1181	1.0021 +1	3.0762	1.2085	0.6881
4480.0	7.6723	6.2224	1.0048	3.0314	1.2108	0.6907
4520.0	7.7927	6.3248	1.0075	2.9850	1.2132	0.6931
4560.0	7.9111	6.4255	1.0101	2.9377	1.2156	0.6950
4600.0	8.0277	6.5243	1.0127	2.8903	1.2179	0.6965
4640.0	8.1424	6.6213	1.0152	2.8436	1.2201	0.6976
4680.0	8.2552	6.7165	1.0176	2.7982	1.2223	0.6982
4720.0	8.3662	6.8101	1.0199	2.7546	1.2243	0.6984
4760.0	8.4756	6.9020	1.0223	2.7135	1.2262	0.6981
4800.0	8.5834	6.9924	1.0245	2.6753	1.2279	0.6974
4840.0	8.6897	7.0815	1.0267	2.6404	1.2294	0.6962
4880.0	8.7946	7.1692	1.0289	2.6091	1.2307	0.6946
4920.0	8.8984	7.2559	1.0310	2.5816	1.2318	0.6926
4960.0	9.0012	7.3416	1.0331	2.5583	1.2326	0.6902
5000.0	9.1032	7.4266	1.0351	2.5393	1.2332	0.6875
5040.0	9.2044	7.5109	1.0371	2.5247	1.2334	0.6844
5080.0	9.3052	7.5947	1.0391	2.5146	1.2334	0.6811
5120.0	9.4057	7.6783	1.0411	2.5091	1.2332	0.6775
5160.0	9.5060	7.7618	1.0430	2.5081	1.2327	0.6737
5200.0	9.6064	7.8453	1.0450	2.5118	1.2319	0.6698
5240.0	9.7070	7.9291	1.0469	2.5201	1.2309	0.6658
5280.0	9.8080	8.0133	1.0488	2.5330	1.2296	0.6616
5320.0	9.9097	8.0981	1.0508	2.5504	1.2282	0.6575
5360.0	1.0012 +4	8.1836	1.0527	2.5725	1.2266	0.6533
5400.0	1.0116	8.2701	1.0546	2.5991	1.2248	0.6491
5440.0	1.0220	8.3576	1.0565	2.6302	1.2228	0.6450
5480.0	1.0326	8.4465	1.0585	2.6658	1.2208	0.6409
5520.0	1.0433	8.5368	1.0604	2.7059	1.2186	0.6369
5560.0	1.0543	8.6286	1.0624	2.7504	1.2163	0.6330
5600.0	1.0654	8.7223	1.0644	2.7994	1.2140	0.6293
5640.0	1.0767	8.8179	1.0664	2.8528	1.2117	0.6257
5680.0	1.0882	8.9156	1.0684	2.9107	1.2093	0.6223
5720.0	1.0999	9.0156	1.0705	2.9729	1.2069	0.6190
5760.0	1.1120	9.1179	1.0726	3.0397	1.2045	0.6159
5800.0	1.1243	9.2229	1.0747	3.1108	1.2021	0.6130
5840.0	1.1369	9.3307	1.0769	3.1864	1.1997	0.6102
5880.0	1.1498	9.4414	1.0791	3.2664	1.1974	0.6077
5920.0	1.1630	9.5552	1.0813	3.3509	1.1951	0.6054
5960.0	1.1766	9.6722	1.0836	3.4399	1.1929	0.6033
6000.0	1.1905	9.7927	1.0859	3.5334	1.1908	0.6014

THERMOPHYSICAL PROPERTIES OF AIR

ISOBAR: 2.0 MPa

T (K)	ρ (kg/m^3)	a (m/s)	μ (Pa·s)	k (J/s·m·K)	$\bar{m}$
288.15	2.4179 +1	340.3	1.7979 −5	0.0249 +0	28.9644
300.0	2.3224	347.2	1.8554	0.0257	28.9644
310.0	2.2475	352.9	1.9030	0.0264	28.9644
320.0	2.1773	358.5	1.9500	0.0271	28.9644
330.0	2.1113	364.1	1.9962	0.0278	28.9644
340.0	2.0492	369.5	2.0419	0.0285	28.9644
350.0	1.9906	374.8	2.0869	0.0292	28.9644
360.0	1.9354	380.1	2.1313	0.0298	28.9644
370.0	1.8830	385.2	2.1751	0.0305	28.9644
380.0	1.8335	390.3	2.2184	0.0312	28.9644
390.0	1.7865	395.3	2.2612	0.0318	28.9644
400.0	1.7418	400.3	2.3035	0.0325	28.9644
410.0	1.6993	405.1	2.3454	0.0331	28.9644
420.0	1.6589	409.9	2.3867	0.0338	28.9644
430.0	1.6203	414.7	2.4276	0.0344	28.9644
440.0	1.5835	419.3	2.4681	0.0350	28.9644
450.0	1.5483	423.9	2.5082	0.0357	28.9644
460.0	1.5146	428.5	2.5479	0.0363	28.9644
470.0	1.4824	433.0	2.5872	0.0370	28.9644
480.0	1.4515	437.4	2.6261	0.0376	28.9644
490.0	1.4219	441.8	2.6647	0.0382	28.9644
500.0	1.3935	446.1	2.7029	0.0388	28.9644
510.0	1.3661	450.4	2.7408	0.0395	28.9644
520.0	1.3399	454.6	2.7783	0.0401	28.9644
530.0	1.3146	458.8	2.8155	0.0407	28.9644
540.0	1.2902	462.9	2.8525	0.0413	28.9644
550.0	1.2668	467.0	2.8891	0.0420	28.9644
560.0	1.2442	471.0	2.9254	0.0426	28.9644
580.0	1.2013	479.0	2.9972	0.0438	28.9644
600.0	1.1612	486.8	3.0680	0.0450	28.9644
620.0	1.1238	494.4	3.1377	0.0463	28.9644
640.0	1.0886	501.9	3.2065	0.0475	28.9644
660.0	1.0556	509.3	3.2743	0.0487	28.9644
680.0	1.0246	516.5	3.3412	0.0499	28.9644
700.0	9.9532 +0	523.6	3.4073	0.0511	28.9644
720.0	9.6768	530.6	3.4726	0.0523	28.9644
740.0	9.4152	537.5	3.5371	0.0535	28.9644
760.0	9.1675	544.3	3.6008	0.0547	28.9644
780.0	8.9324	551.0	3.6639	0.0559	28.9644
800.0	8.7091	557.5	3.7263	0.0571	28.9644
820.0	8.4967	564.1	3.7880	0.0583	28.9644
840.0	8.2944	570.5	3.8490	0.0594	28.9644
860.0	8.1015	576.8	3.9095	0.0606	28.9644
880.0	7.9174	583.1	3.9693	0.0618	28.9644
900.0	7.7414	589.3	4.0286	0.0630	28.9644
920.0	7.5731	595.4	4.0873	0.0641	28.9644
940.0	7.4120	601.4	4.1455	0.0653	28.9644
960.0	7.2576	607.4	4.2032	0.0665	28.9645
980.0	7.1095	613.3	4.2603	0.0676	28.9644
1000.0	6.9673	619.2	4.3170	0.0688	28.9644

THERMOPHYSICAL PROPERTIES OF AIR

ISOBAR: 2.0 MPa

T (K)	h (J/kg)	e (J/kg)	s (J/kg·K)	c_p (J/kg·K)	γ	Pr
288.15	-1.4106 +1	-9.6876 +1	5.9741 +0	1.0052 +0	1.4001	0.7254
300.0	-2.1924 +0	-8.8367	6.0146	1.0055	1.3999	0.7246
310.0	7.8646	-8.1182	6.0476	1.0059	1.3997	0.7240
320.0	1.7926 +1	-7.3994	6.0795	1.0064	1.3994	0.7233
330.0	2.7993	-6.6799	6.1105	1.0070	1.3991	0.7228
340.0	3.8067	-5.9598	6.1406	1.0077	1.3987	0.7222
350.0	4.8148	-5.2389	6.1698	1.0085	1.3982	0.7217
360.0	5.8238	-4.5171	6.1982	1.0095	1.3977	0.7212
370.0	6.8337	-3.7944	6.2259	1.0105	1.3972	0.7207
380.0	7.8447	-3.0707	6.2529	1.0116	1.3966	0.7203
390.0	8.8569	-2.3458	6.2791	1.0127	1.3959	0.7199
400.0	9.8702	-1.6197	6.3048	1.0140	1.3953	0.7195
410.0	1.0885 +2	-8.9229 +0	6.3299	1.0153	1.3945	0.7191
420.0	1.1901	-1.6350	6.3543	1.0168	1.3938	0.7188
430.0	1.2918	5.6674	6.3783	1.0182	1.3930	0.7185
440.0	1.3937	1.2985 +1	6.4017	1.0198	1.3921	0.7182
450.0	1.4958	2.0319	6.4246	1.0214	1.3913	0.7179
460.0	1.5980	2.7669	6.4471	1.0231	1.3904	0.7177
470.0	1.7004	3.5036	6.4691	1.0249	1.3894	0.7175
480.0	1.8030	4.2421	6.4907	1.0267	1.3885	0.7173
490.0	1.9058	4.9825	6.5119	1.0285	1.3875	0.7171
500.0	2.0087	5.7247	6.5327	1.0304	1.3865	0.7170
510.0	2.1119	6.4689	6.5531	1.0324	1.3855	0.7169
520.0	2.2152	7.2150	6.5732	1.0344	1.3845	0.7167
530.0	2.3187	7.9632	6.5929	1.0365	1.3834	0.7167
540.0	2.4225	8.7134	6.6123	1.0385	1.3823	0.7166
550.0	2.5264	9.4658	6.6314	1.0407	1.3813	0.7165
560.0	2.6306	1.0220 +2	6.6502	1.0428	1.3802	0.7165
580.0	2.8396	1.1736	6.6868	1.0473	1.3780	0.7165
600.0	3.0495	1.3260	6.7224	1.0518	1.3757	0.7165
620.0	3.2603	1.4794	6.7570	1.0564	1.3735	0.7165
640.0	3.4721	1.6337	6.7906	1.0611	1.3712	0.7166
660.0	3.6848	1.7890	6.8233	1.0659	1.3689	0.7168
680.0	3.8984	1.9452	6.8552	1.0707	1.3667	0.7169
700.0	4.1131	2.1023	6.8863	1.0755	1.3644	0.7171
720.0	4.3286	2.2605	6.9167	1.0803	1.3622	0.7172
740.0	4.5452	2.4196	6.9464	1.0851	1.3600	0.7174
760.0	4.7627	2.5796	6.9754	1.0900	1.3578	0.7176
780.0	4.9812	2.7406	7.0037	1.0947	1.3557	0.7177
800.0	5.2006	2.9026	7.0315	1.0995	1.3537	0.7178
820.0	5.4210	3.0655	7.0587	1.1042	1.3516	0.7179
840.0	5.6423	3.2294	7.0854	1.1088	1.3496	0.7179
860.0	5.8645	3.3941	7.1115	1.1134	1.3477	0.7179
880.0	6.0876	3.5598	7.1372	1.1178	1.3458	0.7179
900.0	6.3116	3.7264	7.1623	1.1223	1.3440	0.7178
920.0	6.5365	3.8938	7.1871	1.1266	1.3422	0.7176
940.0	6.7622	4.0621	7.2113	1.1308	1.3405	0.7174
960.0	6.9888	4.2312	7.2352	1.1349	1.3389	0.7172
980.0	7.2162	4.4012	7.2586	1.1389	1.3373	0.7169
1000.0	7.4444	4.5719	7.2817	1.1428	1.3357	0.7165

THERMOPHYSICAL PROPERTIES OF AIR

ISOBAR: 2.0 MPa

T (K)	ρ (kg/m^3)		a (m/s)	μ (Pa·s)		k (J/s·m·K)		$\bar{m}$
1020.0	6.8307	+0	625.0	4.3745	−5	0.0697	+0	28.9644
1040.0	6.6993		630.8	4.4309		0.0708		28.9644
1060.0	6.5729		636.5	4.4870		0.0719		28.9644
1080.0	6.4512		642.2	4.5426		0.0730		28.9644
1100.0	6.3339		647.8	4.5978		0.0741		28.9644
1120.0	6.2208		653.3	4.6526		0.0754		28.9644
1140.0	6.1116		658.8	4.7070		0.0765		28.9644
1160.0	6.0063		664.2	4.7610		0.0776		28.9644
1180.0	5.9045		669.6	4.8147		0.0787		28.9644
1200.0	5.8061		674.9	4.8680		0.0797		28.9644
1220.0	5.7109		680.2	4.9209		0.0808		28.9644
1240.0	5.6188		685.5	4.9735		0.0819		28.9644
1260.0	5.5296		690.6	5.0258		0.0830		28.9644
1280.0	5.4432		695.8	5.0778		0.0841		28.9644
1300.0	5.3594		700.9	5.1294		0.0851		28.9644
1320.0	5.2782		705.9	5.1807		0.0862		28.9644
1340.0	5.1995		710.9	5.2318		0.0873		28.9644
1360.0	5.1230		715.9	5.2825		0.0884		28.9644
1380.0	5.0487		720.8	5.3329		0.0894		28.9644
1400.0	4.9766		725.7	5.3831		0.0905		28.9644
1420.0	4.9065		730.6	5.4330		0.0916		28.9644
1440.0	4.8384		735.4	5.4827		0.0927		28.9644
1460.0	4.7721		740.1	5.5320		0.0937		28.9644
1480.0	4.7076		744.9	5.5811		0.0948		28.9644
1500.0	4.6448		749.6	5.6300		0.0959		28.9644
1520.0	4.5837		754.2	5.6786		0.0969		28.9644
1540.0	4.5242		758.8	5.7270		0.0980		28.9644
1560.0	4.4662		763.4	5.7751		0.0991		28.9644
1580.0	4.4097		768.0	5.8231		0.1002		28.9644
1600.0	4.3545		772.5	5.8707		0.1013		28.9644
1620.0	4.3008		777.0	5.9182		0.1023		28.9644
1640.0	4.2483		781.4	5.9654		0.1034		28.9644
1660.0	4.1971		785.9	6.0125		0.1045		28.9643
1680.0	4.1472		790.2	6.0593		0.1056		28.9643
1700.0	4.0984		794.6	6.1059		0.1067		28.9643
1720.0	4.0507		798.9	6.1523		0.1078		28.9643
1740.0	4.0042		803.2	6.1985		0.1089		28.9643
1760.0	3.9587		807.5	6.2445		0.1099		28.9643
1780.0	3.9142		811.7	6.2903		0.1110		28.9642
1800.0	3.8707		816.0	6.3360		0.1121		28.9642
1820.0	3.8281		820.1	6.3814		0.1132		28.9642
1840.0	3.7865		824.3	6.4267		0.1143		28.9641
1860.0	3.7458		828.4	6.4717		0.1154		28.9641
1880.0	3.7059		832.5	6.5166		0.1165		28.9640
1900.0	3.6669		836.6	6.5614		0.1176		28.9639
1920.0	3.6287		840.7	6.6059		0.1187		28.9639
1940.0	3.5913		844.7	6.6503		0.1198		28.9638
1960.0	3.5546		848.7	6.6945		0.1209		28.9637
1980.0	3.5187		852.7	6.7385		0.1220		28.9635
2000.0	3.4835		856.6	6.7824		0.1231		28.9634

THERMOPHYSICAL PROPERTIES OF AIR

ISOBAR: 2.0 MPa

T (K)	h (J/kg)	e (J/kg)	s (J/kg·K)	c_p (J/kg·K)	γ	Pr
1020.0	7.6733 +2	4.7434 +2	7.3043 +0	1.1465 +0	1.3343	0.7185
1040.0	7.9030	4.9156	7.3266	1.1501	1.3329	0.7185
1060.0	8.1333	5.0885	7.3486	1.1537	1.3315	0.7185
1080.0	8.3644	5.2622	7.3702	1.1572	1.3302	0.7185
1100.0	8.5962	5.4365	7.3914	1.1608	1.3288	0.7186
1120.0	8.8288	5.6116	7.4124	1.1643	1.3275	0.7187
1140.0	9.0620	5.7873	7.4330	1.1678	1.3262	0.7188
1160.0	9.2959	5.9638	7.4534	1.1712	1.3249	0.7190
1180.0	9.5305	6.1409	7.4734	1.1747	1.3237	0.7191
1200.0	9.7657	6.3187	7.4932	1.1781	1.3224	0.7192
1220.0	1.0002 +3	6.4973	7.5127	1.1816	1.3212	0.7194
1240.0	1.0238	6.6765	7.5319	1.1850	1.3200	0.7195
1260.0	1.0476	6.8564	7.5509	1.1884	1.3188	0.7197
1280.0	1.0714	7.0369	7.5697	1.1918	1.3176	0.7199
1300.0	1.0952	7.2182	7.5882	1.1952	1.3164	0.7201
1320.0	1.1192	7.4001	7.6064	1.1986	1.3152	0.7202
1340.0	1.1432	7.5827	7.6245	1.2020	1.3140	0.7204
1360.0	1.1673	7.7660	7.6423	1.2054	1.3128	0.7206
1380.0	1.1914	7.9500	7.6599	1.2089	1.3117	0.7208
1400.0	1.2156	8.1347	7.6774	1.2123	1.3105	0.7210
1420.0	1.2399	8.3200	7.6946	1.2157	1.3094	0.7212
1440.0	1.2642	8.5061	7.7116	1.2192	1.3082	0.7214
1460.0	1.2887	8.6928	7.7284	1.2226	1.3071	0.7216
1480.0	1.3131	8.8802	7.7451	1.2261	1.3060	0.7218
1500.0	1.3377	9.0683	7.7616	1.2295	1.3048	0.7220
1520.0	1.3623	9.2571	7.7779	1.2330	1.3037	0.7222
1540.0	1.3870 ·	9.4466	7.7940	1.2365	1.3026	0.7224
1560.0	1.4118	9.6368	7.8100	1.2400	1.3015	0.7225
1580.0	1.4366	9.8277	7.8258	1.2435	1.3004	0.7227
1600.0	1.4615	1.0019 +3	7.8415	1.2470	1.2993	0.7229
1620.0	1.4865	1.0212	7.8570	1.2506	1.2982	0.7231
1640.0	1.5115	1.0405	7.8724	1.2541	1.2971	0.7232
1660.0	1.5367	1.0598	7.8876	1.2577	1.2960	0.7234
1680.0	1.5619	1.0793	7.9027	1.2613	1.2949	0.7236
1700.0	1.5871	1.0988	7.9176	1.2649	1.2939	0.7237
1720.0	1.6125	1.1184	7.9324	1.2685	1.2928	0.7239
1740.0	1.6379	1.1380	7.9471	1.2721	1.2917	0.7240
1760.0	1.6633	1.1578	7.9617	1.2758	1.2906	0.7242
1780.0	1.6889	1.1776	7.9761	1.2794	1.2896	0.7243
1800.0	1.7145	1.1975	7.9904	1.2831	1.2885	0.7245
1820.0	1.7402	1.2174	8.0046	1.2868	1.2875	0.7246
1840.0	1.7660	1.2374	8.0187	1.2906	1.2864	0.7247
1860.0	1.7918	1.2575	8.0327	1.2943	1.2853	0.7249
1880.0	1.8178	1.2777	8.0466	1.2981	1.2843	0.7250
1900.0	1.8438	1.2980	8.0603	1.3020	1.2833	0.7251
1920.0	1.8698	1.3183	8.0740	1.3058	1.2822	0.7252
1940.0	1.8960	1.3387	8.0875	1.3097	1.2812	0.7253
1960.0	1.9222	1.3592	8.1010	1.3137	1.2801	0.7254
1980.0	1.9485	1.3798	8.1143	1.3177	1.2791	0.7255
2000.0	1.9749	1.4004	8.1276	1.3217	1.2780	0.7256

THERMOPHYSICAL PROPERTIES OF AIR

ISOBAR: 2.0 MPa

T (K)	ρ (kg/m^3)	a (m/s)	μ (Pa·s)	k (J/s·m·K)	$\bar{m}$
2040.0	3.4152 +0	864.4	6.8697 -5	0.1253 +0	28.9631
2080.0	3.3494	872.1	6.9564	0.1285	28.9626
2120.0	3.2862	879.8	7.0425	0.1310	28.9620
2160.0	3.2252	887.3	7.1280	0.1336	28.9613
2200.0	3.1665	894.8	7.2129	0.1362	28.9604
2240.0	3.1098	902.1	7.2973	0.1390	28.9593
2280.0	3.0551	909.4	7.3812	0.1418	28.9579
2320.0	3.0023	916.5	7.4645	0.1447	28.9563
2360.0	2.9512	923.6	7.5473	0.1478	28.9542
2400.0	2.9018	930.6	7.6296	0.1510	28.9518
2440.0	2.8539	937.5	7.7115	0.1544	28.9489
2480.0	2.8075	944.3	7.7929	0.1579	28.9454
2520.0	2.7626	951.0	7.8738	0.1617	28.9413
2560.0	2.7190	957.7	7.9544	0.1656	28.9365
2600.0	2.6766	964.3	8.0345	0.1698	28.9309
2640.0	2.6355	970.8	8.1142	0.1743	28.9245
2680.0	2.5955	977.2	8.1935	0.1790	28.9170
2720.0	2.5565	983.6	8.2725	0.1841	28.9084
2760.0	2.5186	989.9	8.3511	0.1894	28.8987
2800.0	2.4817	996.2	8.4294	0.1951	28.8875
2840.0	2.4457	1002.4	8.5073	0.2012	28.8750
2880.0	2.4105	1008.6	8.5850	0.2076	28.8608
2920.0	2.3762	1014.8	8.6624	0.2144	28.8449
2960.0	2.3427	1021.0	8.7395	0.2216	28.8272
3000.0	2.3098	1027.1	8.8164	0.2292	28.8075
3040.0	2.2777	1033.2	8.8931	0.2373	28.7856
3080.0	2.2463	1039.4	8.9695	0.2457	28.7615
3120.0	2.2154	1045.6	9.0458	0.2546	28.7350
3160.0	2.1852	1051.8	9.1219	0.2639	28.7060
3200.0	2.1555	1058.1	9.1978	0.2737	28.6743
3240.0	2.1263	1064.4	9.2737	0.2838	28.6398
3280.0	2.0976	1070.7	9.3493	0.2945	28.6024
3320.0	2.0694	1077.2	9.4250	0.3054	28.5620
3360.0	2.0417	1083.7	9.5005	0.3165	28.5185
3400.0	2.0144	1090.3	9.5761	0.3280	28.4719
3440.0	1.9874	1096.9	9.6515	0.3397	28.4220
3480.0	1.9609	1103.7	9.7270	0.3516	28.3688
3520.0	1.9348	1110.6	9.8025	0.3636	28.3123
3560.0	1.9090	1117.6	9.8781	0.3766	28.2524
3600.0	1.8836	1124.7	9.9537	0.3889	28.1892
3640.0	1.8585	1131.9	1.0029 -4	0.4011	28.1228
3680.0	1.8337	1139.2	1.0105	0.4132	28.0530
3720.0	1.8093	1146.7	1.0181	0.4251	27.9802
3760.0	1.7852	1154.2	1.0257	0.4367	27.9042
3800.0	1.7614	1161.9	1.0333	0.4480	27.8253
3840.0	1.7379	1169.7	1.0409	0.4588	27.7436
3880.0	1.7148	1177.7	1.0485	0.4690	27.6593
3920.0	1.6920	1185.7	1.0561	0.4787	27.5726
3960.0	1.6695	1193.9	1.0638	0.4877	27.4837
4000.0	1.6473	1202.2	1.0714	0.4959	27.3928

THERMOPHYSICAL PROPERTIES OF AIR

ISOBAR: 2.0 MPa

T (K)	h (J/kg)	e (J/kg)	s (J/kg·K)	c_p (J/kg·K)	γ	Pr
2040.0	2.0280 +3	1.4420 +3	8.1539 +0	1.3299 +0	1.2760	0.7258
2080.0	2.0813	1.4838	8.1798	1.3384	1.2739	0.7243
2120.0	2.1350	1.5260	8.2053	1.3472	1.2718	0.7240
2160.0	2.1891	1.5686	8.2306	1.3563	1.2697	0.7236
2200.0	2.2436	1.6115	8.2556	1.3658	1.2675	0.7231
2240.0	2.2984	1.6548	8.2803	1.3758	1.2654	0.7224
2280.0	2.3536	1.6986	8.3047	1.3864	1.2632	0.7216
2320.0	2.4093	1.7427	8.3289	1.3975	1.2610	0.7207
2360.0	2.4654	1.7873	8.3529	1.4093	1.2588	0.7196
2400.0	2.5221	1.8324	8.3767	1.4219	1.2565	0.7184
2440.0	2.5792	1.8779	8.4003	1.4354	1.2542	0.7169
2480.0	2.6369	1.9241	8.4238	1.4498	1.2518	0.7153
2520.0	2.6952	1.9708	8.4471	1.4652	1.2494	0.7134
2560.0	2.7541	2.0181	8.4703	1.4818	1.2469	0.7114
2600.0	2.8138	2.0661	8.4934	1.4996	1.2444	0.7092
2640.0	2.8741	2.1147	8.5164	1.5187	1.2419	0.7068
2680.0	2.9353	2.1642	8.5394	1.5392	1.2393	0.7041
2720.0	2.9973	2.2145	8.5624	1.5613	1.2367	0.7013
2760.0	3.0602	2.2656	8.5854	1.5849	1.2341	0.6983
2800.0	3.1241	2.3177	8.6083	1.6102	1.2314	0.6952
2840.0	3.1890	2.3707	8.6314	1.6372	1.2288	0.6919
2880.0	3.2551	2.4249	8.6545	1.6661	1.2262	0.6885
2920.0	3.3223	2.4801	8.6777	1.6968	1.2235	0.6850
2960.0	3.3909	2.5366	8.7010	1.7294	1.2209	0.6814
3000.0	3.4607	2.5943	8.7244	1.7640	1.2184	0.6778
3040.0	3.5320	2.6533	8.7480	1.8005	1.2159	0.6741
3080.0	3.6048	2.7138	8.7718	1.8390	1.2134	0.6705
3120.0	3.6792	2.7758	8.7958	1.8795	1.2110	0.6669
3160.0	3.7552	2.8393	8.8200	1.9219	1.2087	0.6633
3200.0	3.8329	2.9044	8.8444	1.9662	1.2065	0.6599
3240.0	3.9125	2.9713	8.8692	2.0122	1.2044	0.6566
3280.0	3.9939	3.0398	8.8941	2.0600	1.2024	0.6536
3320.0	4.0773	3.1102	8.9194	2.1094	1.2006	0.6506
3360.0	4.1627	3.1825	8.9450	2.1603	1.1988	0.6478
3400.0	4.2502	3.2566	8.9708	2.2124	1.1972	0.6453
3440.0	4.3397	3.3327	8.9970	2.2657	1.1957	0.6430
3480.0	4.4314	3.4108	9.0235	2.3198	1.1944	0.6409
3520.0	4.5253	3.4909	9.0504	2.3745	1.1932	0.6391
3560.0	4.6214	3.5730	9.0775	2.4297	1.1922	0.6373
3600.0	4.7197	3.6572	9.1050	2.4849	1.1913	0.6360
3640.0	4.8202	3.7433	9.1327	2.5399	1.1905	0.6351
3680.0	4.9229	3.8314	9.1608	2.5942	1.1899	0.6344
3720.0	5.0277	3.9215	9.1891	2.6477	1.1895	0.6341
3760.0	5.1347	4.0136	9.2177	2.6998	1.1891	0.6341
3800.0	5.2437	4.1074	9.2465	2.7502	1.1890	0.6344
3840.0	5.3546	4.2031	9.2756	2.7986	1.1890	0.6350
3880.0	5.4675	4.3004	9.3048	2.8444	1.1891	0.6358
3920.0	5.5822	4.3993	9.3342	2.8874	1.1894	0.6370
3960.0	5.6985	4.4997	9.3638	2.9271	1.1898	0.6385
4000.0	5.8163	4.6014	9.3934	2.9633	1.1903	0.6402

THERMOPHYSICAL PROPERTIES OF AIR

ISOBAR: 2.0 MPa

T (K)	ρ (kg/m^3)	a (m/s)	μ (Pa·s)	k (J/s·m·K)	$\bar{m}$
4040.0	1.6255 +0	1210.5	1.0791 −4	0.5033 +0	27.3002
4080.0	1.6040	1219.0	1.0868	0.5097	27.2062
4120.0	1.5829	1227.6	1.0944	0.5154	27.1111
4160.0	1.5621	1236.3	1.1021	0.5201	27.0152
4200.0	1.5417	1245.1	1.1098	0.5239	26.9187
4240.0	1.5217	1253.9	1.1175	0.5268	26.8220
4280.0	1.5020	1262.8	1.1252	0.5287	26.7253
4320.0	1.4828	1271.8	1.1329	0.5297	26.6290
4360.0	1.4639	1280.8	1.1406	0.5298	26.5333
4400.0	1.4454	1289.9	1.1483	0.5292	26.4386
4440.0	1.4273	1299.0	1.1560	0.5278	26.3450
4480.0	1.4096	1308.1	1.1637	0.5258	26.2528
4520.0	1.3923	1317.2	1.1714	0.5232	26.1623
4560.0	1.3754	1326.4	1.1790	0.5201	26.0735
4600.0	1.3589	1335.5	1.1867	0.5167	25.9867
4640.0	1.3428	1344.5	1.1943	0.5130	25.9021
4680.0	1.3271	1353.6	1.2019	0.5092	25.8196
4720.0	1.3118	1362.6	1.2096	0.5053	25.7394
4760.0	1.2968	1371.5	1.2172	0.5015	25.6615
4800.0	1.2822	1380.3	1.2247	0.4978	25.5860
4840.0	1.2680	1389.1	1.2323	0.4944	25.5129
4880.0	1.2541	1397.7	1.2398	0.4912	25.4421
4920.0	1.2406	1406.2	1.2474	0.4885	25.3737
4960.0	1.2273	1414.6	1.2549	0.4863	25.3075
5000.0	1.2144	1422.9	1.2624	0.4847	25.2434
5040.0	1.2018	1431.0	1.2698	0.4836	25.1815
5080.0	1.1895	1439.0	1.2773	0.4832	25.1216
5120.0	1.1775	1446.8	1.2847	0.4835	25.0635
5160.0	1.1658	1454.5	1.2922	0.4846	25.0072
5200.0	1.1543	1461.9	1.2996	0.4864	24.9526
5240.0	1.1430	1469.2	1.3070	0.4890	24.8994
5280.0	1.1320	1476.4	1.3144	0.4925	24.8477
5320.0	1.1212	1483.3	1.3218	0.4969	24.7972
5360.0	1.1106	1490.2	1.3292	0.5021	24.7478
5400.0	1.1002	1496.8	1.3366	0.5083	24.6994
5440.0	1.0901	1503.3	1.3440	0.5153	24.6518
5480.0	1.0800	1509.7	1.3514	0.5233	24.6050
5520.0	1.0702	1515.9	1.3588	0.5322	24.5587
5560.0	1.0605	1522.0	1.3662	0.5421	24.5129
5600.0	1.0510	1528.0	1.3737	0.5529	24.4673
5640.0	1.0416	1533.9	1.3811	0.5647	24.4220
5680.0	1.0323	1539.7	1.3886	0.5774	24.3768
5720.0	1.0232	1545.5	1.3960	0.5912	24.3315
5760.0	1.0142	1551.1	1.4035	0.6060	24.2861
5800.0	1.0053	1556.8	1.4110	0.6217	24.2404
5840.0	9.9655 −1	1562.4	1.4186	0.6385	24.1943
5880.0	9.8786	1567.9	1.4261	0.6563	24.1477
5920.0	9.7927	1573.5	1.4337	0.6752	24.1006
5960.0	9.7077	1579.0	1.4414	0.6950	24.0527
6000.0	9.6235	1584.6	1.4490	0.7160	24.0041

THERMOPHYSICAL PROPERTIES OF AIR

ISOBAR: 2.0 MPa

T (K)	h (J/kg)	e (J/kg)	s (J/kg·K)	c_p (J/kg·K)	γ	Pr
4040.0	5.9355 +3	4.7042 +3	9.4230 +0	2.9954 +0	1.1910	0.6422
4080.0	6.0559	4.8082	9.4527	3.0234	1.1918	0.6443
4120.0	6.1773	4.9129	9.4823	3.0469	1.1927	0.6468
4160.0	6.2996	5.0184	9.5118	3.0657	1.1938	0.6494
4200.0	6.4225	5.1243	9.5412	3.0796	1.1950	0.6522
4240.0	6.5459	5.2306	9.5705	3.0887	1.1963	0.6551
4280.0	6.6695	5.3371	9.5995	3.0928	1.1977	0.6581
4320.0	6.7932	5.4435	9.6282	3.0921	1.1992	0.6612
4360.0	6.9168	5.5497	9.6567	3.0866	1.2008	0.6643
4400.0	7.0401	5.6554	9.6849	3.0766	1.2024	0.6674
4440.0	7.1629	5.7607	9.7127	3.0622	1.2042	0.6705
4480.0	7.2850	5.8652	9.7400	3.0440	1.2060	0.6736
4520.0	7.4063	5.9689	9.7670	3.0221	1.2079	0.6765
4560.0	7.5267	6.0717	9.7935	2.9971	1.2098	0.6792
4600.0	7.6461	6.1733	9.8196	2.9694	1.2118	0.6818
4640.0	7.7643	6.2738	9.8452	2.9395	1.2138	0.6842
4680.0	7.8812	6.3731	9.8703	2.9079	1.2157	0.6863
4720.0	7.9969	6.4712	9.8949	2.8752	1.2177	0.6881
4760.0	8.1112	6.5679	9.9190	2.8418	1.2196	0.6896
4800.0	8.2242	6.6634	9.9426	2.8083	1.2215	0.6908
4840.0	8.3359	6.7575	9.9658	2.7751	1.2233	0.6917
4880.0	8.4462	6.8504	9.9885	2.7427	1.2250	0.6921
4920.0	8.5553	6.9421	1.0011 +1	2.7116	1.2266	0.6922
4960.0	8.6632	7.0326	1.0033	2.6820	1.2281	0.6920
5000.0	8.7699	7.1220	1.0054	2.6544	1.2294	0.6913
5040.0	8.8756	7.2104	1.0075	2.6291	1.2306	0.6903
5080.0	8.9803	7.2978	1.0096	2.6064	1.2316	0.6889
5120.0	9.0841	7.3845	1.0116	2.5864	1.2324	0.6872
5160.0	9.1872	7.4705	1.0136	2.5695	1.2331	0.6852
5200.0	9.2897	7.5559	1.0156	2.5557	1.2335	0.6828
5240.0	9.3917	7.6408	1.0176	2.5452	1.2337	0.6802
5280.0	9.4934	7.7254	1.0195	2.5381	1.2337	0.6773
5320.0	9.5948	7.8098	1.0214	2.5345	1.2335	0.6742
5360.0	9.6962	7.8942	1.0233	2.5345	1.2331	0.6709
5400.0	9.7976	7.9786	1.0252	2.5381	1.2325	0.6674
5440.0	9.8993	8.0633	1.0271	2.5453	1.2317	0.6638
5480.0	1.0001 +4	8.1483	1.0289	2.5563	1.2308	0.6602
5520.0	1.0104	8.2338	1.0308	2.5709	1.2296	0.6564
5560.0	1.0207	8.3199	1.0327	2.5892	1.2284	0.6526
5600.0	1.0311	8.4068	1.0345	2.6113	1.2269	0.6488
5640.0	1.0416	8.4946	1.0364	2.6370	1.2254	0.6450
5680.0	1.0522	8.5834	1.0383	2.6665	1.2237	0.6412
5720.0	1.0629	8.6734	1.0401	2.6996	1.2220	0.6374
5760.0	1.0738	8.7648	1.0420	2.7364	1.2201	0.6338
5800.0	1.0848	8.8576	1.0439	2.7769	1.2182	0.6302
5840.0	1.0960	8.9520	1.0459	2.8211	1.2163	0.6267
5880.0	1.1074	9.0481	1.0478	2.8689	1.2143	0.6234
5920.0	1.1190	9.1461	1.0498	2.9204	1.2123	0.6201
5960.0	1.1308	9.2461	1.0518	2.9756	1.2102	0.6171
6000.0	1.1428	9.3483	1.0538	3.0344	1.2082	0.6141

THERMOPHYSICAL PROPERTIES OF AIR

ISOBAR: 5.0 MPa

T (K)	ρ (kg/m^3)	a (m/s)	μ (Pa·s)	k (J/s·m·K)	$\bar{m}$
288.15	6.0448 +1	340.3	1.7979 −5	0.0249 +0	28.9644
300.0	5.8061	347.2	1.8554	0.0257	28.9644
310.0	5.6188	352.9	1.9030	0.0264	28.9644
320.0	5.4432	358.5	1.9500	0.0271	28.9644
330.0	5.2782	364.1	1.9962	0.0278	28.9644
340.0	5.1230	369.5	2.0419	0.0285	28.9644
350.0	4.9766	374.8	2.0869	0.0292	28.9644
360.0	4.8384	380.1	2.1313	0.0298	28.9644
370.0	4.7076	385.2	2.1751	0.0305	28.9644
380.0	4.5837	390.3	2.2184	0.0312	28.9644
390.0	4.4662	395.3	2.2612	0.0318	28.9644
400.0	4.3545	400.3	2.3035	0.0325	28.9644
410.0	4.2483	405.1	2.3454	0.0331	28.9644
420.0	4.1472	409.9	2.3867	0.0338	28.9644
430.0	4.0507	414.7	2.4276	0.0344	28.9644
440.0	3.9587	419.3	2.4681	0.0350	28.9644
450.0	3.8707	423.9	2.5082	0.0357	28.9644
460.0	3.7866	428.5	2.5479	0.0363	28.9644
470.0	3.7060	433.0	2.5872	0.0370	28.9644
480.0	3.6288	437.4	2.6261	0.0376	28.9644
490.0	3.5547	441.8	2.6647	0.0382	28.9644
500.0	3.4836	446.1	2.7029	0.0388	28.9644
510.0	3.4153	450.4	2.7408	0.0395	28.9644
520.0	3.3497	454.6	2.7783	0.0401	28.9644
530.0	3.2864	458.8	2.8155	0.0407	28.9644
540.0	3.2256	462.9	2.8525	0.0413	28.9644
550.0	3.1669	467.0	2.8891	0.0420	28.9644
560.0	3.1104	471.0	2.9254	0.0426	28.9644
580.0	3.0031	479.0	2.9972	0.0438	28.9644
600.0	2.9030	486.8	3.0680	0.0450	28.9644
620.0	2.8094	494.4	3.1377	0.0463	28.9644
640.0	2.7216	501.9	3.2065	0.0475	28.9644
660.0	2.6391	509.3	3.2743	0.0487	28.9644
680.0	2.5615	516.5	3.3412	0.0499	28.9644
700.0	2.4883	523.6	3.4073	0.0511	28.9644
720.0	2.4192	530.6	3.4726	0.0523	28.9644
740.0	2.3538	537.5	3.5371	0.0535	28.9644
760.0	2.2919	544.3	3.6008	0.0547	28.9644
780.0	2.2331	551.0	3.6639	0.0559	28.9644
800.0	2.1773	557.5	3.7263	0.0571	28.9644
820.0	2.1242	564.1	3.7880	0.0583	28.9644
840.0	2.0736	570.5	3.8490	0.0594	28.9644
860.0	2.0254	576.8	3.9095	0.0606	28.9644
880.0	1.9793	583.1	3.9693	0.0618	28.9644
900.0	1.9354	589.3	4.0286	0.0630	28.9644
920.0	1.8933	595.4	4.0873	0.0641	28.9644
940.0	1.8530	601.4	4.1455	0.0653	28.9644
960.0	1.8144	607.4	4.2032	0.0665	28.9645
980.0	1.7774	613.3	4.2603	0.0676	28.9644
1000.0	1.7418	619.2	4.3170	0.0688	28.9644

THERMOPHYSICAL PROPERTIES OF AIR

ISOBAR: 5.0 MPa

T (K)	h (J/kg)	e (J/kg)	s (J/kg·K)	c_p (J/kg·K)	γ	Pr
288.15	-1.4106 +1	-9.6876 +1	5.7109 +0	1.0052 +0	1.4001	0.7254
300.0	-2.1924 +0	-8.8367	5.7514	1.0055	1.3999	0.7246
310.0	7.8646	-8.1182	5.7844	1.0059	1.3997	0.7240
320.0	1.7926 +1	-7.3994	5.8163	1.0064	1.3994	0.7233
330.0	2.7993	-6.6799	5.8473	1.0070	1.3991	0.7228
340.0	3.8067	-5.9598	5.8774	1.0077	1.3987	0.7222
350.0	4.8148	-5.2389	5.9066	1.0085	1.3982	0.7217
360.0	5.8238	-4.5171	5.9350	1.0095	1.3977	0.7212
370.0	6.8337	-3.7944	5.9627	1.0105	1.3972	0.7207
380.0	7.8447	-3.0707	5.9897	1.0116	1.3966	0.7203
390.0	8.8569	-2.3458	6.0159	1.0127	1.3959	0.7199
400.0	9.8702	-1.6197	6.0416	1.0140	1.3953	0.7195
410.0	1.0885 +2	-8.9229 +0	6.0667	1.0153	1.3945	0.7191
420.0	1.1901	-1.6350	6.0911	1.0168	1.3938	0.7188
430.0	1.2918	5.6674	6.1151	1.0182	1.3930	0.7185
440.0	1.3937	1.2985 +1	6.1385	1.0198	1.3921	0.7182
450.0	1.4958	2.0319	6.1614	1.0214	1.3913	0.7179
460.0	1.5980	2.7669	6.1839	1.0231	1.3904	0.7177
470.0	1.7004	3.5036	6.2059	1.0249	1.3894	0.7175
480.0	1.8030	4.2421	6.2275	1.0267	1.3885	0.7173
490.0	1.9058	4.9825	6.2487	1.0285	1.3875	0.7171
500.0	2.0087	5.7247	6.2695	1.0304	1.3865	0.7170
510.0	2.1119	6.4689	6.2899	1.0324	1.3855	0.7169
520.0	2.2152	7.2150	6.3100	1.0344	1.3845	0.7167
530.0	2.3187	7.9632	6.3297	1.0365	1.3834	0.7167
540.0	2.4225	8.7134	6.3491	1.0385	1.3823	0.7166
550.0	2.5264	9.4658	6.3682	1.0407	1.3813	0.7165
560.0	2.6306	1.0220 +2	6.3870	1.0428	1.3802	0.7165
580.0	2.8396	1.1736	6.4236	1.0473	1.3780	0.7165
600.0	3.0495	1.3260	6.4592	1.0518	1.3757	0.7165
620.0	3.2603	1.4794	6.4938	1.0564	1.3735	0.7165
640.0	3.4721	1.6337	6.5274	1.0611	1.3712	0.7166
660.0	3.6848	1.7890	6.5601	1.0659	1.3689	0.7168
680.0	3.8984	1.9452	6.5920	1.0707	1.3667	0.7169
700.0	4.1131	2.1023	6.6231	1.0755	1.3644	0.7171
720.0	4.3286	2.2605	6.6535	1.0803	1.3622	0.7172
740.0	4.5452	2.4196	6.6831	1.0851	1.3600	0.7174
760.0	4.7627	2.5796	6.7122	1.0900	1.3573	0.7176
780.0	4.9812	2.7406	6.7405	1.0947	1.3557	0.7177
800.0	5.2006	2.9026	6.7683	1.0995	1.3537	0.7178
820.0	5.4210	3.0655	6.7955	1.1042	1.3516	0.7179
840.0	5.6423	3.2294	6.8222	1.1088	1.3496	0.7179
860.0	5.8645	3.3941	6.8483	1.1134	1.3477	0.7179
880.0	6.0876	3.5598	6.8740	1.1178	1.3458	0.7179
900.0	6.3116	3.7264	6.8991	1.1223	1.3440	0.7178
920.0	6.5365	3.8938	6.9238	1.1266	1.3422	0.7176
940.0	6.7622	4.0621	6.9481	1.1308	1.3405	0.7174
960.0	6.9888	4.2312	6.9720	1.1349	1.3389	0.7172
980.0	7.2162	4.4012	6.9954	1.1389	1.3373	0.7169
1000.0	7.4444	4.5719	7.0185	1.1428	1.3357	0.7165

THERMOPHYSICAL PROPERTIES OF AIR

ISOBAR: 5.0 MPa

T (K)	ρ (kg/m^3)	a (m/s)	μ (Pa·s)	k (J/s·m·K)	$\bar{m}$
1020.0	1.7077 +1	625.0	4.3745 -5	0.0697 +0	28.9644
1040.0	1.6748	630.8	4.4309	0.0708	28.9644
1060.0	1.6432	636.5	4.4870	0.0719	28.9644
1080.0	1.6128	642.2	4.5426	0.0730	28.9644
1100.0	1.5835	647.8	4.5978	0.0741	28.9644
1120.0	1.5552	653.3	4.6526	0.0754	28.9644
1140.0	1.5279	658.8	4.7070	0.0765	28.9644
1160.0	1.5016	664.2	4.7610	0.0776	28.9644
1180.0	1.4761	669.6	4.8147	0.0787	28.9644
1200.0	1.4515	674.9	4.8680	0.0797	28.9644
1220.0	1.4277	680.2	4.9209	0.0808	28.9644
1240.0	1.4047	685.5	4.9735	0.0819	28.9644
1260.0	1.3824	690.6	5.0258	0.0830	28.9644
1280.0	1.3608	695.8	5.0778	0.0841	28.9644
1300.0	1.3399	700.9	5.1294	0.0851	28.9644
1320.0	1.3196	705.9	5.1807	0.0862	28.9644
1340.0	1.2999	710.9	5.2318	0.0873	28.9644
1360.0	1.2807	715.9	5.2825	0.0884	28.9644
1380.0	1.2622	720.8	5.3329	0.0894	28.9644
1400.0	1.2442	725.7	5.3831	0.0905	28.9644
1420.0	1.2266	730.6	5.4330	0.0916	28.9644
1440.0	1.2096	735.4	5.4827	0.0927	28.9644
1460.0	1.1930	740.1	5.5320	0.0937	28.9644
1480.0	1.1769	744.9	5.5811	0.0948	28.9644
1500.0	1.1612	749.6	5.6300	0.0959	28.9644
1520.0	1.1459	754.2	5.6786	0.0969	28.9644
1540.0	1.1311	758.8	5.7270	0.0980	28.9644
1560.0	1.1165	763.4	5.7751	0.0991	28.9644
1580.0	1.1024	768.0	5.8231	0.1002	28.9644
1600.0	1.0886	772.5	5.8707	0.1013	28.9644
1620.0	1.0752	777.0	5.9182	0.1023	28.9644
1640.0	1.0621	781.4	5.9654	0.1034	28.9644
1660.0	1.0493	785.9	6.0125	0.1045	28.9644
1680.0	1.0368	790.3	6.0593	0.1056	28.9644
1700.0	1.0246	794.6	6.1059	0.1067	28.9643
1720.0	1.0127	798.9	6.1523	0.1078	28.9643
1740.0	1.0010	803.2	6.1985	0.1089	28.9643
1760.0	9.8967 +0	807.5	6.2445	0.1099	28.9643
1780.0	9.7855	811.8	6.2903	0.1110	28.9643
1800.0	9.6767	816.0	6.3360	0.1121	28.9643
1820.0	9.5704	820.2	6.3814	0.1132	28.9643
1840.0	9.4664	824.3	6.4267	0.1143	28.9642
1860.0	9.3646	828.5	6.4717	0.1154	28.9642
1880.0	9.2649	832.6	6.5166	0.1165	28.9642
1900.0	9.1674	836.7	6.5614	0.1176	28.9641
1920.0	9.0719	840.7	6.6059	0.1187	28.9641
1940.0	8.9783	844.7	6.6503	0.1198	28.9640
1960.0	8.8867	848.8	6.6945	0.1209	28.9639
1980.0	8.7969	852.7	6.7385	0.1220	28.9639
2000.0	8.7089	856.7	6.7824	0.1231	28.9638

THERMOPHYSICAL PROPERTIES OF AIR

ISOBAR: 5.0 MPa

T (K)	h (J/kg)	e (J/kg)	s (J/kg·K)	c_p (J/kg·K)	γ	Pr
1020.0	7.6733 +2	4.7434 +2	7.0411 +0	1.1465 +0	1.3343	0.7185
1040.0	7.9030	4.9156	7.0634	1.1501	1.3329	0.7185
1060.0	8.1333	5.0885	7.0854	1.1537	1.3315	0.7185
1080.0	8.3644	5.2622	7.1070	1.1572	1.3302	0.7185
1100.0	8.5962	5.4365	7.1282	1.1608	1.3288	0.7186
1120.0	8.8288	5.6116	7.1492	1.1643	1.3275	0.7187
1140.0	9.0620	5.7873	7.1698	1.1678	1.3262	0.7188
1160.0	9.2959	5.9638	7.1902	1.1712	1.3249	0.7190
1180.0	9.5305	6.1409	7.2102	1.1747	1.3237	0.7191
1200.0	9.7657	6.3187	7.2300	1.1781	1.3224	0.7192
1220.0	1.0002 +3	6.4973	7.2495	1.1816	1.3212	0.7194
1240.0	1.0238	6.6765	7.2687	1.1850	1.3200	0.7195
1260.0	1.0476	6.8564	7.2877	1.1884	1.3188	0.7197
1280.0	1.0714	7.0369	7.3065	1.1918	1.3176	0.7199
1300.0	1.0952	7.2182	7.3250	1.1952	1.3164	0.7201
1320.0	1.1192	7.4001	7.3432	1.1986	1.3152	0.7202
1340.0	1.1432	7.5827	7.3613	1.2020	1.3140	0.7204
1360.0	1.1673	7.7660	7.3791	1.2054	1.3128	0.7206
1380.0	1.1914	7.9500	7.3967	1.2089	1.3117	0.7208
1400.0	1.2156	8.1347	7.4142	1.2123	1.3105	0.7210
1420.0	1.2399	8.3200	7.4314	1.2157	1.3094	0.7212
1440.0	1.2642	8.5060	7.4484	1.2191	1.3082	0.7214
1460.0	1.2887	8.6928	7.4652	1.2226	1.3071	0.7216
1480.0	1.3131	8.8802	7.4819	1.2260	1.3060	0.7218
1500.0	1.3377	9.0683	7.4984	1.2295	1.3049	0.7220
1520.0	1.3623	9.2571	7.5147	1.2330	1.3037	0.7222
1540.0	1.3870	9.4466	7.5308	1.2365	1.3026	0.7224
1560.0	1.4118	9.6368	7.5468	1.2400	1.3015	0.7225
1580.0	1.4366	9.8277	7.5626	1.2435	1.3004	0.7227
1600.0	1.4615	1.0019 +3	7.5783	1.2470	1.2993	0.7229
1620.0	1.4865	1.0212	7.5938	1.2505	1.2982	0.7231
1640.0	1.5115	1.0405	7.6092	1.2540	1.2971	0.7232
1660.0	1.5367	1.0598	7.6244	1.2576	1.2960	0.7234
1680.0	1.5618	1.0793	7.6395	1.2611	1.2950	0.7236
1700.0	1.5871	1.0988	7.6544	1.2647	1.2939	0.7237
1720.0	1.6124	1.1184	7.6692	1.2683	1.2928	0.7239
1740.0	1.6378	1.1380	7.6839	1.2719	1.2918	0.7240
1760.0	1.6633	1.1578	7.6985	1.2755	1.2907	0.7242
1780.0	1.6889	1.1776	7.7129	1.2791	1.2896	0.7243
1800.0	1.7145	1.1974	7.7272	1.2827	1.2886	0.7245
1820.0	1.7402	1.2174	7.7414	1.2864	1.2875	0.7246
1840.0	1.7659	1.2374	7.7555	1.2901	1.2865	0.7247
1860.0	1.7918	1.2575	7.7695	1.2937	1.2855	0.7249
1880.0	1.8177	1.2776	7.7833	1.2974	1.2844	0.7250
1900.0	1.8437	1.2979	7.7971	1.3011	1.2834	0.7251
1920.0	1.8697	1.3182	7.8107	1.3049	1.2824	0.7252
1940.0	1.8959	1.3386	7.8242	1.3086	1.2814	0.7253
1960.0	1.9221	1.3591	7.8377	1.3124	1.2804	0.7254
1980.0	1.9484	1.3796	7.8510	1.3162	1.2794	0.7255
2000.0	1.9747	1.4002	7.8643	1.3200	1.2784	0.7256

THERMOPHYSICAL PROPERTIES OF AIR

ISOBAR: 5.0 MPa

T (K)	ρ (kg/m^3)	a (m/s)	μ (Pa·s)	k (J/s·m·K)	$\bar{m}$
2040.0	8.5381 +0	864.6	6.8697 −5	0.1253 +0	28.9635
2080.0	8.3738	872.3	6.9564	0.1275	28.9633
2120.0	8.2157	880.0	7.0425	0.1306	28.9629
2160.0	8.0634	887.6	7.1280	0.1330	28.9625
2200.0	7.9167	895.1	7.2129	0.1355	28.9619
2240.0	7.7751	902.5	7.2973	0.1380	28.9612
2280.0	7.6385	909.8	7.3811	0.1406	28.9603
2320.0	7.5065	917.1	7.4644	0.1433	28.9592
2360.0	7.3789	924.2	7.5472	0.1460	28.9580
2400.0	7.2556	931.3	7.6295	0.1488	28.9564
2440.0	7.1362	938.3	7.7113	0.1517	28.9546
2480.0	7.0205	945.3	7.7927	0.1548	28.9524
2520.0	6.9085	952.1	7.8736	0.1579	28.9498
2560.0	6.7998	958.9	7.9540	0.1612	28.9467
2600.0	6.6944	965.7	8.0340	0.1647	28.9432
2640.0	6.5920	972.3	8.1137	0.1683	28.9391
2680.0	6.4926	978.9	8.1929	0.1721	28.9344
2720.0	6.3959	985.5	8.2717	0.1760	28.9289
2760.0	6.3019	991.9	8.3502	0.1802	28.9227
2800.0	6.2103	998.4	8.4283	0.1846	28.9156
2840.0	6.1211	1004.7	8.5060	0.1892	28.9076
2880.0	6.0342	1011.1	8.5835	0.1940	28.8986
2920.0	5.9495	1017.3	8.6606	0.1991	28.8885
2960.0	5.8668	1023.6	8.7374	0.2044	28.8772
3000.0	5.7860	1029.8	8.8139	0.2101	28.8645
3040.0	5.7071	1036.0	8.8901	0.2160	28.8505
3080.0	5.6300	1042.2	8.9661	0.2222	28.8350
3120.0	5.5545	1048.4	9.0419	0.2287	28.8180
3160.0	5.4807	1054.6	9.1174	0.2354	28.7992
3200.0	5.4083	1060.8	9.1927	0.2425	28.7787
3240.0	5.3374	1067.0	9.2678	0.2499	28.7563
3280.0	5.2678	1073.2	9.3428	0.2576	28.7320
3320.0	5.1996	1079.4	9.4175	0.2655	28.7056
3360.0	5.1326	1085.7	9.4922	0.2738	28.6771
3400.0	5.0667	1092.0	9.5666	0.2825	28.6464
3440.0	5.0021	1098.4	9.6410	0.2913	28.6134
3480.0	4.9385	1104.8	9.7153	0.3003	28.5780
3520.0	4.8759	1111.2	9.7895	0.3095	28.5403
3560.0	4.8143	1117.8	9.8637	0.3189	28.5000
3600.0	4.7537	1124.3	9.9378	0.3285	28.4573
3640.0	4.6940	1131.0	1.0012 −4	0.3382	28.4120
3680.0	4.6351	1137.7	1.0086	0.3489	28.3642
3720.0	4.5771	1144.5	1.0160	0.3588	28.3138
3760.0	4.5200	1151.4	1.0234	0.3688	28.2608
3800.0	4.4636	1158.4	1.0308	0.3788	28.2053
3840.0	4.4080	1165.5	1.0382	0.3887	28.1473
3880.0	4.3532	1172.6	1.0456	0.3986	28.0868
3920.0	4.2991	1179.9	1.0530	0.4082	28.0239
3960.0	4.2458	1187.2	1.0605	0.4177	27.9587
4000.0	4.1932	1194.6	1.0679	0.4269	27.8912

THERMOPHYSICAL PROPERTIES OF AIR

ISOBAR: 5.0 MPa

T (K)	h (J/kg)	e (J/kg)	s (J/kg·K)	c_p (J/kg·K)	γ	Pr
2040.0	2.0277 +3	1.4417 +3	7.8905 +0	1.3277 +0	1.2764	0.7258
2080.0	2.0809	1.4834	7.9164	1.3356	1.2744	0.7259
2120.0	2.1345	1.5255	7.9419	1.3436	1.2724	0.7247
2160.C	2.1884	1.5679	7.9671	1.3518	1.2704	0.7245
2200.0	2.2427	1.6107	7.9919	1.3603	1.2685	0.7242
2240.0	2.2973	1.6537	8.0165	1.3689	1.2665	0.7239
2280.0	2.3522	1.6972	8.0408	1.3779	1.2646	0.7234
2320.0	2.4075	1.7410	8.0649	1.3872	1.2626	0.7228
2360.0	2.4632	1.7851	8.0887	1.3969	1.2606	0.7221
2400.0	2.5193	1.8297	8.1122	1.4071	1.2586	0.7213
2440.0	2.5757	1.8746	8.1356	1.4177	1.2567	0.7204
2480.0	2.6327	1.9200	8.1587	1.4289	1.2546	0.7193
2520.0	2.6901	1.9658	8.1817	1.4407	1.2526	0.7181
2560.0	2.7479	2.0121	8.2045	1.4531	1.2506	0.7167
2600.0	2.8063	2.0589	8.2271	1.4663	1.2485	0.7152
2640.0	2.8653	2.1063	8.2496	1.4803	1.2464	0.7135
2680.0	2.9248	2.1541	8.2720	1.4950	1.2443	0.7117
2720.0	2.9849	2.2026	8.2942	1.5107	1.2422	0.7098
2760.0	3.0456	2.2517	8.3164	1.5274	1.2401	0.7076
2800.0	3.1071	2.3014	8.3385	1.5451	1.2380	0.7054
2840.0	3.1693	2.3519	8.3606	1.5638	1.2358	0.7029
2880.0	3.2322	2.4030	8.3826	1.5837	1.2337	0.7004
2920.0	3.2960	2.4550	8.4046	1.6048	1.2315	0.6977
2960.0	3.3606	2.5078	8.4265	1.6270	1.2294	0.6949
3000.0	3.4261	2.5614	8.4485	1.6505	1.2273	0.6921
3040.0	3.4927	2.6160	8.4706	1.6753	1.2252	0.6891
3080.0	3.5602	2.6715	8.4926	1.7014	1.2231	0.6861
3120.0	3.6288	2.7280	8.5148	1.7288	1.2211	0.6830
3160.0	3.6985	2.7856	8.5370	1.7575	1.2191	0.6799
3200.0	3.7694	2.8443	8.5593	1.7875	1.2172	0.6768
3240.0	3.8415	2.9041	8.5816	1.8188	1.2153	0.6737
3280.0	3.9149	2.9651	8.6042	1.8514	1.2135	0.6707
3320.0	3.9897	3.0274	8.6268	1.8853	1.2117	0.6677
3360.0	4.0658	3.0909	8.6496	1.9203	1.2100	0.6647
3400.0	4.1433	3.1558	8.6725	1.9564	1.2084	0.6620
3440.0	4.2223	3.2220	8.6956	1.9937	1.2069	0.6593
3480.0	4.3028	3.2897	8.7189	2.0319	1.2055	0.6567
3520.0	4.3849	3.3587	8.7423	2.0709	1.2042	0.6543
3560.0	4.4685	3.4292	8.7660	2.1108	1.2030	0.6520
3600.0	4.5537	3.5012	8.7898	2.1513	1.2019	0.6499
3640.0	4.6406	3.5747	8.8138	2.1923	1.2009	0.6480
3680.0	4.7291	3.6497	8.8380	2.2337	1.2000	0.6458
3720.0	4.8193	3.7262	8.8623	2.2753	1.1992	0.6442
3760.0	4.9112	3.8042	8.8869	2.3170	1.1985	0.6429
3800.0	5.0047	3.8837	8.9116	2.3584	1.1979	0.6418
3840.0	5.0998	3.9648	8.9365	2.3996	1.1975	0.6409
3880.0	5.1966	4.0473	8.9616	2.4402	1.1971	0.6402
3920.0	5.2950	4.1312	8.9869	2.4800	1.1969	0.6397
3960.0	5.3950	4.2166	9.0122	2.5189	1.1968	0.6395
4000.0	5.4965	4.3033	9.0377	2.5566	1.1968	0.6395

THERMOPHYSICAL PROPERTIES OF AIR

ISOBAR: 5.0 MPa

T (K)	ρ (kg/m^3)	a (m/s)	μ (Pa·s)	k (J/s·m·K)	$\bar{m}$
4040.0	4.1413 +0	1202.1	1.0753 -4	0.4358 +0	27.8216
4080.0	4.0902	1209.7	1.0828	0.4444	27.7499
4120.0	4.0397	1217.4	1.0902	0.4526	27.6764
4160.0	3.9900	1225.2	1.0977	0.4603	27.6011
4200.0	3.9410	1233.0	1.1052	0.4676	27.5241
4240.0	3.8927	1241.0	1.1126	0.4743	27.4458
4280.0	3.8451	1249.0	1.1201	0.4805	27.3662
4320.0	3.7983	1257.1	1.1276	0.4860	27.2855
4360.0	3.7522	1265.3	1.1351	0.4911	27.2039
4400.0	3.7068	1273.5	1.1426	0.4955	27.1215
4440.0	3.6622	1281.8	1.1501	0.4994	27.0387
4480.0	3.6183	1290.1	1.1576	0.5026	26.9555
4520.0	3.5752	1298.5	1.1651	0.5053	26.8723
4560.0	3.5329	1307.0	1.1726	0.5074	26.7890
4600.0	3.4913	1315.5	1.1801	0.5090	26.7061
4640.0	3.4505	1324.0	1.1876	0.5100	26.6235
4680.0	3.4105	1332.5	1.1952	0.5105	26.5415
4720.0	3.3713	1341.1	1.2027	0.5107	26.4603
4760.0	3.3328	1349.6	1.2102	0.5104	26.3800
4800.0	3.2951	1358.2	1.2177	0.5098	26.3007
4840.0	3.2581	1366.7	1.2251	0.5089	26.2226
4880.0	3.2220	1375.2	1.2326	0.5079	26.1457
4920.0	3.1865	1383.7	1.2401	0.5066	26.0702
4960.0	3.1518	1392.2	1.2476	0.5053	25.9961
5000.0	3.1179	1400.6	1.2550	0.5040	25.9234
5040.0	3.0847	1409.0	1.2625	0.5027	25.8523
5080.0	3.0521	1417.3	1.2699	0.5016	25.7828
5120.0	3.0203	1425.5	1.2773	0.5006	25.7148
5160.0	2.9892	1433.6	1.2847	0.4998	25.6485
5200.0	2.9587	1441.7	1.2921	0.4993	25.5837
5240.0	2.9288	1449.7	1.2995	0.4991	25.5204
5280.0	2.8996	1457.5	1.3069	0.4993	25.4586
5320.0	2.8710	1465.3	1.3143	0.5000	25.3984
5360.0	2.8430	1473.0	1.3216	0.5011	25.3395
5400.0	2.8155	1480.5	1.3290	0.5027	25.2820
5440.0	2.7886	1487.9	1.3363	0.5048	25.2258
5480.0	2.7622	1495.2	1.3437	0.5075	25.1708
5520.0	2.7363	1502.4	1.3510	0.5108	25.1170
5560.0	2.7109	1509.5	1.3584	0.5148	25.0642
5600.0	2.6860	1516.4	1.3657	0.5194	25.0124
5640.0	2.6615	1523.2	1.3730	0.5246	24.9615
5680.0	2.6375	1529.9	1.3803	0.5305	24.9114
5720.0	2.6138	1536.5	1.3877	0.5372	24.8620
5760.0	2.5906	1543.0	1.3950	0.5446	24.8132
5800.0	2.5677	1549.4	1.4024	0.5527	24.7649
5840.0	2.5452	1555.7	1.4097	0.5615	24.7171
5880.0	2.5230	1561.9	1.4171	0.5711	24.6696
5920.0	2.5012	1568.1	1.4244	0.5815	24.6224
5960.0	2.4797	1574.1	1.4318	0.5926	24.5754
6000.0	2.4584	1580.1	1.4392	0.6046	24.5284

THERMOPHYSICAL PROPERTIES OF AIR

ISOBAR: 5.0 MPa

T	h	e	s	c_p	γ	Pr
(K)	(J/kg)	(J/kg)	(J/kg·K)	(J/kg·K)		
4040.0	5.5995 +3	4.3914 +3	9.0634 +0	2.5929 +0	1.1969	0.6398
4080.0	5.7039	4.4807	9.0891	2.6276	1.1971	0.6402
4120.0	5.8097	4.5712	9.1149	2.6604	1.1974	0.6409
4160.0	5.9167	4.6628	9.1407	2.6913	1.1978	0.6418
4200.0	6.0250	4.7554	9.1666	2.7199	1.1983	0.6429
4240.0	6.1343	4.8490	9.1925	2.7460	1.1989	0.6442
4280.0	6.2446	4.9434	9.2184	2.7697	1.1997	0.6456
4320.0	6.3558	5.0386	9.2443	2.7906	1.2005	0.6472
4360.0	6.4678	5.1344	9.2701	2.8087	1.2013	0.6490
4400.0	6.5805	5.2307	9.2958	2.8239	1.2023	0.6509
4440.0	6.6937	5.3275	9.3214	2.8361	1.2034	0.6529
4480.0	6.8073	5.4246	9.3469	2.8453	1.2045	0.6551
4520.0	6.9213	5.5218	9.3722	2.8515	1.2057	0.6573
4560.0	7.0354	5.6192	9.3974	2.8547	1.2070	0.6595
4600.0	7.1496	5.7165	9.4223	2.8550	1.2083	0.6618
4640.0	7.2638	5.8138	9.4470	2.8525	1.2097	0.6641
4680.0	7.3778	5.9108	9.4715	2.8474	1.2111	0.6664
4720.0	7.4915	6.0074	9.4957	2.8397	1.2126	0.6687
4760.0	7.6049	6.1037	9.5196	2.8298	1.2141	0.6708
4800.0	7.7179	6.1995	9.5432	2.8177	1.2156	0.6729
4840.0	7.8303	6.2947	9.5666	2.8039	1.2172	0.6748
4880.0	7.9422	6.3893	9.5896	2.7884	1.2187	0.6766
4920.0	8.0534	6.4832	9.6123	2.7716	1.2203	0.6783
4960.0	8.1639	6.5765	9.6346	2.7537	1.2218	0.6797
5000.0	8.2737	6.6690	9.6567	2.7352	1.2233	0.6810
5040.0	8.3827	6.7607	9.6784	2.7161	1.2247	0.6820
5080.0	8.4910	6.8517	9.6998	2.6969	1.2261	0.6827
5120.0	8.5985	6.9419	9.7209	2.6778	1.2275	0.6832
5160.0	8.7052	7.0314	9.7416	2.6591	1.2287	0.6834
5200.0	8.8112	7.1201	9.7621	2.6410	1.2299	0.6834
5240.0	8.9165	7.2082	9.7823	2.6238	1.2310	0.6831
5280.0	9.0211	7.2956	9.8022	2.6077	1.2320	0.6825
5320.0	9.1251	7.3824	9.8218	2.5930	1.2329	0.6816
5360.0	9.2286	7.4687	9.8412	2.5797	1.2336	0.6804
5400.0	9.3315	7.5545	9.8603	2.5682	1.2342	0.6789
5440.0	9.4341	7.6398	9.8792	2.5585	1.2347	0.6772
5480.0	9.5362	7.7249	9.8979	2.5508	1.2351	0.6753
5520.0	9.6381	7.8096	9.9165	2.5452	1.2353	0.6731
5560.0	9.7399	7.8943	9.9348	2.5419	1.2354	0.6707
5600.0	9.8415	7.9788	9.9530	2.5409	1.2353	0.6681
5640.0	9.9432	8.0633	9.9711	2.5423	1.2351	0.6653
5680.0	1.0045 +4	8.1479	9.9891	2.5461	1.2347	0.6624
5720.0	1.0147	8.2327	1.0007 +1	2.5525	1.2342	0.6593
5760.0	1.0249	8.3178	1.0025	2.5615	1.2336	0.6562
5800.0	1.0352	8.4033	1.0043	2.5731	1.2329	0.6529
5840.0	1.0455	8.4893	1.0060	2.5874	1.2320	0.6496
5880.0	1.0559	8.5758	1.0078	2.6043	1.2311	0.6462
5920.0	1.0663	8.6631	1.0096	2.6239	1.2300	0.6427
5960.0	1.0769	8.7511	1.0113	2.6463	1.2289	0.6393
6000.0	1.0875	8.8400	1.0131	2.6713	1.2276	0.6359

THERMOPHYSICAL PROPERTIES OF AIR

ISOBAR: 10.0 MPa

T (K)	ρ (kg/m^3)	a (m/s)	μ (Pa·s)	k (J/s·m·K)	$\bar{m}$
288.15	1.1845 +2	340.3	1.7979 −5	0.0249 +0	28.9644
300.0	1.1377	347.2	1.8554	0.0257	28.9644
310.0	1.1010	352.9	1.9030	0.0264	28.9644
320.0	1.0666	358.5	1.9500	0.0271	28.9644
330.0	1.0343	364.1	1.9962	0.0278	28.9644
340.0	1.0038	369.5	2.0419	0.0285	28.9644
350.0	9.7515 +1	374.8	2.0869	0.0292	28.9644
360.0	9.4807	380.1	2.1313	0.0298	28.9644
370.0	9.2244	385.2	2.1751	0.0305	28.9644
380.0	8.9817	390.3	2.2184	0.0312	28.9644
390.0	8.7514	395.3	2.2612	0.0318	28.9644
400.0	8.5326	400.3	2.3035	0.0325	28.9644
410.0	8.3245	405.1	2.3454	0.0331	28.9644
420.0	8.1263	409.9	2.3867	0.0338	28.9644
430.0	7.9373	414.7	2.4276	0.0344	28.9644
440.0	7.7569	419.3	2.4681	0.0350	28.9644
450.0	7.5845	423.9	2.5082	0.0357	28.9644
460.0	7.4197	428.5	2.5479	0.0363	28.9644
470.0	7.2618	433.0	2.5872	0.0370	28.9644
480.0	7.1105	437.4	2.6261	0.0376	28.9644
490.0	6.9654	441.8	2.6647	0.0382	28.9644
500.0	6.8261	446.1	2.7029	0.0388	28.9644
510.0	6.6922	450.4	2.7408	0.0395	28.9644
520.0	6.5635	454.6	2.7783	0.0401	28.9644
530.0	6.4397	458.8	2.8155	0.0407	28.9644
540.0	6.3204	462.9	2.8525	0.0413	28.9644
550.0	6.2055	467.0	2.8891	0.0420	28.9644
560.0	6.0947	471.0	2.9254	0.0426	28.9644
580.0	5.8846	479.0	2.9972	0.0438	28.9644
600.0	5.6884	486.8	3.0680	0.0450	28.9644
620.0	5.5049	494.4	3.1377	0.0463	28.9644
640.0	5.3329	501.9	3.2065	0.0475	28.9644
660.0	5.1713	509.3	3.2743	0.0487	28.9644
680.0	5.0192	516.5	3.3412	0.0499	28.9644
700.0	4.8758	523.6	3.4073	0.0511	28.9644
720.0	4.7403	530.6	3.4726	0.0523	28.9644
740.0	4.6122	537.5	3.5371	0.0535	28.9644
760.0	4.4908	544.3	3.6008	0.0547	28.9644
780.0	4.3757	551.0	3.6639	0.0559	28.9644
800.0	4.2663	557.5	3.7263	0.0571	28.9644
820.0	4.1622	564.1	3.7880	0.0583	28.9644
840.0	4.0631	570.5	3.8490	0.0594	28.9644
860.0	3.9687	576.8	3.9095	0.0606	28.9644
880.0	3.8785	583.1	3.9693	0.0618	28.9644
900.0	3.7923	589.3	4.0286	0.0630	28.9644
920.0	3.7098	595.4	4.0873	0.0641	28.9644
940.0	3.6309	601.4	4.1455	0.0653	28.9644
960.0	3.5553	607.4	4.2032	0.0665	28.9645
980.0	3.4827	613.3	4.2603	0.0676	28.9644
1000.0	3.4130	619.2	4.3170	0.0688	28.9644

THERMOPHYSICAL PROPERTIES OF AIR

ISOBAR: 10.0 MPa

T (K)	h (J/kg)	e (J/kg)	s (J/kg·K)	c_p (J/kg·K)	γ	Pr
288.15	-1.4106 +1	-9.6876 +1	5.5177 +0	1.0052 +0	1.4001	0.7254
300.0	-2.1924 +0	-8.8367	5.5582	1.0055	1.3999	0.7246
310.0	7.8646	-8.1182	5.5912	1.0059	1.3997	0.7240
320.0	1.7926 +1	-7.3994	5.6231	1.0064	1.3994	0.7233
330.0	2.7993	-6.6799	5.6541	1.0070	1.3991	0.7228
340.0	3.8067	-5.9598	5.6842	1.0077	1.3987	0.7222
350.0	4.8148	-5.2389	5.7134	1.0085	1.3982	0.7217
360.0	5.8238	-4.5171	5.7418	1.0095	1.3977	0.7212
370.0	6.8337	-3.7944	5.7695	1.0105	1.3972	0.7207
380.0	7.8447	-3.0707	5.7964	1.0116	1.3966	0.7203
390.0	8.8569	-2.3458	5.8227	1.0127	1.3959	0.7199
400.0	9.8702	-1.6197	5.8484	1.0140	1.3953	0.7195
410.0	1.0885 +2	-8.9229 +0	5.8734	1.0153	1.3945	0.7191
420.0	1.1901	-1.6350	5.8979	1.0168	1.3938	0.7188
430.0	1.2918	5.6674	5.9219	1.0182	1.3930	0.7185
440.0	1.3937	1.2985 +1	5.9453	1.0198	1.3921	0.7182
450.0	1.4958	2.0319	5.9682	1.0214	1.3913	0.7179
460.0	1.5980	2.7669	5.9907	1.0231	1.3904	0.7177
470.0	1.7004	3.5036	6.0127	1.0249	1.3894	0.7175
480.0	1.8030	4.2421	6.0343	1.0267	1.3885	0.7173
490.0	1.9058	4.9825	6.0555	1.0285	1.3875	0.7171
500.0	2.0087	5.7247	6.0763	1.0304	1.3865	0.7170
510.0	2.1119	6.4689	6.0967	1.0324	1.3855	0.7169
520.0	2.2152	7.2150	6.1168	1.0344	1.3845	0.7167
530.0	2.3187	7.9632	6.1365	1.0365	1.3834	0.7167
540.0	2.4225	8.7134	6.1559	1.0385	1.3823	0.7166
550.0	2.5264	9.4658	6.1750	1.0407	1.3813	0.7165
560.0	2.6306	1.0220 +2	6.1937	1.0428	1.3802	0.7165
580.0	2.8396	1.1736	6.2304	1.0473	1.3780	0.7165
600.0	3.0495	1.3260	6.2660	1.0518	1.3757	0.7165
620.0	3.2603	1.4794	6.3006	1.0564	1.3735	0.7165
640.0	3.4721	1.6337	6.3342	1.0611	1.3712	0.7166
660.0	3.6848	1.7890	6.3669	1.0659	1.3689	0.7168
680.0	3.8984	1.9452	6.3988	1.0707	1.3667	0.7169
700.0	4.1131	2.1023	6.4299	1.0755	1.3644	0.7171
720.0	4.3286	2.2605	6.4603	1.0803	1.3622	0.7172
740.0	4.5452	2.4196	6.4899	1.0851	1.3600	0.7174
760.0	4.7627	2.5796	6.5189	1.0900	1.3578	0.7176
780.0	4.9812	2.7406	6.5473	1.0947	1.3557	0.7177
800.0	5.2006	2.9026	6.5751	1.0995	1.3537	0.7178
820.0	5.4210	3.0655	6.6023	1.1042	1.3516	0.7179
840.0	5.6423	3.2294	6.6289	1.1088	1.3496	0.7179
860.0	5.8645	3.3941	6.6551	1.1134	1.3477	0.7179
880.0	6.0876	3.5598	6.6807	1.1178	1.3458	0.7179
900.0	6.3116	3.7264	6.7059	1.1223	1.3440	0.7178
920.0	6.5365	3.8938	6.7306	1.1266	1.3422	0.7176
940.0	6.7622	4.0621	6.7549	1.1308	1.3405	0.7174
960.0	6.9888	4.2312	6.7788	1.1349	1.3389	0.7172
980.0	7.2162	4.4012	6.8022	1.1389	1.3373	0.7169
1000.0	7.4444	4.5719	6.8252	1.1428	1.3357	0.7165

THERMOPHYSICAL PROPERTIES OF AIR

ISOBAR: 10.0 MPa

T (K)	ρ (kg/m^3)	a (m/s)	μ (Pa·s)	k (J/s·m·K)	$\bar{m}$
1020.0	3.3461 +1	625.0	4.3745 −5	0.0697 +0	28.9644
1040.0	3.2818	630.8	4.4309	0.0708	28.9644
1060.0	3.2198	636.5	4.4870	0.0719	28.9644
1080.0	3.1602	642.2	4.5426	0.0730	28.9644
1100.0	3.1028	647.8	4.5978	0.0741	28.9644
1120.0	3.0474	653.3	4.6526	0.0754	28.9644
1140.0	2.9939	658.8	4.7070	0.0765	28.9644
1160.0	2.9423	664.2	4.7610	0.0776	28.9644
1180.0	2.8924	669.6	4.8147	0.0787	28.9644
1200.0	2.8442	674.9	4.8680	0.0797	28.9644
1220.0	2.7976	680.2	4.9209	0.0808	28.9644
1240.0	2.7525	685.5	4.9735	0.0819	28.9644
1260.0	2.7088	690.6	5.0258	0.0830	28.9644
1280.0	2.6664	695.8	5.0778	0.0841	28.9644
1300.0	2.6254	700.9	5.1294	0.0851	28.9644
1320.0	2.5856	705.9	5.1807	0.0862	28.9644
1340.0	2.5470	710.9	5.2318	0.0873	28.9644
1360.0	2.5096	715.9	5.2825	0.0884	28.9644
1380.0	2.4732	720.8	5.3329	0.0894	28.9644
1400.0	2.4379	725.7	5.3831	0.0905	28.9644
1420.0	2.4035	730.6	5.4330	0.0916	28.9644
1440.0	2.3702	735.4	5.4827	0.0927	28.9644
1460.0	2.3377	740.1	5.5320	0.0937	28.9644
1480.0	2.3061	744.9	5.5811	0.0948	28.9644
1500.0	2.2754	749.6	5.6300	0.0959	28.9644
1520.0	2.2454	754.2	5.6786	0.0969	28.9644
1540.0	2.2163	758.8	5.7270	0.0980	28.9644
1560.0	2.1878	763.4	5.7751	0.0991	28.9644
1580.0	2.1602	768.0	5.8231	0.1002	28.9644
1600.0	2.1331	772.5	5.8707	0.1013	28.9644
1620.0	2.1068	777.0	5.9182	0.1023	28.9644
1640.0	2.0811	781.4	5.9654	0.1034	28.9644
1660.0	2.0560	785.9	6.0125	0.1045	28.9644
1680.0	2.0316	790.3	6.0593	0.1056	28.9644
1700.0	2.0077	794.6	6.1059	0.1067	28.9644
1720.0	1.9843	799.0	6.1523	0.1078	28.9644
1740.0	1.9615	803.3	6.1985	0.1089	28.9643
1760.0	1.9392	807.5	6.2445	0.1099	28.9643
1780.0	1.9174	811.8	6.2903	0.1110	28.9643
1800.0	1.8961	816.0	6.3360	0.1121	28.9643
1820.0	1.8753	820.2	6.3814	0.1132	28.9643
1840.0	1.8549	824.3	6.4267	0.1143	28.9643
1860.0	1.8350	828.5	6.4717	0.1154	28.9642
1880.0	1.8154	832.6	6.5166	0.1165	28.9642
1900.0	1.7963	836.7	6.5614	0.1176	28.9642
1920.0	1.7776	840.7	6.6059	0.1187	28.9642
1940.0	1.7593	844.8	6.6503	0.1198	28.9641
1960.0	1.7413	848.8	6.6945	0.1209	28.9641
1980.0	1.7237	852.8	6.7385	0.1220	28.9640
2000.0	1.7065	856.8	6.7824	0.1231	28.9639

THERMOPHYSICAL PROPERTIES OF AIR

ISOBAR: 10.0 MPa

T (K)	h (J/kg)	e (J/kg)	s (J/kg·K)	c_p (J/kg·K)	γ	Pr
1020.0	7.6733 +2	4.7434 +2	6.8479 +0	1.1465 +0	1.3343	0.7185
1040.0	7.9030	4.9156	6.8702	1.1501	1.3329	0.7185
1060.0	8.1333	5.0885	6.8921	1.1537	1.3315	0.7185
1080.0	8.3644	5.2622	6.9137	1.1572	1.3302	0.7185
1100.0	8.5962	5.4365	6.9350	1.1608	1.3288	0.7186
1120.0	8.8288	5.6116	6.9560	1.1643	1.3275	0.7187
1140.0	9.0620	5.7873	6.9766	1.1678	1.3262	0.7188
1160.0	9.2959	5.9638	6.9969	1.1712	1.3249	0.7190
1180.0	9.5305	6.1409	7.0170	1.1747	1.3237	0.7191
1200.0	9.7657	6.3187	7.0368	1.1781	1.3224	0.7192
1220.0	1.0002 +3	6.4973	7.0563	1.1816	1.3212	0.7194
1240.0	1.0238	6.6765	7.0755	1.1850	1.3200	0.7195
1260.0	1.0476	6.8564	7.0945	1.1884	1.3188	0.7197
1280.0	1.0714	7.0369	7.1132	1.1918	1.3176	0.7199
1300.0	1.0952	7.2182	7.1317	1.1952	1.3164	0.7201
1320.0	1.1192	7.4001	7.1500	1.1986	1.3152	0.7202
1340.0	1.1432	7.5827	7.1681	1.2020	1.3140	0.7204
1360.0	1.1673	7.7660	7.1859	1.2054	1.3128	0.7206
1380.0	1.1914	7.9500	7.2035	1.2089	1.3117	0.7208
1400.0	1.2156	8.1347	7.2209	1.2123	1.3105	0.7210
1420.0	1.2399	8.3200	7.2382	1.2157	1.3094	0.7212
1440.0	1.2642	8.5060	7.2552	1.2191	1.3082	0.7214
1460.0	1.2887	8.6928	7.2720	1.2226	1.3071	0.7216
1480.0	1.3131	8.8802	7.2887	1.2260	1.3060	0.7218
1500.0	1.3377	9.0683	7.3052	1.2295	1.3049	0.7220
1520.0	1.3623	9.2571	7.3215	1.2330	1.3037	0.7222
1540.0	1.3870	9.4466	7.3376	1.2364	1.3026	0.7224
1560.0	1.4118	9.6368	7.3536	1.2399	1.3015	0.7225
1580.0	1.4366	9.8276	7.3694	1.2434	1.3004	0.7227
1600.0	1.4615	1.0019 +3	7.3851	1.2469	1.2993	0.7229
1620.0	1.4865	1.0212	7.4006	1.2505	1.2982	0.7231
1640.0	1.5115	1.0405	7.4159	1.2540	1.2971	0.7232
1660.0	1.5367	1.0598	7.4312	1.2575	1.2961	0.7234
1680.0	1.5618	1.0793	7.4462	1.2611	1.2950	0.7236
1700.0	1.5871	1.0988	7.4612	1.2646	1.2939	0.7237
1720.0	1.6124	1.1184	7.4760	1.2682	1.2928	0.7239
1740.0	1.6378	1.1380	7.4907	1.2718	1.2918	0.7240
1760.0	1.6633	1.1577	7.5052	1.2754	1.2907	0.7242
1780.0	1.6888	1.1775	7.5197	1.2790	1.2897	0.7243
1800.0	1.7145	1.1974	7.5340	1.2826	1.2886	0.7245
1820.0	1.7401	1.2173	7.5482	1.2862	1.2876	0.7246
1840.0	1.7659	1.2374	7.5622	1.2898	1.2866	0.7247
1860.0	1.7917	1.2574	7.5762	1.2934	1.2855	0.7249
1880.0	1.8176	1.2776	7.5901	1.2971	1.2845	0.7250
1900.0	1.8436	1.2978	7.6038	1.3007	1.2835	0.7251
1920.0	1.8697	1.3181	7.6174	1.3044	1.2825	0.7252
1940.0	1.8958	1.3385	7.6310	1.3081	1.2815	0.7253
1960.0	1.9220	1.3590	7.6444	1.3118	1.2805	0.7254
1980.0	1.9483	1.3795	7.6578	1.3155	1.2795	0.7255
2000.0	1.9746	1.4001	7.6710	1.3192	1.2785	0.7256

THERMOPHYSICAL PROPERTIES OF AIR

ISOBAR: 10.0 MPa

T (K)	ρ (kg/m^3)	a (m/s)	μ (Pa·s)	k (J/s·m·K)	$\bar{m}$
2040.0	1.6730 +1	864.6	6.8697 −5	0.1253 +0	28.9638
2080.0	1.6408	872.4	6.9564	0.1275	28.9636
2120.0	1.6099	880.1	7.0424	0.1297	28.9633
2160.0	1.5800	887.7	7.1279	0.1319	28.9630
2200.0	1.5513	895.2	7.2129	0.1351	28.9626
2240.0	1.5236	902.7	7.2972	0.1375	28.9621
2280.0	1.4968	910.0	7.3811	0.1400	28.9615
2320.0	1.4710	917.3	7.4644	0.1425	28.9607
2360.0	1.4460	924.5	7.5471	0.1451	28.9598
2400.0	1.4218	931.7	7.6294	0.1477	28.9587
2440.0	1.3984	938.8	7.7112	0.1505	28.9574
2480.0	1.3758	945.8	7.7926	0.1532	28.9558
2520.0	1.3539	952.7	7.8734	0.1561	28.9540
2560.0	1.3326	959.6	7.9538	0.1591	28.9518
2600.0	1.3120	966.4	8.0338	0.1621	28.9492
2640.0	1.2920	973.1	8.1134	0.1653	28.9463
2680.0	1.2726	979.8	8.1926	0.1686	28.9429
2720.0	1.2537	986.4	8.2713	0.1720	28.9390
2760.0	1.2353	993.0	8.3497	0.1756	28.9346
2800.0	1.2175	999.5	8.4277	0.1793	28.9295
2840.0	1.2001	1006.0	8.5054	0.1832	28.9238
2880.0	1.1832	1012.4	8.5827	0.1872	28.9173
2920.0	1.1667	1018.8	8.6596	0.1914	28.9100
2960.0	1.1506	1025.1	8.7363	0.1959	28.9019
3000.0	1.1349	1031.4	8.8126	0.2005	28.8928
3040.0	1.1195	1037.7	8.8887	0.2053	28.8828
3080.0	1.1046	1043.9	8.9644	0.2103	28.8716
3120.0	1.0900	1050.2	9.0399	0.2155	28.8593
3160.0	1.0757	1056.4	9.1152	0.2210	28.8458
3200.0	1.0617	1062.6	9.1901	0.2267	28.8310
3240.0	1.0480	1068.8	9.2649	0.2326	28.8148
3280.0	1.0346	1075.0	9.3394	0.2387	28.7972
3320.0	1.0214	1081.2	9.4138	0.2450	28.7780
3360.0	1.0085	1087.4	9.4879	0.2516	28.7573
3400.0	9.9588 +0	1093.6	9.5619	0.2584	28.7349
3440.0	9.8348	1099.8	9.6357	0.2655	28.7108
3480.0	9.7130	1106.1	9.7094	0.2727	28.6849
3520.0	9.5933	1112.4	9.7829	0.2803	28.6572
3560.0	9.4757	1118.7	9.8563	0.2879	28.6276
3600.0	9.3601	1125.1	9.9296	0.2957	28.5960
3640.0	9.2464	1131.5	1.0003 −4	0.3036	28.5625
3680.0	9.1345	1137.9	1.0076	0.3117	28.5269
3720.0	9.0243	1144.4	1.0149	0.3199	28.4893
3760.0	8.9159	1151.0	1.0222	0.3291	28.4496
3800.0	8.8091	1157.6	1.0295	0.3376	28.4078
3840.0	8.7039	1164.3	1.0368	0.3461	28.3640
3880.0	8.6002	1171.0	1.0441	0.3547	28.3180
3920.0	8.4980	1177.8	1.0514	0.3632	28.2700
3960.0	8.3972	1184.7	1.0587	0.3718	28.2199
4000.0	8.2979	1191.6	1.0660	0.3803	28.1677

THERMOPHYSICAL PROPERTIES OF AIR

ISOBAR: 10.0 MPa

T	h	e	s	c_p	γ	Pr
(K)	(J/kg)	(J/kg)	(J/kg·K)	(J/kg·K)		
2040.0	2.0275 +3	1.4415 +3	7.6972 +0	1.3267 +0	1.2766	0.7258
2080.0	2.0807	1.4832	7.7230	1.3342	1.2746	0.7259
2120.0	2.1343	1.5253	7.7485	1.3419	1.2727	0.7260
2160.0	2.1881	1.5676	7.7737	1.3496	1.2708	0.7262
2200.0	2.2422	1.6103	7.7985	1.3575	1.2689	0.7248
2240.0	2.2967	1.6532	7.8230	1.3656	1.2671	0.7246
2280.0	2.3515	1.6965	7.8473	1.3738	1.2652	0.7242
2320.0	2.4066	1.7401	7.8712	1.3822	1.2634	0.7238
2360.0	2.4621	1.7840	7.8949	1.3909	1.2616	0.7234
2400.0	2.5179	1.8283	7.9184	1.3998	1.2597	0.7228
2440.0	2.5740	1.8730	7.9416	1.4090	1.2579	0.7221
2480.0	2.6306	1.9180	7.9646	1.4186	1.2561	0.7213
2520.0	2.6875	1.9634	7.9874	1.4286	1.2543	0.7204
2560.0	2.7449	2.0092	8.0100	1.4390	1.2524	0.7194
2600.0	2.8027	2.0554	8.0324	1.4499	1.2506	0.7183
2640.0	2.8609	2.1021	8.0546	1.4613	1.2488	0.7171
2680.0	2.9196	2.1492	8.0766	1.4732	1.2470	0.7157
2720.0	2.9788	2.1968	8.0986	1.4858	1.2451	0.7142
2760.0	3.0385	2.2448	8.1203	1.4990	1.2433	0.7126
2800.0	3.0987	2.2934	8.1420	1.5129	1.2414	0.7109
2840.0	3.1595	2.3426	8.1636	1.5275	1.2396	0.7090
2880.0	3.2209	2.3923	8.1850	1.5429	1.2378	0.7070
2920.0	3.2829	2.4426	8.2064	1.5591	1.2359	0.7049
2960.0	3.3456	2.4936	8.2278	1.5761	1.2341	0.7027
3000.0	3.4090	2.5452	8.2490	1.5940	1.2323	0.7004
3040.0	3.4732	2.5975	8.2703	1.6128	1.2305	0.6980
3080.0	3.5381	2.6505	8.2915	1.6326	1.2287	0.6955
3120.0	3.6038	2.7043	8.3127	1.6532	1.2269	0.6929
3160.0	3.6704	2.7589	8.3339	1.6748	1.2252	0.6903
3200.0	3.7378	2.8143	8.3551	1.6974	1.2235	0.6876
3240.0	3.8062	2.8706	8.3763	1.7209	1.2218	0.6849
3280.0	3.8755	2.9278	8.3976	1.7454	1.2202	0.6822
3320.0	3.9458	2.9860	8.4189	1.7708	1.2186	0.6795
3360.0	4.0172	3.0451	8.4403	1.7972	1.2171	0.6768
3400.0	4.0896	3.1051	8.4617	1.8244	1.2156	0.6741
3440.0	4.1631	3.1663	8.4832	1.8526	1.2142	0.6715
3480.0	4.2378	3.2284	8.5048	1.8815	1.2129	0.6689
3520.0	4.3137	3.2917	8.5264	1.9113	1.2116	0.6665
3560.0	4.3907	3.3561	8.5482	1.9418	1.2104	0.6640
3600.0	4.4690	3.4216	8.5701	1.9729	1.2093	0.6617
3640.0	4.5486	3.4883	8.5921	2.0047	1.2082	0.6595
3680.0	4.6294	3.5561	8.6141	2.0369	1.2073	0.6574
3720.0	4.7115	3.6251	8.6363	2.0697	1.2064	0.6555
3760.0	4.7950	3.6954	8.6586	2.1027	1.2056	0.6532
3800.0	4.8797	3.7668	8.6811	2.1360	1.2049	0.6515
3840.0	4.9659	3.8395	8.7036	2.1695	1.2042	0.6499
3880.0	5.0533	3.9133	8.7263	2.2030	1.2037	0.6485
3920.0	5.1421	3.9884	8.7490	2.2363	1.2033	0.6473
3960.0	5.2322	4.0647	8.7719	2.2695	1.2029	0.6462
4000.0	5.3236	4.1422	8.7949	2.3023	1.2026	0.6454

THERMOPHYSICAL PROPERTIES OF AIR

ISOBAR: 10.0 MPa

T (K)	ρ (kg/m^3)	a (m/s)	μ (Pa·s)	k (J/s·m·K)	$\bar{m}$
4040.0	8.2000 +0	1198.6	1.0733 -4	0.3887 +0	28.1136
4080.0	8.1034	1205.7	1.0806	0.3969	28.0575
4120.0	8.0081	1212.9	1.0879	0.4051	27.9995
4160.0	7.9142	1220.1	1.0952	0.4130	27.9397
4200.0	7.8215	1227.4	1.1025	0.4207	27.8781
4240.0	7.7302	1234.7	1.1098	0.4282	27.8149
4280.0	7.6401	1242.2	1.1171	0.4353	27.7501
4320.0	7.5513	1249.7	1.1244	0.4422	27.6839
4360.0	7.4637	1257.2	1.1318	0.4487	27.6162
4400.0	7.3774	1264.9	1.1391	0.4548	27.5474
4440.0	7.2924	1272.6	1.1464	0.4606	27.4774
4480.0	7.2086	1280.3	1.1538	0.4660	27.4064
4520.0	7.1261	1288.2	1.1611	0.4710	27.3345
4560.0	7.0448	1296.0	1.1685	0.4756	27.2619
4600.0	6.9648	1303.9	1.1758	0.4798	27.1887
4640.0	6.8861	1311.9	1.1832	0.4836	27.1151
4680.0	6.8086	1319.9	1.1906	0.4869	27.0411
4720.0	6.7324	1328.0	1.1979	0.4899	26.9670
4760.0	6.6574	1336.0	1.2053	0.4925	26.8927
4800.0	6.5837	1344.1	1.2127	0.4946	26.8186
4840.0	6.5113	1352.3	1.2200	0.4965	26.7447
4880.0	6.4402	1360.4	1.2274	0.4980	26.6711
4920.0	6.3703	1368.6	1.2348	0.4992	26.5979
4960.0	6.3017	1376.7	1.2421	0.5001	26.5253
5000.0	6.2343	1384.9	1.2495	0.5009	26.4533
5040.0	6.1681	1393.0	1.2568	0.5014	26.3820
5080.0	6.1032	1401.1	1.2642	0.5018	26.3116
5120.0	6.0395	1409.3	1.2715	0.5020	26.2420
5160.0	5.9771	1417.3	1.2789	0.5023	26.1734
5200.0	5.9158	1425.4	1.2862	0.5025	26.1058
5240.0	5.8556	1433.4	1.2936	0.5027	26.0392
5280.0	5.7966	1441.4	1.3009	0.5030	25.9737
5320.0	5.7388	1449.3	1.3082	0.5034	25.9093
5360.0	5.6821	1457.1	1.3155	0.5039	25.8460
5400.0	5.6264	1464.9	1.3228	0.5047	25.7839
5440.0	5.5718	1472.6	1.3301	0.5057	25.7228
5480.0	5.5183	1480.3	1.3374	0.5069	25.6629
5520.0	5.4657	1487.8	1.3447	0.5085	25.6040
5560.0	5.4141	1495.3	1.3520	0.5104	25.5462
5600.0	5.3635	1502.7	1.3593	0.5127	25.4894
5640.0	5.3138	1510.0	1.3665	0.5153	25.4337
5680.0	5.2650	1517.3	1.3738	0.5184	25.3788
5720.0	5.2171	1524.4	1.3811	0.5220	25.3249
5760.0	5.1700	1531.5	1.3883	0.5261	25.2718
5800.0	5.1237	1538.4	1.3956	0.5306	25.2195
5840.0	5.0782	1545.3	1.4028	0.5357	25.1680
5880.0	5.0335	1552.1	1.4101	0.5413	25.1171
5920.0	4.9895	1558.8	1.4174	0.5475	25.0668
5960.0	4.9462	1565.4	1.4247	0.5542	25.0171
6000.0	4.9035	1571.9	1.4319	0.5616	24.9679

THERMOPHYSICAL PROPERTIES OF AIR

ISOBAR: 10.0 MPa

T (K)	h (J/kg)	e (J/kg)	s (J/kg·K)	c_p (J/kg·K)	γ	Pr
4040.0	5.4164 +3	4.2208 +3	8.8180 +0	2.3346 +0	1.2025	0.6447
4080.0	5.5104	4.3005	8.8411	2.3664	1.2024	0.6442
4120.0	5.6057	4.3814	8.8644	2.3973	1.2024	0.6438
4160.0	5.7022	4.4634	8.8877	2.4274	1.2025	0.6437
4200.0	5.7999	4.5464	8.9110	2.4564	1.2026	0.6437
4240.0	5.8987	4.6304	8.9344	2.4843	1.2029	0.6439
4280.0	5.9986	4.7154	8.9579	2.5109	1.2032	0.6443
4320.0	6.0995	4.8012	8.9814	2.5361	1.2037	0.6449
4360.0	6.2015	4.8879	9.0049	2.5597	1.2042	0.6456
4400.0	6.3043	4.9754	9.0283	2.5817	1.2048	0.6463
4440.0	6.4080	5.0636	9.0518	2.6019	1.2054	0.6474
4480.0	6.5124	5.1524	9.0752	2.6203	1.2061	0.6485
4520.0	6.6176	5.2418	9.0986	2.6369	1.2069	0.6498
4560.0	6.7233	5.3317	9.1219	2.6514	1.2078	0.6511
4600.0	6.8297	5.4220	9.1451	2.6639	1.2087	0.6526
4640.0	6.9364	5.5127	9.1682	2.6744	1.2097	0.6541
4680.0	7.0436	5.6036	9.1912	2.6829	1.2107	0.6558
4720.0	7.1510	5.6948	9.2141	2.6893	1.2118	0.6574
4760.0	7.2587	5.7861	9.2368	2.6937	1.2129	0.6591
4800.0	7.3665	5.8774	9.2593	2.6962	1.2141	0.6608
4840.0	7.4744	5.9687	9.2817	2.6968	1.2153	0.6626
4880.0	7.5822	6.0599	9.3039	2.6956	1.2166	0.6643
4920.0	7.6900	6.1510	9.3259	2.6927	1.2178	0.6659
4960.0	7.7976	6.2419	9.3477	2.6883	1.2191	0.6675
5000.0	7.9050	6.3325	9.3692	2.6824	1.2204	0.6691
5040.0	8.0122	6.4228	9.3906	2.6752	1.2217	0.6705
5080.0	8.1190	6.5127	9.4117	2.6669	1.2230	0.6718
5120.0	8.2255	6.6023	9.4326	2.6576	1.2243	0.6730
5160.0	8.3316	6.6914	9.4532	2.6476	1.2255	0.6740
5200.0	8.4373	6.7801	9.4736	2.6369	1.2268	0.6749
5240.0	8.5426	6.8683	9.4938	2.6258	1.2280	0.6756
5280.0	8.6474	6.9561	9.5137	2.6145	1.2292	0.6761
5320.0	8.7517	7.0434	9.5334	2.6031	1.2303	0.6764
5360.0	8.8556	7.1302	9.5529	2.5918	1.2314	0.6765
5400.0	8.9591	7.2166	9.5721	2.5808	1.2324	0.6764
5440.0	9.0621	7.3026	9.5911	2.5704	1.2333	0.6760
5480.0	9.1647	7.3881	9.6099	2.5605	1.2342	0.6755
5520.0	9.2670	7.4733	9.6285	2.5514	1.2349	0.6747
5560.0	9.3689	7.5581	9.6469	2.5433	1.2356	0.6736
5600.0	9.4704	7.6426	9.6651	2.5362	1.2362	0.6724
5640.0	9.5718	7.7268	9.6831	2.5303	1.2367	0.6709
5680.0	9.6729	7.8108	9.7010	2.5257	1.2371	0.6692
5720.0	9.7738	7.8947	9.7187	2.5225	1.2374	0.6673
5760.0	9.8747	7.9784	9.7363	2.5209	1.2376	0.6652
5800.0	9.9755	8.0621	9.7537	2.5209	1.2377	0.6630
5840.0	1.0076 +4	8.1458	9.7710	2.5225	1.2377	0.6605
5880.0	1.0177	8.2296	9.7883	2.5259	1.2376	0.6580
5920.0	1.0278	8.3136	9.8054	2.5311	1.2374	0.6552
5960.0	1.0380	8.3977	9.8225	2.5382	1.2371	0.6524
6000.0	1.0482	8.4822	9.8395	2.5472	1.2367	0.6494

THERMOPHYSICAL PROPERTIES OF AIR

ISOBAR: 20.0 MPa

T (K)	ρ (kg/m^3)	a (m/s)	μ (Pa·s)	k (J/s·m·K)	$\bar{m}$
288.15	2.4179 +2	340.3	1.7979 −5	0.0249 +0	28.9644
300.0	2.3224	347.2	1.8554	0.0257	28.9644
310.0	2.2475	352.9	1.9030	0.0264	28.9644
320.0	2.1773	358.5	1.9500	0.0271	28.9644
330.0	2.1113	364.1	1.9962	0.0278	28.9644
340.0	2.0492	369.5	2.0419	0.0285	28.9644
350.0	1.9906	374.8	2.0869	0.0292	28.9644
360.0	1.9354	380.1	2.1313	0.0298	28.9644
370.0	1.8830	385.2	2.1751	0.0305	28.9644
380.0	1.8335	390.3	2.2184	0.0312	28.9644
390.0	1.7865	395.3	2.2612	0.0318	28.9644
400.0	1.7418	400.3	2.3035	0.0325	28.9644
410.0	1.6993	405.1	2.3454	0.0331	28.9644
420.0	1.6589	409.9	2.3867	0.0338	28.9644
430.0	1.6203	414.7	2.4276	0.0344	28.9644
440.0	1.5835	419.3	2.4681	0.0350	28.9644
450.0	1.5483	423.9	2.5082	0.0357	28.9644
460.0	1.5146	428.5	2.5479	0.0363	28.9644
470.0	1.4824	433.0	2.5872	0.0370	28.9644
480.0	1.4515	437.4	2.6261	0.0376	28.9644
490.0	1.4219	441.8	2.6647	0.0382	28.9644
500.0	1.3935	446.1	2.7029	0.0388	28.9644
510.0	1.3661	450.4	2.7408	0.0395	28.9644
520.0	1.3399	454.6	2.7783	0.0401	28.9644
530.0	1.3146	458.8	2.8155	0.0407	28.9644
540.0	1.2902	462.9	2.8525	0.0413	28.9644
550.0	1.2668	467.0	2.8891	0.0420	28.9644
560.0	1.2442	471.0	2.9254	0.0426	28.9644
580.0	1.2013	479.0	2.9972	0.0438	28.9644
600.0	1.1612	486.8	3.0680	0.0450	28.9644
620.0	1.1238	494.4	3.1377	0.0463	28.9644
640.0	1.0886	501.9	3.2065	0.0475	28.9644
660.0	1.0556	509.3	3.2743	0.0487	28.9644
680.0	1.0246	516.5	3.3412	0.0499	28.9644
700.0	9.9532 +1	523.6	3.4073	0.0511	28.9644
720.0	9.6768	530.6	3.4726	0.0523	28.9644
740.0	9.4152	537.5	3.5371	0.0535	28.9644
760.0	9.1675	544.3	3.6008	0.0547	28.9644
780.0	8.9324	551.0	3.6639	0.0559	28.9644
800.0	8.7091	557.5	3.7263	0.0571	28.9644
820.0	8.4967	564.1	3.7880	0.0583	28.9644
840.0	8.2944	570.5	3.8490	0.0594	28.9644
860.0	8.1015	576.8	3.9095	0.0606	28.9644
880.0	7.9174	583.1	3.9693	0.0618	28.9644
900.0	7.7414	589.3	4.0286	0.0630	28.9644
920.0	7.5731	595.4	4.0873	0.0641	28.9644
940.0	7.4120	601.4	4.1455	0.0653	28.9644
960.0	7.2576	607.4	4.2032	0.0665	28.9645
980.0	7.1095	613.3	4.2603	0.0676	28.9644
1000.0	6.9673	619.2	4.3170	0.0688	28.9644

THERMOPHYSICAL PROPERTIES OF AIR

ISOBAR: 20.0 MPa

T (K)	h (J/kg)	e (J/kg)	s (J/kg·K)	c_p (J/kg·K)	γ	Pr
288.15	-1.4106 +1	-9.6876 +1	5.3127 +0	1.0052 +0	1.4001	0.7254
300.0	-2.1924 +0	-8.8367	5.3532	1.0055	1.3999	0.7246
310.0	7.8646	-8.1182	5.3862	1.0059	1.3997	0.7240
320.0	1.7926 +1	-7.3994	5.4181	1.0064	1.3994	0.7233
330.0	2.7993	-6.6799	5.4491	1.0070	1.3991	0.7228
340.0	3.8067	-5.9598	5.4792	1.0077	1.3987	0.7222
350.0	4.8148	-5.2389	5.5084	1.0085	1.3982	0.7217
360.0	5.8238	-4.5171	5.5368	1.0095	1.3977	0.7212
370.0	6.8337	-3.7944	5.5645	1.0105	1.3972	0.7207
380.0	7.8447	-3.0707	5.5914	1.0116	1.3966	0.7203
390.0	8.8569	-2.3458	5.6177	1.0127	1.3959	0.7199
400.0	9.8702	-1.6197	5.6434	1.0140	1.3953	0.7195
410.0	1.0885 +2	-8.9229 +0	5.6684	1.0153	1.3945	0.7191
420.0	1.1901	-1.6350	5.6929	1.0168	1.3938	0.7188
430.0	1.2918	5.6674	5.7169	1.0182	1.3930	0.7185
440.0	1.3937	1.2985 +1	5.7403	1.0198	1.3921	0.7182
450.0	1.4958	2.0319	5.7632	1.0214	1.3913	0.7179
460.0	1.5980	2.7669	5.7857	1.0231	1.3904	0.7177
470.0	1.7004	3.5036	5.8077	1.0249	1.3894	0.7175
480.0	1.8030	4.2421	5.8293	1.0267	1.3885	0.7173
490.0	1.9058	4.9825	5.8505	1.0285	1.3875	0.7171
500.0	2.0087	5.7247	5.8713	1.0304	1.3865	0.7170
510.0	2.1119	6.4689	5.8917	1.0324	1.3855	0.7169
520.0	2.2152	7.2150	5.9118	1.0344	1.3845	0.7167
530.0	2.3187	7.9632	5.9315	1.0365	1.3834	0.7167
540.0	2.4225	8.7134	5.9509	1.0385	1.3823	0.7166
550.0	2.5264	9.4658	5.9700	1.0407	1.3813	0.7165
560.0	2.6306	1.0220 +2	5.9888	1.0428	1.3802	0.7165
580.0	2.8396	1.1736	6.0254	1.0473	1.3780	0.7165
600.0	3.0495	1.3260	6.0610	1.0518	1.3757	0.7165
620.0	3.2603	1.4794	6.0956	1.0564	1.3735	0.7165
640.0	3.4721	1.6337	6.1292	1.0611	1.3712	0.7166
660.0	3.6848	1.7890	6.1619	1.0659	1.3689	0.7168
680.0	3.8984	1.9452	6.1938	1.0707	1.3667	0.7169
700.0	4.1131	2.1023	6.2249	1.0755	1.3644	0.7171
720.0	4.3286	2.2605	6.2553	1.0803	1.3622	0.7172
740.0	4.5452	2.4196	6.2849	1.0851	1.3600	0.7174
760.0	4.7627	2.5796	6.3139	1.0900	1.3578	0.7176
780.0	4.9812	2.7406	6.3423	1.0947	1.3557	0.7177
800.0	5.2006	2.9026	6.3701	1.0995	1.3537	0.7178
820.0	5.4210	3.0655	6.3973	1.1042	1.3516	0.7179
840.0	5.6423	3.2294	6.4240	1.1088	1.3496	0.7179
860.0	5.8645	3.3941	6.4501	1.1134	1.3477	0.7179
880.0	6.0876	3.5598	6.4758	1.1178	1.3458	0.7179
900.0	6.3116	3.7264	6.5009	1.1223	1.3440	0.7178
920.0	6.5365	3.8938	6.5256	1.1266	1.3422	0.7176
940.0	6.7622	4.0621	6.5499	1.1308	1.3405	0.7174
960.0	6.9888	4.2312	6.5738	1.1349	1.3389	0.7172
980.0	7.2162	4.4012	6.5972	1.1389	1.3373	0.7169
1000.0	7.4444	4.5719	6.6203	1.1428	1.3357	0.7165

THERMOPHYSICAL PROPERTIES OF AIR

ISOBAR: 20.0 MPa

T (K)	ρ (kg/m^3)	a (m/s)	μ (Pa·s)	k (J/s·m·K)	$\overline{m}$
1020.0	6.8307 +1	625.0	4.3745 −5	0.0697 +0	28.9644
1040.0	6.6993	630.8	4.4309	0.0708	28.9644
1060.0	6.5729	636.5	4.4870	0.0719	28.9644
1080.0	6.4512	642.2	4.5426	0.0730	28.9644
1100.0	6.3339	647.8	4.5978	0.0741	28.9644
1120.0	6.2208	653.3	4.6526	0.0754	28.9644
1140.0	6.1116	658.8	4.7070	0.0765	28.9644
1160.0	6.0063	664.2	4.7610	0.0776	28.9644
1180.0	5.9045	669.6	4.8147	0.0787	28.9644
1200.0	5.8061	674.9	4.8680	0.0797	28.9644
1220.0	5.7109	680.2	4.9209	0.0808	28.9644
1240.0	5.6188	685.5	4.9735	0.0819	28.9644
1260.0	5.5296	690.6	5.0258	0.0830	28.9644
1280.0	5.4432	695.8	5.0778	0.0841	28.9644
1300.0	5.3594	700.9	5.1294	0.0851	28.9644
1320.0	5.2782	705.9	5.1807	0.0862	28.9644
1340.0	5.1995	710.9	5.2318	0.0873	28.9644
1360.0	5.1230	715.9	5.2825	0.0884	28.9644
1380.0	5.0487	720.8	5.3329	0.0894	28.9644
1400.0	4.9766	725.7	5.3831	0.0905	28.9644
1420.0	4.9065	730.6	5.4330	0.0916	28.9644
1440.0	4.8384	735.4	5.4827	0.0927	28.9644
1460.0	4.7721	740.1	5.5320	0.0937	28.9644
1480.0	4.7076	744.9	5.5811	0.0948	28.9644
1500.0	4.6448	749.6	5.6300	0.0959	28.9644
1520.0	4.5837	754.2	5.6786	0.0969	28.9644
1540.0	4.5242	758.8	5.7270	0.0980	28.9644
1560.0	4.4662	763.4	5.7751	0.0991	28.9644
1580.0	4.4097	768.0	5.8231	0.1002	28.9644
1600.0	4.3545	772.5	5.8707	0.1013	28.9644
1620.0	4.3008	777.0	5.9182	0.1023	28.9644
1640.0	4.2483	781.4	5.9654	0.1034	28.9644
1660.0	4.1972	785.9	6.0125	0.1045	28.9644
1680.0	4.1472	790.3	6.0593	0.1056	28.9644
1700.0	4.0984	794.6	6.1059	0.1067	28.9644
1720.0	4.0507	799.0	6.1523	0.1078	28.9644
1740.0	4.0042	803.3	6.1985	0.1089	28.9644
1760.0	3.9587	807.5	6.2445	0.1099	28.9643
1780.0	3.9142	811.8	6.2903	0.1110	28.9643
1800.0	3.8707	816.0	6.3360	0.1121	28.9643
1820.0	3.8282	820.2	6.3814	0.1132	28.9643
1840.0	3.7866	824.4	6.4267	0.1143	28.9643
1860.0	3.7458	828.5	6.4717	0.1154	28.9643
1880.0	3.7060	832.6	6.5166	0.1165	28.9643
1900.0	3.6670	836.7	6.5614	0.1176	28.9642
1920.0	3.6288	840.8	6.6059	0.1187	28.9642
1940.0	3.5914	844.8	6.6503	0.1198	28.9642
1960.0	3.5547	848.8	6.6945	0.1209	28.9642
1980.0	3.5188	852.8	6.7385	0.1220	28.9641
2000.0	3.4836	856.8	6.7824	0.1231	28.9641

THERMOPHYSICAL PROPERTIES OF AIR

ISOBAR: 20.0 MPa

T (K)	h (J/kg)	e (J/kg)	s (J/kg·K)	c_p (J/kg·K)	γ	Pr
1020.0	7.6733 +2	4.7434 +2	6.6429 +0	1.1465 +0	1.3343	0.7185
1040.0	7.9030	4.9156	6.6652	1.1501	1.3329	0.7185
1060.0	8.1333	5.0885	6.6872	1.1537	1.3315	0.7185
1080.0	8.3644	5.2622	6.7088	1.1572	1.3302	0.7185
1100.0	8.5962	5.4365	6.7300	1.1608	1.3288	0.7186
1120.0	8.8288	5.6116	6.7510	1.1643	1.3275	0.7187
1140.0	9.0620	5.7873	6.7716	1.1678	1.3262	0.7188
1160.0	9.2959	5.9638	6.7919	1.1712	1.3249	0.7190
1180.0	9.5305	6.1409	6.8120	1.1747	1.3237	0.7191
1200.0	9.7657	6.3187	6.8318	1.1781	1.3224	0.7192
1220.0	1.0002 +3	6.4973	6.8513	1.1816	1.3212	0.7194
1240.0	1.0238	6.6765	6.8705	1.1850	1.3200	0.7195
1260.0	1.0476	6.8564	6.8895	1.1884	1.3188	0.7197
1280.0	1.0714	7.0369	6.9082	1.1918	1.3176	0.7199
1300.0	1.0952	7.2182	6.9267	1.1952	1.3164	0.7201
1320.0	1.1192	7.4001	6.9450	1.1986	1.3152	0.7202
1340.0	1.1432	7.5827	6.9631	1.2020	1.3140	0.7204
1360.0	1.1673	7.7660	6.9809	1.2054	1.3128	0.7206
1380.0	1.1914	7.9500	6.9985	1.2089	1.3117	0.7208
1400.0	1.2156	8.1347	7.0159	1.2123	1.3105	0.7210
1420.0	1.2399	8.3200	7.0332	1.2157	1.3094	0.7212
1440.0	1.2642	8.5060	7.0502	1.2191	1.3082	0.7214
1460.0	1.2887	8.6928	7.0670	1.2226	1.3071	0.7216
1480.0	1.3131	8.8802	7.0837	1.2260	1.3060	0.7218
1500.0	1.3377	9.0683	7.1002	1.2295	1.3049	0.7220
1520.0	1.3623	9.2571	7.1165	1.2330	1.3037	0.7222
1540.0	1.3870	9.4466	7.1326	1.2364	1.3026	0.7224
1560.0	1.4118	9.6367	7.1486	1.2399	1.3015	0.7225
1580.0	1.4366	9.8276	7.1644	1.2434	1.3004	0.7227
1600.0	1.4615	1.0019 +3	7.1801	1.2469	1.2993	0.7229
1620.0	1.4865	1.0211	7.1956	1.2504	1.2982	0.7231
1640.0	1.5115	1.0404	7.2110	1.2540	1.2971	0.7232
1660.0	1.5366	1.0598	7.2262	1.2575	1.2961	0.7234
1680.0	1.5618	1.0793	7.2413	1.2610	1.2950	0.7236
1700.0	1.5871	1.0988	7.2562	1.2646	1.2939	0.7237
1720.0	1.6124	1.1184	7.2710	1.2681	1.2929	0.7239
1740.0	1.6378	1.1380	7.2857	1.2717	1.2918	0.7240
1760.0	1.6633	1.1577	7.3002	1.2753	1.2907	0.7242
1780.0	1.6888	1.1775	7.3147	1.2788	1.2897	0.7243
1800.0	1.7144	1.1974	7.3290	1.2824	1.2887	0.7245
1820.0	1.7401	1.2173	7.3432	1.2860	1.2876	0.7246
1840.0	1.7659	1.2373	7.3572	1.2896	1.2866	0.7247
1860.0	1.7917	1.2574	7.3712	1.2932	1.2856	0.7249
1880.0	1.8176	1.2776	7.3851	1.2968	1.2846	0.7250
1900.0	1.8436	1.2978	7.3988	1.3004	1.2836	0.7251
1920.0	1.8696	1.3181	7.4124	1.3040	1.2826	0.7252
1940.0	1.8957	1.3385	7.4260	1.3077	1.2816	0.7253
1960.0	1.9219	1.3589	7.4394	1.3113	1.2806	0.7254
1980.0	1.9482	1.3794	7.4527	1.3149	1.2796	0.7255
2000.0	1.9745	1.4000	7.4660	1.3186	1.2786	0.7256

THERMOPHYSICAL PROPERTIES OF AIR

ISOBAR: 20.0 MPa

T (K)	ρ (kg/m^3)	a (m/s)	μ (Pa·s)	k (J/s·m·K)	$\bar{m}$
2040.0	3.4153 +1	864.7	6.8697 −5	0.1253 +0	28.9640
2080.0	3.3496	872.5	6.9564	0.1275	28.9638
2120.0	3.2864	880.2	7.0424	0.1297	28.9636
2160.0	3.2255	887.8	7.1279	0.1319	28.9634
2200.0	3.1668	895.3	7.2128	0.1341	28.9631
2240.0	3.1102	902.8	7.2972	0.1372	28.9628
2280.0	3.0556	910.2	7.3810	0.1396	28.9623
2320.0	3.0029	917.5	7.4643	0.1420	28.9618
2360.0	2.9519	924.8	7.5471	0.1444	28.9612
2400.0	2.9026	932.0	7.6294	0.1469	28.9604
2440.0	2.8550	939.1	7.7112	0.1495	28.9595
2480.0	2.8088	946.1	7.7925	0.1521	28.9584
2520.0	2.7641	953.1	7.8733	0.1547	28.9571
2560.0	2.7208	960.1	7.9537	0.1574	28.9556
2600.0	2.6787	966.9	8.0337	0.1602	28.9538
2640.0	2.6380	973.7	8.1132	0.1631	28.9517
2680.0	2.5984	980.5	8.1923	0.1660	28.9494
2720.0	2.5599	987.2	8.2710	0.1690	28.9466
2760.0	2.5226	993.8	8.3494	0.1721	28.9435
2800.0	2.4862	1000.4	8.4273	0.1754	28.9400
2840.0	2.4509	1007.0	8.5049	0.1787	28.9359
2880.0	2.4164	1013.5	8.5821	0.1821	28.9314
2920.0	2.3829	1019.9	8.6590	0.1857	28.9263
2960.0	2.3502	1026.3	8.7355	0.1894	28.9206
3000.0	2.3184	1032.7	8.8117	0.1932	28.9142
3040.0	2.2873	1039.1	8.8876	0.1971	28.9071
3080.0	2.2570	1045.4	8.9632	0.2012	28.8993
3120.0	2.2274	1051.7	9.0384	0.2055	28.8906
3160.0	2.1985	1057.9	9.1135	0.2099	28.8811
3200.0	2.1702	1064.2	9.1882	0.2145	28.8706
3240.0	2.1426	1070.4	9.2627	0.2192	28.8592
3280.0	2.1155	1076.6	9.3369	0.2241	28.8467
3320.0	2.0891	1082.8	9.4109	0.2292	28.8331
3360.0	2.0631	1089.0	9.4847	0.2344	28.8184
3400.0	2.0377	1095.2	9.5583	0.2398	28.8025
3440.0	2.0129	1101.3	9.6316	0.2454	28.7854
3480.0	1.9884	1107.5	9.7048	0.2511	28.7669
3520.0	1.9645	1113.7	9.7778	0.2570	28.7471
3560.0	1.9410	1120.0	9.8506	0.2631	28.7259
3600.0	1.9179	1126.2	9.9233	0.2695	28.7033
3640.0	1.8952	1132.4	9.9958	0.2758	28.6791
3680.0	1.8730	1138.7	1.0068 −4	0.2823	28.6535
3720.0	1.8511	1145.0	1.0140	0.2889	28.6263
3760.0	1.8295	1151.3	1.0213	0.2956	28.5975
3800.0	1.8083	1157.7	1.0285	0.3024	28.5671
3840.0	1.7875	1164.1	1.0357	0.3101	28.5351
3880.0	1.7670	1170.5	1.0429	0.3172	28.5014
3920.0	1.7468	1177.0	1.0501	0.3244	28.4660
3960.0	1.7269	1183.6	1.0572	0.3316	28.4290
4000.0	1.7073	1190.1	1.0644	0.3388	28.3903

THERMOPHYSICAL PROPERTIES OF AIR

ISOBAR: 20.0 MPa

T	h	e	s	c_p	γ	Pr
(K)	(J/kg)	(J/kg)	(J/kg·K)	(J/kg·K)		
2040.0	2.0274 +3	1.4414 +3	7.4921 +0	1.3259 +0	1.2767	0.7258
2080.0	2.0806	1.4831	7.5180	1.3332	1.2748	0.7259
2120.0	2.1341	1.5251	7.5434	1.3406	1.2730	0.7260
2160.0	2.1878	1.5674	7.5686	1.3480	1.2711	0.7262
2200.0	2.2419	1.6099	7.5934	1.3555	1.2693	0.7263
2240.0	2.2963	1.6528	7.6179	1.3630	1.2675	0.7251
2280.0	2.3510	1.6960	7.6420	1.3707	1.2657	0.7249
2320.0	2.4059	1.7395	7.6659	1.3784	1.2640	0.7246
2360.0	2.4612	1.7832	7.6896	1.3863	1.2623	0.7243
2400.0	2.5168	1.8273	7.7129	1.3943	1.2606	0.7239
2440.0	2.5728	1.8718	7.7361	1.4025	1.2589	0.7234
2480.0	2.6290	1.9165	7.7589	1.4109	1.2572	0.7229
2520.0	2.6856	1.9616	7.7816	1.4195	1.2555	0.7223
2560.0	2.7426	2.0070	7.8040	1.4284	1.2539	0.7215
2600.0	2.7999	2.0528	7.8262	1.4376	1.2522	0.7207
2640.0	2.8576	2.0989	7.8482	1.4470	1.2506	0.7198
2680.0	2.9157	2.1455	7.8701	1.4568	1.2490	0.7189
2720.0	2.9742	2.1924	7.8917	1.4670	1.2474	0.7178
2760.0	3.0331	2.2397	7.9132	1.4776	1.2458	0.7166
2800.0	3.0924	2.2874	7.9346	1.4887	1.2442	0.7153
2840.0	3.1522	2.3356	7.9558	1.5001	1.2426	0.7139
2880.0	3.2124	2.3842	7.9768	1.5121	1.2410	0.7124
2920.0	3.2731	2.4333	7.9978	1.5246	1.2394	0.7108
2960.0	3.3344	2.4828	8.0186	1.5377	1.2379	0.7091
3000.0	3.3962	2.5329	8.0393	1.5513	1.2363	0.7073
3040.0	3.4585	2.5835	8.0600	1.5655	1.2348	0.7055
3080.0	3.5214	2.6347	8.0805	1.5804	1.2333	0.7035
3120.0	3.5849	2.6864	8.1010	1.5958	1.2318	0.7015
3160.0	3.6491	2.7388	8.1215	1.6120	1.2303	0.6994
3200.0	3.7139	2.7917	8.1418	1.6287	1.2288	0.6972
3240.0	3.7794	2.8453	8.1622	1.6462	1.2274	0.6950
3280.0	3.8456	2.8996	8.1825	1.6643	1.2260	0.6927
3320.0	3.9125	2.9545	8.2028	1.6831	1.2246	0.6904
3360.0	3.9803	3.0102	8.2230	1.7025	1.2233	0.6881
3400.0	4.0488	3.0666	8.2433	1.7226	1.2220	0.6858
3440.0	4.1181	3.1238	8.2636	1.7434	1.2208	0.6834
3480.0	4.1882	3.1817	8.2839	1.7648	1.2196	0.6811
3520.0	4.2593	3.2405	8.3041	1.7868	1.2184	0.6787
3560.0	4.3312	3.3001	8.3245	1.8094	1.2173	0.6764
3600.0	4.4040	3.3605	8.3448	1.8326	1.2162	0.6743
3640.0	4.4778	3.4218	8.3652	1.8563	1.2152	0.6721
3680.0	4.5525	3.4840	8.3856	1.8805	1.2143	0.6699
3720.0	4.6282	3.5471	8.4061	1.9051	1.2134	0.6679
3760.0	4.7049	3.6110	8.4266	1.9302	1.2126	0.6659
3800.0	4.7827	3.6759	8.4471	1.9556	1.2119	0.6640
3840.0	4.8614	3.7418	8.4678	1.9813	1.2112	0.6616
3880.0	4.9412	3.8085	8.4884	2.0073	1.2105	0.6599
3920.0	5.0220	3.8763	8.5091	2.0334	1.2100	0.6582
3960.0	5.1038	3.9449	8.5299	2.0596	1.2095	0.6567
4000.0	5.1868	4.0145	8.5507	2.0859	1.2091	0.6553

THERMOPHYSICAL PROPERTIES OF AIR

ISOBAR: 20.0 MPa

T (K)	ρ (kg/m^3)	a (m/s)	μ (Pa·s)	k (J/s·m·K)	$\bar{m}$
4040.0	1.6880 +1	1196.7	1.0716 −4	0.3461 +0	28.3499
4080.0	1.6690	1203.4	1.0787	0.3533	28.3079
4120.0	1.6502	1210.1	1.0859	0.3606	28.2642
4160.0	1.6317	1216.9	1.0931	0.3678	28.2189
4200.0	1.6135	1223.7	1.1002	0.3749	28.1720
4240.0	1.5955	1230.6	1.1074	0.3820	28.1235
4280.0	1.5778	1237.5	1.1146	0.3889	28.0735
4320.0	1.5603	1244.5	1.1217	0.3957	28.0221
4360.0	1.5431	1251.5	1.1289	0.4024	27.9692
4400.0	1.5261	1258.6	1.1361	0.4090	27.9149
4440.0	1.5093	1265.8	1.1433	0.4153	27.8593
4480.0	1.4928	1273.0	1.1504	0.4215	27.8025
4520.0	1.4765	1280.2	1.1576	0.4273	27.7445
4560.0	1.4604	1287.5	1.1648	0.4330	27.6854
4600.0	1.4446	1294.9	1.1720	0.4384	27.6253
4640.0	1.4290	1302.3	1.1792	0.4437	27.5643
4680.0	1.4136	1309.7	1.1864	0.4486	27.5024
4720.0	1.3984	1317.2	1.1936	0.4533	27.4398
4760.0	1.3835	1324.7	1.2008	0.4578	27.3765
4800.0	1.3687	1332.3	1.2080	0.4619	27.3126
4840.0	1.3542	1339.9	1.2152	0.4658	27.2483
4880.0	1.3399	1347.5	1.2224	0.4694	27.1836
4920.0	1.3259	1355.2	1.2296	0.4728	27.1186
4960.0	1.3120	1362.9	1.2368	0.4759	27.0533
5000.0	1.2984	1370.6	1.2440	0.4787	26.9880
5040.0	1.2849	1378.3	1.2512	0.4813	26.9227
5080.0	1.2717	1386.0	1.2585	0.4837	26.8574
5120.0	1.2587	1393.8	1.2657	0.4859	26.7922
5160.0	1.2460	1401.5	1.2729	0.4880	26.7273
5200.0	1.2334	1409.3	1.2801	0.4898	26.6626
5240.0	1.2210	1417.1	1.2873	0.4916	26.5983
5280.0	1.2089	1424.8	1.2945	0.4932	26.5345
5320.0	1.1969	1432.5	1.3017	0.4948	26.4711
5360.0	1.1851	1440.3	1.3089	0.4963	26.4082
5400.0	1.1736	1448.0	1.3161	0.4978	26.3459
5440.0	1.1622	1455.6	1.3233	0.4992	26.2842
5480.0	1.1511	1463.3	1.3305	0.5007	26.2232
5520.0	1.1401	1470.9	1.3377	0.5023	26.1628
5560.0	1.1293	1478.5	1.3449	0.5040	26.1031
5600.0	1.1187	1486.0	1.3521	0.5058	26.0442
5640.0	1.1083	1493.5	1.3593	0.5077	25.9859
5680.0	1.0981	1501.0	1.3665	0.5098	25.9284
5720.0	1.0880	1508.4	1.3737	0.5121	25.8716
5760.0	1.0781	1515.7	1.3808	0.5147	25.8156
5800.0	1.0684	1523.0	1.3880	0.5175	25.7602
5840.0	1.0588	1530.3	1.3952	0.5206	25.7056
5880.0	1.0494	1537.4	1.4024	0.5239	25.6516
5920.0	1.0401	1544.5	1.4095	0.5276	25.5983
5960.0	1.0310	1551.6	1.4167	0.5317	25.5456
6000.0	1.0221	1558.6	1.4239	0.5361	25.4935

THERMOPHYSICAL PROPERTIES OF AIR

ISOBAR: 20.0 MPa

T (K)	h (J/kg)	e (J/kg)	s (J/kg·K)	c_p (J/kg·K)	γ	Pr
4040.0	5.2707 +3	4.0851 +3	8.5716 +0	2.1121 +0	1.2088	0.6540
4080.0	5.3557	4.1566	8.5926	2.1382	1.2085	0.6528
4120.0	5.4418	4.2290	8.6136	2.1641	1.2083	0.6517
4160.0	5.5288	4.3023	8.6346	2.1898	1.2081	0.6508
4200.0	5.6169	4.3766	8.6557	2.2151	1.2081	0.6501
4240.0	5.7061	4.4517	8.6768	2.2400	1.2081	0.6494
4280.0	5.7961	4.5277	8.6979	2.2643	1.2082	0.6489
4320.0	5.8872	4.6045	8.7191	2.2881	1.2083	0.6486
4360.0	5.9792	4.6822	8.7403	2.3112	1.2085	0.6483
4400.0	6.0721	4.7607	8.7615	2.3335	1.2088	0.6482
4440.0	6.1658	4.8399	8.7827	2.3550	1.2091	0.6483
4480.0	6.2605	4.9198	8.8039	2.3755	1.2095	0.6484
4520.0	6.3559	5.0004	8.8251	2.3951	1.2100	0.6486
4560.0	6.4521	5.0817	8.8463	2.4137	1.2105	0.6491
4600.0	6.5490	5.1636	8.8675	2.4311	1.2111	0.6496
4640.0	6.6465	5.2460	8.8886	2.4473	1.2117	0.6502
4680.0	6.7447	5.3289	8.9097	2.4624	1.2124	0.6509
4720.0	6.8435	5.4124	8.9307	2.4762	1.2131	0.6517
4760.0	6.9428	5.4962	8.9516	2.4887	1.2139	0.6526
4800.0	7.0426	5.5804	8.9725	2.4999	1.2147	0.6536
4840.0	7.1428	5.6649	8.9933	2.5099	1.2156	0.6546
4880.0	7.2434	5.7498	9.0140	2.5185	1.2165	0.6556
4920.0	7.3442	5.8348	9.0346	2.5257	1.2175	0.6567
4960.0	7.4454	5.9200	9.0551	2.5317	1.2185	0.6578
5000.0	7.5468	6.0054	9.0754	2.5365	1.2195	0.6590
5040.0	7.6483	6.0908	9.0956	2.5400	1.2205	0.6601
5080.0	7.7499	6.1762	9.1157	2.5423	1.2216	0.6612
5120.0	7.8517	6.2617	9.1357	2.5435	1.2226	0.6623
5160.0	7.9534	6.3472	9.1555	2.5437	1.2237	0.6634
5200.0	8.0552	6.4325	9.1751	2.5428	1.2248	0.6644
5240.0	8.1568	6.5178	9.1946	2.5411	1.2259	0.6653
5280.0	8.2584	6.6029	9.2139	2.5385	1.2270	0.6661
5320.0	8.3599	6.6878	9.2331	2.5353	1.2281	0.6669
5360.0	8.4612	6.7726	9.2520	2.5314	1.2292	0.6675
5400.0	8.5624	6.8571	9.2708	2.5270	1.2303	0.6681
5440.0	8.6634	6.9414	9.2895	2.5221	1.2313	0.6684
5480.0	8.7642	7.0255	9.3079	2.5170	1.2323	0.6687
5520.0	8.8648	7.1093	9.3262	2.5117	1.2333	0.6688
5560.0	8.9651	7.1929	9.3443	2.5063	1.2343	0.6687
5600.0	9.0653	7.2763	9.3623	2.5010	1.2352	0.6685
5640.0	9.1652	7.3594	9.3801	2.4958	1.2361	0.6681
5680.0	9.2649	7.4423	9.3977	2.4908	1.2369	0.6675
5720.0	9.3645	7.5250	9.4151	2.4862	1.2377	0.6668
5760.0	9.4638	7.6075	9.4325	2.4821	1.2384	0.6658
5800.0	9.5630	7.6898	9.4496	2.4785	1.2391	0.6647
5840.0	9.6621	7.7719	9.4666	2.4755	1.2397	0.6634
5880.0	9.7611	7.8539	9.4835	2.4733	1.2402	0.6619
5920.0	9.8600	7.9359	9.5003	2.4719	1.2407	0.6603
5960.0	9.9589	8.0177	9.5169	2.4714	1.2411	0.6585
6000.0	1.0058 +4	8.0996	9.5335	2.4719	1.2414	0.6565

THERMOPHYSICAL PROPERTIES OF AIR

ISOBAR: 100.0 MPa

T (K)	ρ (kg/m^3)	a (m/s)	μ (Pa·s)	k (J/s·m·K)	$\bar{m}$
288.15	1.2090 +3	340.3	1.7979 -5	0.0249 +0	28.9644
300.0	1.1612	347.2	1.8554	0.0257	28.9644
310.0	1.1238	352.9	1.9030	0.0264	28.9644
320.0	1.0886	358.5	1.9500	0.0271	28.9644
330.0	1.0556	364.1	1.9962	0.0278	28.9644
340.0	1.0246	369.5	2.0419	0.0285	28.9644
350.0	9.9532 +2	374.8	2.0869	0.0292	28.9644
360.0	9.6768	380.1	2.1313	0.0298	28.9644
370.0	9.4152	385.2	2.1751	0.0305	28.9644
380.0	9.1675	390.3	2.2184	0.0312	28.9644
390.0	8.9324	395.3	2.2612	0.0318	28.9644
400.0	8.7091	400.3	2.3035	0.0325	28.9644
410.0	8.4967	405.1	2.3454	0.0331	28.9644
420.0	8.2944	409.9	2.3867	0.0338	28.9644
430.0	8.1015	414.7	2.4276	0.0344	28.9644
440.0	7.9174	419.3	2.4681	0.0350	28.9644
450.0	7.7414	423.9	2.5082	0.0357	28.9644
460.0	7.5731	428.5	2.5479	0.0363	28.9644
470.0	7.4120	433.0	2.5872	0.0370	28.9644
480.0	7.2576	437.4	2.6261	0.0376	28.9644
490.0	7.1095	441.8	2.6647	0.0382	28.9644
500.0	6.9673	446.1	2.7029	0.0388	28.9644
510.0	6.8307	450.4	2.7408	0.0395	28.9644
520.0	6.6993	454.6	2.7783	0.0401	28.9644
530.0	6.5729	458.8	2.8155	0.0407	28.9644
540.0	6.4512	462.9	2.8525	0.0413	28.9644
550.0	6.3339	467.0	2.8891	0.0420	28.9644
560.0	6.2208	471.0	2.9254	0.0426	28.9644
580.0	6.0063	479.0	2.9972	0.0438	28.9644
600.0	5.8061	486.8	3.0680	0.0450	28.9644
620.0	5.6188	494.4	3.1377	0.0463	28.9644
640.0	5.4432	501.9	3.2065	0.0475	28.9644
660.0	5.2782	509.3	3.2743	0.0487	28.9644
680.0	5.1230	516.5	3.3412	0.0499	28.9644
700.0	4.9766	523.6	3.4073	0.0511	28.9644
720.0	4.8384	530.6	3.4726	0.0523	28.9644
740.0	4.7076	537.5	3.5371	0.0535	28.9644
760.0	4.5837	544.3	3.6008	0.0547	28.9644
780.0	4.4662	551.0	3.6639	0.0559	28.9644
800.0	4.3545	557.5	3.7263	0.0571	28.9644
820.0	4.2483	564.1	3.7880	0.0583	28.9644
840.0	4.1472	570.5	3.8490	0.0594	28.9644
860.0	4.0507	576.8	3.9095	0.0606	28.9644
880.0	3.9587	583.1	3.9693	0.0618	28.9644
900.0	3.8707	589.3	4.0286	0.0630	28.9644
920.0	3.7866	595.4	4.0873	0.0641	28.9644
940.0	3.7060	601.4	4.1455	0.0653	28.9644
960.0	3.6288	607.4	4.2032	0.0665	28.9645
980.0	3.5547	613.3	4.2603	0.0676	28.9644
1000.0	3.4836	619.2	4.3170	0.0688	28.9644

THERMOPHYSICAL PROPERTIES OF AIR

ISOBAR: 100.0 MPa

288.15	− 1.4106 +1	−9.6876 +1	4.8504 +0	1.0052 +0	1.4001	0.725
300.0	− 2.1924 +0	−8.8367	4.8909	1.0055	1.3999	0.724
310.0	7.8646	−8.1182	4.9239	1.0059	1.3997	0.724
320.0	1.7926 +1	−7.3994	4.9558	1.0064	1.3994	0.723
330.0	2.7993	−6.6799	4.9868	1.0070	1.3991	0.722
340.0	3.8067	−5.9598	5.0169	1.0077	1.3987	0.722
350.0	4.8148	−5.2389	5.0461	1.0085	1.3982	0.721
360.0	5.8238	−4.5171	5.0745	1.0095	1.3977	0.721
370.0	6.8337	−3.7944	5.1022	1.0105	1.3972	0.720
380.0	7.8447	−3.0707	5.1291	1.0116	1.3966	0.720
390.0	8.8569	−2.3458	5.1554	1.0127	1.3959	0.7199
400.0	9.8702	−1.6197	5.1811	1.0140	1.3953	0.7195
410.0	1.0885 +2	−8.9229 +0	5.2061	1.0153	1.3945	0.7191
420.0	1.1901	−1.6350	5.2306	1.0168	1.3938	0.7188
430.0	1.2918	5.6674	5.2546	1.0182	1.3930	0.7185
440.0	1.3937	1.2985 +1	5.2780	1.0198	1.3921	0.7182
450.0	1.4958	2.0319	5.3009	1.0214	1.3913	0.7179
460.0	1.5980	2.7669	5.3234	1.0231	1.3904	0.7177
470.0	1.7004	3.5036	5.3454	1.0249	1.3894	0.7175
480.0	1.8030	4.2421	5.3670	1.0267	1.3885	0.7173
490.0	1.9058	4.9825	5.3882	1.0285	1.3875	0.7171
500.0	2.0087	5.7247	5.4090	1.0304	1.3865	0.7170
510.0	2.1119	6.4689	5.4294	1.0324	1.3855	0.7169
520.0	2.2152	7.2150	5.4495	1.0344	1.3845	0.7167
530.0	2.3187	7.9632	5.4692	1.0365	1.3834	0.7167
540.0	2.4225	8.7134	5.4886	1.0385	1.3823	0.7166
550.0	2.5264	9.4658	5.5077	1.0407	1.3813	0.7165
560.0	2.6306	1.0220 +2	5.5264	1.0428	1.3802	0.7165
580.0	2.8396	1.1736	5.5631	1.0473	1.3780	0.7165
600.0	3.0495	1.3260	5.5987	1.0518	1.3757	0.7165
620.0	3.2603	1.4794	5.6333	1.0564	1.3735	0.7165
640.0	3.4721	1.6337	5.6669	1.0611	1.3712	0.7166
660.0	3.6848	1.7890	5.6996	1.0659	1.3689	0.7168
680.0	3.8984	1.9452	5.7315	1.0707	1.3667	0.7169
700.0	4.1131	2.1023	5.7626	1.0755	1.3644	0.7171
720.0	4.3286	2.2605	5.7930	1.0803	1.3622	0.7172
740.0	4.5452	2.4196	5.8226	1.0851	1.3600	0.7174
760.0	4.7627	2.5796	5.8516	1.0900	1.3578	0.7176
780.0	4.9812	2.7406	5.8800	1.0947	1.3557	0.7177
800.0	5.2006	2.9026	5.9078	1.0995	1.3537	0.7178
820.0	5.4210	3.0655	5.9350	1.1042	1.3516	0.7179
840.0	5.6423	3.2294	5.9617	1.1088	1.3496	0.7179
860.0	5.8645	3.3941	5.9878	1.1134	1.3477	0.7179
880.0	6.0876	3.5598	6.0134	1.1178	1.3458	0.7179
900.0	6.3116	3.7264	6.0386	1.1223	1.3440	0.7178
920.0	6.5365	3.8938	6.0633	1.1266	1.3422	0.7176
940.0	6.7622	4.0621	6.0876	1.1308	1.3405	0.7174
960.0	6.9888	4.2312	6.1115	1.1349	1.3389	0.7172
980.0	7.2162	4.4012	6.1349	1.1389	1.3373	0.7169
1000.0	7.4444	4.5719	6.1579	1.1428	1.3357	0.7165

THERMOPHYSICAL PROPERTIES OF AIR

ISOBAR: 100.0 MPa

T (K)	ρ (kg/m^3)	a (m/s)	μ (Pa·s)	k (J/s·m·K)	$\bar{m}$
1020.0	3.4153 +2	625.0	4.3745 -5	0.0697 +0	28.9644
1040.0	3.3497	630.8	4.4309	0.0708	28.9644
1060.0	3.2864	636.5	4.4870	0.0719	28.9644
1080.0	3.2256	642.2	4.5426	0.0730	28.9644
1100.0	3.1669	647.8	4.5978	0.0741	28.9644
1120.0	3.1104	653.3	4.6526	0.0754	28.9644
114C.0	3.0558	658.8	4.7070	0.0765	28.9644
1160.0	3.0031	664.2	4.7610	0.0776	28.9644
1180.0	2.9522	669.6	4.8147	0.0787	28.9644
1200.0	2.9030	674.9	4.8680	0.0797	28.9644
1220.0	2.8554	680.2	4.9209	0.0808	28.9644
1240.0	2.8094	685.5	4.9735	0.0819	28.9644
1260.0	2.7648	690.6	5.0258	0.0830	28.9644
1280.0	2.7216	695.8	5.0778	0.0841	28.9644
1300.0	2.6797	700.9	5.1294	0.0851	28.9644
1320.0	2.6391	705.9	5.1807	0.0862	28.9644
1340.0	2.5997	710.9	5.2318	0.0873	28.9644
1360.0	2.5615	715.9	5.2825	0.0884	28.9644
1380.0	2.5244	720.8	5.3329	0.0894	28.9644
1400.0	2.4883	725.7	5.3831	0.0905	28.9644
1420.0	2.4533	730.6	5.4330	0.0916	28.9644
1440.0	2.4192	735.4	5.4827	0.0927	28.9644
1460.0	2.3861	740.1	5.5320	0.0937	28.9644
1480.0	2.3538	744.9	5.5811	0.0948	28.9644
1500.0	2.3224	749.6	5.6300	0.0959	28.9644
1520.0	2.2919	754.2	5.6786	0.0969	28.9644
1540.0	2.2621	758.8	5.7270	0.0980	28.9644
1560.0	2.2331	763.4	5.7751	0.0991	28.9644
1580.0	2.2048	768.0	5.8231	0.1002	28.9644
1600.0	2.1773	772.5	5.8707	0.1013	28.9644
1620.0	2.1504	777.0	5.9182	0.1023	28.9644
1640.0	2.1242	781.5	5.9654	0.1034	28.9644
1660.0	2.0986	785.9	6.0125	0.1045	28.9644
1680.0	2.0736	790.3	6.0593	0.1056	28.9644
1700.0	2.0492	794.6	6.1059	0.1067	28.9644
1720.0	2.0254	799.0	6.1523	0.1078	28.9644
1740.0	2.0021	803.3	6.1985	0.1089	28.9644
176C.0	1.9793	807.5	6.2445	0.1099	28.9644
1780.0	1.9571	811.8	6.2903	0.1110	28.9644
1800.0	1.9354	816.0	6.3360	0.1121	28.9644
1820.0	1.9141	820.2	6.3814	0.1132	28.9644
1840.0	1.8933	824.4	6.4267	0.1143	28.9644
1860.0	1.8729	828.5	6.4717	0.1154	28.9643
1880.0	1.8530	832.6	6.5166	0.1165	28.9643
1900.0	1.8335	836.7	6.5614	0.1176	28.9643
1920.0	1.8144	840.8	6.6059	0.1187	28.9643
1940.0	1.7957	844.8	6.6503	0.1198	28.9643
1960.0	1.7774	848.9	6.6945	0.1209	28.9643
1980.0	1.7594	852.9	6.7385	0.1220	28.9643
2000.0	1.7418	856.8	6.7824	0.1231	28.9643

THERMOPHYSICAL PROPERTIES OF AIR

ISOBAR: 100.0 MPa

T (K)	h (J/kg)	e (J/kg)	s (J/kg·K)	c_p (J/kg·K)	γ	Pr
1020.0	7.6733 +2	4.7434 +2	6.1806 +0	1.1465 +0	1.3343	0.7185
1040.0	7.9030	4.9156	6.2029	1.1501	1.3329	0.7185
1060.0	8.1333	5.0885	6.2248	1.1537	1.3315	0.7185
1080.0	8.3644	5.2622	6.2465	1.1572	1.3302	0.7185
1100.0	8.5962	5.4365	6.2677	1.1608	1.3288	0.7186
1120.0	8.8288	5.6116	6.2887	1.1643	1.3275	0.7187
1140.0	9.0620	5.7873	6.3093	1.1678	1.3262	0.7188
1160.0	9.2959	5.9638	6.3296	1.1712	1.3249	0.7190
1180.0	9.5305	6.1409	6.3497	1.1747	1.3237	0.7191
1200.0	9.7657	6.3187	6.3695	1.1781	1.3224	0.7192
1220.0	1.0002 +3	6.4973	6.3890	1.1816	1.3212	0.7194
1240.0	1.0238	6.6765	6.4082	1.1850	1.3200	0.7195
1260.0	1.0476	6.8564	6.4272	1.1884	1.3188	0.7197
1280.0	1.0714	7.0369	6.4459	1.1918	1.3176	0.7199
1300.0	1.0952	7.2182	6.4644	1.1952	1.3164	0.7201
1320.0	1.1192	7.4001	6.4827	1.1986	1.3152	0.7202
1340.0	1.1432	7.5827	6.5008	1.2020	1.3140	0.7204
1360.0	1.1673	7.7660	6.5186	1.2054	1.3128	0.7206
1380.0	1.1914	7.9500	6.5362	1.2088	1.3117	0.7208
1400.0	1.2156	8.1347	6.5536	1.2123	1.3105	0.7210
1420.0	1.2399	8.3200	6.5709	1.2157	1.3094	0.7212
1440.0	1.2642	8.5060	6.5879	1.2191	1.3082	0.7214
1460.0	1.2887	8.6928	6.6047	1.2226	1.3071	0.7216
1480.0	1.3131	8.8802	6.6214	1.2260	1.3060	0.7218
1500.0	1.3377	9.0683	6.6379	1.2295	1.3049	0.7220
1520.0	1.3623	9.2571	6.6542	1.2329	1.3037	0.7222
1540.0	1.3870	9.4466	6.6703	1.2364	1.3026	0.7224
1560.0	1.4118	9.6367	6.6863	1.2399	1.3015	0.7225
1580.0	1.4366	9.8276	6.7021	1.2434	1.3004	0.7227
1600.0	1.4615	1.0019 +3	6.7178	1.2469	1.2993	0.7229
1620.0	1.4865	1.0211	6.7333	1.2504	1.2982	0.7231
1640.0	1.5115	1.0404	6.7486	1.2539	1.2972	0.7232
1660.0	1.5366	1.0598	6.7639	1.2574	1.2961	0.7234
1680.0	1.5618	1.0793	6.7789	1.2610	1.2950	0.7236
1700.0	1.5871	1.0988	6.7939	1.2645	1.2939	0.7237
1720.0	1.6124	1.1183	6.8087	1.2680	1.2929	0.7239
1740.0	1.6378	1.1380	6.8234	1.2716	1.2918	0.7240
1760.0	1.6633	1.1577	6.8379	1.2751	1.2908	0.7242
1780.0	1.6888	1.1775	6.8524	1.2787	1.2897	0.7243
1800.0	1.7144	1.1974	6.8667	1.2822	1.2887	0.7245
1820.0	1.7401	1.2173	6.8809	1.2858	1.2877	0.7246
1840.0	1.7659	1.2373	6.8949	1.2894	1.2867	0.7247
1860.0	1.7917	1.2574	6.9089	1.2929	1.2856	0.7249
1880.0	1.8176	1.2775	6.9227	1.2965	1.2846	0.7250
1900.0	1.8435	1.2978	6.9365	1.3000	1.2836	0.7251
1920.0	1.8696	1.3181	6.9501	1.3036	1.2827	0.7252
1940.0	1.8957	1.3384	6.9636	1.3071	1.2817	0.7253
1960.0	1.9219	1.3588	6.9770	1.3107	1.2807	0.7254
1980.0	1.9481	1.3793	6.9904	1.3142	1.2797	0.7255
2000.0	1.9744	1.3999	7.0036	1.3178	1.2788	0.7256

THERMOPHYSICAL PROPERTIES OF AIR

ISOBAR: 100.0 MPa

T (K)	ρ (kg/m^3)	a (m/s)	μ (Pa·s)	k (J/s·m·K)	$\bar{m}$
2040.0	1.7077 +2	864.7	6.8697 −5	0.1253 +0	28.9642
2080.0	1.6748	872.5	6.9564	0.1275	28.9641
2120.0	1.6432	880.3	7.0425	0.1297	28.9641
2160.0	1.6128	887.9	7.1279	0.1319	28.9640
2200.0	1.5834	895.5	7.2128	0.1341	28.9638
2240.0	1.5552	903.0	7.2972	0.1363	28.9637
2280.0	1.5279	910.4	7.3809	0.1385	28.9635
2320.0	1.5015	917.8	7.4642	0.1407	28.9632
2360.0	1.4760	925.1	7.5469	0.1429	28.9630
2400.0	1.4514	932.3	7.6293	0.1459	28.9626
2440.0	1.4276	939.5	7.7111	0.1482	28.9622
2480.0	1.4046	946.6	7.7924	0.1506	28.9617
2520.0	1.3822	953.7	7.8732	0.1529	28.9611
2560.0	1.3606	960.7	7.9535	0.1553	28.9605
2600.0	1.3396	967.6	8.0335	0.1577	28.9597
2640.0	1.3193	974.5	8.1130	0.1602	28.9587
2680.0	1.2996	981.4	8.1920	0.1626	28.9577
2720.0	1.2804	988.2	8.2707	0.1651	28.9565
2760.0	1.2618	995.0	8.3489	0.1677	28.9551
2800.0	1.2437	1001.7	8.4268	0.1702	28.9535
2840.0	1.2261	1008.3	8.5042	0.1728	28.9517
2880.0	1.2090	1015.0	8.5813	0.1755	28.9496
2920.0	1.1923	1021.5	8.6581	0.1782	28.9473
2960.0	1.1761	1028.1	8.7344	0.1809	28.9448
3000.0	1.1603	1034.6	8.8105	0.1837	28.9419
3040.0	1.1449	1041.1	8.8862	0.1865	28.9387
3080.0	1.1299	1047.5	8.9615	0.1894	28.9352
3120.0	1.1153	1053.9	9.0365	0.1924	28.9313
3160.0	1.1010	1060.2	9.1113	0.1954	28.9270
3200.0	1.0871	1066.6	9.1857	0.1985	28.9223
3240.0	1.0734	1072.9	9.2598	0.2016	28.9171
3280.0	1.0601	1079.2	9.3336	0.2049	28.9114
3320.0	1.0471	1085.4	9.4072	0.2082	28.9053
3360.0	1.0344	1091.7	9.4805	0.2115	28.8986
3400.0	1.0220	1097.9	9.5535	0.2150	28.8913
3440.0	1.0099	1104.1	9.6262	0.2185	28.8835
3480.0	9.9796 +1	1110.3	9.6988	0.2221	28.8750
3520.0	9.8631	1116.4	9.7710	0.2258	28.8660
3560.0	9.7489	1122.6	9.8431	0.2295	28.8562
3600.0	9.6371	1128.7	9.9149	0.2333	28.8457
3640.0	9.5275	1134.8	9.9865	0.2372	28.8346
3680.0	9.4201	1141.0	1.0058 −4	0.2412	28.8226
3720.0	9.3147	1147.1	1.0129	0.2453	28.8100
3760.0	9.2113	1153.2	1.0200	0.2494	28.7965
3800.0	9.1098	1159.3	1.0271	0.2536	28.7822
3840.0	9.0102	1165.4	1.0342	0.2578	28.7670
3880.0	8.9123	1171.6	1.0412	0.2623	28.7511
3920.0	8.8162	1177.7	1.0482	0.2667	28.7342
3960.0	8.7218	1183.8	1.0553	0.2712	28.7164
4000.0	8.6289	1189.9	1.0623	0.2757	28.6978

THERMOPHYSICAL PROPERTIES OF AIR

ISOBAR: 100.0 MPa

T (K)	h (J/kg)	e (J/kg)	s (J/kg·K)	c_p (J/kg·K)	γ	Pr
2040.0	2.0273 +3	1.4413 +3	7.0298 +0	1.3248 +0	1.2769	0.7258
2080.0	2.0804	1.4829	7.0556	1.3319	1.2751	0.7259
2120.0	2.1338	1.5248	7.0810	1.3389	1.2733	0.7260
2160.0	2.1875	1.5671	7.1061	1.3458	1.2715	0.7252
2200.0	2.2415	1.6095	7.1308	1.3528	1.2697	0.7263
2240.0	2.2957	1.6523	7.1553	1.3597	1.2681	0.7264
2280.0	2.3503	1.6953	7.1794	1.3666	1.2664	0.7264
2320.0	2.4051	1.7386	7.2032	1.3735	1.2648	0.7265
2360.0	2.4601	1.7822	7.2268	1.3804	1.2632	0.7266
2400.0	2.5155	1.8261	7.2500	1.3872	1.2616	0.7254
2440.0	2.5711	1.8702	7.2730	1.3941	1.2601	0.7252
2480.0	2.6270	1.9146	7.2957	1.4009	1.2586	0.7249
2520.0	2.6832	1.9593	7.3182	1.4078	1.2572	0.7247
2560.0	2.7397	2.0042	7.3404	1.4147	1.2557	0.7243
2600.0	2.7964	2.0494	7.3624	1.4216	1.2544	0.7240
2640.0	2.8534	2.0949	7.3842	1.4286	1.2530	0.7235
2680.0	2.9107	2.1407	7.4057	1.4357	1.2517	0.7231
2720.0	2.9682	2.1867	7.4270	1.4428	1.2504	0.7225
2760.0	3.0261	2.2330	7.4482	1.4500	1.2491	0.7220
2800.0	3.0842	2.2796	7.4691	1.4573	1.2478	0.7213
2840.0	3.1427	2.3265	7.4898	1.4647	1.2466	0.7206
2880.0	3.2014	2.3737	7.5103	1.4723	1.2454	0.7199
2920.0	3.2605	2.4212	7.5307	1.4799	1.2442	0.7191
2960.0	3.3198	2.4690	7.5509	1.4878	1.2431	0.7182
3000.0	3.3795	2.5171	7.5709	1.4958	1.2420	0.7173
3040.0	3.4395	2.5655	7.5908	1.5040	1.2408	0.7163
3080.0	3.4998	2.6142	7.6105	1.5124	1.2398	0.7153
3120.0	3.5605	2.6632	7.6301	1.5210	1.2387	0.7142
3160.0	3.6215	2.7126	7.6495	1.5298	1.2377	0.7131
3200.0	3.6829	2.7623	7.6688	1.5389	1.2366	0.7119
3240.0	3.7446	2.8124	7.6880	1.5482	1.2356	0.7107
3280.0	3.8067	2.8628	7.7070	1.5577	1.2347	0.7094
3320.0	3.8692	2.9136	7.7260	1.5675	1.2337	0.7080
3360.0	3.9321	2.9648	7.7448	1.5776	1.2328	0.7066
3400.0	3.9954	3.0163	7.7635	1.5879	1.2319	0.7052
3440.0	4.0592	3.0683	7.7822	1.5984	1.2310	0.7038
3480.0	4.1233	3.1206	7.8007	1.6093	1.2302	0.7023
3520.0	4.1879	3.1734	7.8192	1.6204	1.2293	0.7008
3560.0	4.2530	3.2265	7.8375	1.6318	1.2285	0.6992
3600.0	4.3185	3.2801	7.8558	1.6434	1.2278	0.6977
3640.0	4.3844	3.3341	7.8741	1.6553	1.2270	0.6961
3680.0	4.4509	3.3886	7.8922	1.6675	1.2263	0.6945
3720.0	4.5178	3.4435	7.9103	1.6799	1.2256	0.6929
3760.0	4.5853	3.4989	7.9283	1.6925	1.2250	0.6913
3800.0	4.6532	3.5548	7.9463	1.7054	1.2244	0.6897
3840.0	4.7217	3.6111	7.9642	1.7185	1.2238	0.6881
3880.0	4.7907	3.6679	7.9821	1.7318	1.2233	0.6866
3920.0	4.8603	3.7252	8.0000	1.7453	1.2227	0.6850
3960.0	4.9303	3.7830	8.0177	1.7590	1.2223	0.6835
4000.0	5.0010	3.8413	8.0355	1.7729	1.2218	0.6820

THERMOPHYSICAL PROPERTIES OF AIR

ISOBAR: 100.0 MPa

T (K)	ρ (kg/m^3)	a (m/s)	μ (Pa·s)	k (J/s·m·K)	$\bar{m}$
4040.0	8.5377 +1	1196.1	1.0693 −4	0.2802 +0	28.6782
4080.0	8.4479	1202.3	1.0762	0.2857	28.6576
4120.0	8.3596	1208.4	1.0832	0.2904	28.6362
4160.0	8.2727	1214.6	1.0902	0.2952	28.6137
4200.0	8.1873	1220.8	1.0971	0.3001	28.5903
4240.0	8.1031	1227.0	1.1041	0.3049	28.5659
4280.0	8.0202	1233.3	1.1110	0.3098	28.5406
4320.0	7.9387	1239.5	1.1179	0.3148	28.5143
4360.0	7.8583	1245.8	1.1248	0.3197	28.4870
4400.0	7.7791	1252.1	1.1317	0.3247	28.4587
4440.0	7.7011	1258.4	1.1386	0.3296	28.4295
4480.0	7.6243	1264.8	1.1455	0.3346	28.3993
4520.0	7.5485	1271.1	1.1524	0.3395	28.3681
4560.0	7.4738	1277.5	1.1592	0.3445	28.3360
4600.0	7.4002	1283.9	1.1661	0.3494	28.3030
4640.0	7.3276	1290.4	1.1730	0.3543	28.2690
4680.0	7.2560	1296.8	1.1799	0.3591	28.2342
4720.0	7.1854	1303.3	1.1867	0.3640	28.1984
4760.0	7.1158	1309.9	1.1936	0.3687	28.1618
4800.0	7.0471	1316.4	1.2004	0.3735	28.1244
4840.0	6.9794	1323.0	1.2073	0.3781	28.0861
4880.0	6.9125	1329.6	1.2141	0.3828	28.0471
4920.0	6.8466	1336.2	1.2210	0.3873	28.0073
4960.0	6.7815	1342.9	1.2278	0.3917	27.9667
5000.0	6.7174	1349.5	1.2347	0.3961	27.9255
5040.0	6.6540	1356.2	1.2415	0.4004	27.8835
5080.0	6.5916	1363.0	1.2484	0.4047	27.8409
5120.0	6.5299	1369.7	1.2552	0.4089	27.7978
5160.0	6.4691	1376.5	1.2621	0.4130	27.7540
5200.0	6.4091	1383.3	1.2689	0.4170	27.7097
5240.0	6.3499	1390.1	1.2758	0.4209	27.6648
5280.0	6.2914	1397.0	1.2826	0.4248	27.6195
5320.0	6.2338	1403.8	1.2895	0.4286	27.5738
5360.0	6.1769	1410.7	1.2963	0.4323	27.5276
5400.0	6.1208	1417.6	1.3032	0.4359	27.4811
5440.0	6.0654	1424.5	1.3100	0.4394	27.4342
5480.0	6.0108	1431.5	1.3169	0.4429	27.3870
5520.0	5.9569	1438.4	1.3237	0.4463	27.3396
5560.0	5.9038	1445.4	1.3306	0.4497	27.2919
5600.0	5.8513	1452.4	1.3375	0.4530	27.2441
5640.0	5.7996	1459.4	1.3443	0.4563	27.1960
5680.0	5.7485	1466.4	1.3512	0.4595	27.1478
5720.0	5.6982	1473.4	1.3580	0.4627	27.0996
5760.0	5.6485	1480.4	1.3649	0.4658	27.0512
5800.0	5.5995	1487.4	1.3717	0.4690	27.0027
5840.0	5.5512	1494.5	1.3786	0.4721	26.9543
5880.0	5.5035	1501.5	1.3855	0.4753	26.9058
5920.0	5.4565	1508.6	1.3923	0.4784	26.8574
5960.0	5.4101	1515.6	1.3992	0.4816	26.8090
6000.0	5.3643	1522.7	1.4061	0.4848	26.7606

THERMOPHYSICAL PROPERTIES OF AIR

ISOBAR: 100.0 MPa

T (K)	h (J/kg)	e (J/kg)	s (J/kg·K)	c_p (J/kg·K)	γ	Pr
4040.0	5.0722 +3	3.9001 +3	8.0532 +0	1.7869 +0	1.2214	0.6805
4080.0	5.1439	3.9594	8.0709	1.8010	1.2211	0.6785
4120.0	5.2163	4.0192	8.0885	1.8152	1.2207	0.6770
4160.0	5.2892	4.0796	8.1061	1.8295	1.2205	0.6756
4200.0	5.3626	4.1404	8.1237	1.8439	1.2202	0.6742
4240.0	5.4367	4.2017	8.1412	1.8584	1.2200	0.6728
4280.0	5.5113	4.2636	8.1588	1.8728	1.2198	0.6715
4320.0	5.5865	4.3260	8.1762	1.8873	1.2197	0.6703
4360.0	5.6623	4.3889	8.1937	1.9018	1.2196	0.6691
4400.0	5.7386	4.4523	8.2111	1.9162	1.2196	0.6679
4440.0	5.8156	4.5162	8.2285	1.9305	1.2195	0.6668
4480.0	5.8931	4.5806	8.2459	1.9447	1.2196	0.6658
4520.0	5.9711	4.6455	8.2633	1.9588	1.2196	0.6648
4560.0	6.0498	4.7109	8.2806	1.9728	1.2198	0.6639
4600.0	6.1290	4.7767	8.2979	1.9866	1.2199	0.6630
4640.0	6.2087	4.8431	8.3151	2.0002	1.2201	0.6622
4680.0	6.2890	4.9099	8.3324	2.0136	1.2203	0.6615
4720.0	6.3698	4.9771	8.3496	2.0267	1.2206	0.6608
4760.0	6.4511	5.0448	8.3667	2.0396	1.2209	0.6602
4800.0	6.5330	5.1130	8.3838	2.0522	1.2212	0.6596
4840.0	6.6153	5.1815	8.4009	2.0644	1.2216	0.6591
4880.0	6.6981	5.2505	8.4180	2.0764	1.2220	0.6586
4920.0	6.7814	5.3198	8.4350	2.0880	1.2224	0.6582
4960.0	6.8651	5.3896	8.4519	2.0992	1.2229	0.6578
5000.0	6.9493	5.4596	8.4688	2.1100	1.2234	0.6575
5040.0	7.0339	5.5301	8.4857	2.1205	1.2239	0.6572
5080.0	7.1190	5.6008	8.5025	2.1305	1.2245	0.6570
5120.0	7.2044	5.6719	8.5192	2.1400	1.2251	0.6567
5160.0	7.2902	5.7433	8.5359	2.1492	1.2257	0.6566
5200.0	7.3763	5.8150	8.5525	2.1579	1.2264	0.6564
5240.0	7.4628	5.8869	8.5691	2.1661	1.2271	0.6563
5280.0	7.5496	5.9591	8.5856	2.1739	1.2278	0.6562
5320.0	7.6367	6.0315	8.6020	2.1812	1.2285	0.6561
5360.0	7.7241	6.1041	8.6184	2.1881	1.2293	0.6560
5400.0	7.8117	6.1769	8.6347	2.1945	1.2300	0.6559
5440.0	7.8996	6.2498	8.6509	2.2005	1.2308	0.6558
5480.0	7.9878	6.3230	8.6671	2.2059	1.2317	0.6557
5520.0	8.0761	6.3963	8.6831	2.2110	1.2325	0.6555
5560.0	8.1646	6.4697	8.6991	2.2156	1.2334	0.6554
5600.0	8.2533	6.5432	8.7150	2.2198	1.2343	0.6552
5640.0	8.3422	6.6168	8.7308	2.2236	1.2352	0.6550
5680.0	8.4312	6.6905	8.7465	2.2270	1.2361	0.6547
5720.0	8.5204	6.7642	8.7622	2.2301	1.2370	0.6544
5760.0	8.6096	6.8381	8.7777	2.2328	1.2379	0.6540
5800.0	8.6990	6.9119	8.7932	2.2351	1.2389	0.6536
5840.0	8.7884	6.9858	8.8086	2.2372	1.2398	0.6531
5880.0	8.8780	7.0597	8.8238	2.2389	1.2408	0.6525
5920.0	8.9675	7.1336	8.8390	2.2405	1.2417	0.6519
5960.0	9.0572	7.2075	8.8541	2.2417	1.2427	0.6511
6000.0	9.1469	7.2815	8.8691	2.2428	1.2437	0.6503

THERMOPHYSICAL PROPERTIES OF AIR

ISOBAR: 600.0 MPa

T (K)	ρ (kg/m^3)	a (m/s)	μ (Pa·s)	k (J/s·m·K)	$\bar{m}$
288.15	7.2538 +3	340.3	1.7979 −5	0.0249 +0	28.9644
300.0	6.9673	347.2	1.8554	0.0257	28.9644
310.0	6.7425	352.9	1.9030	0.0264	28.9644
320.0	6.5318	358.5	1.9500	0.0271	28.9644
330.0	6.3339	364.1	1.9962	0.0278	28.9644
340.0	6.1476	369.5	2.0419	0.0285	28.9644
350.0	5.9719	374.8	2.0869	0.0292	28.9644
360.0	5.8061	380.1	2.1313	0.0298	28.9644
370.0	5.6491	385.2	2.1751	0.0305	28.9644
380.0	5.5005	390.3	2.2184	0.0312	28.9644
390.0	5.3594	395.3	2.2612	0.0318	28.9644
400.0	5.2255	400.3	2.3035	0.0325	28.9644
410.0	5.0980	405.1	2.3454	0.0331	28.9644
420.0	4.9766	409.9	2.3867	0.0338	28.9644
430.0	4.8609	414.7	2.4276	0.0344	28.9644
440.0	4.7504	419.3	2.4681	0.0350	28.9644
450.0	4.6448	423.9	2.5082	0.0357	28.9644
460.0	4.5439	428.5	2.5479	0.0363	28.9644
470.0	4.4472	433.0	2.5872	0.0370	28.9644
480.0	4.3545	437.4	2.6261	0.0376	28.9644
490.0	4.2657	441.8	2.6647	0.0382	28.9644
500.0	4.1804	446.1	2.7029	0.0388	28.9644
510.0	4.0984	450.4	2.7408	0.0395	28.9644
520.0	4.0196	454.6	2.7783	0.0401	28.9644
530.0	3.9437	458.8	2.8155	0.0407	28.9644
540.0	3.8707	462.9	2.8525	0.0413	28.9644
550.0	3.8003	467.0	2.8891	0.0420	28.9644
560.0	3.7325	471.0	2.9254	0.0426	28.9644
580.0	3.6038	479.0	2.9972	0.0438	28.9644
600.0	3.4836	486.8	3.0680	0.0450	28.9644
620.0	3.3713	494.4	3.1377	0.0463	28.9644
640.0	3.2659	501.9	3.2065	0.0475	28.9644
660.0	3.1669	509.3	3.2743	0.0487	28.9644
680.0	3.0738	516.5	3.3412	0.0499	28.9644
700.0	2.9860	523.6	3.4073	0.0511	28.9644
720.0	2.9030	530.6	3.4726	0.0523	28.9644
740.0	2.8246	537.5	3.5371	0.0535	28.9644
760.0	2.7502	544.3	3.6008	0.0547	28.9644
780.0	2.6797	551.0	3.6639	0.0559	28.9644
800.0	2.6127	557.5	3.7263	0.0571	28.9644
820.0	2.5490	564.1	3.7880	0.0583	28.9644
840.0	2.4883	570.5	3.8490	0.0594	28.9644
860.0	2.4304	576.8	3.9095	0.0606	28.9644
880.0	2.3752	583.1	3.9693	0.0618	28.9644
900.0	2.3224	589.3	4.0286	0.0630	28.9644
920.0	2.2719	595.4	4.0873	0.0641	28.9644
940.0	2.2236	601.4	4.1455	0.0653	28.9644
960.0	2.1773	607.4	4.2032	0.0665	28.9645
980.0	2.1328	613.3	4.2603	0.0676	28.9644
1000.0	2.0902	619.2	4.3170	0.0688	28.9644

THERMOPHYSICAL PROPERTIES OF AIR

ISOBAR: 600.0 MPa

T (K)	h (J/kg)		e (J/kg)		s (J/kg·K)		c_p (J/kg·K)		γ	Pr
288.15	−1.4106	+1	−9.6876	+1	4.3357	+0	1.0052	+0	1.4001	0.7254
300.0	−2.1924	+0	−8.8367		4.3762		1.0055		1.3999	0.7246
310.0	7.8646		−8.1182		4.4092		1.0059		1.3997	0.7240
320.0	1.7926	+1	−7.3994		4.4411		1.0064		1.3994	0.7233
330.0	2.7993		−6.6799		4.4721		1.0070		1.3991	0.7228
340.0	3.8067		−5.9598		4.5022		1.0077		1.3987	0.7222
350.0	4.8148		−5.2389		4.5314		1.0085		1.3982	0.7217
360.0	5.8238		−4.5171		4.5598		1.0095		1.3977	0.7212
370.0	6.8337		−3.7944		4.5875		1.0105		1.3972	0.7207
380.0	7.8447		−3.0707		4.6145		1.0116		1.3966	0.7203
390.0	8.8569		−2.3458		4.6407		1.0127		1.3959	0.7199
400.0	9.8702		−1.6197		4.6664		1.0140		1.3953	0.7195
410.0	1.0885	+2	−8.9229	+0	4.6915		1.0153		1.3945	0.7191
420.0	1.1901		−1.6350		4.7159		1.0168		1.3938	0.7188
430.0	1.2918		5.6674		4.7399		1.0182		1.3930	0.7185
440.0	1.3937		1.2985	+1	4.7633		1.0198		1.3921	0.7182
450.0	1.4958		2.0319		4.7862		1.0214		1.3913	0.7179
460.0	1.5980		2.7669		4.8087		1.0231		1.3904	0.7177
470.0	1.7004		3.5036		4.8307		1.0249		1.3894	0.7175
480.0	1.8030		4.2421		4.8523		1.0267		1.3885	0.7173
490.0	1.9058		4.9825		4.8735		1.0285		1.3875	0.7171
500.0	2.0087		5.7247		4.8943		1.0304		1.3865	0.7170
510.0	2.1119		6.4689		4.9147		1.0324		1.3855	0.7169
520.0	2.2152		7.2150		4.9348		1.0344		1.3845	0.7167
530.0	2.3187		7.9632		4.9545		1.0365		1.3834	0.7167
540.0	2.4225		8.7134		4.9739		1.0385		1.3823	0.7166
550.0	2.5264		9.4658		4.9930		1.0407		1.3813	0.7165
560.0	2.6306		1.0220	+2	5.0118		1.0428		1.3802	0.7165
580.0	2.8396		1.1736		5.0484		1.0473		1.3780	0.7165
600.0	3.0495		1.3260		5.0840		1.0518		1.3757	0.7165
620.0	3.2603		1.4794		5.1186		1.0564		1.3735	0.7165
640.0	3.4721		1.6337		5.1522		1.0611		1.3712	0.7166
660.0	3.6848		1.7890		5.1849		1.0659		1.3689	0.7168
680.0	3.8984		1.9452		5.2168		1.0707		1.3667	0.7169
700.0	4.1131		2.1023		5.2479		1.0755		1.3644	0.7171
720.0	4.3286		2.2605		5.2783		1.0803		1.3622	0.7172
740.0	4.5452		2.4196		5.3079		1.0851		1.3600	0.7174
760.0	4.7627		2.5796		5.3370		1.0900		1.3578	0.7176
780.0	4.9812		2.7406		5.3653		1.0947		1.3557	0.7177
800.0	5.2006		2.9026		5.3931		1.0995		1.3537	0.7178
820.0	5.4210		3.0655		5.4203		1.1042		1.3516	0.7179
840.0	5.6423		3.2294		5.4470		1.1088		1.3496	0.7179
860.0	5.8645		3.3941		5.4731		1.1134		1.3477	0.7179
880.0	6.0876		3.5598		5.4988		1.1178		1.3458	0.7179
900.0	6.3116		3.7264		5.5239		1.1223		1.3440	0.7178
920.0	6.5365		3.8938		5.5486		1.1266		1.3422	0.7176
940.0	6.7622		4.0621		5.5729		1.1308		1.3405	0.7174
960.0	6.9888		4.2312		5.5968		1.1349		1.3389	0.7172
980.0	7.2162		4.4012		5.6202		1.1389		1.3373	0.7169
1000.0	7.4444		4.5719		5.6433		1.1428		1.3357	0.7165

THERMOPHYSICAL PROPERTIES OF AIR

ISOBAR: 600.0 MPa

T (K)	ρ (kg/m^3)	a (m/s)	μ (Pa·s)	k (J/s·m·K)	$\bar{m}$
1020.0	2.0492 +3	625.0	4.3745 -5	0.0697 +0	28.9644
1040.0	2.0098	630.8	4.4309	0.0708	28.9644
1060.0	1.9719	636.5	4.4870	0.0719	28.9644
1080.0	1.9354	642.2	4.5426	0.0730	28.9644
1100.0	1.9002	647.8	4.5978	0.0741	28.9644
1120.0	1.8662	653.3	4.6526	0.0754	28.9644
1140.0	1.8335	658.8	4.7070	0.0765	28.9644
1160.0	1.8019	664.2	4.7610	0.0776	28.9644
1180.0	1.7713	669.6	4.8147	0.0787	28.9644
1200.0	1.7418	674.9	4.8680	0.0797	28.9644
1220.0	1.7133	680.2	4.9209	0.0808	28.9644
1240.0	1.6856	685.5	4.9735	0.0819	28.9644
1260.0	1.6589	690.6	5.0258	0.0830	28.9644
1280.0	1.6330	695.8	5.0778	0.0841	28.9644
1300.0	1.6078	700.9	5.1294	0.0851	28.9644
1320.0	1.5835	705.9	5.1807	0.0862	28.9644
1340.0	1.5598	710.9	5.2318	0.0873	28.9644
1360.0	1.5369	715.9	5.2825	0.0884	28.9644
1380.0	1.5146	720.8	5.3329	0.0894	28.9644
1400.0	1.4930	725.7	5.3831	0.0905	28.9644
1420.0	1.4720	730.6	5.4330	0.0916	28.9644
1440.0	1.4515	735.4	5.4827	0.0927	28.9644
1460.0	1.4316	740.1	5.5320	0.0937	28.9644
1480.0	1.4123	744.9	5.5811	0.0948	28.9644
1500.0	1.3935	749.6	5.6300	0.0959	28.9644
1520.0	1.3751	754.2	5.6786	0.0969	28.9644
1540.0	1.3573	758.8	5.7270	0.0980	28.9644
1560.0	1.3399	763.4	5.7751	0.0991	28.9644
1580.0	1.3229	768.0	5.8231	0.1002	28.9644
1600.0	1.3064	772.5	5.8707	0.1013	28.9644
1620.0	1.2902	777.0	5.9182	0.1023	28.9644
1640.0	1.2745	781.5	5.9654	0.1034	28.9644
1660.0	1.2591	785.9	6.0125	0.1045	28.9644
1680.0	1.2442	790.3	6.0593	0.1056	28.9644
1700.0	1.2295	794.6	6.1059	0.1067	28.9644
1720.0	1.2152	799.0	6.1523	0.1078	28.9644
1740.0	1.2013	803.3	6.1985	0.1089	28.9644
1760.0	1.1876	807.5	6.2445	0.1099	28.9644
1780.0	1.1743	811.8	6.2903	0.1110	28.9644
1800.0	1.1612	816.0	6.3360	0.1121	28.9644
1820.0	1.1485	820.2	6.3814	0.1132	28.9644
1840.0	1.1360	824.4	6.4267	0.1143	28.9644
1860.0	1.1238	828.5	6.4717	0.1154	28.9644
1880.0	1.1118	832.6	6.5166	0.1165	28.9644
1900.0	1.1001	836.7	6.5614	0.1176	28.9644
1920.0	1.0886	840.8	6.6059	0.1187	28.9644
1940.0	1.0774	844.9	6.6503	0.1198	28.9644
1960.0	1.0664	848.9	6.6945	0.1209	28.9644
1980.0	1.0556	852.9	6.7386	0.1220	28.9644
2000.0	1.0451	856.9	6.7824	0.1231	28.9643

THERMOPHYSICAL PROPERTIES OF AIR

ISOBAR: 600.0 MPa

T (K)	h (J/kg)	e (J/kg)	s (J/kg·K)	c_p (J/kg·K)	γ	Pr
1020.0	7.6733 +2	4.7434 +2	5.6659 +0	1.1465 +0	1.3343	0.7185
1040.0	7.9030	4.9156	5.6882	1.1501	1.3329	0.7185
1060.0	8.1333	5.0885	5.7102	1.1537	1.3315	0.7185
1080.0	8.3644	5.2622	5.7318	1.1572	1.3302	0.7185
1100.0	8.5962	5.4365	5.7530	1.1608	1.3288	0.7186
1120.0	8.8288	5.6116	5.7740	1.1643	1.3275	0.7187
1140.0	9.0620	5.7873	5.7946	1.1678	1.3262	0.7188
1160.0	9.2959	5.9638	5.8150	1.1712	1.3249	0.7190
1180.0	9.5305	6.1409	5.8350	1.1747	1.3237	0.7191
1200.0	9.7657	6.3187	5.8548	1.1781	1.3224	0.7192
1220.0	1.0002 +3	6.4973	5.8743	1.1816	1.3212	0.7194
1240.0	1.0238	6.6765	5.8935	1.1850	1.3200	0.7195
1260.0	1.0476	6.8564	5.9125	1.1884	1.3188	0.7197
1280.0	1.0714	7.0369	5.9313	1.1918	1.3176	0.7199
1300.0	1.0952	7.2182	5.9498	1.1952	1.3164	0.7201
1320.0	1.1192	7.4001	5.9680	1.1986	1.3152	0.7202
1340.0	1.1432	7.5827	5.9861	1.2020	1.3140	0.7204
1360.0	1.1673	7.7660	6.0039	1.2054	1.3128	0.7206
1380.0	1.1914	7.9500	6.0215	1.2088	1.3117	0.7208
1400.0	1.2156	8.1347	6.0390	1.2123	1.3105	0.7210
1420.0	1.2399	8.3200	6.0562	1.2157	1.3094	0.7212
1440.0	1.2642	8.5060	6.0732	1.2191	1.3082	0.7214
1460.0	1.2887	8.6928	6.0900	1.2226	1.3071	0.7216
1480.0	1.3131	8.8802	6.1067	1.2260	1.3060	0.7218
1500.0	1.3377	9.0683	6.1232	1.2295	1.3049	0.7220
1520.0	1.3623	9.2571	6.1395	1.2329	1.3037	0.7222
1540.0	1.3870	9.4465	6.1556	1.2364	1.3026	0.7224
1560.0	1.4118	9.6367	6.1716	1.2399	1.3015	0.7225
1580.0	1.4366	9.8276	6.1874	1.2434	1.3004	0.7227
1600.0	1.4615	1.0019 +3	6.2031	1.2469	1.2993	0.7229
1620.0	1.4865	1.0211	6.2186	1.2504	1.2982	0.7231
1640.0	1.5115	1.0404	6.2340	1.2539	1.2972	0.7232
1660.0	1.5366	1.0598	6.2492	1.2574	1.2961	0.7234
1680.0	1.5618	1.0792	6.2643	1.2609	1.2950	0.7236
1700.0	1.5871	1.0988	6.2792	1.2645	1.2939	0.7237
1720.0	1.6124	1.1183	6.2940	1.2680	1.2929	0.7239
1740.0	1.6378	1.1380	6.3087	1.2715	1.2918	0.7240
1760.0	1.6633	1.1577	6.3232	1.2751	1.2908	0.7242
1780.0	1.6888	1.1775	6.3377	1.2786	1.2898	0.7243
1800.0	1.7144	1.1974	6.3520	1.2822	1.2887	0.7245
1820.0	1.7401	1.2173	6.3662	1.2857	1.2877	0.7246
1840.0	1.7658	1.2373	6.3802	1.2892	1.2867	0.7247
1860.0	1.7917	1.2574	6.3942	1.2928	1.2857	0.7249
1880.0	1.8176	1.2775	6.4080	1.2963	1.2847	0.7250
1900.0	1.8435	1.2977	6.4218	1.2998	1.2837	0.7251
1920.0	1.8695	1.3180	6.4354	1.3034	1.2827	0.7252
1940.0	1.8956	1.3384	6.4489	1.3069	1.2817	0.7253
1960.0	1.9218	1.3588	6.4624	1.3104	1.2808	0.7254
1980.0	1.9481	1.3793	6.4757	1.3139	1.2798	0.7255
2000.0	1.9744	1.3999	6.4889	1.3174	1.2789	0.7256

THERMOPHYSICAL PROPERTIES OF AIR

ISOBAR: 600.0 MPa

T (K)	ρ (kg/m^3)	a (m/s)	μ (Pa·s)	k (J/s·m·K)	$\bar{m}$
2040.0	1.0246 +3	864.8	6.8697 −5	0.1253 +0	28.9643
2080.0	1.0049	872.6	6.9564	0.1275	28.9643
2120.0	9.8593 +2	880.3	7.0425	0.1297	28.9643
2160.0	9.6767	888.0	7.1279	0.1319	28.9642
2200.0	9.5008	895.6	7.2128	0.1341	28.9642
2240.0	9.3311	903.1	7.2972	0.1363	28.9641
2280.0	9.1674	910.5	7.3810	0.1385	28.9640
2320.0	9.0093	917.9	7.4642	0.1407	28.9640
2360.0	8.8565	925.2	7.5470	0.1429	28.9638
2400.0	8.7089	932.5	7.6292	0.1451	28.9637
2440.0	8.5661	939.7	7.7110	0.1472	28.9635
2480.0	8.4278	946.9	7.7922	0.1494	28.9633
2520.0	8.2940	954.0	7.8730	0.1515	28.9631
2560.0	8.1643	961.0	7.9535	0.1543	28.9628
2600.0	8.0386	968.0	8.0334	0.1566	28.9625
2640.0	7.9167	974.9	8.1128	0.1588	28.9621
2680.0	7.7985	981.8	8.1919	0.1610	28.9617
2720.0	7.6836	988.7	8.2705	0.1633	28.9612
2760.0	7.5721	995.5	8.3487	0.1655	28.9607
2800.0	7.4638	1002.3	8.4265	0.1678	28.9600
2840.0	7.3585	1009.0	8.5039	0.1700	28.9593
2880.0	7.2561	1015.7	8.5810	0.1723	28.9585
2920.0	7.1565	1022.4	8.6576	0.1745	28.9575
2960.0	7.0595	1029.0	8.7339	0.1768	28.9565
3000.0	6.9651	1035.5	8.8099	0.1791	28.9553
3040.0	6.8731	1042.1	8.8855	0.1814	28.9540
3080.0	6.7835	1048.6	8.9607	0.1837	28.9526
3120.0	6.6962	1055.1	9.0356	0.1860	28.9510
3160.0	6.6110	1061.5	9.1102	0.1884	28.9492
3200.0	6.5280	1067.9	9.1844	0.1907	28.9473
3240.0	6.4469	1074.3	9.2584	0.1931	28.9452
3280.0	6.3678	1080.6	9.3320	0.1955	28.9429
3320.0	6.2905	1087.0	9.4054	0.1979	28.9403
3360.0	6.2150	1093.3	9.4784	0.2003	28.9376
3400.0	6.1413	1099.5	9.5511	0.2027	28.9346
3440.0	6.0692	1105.8	9.6236	0.2052	28.9314
3480.0	5.9987	1112.0	9.6958	0.2077	28.9279
3520.0	5.9298	1118.2	9.7677	0.2102	28.9242
3560.0	5.8623	1124.4	9.8394	0.2127	28.9201
3600.0	5.7963	1130.5	9.9108	0.2152	28.9158
3640.0	5.7317	1136.6	9.9819	0.2178	28.9112
3680.0	5.6684	1142.7	1.0053 −4	0.2204	28.9062
3720.0	5.6065	1148.8	1.0124	0.2230	28.9009
3760.0	5.5457	1154.9	1.0194	0.2257	28.8953
3800.0	5.4862	1161.0	1.0264	0.2284	28.8893
3840.0	5.4279	1167.0	1.0334	0.2311	28.8830
3880.0	5.3707	1173.1	1.0404	0.2338	28.8763
3920.0	5.3146	1179.1	1.0473	0.2366	28.8692
3960.0	5.2595	1185.1	1.0543	0.2393	28.8617
4000.0	5.2055	1191.1	1.0612	0.2421	28.8538

THERMOPHYSICAL PROPERTIES OF AIR

ISOBAR: 600.0 MPa

T (K)	h (J/kg)	e (J/kg)	s (J/kg·K)	c_p (J/kg·K)	γ	Pr
2040.0	2.0272 +3	1.4412 +3	6.5151 +0	1.3243 +0	1.2770	0.7258
2080.0	2.0803	1.4828	6.5408	1.3312	1.2752	0.7259
2120.0	2.1337	1.5247	6.5663	1.3381	1.2734	0.7260
2160.0	2.1874	1.5669	6.5913	1.3448	1.2717	0.7262
2200.0	2.2413	1.6093	6.6161	1.3515	1.2700	0.7263
2240.0	2.2955	1.6520	6.6405	1.3582	1.2683	0.7264
2280.0	2.3499	1.6950	6.6646	1.3647	1.2667	0.7264
2320.0	2.4047	1.7382	6.6884	1.3712	1.2651	0.7265
2360.0	2.4596	1.7817	6.7119	1.3776	1.2636	0.7266
2400.0	2.5149	1.8255	6.7351	1.3839	1.2621	0.7266
2440.0	2.5703	1.8694	6.7580	1.3901	1.2607	0.7266
2480.0	2.6261	1.9137	6.7806	1.3962	1.2593	0.7266
2520.0	2.6820	1.9581	6.8030	1.4022	1.2580	0.7266
2560.0	2.7382	2.0029	6.8252	1.4082	1.2567	0.7257
2600.0	2.7947	2.0478	6.8470	1.4140	1.2554	0.7255
2640.0	2.8514	2.0930	6.8687	1.4198	1.2542	0.7254
2680.0	2.9083	2.1384	6.8901	1.4255	1.2530	0.7251
2720.0	2.9654	2.1840	6.9112	1.4312	1.2518	0.7249
2760.0	3.0228	2.2299	6.9322	1.4367	1.2507	0.7247
2800.0	3.0803	2.2759	6.9529	1.4422	1.2497	0.7244
2840.0	3.1382	2.3222	6.9734	1.4477	1.2486	0.7241
2880.0	3.1962	2.3687	6.9937	1.4531	1.2476	0.7237
2920.0	3.2544	2.4154	7.0137	1.4585	1.2467	0.7234
2960.0	3.3128	2.4624	7.0336	1.4638	1.2457	0.7230
3000.0	3.3715	2.5095	7.0533	1.4691	1.2448	0.7225
3040.0	3.4304	2.5568	7.0728	1.4744	1.2440	0.7221
3080.0	3.4895	2.6044	7.0921	1.4796	1.2431	0.7216
3120.0	3.5487	2.6521	7.1112	1.4849	1.2423	0.7211
3160.0	3.6082	2.7001	7.1302	1.4901	1.2415	0.7206
3200.0	3.6679	2.7482	7.1490	1.4954	1.2408	0.7200
3240.0	3.7279	2.7966	7.1676	1.5007	1.2401	0.7195
3280.0	3.7880	2.8451	7.1860	1.5059	1.2394	0.7189
3320.0	3.8483	2.8939	7.2043	1.5112	1.2387	0.7182
3360.0	3.9089	2.9429	7.2224	1.5166	1.2380	0.7176
3400.0	3.9697	2.9920	7.2404	1.5220	1.2374	0.7169
3440.0	4.0307	3.0414	7.2582	1.5274	1.2368	0.7162
3480.0	4.0919	3.0910	7.2759	1.5328	1.2362	0.7154
3520.0	4.1533	3.1408	7.2935	1.5383	1.2357	0.7147
3560.0	4.2149	3.1908	7.3109	1.5439	1.2352	0.7139
3600.0	4.2768	3.2410	7.3282	1.5495	1.2347	0.7131
3640.0	4.3389	3.2914	7.3453	1.5551	1.2342	0.7123
3680.0	4.4012	3.3420	7.3624	1.5609	1.2337	0.7115
3720.0	4.4638	3.3928	7.3793	1.5667	1.2333	0.7106
3760.0	4.5265	3.4439	7.3960	1.5725	1.2329	0.7098
3800.0	4.5896	3.4952	7.4127	1.5784	1.2325	0.7089
3840.0	4.6528	3.5467	7.4293	1.5844	1.2321	0.7080
3880.0	4.7163	3.5984	7.4457	1.5904	1.2317	0.7071
3920.0	4.7801	3.6503	7.4621	1.5965	1.2314	0.7061
3960.0	4.8440	3.7025	7.4783	1.6027	1.2311	0.7052
4000.0	4.9083	3.7549	7.4945	1.6089	1.2308	0.7043

THERMOPHYSICAL PROPERTIES OF AIR

ISOBAR: 600.0 MPa

T (K)	ρ (kg/m^3)	a (m/s)	μ (Pa·s)	k (J/s·m·K)	$\overline{m}$
4040.0	5.1525 +2	1197.1	1.0681 −4	0.2450 +0	28.8455
4080.0	5.1004	1203.0	1.0750	0.2478	28.8368
4120.0	5.0493	1209.0	1.0818	0.2512	28.8276
4160.0	4.9991	1215.0	1.0887	0.2541	28.8180
4200.0	4.9497	1220.9	1.0955	0.2566	28.8079
4240.0	4.9013	1226.8	1.1023	0.2595	28.7974
4280.0	4.8536	1232.8	1.1091	0.2625	28.7864
4320.0	4.8067	1238.7	1.1159	0.2662	28.7750
4360.0	4.7607	1244.7	1.1226	0.2693	28.7630
4400.0	4.7154	1250.6	1.1294	0.2723	28.7506
4440.0	4.6708	1256.5	1.1361	0.2755	28.7377
4480.0	4.6269	1262.4	1.1428	0.2786	28.7243
4520.0	4.5837	1268.4	1.1495	0.2817	28.7104
4560.0	4.5413	1274.3	1.1562	0.2849	28.6960
4600.0	4.4994	1280.2	1.1629	0.2881	28.6812
4640.0	4.4583	1286.2	1.1696	0.2913	28.6658
4680.0	4.4177	1292.1	1.1762	0.2945	28.6499
4720.0	4.3778	1298.1	1.1829	0.2977	28.6335
4760.0	4.3384	1304.0	1.1895	0.3010	28.6166
4800.0	4.2996	1310.0	1.1961	0.3042	28.5992
4840.0	4.2614	1315.9	1.2027	0.3075	28.5813
4880.0	4.2238	1321.9	1.2093	0.3107	28.5629
4920.0	4.1867	1327.8	1.2159	0.3140	28.5440
4960.0	4.1501	1333.8	1.2225	0.3173	28.5245
5000.0	4.1140	1339.8	1.2291	0.3206	28.5046
5040.0	4.0784	1345.8	1.2356	0.3239	28.4843
5080.0	4.0434	1351.8	1.2422	0.3271	28.4634
5120.0	4.0088	1357.8	1.2487	0.3304	28.4420
5160.0	3.9746	1363.9	1.2553	0.3337	28.4202
5200.0	3.9410	1369.9	1.2618	0.3370	28.3979
5240.0	3.9077	1376.0	1.2684	0.3403	28.3751
5280.0	3.8750	1382.0	1.2749	0.3436	28.3518
5320.0	3.8426	1388.1	1.2814	0.3468	28.3282
5360.0	3.8107	1394.2	1.2879	0.3501	28.3040
5400.0	3.7792	1400.3	1.2944	0.3534	28.2795
5440.0	3.7481	1406.4	1.3009	0.3566	28.2545
5480.0	3.7174	1412.6	1.3074	0.3599	28.2290
5520.0	3.6870	1418.7	1.3139	0.3632	28.2032
5560.0	3.6571	1424.9	1.3204	0.3664	28.1770
5600.0	3.6276	1431.1	1.3269	0.3696	28.1503
5640.0	3.5984	1437.3	1.3334	0.3729	28.1233
5680.0	3.5696	1443.5	1.3398	0.3761	28.0959
5720.0	3.5411	1449.7	1.3463	0.3793	28.0681
5760.0	3.5130	1456.0	1.3528	0.3826	28.0400
5800.0	3.4852	1462.3	1.3593	0.3858	28.0115
5840.0	3.4578	1468.6	1.3657	0.3890	27.9827
5880.0	3.4307	1474.9	1.3722	0.3922	27.9536
5920.0	3.4039	1481.2	1.3787	0.3955	27.9241
5960.0	3.3775	1487.6	1.3851	0.3987	27.8943
6000.0	3.3513	1494.0	1.3916	0.4019	27.8643

THERMOPHYSICAL PROPERTIES OF AIR

ISOBAR: 600.0 MPa

T	h	e	s	c_p	γ	Pr
(K)	(J/kg)	(J/kg)	(J/kg·K)	(J/kg·K)		
4040.0	4.9728 +3	3.8075 +3	7.5105 +0	1.6152 +0	1.2305	0.7033
4080.0	5.0375	3.8603	7.5264	1.6215	1.2303	0.7024
4120.0	5.1025	3.9134	7.5423	1.6279	1.2301	0.7011
4160.0	5.1677	3.9667	7.5580	1.6343	1.2299	0.7001
4200.0	5.2332	4.0202	7.5737	1.6408	1.2297	0.6996
4240.0	5.2990	4.0740	7.5893	1.6473	1.2295	0.6986
4280.0	5.3650	4.1280	7.6048	1.6539	1.2294	0.6976
4320.0	5.4313	4.1822	7.6202	1.6605	1.2293	0.6961
4360.0	5.4979	4.2367	7.6356	1.6671	1.2292	0.6951
4400.0	5.5647	4.2914	7.6508	1.6737	1.2291	0.6941
4440.0	5.6317	4.3463	7.6660	1.6804	1.2291	0.6930
4480.0	5.6991	4.4015	7.6811	1.6870	1.2290	0.6920
4520.0	5.7667	4.4569	7.6961	1.6937	1.2290	0.6910
4560.0	5.8346	4.5125	7.7111	1.7004	1.2291	0.6900
4600.0	5.9027	4.5683	7.7259	1.7070	1.2291	0.6890
4640.0	5.9712	4.6244	7.7407	1.7136	1.2292	0.6880
4680.0	6.0398	4.6807	7.7555	1.7202	1.2293	0.6870
4720.0	6.1088	4.7373	7.7702	1.7268	1.2294	0.6860
4760.0	6.1780	4.7941	7.7848	1.7334	1.2295	0.6850
4800.0	6.2474	4.8510	7.7993	1.7398	1.2297	0.6840
4840.0	6.3172	4.9082	7.8138	1.7463	1.2299	0.6831
4880.0	6.3871	4.9657	7.8282	1.7527	1.2301	0.6821
4920.0	6.4574	5.0233	7.8425	1.7590	1.2303	0.6811
4960.0	6.5279	5.0811	7.8568	1.7652	1.2306	0.6801
5000.0	6.5986	5.1392	7.8710	1.7714	1.2309	0.6791
5040.0	6.6696	5.1974	7.8851	1.7774	1.2312	0.6782
5080.0	6.7408	5.2559	7.8992	1.7834	1.2315	0.6772
5120.0	6.8122	5.3145	7.9132	1.7893	1.2319	0.6762
5160.0	6.8839	5.3733	7.9271	1.7950	1.2322	0.6752
5200.0	6.9558	5.4324	7.9410	1.8007	1.2326	0.6742
5240.0	7.0280	5.4915	7.9548	1.8062	1.2331	0.6732
5280.0	7.1003	5.5509	7.9686	1.8116	1.2335	0.6722
5320.0	7.1729	5.6104	7.9823	1.8168	1.2340	0.6712
5360.0	7.2457	5.6701	7.9959	1.8220	1.2345	0.6702
5400.0	7.3187	5.7299	8.0095	1.8269	1.2351	0.6692
5440.0	7.3918	5.7899	8.0230	1.8317	1.2356	0.6681
5480.0	7.4652	5.8501	8.0364	1.8363	1.2362	0.6671
5520.0	7.5387	5.9103	8.0498	1.8408	1.2369	0.6660
5560.0	7.6125	5.9707	8.0631	1.8451	1.2375	0.6649
5600.0	7.6863	6.0312	8.0763	1.8492	1.2382	0.6638
5640.0	7.7604	6.0919	8.0895	1.8532	1.2389	0.6626
5680.0	7.8346	6.1526	8.1026	1.8569	1.2396	0.6615
5720.0	7.9089	6.2134	8.1157	1.8605	1.2404	0.6603
5760.0	7.9834	6.2743	8.1286	1.8638	1.2412	0.6590
5800.0	8.0580	6.3353	8.1415	1.8670	1.2420	0.6578
5840.0	8.1328	6.3964	8.1544	1.8699	1.2429	0.6565
5880.0	8.2076	6.4575	8.1672	1.8727	1.2438	0.6551
5920.0	8.2826	6.5187	8.1799	1.8752	1.2447	0.6537
5960.0	8.3577	6.5800	8.1925	1.8776	1.2457	0.6523
6000.0	8.4328	6.6413	8.2051	1.8797	1.2467	0.6508

THERMOPHYSICAL PROPERTIES OF AIR

ISOBAR: 1000.0 MPa

T (K)	ρ (kg/m^3)	a (m/s)	μ (Pa·s)	k (J/s·m·K)	$\bar{m}$
288.15	1.2090 +4	340.3	1.7979 −5	0.0249 +0	28.9644
300.0	1.1612	347.2	1.8554	0.0257	28.9644
310.0	1.1238	352.9	1.9030	0.0264	28.9644
320.0	1.0886	358.5	1.9500	0.0271	28.9644
330.0	1.0556	364.1	1.9962	0.0278	28.9644
340.0	1.0246	369.5	2.0419	0.0285	28.9644
350.0	9.9532 +3	374.8	2.0869	0.0292	28.9644
360.0	9.6768	380.1	2.1313	0.0298	28.9644
370.0	9.4152	385.2	2.1751	0.0305	28.9644
380.0	9.1675	390.3	2.2184	0.0312	28.9644
390.0	8.9324	395.3	2.2612	0.0318	28.9644
400.0	8.7091	400.3	2.3035	0.0325	28.9644
410.0	8.4967	405.1	2.3454	0.0331	28.9644
420.0	8.2944	409.9	2.3867	0.0338	28.9644
430.0	8.1015	414.7	2.4276	0.0344	28.9644
440.0	7.9174	419.3	2.4681	0.0350	28.9644
450.0	7.7414	423.9	2.5082	0.0357	28.9644
460.0	7.5731	428.5	2.5479	0.0363	28.9644
470.0	7.4120	433.0	2.5872	0.0370	28.9644
480.0	7.2576	437.4	2.6261	0.0376	28.9644
490.0	7.1095	441.8	2.6647	0.0382	28.9644
500.0	6.9673	446.1	2.7029	0.0388	28.9644
510.0	6.8307	450.4	2.7408	0.0395	28.9644
520.0	6.6993	454.6	2.7783	0.0401	28.9644
530.0	6.5729	458.8	2.8155	0.0407	28.9644
540.0	6.4512	462.9	2.8525	0.0413	28.9644
550.0	6.3339	467.0	2.8891	0.0420	28.9644
560.0	6.2208	471.0	2.9254	0.0426	28.9644
580.0	6.0063	479.0	2.9972	0.0438	28.9644
600.0	5.8061	486.8	3.0680	0.0450	28.9644
620.0	5.6188	494.4	3.1377	0.0463	28.9644
640.0	5.4432	501.9	3.2065	0.0475	28.9644
660.0	5.2782	509.3	3.2743	0.0487	28.9644
680.0	5.1230	516.5	3.3412	0.0499	28.9644
700.0	4.9766	523.6	3.4073	0.0511	28.9644
720.0	4.8384	530.6	3.4726	0.0523	28.9644
740.0	4.7076	537.5	3.5371	0.0535	28.9644
760.0	4.5837	544.3	3.6008	0.0547	28.9644
780.0	4.4662	551.0	3.6639	0.0559	28.9644
800.0	4.3545	557.5	3.7263	0.0571	28.9644
820.0	4.2483	564.1	3.7880	0.0583	28.9644
840.0	4.1472	570.5	3.8490	0.0594	28.9644
860.0	4.0507	576.8	3.9095	0.0606	28.9644
880.0	3.9587	583.1	3.9693	0.0618	28.9644
900.0	3.8707	589.3	4.0286	0.0630	28.9644
920.0	3.7866	595.4	4.0873	0.0641	28.9644
940.0	3.7060	601.4	4.1455	0.0653	28.9644
960.0	3.6288	607.4	4.2032	0.0665	28.9645
980.0	3.5547	613.3	4.2603	0.0676	28.9644
1000.0	3.4836	619.2	4.3170	0.0688	28.9644

THERMOPHYSICAL PROPERTIES OF AIR

ISOBAR: 1000.0 MPa

T (K)	h (J/kg)	e (J/kg)	s (J/kg·K)	c_p (J/kg·K)	γ	Pr
288.15	-1.4106 +1	-9.6876 +1	4.1890 +0	1.0052 +0	1.4001	0.7254
300.0	-2.1924 +0	-8.8367	4.2295	1.0055	1.3999	0.7246
310.0	7.8646	-8.1182	4.2624	1.0059	1.3997	0.7240
320.0	1.7926 +1	-7.3994	4.2944	1.0064	1.3994	0.7233
330.0	2.7993	-6.6799	4.3254	1.0070	1.3991	0.7228
340.0	3.8067	-5.9598	4.3554	1.0077	1.3987	0.7222
350.0	4.8148	-5.2389	4.3847	1.0085	1.3982	0.7217
360.0	5.8238	-4.5171	4.4131	1.0095	1.3977	0.7212
370.0	6.8337	-3.7944	4.4408	1.0105	1.3972	0.7207
380.0	7.8447	-3.0707	4.4677	1.0116	1.3966	0.7203
390.0	8.8569	-2.3458	4.4940	1.0127	1.3959	0.7199
400.0	9.8702	-1.6197	4.5197	1.0140	1.3953	0.7195
410.0	1.0885 +2	-8.9229 +0	4.5447	1.0153	1.3945	0.7191
420.0	1.1901	-1.6350	4.5692	1.0168	1.3938	0.7188
430.0	1.2918	5.6674	4.5931	1.0182	1.3930	0.7185
440.0	1.3937	1.2985 +1	4.6166	1.0198	1.3921	0.7182
450.0	1.4958	2.0319	4.6395	1.0214	1.3913	0.7179
460.0	1.5980	2.7669	4.6620	1.0231	1.3904	0.7177
470.0	1.7004	3.5036	4.6840	1.0249	1.3894	0.7175
480.0	1.8030	4.2421	4.7056	1.0267	1.3885	0.7173
490.0	1.9058	4.9825	4.7268	1.0285	1.3875	0.7171
500.0	2.0087	5.7247	4.7476	1.0304	1.3865	0.7170
510.0	2.1119	6.4689	4.7680	1.0324	1.3855	0.7169
520.0	2.2152	7.2150	4.7881	1.0344	1.3845	0.7167
530.0	2.3187	7.9632	4.8078	1.0365	1.3834	0.7167
540.0	2.4225	8.7134	4.8272	1.0385	1.3823	0.7166
550.0	2.5264	9.4658	4.8463	1.0407	1.3813	0.7165
560.0	2.6306	1.0220 +2	4.8650	1.0428	1.3802	0.7165
580.0	2.8396	1.1736	4.9017	1.0473	1.3780	0.7165
600.0	3.0495	1.3260	4.9373	1.0518	1.3757	0.7165
620.0	3.2603	1.4794	4.9719	1.0564	1.3735	0.7165
640.0	3.4721	1.6337	5.0055	1.0611	1.3712	0.7166
660.0	3.6848	1.7890	5.0382	1.0659	1.3689	0.7168
680.0	3.8984	1.9452	5.0701	1.0707	1.3667	0.7169
700.0	4.1131	2.1023	5.1012	1.0755	1.3644	0.7171
720.0	4.3286	2.2605	5.1315	1.0803	1.3622	0.7172
740.0	4.5452	2.4196	5.1612	1.0851	1.3600	0.7174
760.0	4.7627	2.5796	5.1902	1.0900	1.3578	0.7176
780.0	4.9812	2.7406	5.2186	1.0947	1.3557	0.7177
800.0	5.2006	2.9026	5.2464	1.0995	1.3537	0.7178
820.0	5.4210	3.0655	5.2736	1.1042	1.3516	0.7179
840.0	5.6423	3.2294	5.3002	1.1088	1.3496	0.7179
860.0	5.8645	3.3941	5.3264	1.1134	1.3477	0.7179
880.0	6.0876	3.5598	5.3520	1.1178	1.3458	0.7179
900.0	6.3116	3.7264	5.3772	1.1223	1.3440	0.7178
920.0	6.5365	3.8938	5.4019	1.1266	1.3422	0.7176
940.0	6.7622	4.0621	5.4262	1.1308	1.3405	0.7174
960.0	6.9888	4.2312	5.4500	1.1349	1.3389	0.7172
980.0	7.2162	4.4012	5.4735	1.1389	1.3373	0.7169
1000.0	7.4444	4.5719	5.4965	1.1428	1.3357	0.7165

THERMOPHYSICAL PROPERTIES OF AIR

ISOBAR: 1000.0 MPa

T (K)	ρ (kg/m^3)	a (m/s)	μ (Pa·s)	k (J/s·m·K)	$\bar{m}$
1020.0	3.4153 +3	625.0	4.3745 −5	0.0697 +0	28.9644
1040.0	3.3497	630.8	4.4309	0.0708	28.9644
1060.0	3.2864	636.5	4.4870	0.0719	28.9644
1080.0	3.2256	642.2	4.5426	0.0730	28.9644
1100.0	3.1669	647.8	4.5978	0.0741	28.9644
1120.0	3.1104	653.3	4.6526	0.0754	28.9644
1140.0	3.0558	658.8	4.7070	0.0765	28.9644
1160.0	3.0031	664.2	4.7610	0.0776	28.9644
1180.0	2.9522	669.6	4.8147	0.0787	28.9644
1200.0	2.9030	674.9	4.8680	0.0797	28.9644
1220.0	2.8554	680.2	4.9209	0.0808	28.9644
1240.0	2.8094	685.5	4.9735	0.0819	28.9644
1260.0	2.7648	690.6	5.0258	0.0830	28.9644
1280.0	2.7216	695.8	5.0778	0.0841	28.9644
1300.0	2.6797	700.9	5.1294	0.0851	28.9644
1320.0	2.6391	705.9	5.1807	0.0862	28.9644
1340.0	2.5997	710.9	5.2318	0.0873	28.9644
1360.0	2.5615	715.9	5.2825	0.0884	28.9644
1380.0	2.5244	720.8	5.3329	0.0894	28.9644
1400.0	2.4883	725.7	5.3831	0.0905	28.9644
1420.0	2.4533	730.6	5.4330	0.0916	28.9644
1440.0	2.4192	735.4	5.4827	0.0927	28.9644
1460.0	2.3861	740.1	5.5320	0.0937	28.9644
1480.0	2.3538	744.9	5.5811	0.0948	28.9644
1500.0	2.3224	749.6	5.6300	0.0959	28.9644
1520.0	2.2919	754.2	5.6786	0.0969	28.9644
1540.0	2.2621	758.8	5.7270	0.0980	28.9644
1560.0	2.2331	763.4	5.7751	0.0991	28.9644
1580.0	2.2048	768.0	5.8231	0.1002	28.9644
1600.0	2.1773	772.5	5.8707	0.1013	28.9644
1620.0	2.1504	777.0	5.9182	0.1023	28.9644
1640.0	2.1242	781.5	5.9654	0.1034	28.9644
1660.0	2.0986	785.9	6.0125	0.1045	28.9644
1680.0	2.0736	790.3	6.0593	0.1056	28.9644
1700.0	2.0492	794.6	6.1059	0.1067	28.9644
1720.0	2.0254	799.0	6.1523	0.1078	28.9644
1740.0	2.0021	803.3	6.1985	0.1089	28.9644
1760.0	1.9793	807.5	6.2445	0.1099	28.9644
1780.0	1.9571	811.8	6.2903	0.1110	28.9644
1800.0	1.9354	816.0	6.3360	0.1121	28.9644
1820.0	1.9141	820.2	6.3814	0.1132	28.9644
1840.0	1.8933	824.4	6.4267	0.1143	28.9644
1860.0	1.8729	828.5	6.4717	0.1154	28.9644
1880.0	1.8530	832.6	6.5166	0.1165	28.9644
1900.0	1.8335	836.7	6.5614	0.1176	28.9644
1920.0	1.8144	840.8	6.6059	0.1187	28.9644
1940.0	1.7957	844.9	6.6503	0.1198	28.9644
1960.0	1.7774	848.9	6.6945	0.1209	28.9644
1980.0	1.7594	852.9	6.7386	0.1220	28.9644
2000.0	1.7418	856.9	6.7824	0.1231	28.9644

THERMOPHYSICAL PROPERTIES OF AIR

ISOBAR: 1000.0 MPa

T	h	e	s	c_p	γ	Pr
(K)	(J/kg)	(J/kg)	(J/kg·K)	(J/kg·K)		
1020.0	7.6733 +2	4.7434 +2	5.5192 +0	1.1465 +0	1.3343	0.7185
1040.0	7.9030	4.9156	5.5415	1.1501	1.3329	0.7185
1060.0	8.1333	5.0885	5.5634	1.1537	1.3315	0.7185
1080.0	8.3644	5.2622	5.5850	1.1572	1.3302	0.7185
1100.0	8.5962	5.4365	5.6063	1.1608	1.3288	0.7186
1120.0	8.8288	5.6116	5.6272	1.1643	1.3275	0.7187
1140.0	9.0620	5.7873	5.6479	1.1678	1.3262	0.7188
1160.0	9.2959	5.9638	5.6682	1.1712	1.3249	0.7190
1180.0	9.5305	6.1409	5.6883	1.1747	1.3237	0.7191
1200.0	9.7657	6.3187	5.7080	1.1781	1.3224	0.7192
1220.0	1.0002 +3	6.4973	5.7276	1.1816	1.3212	0.7194
1240.0	1.0238	6.6765	5.7468	1.1850	1.3200	0.7195
1260.0	1.0476	6.8564	5.7658	1.1884	1.3188	0.7197
1280.0	1.0714	7.0369	5.7845	1.1918	1.3176	0.7199
1300.0	1.0952	7.2182	5.8030	1.1952	1.3164	0.7201
1320.0	1.1192	7.4001	5.8213	1.1986	1.3152	0.7202
1340.0	1.1432	7.5827	5.8393	1.2020	1.3140	0.7204
1360.0	1.1673	7.7660	5.8572	1.2054	1.3128	0.7206
1380.0	1.1914	7.9500	5.8748	1.2088	1.3117	0.7208
1400.0	1.2156	8.1347	5.8922	1.2123	1.3105	0.7210
1420.0	1.2399	8.3200	5.9094	1.2157	1.3094	0.7212
1440.0	1.2642	8.5060	5.9265	1.2191	1.3082	0.7214
1460.0	1.2887	8.6928	5.9433	1.2226	1.3071	0.7216
1480.0	1.3131	8.8802	5.9600	1.2260	1.3060	0.7218
1500.0	1.3377	9.0683	5.9764	1.2295	1.3049	0.7220
1520.0	1.3623	9.2571	5.9928	1.2329	1.3037	0.7222
1540.0	1.3870	9.4466	6.0089	1.2364	1.3026	0.7224
1560.0	1.4118	9.6367	6.0249	1.2399	1.3015	0.7225
1580.0	1.4366	9.8276	6.0407	1.2434	1.3004	0.7227
1600.0	1.4615	1.0019 +3	6.0564	1.2469	1.2993	0.7229
1620.0	1.4865	1.0211	6.0719	1.2504	1.2982	0.7231
1640.0	1.5115	1.0404	6.0872	1.2539	1.2972	0.7232
1660.0	1.5366	1.0598	6.1024	1.2574	1.2961	0.7234
1680.0	1.5618	1.0792	6.1175	1.2609	1.2950	0.7236
1700.0	1.5871	1.0988	6.1325	1.2645	1.2939	0.7237
1720.0	1.6124	1.1183	6.1473	1.2680	1.2929	0.7239
1740.0	1.6378	1.1380	6.1620	1.2715	1.2918	0.7240
1760.0	1.6633	1.1577	6.1765	1.2751	1.2908	0.7242
1780.0	1.6888	1.1775	6.1909	1.2786	1.2898	0.7243
1800.0	1.7144	1.1974	6.2052	1.2821	1.2887	0.7245
1820.0	1.7401	1.2173	6.2194	1.2857	1.2877	0.7246
1840.0	1.7658	1.2373	6.2335	1.2892	1.2867	0.7247
1860.0	1.7917	1.2574	6.2475	1.2928	1.2857	0.7249
1880.0	1.8175	1.2775	6.2613	1.2963	1.2847	0.7250
1900.0	1.8435	1.2977	6.2750	1.2998	1.2837	0.7251
1920.0	1.8695	1.3180	6.2887	1.3033	1.2827	0.7252
1940.0	1.8956	1.3384	6.3022	1.3068	1.2817	0.7253
1960.0	1.9218	1.3588	6.3156	1.3103	1.2808	0.7254
1980.0	1.9481	1.3793	6.3289	1.3138	1.2798	0.7255
2000.0	1.9744	1.3999	6.3422	1.3173	1.2789	0.7256

THERMOPHYSICAL PROPERTIES OF AIR

ISOBAR: 1000.0 MPa

T (K)	ρ (kg/m^3)	a (m/s)	μ (Pa·s)	k (J/s·m·K)	$\bar{m}$
2040.0	1.7077 +3	864.8	6.8697 −5	0.1253 +0	28.9644
2080.0	1.6748	872.6	6.9564	0.1275	28.9643
2120.0	1.6432	880.3	7.0425	0.1297	28.9643
2160.0	1.6128	888.0	7.1279	0.1319	28.9643
2200.0	1.5835	895.6	7.2128	0.1341	28.9642
2240.0	1.5552	903.1	7.2972	0.1363	28.9642
2280.0	1.5279	910.5	7.3810	0.1385	28.9641
2320.0	1.5016	917.9	7.4642	0.1407	28.9641
2360.0	1.4761	925.3	7.5470	0.1429	28.9640
2400.0	1.4515	932.5	7.6292	0.1451	28.9639
2440.0	1.4277	939.7	7.7110	0.1472	28.9638
2480.0	1.4047	946.9	7.7922	0.1494	28.9636
2520.0	1.3823	954.0	7.8730	0.1515	28.9634
2560.0	1.3607	961.0	7.9534	0.1536	28.9632
2600.0	1.3398	968.0	8.0334	0.1564	28.9630
2640.0	1.3195	975.0	8.1128	0.1586	28.9627
2680.0	1.2998	981.9	8.1919	0.1608	28.9624
2720.0	1.2806	988.8	8.2705	0.1630	28.9620
2760.0	1.2621	995.6	8.3487	0.1652	28.9616
2800.0	1.2440	1002.4	8.4265	0.1674	28.9611
2840.0	1.2265	1009.1	8.5039	0.1696	28.9605
2880.0	1.2094	1015.8	8.5809	0.1718	28.9599
2920.0	1.1928	1022.5	8.6576	0.1740	28.9592
2960.0	1.1767	1029.1	8.7339	0.1762	28.9583
3000.0	1.1609	1035.7	8.8098	0.1784	28.9574
3040.0	1.1456	1042.2	8.8853	0.1806	28.9564
3080.0	1.1307	1048.8	8.9606	0.1828	28.9553
3120.0	1.1162	1055.2	9.0355	0.1850	28.9541
3160.0	1.1020	1061.7	9.1100	0.1873	28.9528
3200.0	1.0881	1068.1	9.1842	0.1895	28.9513
3240.0	1.0746	1074.5	9.2582	0.1917	28.9496
3280.0	1.0615	1080.9	9.3318	0.1940	28.9478
3320.0	1.0486	1087.2	9.4051	0.1962	28.9459
3360.0	1.0361	1093.5	9.4781	0.1985	28.9438
3400.0	1.0238	1099.8	9.5508	0.2008	28.9415
3440.0	1.0118	1106.0	9.6232	0.2031	28.9390
3480.0	1.0001	1112.3	9.6953	0.2054	28.9363
3520.0	9.8861 +2	1118.5	9.7672	0.2077	28.9334
3560.0	9.7740	1124.7	9.8388	0.2100	28.9302
3600.0	9.6642	1130.8	9.9101	0.2124	28.9269
3640.0	9.5569	1137.0	9.9812	0.2147	28.9233
3680.0	9.4517	1143.1	1.0052 −4	0.2171	28.9195
3720.0	9.3488	1149.2	1.0123	0.2195	28.9154
3760.0	9.2479	1155.2	1.0193	0.2219	28.9110
3800.0	9.1491	1161.3	1.0263	0.2243	28.9064
3840.0	9.0523	1167.3	1.0333	0.2268	28.9014
3880.0	8.9573	1173.4	1.0403	0.2292	28.8962
3920.0	8.8642	1179.4	1.0472	0.2317	28.8907
3960.0	8.7729	1185.4	1.0541	0.2342	28.8849
4000.0	8.6833	1191.3	1.0610	0.2367	28.8787

THERMOPHYSICAL PROPERTIES OF AIR

ISOBAR: 1000.0 MPa

T	h	e	s	c_p	γ	Pr
(K)	(J/kg)	(J/kg)	(J/kg·K)	(J/kg·K)		
2040.0	2.0272 +3	1.4412 +3	6.3683 +0	1.3243 +0	1.2770	0.7258
2080.0	2.0803	1.4828	6.3941	1.3311	1.2752	0.7259
2120.0	2.1337	1.5247	6.4195	1.3379	1.2734	0.7260
2160.0	2.1873	1.5669	6.4446	1.3447	1.2717	0.7262
2200.0	2.2413	1.6093	6.4693	1.3514	1.2700	0.7263
2240.0	2.2954	1.6520	6.4937	1.3579	1.2683	0.7264
2280.0	2.3499	1.6950	6.5178	1.3644	1.2667	0.7264
2320.0	2.4046	1.7382	6.5416	1.3708	1.2652	0.7265
2360.0	2.4596	1.7816	6.5651	1.3771	1.2637	0.7266
2400.0	2.5148	1.8254	6.5883	1.3833	1.2622	0.7266
2440.0	2.5702	1.8693	6.6112	1.3895	1.2608	0.7266
2480.0	2.6259	1.9135	6.6339	1.3955	1.2594	0.7266
2520.0	2.6819	1.9580	6.6562	1.4014	1.2581	0.7266
2560.0	2.7380	2.0026	6.6783	1.4072	1.2568	0.7266
2600.0	2.7944	2.0476	6.7002	1.4129	1.2555	0.7258
2640.0	2.8511	2.0927	6.7218	1.4185	1.2543	0.7256
2680.0	2.9079	2.1380	6.7432	1.4240	1.2532	0.7255
2720.0	2.9650	2.1836	6.7643	1.4294	1.2521	0.7253
2760.0	3.0223	2.2294	6.7852	1.4347	1.2510	0.7251
2800.0	3.0797	2.2754	6.8059	1.4399	1.2499	0.7249
2840.0	3.1375	2.3216	6.8264	1.4451	1.2489	0.7246
2880.0	3.1954	2.3680	6.8466	1.4501	1.2480	0.7243
2920.0	3.2535	2.4145	6.8667	1.4551	1.2470	0.7240
2960.0	3.3118	2.4613	6.8865	1.4601	1.2461	0.7237
3000.0	3.3703	2.5083	6.9061	1.4649	1.2453	0.7234
3040.0	3.4290	2.5555	6.9256	1.4698	1.2444	0.7230
3080.0	3.4878	2.6028	6.9448	1.4745	1.2436	0.7227
3120.0	3.5469	2.6504	6.9639	1.4793	1.2429	0.7223
3160.0	3.6062	2.6981	6.9827	1.4839	1.2421	0.7218
3200.0	3.6656	2.7460	7.0014	1.4886	1.2414	0.7214
3240.0	3.7253	2.7941	7.0199	1.4932	1.2408	0.7209
3280.0	3.7851	2.8424	7.0383	1.4979	1.2401	0.7204
3320.0	3.8451	2.8908	7.0565	1.5025	1.2395	0.7199
3360.0	3.9053	2.9394	7.0745	1.5070	1.2389	0.7194
3400.0	3.9657	2.9882	7.0924	1.5116	1.2383	0.7189
3440.0	4.0262	3.0372	7.1101	1.5162	1.2378	0.7183
3480.0	4.0870	3.0864	7.1276	1.5208	1.2372	0.7177
3520.0	4.1479	3.1357	7.1450	1.5254	1.2367	0.7171
3560.0	4.2090	3.1852	7.1623	1.5300	1.2363	0.7165
3600.0	4.2703	3.2349	7.1794	1.5347	1.2358	0.7159
3640.0	4.3318	3.2847	7.1964	1.5393	1.2354	0.7152
3680.0	4.3934	3.3347	7.2133	1.5440	1.2350	0.7145
3720.0	4.4553	3.3849	7.2300	1.5487	1.2346	0.7138
3760.0	4.5173	3.4353	7.2466	1.5535	1.2342	0.7131
3800.0	4.5796	3.4858	7.2630	1.5582	1.2339	0.7124
3840.0	4.6420	3.5366	7.2794	1.5630	1.2335	0.7117
3880.0	4.7046	3.5875	7.2956	1.5679	1.2332	0.7110
3920.0	4.7674	3.6385	7.3117	1.5727	1.2329	0.7102
3960.0	4.8304	3.6898	7.3277	1.5776	1.2327	0.7094
4000.0	4.8936	3.7412	7.3436	1.5826	1.2324	0.7087

THERMOPHYSICAL PROPERTIES OF AIR

ISOBAR: 1000.0 MPa

T (K)	ρ (kg/m^3)	a (m/s)	μ (Pa·s)	k (J/s·m·K)	$\bar{m}$
4040.0	8.5954 +2	1197.3	1.0679 −4	0.2392 +0	28.8723
4080.0	8.5092	1203.3	1.0748	0.2418	28.8655
4120.0	8.4245	1209.2	1.0816	0.2443	28.8583
4160.0	8.3413	1215.1	1.0884	0.2473	28.8508
4200.0	8.2596	1221.1	1.0952	0.2500	28.8430
4240.0	8.1794	1227.0	1.1020	0.2526	28.8348
4280.0	8.1005	1232.9	1.1088	0.2553	28.8262
4320.0	8.0230	1238.8	1.1155	0.2574	28.8172
4360.0	7.9468	1244.6	1.1223	0.2601	28.8079
4400.0	7.8719	1250.5	1.1290	0.2634	28.7982
4440.0	7.7983	1256.4	1.1357	0.2662	28.7880
4480.0	7.7258	1262.3	1.1424	0.2689	28.7775
4520.0	7.6545	1268.1	1.1491	0.2717	28.7666
4560.0	7.5844	1274.0	1.1557	0.2745	28.7553
4600.0	7.5154	1279.9	1.1624	0.2773	28.7436
4640.0	7.4475	1285.7	1.1690	0.2801	28.7315
4680.0	7.3806	1291.6	1.1756	0.2829	28.7190
4720.0	7.3148	1297.4	1.1822	0.2858	28.7061
4760.0	7.2499	1303.3	1.1888	0.2886	28.6927
4800.0	7.1861	1309.2	1.1954	0.2915	28.6790
4840.0	7.1231	1315.0	1.2020	0.2944	28.6648
4880.0	7.0612	1320.9	1.2085	0.2973	28.6502
4920.0	7.0001	1326.8	1.2151	0.3002	28.6352
4960.0	6.9399	1332.6	1.2216	0.3031	28.6198
5000.0	6.8806	1338.5	1.2281	0.3060	28.6040
5040.0	6.8221	1344.4	1.2346	0.3089	28.5878
5080.0	6.7644	1350.3	1.2411	0.3118	28.5711
5120.0	6.7076	1356.2	1.2476	0.3147	28.5541
5160.0	6.6515	1362.1	1.2541	0.3177	28.5366
5200.0	6.5962	1368.0	1.2606	0.3206	28.5188
5240.0	6.5417	1373.9	1.2671	0.3236	28.5005
5280.0	6.4879	1379.8	1.2735	0.3265	28.4819
5320.0	6.4348	1385.7	1.2800	0.3295	28.4628
5360.0	6.3824	1391.7	1.2864	0.3324	28.4434
5400.0	6.3307	1397.7	1.2929	0.3354	28.4236
5440.0	6.2797	1403.6	1.2993	0.3383	28.4034
5480.0	6.2294	1409.6	1.3057	0.3413	28.3828
5520.0	6.1797	1415.6	1.3121	0.3443	28.3619
5560.0	6.1306	1421.6	1.3186	0.3472	28.3406
5600.0	6.0821	1427.6	1.3250	0.3502	28.3189
5640.0	6.0343	1433.7	1.3314	0.3532	28.2969
5680.0	5.9871	1439.7	1.3378	0.3561	28.2745
5720.0	5.9404	1445.8	1.3442	0.3591	28.2518
5760.0	5.8944	1451.9	1.3505	0.3621	28.2288
5800.0	5.8489	1458.0	1.3569	0.3651	28.2054
5840.0	5.8039	1464.1	1.3633	0.3681	28.1817
5880.0	5.7595	1470.3	1.3697	0.3710	28.1577
5920.0	5.7157	1476.5	1.3760	0.3740	28.1333
5960.0	5.6724	1482.7	1.3824	0.3770	28.1087
6000.0	5.6295	1488.9	1.3888	0.3801	28.0838

THERMOPHYSICAL PROPERTIES OF AIR

ISOBAR: 1000.0 MPa

T (K)	h (J/kg)	e (J/kg)	s (J/kg·K)	c_p (J/kg·K)	γ	Pr
4040.0	4.9570 +3	3.7929 +3	7.3593 +0	1.5875 +0	1.2322	0.7079
4080.0	5.0206	3.8447	7.3750	1.5925	1.2320	0.7071
4120.0	5.0844	3.8966	7.3906	1.5976	1.2318	0.7063
4160.0	5.1484	3.9488	7.4060	1.6026	1.2316	0.7052
4200.0	5.2127	4.0011	7.4214	1.6077	1.2315	0.7044
4240.0	5.2771	4.0537	7.4366	1.6128	1.2313	0.7035
4280.0	5.3417	4.1064	7.4518	1.6179	1.2312	0.7027
4320.0	5.4065	4.1592	7.4669	1.6231	1.2311	0.7023
4360.0	5.4715	4.2123	7.4819	1.6282	1.2311	0.7015
4400.0	5.5368	4.2656	7.4968	1.6334	1.2310	0.7000
4440.0	5.6022	4.3190	7.5116	1.6386	1.2310	0.6991
4480.0	5.6678	4.3726	7.5263	1.6438	1.2310	0.6983
4520.0	5.7337	4.4264	7.5409	1.6489	1.2310	0.6974
4560.0	5.7998	4.4804	7.5555	1.6541	1.2310	0.6964
4600.0	5.8660	4.5345	7.5699	1.6593	1.2311	0.6955
4640.0	5.9325	4.5889	7.5843	1.6644	1.2311	0.6946
4680.0	5.9992	4.6434	7.5986	1.6696	1.2312	0.6937
4720.0	6.0661	4.6981	7.6129	1.6747	1.2313	0.6928
4760.0	6.1332	4.7529	7.6270	1.6798	1.2315	0.6918
4800.0	6.2004	4.8079	7.6411	1.6848	1.2316	0.6909
4840.0	6.2679	4.8631	7.6551	1.6898	1.2318	0.6900
4880.0	6.3356	4.9185	7.6690	1.6948	1.2320	0.6890
4920.0	6.4035	4.9740	7.6829	1.6997	1.2322	0.6880
4960.0	6.4716	5.0297	7.6967	1.7046	1.2324	0.6871
5000.0	6.5399	5.0855	7.7104	1.7093	1.2327	0.6861
5040.0	6.6084	5.1415	7.7240	1.7141	1.2330	0.6851
5080.0	6.6770	5.1977	7.7376	1.7187	1.2333	0.6841
5120.0	6.7458	5.2540	7.7511	1.7233	1.2336	0.6831
5160.0	6.8149	5.3104	7.7645	1.7278	1.2340	0.6821
5200.0	6.8841	5.3670	7.7779	1.7322	1.2344	0.6810
5240.0	6.9534	5.4238	7.7912	1.7365	1.2348	0.6800
5280.0	7.0230	5.4806	7.8044	1.7406	1.2352	0.6789
5320.0	7.0927	5.5376	7.8175	1.7447	1.2357	0.6778
5360.0	7.1626	5.5947	7.8306	1.7487	1.2362	0.6767
5400.0	7.2326	5.6519	7.8436	1.7525	1.2367	0.6756
5440.0	7.3028	5.7093	7.8566	1.7563	1.2372	0.6745
5480.0	7.3731	5.7667	7.8695	1.7598	1.2378	0.6733
5520.0	7.4436	5.8243	7.8823	1.7633	1.2384	0.6721
5560.0	7.5141	5.8819	7.8950	1.7666	1.2390	0.6708
5600.0	7.5849	5.9396	7.9077	1.7698	1.2396	0.6696
5640.0	7.6557	5.9974	7.9203	1.7728	1.2403	0.6683
5680.0	7.7267	6.0553	7.9328	1.7756	1.2410	0.6670
5720.0	7.7978	6.1133	7.9453	1.7783	1.2418	0.6656
5760.0	7.8690	6.1713	7.9577	1.7808	1.2425	0.6642
5800.0	7.9402	6.2294	7.9700	1.7831	1.2433	0.6628
5840.0	8.0116	6.2875	7.9823	1.7853	1.2442	0.6613
5880.0	8.0831	6.3456	7.9945	1.7873	1.2451	0.6597
5920.0	8.1546	6.4038	8.0066	1.7891	1.2460	0.6581
5960.0	8.2262	6.4621	8.0187	1.7907	1.2469	0.6565
6000.0	8.2978	6.5203	8.0307	1.7921	1.2479	0.6548

TABLE H.1

SELECTED VALUES

U.S. STANDARD ATMOSPHERE, 1976

SI UNITS

PROPERTIES OF THE U.S. STANDARD ATMOSPHERE, 1976
SI UNITS

Geometric Altitude H(m)	Geopotential Altitude z(m)	Temperature T(K)	Pressure Ratio p/p_a	Density Ratio ρ/ρ_a	Particle Number Density ν(m^{-3})
-500	-500	291.400	1.0607 +0	1.0489 +0	2.6715 +25
-200	-200	289.450	1.0239	1.0193	2.5962
0	0	288.150	1.0000	1.0000	2.5470
200	200	286.850	9.7651 -1	9.8094 -1	2.4984
400	400	285.550	9.5348	9.6216	2.4506
600	600	284.250	9.3088	9.4366	2.4035
800	800	282.951	9.0873	9.2543	2.3571
1000	1000	281.651	8.8700	9.0748	2.3113
1200	1200	280.351	8.6570	8.8979	2.2663
1400	1400	279.052	8.4482	8.7237	2.2219
1600	1600	277.753	8.2435	8.5521	2.1782
1800	1799	276.435	8.0428	8.3632	2.1352
2000	1999	275.154	7.8461	8.2168	2.0928
2200	2199	273.855	7.6534	8.0529	2.0511
2400	2399	272.556	7.4645	7.8916	2.0100
2600	2599	271.257	7.2794	7.7328	1.9695
2800	2799	269.958	7.0980	7.5764	1.9297
3000	2999	268.659	6.9204	7.4225	1.8905
3200	3198	267.360	6.7463	7.2710	1.8519
3400	3398	266.062	6.5759	7.1219	1.8139
3600	3598	264.763	6.4089	6.9751	1.7765
3800	3798	263.465	6.2454	6.8307	1.7398
4000	3997	262.166	6.0854	6.6885	1.7036
4200	4197	260.868	5.9286	6.5487	1.6679
4400	4397	259.570	5.7752	6.4111	1.6329
4600	4597	258.272	5.6250	6.2758	1.5984
4800	4796	256.974	5.4780	6.1426	1.5645
5000	4996	255.676	5.3341	6.0117	1.5312
5200	5196	254.378	5.1933	5.8829	1.4983
5400	5395	253.080	5.0556	5.7562	1.4661
5600	5595	251.782	4.9208	5.6316	1.4344
5800	5795	250.484	4.7889	5.5091	1.4032
6000	5994	249.187	4.6600	5.3887	1.3725
6200	6194	247.889	4.5338	5.2703	1.3423
6400	6394	246.592	4.4105	5.1539	1.3127
6600	6593	245.294	4.2899	5.0394	1.2835
6800	6793	243.997	4.1720	4.9270	1.2549
7000	6992	242.700	4.0567	4.8165	1.2267
7200	7192	241.403	3.9441	4.7079	1.1991
7400	7391	240.106	3.8339	4.6012	1.1719
7600	7591	238.809	3.7263	4.4963	1.1452
7800	7790	237.512	3.6212	4.3933	1.1190
8000	7990	236.215	3.5185	4.2921	1.0932
8500	8489	232.974	3.2720	4.0470	1.0308
9000	8987	229.733	3.0397	3.8128	9.7110 +24
9500	9486	226.492	2.8210	3.5891	9.1413
10000	9984	223.252	2.6153	3.3756	8.5976

PROPERTIES OF THE U.S. STANDARD ATMOSPHERE, 1976
SI UNITS

Geometric Altitude	Particle Mean-Free Path Length	Particle Mean-Free Speed	Speed of Sound	Dynamic Viscosity Ratio	Thermal Conductivity Ratio
H(m)	λ(m)	C(m/s)	a(m/s)	μ/μ_a	k/k_a
−500	6.3240 −8	461.53	342.21	1.0087 +0	1.0100 +0
−200	6.5074	459.98	341.06	1.0035	1.0040
0	6.6332	458.94	340.29	1.0000	1.0000
200	6.7621	457.91	339.53	9.9649 −1	9.9596 −1
400	6.8941	456.87	338.76	9.9297	9.9191
600	7.0293	455.83	337.98	9.8945	9.8787
800	7.1677	454.79	337.21	9.8591	9.8382
1000	7.3095	453.74	336.43	9.8237	9.7976
1200	7.4548	452.69	335.66	9.7883	9.7570
1400	7.6037	451.64	334.88	9.7527	9.7163
1600	7.7562	450.59	334.10	9.7171	9.6756
1800	7.9126	449.53	333.32	9.6814	9.6348
2000	8.0728	448.48	332.53	9.6456	9.5940
2200	8.2371	447.42	331.75	9.6098	9.5531
2400	8.4054	446.35	330.96	9.5739	9.5122
2600	8.5781	445.29	330.17	9.5379	9.4712
2800	8.7551	444.22	329.38	9.5018	9.4302
3000	8.9367	443.15	328.58	9.4656	9.3891
3200	9.1229	442.08	327.79	9.4294	9.3480
3400	9.3139	441.00	326.99	9.3930	9.3068
3600	9.5099	439.93	326.19	9.3566	9.2656
3800	9.7110	438.85	325.39	9.3201	9.2244
4000	9.9173	437.76	324.59	9.2836	9.1830
4200	1.0129 −7	436.68	323.78	9.2469	9.1417
4400	1.0346	435.59	322.98	9.2102	9.1003
4600	1.0570	434.50	322.17	9.1733	9.0588
4800	1.0799	433.41	321.36	9.1364	9.0173
5000	1.1034	432.31	320.55	9.0995	8.9757
5200	1.1276	431.21	319.73	9.0624	8.9341
5400	1.1524	430.11	318.91	9.0252	8.8924
5600	1.1779	429.01	318.10	8.9880	8.8507
5800	1.2040	427.90	317.27	8.9507	8.8089
6000	1.2310	426.79	316.45	8.9133	8.7671
6200	1.2586	425.68	315.63	8.8758	8.7253
6400	1.2870	424.56	314.80	8.8382	8.6833
6600	1.3163	423.44	313.97	8.8005	8.6414
6800	1.3463	422.32	313.14	8.7627	8.5994
7000	1.3772	421.20	312.31	8.7249	8.5573
7200	1.4090	420.07	311.47	8.6869	8.5152
7400	1.4416	418.94	310.63	8.6489	8.4730
7600	1.4753	417.81	309.79	8.6108	8.4308
7800	1.5098	416.67	308.95	8.5726	8.3885
8000	1.5454	415.53	308.11	8.5343	8.3462
8500	1.6390	412.67	305.98	8.4381	8.2402
9000	1.7397	409.79	303.85	8.3414	8.1339
9500	1.8482	406.89	301.70	8.2441	8.0273
10000	1.9651	403.97	299.53	8.1461	7.9203

PROPERTIES OF THE U.S. STANDARD ATMOSPHERE, 1976
SI UNITS

Geometric Altitude H(m)	Geopotential Altitude z(m)	Temperature T(K)	Pressure Ratio p/p_a	Density Ratio ρ/ρ_a	Particle Number Density ν(m^{-3})
11000	10981	216.774	2.2403 −1	2.9780 −1	7.5848 +24
12000	11977	216.650	1.9145	2.5464	6.4857
13000	12973	216.650	1.6362	2.1763	5.5430
14000	13969	216.650	1.3985	1.8601	4.7375
15000	14965	216.650	1.1953	1.5898	4.0493
16000	15960	216.650	1.0217	1.3589	3.4612
17000	16955	216.650	8.7340 −2	1.1616	2.9587
18000	17949	216.650	7.4663	9.9304 −2	2.5292
19000	18943	216.650	6.3829	8.4894	2.1622
20000	19937	216.650	5.4570	7.2580	1.8486
21000	20931	217.581	4.6671	6.1808	1.5742
22000	21924	218.574	3.9945	5.2661	1.3413
23000	22917	219.567	3.4215	4.4903	1.1437
24000	23910	220.560	2.9328	3.8317	9.7591 +23
25000	24902	221.552	2.5158	3.2722	8.3341
26000	25894	222.544	2.1597	2.7965	7.1225
27000	26886	223.536	1.8553	2.3917	6.0916
28000	27877	224.527	1.5950	2.0470	5.2138
29000	28868	225.518	1.3722	1.7533	4.4657
30000	29859	226.509	1.1813	1.5029	3.8278
32000	31840	228.490	8.7743 −3	1.1065	2.8183
34000	33819	237.743	6.5473	8.0714 −3	2.0558
36000	35797	239.282	4.9200	5.9248	1.5090
38000	37774	244.818	3.7220	4.3809	1.1158
40000	39750	250.350	2.8338	3.2618	8.3077 +22
42000	41724	255.878	2.1709	2.4447	6.2266
44000	43698	261.403	1.6728	1.8440	4.6965
46000	45669	266.925	1.2962	1.3993	3.5640
48000	47640	270.650	1.0095	1.0749	2.7376
50000	49610	270.650	7.8735 −4	8.3827 −4	2.1351
52000	51578	269.031	6.1401	6.5765	1.6750
54000	53545	263.524	4.7705	5.2164	1.3286
56000	55511	258.019	3.6873	4.1180	1.0488
58000	57476	252.518	2.8348	3.2348	8.2390 +21
60000	59439	247.021	2.1671	2.5280	6.4387
62000	61401	241.527	1.6470	1.9650	5.0048
64000	63362	236.036	1.2441	1.5188	3.8683
66000	65322	230.549	9.3372 −5	1.1670	2.9723
68000	67280	225.065	6.9607	8.9118 −5	2.2698
70000	69238	219.585	5.1526	6.7616	1.7222
72000	71194	214.263	3.7861	5.0917	1.2968
74000	73148	210.353	2.7642	3.7866	9.6443 +20
76000	75102	206.446	2.0067	2.8009	7.1338
78000	77054	202.541	1.4481	2.0603	5.2475
80000	79006	198.639	1.0387	1.5068	3.8378
82000	80956	194.739	7.4028 −6	1.0954	2.7899
84000	82904	190.841	5.2410	7.9134 −6	2.0155
85500	84365	187.920	4.0269	6.1747	1.5727

PROPERTIES OF THE U.S. STANDARD ATMOSPHERE, 1976
SI UNITS

Geometric Altitude	Particle Mean-Free Path Length	Particle Mean-Free Speed	Speed of Sound	Dynamic Viscosity Ratio	Thermal Conductivity Ratio
H(m)	λ(m)	C(m/s)	a(m/s)	μ/μ_a	k/k_a
11000	2.2274 -7	398.07	295.15	7.9485 -1	7.7055 -1
12000	2.6049	397.95	295.07	7.9447	7.7014
13000	3.0479	397.95	295.07	7.9447	7.7014
14000	3.5662	397.95	295.07	7.9447	7.7014
15000	4.1723	397.95	295.07	7.9447	7.7014
16000	4.8812	397.95	295.07	7.9447	7.7014
17000	5.7102	397.95	295.07	7.9447	7.7014
18000	6.6797	397.95	295.07	7.9447	7.7014
19000	7.8135	397.95	295.07	7.9447	7.7014
20000	9.1393	397.95	295.07	7.9447	7.7014
21000	1.0732 -6	398.81	295.70	7.9732	7.7324
22000	1.2596	399.72	296.38	8.0037	7.7653
23000	1.4772	400.62	297.05	8.0340	7.7983
24000	1.7312	401.53	297.72	8.0643	7.8312
25000	2.0272	402.43	298.39	8.0945	7.8641
26000	2.3720	403.33	299.06	8.1247	7.8969
27000	2.7734	404.23	299.72	8.1547	7.9297
28000	3.2404	405.12	300.39	8.1847	7.9624
29000	3.7832	406.01	301.05	8.2147	7.9951
30000	4.4137	406.91	301.71	8.2446	8.0278
32000	5.9945	408.68	303.02	8.3041	8.0930
34000	8.2182	413.35	306.49	8.4610	8.2654
36000	1.1196 -5	418.22	310.10	8.6247	8.4462
38000	1.5141	423.03	313.67	8.7866	8.6259
40000	2.0336	427.78	317.19	8.9468	8.8046
42000	2.7133	432.48	320.67	9.1052	8.9822
44000	3.5973	437.13	324.12	9.2620	9.1587
46000	4.7404	441.72	327.52	9.4172	9.3342
48000	6.1713	444.79	329.80	9.5210	9.4521
50000	7.9130	444.79	329.80	9.5210	9.4521
52000	1.0086 -4	443.46	328.81	9.4760	9.4009
54000	1.2716	438.90	325.43	9.3218	9.2262
56000	1.6108	434.29	322.01	9.1662	9.0507
58000	2.0506	429.63	318.56	9.0091	8.8744
60000	2.6239	424.93	315.07	8.8506	8.6972
62000	3.3757	420.18	311.55	8.6906	8.5192
64000	4.3675	415.38	307.99	8.5290	8.3404
66000	5.6840	410.52	304.39	8.3658	8.1607
68000	7.4432	405.61	300.75	8.2010	7.9802
70000	9.8102	400.64	297.06	8.0346	7.7989
72000	1.3027 -3	395.75	293.44	7.8712	7.6219
74000	1.7518	392.13	290.75	7.7501	7.4914
76000	2.3682	388.47	288.04	7.6280	7.3604
78000	3.2195	384.78	285.30	7.5051	7.2291
80000	4.4022	381.05	282.54	7.3813	7.0975
82000	6.0556	377.29	279.75	7.2566	6.9654
84000	8.3822	373.50	276.94	7.1309	6.8330
85500	1.0743 -2	370.63	274.81	7.0360	6.7335

TABLE H.2

SELECTED VALUES

U.S. STANDARD ATMOSPHERE, 1976

ENGINEERING UNITS

PROPERTIES OF THE U.S. STANDARD ATMOSPHERE, 1976
ENGINEERING UNITS

Geometric Altitude H(ft)	Geopotential Altitude z(ft)	Temperature T(°R)	Pressure Ratio p/p_a	Density Ratio ρ/ρ_a	Gravity Ratio g/g_a
-1000	-1000	522.236	1.0366 +0	1.0296 +0	1.0001
-500	-500	520.452	1.0182	1.0147	1.0000
0	0	518.669	1.0000	1.0000	1.0000
500	500	516.886	9.8206 -1	9.8545 -1	1.0000
1000	1000	515.104	9.6438	9.7107	0.9999
1500	1500	513.320	9.4697	9.5684	0.9999
2000	2000	511.538	9.2981	9.4278	0.9998
2500	2500	509.754	9.1291	9.2887	0.9998
3000	3000	507.972	8.9625	9.1513	0.9997
3500	3499	506.190	8.7984	9.0154	0.9997
4000	3999	504.408	8.6368	8.8811	0.9996
4500	4499	502.625	8.4776	8.7483	0.9996
5000	4999	500.843	8.3208	8.6170	0.9995
5500	5499	499.061	8.1664	8.4873	0.9995
6000	5998	497.279	8.0142	8.3590	0.9994
6500	6498	495.496	7.8644	8.2323	0.9994
7000	6998	491.932	7.5717	7.9832	0.9993
8000	7997	490.150	7.4286	7.8609	0.9992
8500	8497	488.370	7.2878	7.7400	0.9992
9000	8996	486.588	7.1492	7.6206	0.9991
9500	9496	484.806	7.0127	7.5025	0.9991
10000	9995	483.025	6.8783	7.3859	0.9990
11000	10994	479.462	6.6158	7.1568	0.9989
12000	11993	475.900	6.3615	6.9333	0.9989
13000	12992	472.337	6.1152	6.7151	0.9988
14000	13991	468.777	5.8767	6.5022	0.9987
15000	14989	465.215	5.6458	6.2946	0.9986
16000	15988	461.655	5.4224	6.0921	0.9985
17000	16986	458.094	5.2001	5.8946	0.9984
18000	17984	454.534	4.9970	5.7021	0.9983
19000	18983	450.973	4.7947	5.5144	0.9982
20000	19981	447.415	4.5991	5.3316	0.9981
22000	21997	440.296	4.2273	4.9798	0.9979
24000	23972	433.180	3.8803	4.6462	0.9977
26000	25968	426.065	3.5568	4.3300	0.9975
28000	27962	418.951	3.2556	4.0305	0.9973
30000	29957	411.838	2.9754	3.7473	0.9971
32000	31951	404.728	2.7151	3.4795	0.9969
34000	33945	397.618	2.4736	3.2267	0.9967
36000	35938	390.510	2.2498	2.9883	0.9966
38000	37931	389.970	2.0443	2.7191	0.9964
40000	39923	389.970	1.8576	2.4708	0.9962
42000	41916	389.970	1.6880	2.2452	0.9960
44000	43907	389.970	1.5339	2.0402	0.9958
46000	45899	389.970	1.3939	1.8540	0.9956
48000	47890	389.970	1.2667	1.6848	0.9954
50000	49880	389.970	1.1511	1.5311	0.9952

PROPERTIES OF THE U.S. STANDARD ATMOSPHERE, 1976

ENGINEERING UNITS

Geometric Altitude H(ft)	Particle Number Density ν(ft^{-3})	Particle Mean-Free Path Length λ(ft)	Speed of Sound a(ft/sec)	Dynamic Viscosity Ratio μ/μ_a	Thermal Conductivity Ratio k/k_a
-1000	7.4255 +23	2.1137 -7	1120.3	1.0053 +0	1.0061 +0
-500	7.3182	2.1447	1118.4	1.0026	1.0030
0	7.2123	2.1762	1116.4	1.0000	1.0000
500	7.1072	2.2084	1114.5	9.9732 -1	9.9692 -1
1000	7.0036	2.2411	1112.6	9.9464	9.9384
1500	6.9011	2.2744	1110.7	9.9196	9.9076
2000	6.7994	2.3083	1108.8	9.8928	9.8767
2500	6.6992	2.3429	1106.8	9.8659	9.8459
3000	6.6001	2.3781	1104.9	9.8389	9.8150
3500	6.5021	2.4139	1103.0	9.8119	9.7840
4000	6.4053	2.4505	1101.0	9.7849	9.7531
4500	6.3096	2.4876	1099.1	9.7578	9.7221
5000	6.2147	2.5255	1097.1	9.7307	9.6910
5500	6.1213	2.5641	1095.1	9.7035	9.6600
6000	6.0287	2.6035	1093.2	9.6763	9.6289
6500	5.9372	2.6436	1091.2	9.6490	9.5978
7000	5.8469	2.6844	1089.3	9.6217	9.5667
7500	5.7577	2.7260	1087.3	9.5944	9.5355
8000	5.6696	2.7684	1085.4	9.5670	9.5043
8500	5.5824	2.8117	1083.3	9.5395	9.4731
9000	5.4960	2.8558	1081.4	9.5120	9.4419
9500	5.4111	2.9007	1079.4	9.4845	9.4106
10000	5.3270	2.9465	1077.4	9.4569	9.3793
11000	5.1616	3.0408	1073.4	9.4016	9.3166
12000	5.0005	3.1389	1069.4	9.3461	9.2537
13000	4.8430	3.2408	1065.4	9.2904	9.1908
14000	4.6896	3.3468	1061.4	9.2346	9.1278
15000	4.5398	3.4573	1057.4	9.1785	9.0646
16000	4.3936	3.5722	1053.3	9.1223	9.0013
17000	4.2512	3.6919	1049.3	9.0658	8.9379
18000	4.1125	3.8166	1045.1	9.0092	8.8744
19000	3.9771	3.9465	1041.0	8.9523	8.8108
20000	3.8451	4.0817	1036.9	8.8953	8.7470
22000	3.5914	4.3701	1028.6	8.7806	8.6192
24000	3.2944	4.6841	1020.3	8.6650	8.4909
26000	3.1228	5.0259	1011.9	8.5487	8.3621
28000	2.9070	5.3993	1003.4	8.4315	8.2329
30000	2.7026	5.8074	994.9	8.3134	8.1032
32000	2.5095	6.2546	986.2	8.1945	7.9730
34000	2.3272	6.7444	977.5	8.0746	7.8424
36000	2.1552	7.2825	968.7	7.9539	7.7114
38000	1.9611	8.0036	968.1	7.9447	7.7014
40000	1.7820	8.8081	968.1	7.9447	7.7014
42000	1.6193	9.6929	968.1	7.9447	7.7014
44000	1.4715	1.0667 -6	968.1	7.9447	7.7014
46000	1.3371	1.1738	968.1	7.9447	7.7014
48000	1.2151	1.2917	968.1	7.9447	7.7014
50000	1.1042	1.4214	968.1	7.9447	7.7014

PROPERTIES OF THE U.S. STANDARD ATMOSPHERE, 1976
ENGINEERING UNITS

Geometric Altitude	Geopotential Altitude	Temperature	Pressure Ratio	Density Ratio	Gravity Ratio
H(ft)	z(ft)	T(°R)	p/p_a	ρ/ρ_a	g/g_a
50000	49880	389.970	1.1511 −1	1.5311 −1	0.9952
55000	54855	389.970	9.0633 −2	1.2055	0.9947
60000	59828	389.970	7.1366	9.4919 −2	0.9943
65000	64798	389.970	5.6201	7.4750	0.9938
70000	69766	392.245	4.4289	5.8565	0.9933
75000	74731	394.970	3.4963	4.5914	0.9928
80000	79694	397.694	2.7649	3.6060	0.9924
85000	84655	400.415	2.1902	2.8371	0.9919
90000	89613	403.135	1.7379	2.2360	0.9914
95000	94569	405.855	1.6596	2.1325	0.9910
100000	99523	408.571	1.0997	1.3960	0.9905
105000	104474	411.289	8.7691 −3	1.1059	0.9900
110000	109423	418.385	7.0112	8.6918 −3	0.9895
115000	114369	425.982	5.6288	6.8536	0.9891
120000	119313	433.578	4.5370	5.4275	0.9886
125000	124255	441.169	3.6711	4.3160	0.9881
130000	129195	448.758	2.9815	3.4460	0.9876
135000	134132	456.341	2.4301	2.7620	0.9872
140000	139066	463.923	1.9875	2.2221	0.9867
145000	143999	471.499	1.6311	1.7943	0.9862
150000	148929	479.073	1.3429	1.4539	0.9858
155000	153856	486.642	1.1091	1.1821	0.9853
160000	158782	487.170	9.1763 −4	9.7697 −4	0.9848
165000	163705	487.170	7.5930	8.0440	0.9844
170000	168625	485.168	6.2827	6.7166	0.9839
175000	173544	477.614	5.1877	5.6337	0.9834
180000	178460	470.061	4.2709	4.7126	0.9830
185000	183373	462.513	3.5054	3.9311	0.9825
190000	188285	454.968	2.8681	3.2697	0.9820
195000	193194	447.427	2.3390	2.7114	0.9816
200000	198100	439.889	1.9011	2.2416	0.9811
205000	203004	432.356	1.5398	1.8472	0.9806
210000	207906	424.825	1.2426	1.5172	0.9802
215000	212806	417.299	9.9916 −5	1.2419	0.9797
220000	217703	409.775	8.0026	1.0129	0.9792
225000	222598	402.257	6.3840	8.2316 −5	0.9788
230000	227491	394.740	5.0716	6.6639	0.9783
235000	232381	387.227	4.0117	5.3735	0.9778
240000	237269	381.618	3.1608	4.2959	0.9774
245000	242155	376.257	2.4811	3.4218	0.9769
250000	247039	370.899	1.9428	2.7169	0.9764
255000	251920	365.544	1.5154	2.1502	0.9760
260000	256798	360.190	1.1778	1.6961	0.9755
265000	261675	354.839	9.1211 −6	1.3332	0.9751
270000	266549	349.491	7.0369	1.0443	0.9746
275000	271421	344.145	5.4079	8.1505 −6	0.9741
280000	276290	338.801	4.1395	6.3372	0.9737

PROPERTIES OF THE U.S. STANDARD ATMOSPHERE, 1976
ENGINEERING UNITS

Geometric Altitude H(ft)	Particle Number Density ν(ft^{-3})	Particle Mean-Free Path Length λ(ft)	Speed of Sound a(ft/sec)	Dynamic Viscosity Ratio μ/μ_a	Thermal Conductivity Ratio k/k_a
50000	1.1042 +23	1.4214 −6	968.1	7.9447 −1	7.7014 −1
55000	8.6998 +22	1.8053	968.1	7.9447	7.7014
60000	6.8459	2.2927	968.1	7.9447	7.7014
65000	5.3912	2.9114	968.1	7.9447	7.7014
70000	4.2237	3.7159 −6	970.9	7.9835	7.7434
75000	3.3114	4.7398	974.3	8.0298	7.7937
80000	2.6007	6.0351	977.6	8.0759	7.8438
85000	2.0462	7.6709	980.9	8.1219	7.8939
90000	1.6126	9.7329 −5	984.3	8.1677	7.9438
95000	1.2732	1.2328	987.6	8.2134	8.9927
100000	1.0069	1.5589	990.9	8.2589	8.0435
105000	7.9757 +21	1.9679	994.2	8.3042	8.0931
110000	7.6316	2.5038	1002.7	8.4221	8.2226
115000	4.9430	3.1754	1011.8	8.5473	8.3606
120000	3.9145	4.0098	1020.8	8.6715	8.4981
125000	3.1129	5.0423	1029.7	8.7947	8.6349
130000	2.4853	6.3153	1038.5	9.9168	8.7711
135000	1.9920	7.8793	1047.2	9.0379	9.9067
140000	1.6027	9.7933	1055.9	9.1581	9.0416
145000	1.2941	1.2129 −4	1064.5	9.2773	9.1760
150000	1.0486	1.4969	1073.0	9.3955	9.3097
155000	8.5256 +20	1.8410	1081.4	9.5129	9.4428
160000	7.0461	2.2276	1082.0	9.5210	9.4521
165000	25.8304	2.6921	1082.0	9.5210	9.4521
170000	4.8442	3.2401	1079.8	9.4901	9.4169
175000	4.0632	3.8629	1071.4	9.3728	9.2840
180000	3.3989	4.6178	1062.9	9.2547	9.1505
185000	2.8351	5.5361	1054.3	9.1358	9.0166
190000	2.3582	6.6558	1045.6	9.0161	8.8822
195000	1.9556	8.0262	1036/9	8.8955	8.7473
200000	1.6167	9.7087	1028.2	8.7740	8.6119
205000	1.3323	1.1781 −3	1019.3	8.6516	8.4760
210000	1.0942	1.4344	1010.4	8.5383	8.3396
215000	8.9569 +19	1.7524	1001.4	8.4041	8.2028
220000	7.3055	2.1485	992.4	8.2790	8.0655
225000	5.9369	2.6438	983.2	8.1529	7.9277
230000	4.8062	3.2657	974.0	8.0259	7.7894
235000	3.8754	4.0499	964.7	7.8978	7.6507
240000	3.0984	5.0659	957.6	7.8015	7.5468
245000	2.4678	6.3599	950.9	7.7089	7.4472
250000	1.9595	8.0102	944.1	7.6158	7.3473
255000	1.5508	1.0121 −2	937.3	7.5222	7.2473
260000	1.2232	1.2831	930.4	7.4280	7.1470
265000	9.6156 +18	1.6323	923.5	7.3333	7.0465
270000	7.5320	2.0839	916.5	7.2381	6.9458
275000	5.8783	2.6701	909.4	7.1423	6.8449
280000	4.5706	3.4341	902.3	7.0459	6.7438

TABLE H.3

SELECTED VALUES

U.S. STANDARD ATMOSPHERE, 1976

SI UNITS

PROPERTIES OF THE U.S. STANDARD ATMOSPHERE, 1976

ABOVE THE RE-ENTRY ALTITUDE, SI UNITS

Geometric Altitude H(m)	Temperature T(K)	Pressure Ratio p/p_a	Density Ratio ρ/ρ_a	Particle Number Density $\nu(\mathrm{m}^{-3})$	Particle Mean-Free Speed C(m/s)	Particle Mean-Free Path-Length λ(m)
86000	186.87	3.6850 −6	5.680 −6	1.447 +20	369.7	1.17 −2
90000	186.87	1.8119	2.789	7.116 +19	369.9	2.37
95000	188.42	7.4973 −7	1.137	2.920	372.6	5.79
100000	195.08	3.1593	4.575 −7	1.189	381.4	1.42 −1
110000	240.00	7.0113 −8	7.925 −8	2.114 +18	431.7	7.88
120000	360.00	2.5050	1.814	5.107 +17	539.3	3.31 +0
130000	469.27	1.2341	6.655 −9	1.930	625.0	8.8
140000	559.63	7.1087 −9	3.128	9.322 +16	691.9	1.8 +1
150000	634.39	4.4828	1.694	5.186	746.5	3.3
160000	696.29	2.9997	1.007	3.162	792.2	5.3
170000	747.57	2.0933	6.380 −10	2.055	831.3	8.2
180000	790.08	1.5072	4.240	1.400	865.3	1.2 +2
190000	825.31	1.1118	2.923	9.887 +15	895.1	1.7
200000	854.56	8.3628 −10	2.074	7.182	921.6	2.4
210000	878.84	6.3910	1.507	5.337	945.2	3.2
220000	899.01	4.9494	1.116	4.040	966.5	4.2
230000	915.78	3.8763	8.402 −11	3.106	985.8	5.4
240000	929.73	3.0653	6.415	2.420	1003.2	7.0
250000	941.33	2.4443	4.957	1.906	1019.1	8.9
260000	950.99	1.9634	3.871	1.515	1033.5	1.1 +3
270000	959.09	1.5872	3.052	1.215	1046.7	1.4
280000	965.75	1.2905	2.425	9.807 +14	1058.7	1.7
290000	971.34	1.0545	1.941	7.967	1069.7	2.1
300000	976.01	8.6557 −11	1.564	6.509	1079.7	2.6
320000	983.16	5.9014	1.032	4.405	1097.4	3.8
340000	988.15	4.0779	6.941 −12	3.029	1112.4	5.6
360000	991.65	2.8501	4.739	2.109	1125.5	8.0
380000	994.10	2.0117	3.276	1.485	1137.4	1.1 +4
400000	995.83	1.4328	2.288	1.056	1148.5	1.6
420000	997.04	1.0291	1.612	7.575 +13	1159.6	2.2
440000	997.90	7.4529 −12	1.144	5.481	1171.2	3.1
460000	998.50	5.4434	8.180 −13	4.001	1183.9	4.2
480000	998.93	4.0111	5.884	2.947	1198.3	5.7
500000	999.67	1.4939	1.946	1.097	1271.5	1.5 +5
600000	999.85	8.1056 −13	9.279 −14	5.950 +12	1356.4	2.8
650000	999.93	4.8226	4.663	3.540	1476.0	4.8
700000	999.97	3.1491	2.506	2.311	1627.0	7.3
750000	999.99	2.2303	1.460	1.637	1793.9	1.0 +6
800000	999.99	1.6813	9.272 −15	1.234	1954.3	1.4
850000	1000.00	1.3240	6.387	9.717 +11	2089.6	1.7
900000	1000.00	1.0731	4.701	7.876	2192.6	2.1
950000	1000.00	8.8642 −14	3.635	6.505	2266.4	2.6
1000000	1000.00	7.4155	2.907	5.442	2318.1	3.1